Navigation: Science a

The series Navigation: Science and Technology (NST) presents new developments and advances in various aspects of navigation - from land navigation, marine navigation, aeronautic navigation, to space navigation; and from basic theories, mechanisms, to modern techniques. It publishes monographs, edited volumes, lecture notes and professional books on topics relevant to navigation - quickly, up to date and with a high quality. A special focus of the series is the technologies of the Global Navigation Satellite Systems (GNSSs), as well as the latest progress made in the existing systems (GPS, BDS, Galileo, GLONASS, etc.). To help readers keep abreast of the latest advances in the field, the key topics in NST include but are not limited to:

- Satellite Navigation Signal Systems
- GNSS Navigation Applications
- Position Determination
- Navigational Instrument
- Atomic Clock Technique and Time-Frequency System
- X-ray Pulsar-based Navigation and Timing
- Test and Evaluation
- User Terminal Technology
- Navigation in Space
- New Theories and Technologies of Navigation
- Policies and Standards

More information about this series at http://www.springer.com/series/15704

Ian Sharp · Kegen Yu

Wireless Positioning: Principles and Practice

Springer

Ian Sharp
CSIRO ICT Centre
Marsfield, NSW
Australia

Kegen Yu
China University of Mining
and Technology
Xuzhou
China

ISSN 2522-0454 ISSN 2522-0462 (electronic)
Navigation: Science and Technology
ISBN 978-981-13-4240-0 ISBN 978-981-10-8791-2 (eBook)
https://doi.org/10.1007/978-981-10-8791-2

Printed on acid-free paper

This Springer imprint is published by the registered company Springer Nature Singapore Pte Ltd.
part of Springer Nature
The registered company address is: 152 Beach Road, #21-01/04 Gateway East, Singapore 189721, Singapore

Contents

Chapter 1
Introduction

In recent years positioning technology has become an important adjunct to modern life. In the past such technology was restricted to large professional applications such as the navigation of ships and aircraft or land surveying, However, with the advent of modern electronic devices such as smart phones, the ready availability of positioning information has meant that it becomes an important requirement to ordinary peoples daily activities, be it navigating in a car, finding a nearby restaurant, recording the location of a picture snapped on a smart phone, or tracking your daily run on your smart watch. While the global positioning system (GPS) was originally developed for military purposes, the technology has been made accessible for civilian use in the 21st century. With the development of single-chip GPS receivers, positioning has become important to a wide range of fields such as industry, agriculture, national defense, aeronautics and astronautics, space exploration, and the various aspects of modern society.

In the literature there are a large number of reference books on the topic, dealing with local positioning, global positioning, or both. The topic of most of these books is more related to the science and mathematical algorithms associated with determining a position given data such as the range to nearby beacons. For example, the development of GPS was undertaken by the US military, and in particular the development of the complex and expensive satellite network. The civilian aspects have been restricted to developing the cheap GPS receiver hardware, and developing the mathematical algorithms and enhanced signal processing required for improved GPS performance. Such research has resulted in performance accuracies far better than the original designers of the GPS. However, the development of terrestrial positioning system, such as indoor positioning has been largely restricted to investigating the science of indoor radio propagation, and developing algorithms that mitigate the effects of the complex indoor propagation environment. What has been largely missing in this research effort is the engineering required for cheap effective hardware for terrestrial applications which GPS cannot satisfy. Thus this book concentrates on the fundamental principles and engineering practice of wireless positioning with a focus on non-GNSS (global navigation satellite system)

I. Sharp and K. Yu, *Wireless Positioning: Principles and Practice*, Navigation: Science and Technology, https://doi.org/10.1007/978-981-10-8791-2_1

local radio positioning systems. In particular, significant emphasis is placed on *engineering* practice rather than the more *science* aspects, this being the main distinguishing feature between this book and many other books. Further, the book draws decades of experience in developing actual terrestrial-based radiolocation hardware for a variety of applications, and data from these systems have been liberally used to illustrate the performance of real hardware in real-world situations. Indeed, the book provides a guide to the complete process of developing terrestrial positioning systems, from the initial functional specifications, through to system testing, and the performance evaluation in a wide variety of applications.

In this introduction chapter, a number of basic concepts in wireless positioning are described, which may be helpful to readers who are not familiar with wireless positioning or who are not clear with some terminologies in this field. Only basic explanations are presented and more details and in-depth studies may be found in the following main chapters. The overview of the book is then provided so that readers can have a quick guideline about the book structure and select the contents of interest to read.

1.1 Basics of Wireless Positioning

In wireless positioning, the terminologies *positioning* and *localization* are often used interchangeably in the literature. The former term may mainly deal with determination of position of oneself, while the latter might be more related to finding the position of another object (such as a person, an animal, a robot, a ship, a submarine, an aircraft, a missile, or an unmanned aerial vehicle) or a group of such objects. Object tracking can be considered as localizing or locating an object or multiple objects continuously over a period of time. The position information of the located object(s) may be used by other objects, and thus further aid position determination. On the other hand, navigation is defined as determining one's own position and using other relevant data to determine a path to reach a destination efficiently.

A *wireless positioning system* typically refers to be any positioning system that makes use of wireless transmission channels (in air or water) to transfer and/or receive signals. There are a number of such signals widely employed for positioning, including radio, acoustic, light and magnetic signals. Positioning can be enabled through mutual communications between two terminals, which is defined as a *two-way* communications mode. However, the most popular positioning approach is *broadcasting* by one or more transmitters, such as GNSS satellites, which is a *one-way* communications mode. The signal selection for a particular requirement depends on several factors, such as the application scenario, the performance requirements (such as accuracy and coverage), and the cost.

Positioning systems may be further divided into two categories, one for local positioning and the other for global positioning. A local positioning system provides a positioning service over a limited specific area, such as a building, a campus, or a city. Global positioning is enabled by use of satellites to provide services with global coverage. Currently, only US Department of Defense global positioning system (GPS) and Russian global navigation satellite system (GLONASS) provide global coverage services, while China's BeiDou navigation system (BDS) and the European Galileo navigation system will have global coverage in a few years.

In a local positioning system, terminologies like *mobile station* and *base station* are often used. A mobile station is the object to be located, which can be a person with a mobile phone or a positioning tag, an animal with an attached tracking device, or a vehicle. A base station can be associated with a mobile phone cellular network, WiFi (also called an *access point*), or a fixed ultra-wideband device in an indoor environment. In GNSS, a satellite may be called a base station, while any mobile object that carries a GNSS receiver (a chip in a smart phone) may be called a mobile station.

Mobile stations, "fixed" base stations (also sometimes referred to as *anchor nodes*), or both can incorporate receivers to estimate one or more signal parameters (such as time-of-arrival (TOA), angle-of-arrival (AOA), carrier phase, or signal strength). TOA measurements may be used to estimate the transmission time between a transmitter and a receiver (and hence determine distance based on a known signal propagation speed) if the clocks in the two communication sides are synchronized. Alternatively, *pseudorange* measurements can be made if the receiver has a free-running clock but all the transmitters are synchronized in time, or have known time offsets.

Signal Strength may also be used to estimate propagation distances, using a propagation path loss (or power attenuation) model. Such a model is useful in a *free-space* environment. For example, on a very large scale, astronomers use this principle to estimate distance to particular celestial objects, such as Type I supernovas which have a known fixed source power and close to free-space propagation of the electromagnetic radiation. For terrestrial applications outdoors with no path blockage free-space conditions can approximately exist, but ground reflections can significantly affect the propagation loss. For indoor locations typically non-line-of-sight propagation conditions exist, so that the signal attenuation will be greater, and vary statistically spatially. Thus modeling the loss as a function of propagation distance is difficult, so that at best rather inaccurate distance measurements are possible. However, this signal variability can be used as an advantage by generating a signal strength *fingerprint* at many locations in the coverage area, for example with WiFi access points. A mobile position can be estimated by matching the signal strength of signals from multiple access points to a database of fingerprints from a site survey.

Parametric positioning uses a group of such distance or pseudorange measurements in equations to determine the position coordinates of the mobile device. Alternatively, *non-parametric positioning* does not involve solving such measurement equations, but simply judges if the object is within the vicinity, such as in a particular room, by matching between database and real-time measurements. Alternatively, the method utilizes the difference between the database and real-time measurements to weight the positions of reference points to generate a position estimate of the target.

In the detection of intrusions or burglaries, positioning needs to be realized without involving any communications between the target (intruder) and the base stations. This is often termed as *device free* localization. In a monitoring area, one or more devices continuously transmit radio signals, while a group of receivers capture the scattered signals from the room and objects in the room, and measure signal characteristics such as the received signal strength (RSS). In the absence of any intrusion, the RSS has a specific (static) distribution pattern over the localization area. In the presence of one or more intrusions the distribution pattern would change perceptibly. By analyzing for any change in an otherwise static signal pattern, possible intrusions can be detected. This is also the principle of infrared burglar alarm systems.

In some cases, the signal is not deliberately designed but is available naturally, such as geomagnetic or gravitational fields. Although these two geo-fields vary continuously over time and space, the time variations are usually rather small, even over periods of many months or years. Thus it is necessary to record the field strength at many points in the area of interest, and establish a database with the field strengths at these locations in advance. For detection of a major event (such as a land slide or volcanic activity), the real-time measured field data are compared with the recorded field data in the database, to obtain the best match and hence provide a position estimate. This is the basic principle for all database based positioning methods.

Another situation of interest is when a signal can be produced due to physical movement or orientation variation of a mobile target, including but not limited to acceleration, speed, heading angle, and turning rate. Inertial sensors (such as accelerators and gyroscopes) and magnetometers are usually used to measure these parameters. Typically, in modern implementation these accelerators and gyroscopes are integrated into a small solid-state sensor packages, which collectively is termed an *inertial measurement unit* (IMU). An IMU, or a group of IMUs in some cases, is an essential part of an *inertial navigation system* (INS). Due to the great advances in *microelectromechanica*l system (MEMS), modern INS equipment can be manufactured using a tiny chip which can be easily embedded into small devices such as mobile phones. The principle of such an INS is that integration of the acceleration produces a velocity and a further integration provides a position estimate. Angular data from integrating the gyroscope data are also required, as the accelerometer measurements include the acceleration due to the Earth's gravity, so that angular information is required for correct interpretation of the accelerometer data. Such integration processes only provide relative movement, and also small sensor measurement offsets result in a slow drift in position, even when stationary. Thus an INS is rarely used independently, but is integrated with another positioning technology.

In an open outdoor environment, GNSS is the first and probably the best choice for positioning and navigation. In fact, modern smart phones now almost always include an embedded GNSS chip, and the associated navigation software (App) to find the phone's location on a map or a path to a nominated destination. However, when walking from an outdoor open space into a large building or a dense forest, GNSS signals are blocked, so the positioning system will stop working. To deal with this situation and to enable the so-called *seamless positioning*, multiple positioning technologies need to be integrated. For example, an integrated GNSS/INS system is able to provide valid position information even if the GNSS signals are blocked for a short period of time. Alternatively, after moving from outdoors to indoors, a local positioning system can be activated to replace GNSS indoors. In a complex indoor environment, integration of multiple technologies is also helpful, and often necessary to achieve the desired accuracy, reliability, and coverage.

Using the above techniques positioning can be realized on the ground, underground, under water, in the air, or in the deep space. Underground positioning, such as for tracking miners, equipment and vehicles in an underground mine, is similar to positioning in a large building. However, the task is more difficult because of the adverse underground mining environment. Also, radio signals are usually not used for underwater communications due to the very short transmission range, particularly in salty water. Thus for underwater applications acoustic signals are almost universally used for communications and positioning; for instance, sonar is often used for detecting underwater vehicles and measuring water depth. Even with continuous improvements in such technologies, underwater vehicles are often equipped with INS to enable navigation over long distances, on or under the sea or lake surface without any aid from other equipment. Breakthroughs in *quantum gyroscope* technology will revolutionize navigation especially for underwater applications due to the very small measurement errors, which mean the INS integration process results in much smaller errors than older technologies. In the air, similar to the open space on the ground, GNSS is the best option for positioning and navigation of aircraft, UAVs, and missiles. In the deep space where GNSS is not applicable, positioning and navigation rely on locating reference objects such as bright stars, pulsars or quasars for accurate time reference.

1.2 Overview of the Book

This book consists of two parts. Part I (from Chaps. 2 to 11) is on the engineering and design of positioning systems, while Part II (from Chaps. 12 to 18) provides information on recent developments in positioning systems.

The first part of the book is intended to provide readers with the engineering elements necessary for designing, implementation and testing of terrestrial positioning systems. Much of the material is based on the experience in developing a number of such systems, and much of the data and graphical information in the book is associated with the performance of actual systems designed and developed using methods described in the Part I chapters.

Due to the widespread use of the Global Positioning System in a wide variety of devices, much of the literature on location systems has focused on the performance of GPS in a variety of applications. As GPS was developed by the US Department of Defense for military use, the satellite component of the GPS is not discussed in the civilian literature, so that the available literature on radiolocation is largely restricted to the receiver functions only. However, an independent positioning system, particularly for indoor operation, requires the design and development of the complete system, including the radio transmitter and receiver hardware, signal processing, position determination algorithms, and software for both real-time control and user applications.

While much of the engineering literature is devoted to position determination assuming that suitable measurements such as ranging data are available, in any system development project this component represents a negligible fraction of the development effort. Although position determination is important, and the topic has been covered extensively elsewhere (including another book[1] by the authors of this book), this topic is not dealt with comprehensively in this book; instead the focus of Part I of the book is on the processes and design techniques associated with the development of a new system with specific requirements.

The chapters in Part I of this book are broadly organized according to the tasks required to engineer a new positioning system, with a strong emphasis on radiolocation technology in difficult operating environments, such as inside buildings. Chapter 2 provides a broad guideline to the system engineering development processes, with specific details in subsequent chapters. These chapters are aimed at providing practical information to assist the development process, with many examples from the development of actual positioning systems. In addition to the details of the various subsystems, chapters are also devoted to system testing and applications of positioning systems. While the emphasis is on the developing of the radiolocation components, chapters also include the use of sensors to improve positioning accuracy, and the use of WiFi infrastructure for low-accuracy position determination in buildings.

It is suggested that Chap. 2 is first read to get an overview of the other chapters, and the subsequent chapters used as required to provide useful information on various specific aspects of positioning systems. The intention is to provide useful methods and data that could assist in the development of a system, and the expected performance in real-world environments.

Part II of the book provides results from recent research on positioning systems (mainly indoors), and is more theoretical in nature than Part I. While the material in this part is appropriate for the design and development of positioning systems, the information is more related to the *science* of positioning rather than the *engineering* aspects.

The topics in Part II have been chosen to provide information that is related to more recent developments, or providing more detailed background information of a more theoretical and mathematical nature. For example, in Section I reference is made in several chapters to the bias errors in time-of-arrival measurements, and its

[1]Ground-based Wireless Positioning, Kegen Yu, Ian Sharp, Jay Guo, Wiley Press 2009.

impact on design and performance. The data presented in Part I is based on measurements and empirical models. In Chap. 13 in Part II the causes of these bias errors are analyzed, based on more detailed models which relate more closely to the science of radio propagation; the results show that there are good theoretical reasons for the measured results, based on more basic parameters of the bias problem. In other cases, such as receiver signal strength positioning, new techniques based on detailed measurements allow improved accuracy of an old well-known technique. Similarly, the well-known concept of geometrical dilution of precession (GDOP) has been extended to allow statistical performance estimates to be made in modern ad hoc networks, where small mobile nodes form the basis of position determination rather than the traditional anchor nodes system for which GDOP was originally developed.

Fusion of measurements from different sensors and integration of multiple systems is often appropriate for both local and global positioning in complex environments. Also, similar to wireless communications, antenna array can be exploited in a positioning system either for direction finding or for position determination. Two chapters in Part II are presented to cover the issues of data fusion and system integration as well as antenna arrays. The information provided in these chapters may be useful for readers in the understanding and perhaps the building of an advanced positioning system.

It is suggested that the individual topics and chapters in Part II of the book are read to obtain more in-depth information on the specific topics. Where appropriate, these detailed topics are referenced in Part I, but knowledge of these topics in Part II is not required for the understanding of Part I.

Chapter 2
Designing Positioning Systems

2.1 Introduction

A key purpose of this book is to provide information pertinent to the development of terrestrial positioning system, particularly radio-based technology. While much of the literature on positioning is associated with methods of determining a position based on measured data from various types of sensors, this chapter provides an introduction to the engineering required to achieve this aim, and the methods of developing a system that meets a particular set of requirements. While implementation details are important in any such development, the emphasis in this chapter is on the *System Engineering* processes that relate specifically to the development of a completely new positioning system. Such a system, particularly when based on radiolocation, involves many discipline areas, including hardware associated with radio transmitters and receivers, analog and in particular digital signal processing, as well as software associated with the real-time processing of data and the application software on a wide range of devices. Coordinating the development in all these areas is the Systems Engineering task.

The broad principles of Systems Engineering as applied to project managements and product development are essentially universal, and thus can be applied to positioning systems projects. These principles mean that a project will progress through developments phases, starting with a requirements analysis, both from the customer or user perspective as well as from the developer's point of view. This analysis will result in a *Requirements Specification*, which will guide the design, development, testing and manufacture phases. While all such aspects are important, the focus of this book is on the initial design and prototype development of position systems, and the methods and research required to produce a new innovative design. In this chapter only a broad overview of these processes will be given, with subsequent chapters providing details in particularly important areas for position

I. Sharp and K. Yu, *Wireless Positioning: Principles and Practice*, Navigation: Science and Technology, https://doi.org/10.1007/978-981-10-8791-2_2

systems. As position determination methods are available in detail elsewhere (Yu et al. 2009), these topics are only covered briefly in this book, while other topics rarely covered in the literature are covered in some detail in subsequent chapters. In particular, the more detailed coverage includes the topics:

- Radio Link Budgets, and the impact on the overall design
- Time and frequency control for precision performance
- System testing, particular measuring positional accuracy statistics in the normal operating environment
- Adaptation of existing technology for positioning systems to minimize development effort and product costs.

2.2 Overview of Positioning Systems Architecture

Although there are many types of positioning systems and underlying technology, the basic architecture can be summarized into just a few general cases. The two main categories can be classed as either "navigation" or "tracking", each with their specific characteristics and areas of application. Thus before considering the architecture of specific systems, it is useful to briefly summarize the characteristics of these two types of positioning systems. The following will be restricted to radiolocation, but much of these characteristics apply to other technologies, such as ultrasonics and sonar.

2.2.1 Navigation

A navigation system can be loosely defined as a technological method of determining the position of a mobile device, where the positional data is used in the mobile. Common examples include the navigation of aircraft, the in-car navigation using the Global Positioning System (GPS), and in recent years the use of mobile phones and similar such devices for personal navigation. The size of the operating domain can vary widely, from a few tens of meters (say within a building) to global. The most commonly known navigation system with widespread use and applications is the GPS, which has global coverage for mainly outdoor operation. The US military operated GPS can be classified as a Global Navigation Satellite System (GNSS), and is but one of several, others being GLONASS (Russia), Galileo (European Union) and BeiDou (China). The satellite infrastructure is built and maintained typically by a nation's military, but is also available for civilian modes of operation. Thus the civilian design function is limited to the mobile device, and most of the other technological and operational characteristics are beyond the scope of civilian designers.

The following summarizes the characteristics of radio navigation systems:

1. A common characteristic of radio navigation is that "fixed" nodes transmit a radio signal, with a receiver in the mobile device. The "fixed" node is at a known (to the receiver) location, but does not necessarily mean stationary. For example, a satellite with known orbital characteristics relative to the earth can be considered "fixed" for navigation in the environment of the earth (both on the surface of the earth and in nearby space).
2. Because the "fixed" nodes transmit and the mobile nodes receive, this broadcasting mode implies unlimited capacity—any mobile node with a suitable receiver and data processing capability can determine its location, and hence can use the derived location for any appropriate navigation task. For example, GPS receivers are commonly in "smart" mobile phones, with software (Apps) providing a wide range of applications where the positional data are important for the App data processing (where is the nearest restaurant?).
3. A key characteristic of this type of radiolocation is that the signal and other data processing (which can be quite complex) is in the mobile device, although increasingly this processing is being transferred to the "Internet Cloud" is some cases. In the past the cost, size and power requirements severely limited the application of the technology (to say ships and aircraft), but with the rapid development of microelectronics this is no longer a major limitation, so that for example, GPS/GLONASS navigation is available in smart mobile phones. However, the power requirements may still be a limitation in some future applications, such indoor tracking of people wearing "smart badges".

The availability of radio spectrum can constrain possible applications. Military-based systems all use dedicated spectrum for their operation, but any civilian-type system typically will be limited to the Industrial, Scientific and Medical (ISM) bands (Federal Communications Commission, Code of Federal Regulations, Part 15) which allow unlicensed operations, but must be shared with other uses/users. To avoid mutual interference, the radio transmissions in ISM bands are restricted to either frequency hopping or direct sequence spread-spectrum (DSSS). Fortunately, DSSS is ideal for positioning systems. More recently, an alternative transmission method, namely Ultra-wideband (UWB) (Fontana et al. 2003; Yu et al. 2006; Fontana 2004) has also become available, but to limit interference to other radio-based systems within the UWB frequency band (3.1–10.6 GHz) the allowable transmitter power is very low (−41.3 dBm/MHz), so that the operating range is small (typically less than 10–15 m), severely restricting possible applications. Thus most of the common navigation system use spread-spectrum radio transmissions, and typically use code division multiple access (CDMA) to minimize mutual interference between the fixed node transmissions. However, other multiple access methods, such a frequency division multiple access (FDMA) as used by GLONASS, and time-division multiple access (TDMA) could be used.

2.2.2 *Tracking*

A tracking system can be defined as the determination of the location of a mobile object by another party, with or without the cooperation of object being tracked. A common example is radar, typically used for tracking aircraft, and is separate from the systems used for navigation by the pilots. In typical applications the location of an object is important information required by tracking party, but is not required by the mobile party.

The following summarizes radio tracking systems:

1. A common characteristic of radio tracking is that mobile nodes transmit a radio signal, with the receivers in the "fixed" nodes. In the case of radar, the fixed node also transmits a high-powered signal, but the scattering off the mobile device is the involuntary transmission from the mobile device used for tracking. More commonly in radiolocation tracking systems, the mobile has a radio transmitter whose transmissions are received at one or more fixed receivers.
2. The signal processing for the reception of the signal from the mobile, and the data processing required to determine position can be quite complex, but being in the fixed nodes the constraints on processing capability, size, and power requirements are usually not a constraint on the design, unlike the navigation case.
3. Because simultaneous transmissions from mobile devices can cause mutual interference, some form of multiple access to the radio frequency band is required. While any of the various techniques[1] (CDMA, FDMA, TDMA) used in navigation systems could be used in tracking systems, there are further constraints on terrestrial tracking systems that do not apply to GNSS. In particular, for terrestrial navigation systems, as a mobile (receiving) device can be close to one of the transmitting nodes (base station), the high signal from this base station effectively "jams" the mobile from receiving (weak) signals from other more remote transmitting base stations. This type of interference is called the near-far affect, and plays a crucial role in the design of terrestrial positioning systems. Note that there is a crucial difference between receiving data communications (which only needs to communicate with one base station at a time), and positioning systems which require signals simultaneously with multiple base stations. Because of this problem, terrestrial positioning systems tend to operate as a tracking system, namely the transmitter is in the mobile. In this case, the near-far affect rules out the use of CDMA, as the high received signal from a mobile near a base station will jam the transmissions from other more distant mobile nodes. Further, the requirement for large bandwidths to obtain the

---

[1]CDMA (code division multiple access), FDMA (frequency division multiple access), TDMA (time division multiple access).

required positional accuracy for time-of-arrival systems severely restricts the use of FDMA as the multiple access method, as there will be a limited number of available wideband channels and hence a limited number of mobile nodes that can transmit simultaneously. Thus tracking systems tend to use TDMA for access control to the radio spectrum.
4. Because of the use of TDMA, and with transmitters in the mobile device, a tracking system will have a limited capacity, unlike navigation systems. In this respect tracking systems are similar to mobile phone systems. However, terrestrial tracking systems have the additional constraint on the design of requiring simultaneous communications with multiple base stations, which rules out CDMA as often used in mobile phone systems.
5. For terrestrial tracking systems, the designer is tasked with designing both the "fixed" infrastructure (both hardware and software), as well as the mobile device and the signal protocols. These protocols will include both those associated with the radiolocation functions, as well as data communications between the mobile and fixed nodes, and inter-nodal communications. This task is in stark contrast with (say) designing a navigation/tracking application using GNSS. Although the design task is greater, there exists the opportunity to tailor the design to the needs of a particular application, such as improved accuracy or providing indoor operation, where a GNSS solution may be inadequate due to severe signal attenuation and multipath effects.

2.2.3 Mesh Positioning Systems

The previous subsections described the characteristics of the two main generic types of (radio) positioning systems. A common characteristic of these systems is the subdivision of functional elements into "fixed" and mobile nodes. Further, these systems have centralized control and management. For example, the US military control and manage the GPS satellite system, and for a tracking system the system's owner will be responsible for the fixed nodes, communications networking, and the management of the tracked positional data. However, with advances in microelectronics the cost and size of the positioning systems hardware modules are decreasing, so that the distinction between the "fixed" (base stations) and "mobile" assets is becoming blurred. Thus in a Mesh Positioning System (MPS) there is no such distinction between node types. These MPS nodes will have generic functions such as transmitting/receiving (both for position determination and data transmission), and suitable signal processing and data processing functions. In such a system, each node will communicate with its surrounding nodes to share information, but in general there is no centralized control and management. These basic resources can be configured into the functionality of the more classical navigation and tracking systems, but other configurations are also possible. For example, consider the tracking of people in a building. While there will be a few "fixed" nodes attached to the building, most of the nodes will be attached to people. The system could be used both for tracking the people and for sending/receiving information to the people. While the position of the mobile nodes is not originally known, if there is an adequate number of fixed nodes, some of the mobile node positions can be determined, which in turn can be used to determine the location of more nodes. By this method the position of all nodes in the network can be determined; indeed (if necessary) a system-wide (centralized) optimisation procedure can be used to provide even better positional estimates, and to continually update the position estimates of all nodes, which can be communicated to all nodes in the network. These positional data can be used either for navigation (say to find your way around a building to locate a particular room), or for tracking as the fixed nodes can be connected to a local area network within the building, and thus to a centralized monitoring system.

Apart from the elimination of the distinction between node types, a Mesh network implements position determination by a somewhat different method. While a classic positioning system requires a time synchronization subsystem (TSS), a Mesh Positioning system can be designed without the need for any time synchronization at all. Indeed, each node in the network can have a free-running local oscillator which has both a (slightly) different frequency typical of crystal oscillators (say 1–10 ppm variation between individual crystals), and time is local to each node. In its simplest implementation, each node transmits periodically, but with some random variation in the update period. The transmission period will be much less than the average period between transmissions, allowing for many nodes in the network. When not transmitting, the node "listens" for transmissions from other surrounding nodes, and records the TOA (and identity of the transmitting node) of any transmissions. These data are included in any transmissions, so that the surrounding nodes also share this information. With the proper design to the protocol and appropriate signal processing it can be shown that the ranges between nodes can be determined using just the TOA data, and thus the positions of the nodes can be determined. With purely random time of transmission Aloha protocol (Abramson 1970) the efficiency of the use of the available time is only 1/2e (or 16.8%); using slotted Aloha this can be doubled to 33.6%. More sophisticated protocols, such a Carrier Sense Multiple Access (CSMA) can increase the efficiency to near 100%.

Thus by providing nodes with both transmit and receive functions, together with data communications capability, a Mesh Network positioning system can provide similar capability to the classical position systems, but with a greatly simplified design. Because of this simplicity, other positioning applications (other than GNSS-based) can be developed, including indoor operation and enhanced positional accuracy when compared with GNSS.

2.3 Overview of Design Procedures

This section provides an overview of the design processes required for developing a radiolocation system with particular requirements associated with an application that makes off-the-shelf solutions impossible. The subsections briefly describe the major components of the design process, and the associated functional elements of typical systems. These introductory comments are expanded in subsequent chapters.

2.3.1 Functional Requirements

The start of the process of developing a new positioning system begins with specifying the general requirements. Such requirements are important, as there is not a "one solution fits all", particularly for specialist applications. Quite typically the customer will have tried to find an existing technology, such as one based on GPS, but actual performance of existing technologies has proved inadequate. For example, GPS does not function satisfactorily indoors, or the required accuracy cannot be achieved by GPS. In a research environment, the "customer" may be the system developer. In this situation, the system developer perceives a problem with existing solutions, but has ideas how new technology may overcome these limitations. In any case the starting point in developing a new system is a broad "customer" specification, which is the input to the much more detailed *Requirements Specification* which will guide the detailed development of the new system.

The following subsections describe the general engineering development process, but with particular emphasis on the design of custom radiolocation systems.

2.3.1.1 Customer Specifications

It is useful to consider the typical characteristics of customer specifications, as these are often radically different from the technical Requirements Specification use for the system design. In particular, the customer will be focused on the application of the positioning technology, and has little interest or knowledge of positioning technology. From the particular requirements of the application initial rough estimates of the operating environment, positional accuracy, mobile size and battery life will be provided. However, probably the most important requirement is usually the cost of a potential solution. As GPS is now a mature technology, the cost of devices employing GPS chip receivers is very low, so that expectations of the customer is usually overly optimistic regarding other custom solutions. Thus for example, a technically satisfactory but high cost solution will usually not be an appropriate response to solve the customer's application requirements.

A further cautionary note on the interactions between a customer and the technical designers is that often the understanding of concepts such as "accuracy" is very different. In particular, while statistical performance measurements are used extensively by technical people, this concept is typically poorly understood by users of positioning technology. Much of the disappointment of users of positioning technology centers around this concept. Consider a specific example, where the customer's application is to locate people to a particular room in a building—in this case the customer may consider "accuracy" as a fraction of the size of a room, say 1–2 m. It is important to note here that the customer will interpret this as the *maximum* expected positional error, not some statistical measure of positional accuracy. While the customer may understand that the actual errors will vary, an error of (say) 10 m would not be considered satisfactory, even if it occurred infrequently, say 1 in 1000 positions. If the actual system did exhibit this accuracy characteristic, then the customer may consider the system "unusable" even if 99% of positions were within 1 m. Thus from a functional designer's perspective simply defining "accuracy" as a root-mean-squared error of 1 m is not appropriate. In particular, it must be understood by the designer in this case that it is better to have a "no position" than an error of 10 m. Thus in general it is important for the system designer to clearly understand the requirements from the customer's perspective, and to translate these requirements into a technical function specification.

2.3.1.2 Technical Requirements Specification

The purpose of the *Technical Requirements Specification* is to accurately translate the customer's requirements into technical specifications which will guide the design of the system, typically described in the overall *Design Specification* and other more detailed subsystem design specifications. Although the Technical Requirements Specification must be based on the customer's requirements, the detailed specifications will also include a general technical design concept using specific technology. Thus the Functional and Design Specifications are typically managed with feedback, as detailed design may show that all the required functional requirements cannot be met without some modification. Any modifications to the Technical Requirements Specification will also need to be approved by the customer. This procedure is important for avoiding costly changes late in the product design and testing.

The technical requirements specification will include the main (simplified) customer's requirements, but also will include much more detailed requirements. For example, the specification would include more detailed information on the radio range, details of base stations, computer interfaces to the positioning hardware, data rates, application programs and the associated user interfaces, mobile device characteristics (size, battery life, sensors, operator interface (if any), transmitter power etc.), radio transmission characteristics (frequency, bandwidth, modulation, multiple access technique), and the details about the positioning (accuracy, number of users, update period).

2.3.2 *Selection of Radio Frequency and Bandwidth*

For a radiolocation system, the key characteristics which define the overall performance are the radio frequency (RF) and the associated bandwidth. In particular, the radio frequency has a large impact on the radio range, while the bandwidth defines the positional accuracy; both of these must be defined in the functional specification. However, the choice of frequency is also severely constrained by regulatory authorities such as the FCC in the US (FCC 2013), which will define specific frequencies for particular applications, as well as other constraints such as the bandwidth, transmitter (EIRP) power and modulation techniques. While some radiolocation technologies (in particular GPS and aircraft navigational aids) have special frequency bands allocated, in general other niche radiolocation applications must use specific bands (ISM) which must be shared with other users. Often the only practical choice is to adopt one of the ISM bands, in particular the 900 MHz, 2.4 and 5.8 GHz bands, as the cost of licensing specific bands is usually prohibitive for all but the must common commercial uses (such as the spectrum licensed for mobile phones).

The choice of the appropriate frequency to meet all the requirements in the functional specification is not simple, as it requires a tradeoff between various desirable characteristics. In particular, increasing the frequency will reduce the background (interference) noise, increase the available bandwidth, and result in smaller more efficient antennas, all of which are desirable characteristics. However, increasing frequency will also increase the free-space propagation loss, increase the scattering and penetration losses (such as through walls), and at very high frequencies (>10 GHz) result in more expensive electronics. A design analysis is required to determine the optimum solution for a particular application, with many of the key elements defined by a Link Budget (see Sect. 2.3.5 and Chap. 9).

For radiolocation, a key component in the functional specification is the positional accuracy, which is directly related to the available bandwidth. As the bandwidth typically will increase with the radio frequency (UWB being a notable exception which does not have a defined RF), choosing a high RF is a natural choice, but as explained in the previous paragraph this also has detrimental effects. The bandwidth is important as it defines an effective time resolution of the associated "pulse" used for measuring the range. The range measurement errors are essentially inversely proportional to RF channel width (or equivalently proportional to the pulse rise-time), both for the effects of random receiver noise and multipath propagation. However, wide bandwidths also imply fast electronics for the associated signal processing, which in turn requires more power and hence a larger battery. Thus again a design analysis is required to select the best frequency for a given application.

2.3.3 Radio Propagation Characteristics in the Operating Environment

The characteristics of radio propagation in the operating environment has a large effect on the performance of a radiolocation system. In the design of a system, two major effects need to be considered. The first effect is the variation in the propagation loss as a function of position (range). The loss characteristic will determine the operating range, and other subsidiary effects such as the probability of successfully measuring the range. The second effect is the propagation delay excess, which determines the errors in the measured range. While these two effects are important in the design of any radiolocation system, understanding and accommodating the consequences on performance is particularly important for indoor systems where the degrading effects are largest.

2.3.3.1 Loss Excess

The propagation loss between the transmitter and the receiver can be considered on three different scales. The first or large-scale case is line-of-sight (LOS) propagation where the loss is usually considered to be close to the free-space loss, although in practice there will always be surfaces (such as the ground or ceilings) present which can affect the loss. Free-space loss is only a function of the range measured in wavelengths, and thus will have a dominant input to the selection of the RF described in Sect. 2.3.2. In any case, the free-space loss as a function of range represents a baseline for the measurement of loss excess in all environments, including the non line-of-sight (NLOS) case. In the NLOS case, this large-scale loss excess is defined as smoothly increasing with range (Yu et al. 2009), but the exact characteristics depends on the operating frequency (more at higher frequencies) and the nature of the NLOS operating environment. The second or medium-scale signal variation that needs to be considered is when there is significant multipath interference (including the often forgotten ground reflections). When there are significant multipath interference signals, the signal strength when measured (averaged) over a scale of the order of 10–50 wavelength exhibits a log-normal random statistical distribution (Johnson et al. 1994) with a standard deviation (STD) of the order of 6–10 dB. The third or small-scale case occurs over scales of the order of a wavelength, and can exhibit very rapid changes in the signal amplitude in severe NLOS multipath environments. These Rayleigh fades can have a severe affect on data communications, but generally have less severe consequences on positioning systems. The reason for this is that the signal design aims at allowing the TOA to be measured using just the leading edge of the signal. When determining the range using link budget calculations (see Sect. 2.3.5 following), these statistical variations in the signal strength need to be taken into account, so that there will be statistical variation in the maximum range. Thus for example when planning the number of base stations required to cover a given area, some redundancy in the number of receiving base stations (above that nominally required to determine the position from range measurements) needs to be included in any plan.

2.3.3.2 Delay Excess

The basic concept of radiolocation based on time-of-flight measurements is to use the propagation delay between the transmitter and receiver as a proxy for the range given a known propagation speed. Further it is known that in free-space radiowaves propagate in a straight-line, so the delay can be easily converted to a Euclidean distance. However, in all actual operating environments, and in particular indoor environments there will be delay excess (as measured by the receiver) relative to the ideal straight-line case. As a consequence, the range as determined from the propagation delay will have measurement errors. In addition to these propagation-related range errors, errors will also be associated with random receiver noise, but typically with a sufficiently high signal-to-noise ratio (SNR) these effects are relatively insignificant. In the LOS case with some small additional multipath signals the range errors result in position errors that can be approximated by a spatial Gaussian noise with zero mean (Yu et al. 2009, Sect. 2.4), and thus have consequences similar to receiver (temporal) noise (zero mean range error). However, for the NLOS case experienced in indoor environments, there is an additional positive bias error associated with the signals scattered along the path between the transmitter and the receiver. However, it has been shown through measurements (Alavi and Pahlavan 2003, 2006; Alsindi et al. 2009) that typically these errors can be reasonably well accounted for by a bias function which is linear with range. With appropriate techniques (Sharp and Yu 2013) this linear function can be estimated, so after removal of the linear-range bias error the residual errors will have an approximately Gaussian statistical distribution with zero mean, similar to the LOS case. However, a small proportion of measurements (typically less than 10% and often less than 5% (Sharp and Yu 2016) do not fit this linear model, and will have much larger delay excesses associated with the propagation path. For accurate measurement of position such "rogue" measurements should be eliminated as part of the position determination process—refer to Sect. 2.3.8 for more details.

2.3.4 Time-of-Arrival Measurements

Time-of-Arrival measurements are the key concept in all radiolocation systems which use the propagation delay of the radiowaves for position determination. A TOA measurement does not directly lead to range measurements, but provides a framework associated with many other subsystems which when combined allow the range between nodes to be determined, and hence the position. TOA measurements are also directly involved in subsystems such as frequency and time synchronization, the establishing of the time division multiple access protocol (time slots allocated to nodes for data transmissions as well as TOA-related transmissions), round-trip-time (RTT) measurements for determining ranges, and internal delay calibrations which allows RTT measurements to be converted to ranges. Because all these functions are related directly to TOA measurements, the accuracy of the TOA directly effects many aspects of the performance, in addition to the positional accuracy of the system.

A TOA measurement can be defined as the time a radio "pulse" is detected by a radio receiver, *relative* to a clock in the receiver. Note by definition the TOA does not contain explicitly any knowledge of the time of the original transmission of the pulse, and hence does not directly provide any information on the propagation delay. Note also that the local clock may not have any knowledge of *absolute* time, or even time in other nodes. Further, the clock frequency will have an accuracy defined by the local reference oscillator, and will have frequency offset errors typically in the range of 1–10 ppm in simple mobile devices. The "pulse" for TOA measurements is typically not directly associated with RF energy in a time-domain signal, but rather is associated with an output from a signal processor in the baseband of the receiver. While UWB technology does use pulses of RF energy, more typically the RF energy is modulated into a spread-spectrum format, with receiver signal processing (correlation) generating the "pulse". For example, all GNSS (and GPS in particular) use spread-spectrum modulation. For terrestrial systems the only readily available RF spectrum with sufficient bandwidth is in the ISM bands, which mandate the use of spread-spectrum modulation to minimize the effects of mutual interference between (other) users operating within band. Thus terrestrial radiolocation systems will usually operate in an ISM band, and thus use spread-spectrum modulation. However, from the perspective of TOA measurements, the origin of the "pulse" is unimportant.

The TOA is associated with the pulse, but because of the finite signal bandwidth the pulse will have a rise-time given approximately by

$$\tau_{rise} \approx \frac{1}{BW_{3dB}} \approx \frac{1.5}{BW_{chan}} \tag{2.1}$$

where BW_{3dB} is the baseband signal bandwidth, and BW_{chan} is the bandwidth of the RF signal as limited by the radio receiver filters. However, for practical implementation the TOA is not associated in a "fuzzy" manner to a precision of the order of the pulse rise-time, as this time resolution is too coarse for useful position determination. For example, the GPS 3 dB bandwidth is about 1 MHz (civilian mode), which according to (2.1) results in a rise-time of about 1 microsecond (or equivalent to 300 m). In practice a specific point on the pulse, called the *epoch* of the pulse, is defined, and the TOA is defined by the time of the signal epoch as measured by the local clock in the node. The location of the epoch on the pulse is arbitrary, but must be consistent in all nodes. While the peak of the pulse may seem the logical choice, the precision of the determination of the epoch is affected by corruption of the pulse, in particular system (random) noise, and the effects of multipath propagation; the latter effect is particularly important in terrestrial positioning systems. In general, the *precision* of the TOA measurement is proportional to the derivative of the pulse shape, so the peak is a poor choice as for a bandlimited pulse the peak has a zero derivative. A better choice is two points either side of the peak, with the epoch defined by the mean of these two points. However, in a severe multipath environment even this choice is poor, as multipath corruption of the pulse can result in the pulse peak being far from the first detectable signal (associated with the TOA of the straight-line signal path). In such situations the epoch should be defined on the leading edge of the pulse using some algorithm which is related to the shape of the nominal pulse. Note for accurate TOA measurements the TOA should be largely independent of factors such as the pulse amplitude, pulse signal-to-noise ratio, and if possible multipath signals. With such algorithms the *precision* of the TOA measurement can be of the order of 1% of the pulse rise-time for random receiver noise, and of the order of 10% of the pulse rise-time in a severe multipath environment. However, note that there is a difference between "precision" and "accuracy". In particular, in a severe non-line-of-sight case the direct (straight-line) signal may be so severely attenuated that it is below the pulse noise floor. In such a case no detection algorithm can accurately measure the "true" TOA, but the algorithm should be able to measure the TOA of the first *detectable* signal above the noise floor to a precision indicated above.

Another complication with deriving range from TOA measurements is that the propagation delay between nodes is but one component of the total delay. In particular, the total delay includes equipment delays (both in the baseband and RF sections) of both the transmitter and the receiver. Thus any practical system must include in the design either a measurement of these delays, or a method of compensating for the delays. Indeed, for a terrestrial system the total delay from transmitter to receiver is usually dominated by the equipment delays, so that the positional accuracy can be dominated by inaccuracies in correcting/compensating for these delays. One common approach is to synchronize clocks in the transmitting or receiving nodes depending on the particular architecture. Such clock synchronization will in turn utilize TOA measurements. In such cases of time synchronized nodes it can be shown that the TOA measurement is equivalent to the propagation delay plus an unknown offset time common to all measurements. When the TOA is converted to an equivalent distance measurement, the range plus offset is referred to as a pseudorange. This unknown offset can be determined as part of the position determination process, thus effectively allowing the range to be determined from the TOA measurements. More details of these processes is given in Chap. 4.

A further complication with the use of direct-sequence spread-spectrum modulation (such as used for example by GPS) is the finite length of the associate pseudo-random code (or pn-code). For example, with GPS the pn-code length (and hence the rise-time) is about 1 ms, and the pn-code repeats at this period. Thus the TOA measurement is limited to a maximum of this period, which is equivalent to about 300 km. As this distance is much smaller than the distance to satellites, some other method (Alavi and Pahlavan 2003, 2006; Alsindi et al. 2009) is required to determine the range from the TOA measurement, which will be an integer times the pn-code period, plus the TOA measurement. However, for terrestrial systems the ranges are much shorter, so this range ambiguity problem can usually be avoided by the appropriate selection of the pn-code length.

2.3.5 *Link Budget*

A Link Budget provides a summary of the performance of a radio path between a transmitter and a receiver, and in particular lists the gains and losses along the path. From these data, the receiver output SNR can be estimated for a given link, but a signal attenuation model is also required to estimate the propagation losses at specified ranges. The link budget is thus an important tool in the design of radiolocation systems. The output from the Link Budget studies can be compared with the technical requirements specification, and if necessary the design is modified. Typically, this is an iterative process, with various design tradeoffs made to ensure all of the requirements are met.

The Link Budget calculations will require inputs from other subsystem models, including the transmitter modulation, antenna designs, propagation models, receiver signal processing performance modeling, data decoding and position determination. Many of these processes will require separate detailed analysis, either using existing models or developing new models where innovative research and design is necessary to meet the functional requirements. Only when the outputs from these various models are input into the Link Budget can the overall performance of the system be assessed. Once a satisfactory overall design is confirmed, the parameters and concepts from the modeling can be translated into the *Design Specifications* of the various subsystems, which can encompass the radio analog hardware, digital processing hardware, firmware and computer software. Thus the Link Budget is the key integrating tool for allowing the overall design to proceed.

The Link Budget design process will need to be performed many times, depending on the type of link and the various types of transmitting and receiving nodes. In a radiolocation system, both the range determining (such as via TOA or receiver signal strength) and data links need to be analyzed, including links between base stations, from base stations to mobile nodes and from mobile nodes to base stations. In the case of Mesh Networks where all the nodes are typically the same only one inter-node link type needs to be considered. Thus the design process can be quite involved, as all these Link Budgets must meet the required specifications if the system as a whole is to function satisfactorily. In the end the performance is defined by the "weakest link".

In the design of the overall system, it is typical that tradeoffs between the various subsystems are required; these tradeoffs can be facilitated by the use of the Link Budget. For example, for a specified range, the propagation model will define an allowable loss for a given range. As the propagation environment for a given application is fixed, this design loss has to be met by a combination of transmitter RF power, transmitter/receiver antenna gains, the radio receiver noise figure, and the receiver *Process Gain*. The Process Gain of a block is defined by the increased output SNR compared with the input SNR. Typically, the Process Gain is increased by integrating the input signal coherently over time. If the transmitter is in a mobile device, increasing the transmitter power will increase the available loss, but will also reduce the battery life, or require a larger battery which will increase the weight and physical size of the mobile unit; these changes may exceed the requirements in the functional specification. Alternatively, the available loss can be improved by increasing the receiver Process Gain, but this will reduce the receiver throughput, either the data rate or the number of positions per second that an be accommodated by the system in a TMDA system. Finally, the available loss can be improved by increasing the antenna gains, but this is usually only viable in the base stations due to size limitations in mobile devices. Thus in summary, some type of tradeoff may be required to obtain the best possible solution.

The type of Link Budget tradeoffs described above may result in no satisfactory solution given the propagation losses and the desired distances between nodes. For example, the number of fixed nodes required to give adequate coverage may be logistically or financially prohibitive. In such cases the Link Budget can be used to estimate the maximum range given the best compromise of the other parameters. This available loss then defines the operating range of the links, and hence the node spatial density required for satisfactory operation. While this may result in a satisfactory technical solution, the increased node density (particularly fixed base stations) will increase the cost of the system, and the required infrastructure may not be logistically feasible. Thus again the Link Budget output provides useful information about the overall design and deployment of a radiolocation system. Another consideration relating to the spatial node density is the different requirements for the ranging and data links from the mobile unit to base stations. In particular, for position determination at least three (and preferably five after allowing for some redundancy) range measurements are required. In contrast, only one data reception base station is required for data transmissions from the mobile. As the density of base stations is defined by the position determination requirement, there will be a redundancy of receiving base stations in the data communications from the mobile. This redundancy allows a higher data rate (lower Process Gain) to be used, with the statistical performance of data reception determined by the statistical variation in the propagation loss as specified by the Link Budget.

Thus in summary, the Link Budget is a design tool which allows the overall performance parameters of the system (and subsystems) to be determined, and iterative design changes made to optimize the performance and met the functional requirements. For further details on Link Budgets, refer to Chap. 9.

2.3.6 Time and Frequency Synchronization

For a radiolocation system based on using the propagation time of a radio signal as a proxy for distance measurements, it is clear that the prime measurement is associated with time, namely the time-of-flight (TOF) between the transmitter and the receiver. As a direct consequence, the accuracy of time measurements defines the accuracy of the range measurements between nodes, and hence the computed position. The "clock" in a node typically will be based on a (crystal) oscillator which is also used for other functions in the node in addition to TOF measurements, such as in generating the transmitter baseband modulated signals (such as DSSS), providing a reference source for the RF generator for the transmitter RF, and local oscillator frequencies associated with the radio "mixing" down-converting functions in the receiver. By locking all the node clock-related functions to the one clock, this one clock will automatically provide a common frequency reference. This design concept can also be exploited for controlling the node clock frequency through a suitable feedback mechanism based on the received radio frequency from another node. These functions are described in detail in Chap. 5.

The basic accuracy of (cheap) crystal oscillators is of the order of 10 ppm, which is equivalent to a range error of just 1 m in 100 km, or 1 cm/km. For most ground-based tracking systems this accuracy is more than adequate, but as will be shown other functions required for accuracy position determination need much more accurate clocks.

Consider a conceptual design for measuring the range between two nodes whose clocks are synchronised in time. The time of the transmission and the time of reception are recorded, with the differential time the TOF; data communications between the nodes can be used to share the recorded times in the nodes. The distance between the nodes can then be calculated assuming there is straight-line propagation. Thus accurate time synchronization is the key to range measurements.

Before considering radio-only ranging methods, it is useful to consider a TOF method which can be implemented with cheap crystal oscillators with minimal complexity. In particular, consider a dual propagation method (Nissanka et al. 2000), one using radio transmissions, and one using sound (ultrasonics) propagation. As the propagation of radiowaves is about a million times faster than sound propagation, the reception of the radio signal can be considered virtually instantaneous when compared with the audio signal. Thus by using the local clock for recording the TOA of the radio signal and the TOA of the audio signal, the range can be estimated directly from the differential time.[2] However, due to the nature of sound propagation, application of this concept is limited to LOS propagation and ranges up to about 10–20 m; thus the following discussion will be limited to radio-only solutions.

The key to measuring the TOF by radio propagation only this the synchronization of the clocks in the nodes. Because of the very fast speed of propagation of radiowaves (0.3 m/ns), this synchronization must be very accurate indeed. For example, a 1 ns error represents a 0.3 m range error. Further, a 1 ns synchronization error would occur after only in 0.1 ms of elapsed time if the differential clock

[2]Another interesting application of this concept is the informal determination of the range of a lightning strike. By measuring the time from the lightning to hearing the associated thunder, the range can be calculated. In particular, the range in kilometers is approximately the differential time in seconds divided by three.

frequencies were only accurate to 10 ppm, as for cheap crystal oscillators. Thus it is clear that some form of clock synchronization (or at least determining the differential drift in the clocks) is required for a radio-based TOF measurement.

Clock synchronization between stationary nodes is feasible by inverting the range measurement concept. In particular, if the locations of the stationary nodes are accurately known (by some independent surveying technique) such that the internodal ranges can be calculated, then TOA measurements provide a direct method of estimating the differential time and frequency between the clocks in the nodes. Such a method requires that the TOA measurements accurately represent the arrival time of the radio signal, and that the propagation path is a straight line. While such a technique is feasible for outdoor applications, multipath propagation makes such a technique problematical indoors. Further, while such a scheme may be feasible for stationary nodes, clock synchronization in a mobile node is very difficult, so that practical implementations typically will not require clock synchronization in mobile devices.

For classical navigation/tracking systems (such as GNSS), the implementation will include a Time Synchronization Subsystem (TSS), so that the clocks in the "fixed" nodes can be synchronized in time; this synchronization will also include the clock frequency and possibly the clock phase. However, time synchronization with mobile nodes is not performed. As a consequence, it can be shown that the TOA measured in the receiving nodes will include both the propagation delay between the transmitter and the receiver, as well as a time offset (unknown to the receiver) between the time in the synchronized nodes and the time in the mobile node. A further complication is that this offset time will also include the delays in both the transmitter and the receiver, which are also not known with any precision (it can vary from node to node). Thus after converting the propagation delay to an equivalent distance using the known propagation speed, the TOA is effectively a pseudorange, incorporating both the range between the transmitter and receiver and an unknown offset which is common to all the pseudorange measurements. To achieve this aim, the TSS must be carefully designed to appropriately account for the equipment delays in the various nodes, as well as achieving accurate time synchronization, often with large distances between the nodes. As a consequence, much of the design effort in a classical positioning system must be directed to establishing and maintaining accurate time synchronization throughout the network of fixed nodes. Any error in the time synchronization will be directly reflected in errors in the computed position.

Because of the complexity and difficulty of implementation of clock synchronization, an alternative approach which does not require full clock synchronization is more practical in some situations, and in particular in indoor tracking systems. The ranging method is based on measuring the round-trip-time (RTT) between two nodes by utilising the TOA measurements in both nodes. However, the method still requires frequency synchronization (rather than full time synchronization) for the method to be accurate, due to the clock drift effect described previously. However, unlike time synchronization, multipath effects do not affect the accuracy of the method, and thus this technique is particularly useful for indoor positioning systems. However, the measured RTT includes the delays through the equipment (which can be much greater than the propagation delays), so the RTT method also requires a method of accurately measuring these internal delays.

For the details of time a frequency synchronization, refer to Chap. 5. The calibration of internal delays in described in Chap. 11.

2.3.7 Data Communications Requirements

While the main focus of a positioning system is on the aspects associated with determining positions, data communications are required to allow the processes associated with the position determination to work. However, the data requirements in support of position determination are modest, so that only low bit-rate communications are necessary. Further, in some cases the devices will also include existing data communications such as WiFi or Bluetooth, so that the data functions can utilize this method for any necessary data communications in a network of nodes. However, even if such additional data communications infrastructure already exists, the smooth operation of the positioning functionality necessitates an independent data capability.

The signal protocol for a system typically is based on a TDMA structure required for the positioning function. Further, data encoding can be based on the same modulation scheme (such as spread-spectrum) used for TOA measurements, but with further encoding such as phase modulation of the pn-code symbol. The advantage of such a design is that the same basic signal processing can be used for the data transmission as well as the TOA measurement, so that the associated process gain allows the transmitter power to be kept low. Thus the correlation process which results in an output "pulse" can provide dual functions, namely the magnitude of the pulse is used for the TOA determination, and the phase of the pulse for data encoding. As the TOA function is designed to counter multipath effects such as signal fading, these benefits also flow through to the data transmission function.

The following provides the outline of a typical implementation. The basic spread-spectrum signal protocol defined above is organized into time slots, in order to implement the position location functions. These slots can also include a provision for data frames. There are four main functions that the system must perform:

1. The first type of transmission is the spread-spectrum signal transmitted by the mobile units, reception of which at the Base Stations allows the position of the mobile to be determined. Because of the timing accuracy requirements these transmissions should use a high chip rate consistent with the position accuracy requirements.
2. The second type of transmissions are data from the mobile units to the base stations. Because the requirements for the data transmission function are less stringent, lower chip-rate spread-spectrum transmissions can be used. This

would reduce the associated process gain, but as the data only needs to be received at one base station (as opposed to many base stations for TOA measurements), the lower process gain would not reduce the performance of the overall system. The lower chip rate also results in lower power consumption, increasing battery life.

3. The third type of transmissions are data from each base stations to a Master Station. A Master Station is the location where the positioning system is managed, and in a tracking system where the positions are determined from the data received from the base stations. These transmissions relay the mobile data (including from various sensors in the mobile unit) and the mobile tracking data (derived in the Base Station) to the Master Station. Because there can be a large number of mobile units being tracked, these data rates need to be much higher than the separate transmissions from each mobile. Thus a smaller length pn-code would be used for these transmissions, resulting in a lower process gain. However, the base stations will have higher gain antennas, transmitter power and larger battery capacity then the mobile units, so overall the base station to Master Station link should be more reliable than the mobile to base station link.
4. The fourth type of transmissions are those associated with the broadcast messages. These messages are transmitted (solely) by the Master Station to both the Base Stations and mobiles. The broadcast messages typically are used to transmit control commands. Because the broadcast messages are only transmitted by the Master Station, the transmission should be at a slow rate resulting in needing a much higher process gain than for all other types of data transmission on the system.

While the details of data packets may vary with particular requirements, the following is typical for a system. It can be observed most of the data are not associated with position determination. In particular, modern mobile units will include additional sensors, as well as data monitoring the performance of the mobile unit. In the following example the number of bits in the data frame is 200. These data bits are allocated as follows:

1.	Mobile ID	16 bits
2.	3-axis accelerometer	36 bits (3 axes × 12 bits). 1 bit = 1 mg
3.	3-axis magnetometer	36 bits (3 axes × 12 bits). 1 bit = 0.5 mG
4.	Vertical axis rate-gyro	12 bits. 1 bit = 0.01 degrees/second
5.	Mobile receiver power	8 bits. 1 bit = 0.5 dB
6.	Mobile SNR	8 bits. 1 bit = 0.2 dB, range 0–51 dB
7.	VCXO voltage	8 bits. Used as a check for ageing of the crystal
8.	Block CRC16	16 bits. The cyclical redundancy check
9.	Spare	60 bits. Available for other functions
Total		200 bits

2.3.8 Position Determination

The key and final important subsystem in a radiolocation system is the determination of the position of the mobile node, based on either range or pseudorange data. Much of the literature focuses on just this process, but in the design of a system the position determination is usually rather trivial. Typically, the position determination can be implemented in just a few pages of code, particularly for outdoor systems. The position determination indoors is more complex, due to the measurement errors largely associated with the NLOS multipath environment. As the topic of position determination is much covered in the literature and in books (Yu et al. 2009), this book only has a limited coverage of position determination not covered elsewhere, with particular emphasis on the indoor case (see Chap. 6 for TOA positioning, and Chap. 8 for WiFi positioning). Other methods associated with position determination using sensors such as accelerometers and rate-gyroscopes are described in Chap. 7, and new methods for receiver signal strength positioning in Chap. 15.

The broad basis of position determination can be best understood by simple geometric interpretations. In the 2D range case, the position can be found by the intersection of circles centered at each receiving node and a radius as measured by the radio hardware and signal processing subsystems. For 3D positioning the circles can be generalized to spheres. For the pseudorange case the position is found at the intersections of hyperbolas; such systems are sometimes called hyperbolic positioning systems. In both cases at least three measurements are required to determine a position fix without ambiguity. For both the range and pseudorange cases without measurement errors analytical solutions are available with little computational difficulty. Also observe that the position determination process is independent of the underlying technology, so that for example it could apply to ultrasonics-based ranging, or radio methods using signal strength for range determination.

While position determination with no measurement errors presents little difficulty, the situation is far more complex when there are measurement errors; this is especially true for indoor NLOS systems. Measurement errors will include both random and systematic biases. In such cases the key to improving the determined position is redundancy in the measured data, that is more than three range/ pseudorange measurements are required. Such methods are described in detail elsewhere (Sathyan et al. 2011). However, these methods can be computationally intensive and may not be suitable for real-time systems. A common feature of typical methods is the assumption of straight-line radiowave propagation, which is not true for indoor NLOS environments; this difficult particular case is the focus of the methods described in Chap. 6. In particular, improved accuracy can be obtained if a more complex propagation model is adopted which does not necessarily assume straight-line propagation. As each indoor environment is unique and can vary over time, the only practical solution is for a subsystem to continually determine a model for the systematic errors throughout the coverage area. The position determination process thus can partially correct bias errors in the raw measured range/pseudorange data, leaving only random errors with near zero biases. Thus this process can convert the raw range/pseudorange errors to an equivalent random error straight-line case, allowing standard position determination methods to be used. By this process the performance indoors can approach that obtained for an outdoors system with the same standard deviation of the random errors.

While the system design may nominally result in either range or pseudorange measurements, often the range measurements are in fact pseudoranges. The reason for this is that range measurements will include some uncertainty at the mobile end of the transmitter-receiver link in essentially all technological solutions. While the "fixed" structure can have well defined and calibrated hardware parameters (such as internal delays for a TOA-based system, or the effective radiated power in a signal-strength based system), the mobile device usually is not calibrated and thus there is some uncertainty in the mobile device parameters required for range measurements. For a pseudorange measurements these uncertainties are incorporated into the unknown pseudorange offset (common to all measurements as the mobile node is the common factor in all range measurements). Thus while the measurements are nominally ranges, better more accurate positions usually can be obtained if range measurements are treated pseudoranges.

From the above description, and the discussions in previous subsections, it is clear that the position determination process is a complex interaction of many other processes and subsystems, with interactions between the subsystems and feedback loops. These processes are summarized in Fig. 2.1. The design and operation of these subsystems and their interactions is the main topics of subsequent chapters.

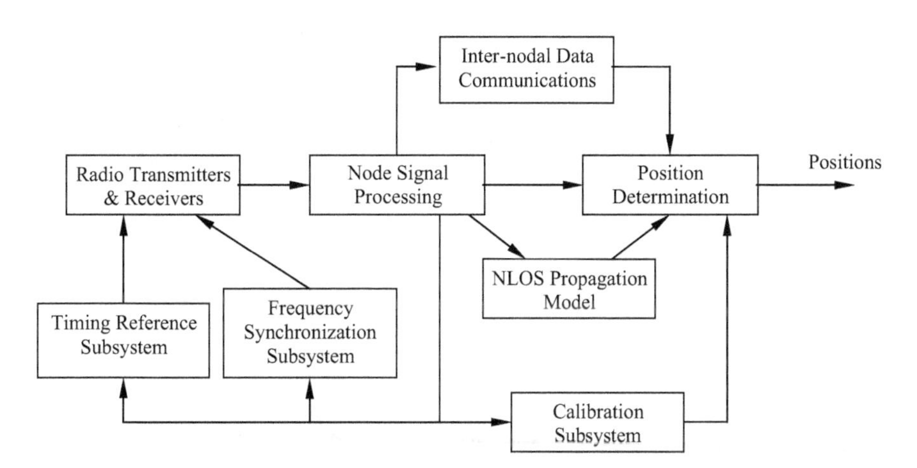

Fig. 2.1 Summary block diagram of the position determination processes and their interactions

2.3.9 *Auxiliary Sensors Requirements*

The main emphasis in the design of a radiolocation system is the method of determining positions based on some property of a radio signal. However, modern devices usually include other sensors which can be used in applications. These sensors can be related directly to the radio system itself, or other auxiliary sensors. For the radio the sensors typically could include measurement of the receiver signal strength (RSS) and possibly the receiver signal-to-noise ratio (SNR). Of particular interest to positioning are inertial sensors and the receiver signal strength indicator (RSSI), as these sensors can provide data that can be used for position determination, either as a standalone function, or to assist the radiolocation system. Inertial sensors can be used in *dead reckoning* position determination, either by integration of accelerometer and rate gyroscope data, or by detecting the gait of a person walking. Such positioning methods provide good short-term accuracy, and can be used to enhance the radiolocation positioning. See Chap. 7 for details on sensor positioning.

In addition to sensors associated with the radio other sensors that can be utilized for position determination. These additional non-radio sensors could include, but not be limited to, inertial sensors (accelerometer and rate-gyroscope), and a compass (magnetometers). The inertial and compass sensors are available as small 3-axis solid-state packages.

Other sensors commonly available that are not associated directly with position determination include temperature sensors and battery monitoring (voltage, state of charge). The temperature sensor would normally be used to check if the temperature is within the operational range of the equipment. Another possible function is to provide an input to a temperature compensation feedback loop to correct for the frequency of the local oscillator. This is important in ensuring accurate radio frequencies and for precision time measurements required for accurate TOA-based range measurements.

Possible additional sensors (or interfaces to them) are related to biological applications such as a heart rate monitoring. Further, generic interfaces such as Bluetooth or USB can provide interfaces to other external devices, which may also be included in the design of the mobile device.

2.4 Tracking Systems Design

The design of a complete tracking system is a complex undertaking, as it involves many aspects which need to be coordinated into a coherent whole. For a terrestrial system, which is the focus of consideration in this section, the components include both the mobile and fixed (base station) components. Further, the design will incorporate a wide range of topics and technologies, including radio receivers and transmitters, analog baseband circuits associated with signal modulation and demodulation, digital signal processing (both in hardware such as field programmable gate arrays (FPGA) and digital signal processing (DSP) firmware), computer software associated with the control and management functions, and user interfaces which may include both personal computers (PCs) and mobile devices such a mobile/cellular phones and Tablet devices. This task is further complicated as the design may vary considerably depending on the particular requirements detailed in the Functional Specification. Thus while much of the technology will be common to all radiolocation systems, allowing use of existing components and software, some requirements many necessitate the development of new methods or taking advantage of advancements in technology. Thus before the system and detailed design can commence considerable prior effort is required in studies and performance analysis. Only when these preliminary steps are completed, can a detailed design commence.

The first step in any design will typically commence with a Link Budget analysis—see Chap. 9 for details of this process. The Link Budget will establish some fundamental system parameters, such as the operating radio frequency, signal bandwidth, broad characteristics of the signal protocol and the associated signal processing required, transmitter power, and the expected coverage area of a base station. The examples of this process given in Chap. 9 shows that this Link Budget analysis can result in quite different designs, even when the original functional specifications are quite different. Such an analysis may also prompt further research if no feasible solution is apparent using existing technology. Further, even if the technology is available the cost of the system many be prohibitive, which may foster further research and development of new methods, or using existing technology in a new innovative way. Only when the Link Budget analysis indicates a (at least potentially) viable solution should the detailed design commence. Note that the design and development phase should not involve any research to solve the project's functional requirements, that is the engineering development should not involve scientific research—this should occur before the design phase. Thus the purpose of the system design is to layout a development plan which is largely risk free.

Failure to perform the necessary pre-design phase tasks can often result in increased development time, or even failure of the whole project. To avoid this outcome, it is sometimes necessary to first develop an adaptation of an existing system for testing the proposed concept; these are referred to as Proof of Concept (POC) systems, and often are based on existing technology, but with some enhancements and refinements.

For example, the POC testing could verify that the number of Base Stations required for satisfactory operation is typically no more than a practical limit, so the designers can be confident that the proposed design will cover the operational area with the specified number of base stations. Because of the complexity of a multipath environment, specifying the required signal bandwidth can be difficult. The design should use the minimum possible bandwidth, and no more, as the cost of the equipment increases rapidly with bandwidth. A POC system may initially use a spread-spectrum chip rate of 10 MHz, but testing finds this to be inadequate for multipath rejection, so the bandwidth is increased so that the tracking chip rate is increased to 30 MHz for further testing. Such system testing is thus an important aspect of any development—see Chap. 11 for examples of such testing.

Further testing with the POC system could show that inertial sensors are necessary for consistently accurate position fixing, but that full three-axis sensors are not required, so the design can be restricted to longitudinal and lateral accelerometers and magnetometers and a vertical axis rate-gyro. Because fewer sensors are on the mobile, the quantity of data transmitted from the mobile is reduced, somewhat easing the design of the data transmission link (see Sect. 2.3.7).

Once the overall design is established, perhaps with POC testing, the high level design for the various subsystems can commence. While the subsystems may vary between projects, the typical design would include components associated with the radio transceiver, the digital signal processing, and the firmware and software. These design requirements are intended to guide the detailed design of the subsystems, but not constrain their detailed design. In particular, the interface between subsystems needs to be specified in some detail, so that the complete design will come together with minimal difficult. This approach recognizes that in such complex projects different teams of developers will be involved with each subsystem, and their detailed design is guided by the system design, but importantly outside these constraints the subsystem designs will be the responsibility of each team of specialist. As these subsystem teams would be involved with any prior POC and computer simulation testing, it should be the case that the detailed design can proceed on the basis of the subsystem high level design specifications. Clearly the overall process would involve some of the same concepts outlined for the overall system design, but at a more detailed level, but by this stage of the development there should be little risk in developing each subsystem.

2.5 Mesh Network Design

In Sect. 2.4 the general design principles of designing a positioning system are discussed. This section goes in more detail of an actual design, which provides an illustrative example which could be applied to a wider array of such development projects. The particular example shows the design process of an innovative positioning system design which was intended to achieve high accuracy (better than 1 m) inside buildings, but uses relatively cheap hardware components which were originally developed and used for purposes other than radiolocation. The design is also based on the concept of a *Mesh Network*, where there is no overall distinction between mobile devices and base stations, and the communications between devices provides both data and positioning services.

The basic design concept for this Mesh Network system is that the hardware is largely based on existing off-the shelf components that were developed for other purposes, such as WiFi (IEEE 802.11 technology). Further the design is intended to operated indoors with a positional accuracy of better than 1 m, but with a minimal number of fixed base stations. Such a design goal means that a mesh network of modules should interact without any centralized control. While typical radiolocation designs (like GPS) require accurate time synchronization (such as in the GPS satellites), this ad hoc network is intended to operate with no such accurate clocks, yet achieve an accuracy at least 10 times greater than GPS. The following provides a summary of the design of such a system. This technology called Wireless Ad hoc System for Positioning (WASP) (Sathyan et al. 2011) has been developed through to commercially available products. However, the following only describes a summary of the initial proof of concept system, and provides a template for other similar developments. The following summarizes the main functional components developed especially for this project:

1. The basis of the POC system was to test the idea that a wideband radio system can be synthesized from a number of narrowband measurements, thus obtaining the ranging accuracy of a wideband system while simultaneously obtaining the benefits of the simpler hardware associated with a narrowband radio and lower clock rates in the digital hardware. In particular, the design is based on a chip radio used for WiFi hardware, where the channel bandwidth is 20 MHz. However, to achieve the required positional accuracy simulations showed that a signal bandwidth of about 150 MHz is required. To minimize costs and development time, the synthesis method was adopted for the POC system. For more details on this concept, refer to Chap. 12.
2. Thus the processing of the sub-channel data is the key process in determining the viability of the concept. The sub-channel processing consists of a number of components, from the initial acquisition to the reconstruction of the wideband spectrum. The basic idea is to split a wideband signal into a number of (overlapping) sub-bands, and to sequentially transmit each sub-band data which are received and reconstructed in the receiver. The key for this process is the requirement to correct for the unknown differential phase between each sub-channel transmission, as it is essential that the reconstructed wideband signal is coherent across the band. This new technique needs to be tested in severe multipath environments.
3. The transceiver used is a chip radio operating in the 5.8 GHz band, originally developed for WiFi applications. While this results in a cheap and simple implementation, the chip radio was not designed for accurate (sub-nanosecond) time-of-arrival measurements, thus requiring additional calibration of its operational characteristics.
4. The design assumes each device has an independent, unsynchronized clock. To determine the range between two nodes, the round-trip time (RTT) is measured. The first receiving unit always transmits on its local clock "tick", and second unit always transmits its reply in its clock "tick". The time of the returned signal at the first unit is measured relative to its local clock, allowing the range to be estimated after compensating for the equipment delays.
5. A classical RTT design requires the response to a received signal to be transmitted immediately (or after a known small delay), so that the range can be estimated from the RTT minus any equipment delays. In the mesh network case

with N nodes there are $N(N-1)/2$ pairs, so that if each transmission uses one time slot the number of slots required becomes very large as the number of nodes becomes even modestly large. In practice this would mean a low update rate of range determination, which is not practical for many applications. Thus the proposed design does not use the classical design, but each node merely transmits in its allocated time slot (one of N), but includes the TOA of any received signal from other nodes. Thus the RTT includes the comparatively large time delays associate with multiple slot delays. As the clocks in the nodes are not synchronized this results in large errors in the RTT using the method described in paragraph (4) above, due to the differential clock frequencies. However, by measuring the drift in the RTT data the frequency drift between the two units can be measured, and transmitted in inter-nodal messages, allowing ranges between nodes to be determined. However, the accuracy of such a method needs to be tested.

The above example provides an illustration of a development process where an innovative design is suggested, driven by requirements of performance and commercial considerations such as development and product costs. The concepts outlined in this chapter, and expanded in subsequent chapters in this book, provide a foundation for the development of a wide range of practical positioning systems.

References

Abramson N (1970) The ALOHA system—another alternative for computer communications. In: Proceedings of the 1970 fall joint computer conference. AFIPS Press

Alavi B, Pahlavan K (2003) Modeling of the distance error for indoor geolocation. In: Proceedings of the IEEE wireless communications and networking, pp 668–672, March 2003

Alavi B, Pahlavan K (2006) Modeling of the distance measure error using UWB indoor radio measurement. IEEE Commun Lett 10(4):275–277

Alsindi N, Alavi B, Pahlavan K (2009) Measurement and modelling of ultra wideband TOA-based ranging in indoor multipath environments. IEEE Trans Veh Technol 58(3):1046–1058

FCC (2013) Federal Communications Commission, Code of Federal Regulations, Part 15—Radio frequency devices. Subpart C—Intentional radiators. Section 15.247, Operation within the bands 902–928 MHz, 2400–2483.5 MHz, and 5725–5850 MHz

Fontana R (2004) Recent system applications of short-pulse ultra-wideband (UWB) technology. IEEE Trans Microw Theory Technol 52(9):2087–2104

Fontana R, Richley E, Barney J (2003) Commercialization of an ultra wideband precision asset location system. In: Proceedings of the IEEE conference on UWB systems and technologies, pp 369–373

Johnson N, Kotz S, Balakrishnan N (1994) Lognormal distributions, continuous univariate distributions. In: Wiley series in probability and mathematical statistics: applied probability and statistics, vol 1. Wiley

Nissanka B, Anit C, Hari B (2000) The cricket location-support system. In: Proceedings of the 6th annual international conference on mobile computing and networking (ACM MOBICOM), Boston, MA, USA, pp 32–43, 6–11 August 2000

Sathyan T, Humphrey D, Hedley M (2011) WASP: a system and algorithms for accurate radio localization using low-cost hardware. IEEE Trans Syst Man Cybern Part C 41(2):211–222

Sharp I, Yu K (2013) Enhanced least squares positioning algorithm for indoor positioning. IEEE Trans Mob Comput 12(8):1640–1650
Sharp I, Yu K (2016) Improved indoor range measurements at various signal bandwidths. IEEE Trans Mob Comput 65(6):1364–1373
Yu K, Montillet J, Rabbachin A, Cheong P, Oppermann I (2006) UWB location and tracking for wireless embedded networks. Signal Process 86(9):2153–2171
Yu K, Sharp I, Guo YJ (2009) Ground-based wireless positioning. Wiley-IEEE Press

Chapter 3
Signaling Techniques

3.1 Introduction

Radio positioning systems rely on transmission and reception of radio-frequency signals for target position determination. A specific system makes use of one or more signal parameters, including the time-of-arrival (TOA), the time-difference-of-arrival (TDOA), the received signal strength (RSS), the received power profile, the angle-of-arrival (AOA), and the received signal phase. To achieve efficient signal transmission and/or reception, careful signal design must be performed, which is vital in the design of positioning systems. The choice of signaling methods will greatly affect the performance of positioning systems. When designing signals, a number of issues needs to be considered, including positioning accuracy, power consumption, frequency bandwidth, interference, cost, and policy compliance. This chapter focuses on some of the main aspects of radio signaling concepts and techniques. Specifically, a number of pulse shapes and their spectrum are first studied, which are important particularly when designing ultra-wideband (UWB) positioning systems. Then spectrum spreading techniques are discussed, which are employed by many existing radio positioning systems including global navigation satellite systems (GNSS). Next, modulation techniques for positioning are presented with a focus on UWB modulation techniques, and discussions on single-carrier and multi-carrier options, single-input signal-output (SISO) and multi-input multi-output (MIMO) schemes. Finally, radar and frequency modulated continuous wave (FMCW) technology are briefly studied.

3.2 Pulse Shapes and Energy Spectrum

In wireless communications and positioning, the radio-frequency (RF) signal travels between the transmitter and the receiver through a wireless channel which is band-limited. The signal bandwidth increases as the modulation rate increases.

I. Sharp and K. Yu, *Wireless Positioning: Principles and Practice*, Navigation: Science and Technology, https://doi.org/10.1007/978-981-10-8791-2_3

When the signal bandwidth is larger than the channel bandwidth, the channel will introduce distortion to the signal. To avoid or reduce such distortions, pulse shaping is typically applied to adjust the signal's energy spectrum. Pulse shaping is also essential for systems which operate on frequency bands that governments apply restrictions on, such as UWB systems. This section presents a number of commonly used pulse shapes and their energy spectrums.

3.2.1 Ultra-wideband Spectrum Masks

An emitted signal of either impulse radio or continuous-wave radio is an UWB signal if its instantaneous bandwidth is at least 500 MHz or its fractional bandwidth is at least 20%. UWB signals have a number of desirable characteristics, including low power transmission, very low spectral density, and fine time resolution. Therefore UWB signals tend to be immune to multipath fading and malicious interception. In particular, UWB technology is suited for indoor or short-range positioning, and can achieve centimeter accuracy.

The FCC (Federal Communications Commission) of the United States released rules to restrict the transmission power for UWB equipment to a very low level to avoid interference to other systems that occupy the same or neighboring bands of frequencies such as the global positioning system (GPS), WiMAX systems, wireless LAN, and ground penetrating radar. In particular, specific UWB power emission masks are defined by FCC, ETSI (European Telecommunications Standard Institute) and other organizations. For instance, the FCC-defined UWB emission masks are illustrated in Fig. 3.1 for outdoor hand-held systems and Fig. 3.2 for indoor systems (FCC 2002).

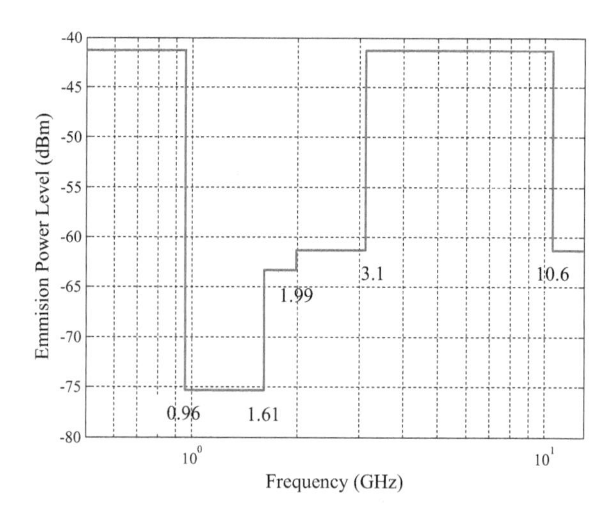

Fig. 3.1 FCC UWB spectrum mask for outdoor hand-held systems

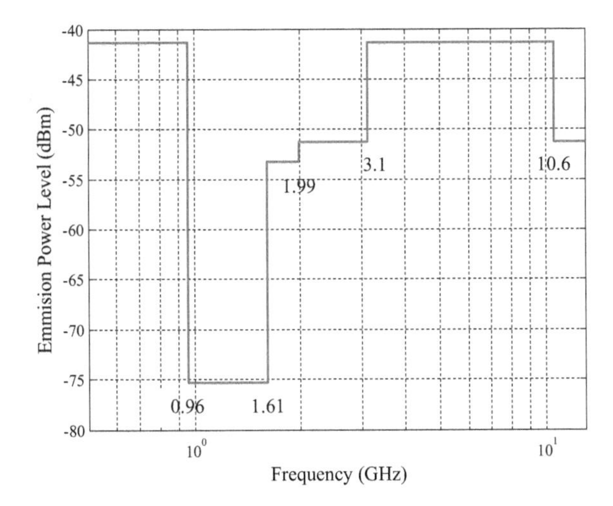

Fig. 3.2 FCC UWB spectrum mask for indoor systems

3.2.2 Rectangular Pulse

Rectangular pulse shaping has commonly been used in narrowband and wideband positioning systems which are based on direct-sequence spread-spectrum (DSSS) technology. The pulse shapes of a single rectangular pulse in time domain and in frequency domain are illustrated in Fig. 3.3. In frequency domain the pulse shape is a sinc function.

The energy spectral density (ESD) of a signal is defined as the square of the magnitude of the continuous Fourier transform of the signal. The ESD of a single rectangular pulse is the squared sinc function. In practice, a low-pass filter is typically employed to adjust the ESD and to restrict the bandwidth of the rectangular pulses.

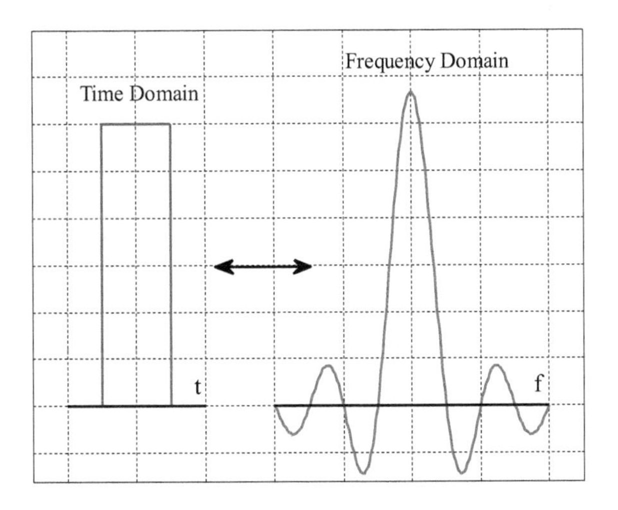

Fig. 3.3 Pulse shapes of a rectangular pulse in time domain and in frequency domain

At the receiver the received signal waveform is correlated with a template waveform generated in the receiver. The output of the correlator when correlating the received rectangular pulse and its shifted version is an approximate triangular pulse waveform. By detecting the leading edge or the peak of the triangular correlation diagram, the signal arrival time can be determined. More details about correlator based signal detection can be found in Chap. 4 of this book and Chap. 3 of Yu et al. (2009).

3.2.3 Gaussian Pulse and Its Derivatives

In the study and practice of UWB communications Gaussian pulse and its derivatives are commonly used pulse shapes. Mathematically, a normalized Gaussian pulse may be described by

$$g_0(t) = \exp\left(-\frac{t^2}{\sigma^2}\right) \tag{3.1}$$

where the pulse duration is determined by σ. The monocycle pulse can be generated as the first derivative of the Gaussian pulses, which is given by

$$g_1(t) = -\frac{2t}{\sigma^2}\exp\left(-\frac{t^2}{\sigma^2}\right) \tag{3.2}$$

Alternatively, the monocycle pulse can also be generated as the second derivative of the Gaussian pulse

$$g_2(t) = -\frac{2}{\sigma^2}\left(1-\frac{2t^2}{\sigma^2}\right)\exp\left(-\frac{t^2}{\sigma^2}\right) \tag{3.3}$$

Figure 3.4 illustrates the Gaussian pulse and its first and second derivatives. Clearly, the first derivative based monocycle pulse consists of two pulses that have a phase difference of 180°, whereas the second derivative based monocycle has one main pulse and two side pulses.

It can be shown that the Fourier transform of the Gaussian pulse is given by

$$G_0(\omega) = \sqrt{\pi}\sigma\exp\left(-\frac{(\sigma\omega)^2}{4}\right) \tag{3.4}$$

Also, the Fourier transform of the first and second derivative based monocycle pulse is respectively given by

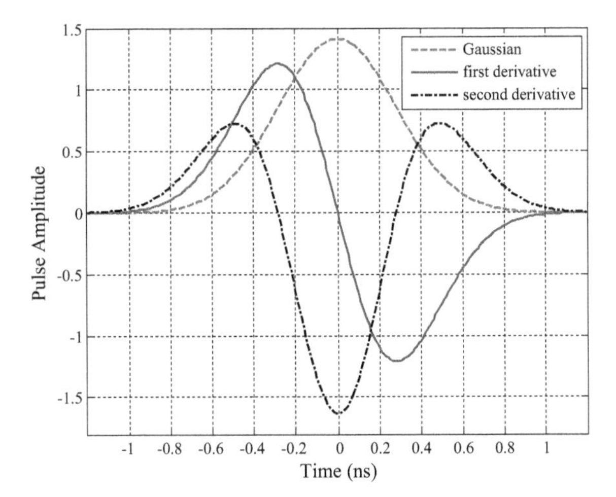

Fig. 3.4 Gaussian pulse and its derivatives

$$G_1(\omega) = j\omega G_0(\omega) = j\sqrt{\pi}\sigma\omega \exp\left(-\frac{(\sigma\omega)^2}{4}\right) \tag{3.5}$$

and

$$G_2(\omega) = (j\omega)^2 G_0(\omega) = -\sqrt{\pi}\sigma\omega^2 \exp\left(-\frac{(\sigma\omega)^2}{4}\right) \tag{3.6}$$

Figure 3.5 shows the normalized energy spectrum of the Gaussian pulse and its first and second derivative when $\sigma = 6.6$ ns. For comparison, the FCC spectrum mask for indoor systems is also included. Note that the emission level of the spectrum mask is the emitted signal power at the antenna in dBm. It is seen that the bandwidth of the Gaussian pulse is the largest, whereas the bandwidth of the second derivative is the smallest. It is clear that the Gaussian pulse and its derivatives are not satisfactory with respect to the rectangular emission masks. That is, they are not suited for direct transmission unless specific modulation or further pulse shaping is employed to comply with the emission masks.

3.2.4 Raised Cosine Pulse

The time-domain expression of the raised cosine pulse is given by

$$x(t) = \frac{\sin(\pi t/T)}{\pi t/T} \frac{\cos(\pi\beta t/T)}{1 - (2\beta t/T)^2} \tag{3.7}$$

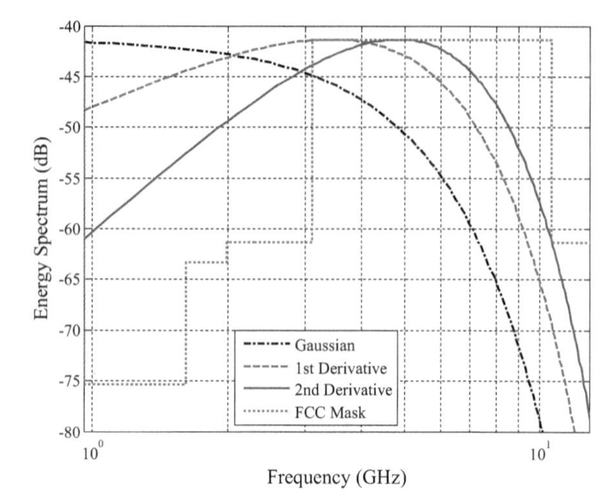

Fig. 3.5 Energy spectrum of single Gaussian pulse and its derivatives

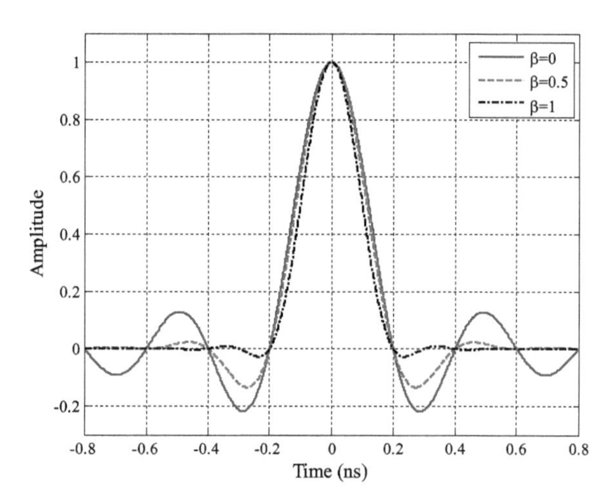

Fig. 3.6 Raised cosine pulse shapes with three different roll-off factors

where T is the pulse duration and β is the roll-off factor. Figure 3.6 shows the shape of the raised cosine pulse in time-domain when T is set at 0.2 ns.

In frequency domain the raised cosine pulse can be described by

$$X(f) = \begin{cases} T, & |f| \leq \frac{1-\beta}{2T} \\ \frac{T}{2}\left(1 + \cos\left(\frac{\pi T}{\beta}\left(|f| - \frac{1-\beta}{2T}\right)\right)\right), & \frac{1-\beta}{2T} < |f| \leq \frac{1+\beta}{2T} \\ 0, & |f| > \frac{1+\beta}{2T} \end{cases} \tag{3.8}$$

The energy spectrum of the raised cosine pulse is shown in Fig. 3.7 when T is set at 0.2 ns.

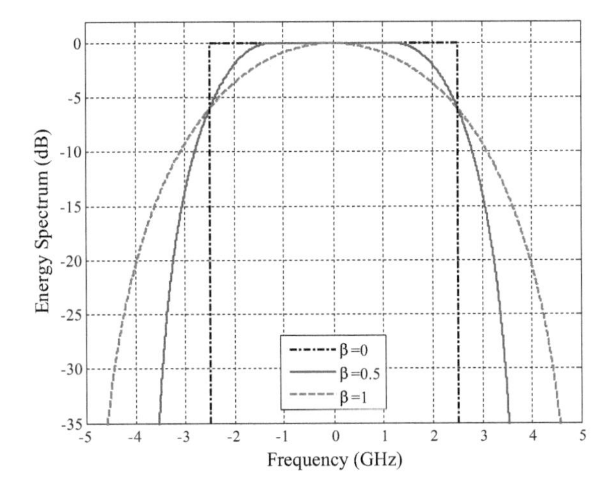

Fig. 3.7 Energy spectrum of raised cosine pulse

It is clear that the raised cosine pulse is a baseband signal. To generate an impulse-radio UWB signal, such a baseband signal needs to be up-converted. That is, the raised cosine pulse can be used to modulate a sinusoidal signal, resulting in

$$y(t) = x(t)\cos(2\pi f_0 t) = \frac{\sin(\pi t/T)}{\pi t/T}\frac{\cos(\pi\beta t/T)}{1-(2\beta t/T)^2}\cos(2\pi f_0 t) \tag{3.9}$$

where the frequency f_0 may be set as the central frequency over the UWB band (3.1–10.6 GHz). Figure 3.8 shows the time-domain pulse shape of the pulse when $f_0 = 6.85$ GHz, $T = 0.2$ ns, and $\beta = 0.5$.

The frequency-domain signal is given by

$$Y(f) = \frac{1}{2}(X(f-f_0)+X(f+f_0)) \tag{3.10}$$

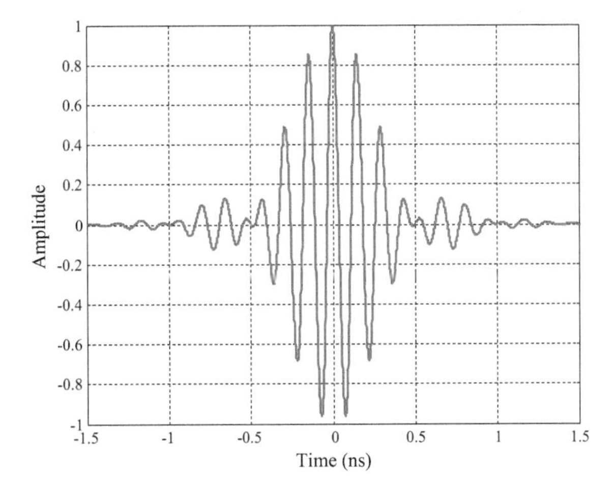

Fig. 3.8 Pulse shape of raised cosine function modulated sinusoidal signal

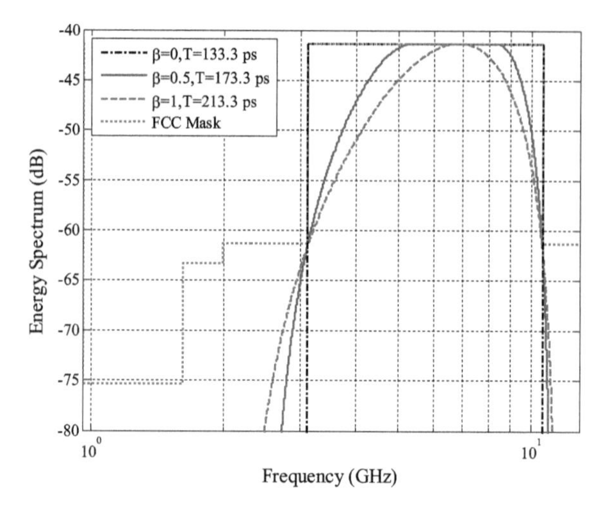

Fig. 3.9 Energy spectrum of raised cosine pulse modulated by a sinusoidal signal

That is, the energy spectrum of the modulated sinusoidal signal is the shifted energy spectrum of the raised cosine pulse, as shown in Fig. 3.9. Clearly, the raised cosine pulse modulated by a sinusoidal signal has s spectral density that has a good match with the required spectrum mask.

3.2.5 *Gaussian Modulated Sinusoidal Pulse*

Although the spectrum of the raised cosine pulse has a good match with the required rectangular spectrum masks, it is difficult to generate for commercial systems. An alternative pulse is the Gaussian modulated sinusoidal pulse which, in time-domain, is represented as

$$y(t) = \exp\left(-\left(\frac{t}{\sigma}\right)^2\right)\cos(2\pi f_0 t) \tag{3.11}$$

In the frequency domain, the signal is described as

$$Y(\omega_0) = \frac{1}{2}\sqrt{\pi}\sigma\left(\exp\left(-\frac{(\sigma(\omega-\omega_0))^2}{4}\right) + \exp\left(-\frac{(\sigma(\omega+\omega_0))^2}{4}\right)\right) \tag{3.12}$$

Figure 3.10 shows the Gaussian modulated sinusoidal pulse shape when $\sigma = 0.13$ ns, while Fig. 3.11 shows the energy spectral density of the signal. Comparatively, the energy spectrum of the Gaussian pulse spreads wider, whereas the energy spectrum of the raised cosine pulse as shown in Fig. 3.9 matches a rectangular mask better in the UWB band. Outside the UWB band, the Gaussian modulated sinusoidal signal has a better match with the mask than the raised cosine

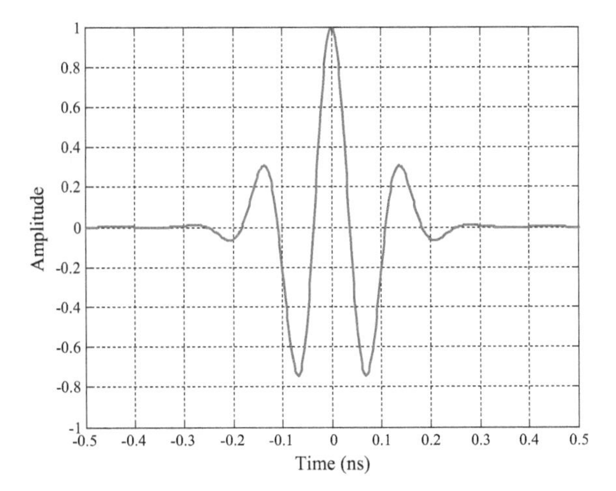

Fig. 3.10 Gaussian modulated sinusoidal pulse

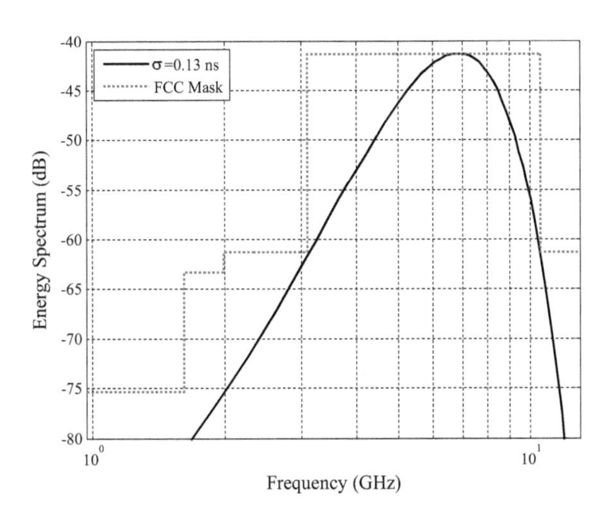

Fig. 3.11 Energy spectrum of the Gaussian modulated sinusoidal pulse

pulse. Because of easy implementation, a Gaussian modulated sinusoidal pulse is a better option in practice.

3.2.6 Chirp Waveform

A chirp, also termed a sweep signal, is a sinusoidal signal whose frequency increases or decreases with time. It is commonly used in many applications like sonar and radar. Chirping techniques are also used in radio positioning, including UWB based systems. Signal bandwidth is limited due to complexity constrains of equipment hardware; thus there is a limit in range resolution that can be achieved in practice. By using a chirp waveform, the signal bandwidth is substantially compressed without reducing the waveform repetition rate, so that digital signal processing can be performed at more practical sample rates to reduce hardware complexity. Equivalently, a larger bandwidth signal can be designed for transmission and reception and therefore improvement in range resolution can be achieved.

Chirp waveforms can be generated by modulating the frequency of a sinusoidal signal, and is described by

$$x(t) = A\sin\left(2\pi\int_0^t f(\tau)d\tau\right) \tag{3.13}$$

where $f(t)$ is the instantaneous frequency of the signal. If the instantaneous frequency is a linear function of time, as given by

$$f(t) = f_0 + \kappa t \tag{3.14}$$

where f_0 is the initial frequency at $t = 0$, and κ is the frequency slope or the chirp rate, then a linear chirp is produced as

$$x(t) = A\sin\left(2\pi\left(f_0 + \frac{\kappa}{2}t\right)t\right) \tag{3.15}$$

Figure 3.12 illustrates the amplitude and frequency of a linear chirp waveform with length T with respect to time.

The Fourier transform of the linear chirp waveform with length T is then given by

$$\mathcal{F}[x(t)] = \int_0^T \sin(\omega_0 t + \pi\kappa t^2)\exp(-j\omega t)dt \tag{3.16}$$

There is no closed-form expression for the above integral and thus numerical integration would be required to calculate the spectrum. Figure 3.13 illustrates the

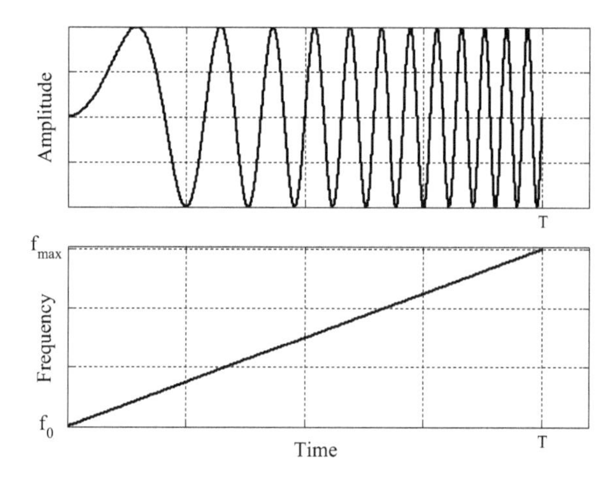

Fig. 3.12 Linear chirp signal amplitude and frequency versus time

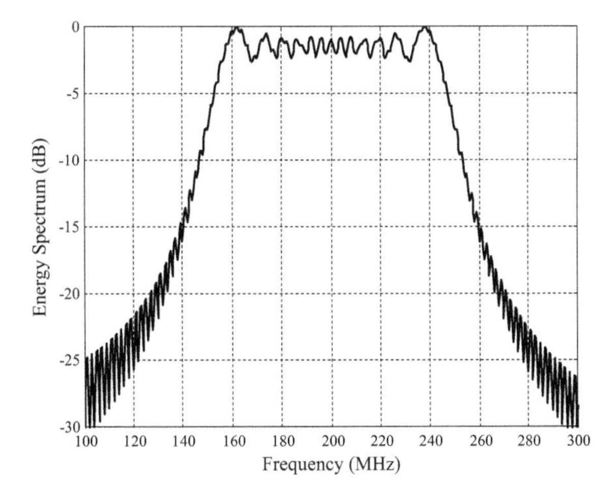

Fig. 3.13 Energy spectrum of a linear chirp signal

energy spectrum of a linear chirp signal which has a bandwidth of 100 MHz. The initial frequency (f_0) is set at 150 MHz, the chirp length (T) is set at 0.5 μs, and the chirp rate (κ) is set at 200 MHz μs^{-1}. It is worth noting that through properly selecting the parameters (chirp rate, frequency f_0 or ω_0, and the pulse length T), a UWB chirp waveform can be generated. However, it is a non-trivial task to design these parameters to efficiently utilize the masked spectrum. Interested readers may do some exercise to design a UWB chirp so that its spectrum has a good match with the required emission masks.

There are other chirp waveforms such as the exponential chirp waveform which is one of the nonlinear chirp waveforms. In an exponential chirp, the signal frequency is an exponential function of time, that is

$$f(t) = f_0 \kappa^t \tag{3.17}$$

Then the modulated signal becomes

$$x(t) = A \sin\left(\frac{2\pi f_0}{\ln(\kappa)}(\kappa^t - 1)\right) \tag{3.18}$$

which by using $\kappa^t = \exp(t \ln \kappa)$, can also be written as

$$x(t) = A \sin\left(2\pi f_0 \tau \left(\exp\left(\frac{t}{\tau}\right) - 1\right)\right) \tag{3.19}$$

Figure 3.14 illustrates the instantaneous frequency and amplitude of the exponential chirp signal, whereas Fig. 3.15 shows an example of the energy spectrum of the exponential chirp. In generating the spectrum, the initial frequency is set at 600 MHz, the chirp length is set at 0.1 μs, and the parameter τ is set at 0.278 μs. Compared with the linear chirp, the exponential chirp is more difficult to generate;

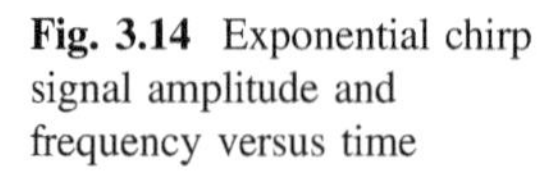

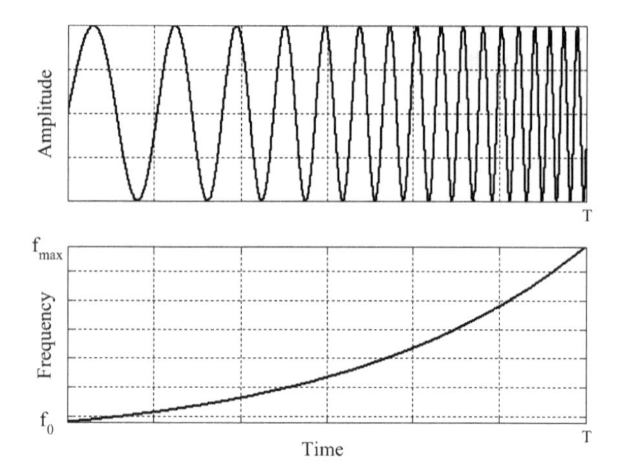

Fig. 3.14 Exponential chirp signal amplitude and frequency versus time

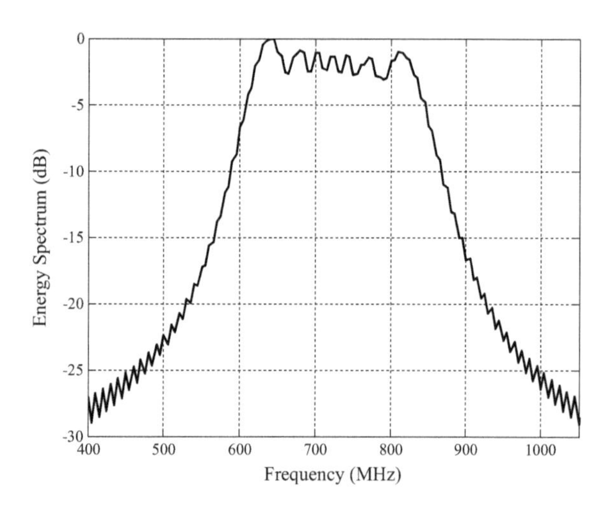

Fig. 3.15 Energy spectrum of an exponential chirp signal

however, the exponential chirp is less sensitive to the effect of Doppler shift. One more chirp waveform, namely the triangular waveform will be dealt with in Sect. 4 when the FMCW radar is studied.

3.3 Modulation Techniques for Positioning

Modulation and demodulation are important for wireless communications as well as for positioning. Positioning system designers need to choose appropriate modulation techniques suited to specific application scenarios to achieve the required objectives such as minimal power consumption, maximal interference resilience, and good accuracy. In this section, five typical UWB modulation techniques are briefly studied since UWB technology is one of the key technologies which are used for short-range positioning. Then properties of single-carrier technology and multi-carrier technology are briefly studied, followed by a discussion on single-antenna systems versus multiple-antenna systems.

3.3.1 *Typical UWB Modulation Techniques*

3.3.1.1 Pulse Position Modulation

Pulse-position modulation (PPM) is a form of signal modulation, encoding the message symbols through placing a pulse in one of a number of possible positions within a symbol period (Win and Scholtz 2002). Theoretically, the number of positions can be infinite; however, it is constrained by multipath fading, noise, and hardware complexity in practice. Figure 3.16 illustrates such a modulation with four different pulse positions within the symbol period, representing four different data symbols. There are a number of advantages associated with PPM. The spectral shaping can be realized in a simple way through time hopping. The system receiver structure has a low complexity and thus the design and implementation is relatively inexpensive. Also, a tradeoff between user population and data rate can be easily made. On the other hand, there are several disadvantages related to PPM. Very precise time base is required to guarantee the precise placement of the extremely short pulses and synchronization becomes a rather difficult issue.

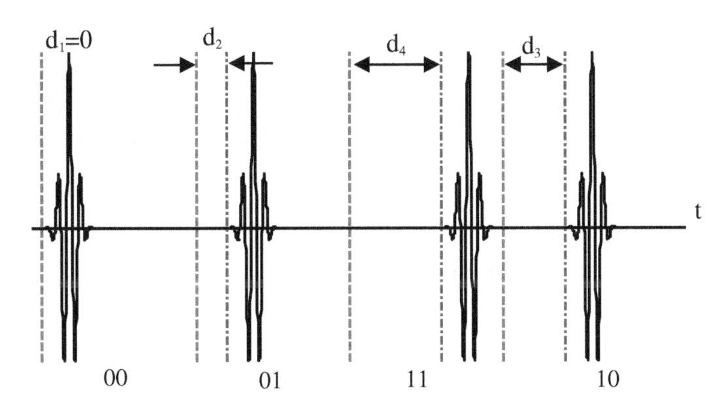

Fig. 3.16 An example of pulse position modulation

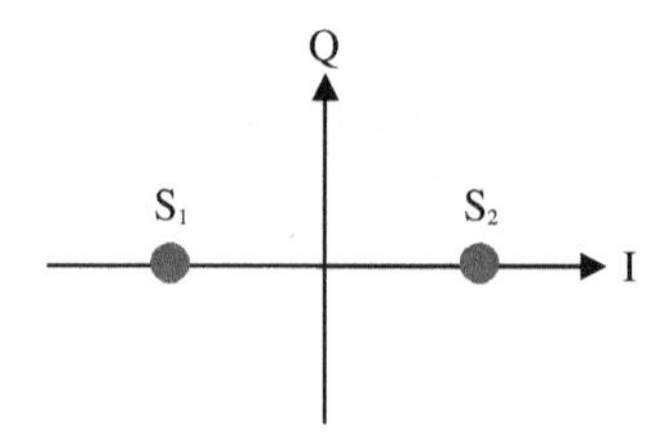

Fig. 3.17 Constellation diagram of bi-phase modulation

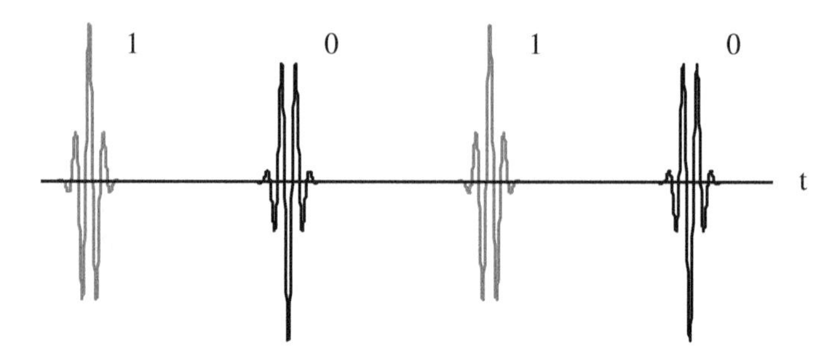

Fig. 3.18 Illustration of bi-phase modulation

3.3.1.2 Bi-phase Modulation

Bi-phase modulation (BPM), also termed binary phase shift keying (BPSK) modulation, is the simplest form of phase shift keying. It uses two phases, separated by 180°, to represent two different data symbols, as illustrated in Fig. 3.17. An example of the BPM is shown in Fig. 3.18. The data are modulated at only one bit per symbol and thus this modulation technique is not a good option for high data rate communications. However, this modulation method is the most robust of all PSK methods. To overcome the phase shift introduced by the fading channels, the data are often differentially encoded in the transmitter and differential detection is employed in the receiver.

3.3.1.3 Pulse Amplitude Modulation

Pulse amplitude modulation (PAM) is the simplest form of pulse modulation. The data to be transmitted are encoded by varying the amplitude of signal pulses. PAM is suited for wired communications such as in the popular Ethernet communication standard. PAM 12 and PAM 8 have also been considered for ten gigabyte Ethernet over copper wire. Figure 3.19 shows a simple example of PAM.

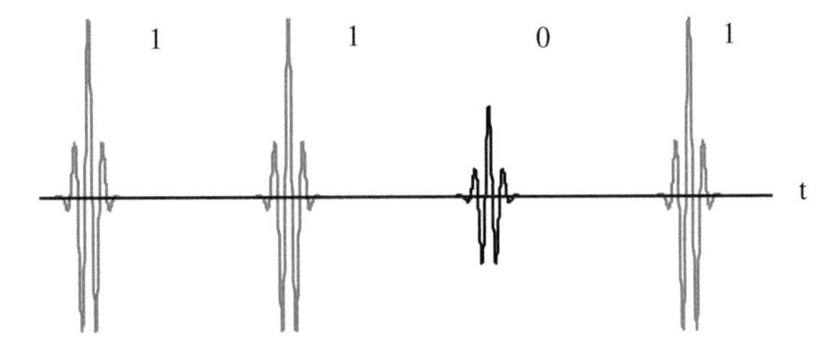

Fig. 3.19 Illustration of pulse amplitude modulation

3.3.1.4 On-Off Keying

On-off keying (OOK) modulation is the simplest form of amplitude shift keying modulation, representing data symbols as the presence or absence of a carrier wave or a pulse. For instance, the presence of a pulse represents binary one, whereas the absence of a pulse represents binary zero. Clearly the key advantage of the OOK modulation is the implementation simplicity, whereas the main disadvantage is the poor energy efficiency. Figure 3.20 shows an example of the OOK modulation.

3.3.1.5 Orthogonal Pulse Modulation

Orthogonal pulse modulation (OPM) is based on a set of pulse shapes which are orthogonal to each other. Each of the pulse shapes corresponds to one of the modulation states. For instance, the modified Hermite pulses of different orders can be employed to form OPM (Michael et al. 2002). The Hermite polynomials are defined as

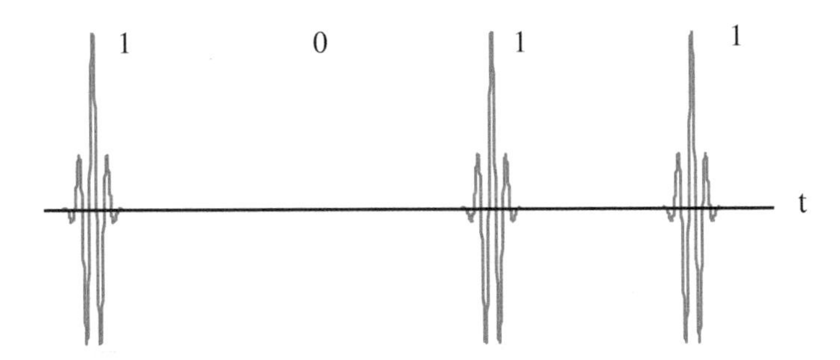

Fig. 3.20 Illustration of on-off keying

$$
\begin{aligned}
h_{e_0}(t) &= 1 \\
h_{e_n}(t) &= (-1)^n e^{\frac{t^2}{2}} \frac{d^n}{dt^n}\left(e^{-\frac{t^2}{2}}\right),\ n = 1,\ 2,\ \ldots
\end{aligned}
\tag{3.20}
$$

which are not orthogonal. However, they can be modified to become orthogonal by multiplying by a weight function as follows (Michael et al. 2002)

$$
\begin{aligned}
h_n(t) &= e^{-\frac{t^2}{4}} h_{e_0}(t) \\
&= (-1)^n e^{\frac{t^2}{4}} \frac{d^n}{dt^n}\left(e^{-\frac{t^2}{2}}\right),\ n = 1,\ 2,\ \ldots
\end{aligned}
\tag{3.21}
$$

By introducing a time-scaling factor τ, the first four modified Hermite polynomials are given by

$$
\begin{aligned}
h_0(t) &= e^{-\frac{t^2}{4\tau^2}} \\
h_1(t) &= \frac{t}{\tau} e^{-\frac{t^2}{4\tau^2}} \\
h_0(t) &= \left(\left(\frac{t}{\tau}\right)^2 - 1\right) e^{-\frac{t^2}{4\tau^2}} \\
h_1(t) &= \frac{t}{\tau}\left(\left(\frac{t}{\tau}\right)^2 - 3\right) e^{-\frac{t^2}{4\tau^2}}
\end{aligned}
\tag{3.22}
$$

Figure 3.21 shows the pulse shape of the modified Hermite pulses versus time. It can be seen that the pulse width for pulses of different orders does not change significantly. Nevertheless, the complexity for successful reception of pulses increases with the order of the polynomial. The Fourier transforms of the first four modified Hermite polynomials can be shown to be

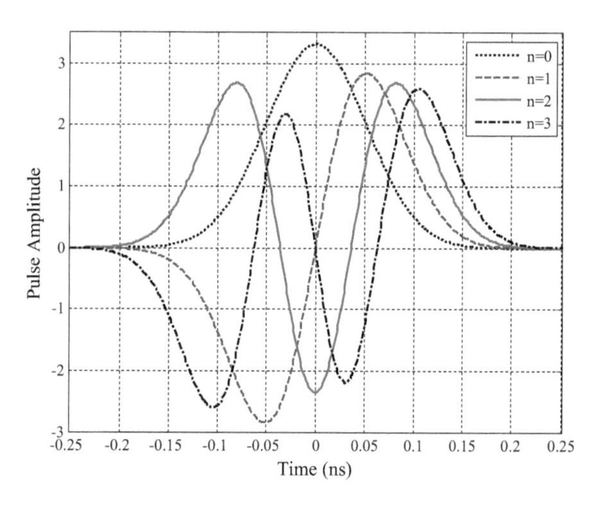

Fig. 3.21 Pulse shape of modified Hermite pulses of orders $n = 0, 1, 2, 3$ normalized to unit energy

$$
\begin{aligned}
H_0(f) &= 2\tau\sqrt{\pi}e^{-(2\pi f\tau)^2} \\
H_1(f) &= (-j4\pi f\tau)2\tau\sqrt{\pi}e^{-(2\pi f\tau)^2} \\
H_2(f) &= (1-(4\pi f\tau)^2)2\tau\sqrt{\pi}e^{-(2\pi f\tau)^2} \\
H_3(f) &= (-j12\pi f\tau + j(4\pi f\tau)^3)2\tau\sqrt{\pi}e^{-(2\pi f\tau)^2}
\end{aligned}
\tag{3.23}
$$

The corresponding energy spectrums of the four polynomials are shown in Fig. 3.22. Clearly, like the Gaussian pulse and its derivatives, the pulses generated from the modified Hermite polynomials cannot be directly used for transmission in UWB systems since they do not comply with the FCC spectrum mask. In fact, the first modified Hermite polynomial is a Gaussian function. Therefore, the modified Hermite pulses need to be further processed such as by modulating a sinusoidal signal, if they are intended for UWB applications.

Figure 3.23 illustrates the modified Hermite polynomials based OPM. Using a set of orthogonal waveforms, an M-ary signaling set can be constructed, so that higher data rate transmission can be realized. In addition orthogonal pulses can be assigned to different users to enable multiuser communications.

3.3.2 *Pulse-Based Systems, Single-Carrier Systems and Multi-carrier Systems*

In this subsection the properties of the pulse-based, single-carrier and multi-carrier systems are briefly studied.

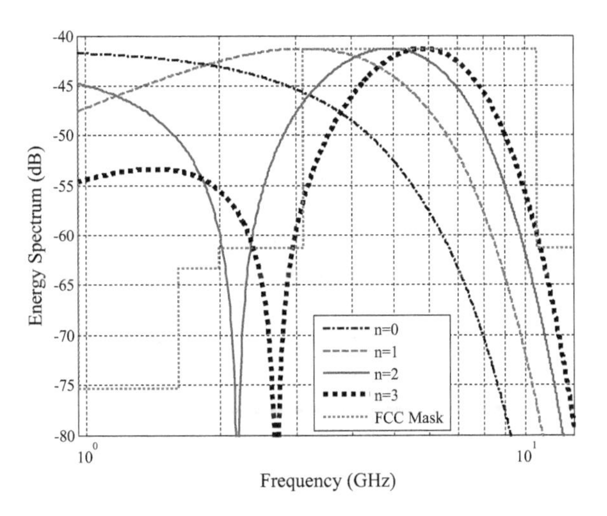

Fig. 3.22 Energy spectrum of modified Hermite pulses

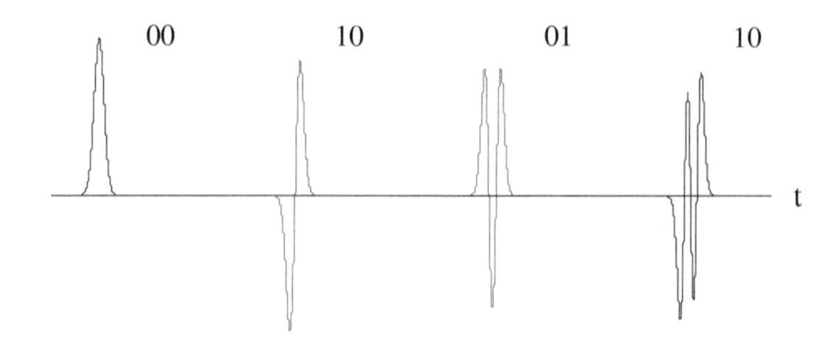

Fig. 3.23 Example of orthogonal pulse modulation

3.3.2.1 Pulse-Based Systems

Pulse-based systems are usually associated with UWB technology. In a pulse-based or impulse-radio UWB system, no carrier is employed to modulate/up-convert the pulses in general. Each transmitted pulse instantaneously occupies the UWB bandwidth. Pulse repetition rates can be either low or very high, depending on specific applications. For instance, pulse-based UWB radars tend to use low repetition rates, typically up to several mega-pulses per second; whereas high data rate communications systems utilize high pulse rates, typically a few giga-pulses per second, thus enabling short-range gigabit-per-second communications systems. Compared to carrier-based systems which are subject to channel fading, pulse-based systems are relatively immune to multipath fading.

3.3.2.2 Single-Carrier Based Systems

In a range of telecommunication standards such as the IS-95, GSM, WCDMA, and TDCDMA, traditional spreading techniques based on single-carrier technology is used. These standards based systems typically employ spreading techniques such as direct-sequence spread spectrum (DSSS) or frequency-hopping spread spectrum (FHSS) technology. The single-carrier based DS-UWB has also been considered as a candidate standard for UWB communications. These spreading-spectrum techniques enable the systems to be resistant to intended or unintended jamming. Also, a single channel can be shared among multiple users and multiuser communications can be enabled through using the DS-CDMA (direct-sequence code-division-multiple-access) or the FH-CDMA technology. It is worth mentioning that three of the four global satellite constellations (GPS, BDS, and Galileo) use the CDMA technology. Although GLONASS uses FDMA technology, it is likely that CDMA will soon supplement GLONASS's FDMA scheme. Since only one carrier is required, the spreading and synchronization are simple and thus the system design is simple. However, the single carrier systems suffer from the sensitivity to non-linear distortion. It is rather difficult to collect significant signal energy by using a single RF chain. Careful code selection/design for multiple accesses is also required to achieve minimal cross-correlation.

3.3.2.3 Multiband Multi-carrier Systems

Multi-band multi-carrier systems employ orthogonal frequency division multiplexing (OFDM) technology, transmitting data on each of a group of sub-bands. Figure 3.24 illustrates how the OFDM symbols are transmitted via frequency-time interleaving in a multi-band OFDM system. Typically, a cyclic prefix (CP) is inserted at the beginning of each OFDM symbol and a guard interval is appended to each OFDM symbol. The CP is used to mitigate multipath effects and make channel estimation easy. The guard interval has been inserted to eliminate the intersymbol interference from the previous symbol and guarantee sufficient time for the transmitter and receiver to switch to the next channel. In multi-band OFDM UWB systems, aggregation of sub-bands of narrow-band carriers must be at least 500 MHz so that access to the UWB spectrum can be permitted under the rules. For instance, in the frequency band of 3.1–10.6 GHz, 14 sub-bands may be defined. Each sub-band is 528 MHz in width and contains certain number (say 128) of modulated OFDM sub-carriers. Comparatively, the OFDM technology is robust against multipath effect due to the long symbol and guard times. Coexistence with other systems is not a difficult issue since frequency notching is relatively easy. In addition, flexible data rates can be readily realized. On the other hand, to realize OFDM oscillators must be very stable and parallel processing of sub-bands is required.

3.3.3 Single-Antenna Systems Versus Multiple-Antenna Systems

In single-antenna systems, both the base station and the mobile terminals use a single antenna for signal transmission or signal reception. The main advantage of

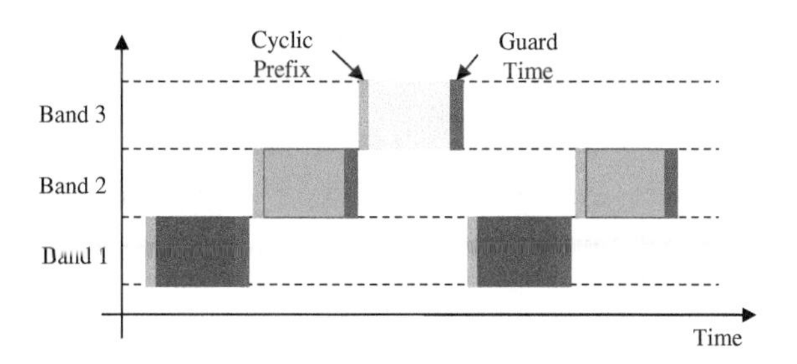

Fig. 3.24 An example of band hopping in multi-band OFDM UWB systems

single-antenna systems is the simplicity in terms of system structure, hardware, and signal processing. On the other hand, single-antenna systems might not perform well in the presence of deep fades when communications are located in rich multipath propagation environments.

In multiple-antenna systems, the base station, the mobile terminal, or both employ multiple antenna elements for signal transmission and/or signal reception. Typically a multiple-antenna system is likely to achieve better communications and positioning performance than its single-antenna counterpart especially in severe signal fading environments. As for positioning, the performance gain comes from the multiple-antenna based beamforming that can significantly improve signal-to-noise-plus-interference ratio. In addition, through processing of the received signals from multiple antenna elements, the angle of arrival of the incoming signal can be estimated, which provides extra useful information to aid position determination. More information on how to use multiple antennas for angle estimation and beamforming can be found in Chap. 18.

At base stations it is feasible to build complex multiple-antenna systems for signal transmission or reception. On the other hand, it may be impractical to equip a mobile terminal such as mobile phone with multiple antennas. Unlike a base station where space is typically not an issue and a powerful processing unit is usually available, a mobile phone does not have a space that is large enough to separate the multiple antennas. The separation distance between a pair of antennas must be greater than half a wavelength of the signal to ensure that the received signals are uncorrelated. For instance, for 2G GSM networks operating in the 900 MHz band, the signal wavelength is about 33 cm. That is, the distance among the antenna elements should be greater than 16.7 cm. Even for 3G GSM operating in the 2100 MHz band, the separation distance should be greater than 7 cm. Clearly, such a dimensional demand is generally troublesome or even impractical for tiny mobile phones. However in the case of millimeter wave communications, a multi-antenna scheme would be feasible to be implemented in mobile phones due to the very short signal wavelength.

3.4 Spread-Spectrum Techniques

By using a spread-spectrum technique, the bandwidth of a signal is deliberately spread in the frequency domain, resulting in a much wider bandwidth. Spread-spectrum signals are highly resistant to narrowband interference and jamming. Further they are difficult to intercept or eavesdrop, and many users can share the same frequency channel with minimal interference. Thus spread spectrum techniques are particularly suited for secure communications and positioning. In this section the three basic spread spectrum techniques are summarized and a number of spreading codes are studied.

3.4.1 *Three Basic Spectrum Spreading Approaches*

3.4.1.1 Direct-Sequence Spread Spectrum

DSSS is a modulation technique widely used in wireless communications. With this technique each information symbol is modulated by a pseudo-random chip sequence of 1 or −1 values. The duration of each chip is much shorter than the symbol duration so that the frequency of the pseudo-random sequence is much higher than that of the original signal. At the receiver the 'de-spreading' process is carried out by correlating the received signal with a copy of the same pseudo-random sequence stored in the receiver. Figures 3.25 and 3.26 show an example of the DSSS modulation in terms of spreading when the second derivative of the Gaussian pulse is modulated by a sequence of 15 pseudo-random chips. A maximum-length sequence is used as the pseudo-random/spreading code. More discussions about the spreading codes will be provided later in this section.

When time-of-arrival (TOA) measurements are employed for position determination, signal bandwidth is an important issue. The accuracy of the TOA measurements is proportional to the signal bandwidth in general; this is nearly always the case where wide-bandwidth signals are considered in the design of positioning systems. This is also the reason why UWB technology has been widely investigated for positioning. On the other hand, a signal with a higher carrier frequency travels a shorter distance, making UWB positioning systems suited for indoor environments with a transmission range typically up to 30–40 m. Thus careful signal design is necessary to accommodate all the requirements including accuracy and transmission range.

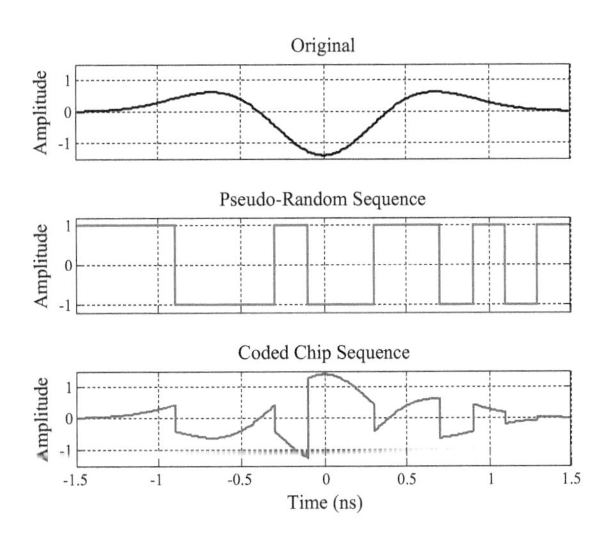

Fig. 3.25 Illustration of direct sequence spreading in time domain

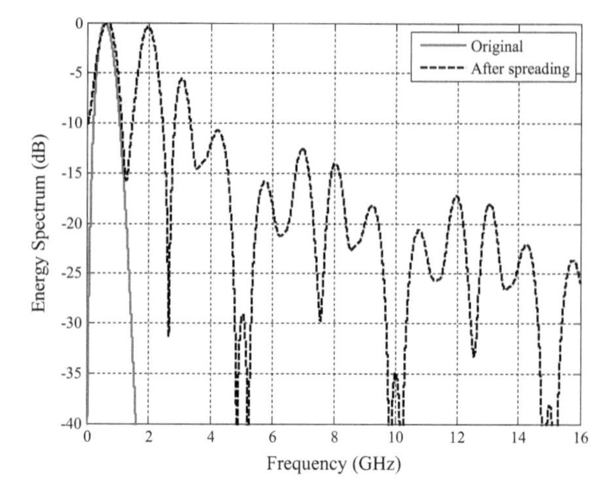

Fig. 3.26 Illustration of direct-sequence spreading in frequency domain

Although direct sequence and frequency hopping are the most commonly used spread-spectrum technology in wireless communications, the direct-sequence technology is typically used for radio positioning systems such as GPS. Chirp spread-spectrum is another technology that has been employed for positioning, which will be discussed later.

3.4.1.2 Frequency-Hopping Spread Spectrum

Frequency-hopping technology achieves the same objective as the direct-sequence technology; however they are two rather different with their own distinctive characteristics. Unlike the DSSS technology, frequency hopping does not spread the signal. Instead, it makes use of different carrier frequency over a given bandwidth at different time period when transmitting radio signals, as illustrated in Fig. 3.27. The carrier frequency is rapidly switched according to a pseudo-random sequence

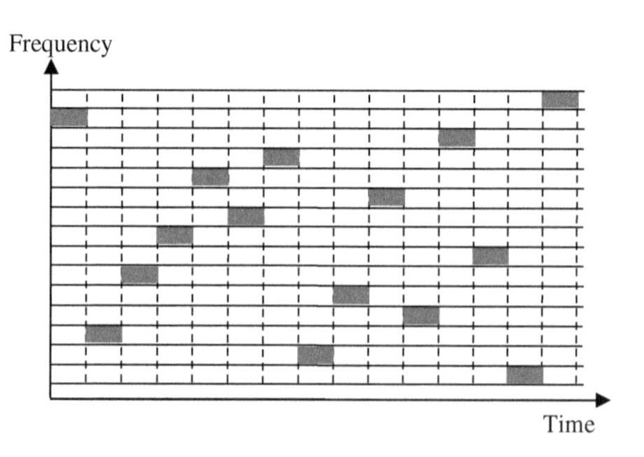

Fig. 3.27 Illustration of frequency hopping

known to both transmitter and receiver. Frequency hopping performs better than direct-sequence technique in multipath propagation environments. In fading channels if a deep fade or null occurs at one frequency, then it is unlikely that a deep fade happens simultaneously at another frequency. Thus rapidly switching the carrier frequency will effectively mitigate the multipath fading effect. As a result frequency-hopping based positioning systems will outperform fixed carrier-frequency positioning systems in multipath fading environments. The main disadvantage of frequency hopping is associated with synchronization. The direct sequence technique only requires synchronization of the timing of the chips, but the frequency hopping technique requires that the transmitter and the receiver are synchronized both in time and in frequency.

3.4.1.3 Chirp Spread Spectrum

Chirp spread-spectrum (CSS) is a technique that uses chirp pulses to encode data symbols. As already studied in Sect. 3.1.6, a chirp is a sinusoidal signal whose frequency increases or decreases over a certain period of time. It is clear that unlike DDSS or FHSS, CSS does not use pseudo-random sequence to modulate the signal pulses. Instead it relies on varying the sinusoidal frequency continuously and typically the frequency is linear with time over a specific duration. To some degree this resembles the FHSS technique which also varies the frequency but in a way of pseudo-random hopping.

Precision ranging was one of the original application areas of the CSS. Although CSS is not currently a favorite technique considered by the IEEE for standardization in the area of precision ranging, some companies have already developed CSS related products such as Nanotron's nanoLOC TRX transceiver which offers robust wireless communication and ranging capabilities so that real-time location can be enabled (Nanotron 2017).

3.4.2 Spreading Codes

Both DSSS technique and FHSS technique require spreading codes which are used to code sequences of chips to form pseudo-random sequences for spreading the signals in the transmitter and for de-spreading the received signal in the receiver. In multiuser communications, each user is assigned with a unique spreading code. To minimize the multiuser interference, the spreading codes should have desirable auto-correlation and cross-correlation properties. In this subsection four different codes are studied.

3.4.2.1 Maximum Length Codes

Maximum/maximal length sequence (MLS), also called m-sequence, can be simply generated using linear feedback shift registers (LFSR). Any LFSR can be represented by a polynomial referred to as generator polynomial. An LFSR represented by a primitive polynomial will produce a m-sequence which is a binary sequence and periodic with length $N = 2^m - 1$ where m is the number of registers. For instance, when using five registers, an m-sequence of length 31 can be generated when the primitive polynomial is given by $G(x) = x^5 + x^3 + 1$. That is, the feedback taps are tap 5 and tap 3, as shown in Fig. 3.28 where $\oplus$ denotes the modulo-2 addition (EX-OR). The output consists of ones and zeros so that they need to be converted to positive and negative ones such as by $1 - 2 \times output$. Note that there are a limited number of primitive polynomials for a given m value ($m \geq 3$). In the case of $m = 5$, two other sets of primitive polynomials are $G(x) = x^5 + x^4 + x^3 + x^2 + 1$ and $G(x) = x^5 + x^4 + x^3 + x^1 + 1$.

The LFSR goes through all $2^m - 1$ states, not including the all-zero state, before repeating. An m-sequence has 2^{m-1} ones and $2^{m-1} - 1$ zeros. The sum of any two distinct shifts of an m-sequence is another shift of the same m-sequence. The periodic autocorrelation function (ACF) of the three sets of the m-sequences of length $N = 31$ is shown in Fig. 3.29. The m-sequences are normalized so that they consist of $\pm\frac{1}{\sqrt{N}}$. In fact the periodic ACF just has two different values given by

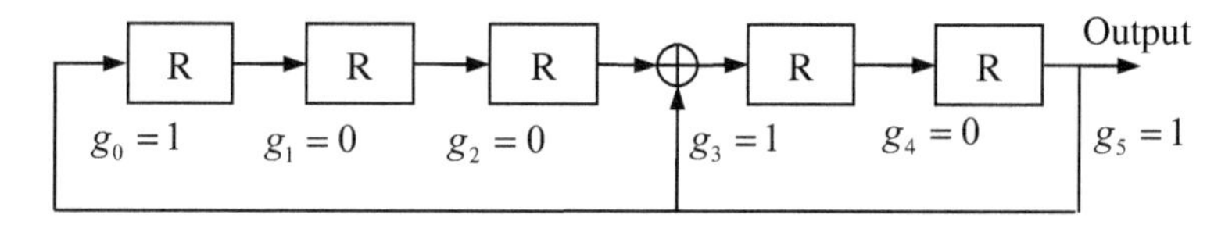

Fig. 3.28 Five-stage LFSR for generating m-sequences of length 15

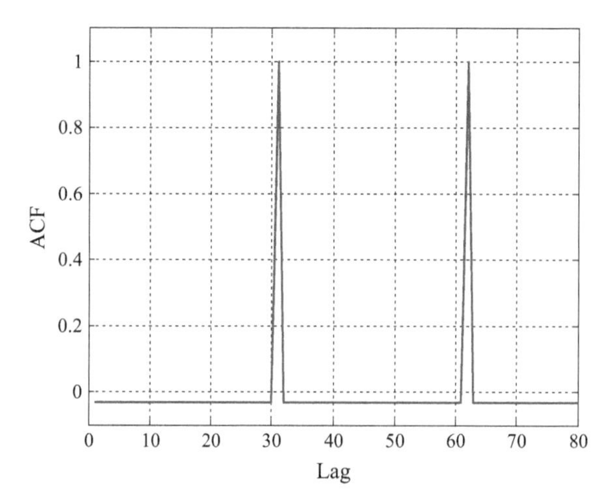

Fig. 3.29 Periodic autocorrelation function of m-sequence of length 31

$$ACF = \begin{cases} 1, & lag = kN,\ k = 0,\ 1,\ 2,\ \ldots \\ -\frac{1}{N}, & \text{other lags} \end{cases} \tag{3.24}$$

Clearly the peak value occurs at lag kN and the other occurs at other lags. When the length (N) of the m-sequence is large the smaller ACF is close to zero.

3.4.2.2 Gold Codes

Gold codes/sequences are useful in CDMA communications. In particular, they can be utilized to permit asynchronous transmissions among multiple users. The receiver can synchronize with the transmitter by using the auto-correlation property of the Gold codes. The well-known global positioning system (GPS) also makes use of Gold codes of length 1023 because they have bounded small cross-correlations, making an equivocal identification possible. Gold codes can be produced by modulo-2 addition of two m-sequences of the same length, yielded by two m-sequence generators as shown in Fig. 3.30. In this case, there are $2^m + 1$ Gold codes in total, generated as follows. The initial states of the first LFSR are set at non-zero and the states of the second LFSR is changed one by one from zero to $2^m - 1$. In addition, the m-sequence generated by the first LFSR is a Gold code.

Alternatively, Gold codes can also be generated from a single LFSR according to the following steps:

(1) Take a m-sequence of length $N = 2^m - 1$, which is called the original m-sequence
(2) Decimate this original m-sequence by a decimation factor q, generating a decimated m-sequence
(3) The decimation factor and the size of the LFSR satisfy the following conditions:

 (a) The LFSR size m is odd or $\mathrm{mod}(m, 4) = 2$

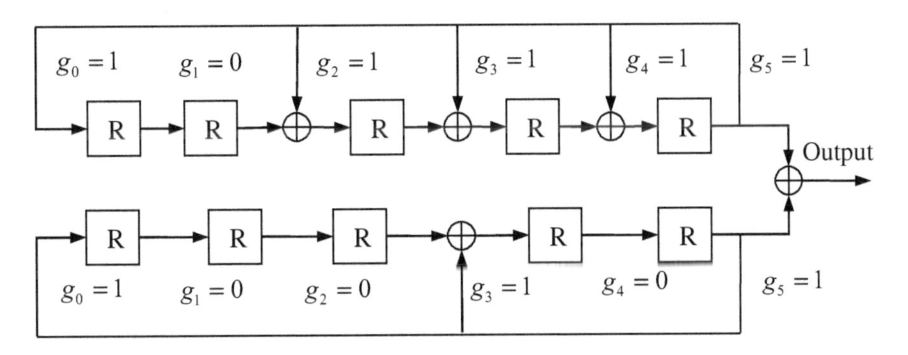

Fig. 3.30 Generation of Gold codes by combining two LFSRs

(b) The decimation factor is odd and either $q = 2^k + 1$ or $q = 2^{2k} - 2^k + 1$ for an integer k

(c) The greatest common divisor of m and k, $\gcd(m, k) = 1$ when m is odd and $\gcd(m, k) = 2$ when $\mathrm{mod}(m, 4) = 2$

(4) Then the original m-sequence and the decimated m-sequence will be a preferred pair of m-sequences which can be combined via modulo-2 addition to form a Gold sequence.

Similar to ordinary m-sequences, Gold codes have a large autocorrelation peak at lag zero. But differently, instead of a two-valued ACF, Gold codes also have small peaks of three distinct values, as shown in Fig. 3.31. Further, Gold codes have three-valued cross-correlation function and the values are

$$\left\{ -\tfrac{1}{N}t(m), \quad -\tfrac{1}{N}, \quad \tfrac{1}{N}(t(m) - 2) \right\} \tag{3.25}$$

where

$$t(m) = \begin{cases} 2^{(m+1)/2} + 1, & \text{for odd } m \\ 2^{(m+2)/2} + 1, & \text{for even } m \end{cases} \tag{3.26}$$

Figure 3.32 shows the cross-correlation of two Gold codes of length 31 and the three values are in accordance with (3.25).

3.4.2.3 Kasami Sequence Codes

Similar to Gold codes, Kasami codes are produced by modulo-2 addition of two distinct code sequences. The first code sequence is an m-sequence generated using a LFSR of size m and the second one is derived from the first one by decimating it by

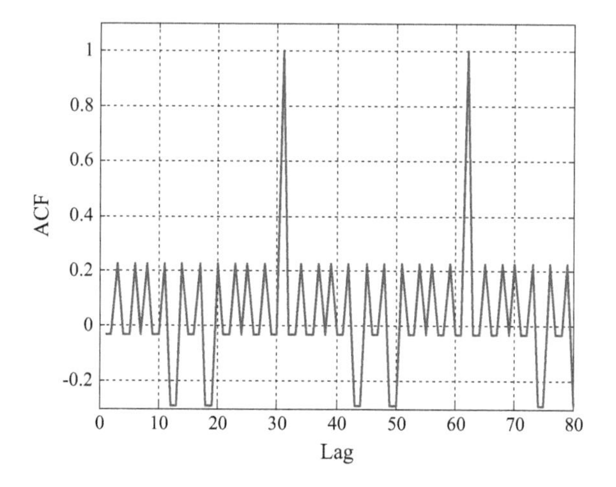

Fig. 3.31 An example of periodic ACF of Gold codes of length 31

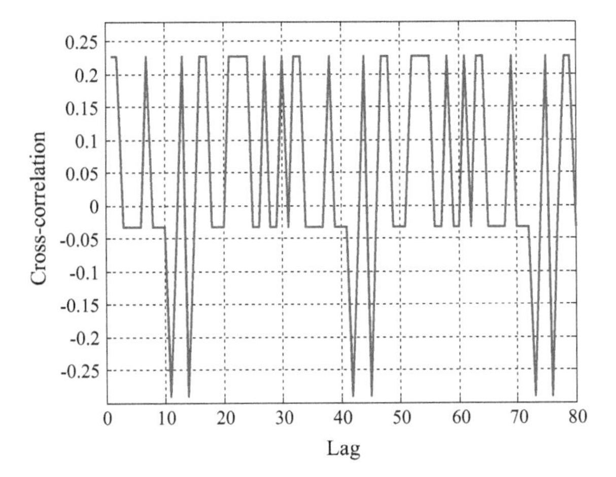

Fig. 3.32 An example of cross-correlation of two Gold codes of length 31

an integer factor $q = 2^{m/2} + 1$ where m is even. It is seen that the length of the resulting sequence from the decimation of the m-sequence of length $2^m - 1$ is $2^{m/2} - 1$. Repeating this sequence $2^{m/2} + 1$ times produces the second sequence of length $2^m - 1$. The modulo-2 addition of the original m-sequence and the second sequence yields a Kasami code. There are $2^{m/2} - 2$ cyclic shifts of the second sequence, which can be used to produce $2^{m/2} - 2$ Kasami codes. Including the original m-sequence, there are totally $2^{m/2}$ (small set of) Kasami codes. Figures 3.33 and 3.34 show examples of the autocorrelation function and the cross-correlation function of the Kasami codes of length 63, respectively. Similar to

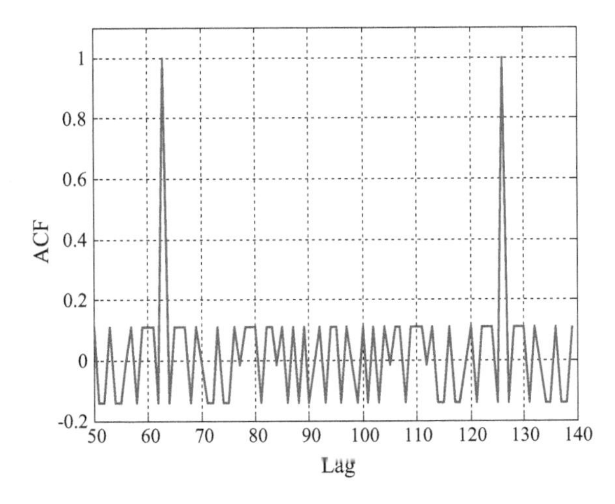

Fig. 3.33 An example of periodic autocorrelation function of Kasami codes of length 63

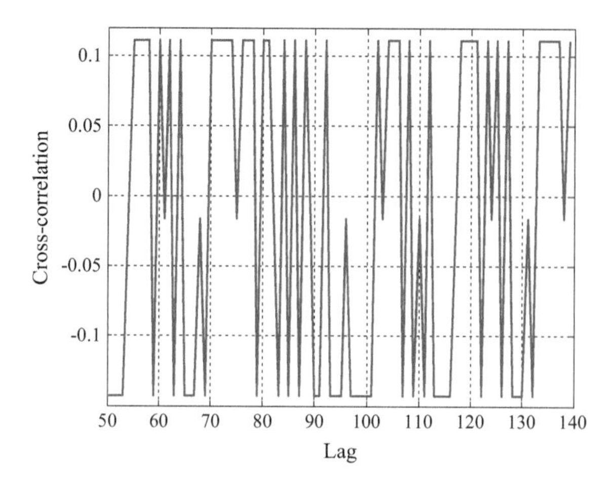

Fig. 3.34 An example of cross-correlation of two Kasami codes of length 63

the Gold codes, the ACF of the Kasami codes has the peak value at lag zero. The other values of the ACF and the values of the cross-correlation are from the set $\left\{-\frac{1}{N}, \quad -\frac{1}{N}(2^{m/2}+1), \quad \frac{1}{N}(2^{m/2}-1)\right\}$.

3.4.2.4 Walsh-Hadamard Codes

Walsh-Hadamard (WH) codes are orthogonal codes when there is no timing mismatch between each pair of codes. The generation of the WH codes is simple and follows a recursive procedure. That is, the Hadmart matrix of order 2 is first defined as

$$\mathbf{H}_2 = \begin{pmatrix} +1 & +1 \\ +1 & -1 \end{pmatrix} \tag{3.27}$$

Then for $M = 2^k$ where k is an integer of greater than one, the Hadamart matrix of order M is constructed recursively according to

$$\mathbf{H}_M = \begin{pmatrix} \mathbf{H}_{M/2} & \mathbf{H}_{M/2} \\ \mathbf{H}_{M/2} & -\mathbf{H}_{M/2} \end{pmatrix} \tag{3.28}$$

It can be seen that the rows of the matrix $\mathbf{H}_M$ form M sequences, namely the WH codes, which are mutually orthogonal to each other. Clearly these codes have zero cross-correlation when they are synchronous. However in the asynchronous case, the cross-correlation varies significantly, depending on which pair of codes are involved. For certain pairs of codes, the cross-correlation of the asynchronous case can be very high, making them unsuitable for asynchronous transmissions. The WH codes do not have a single narrow autocorrelation peak. Also the spreading is over a number of discrete frequency components, instead of over the whole frequency band. In spite of these disadvantages the WH codes are used in the IS-95 systems which use the WH codes based 64-ary orthogonal modulation scheme on the reverse links.

3.5 Radar and FMCW

Radar has been widely used for object identification and tracking. In this section the basic concepts of radar and FMCW technology are briefly studied.

3.5.1 Pulse Radar

Distance measurement can be made by use of pulse radar. A sequence of short radio frequency (RF) pulses is transmitted toward the target such as a vehicle or an airplane. The propagating pulses hit the surface of the target and some of their energy bounce back to the radar station. By measuring the time of flight of the pulses, the range between the radar station and the target can be estimated. Thus the basic usage of pulse radar is for pure time of flight measurements. Pulse radar requires high-power transmission and is complex and expensive in general. The history of pulse radar goes back more than half a century during the Second World War when more than eight countries independently and secretly developed radar systems for detecting airplanes or ships.

Since 1990s pulse radio has re-drawn significant attention especially in the area of UWB technology and applications (Oppermann et al. 2004). One of the main applications of the UWB technology is in assets and people tracking in indoor environments. The fine time-resolution of the UWB signals can be employed to achieve precise localization in rich multipath scenarios (Fontana 2004).

3.5.2 FMCW Radar

Different from pulse radar systems, continuous wave (CW) radar systems transmits electromagnetic waves at all times. Conventional CW radar cannot make distance measurement simply since there is no reference point in the transmitted or returned signal to make time delay measurement. However a time reference mark can be achieved through modulating the frequency in a known manner so that the CW radar can be utilized to measure distance, resulting in the frequency modulated CW (FMCW) radar which is illustrated by Fig. 3.35.

By measuring the frequency of the return signal, the round trip time can be measured and therefore the range can be determined. FMCW radar has a number of advantages over traditional pulse radar. FMCW radar avoids high peak transmission

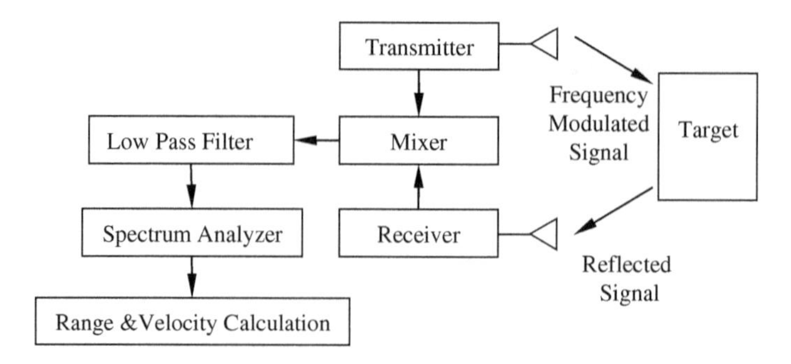

Fig. 3.35 Simplified diagram of FMCW radar

power and in fact the emitted energy is much less than that of pulse radar and even less than a small fraction of that of a typical cell phone. Thus FMCW radar emits no harmful radiation and the design of radio frequency components can be simplified substantially. Another advantage, perhaps the most important one, is the superior resolution achieved from constantly accumulating the return signal energy. One disadvantage of FMCW technology is that its performance is affected by the Doppler frequency caused by the relative movement between the target and the radar system. To reduce the Doppler frequency effect, the amount of frequency deviation should be significantly greater than the expected Doppler shift or the maximum Doppler frequency.

Different frequency modulation techniques can be employed in FMCW technology; however triangular wave frequency modulation is commonly used. The sinusoidal carrier at frequency f_0 is modulated by the triangular shape frequency $f_m(t)$ such that the instantaneous FM signal frequency of the transmitted signal is given by

$$f_i(t) = f_0 + f_m(t) \tag{3.29}$$

The received signal frequency is expressed by

$$f_r(t) = f_0 + f_m(t - \tau) \tag{3.30}$$

where τ is the time delay or the round trip time of the signal. The frequency difference magnitude between the transmitted and the received signal is then given by

$$f_d(t) = |f_m(t - \tau) - f_m(t)| \tag{3.31}$$

In Fig. 3.36 the upper part illustrates the transmitted signal and the received signal frequencies, whereas the lower part shows the difference of the two signal frequencies. In this case the target is stationary or the distance between the target and the radar remains constant. Note that the frequency difference $f_d(t)$ is the

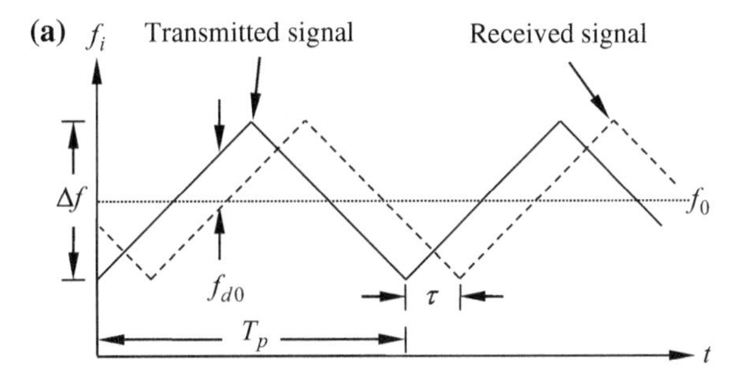

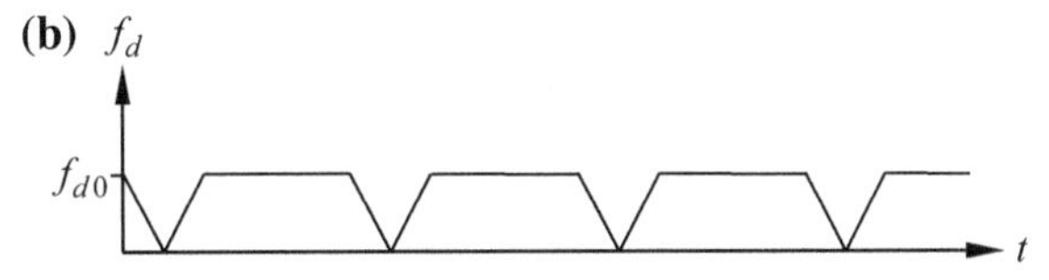

Fig. 3.36 Upper part shows the instantaneous frequencies of the transmitted and received FMCW signals. Lower part shows the difference magnitude of the two frequencies

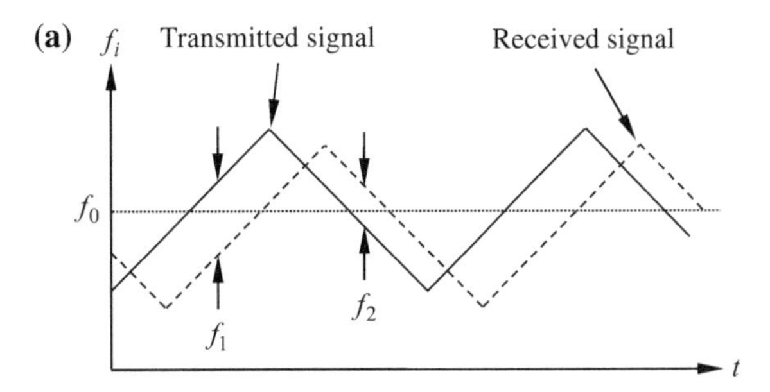

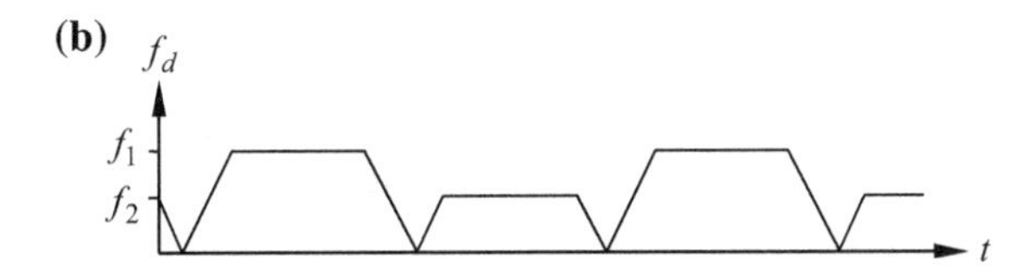

Fig. 3.37 The transmitted and the received signal frequencies and their difference for a moving target

frequency of the output signal at the low pass filter. The low-pass signal is also termed beat signal which is analyzed by the spectrum analyzer to detect the peak frequency (f_{d0}) from the signal spectrum. Given the modulation frequency span Δf and the triangular pulse duration T_p, the peak frequency is determined by

$$f_{d0} = \frac{2\Delta f \tau}{T_p} \tag{3.32}$$

When the target is moving, the frequency difference between the transmitted and received signals is affected by the Doppler frequency shift as shown in Fig. 3.37. Compared to Fig. 3.36, the received signal frequency decreases by an amount that is equal to the Doppler frequency shift. This corresponds to the case where the

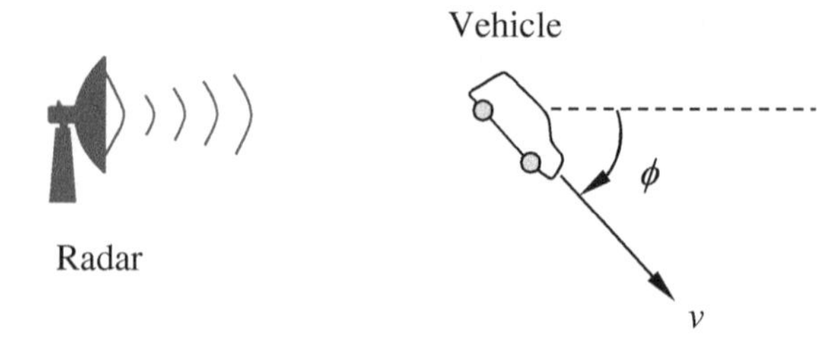

Fig. 3.38 Vehicle is moving away from radar

target is moving away from the radar. The FMCW radar can be used for the determination of both the velocity and the range of the moving target. In Fig. 3.37 the frequency difference over the positive slopes increases, producing

$$f_1 = f_{d0} + f_{Dop} \tag{3.33}$$

where f_{Dop} is the Doppler shift. Suppose that the target is moving with a velocity v at an angle φ as shown in Fig. 3.38. The radial velocity is equal to $v\cos\varphi$ so that the Doppler shift is given by

$$f_{Dop} = \frac{2vf_0 \cos\varphi}{c} \tag{3.34}$$

where c is the speed of signal propagation. However the frequency difference along the negative slopes decreases, resulting in

$$f_2 = f_{d0} - f_{Dop} \tag{3.35}$$

In (3.33) and (3.35) we used the fact that the increment in the positive slope frequency difference equals the decrement in the negative slope frequency difference. Therefore, adding (3.33) and (3.35) produces

$$f_{d0} = \frac{1}{2}(f_1 + f_2) \tag{3.36}$$

whereas subtracting (3.33) from (3.35) produces

$$f_{Dop} = \frac{1}{2}(f_1 - f_2) \tag{3.37}$$

The relationship between the range, the modulation parameters and the peak frequencies is described by

$$R = \frac{cf_{d0}T_p}{4\Delta f} = \frac{c(f_1 + f_2)T_p}{8\Delta f} \tag{3.38}$$

The radial velocity is computed by

$$v_r = v\cos\varphi = \frac{cf_{Dop}}{2f_0} = \frac{c(f_1 - f_2)}{4f_0} \tag{3.39}$$

From (3.39) it is seen that the target velocity v can be calculated only when the angle φ can be estimated. Therefore when the target is moving and thus a Doppler shift is produced, the spectrum analyzer will output two peak frequencies, f_1 and f_2. They are then used to calculate the range and the velocity based on (3.38) and (3.39) where the other parameters are known in advance.

The above formulas for calculating the range and velocity are valid provided that the modulation frequency span Δf is greater than the Doppler shift. In the event that the Doppler shift is greater than the modulation frequency span, (3.38) and (3.39) are changed to

$$R = \frac{cf_{d0}T_p}{4\Delta f} = \frac{c|f_1 - f_2|T_p}{8\Delta f} \tag{3.40}$$

and

$$v_r = v\cos\varphi = \frac{cf_{Dop}}{2f_0} = \begin{cases} \frac{c(f_1+f_2)}{4f_0}, & f_2 > f_1 \\ -\frac{c(f_1+f_2)}{4f_0}, & f_2 < f_1 \end{cases} \tag{3.41}$$

Thus the modulation frequency span should be sufficiently greater than the Doppler shift to avoid the ambiguity in calculating the range and velocity.

3.5.3 FMCW for Network Based Positioning

The basic principle of the FMCW technology can be exploited for assets, people, and vehicle tracking in a wireless network. However there is some distinct differences between these network based tracking systems and the radar systems. A conventional radar system is typically used to detect objects which do not have any communication link with the radar system. Thus such radars rely on measuring the echoes of the signal transmitted by the radar station to detect the target and to measure the range to the target and its velocity. On the other hand, in network based tracking systems, such an echo-based detection strategy is not suitable. One main reason is that there are usually a large number of other objects around the specific target. Typically it is impractical to distinguish the signals reflected by these objects from that reflected by the target of interest. Another reason is that assets, people, and even some vehicles do not have an appropriate surface to effectively scatter the signals so that the energy of the return signal would be too weak to be detectable.

Therefore in network based tracking, the target detects the signals transmitted from the base stations and then re-transmits the received signals back to the base stations.

It is worth mentioning that when the round trip time includes the processing delay in the target, (3.38) should be modified to become

$$R = \frac{cf_{d0}T_p}{4\Delta f} - \frac{1}{2}ct_0 = \frac{c(f_1 + f_2)T_p}{8\Delta f} - \frac{1}{2}ct_0 \tag{3.42}$$

where t_0 is the processing delay which is the time interval between the instant when the signal arrives at the target and the time when the signal is transmitted from the target. Similar change should be made to (3.40) in the presence of processing delay. On the other hand, the target velocity is independent of the round trip time and thus still calculated by using (3.39) or (3.41) in the presence of processing time.

From (3.42) it is seen that when the errors of the peak frequencies f_1 and f_2 are given and denoted by ε_1 and ε_2, respectively, the distance estimation error only depends on the ratio of the pulse duration over the modulation frequency span. That is,

$$\varepsilon_R = \frac{c(\varepsilon_1 + \varepsilon_2)}{8}\frac{T_p}{\Delta f} \tag{3.43}$$

It is clear from (3.43) that one way to reduce the distance estimation error is to decrease the pulse width and/or increase the modulation frequency span. However the selection of these two parameters will generally affect the performance of the peak frequency estimation. Thus in practice, the parameter selection should take account of the properties of the mixer and the low pass filter, preferably by using experimental results.

Figure 3.39 shows the root mean square error (RMSE) of the distance estimation with respect to the standard deviation (STD) of the peak frequency estimation error and the modulation frequency span. The STD of the two peak frequency errors is set at the same value for convenience. The round trip time is set at 20 μs and the distance error is calculated based on (3.43). The frequency estimation error of both f_1 and f_2 is assumed a Gaussian random variable with a zero mean and a STD ranging from 0 to 100 Hz. When the modulation frequency is greater than 30 kHz the maximum distance error is less than 4 m for the given range of the frequency errors. However the distance error can be more than 10 m when the modulation frequency is less than 10 kHz.

Figure 3.40 shows the distance estimation error versus the STD of the peak frequency estimation error and the triangular pulse width when the modulation frequency is set at 20 kHz. At the frequency error of 100 Hz, the minimum and the

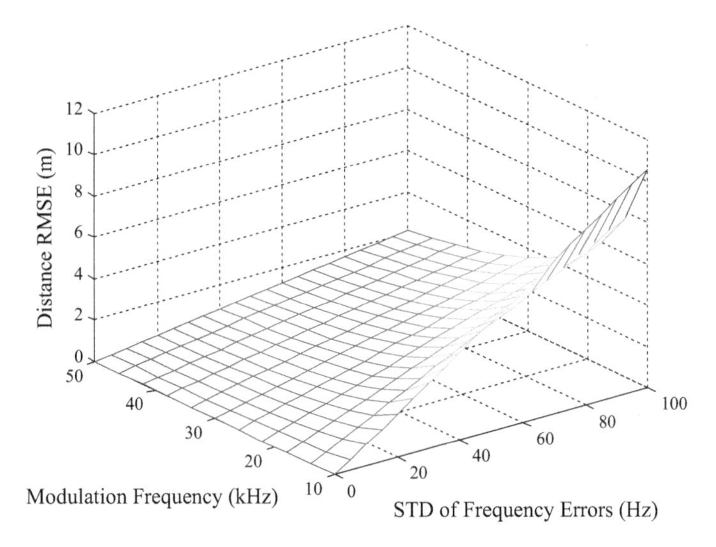

Fig. 3.39 RMSE of distance estimation versus modulation frequency and peak frequency error

maximum distance estimation error is about 2 and 11 m, respectively, over the given range of the pulse width.

The following example shows one basic procedure in the parameter selection for FMCW based ranging. Suppose that the tracking system works in the 2.45 GHz ISM band. When tracking ground vehicles, the potential target maximum velocity may be set at 120 km/h. Then from (3.34) the potential maximum Doppler shift can be calculated to be 544 Hz. Thus a modulation frequency span Δf of 20 kHz will be desirable since it is much larger than the maximum possible Doppler shift, and a large Δf will decrease the ranging error as shown by (3.43) and in Figs. 3.39 and 3.40. Assuming that the maximum radio range of both the base station and the target is 1 km, the round trip time of the radio signal will be 6.667 μs. Adding the transmission delay in the target due to signal processing, which is set at 5 μs, resulting in the round trip time of 11.667 μs. Note that the processing-induced delay will be different among different systems depending on the communication protocol, the software, and the hardware. The delay should be minimized to reduce the round trip time. The triangular pulse width T_p should be greater than the round trip time, but should not be too large to reduce the distance error as indicated by (3.43). Thus T_p is set at 26 μs. In the event that the round trip time varies significantly because of the range variations, the pulse duration may be adaptively selected if permitted. Thus when the estimated range is small, the pulse duration is reduced accordingly to enhance ranging performance.

Table 3.1 shows the design parameters and the corresponding parameters calculated with different velocities and ranges. The positive velocity indicates the target is moving away from the base station, whereas the negative velocity indicates the target is travelling towards the base station. The parameter f_{d0} (peak frequency

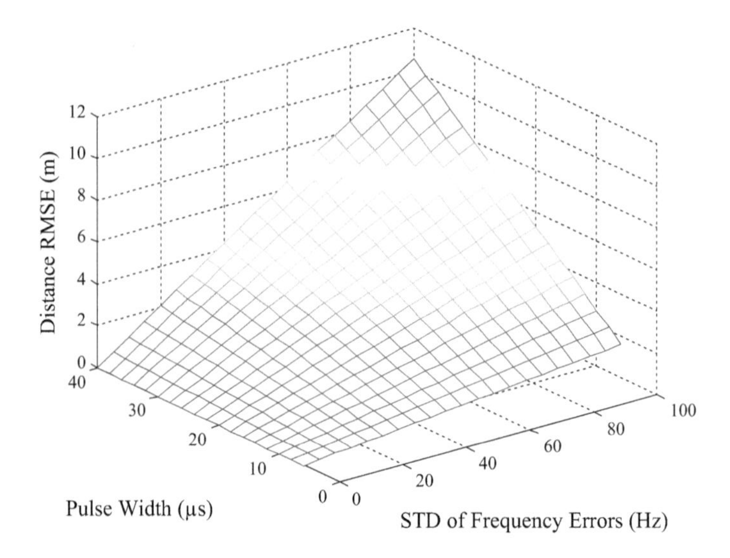

Fig. 3.40 RMSE of distance estimation versus pulse width and peak frequency error

Table 3.1 An example of parameter design for FMCW based ranging

Δf (kHz)	T_p (μs)	R (km)	v_r (km/h)	f_{Dop} (Hz)	f_{d0} (Hz)	f_1 (Hz)	f_2 (Hz)
20	26	1	120	544.4	17949	18493	17404
20	26	1	30	136.1	17949	18085	17813
20	26	1	−120	−544.4	17949	17404	18493
20	26	1	−30	−136.1	17949	17813	18085
20	26	0.1	120	544.4	8718	9262	8174
20	26	0.1	30	136.1	8718	8854	8582
20	26	0.1	−120	−544.4	8718	8174	9262
20	26	0.1	−30	−136.1	8718	8582	8854

for stationary targets) should be greater than the Doppler shift, but smaller than the modulation frequency span in general. This is just an illustrative example and the parameter selection will be dependent on the specific application in practice.

References

FCC (2002) FCC first report and order on ultra-wideband transmission systems, ET Docket 98-153, FCC 02-48, released on 22 Apr 2002

Fontana RJ (2004) Recent system applications of short-pulse ultra-wideband (UWB) technology. IEEE Trans Microw Theory Technol 52(9):2087–2104

Michael LB, Ghavami M, Kohno R (2002) Multiple pulse generator for ultra-wideband communication using Hermite polynomial based orthogonal pulses. In: Proceedings of IEEE conference on ultra wideband systems and technologies, Baltimore, Maryland, USA, pp. 47–52, May 2002

Nanotron (2017). http://www.nanotron.com/EN/PR_ic_modules.php

Oppermann I, Hamalainen M, Iinatti J (2004) UWB theory and applications. Wiley, Chichester

Win MZ, Scholtz RA (2002) Ultra-wide bandwidth time-hopping spread-spectrum impulse radio for wireless multiple-access communications. IEEE Trans Commun 48(4):679–689

Yu K, Sharp I, Guo YJ (2009) Ground-based wireless positioning. Wiley-IEEE Press

Chapter 4
Time-of-Arrival Measurements

4.1 Introduction

For a radiolocation system the position of an object is determined using some characteristic of the radio frequency signal. These characteristics could include the signal strength, the signal phase (for angle of arrival measurements) or the propagation time-of-flight (TOF) from the transmitter to the receiver. As very accurate time synchronization (see Chap. 5 for more details) is not possible in a mobile device, direct measurement of the TOF is not feasible in actual positioning systems, but the time-of-arrival (TOA) of a radio "pulse" can be used for position determination by processing the TOA data from multiple nodes. Clearly the accuracy of such position determination is directly related to the accuracy of the TOA measurement. This chapter provides details of various methods of measuring the TOA, and determining the associated accuracy of the measurement.

4.2 Overview of Time-of-Arrival Measurements

The fundamental measurement for determining positions is the time-of-arrival of a "pulse" at the receiver. The "pulse" for these measurements can be a baseband pulse as in UWB systems, or more commonly a reconstructed "pulse" (using a correlator) in spread-spectrum systems operating in the ISM bands. The details of such signal processing to generate the pulse are not of interest here (see Chap. 3 for common signal processing functions), but rather how this pulse is processed to determine the TOA. Of prime interest is the accuracy of the TOA measurement. There are two main sources of measurement errors, one associated with random noise in the receiver, and the second associated with the effects of multipath propagation.

The relative importance of these two sources of TOA measurement error depends on the particular application. In outdoor applications with LOS propagation the

I. Sharp and K. Yu, *Wireless Positioning: Principles and Practice*, Navigation: Science and Technology, https://doi.org/10.1007/978-981-10-8791-2_4

multipath effects may be small. However, generally the multipath environment is something that cannot be controlled by the system designer, rather the designer must try to mitigate its effects. In contrast, the receiver noise can be managed by the designer, as parameters such as the effective transmitter power (effective isotropic radiated power), the receiver noise figure, and particularly the receiver process gain (for spread-spectrum systems) can greatly effect the receiver noise. Ultimately however, the main affect of noise is to define a lower limit of the signal-to-noise ratio (SNR) for acceptable performance, which in turn will determine the maximum operational range of the radio transmissions. As the maximum operational range effects the design in terms of the coverage area per base station, the indirect affects of system noise may have the biggest impact on the system design, rather than the direct affect on the TOA performance. For more details in this aspect of the design of a radiolocation system, refer to Chap. 9 on Link Budgets.

In indoor environments, the main performance limitation is associated with the multipath effects associated with NLOS propagation, and thus the design typically concentrates on mitigating the effects of multipath propagation, even at the expense of degrading the performance associated with receiver noise. Indeed, the noise and multipath effects tend to be complementary—improving the performance of one tends to degrade the performance of the other. Thus the designer's task is to provide the optimum compromise between these two performance degrading effects.

4.2.1 Overview of TOA Measurements

Given the operating environment, the system designer needs to consider what type of TOA algorithm is appropriate. However, in most designs for tracking systems, the focus is on mitigating the effects of multipath propagation. This topic is considered in detail in Chap. 16, but the following summarizes the main principles.

As described in Sect. 4.2 above, the TOA measurement is assumed to be associated with straight-line propagation between the transmitter and the receiver, although the measurement includes other factors such as the local clocks and internal delays in the radios. However, the TOA algorithm focus is on just one of these, namely attempting to measure the arrival time of the signal associated with the propagation path between the transmitter and the receiver, where "time" is as measured of a clock local to the node. The topic of the use and synchronisation of clocks in nodes is covered in detail in Chap. 5, but the procedures typically rely on the TOA measurements described in this chapter.

To more fully understand this task, consider the measured UWB pulse shown in Fig. 4.3. This figure shows a portion of the pulse near the leading edge. Note the following:

1. Although a single pulse (with a rise-time of about 0.7 ns) was transmitted, the received pulse has many peaks, each associated with a different path from the transmitter to the receiver.

2. The straight-line path (indicated by the D/c) line is not associated with any received pulse. Thus no TOA algorithm could measure the distance D without some error in this case. Mitigating these errors is a function of the position determination algorithms.
3. Before the measured pulse there is a noise floor which must not be mistaken as the signal. Within this noise floor there may be a signal associated with the straight-line path, but it cannot be measured with the presence of the receiver noise. Further, if the noise were greater, the first detectable signal would become embedded in noise, and thus first TOA signal is not detectable. In this case the first detectable pulse would be further delayed relative to the straight-line path. Thus the noise floor can have an indirect effect on the TOA measurement accuracy in a NLOS environment, resulting in "bias" errors. Thus with NLOS measurements, the mean measured TOA will always (ignoring random receiver noise effects) be greater than the straight line TOA.
4. A reasonable approach (as shown in the figure) is to assume the TOA is located at a point where the pulse is clearly above the noise floor. Such approach would result in a measurement with is dependent on the noise floor amplitude.
5. The performance of a peak detection TOA algorithm (as typically used in GPS receivers) would be poor in the multipath environment shown in Fig. 4.1. Thus indoor TOA algorithms should be based on the leading edge of the first detectable signal above the noise floor.
6. The shape of the receiver pulse is highly dependent on the bandwidth of the signal. For example, if the signal bandwidth in Fig. 4.1 is reduced to (say) 500 MHz the pulse rise time would be about 4 ns, and the pulse would be smeared (convolved with) by this wider time-pulse. As a consequence, the first delectable pulse would be at the main peak (near the 20 ns point in Fig. 4.1), and the measured TOA would be delayed considerably. Clearly the best accuracy is achieved with wide bandwidths, but regulations limit the available bandwidth in practical systems. Broadly the TOA measurement errors would be expected to be inversely proportional to the bandwidth of the transmitted signal.

Given the above characteristics of NLOS pulses, the following are guidelines to the desirable characteristics of leading edge algorithms. Ideally the measurement of TOA for a positioning system in a multipath environment should:

1. Scan from the noise region towards the signal region to locate the first detectable signal for further processing.
2. Be independent of the pulse amplitude.
3. Be independent of the RMS noise level before the signal.
4. Should try to minimize both the effects of noise and multipath interference; this typically requires a compromise in performance between these two effects.
5. At high signal-to-noise ratio (SNR) and low multipath, the TOA should approach a constant value, called the *epoch* of the pulse.
6. All nodes in a network should have similar pulse and TOA algorithm characteristics.

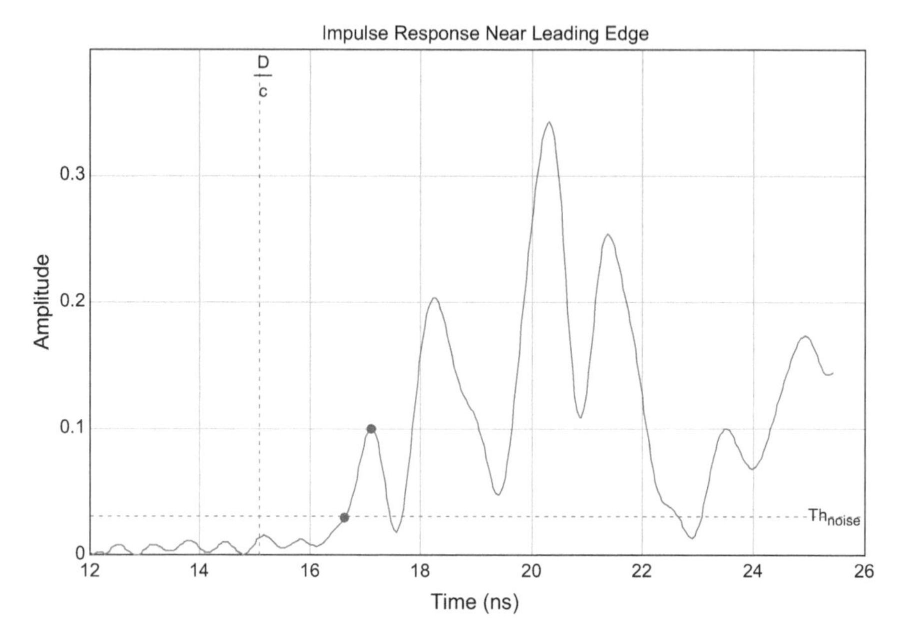

Fig. 4.1 Example of the receiver output pulse in a NLOS environment. The signal is UWB 6–9 GHz, with a range of $D = 4.5$ m through a wall, with the path approximately normal to the wall. In this case the signal epoch is measured where the signal exceeds the noise threshold level, and is shown by a dot. The local peak on the leading edge is also shown by another dot; note that this peak has a much smaller amplitude than the signal peak which is also delayed relative to the measured signal epoch

4.2.2 *Transmitter to Receiver Timing Measurements*

As described in the summary of the architecture of typical radio-based positioning systems (see Chap. 2) a key component is the measurement of the TOA of a radio signal. However, the TOA measurement cannot be used directly to measure the propagation delay, as the total delay from a transmitter to a receiver includes many other delays associated with the radio equipment. Thus before TOA can be used to determine position, it is important to understand the details of the delays through the radio equipment. In the following initial analysis, it will be assumed that the clocks in the transmitter and receiver have the same frequency, but later this restriction will be removed. However, it is important to note that the clocks are not synchronised in time, so that the clocks will have a random offset relative to an absolute time reference.[1] The following analysis is appropriate to all the types of

[1]The applications considered in this book are those that have low speeds relative to nodes in the system. In the case of the GPS system where the transmitters are on satellites travelling at a relatively high speed, the effects associated with Einstein's theory of special and general relativity need to be considered.

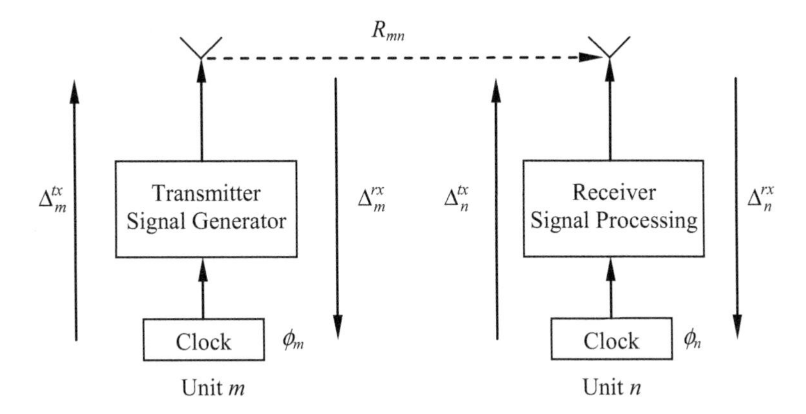

Fig. 4.2 Block diagram showing the major components in the transmission path from the transmitter to the receiver

positioning systems discussed in Chap. 2, namely navigation, tracking and mesh networks.

The TOA measurement encompasses the total transmission delays from the transmitter to the receiver as measured relative to the clocks in the transmitter and the receiver, and includes the delays in the baseband hardware, the radios, and the propagation delays. A block diagram of the transmission path and the associated parameters is shown in Fig. 4.2. Note that transmission can be from node m to node n, or in the reverse direction.

The transmitter baseband electronics has a clock which has a (unknown) offset relative to absolute time. Further, in the following analysis it is assumed that the transmissions are based on a pseudorandom code of finite length, which repeats at the code period, such as used in the GPS. As the receiver measures the TOA relative to this code, the time delay cannot be determined in an absolute sense, as time delays of multiple code lengths cannot be distinguished from one another. However, it will be assumed that the code length is greater than the total propagation delay, so that this ambiguity does not occur.[2] This restriction is typically unimportant for small terrestrial-based systems such as wireless sensor networks, as a code length of the order of (say) 10 μs allows a range of around 3 km. However, as the time measurement has a repetition period, it is appropriate for the clock offset to be referred as a phase offset (ϕ) rather than an (unknown) absolute time offset. The baseband pseudorandom code is used to modulate the RF signal in the radio, which outputs the radio signal from an antenna. This process involves transmitter delays in the radio, filters, and cables, all of which are assumed to be of unknown magnitude. The total delay in the transmitter processes is defined as Δ_m^{tx} for node m. Note that the delays in the radio transmitter, particularly the filters can be considerably greater than the propagation delay between the nodes, and further these

[2]This assumption is not valid to GNSS. The analysis is intended for terrestrial systems only.

delays vary over time. The reason for the variability is mostly associated with filter delays varying due to temperature and ageing effects. Further the length of the cable from the radio to the antenna can vary from node to node, due to particular details of an installation.

Now consider the TOA measurement in more detail. The basic time period in the transmitter and the receiver is the length of the symbol period (T_{pn}). As the code repeats with this period, normal continuous time is not available in the receiver, so that the transmitter and receiver time is related to absolute time by the expression

$$\tau = t \bmod T_{pn} \tag{4.1}$$

As this parameter repeats with a fixed period T_{pn} it is appropriate to consider the time measurements as a phase, rather than time which increases without limit. Further, if the clocks are assumed to be accurately synchronised (see Chap. 5 for details of the frequency synchronisation), then the local time in both the transmitter and the receiver can be defined by a single phase offset determined at $t = 0$. Thus the TOA measurement from m to n can be represented by the expression

$$TOA_{m,n} = \phi_m + \Delta_m^{tx} + D_{mn} + \Delta_n^{rx} - \phi_n \tag{4.2}$$

where ϕ is the clock offset phase $(0 \leq \phi \leq T_{pn})$, and the propagation delays and ranges are defined in a consistent set of units. Thus if distance (R) is defined in meters, and D is the range propagation delay in nanoseconds, then

$$D = R/c \tag{4.3}$$

where c is the propagation speed of 0.2998 m/ns. Thus the TOA measurement involves two parameters in the transmitter, and two in the receiver, but as each module has both transmit and receive functions, each node has three parameters associated with it (the clock phase, the transmitter delay, and the receiver delay).

While (4.2) is perfectly satisfactory for determining the TOA and further data processing, some simplification is possible by adopting a different time reference in each node. In particular, if the clock reference is altered so that the reference point is at the transmitter antenna (rather than the baseband clock), then the TOA expression becomes

$$\begin{aligned} TOA_{m,n} &= (\phi_m - \Delta_m^{tx}) + \Delta_m^{tx} + D_{mn} + \Delta_n^{rx} - (\phi_n - \Delta_n^{tx}) \\ &= \phi_m + D_{mn} + (\Delta_n^{rx} + \Delta_n^{tx}) - \phi_n = \phi_m + D_{mn} + \Delta_n - \phi_n \end{aligned} \tag{4.4}$$

Thus the TOA measurement can be expressed in terms of the clock phases in the transmitter and the receiver, the propagation delay, and the combined transmitter and receiver delays in the *receiving* unit only. As a consequence, only two parameters in each unit need to be defined, namely the local clock phase and the combined radio delays. As the clock phases and radio delays are unknowns to be determined as part of the position determination process, the processing is somewhat simplified, as only two (nuisance) parameters need to be measured instead of three as in the original TOA expression. Thus (4.4) is the basic starting point for the position determination for all types of systems described in Sect. 2.2 in Chap. 2. However, for nodes that both transmit and receive, it useful to consider the TOA measurements associated with of a pair of units, namely m and n. Thus applying (4.4) twice, the resulting two TOA measurements are

$$\begin{aligned} TOA_{m,n} &= \phi_m + D_{mn} + \Delta_n - \phi_n \\ TOA_{n,m} &= \phi_n + D_{mn} + \Delta_m - \phi_m \\ RTT_{m,n} &= \left(TOA_{m,n} + TOA_{n,m}\right)/2 = D_{mn} + (\Delta_m + \Delta_n)/2 \end{aligned} \tag{4.5}$$

Thus by combining two TOA measurements into a single RTT measurement, the clock phase parameters can be eliminated. Thus in this case, the TOA measurement data can be expressed in the simple form

$$RTT_{m,n} = D_{mn} + (\Delta_m + \Delta_n)/2 \approx D_{mn} + \Delta_0 \qquad m < n \tag{4.6}$$

where the measured data only involves the required range measurement and a delay parameter associated with a pair of units. The measurement RTT has a useful physical interpretation, namely the round-trip time from unit m to unit n and back to unit m (divided by two so that the range between the two units appears only once). The second expression in (4.6) equates the measurement with the range plus the radio delay parameter Δ_0, assumed to be the same in each unit. If this delay is assumed to be a known value, then the two TOA measurements combined can be used to estimate the range. However, as described previously, this results in low accuracy measurements due to the variability in the radio delays over time. Alternatively, the measurements can be used to estimate the radio delay parameter, if the two modes are at known physical positions (hence known separation).

4.2.3 Overview of TOA Algorithms

The overview of the characteristics of NLOS propagation and the general requirements for a TOA algorithm provide guidelines for appropriate TOA algorithms. This section provides a summary of some particular algorithms, with particular reference to leading edge algorithms. In particular, the following algorithms will be briefly reviewed; details of the algorithms and their performance are given elsewhere (Yu et al. 2009; Sharp and Yu 2014; Humphrey and Hedley 2008; Dardari et al. 2008; Chehri et al. 2007). However, the following subsections give some general performance formulas appropriate in the presence of Gaussian noise and multipath interference.

There have been many approaches reported in the literature that attempt to improve TOA estimation of bandlimited signals in the presence of noise; these approaches can be summarised at "super-resolution" algorithms. For example, (Humphrey and Hedley 2008, 2009) provide a comparison of several techniques, including the well known MUSIC super-resolution algorithm and an algorithm based on a database of pulse templates in multipath situations, which is shown to be superior to MUSIC. However, these methods are computationally intensive, so a simpler, low computational direct approach based on the characteristics of the leading edge of multipath pulses is suggested in the following subsections. These algorithms are based on the observed characteristics of received pulses in a multipath environment.

4.2.3.1 Peak Algorithm

The concept of the Peak algorithm is simple—locate the peak of the pulse and define this as the epoch of the signal. For an ideal *wideband* signal without multipath interference the peak is sharply defined, so that accurately locating the peak is possible. However, with multipath interference, as illustrated by Fig. 4.1 the signal peak can be a poor estimate of the TOA.

Some form of peak detection is often used in GPS receivers, so that there is much literature on the performance of peak detection (Sharp et al. 2009). However, most of this literature is not particularly relevant to other positioning systems. The reason for this is the peculiar nature of the GPS spread-spectrum (civilian or C/A mode) signal. While the chip rate is only about 1 Mchips/s, the signal channel-width is 20 MHz. This results in a very shape peak (see Fig. 4.3), which has been shown to assist in mitigating the effects of multipath interference. The military (or P code) has a chip rate of about 10 Mchips/s, and the 20 MHz bandwidth was designed to accommodate this signal. At the time the GPS was designed it was believed that only the chip rate was important in characterizing the accuracy of the TOA algorithm, but it was later discovered that the signal bandwidth is also of prime importance. Thus modern GPS receivers have a performance which approaches that of the military mode.

In a typical design the chip rate would be selected to be about half the RF *channel* bandwidth, but the filtering of the wideband binary signal results in the pulse distortion shown in Fig. 4.3, including rounding of the peak and a leading edge "tail". Further filtering results in a distorted pulse with "sidelobes", which restricts detection of small first-arrival signals near the noise threshold. The rounding of the peak makes accurate peak detection difficult in the presence of noise or multipath signals, as these small noise signals can result in large peak TOA detection errors. In general, the accuracy of the TOA peak detection is inversely proportional to the derivative of the pulse shape curve, so that a rounded peak results in low accuracy. Conversely the straight section of the leading edge has maximum accuracy in the presence of noise. For Gaussian type receiver noise the amplitude statistics of the noise are independent of the location on the pulse, but multipath related pulse noise can vary considerably as a function of the delay excess

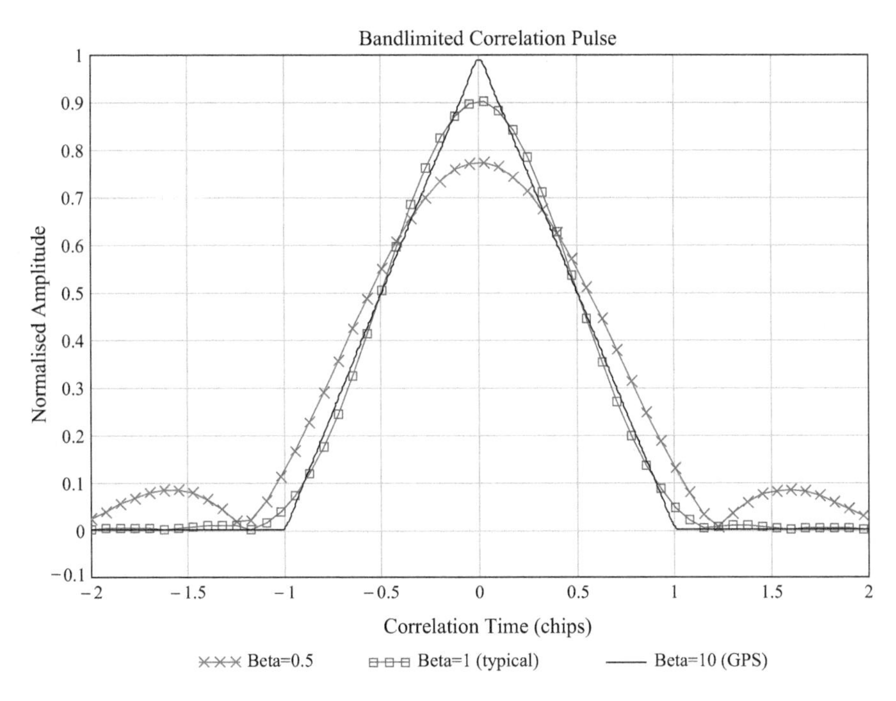

Fig. 4.3 Effect of bandlimiting the signal, where β is the bandwidth normalise to the chip rate, so that $\beta = B\tau_0$, where τ_0 is the chip period and B is the channel-width

of the multipath signal, so that detection at pulse positions away from the peak typically result in better performance.

For the practical implementation of the Peak algorithm usually the peak is not detected, but rather two points near the peak are used, separated by a time differential which is related to the bandwidth parameter β. For example, for GPS a separation of 0.2 chips could be used, while for the typical design with $\beta = 1$ a separation of a quarter of a chip would be appropriate. The peak detection algorithm shifts these two points on the pulse shape until they are of equal amplitude, with the epoch defined at the mid-point. Note that this shifting is made using a feedback loop which adjusts the frequency of the local oscillator in the receiver to vary the sampling times associated with pulse detection. Details of this process are given in Chap. 5. Also for the typical case, due to the sampling theorem for bandlimited signals only points with half-chip spacing are required to fully define the shape of the pulse, so that interpolation of the sampled data (half-chip spacing) can specify all points on the pulse without the need for any additional signal processing by the correlator.

4.2.3.2 Threshold Algorithm

The Threshold algorithm is perhaps the simplest form of leading edge algorithm, with the principles related to the bandlimited pulse shape as outlined in Fig. 4.3. The concept of a TOA Threshold algorithm (Wang et al. 2009; Humphrey and Hedley 2009) is simple—the epoch of the pulse is defined by the position where the leading edge of the (normalized) pulse crosses (exceeds) a specified threshold level. In the case of the nominal pulse without multipath or noise corruption, the threshold amplitude (a_{th}) can be defined by the relationship to the peak (A) of the pulse, namely

$$a_{th} = \alpha A \tag{4.7}$$

here α is a fixed parameter of the algorithm so that the threshold time position (pulse epoch) is independent of the pulse amplitude. The corresponding nominal reference epoch is $\alpha\tau_{pulse}$ (assuming a triangular pulse shape) from which TOA measurement errors can be determined for a multipath pulse. Hence in the ideal case the TOA error will be zero. However, in the presence of receiver noise and multipath signals, defining the threshold level is more difficult. In particular, in a multipath environment, the peak signal can be remote from the leading edge (see Fig. 4.1), so that applying (4.7) will result in a threshold different from that in the error-free case, resulting in measurement errors. Secondly, receiver noise before the leading edge can be falsely interpreted as the leading edge of the pulse. Clearly from the above analysis of the variation in the shape of the leading edge the threshold should be as low as possible, but conversely should be well above the noise level. Further, the multipath "noise" will result in noise-related errors in determining the peak amplitude (A), which in turn will result in errors in defining the threshold level and consequentially the pulse epoch. Based on this discussion, the proposed Threshold algorithm is summarized as follows:

1. Using the position of the pulse peak amplitude as a time reference guide, the noise section of the pulse (well before the peak) is located, and the RMS noise (σ_n) determined. An initial noise threshold is defined as $k\sigma_n$, where k is typically 3 or 4 to minimise false detection of the pulse leading edge signal.
2. The time position where the pulse leading edge exceeds the noise threshold is determined. To minimize false detection, a minimum of three adjacent samples should be checked, and the search continues if all three do not exceed the threshold.
3. Due to the curved nature of the multipath leading edge "tail" (see Fig. 4.3), this position will be somewhat greater in time than the "true" start of the leading edge, and will also be a function of the pulse SNR.
4. Using this noise threshold location on the leading edge, define a position in the measured data advanced by the known rise-time of the nominal pulse. The peak of the pulse within this time interval is located, and defined as the nominal peak

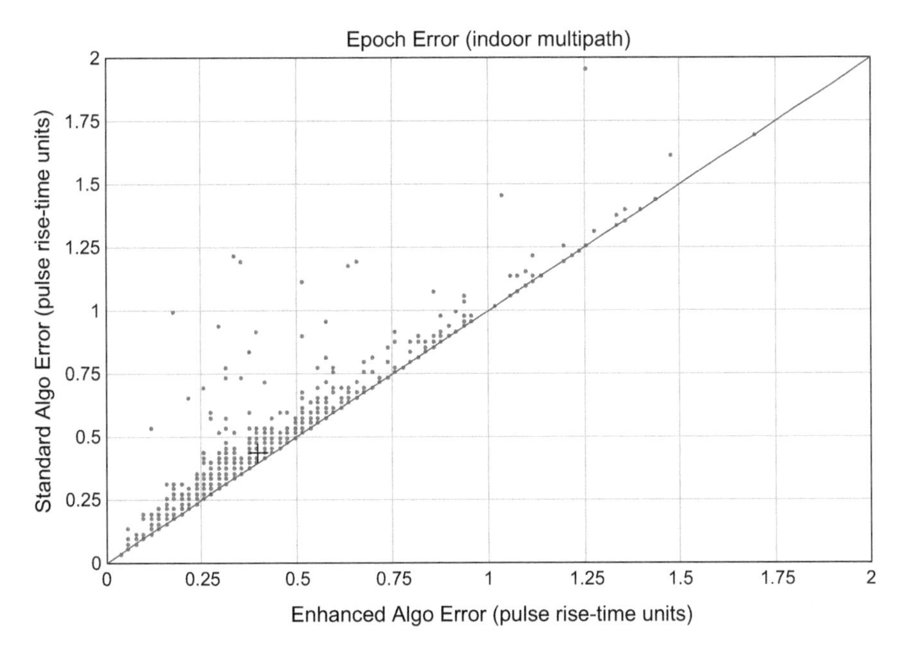

Fig. 4.4 Comparison of the standard threshold algorithm and enhanced algorithm TOA performance. The line shows that the standard algorithm has errors greater or equal to the enhanced algorithm. The mean performance of both algorithms is shown by a cross (at about an error of 0.4 pulse rise-time units). The standard of errors for both algorithms is quite similar, being about 0.25 in normalized units

A for use in (4.7). Note that $a_{th} > k\sigma_n$ is assumed to hold; otherwise define the threshold amplitude as $a_{th} = k\sigma_n$.

5. The scan as in step 2 is repeated with the updated threshold level to locate the pulse epoch.

In Fig. 4.4, the performance of the enhanced threshold algorithm described above is compared with the standard algorithm where the threshold is adjusted using the pulse peak. The results are based on a multipath simulation with 30 multipath signals with the amplitudes having Rayleigh statistics, and a SNR of 30 dB. The results show that the enhanced algorithm has less high-magnitude errors, but there is considerable scatter of errors in both cases. Note also there is a considerable bias error, 0.44 for the standard algorithm, and 0.40 for the enhanced.

4.2.3.3 Ratio Algorithm

An ideal TOA algorithm would be unaffected by the amplitude of the pulse, or by other effects such as multipath interference. As shown the Threshold algorithm is affected by both of these, and thus an alternative type of algorithm may have a superior performance. In particular, an ideal algorithm would be one that was insensitive to the amplitude of the signal in determining the TOA, but rather was based on the *shape* of the leading edge. Also the algorithm should be minimally affected by multipath signals. One such algorithm is the Ratio algorithm, the characteristics of which are now summarized; for more details, refer to (Sharp et al. 2009, Sect. 4.6; Sharp and Yu 2014).

As the pulse will be bandlimited so that typically there is only two measurements in the sampled pulse data, an algorithm should ideally only use two samples from the leading edge. However, because of the sampling theorem, if required the complete pulse shape can be reconstructed from the sampled data. To avoid using the samples directly, the Ratio algorithm uses the ratio of the two samples to define the epoch of the pulse. Clearly, the ratio (ρ) is unaffected by the amplitude of the pulse, so errors associated with defining the signal amplitude (particularly in the case of multiple peaks as shown in Fig. 4.1) are avoided. The ratio of two points on the interpolated leading edge as close as possible to where the pulse emerges from the noise will be close to the best possible estimate of the TOA. The location of these two points determines the performance of the algorithm. If point #1 has a low amplitude and point #2 has a larger amplitude but a relatively small separation from the first point, then only data near the first detectable signal are used thus minimizing errors due to multipath interference. However, as the points become closer together the ratio is more affected be random noise; thus there is a trade-off between multipath and random noise performance. The optimum parameters will depend on the particular application, and the relative performance of these two corrupting effects.

For a practical implementation, the pulse is scanned from the noise section of the pulse until a threshold noise level is exceeded. The ratio of the two amplitudes will then tend to increase from a low value as the pulse is scanned. The shape of this ratio as a function of the scan time for the ideal pulse is a known shape which is directly related to the pulse shape only. The epoch of the pulse can be defined when the ratio is at some set point, namely $\rho = \rho_0$. A feedback loop can be devised such that this set point is maintained by adjusting the local clock frequency as described in Chap. 5. This procedure ensures that the epoch of the local pn-code closely matches that of the TOA of the received signal.

4.2.3.4 Projection Algorithm

The TOA measurements described previously while simple to implement do not specifically address the problems associated with severe NLOS conditions commonly experienced indoors. While some improvements in the Threshold and Ratio methods result in superior performance compared with simpler algorithms (such as Peak and fixed threshold), even the improved algorithms fail when the first signal detectable is small compared with stronger but more delayed multipath signals. To consider this problem in more detail it is useful to examine in some detail examples of measured pulse data, and further the effect of bandlimiting the signal. In the following example, the data were measured with a bandwidth of 2–6 GHz, and then the performance at lower bandwidth computed by convolving the original UWB data with a narrowband pulse with a specified pulse rise-time. In the following example the narrowband signal has a pulse rise-time of 25 ns, which corresponds to a bandwidth of 40 MHz, and the signal could thus be transmitted in the 2.4 GHz ISM band which has a channel band of 80 MHz.

The data were originally processed using the standard fixed threshold method, but the performance was poor as the bandwidth decreased. Thus the leading-edge pulse data was analysed to attempt to discover the reasons for the poor performance. A typical problem can be observed in the difficult example case shown in Fig. 4.5, where noise can mask the initial narrowband signal, resulting in measurement bias errors; clearly this effect worsens as the SNR reduces. For example, in Fig. 4.5a, at high SNR (as in the figure) the long leading-edge "tongue" can be detected, so that the TOA error is small (measured relative to the leading edge of the nominal triangular pulse). However, as the SNR reduces, this low-amplitude section will be buried in the noise, resulting in this case with an error of about 20 ns. Thus it is clear for good NLOS indoor performance the TOA measurements must operate at high SNR; the tradeoff for good NLOS performance at relatively low bandwidths (say less than 200 MHz) is that the SNR must be relatively high when compared with LOS situations. Thus a SNR of at least 30 dB is used in practical positioning systems, such as WASP. Such high SNR figures can be obtained by applying process gain in typical spread-spectrum systems operating in ISM bands without having to increase the transmitter power, which should be as low as possible for mobile applications.

The proposed algorithm first detects the position where the signal emerges from the noise, and then projects backwards in time to estimate the "epoch" position buried in the noise. This process is best explained with an example, such as shown in Fig. 4.5b, which is a zoomed-in version of Fig. 4.5a. The first process is to determine the noise level before the signal in the impulse response. The threshold is (say) set at 1.2 times the peak of the noise in the noise preamble region (negative time in Fig. 4.6b). The 1.2 figure is selected to allow a small extra margin in the noise estimate, as there is a small probability the noise estimate in the preamble is exceeded by noise in the signal section, which would lead to a false signal detection. For a signal of peak amplitude A, the noise threshold level is defined as $Th_{noise} = \alpha A$. The point where the pulse exceeds this threshold (circular dot in Fig. 4.5b) is the preliminary estimate of the epoch. At this point, the slope of the pulse is also estimated. Starting at this preliminary epoch estimation, the pulse is traced backwards in time while the following conditions apply:

1. The pulse amplitude is greater than $\beta = 0.3$ times the noise threshold (Th_{noise}).
2. The slope of the pulse is greater than $s_{\min} = 0.3$ times the nominal (triangular) slope of the leading edge of the pulse. See Sect. 4.2.5 for details.

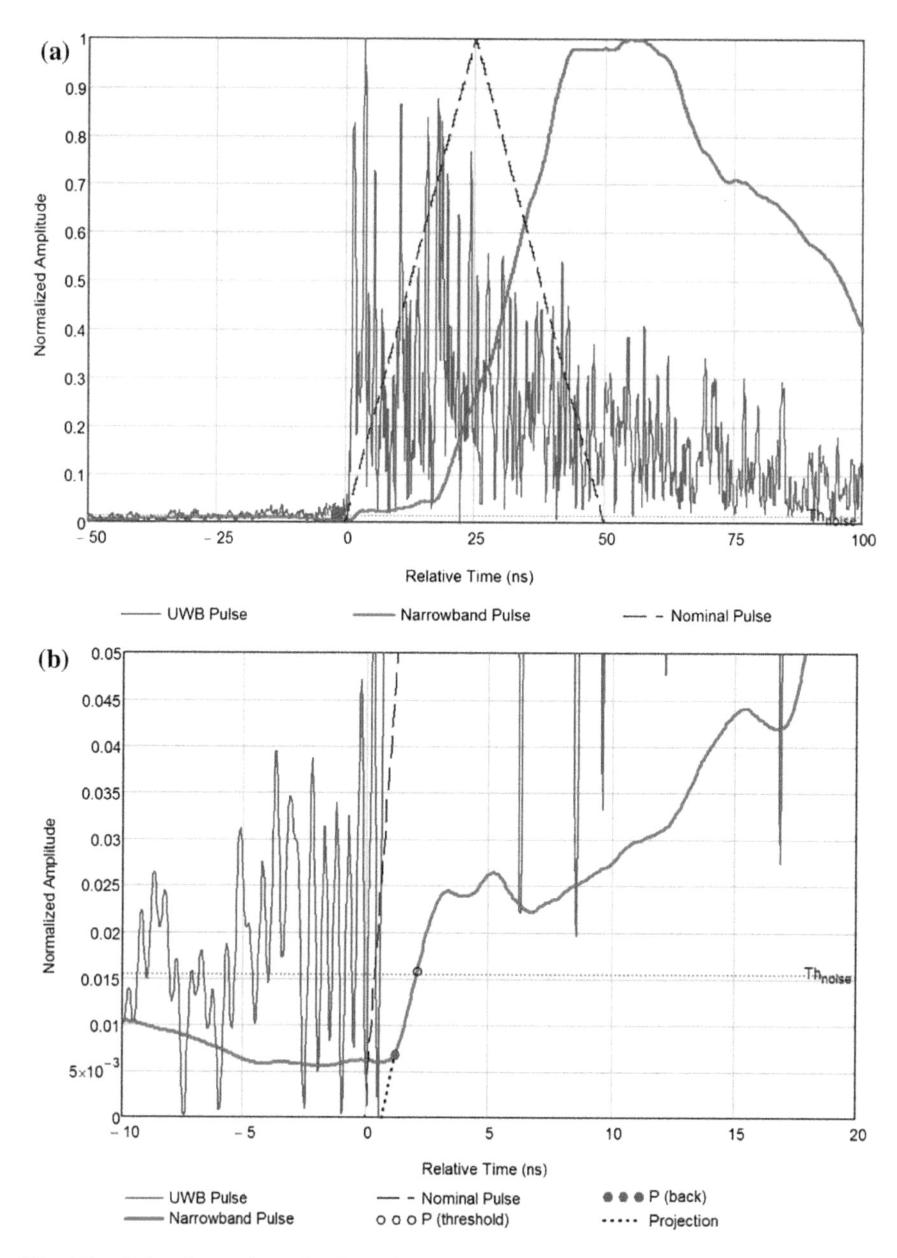

Fig. 4.5 **a** Pulse shape with a rise-time of 25 ns (thick line) The UWB pulse is also shown (thin line). The nominal pulse is the narrowband triangular (dashed) transmitted pulse. **b** Zoomed-in section of the leading edge of the pulse in Fig. 4.5a. The thin (noisy) line is the UWB pulse, and the thick line is the bandlimited pulse (rise-time 25 ns) whose TOA is to be measured. The purpose of the dots is explained in the text

These two conditions minimize the possibility of the back-tracing process using (by mistake) the noise section before the pulse, but allows the epoch to be located at signal amplitudes below the noise threshold level.

Finally, using this second point (dot) as the starting point (see Fig. 4.5b), and the previously estimated slope of the pulse near the threshold, the pulse is projected to zero amplitude. This zero-amplitude intersection point is then considered to be the signal epoch. If the above process is not followed in this case, it is clear that the initial epoch position would be very sensitive to the noise threshold level. Note also that this procedure correctly predicts the location of the epoch in the absence of multipath signals when the pulse is (nearly) triangular in shape.

4.2.4 Theoretical Gaussian Noise Performance

The performance of TOA algorithms in the presence of random (Gaussian) noise is briefly reviewed in this section. Rather than analyse each algorithm (such as those defined in Sect. 2.3.2), the general characteristics of all algorithms are presented. In this analysis it is convenient to assume the nominal pulse is normalized to unit amplitude, but corrupted with random noise with zero mean and a standard deviation such that the signal-to-noise ratio is $\gamma = 1/\sigma_n^2$. Further, the ideal pulse will be typically be approximately triangular in shape, as shown in Fig. 4.3, with a rise-time $\left(\tau_{pulse}\right)$. Using techniques described in (Sharp et al. 2009) it can be shown that the standard deviation in the TOA error can be approximately expressed in the form

$$\sigma_\varepsilon = \frac{K_{algo}\tau_{pulse}}{\sqrt{\gamma}} \tag{4.8}$$

where K_{algo} is a dimensionless constant which depends on the implementation details of the particular algorithm; typical values are in the range $0.7 \le K_{algo} \le 1.5$.

It is informative to compare the expression (4.8) with the Cramer-Rao Lower Bound (CRLB) (Sharp and Yu 2013). For a pulse with a signal bandwidth B (not the channel bandwidth), the error STD is constrained by the lower bound

$$\sigma_{CRLB} = \frac{1}{\sqrt{8}\pi B\sqrt{\gamma}} \tag{4.9}$$

Note in this case the SNR is measured after the correlation process, which enhances the baseband SNR, and hence reduces the errors. The ratio of the TOA performance to the CRLB is thus

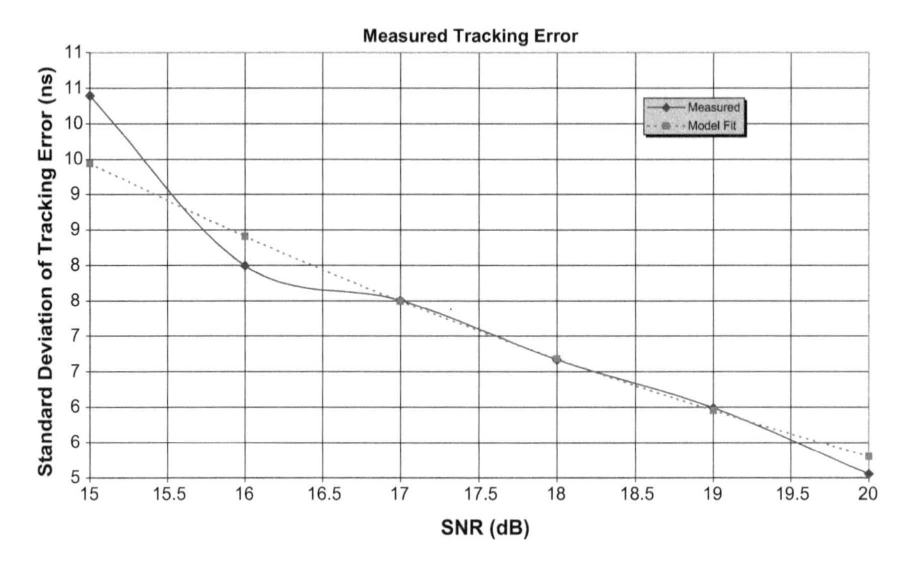

Fig. 4.6 Measured tracking error as a function of SNR. Chip period 25 ns

$$\sigma_{\varepsilon}/\sigma_{CRLB} = K_{algo}\sqrt{8}\pi\left(\tau_{pulse}B\right) \approx K_{algo}\sqrt{8}\pi \tag{4.10}$$

where typically $\tau_{pulse}B \approx 1$ has been invoked in the final expression in (4.10). Note also that (4.10) is independent of the SNR. Thus leading edge algorithms, which are optimized to reduce the effects of multipath signals, are far from the lower bound with Gaussian noise corruption. However, if the SNR is sufficiently high (say 30 dB), the standard deviation of the TOA is only about 3% of the pulse rise-time, which is much smaller than typical errors associated with multipath interference. This is shown in Fig. 4.6 for the Threshold algorithm (threshold set to 25% of pulse-rise-time), and more generally as considered in the next section. In most designs for indoor systems, the limiting factor is multipath interference, not random receiver noise.

4.2.5 TOA Algorithm Multipath Performance

This subsection provides a theoretical analysis of the performance of a simple threshold detection algorithm (Sect. 4.2.3.2) and the performance of the proposed enhanced algorithm using back projection (Sect. 4.2.3.4).

4.2.5.1 Simple Leading-Edge Algorithm Performance

The TOA estimation with a simple threshold algorithm is based on detecting the first signal above a threshold determined by the noise before the measured signal pulse. The case of a NLOS multipath signal with no detectable direct signal has been analysed in detail in Chap. 13, Sect. 13.6, so that only a summary will be given here.

In a multipath environment, the shape of the leading edge will vary according to the multipath signals present in the leading edge. Thus the shape of the leading edge will be statistically random. However, it is shown in (Sharp and Yu 2013) that the *average* shape of the leading edge in a NLOS multipath environment when the transmitted pulse is triangular is in normalized form

$$P_{mp}(\tau) = \tau^{3/2} \qquad (0 \leq \tau \leq 1) \tag{4.11}$$

where the time (τ) is normalized by the pulse rise-time, and the amplitude at the nominal peak delay is normalized to unity. The shape of the nominal triangular pulse and the multipath leading edge is shown in Fig. 4.7; also shown is noise before the pulse and the corresponding threshold level.

A simple estimate of the "average" TOA performance in a multipath environment can be determined using the average pulse shape, and the correction described above. Thus the error estimate ε_0 is given by the time delay between the curved multipath and triangular pulses, namely

$$\varepsilon_0 = \left[Th_{noise}^{2/3} - Th_{noise} \right] \tau_{pulse} \tag{4.12}$$

For example, the threshold amplitude used in the WASP system (5.8 GHz ISM band, 125 MHz channel width) is fixed at $\alpha = 0.125$ (Sathyan et al. 2011), and the pulse rise-time is 13.5 ns, so using (4.12) the average bias error will be $0.125\tau_{pulse}$, about 1.6 ns, or equivalent to about 0.5 m. This performance figure can be used to compare with the performance of the proposed enhanced TOA algorithm.

4.2.5.2 Enhanced Leading-Edge Algorithm Performance

The methodology of determining the performance of the simple threshold algorithm can be applied to the proposed "enhanced" algorithm described above. Initially the analysis is again based on the average pulse shape, as shown in Fig. 4.7. In particular, the algorithm defines three points on the multipath pulse leading edge, namely the threshold point[3] P_1 (dot), a second point P_2 (circle) which is used in the

[3]For simplicity, the symbol P is used to both identify the various points used in the algorithm, and the amplitude of the signal (italic symbol) at that point. The subscripts 1, 2 and 3 refer to the three points shown in Fig. 4.7, and their associated parameters.

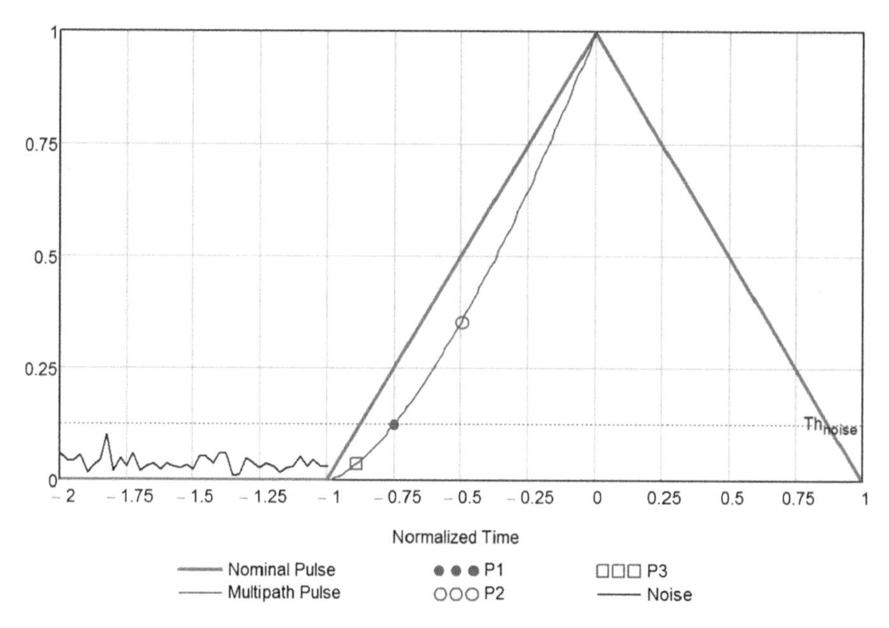

Fig. 4.7 Geometry of the pulses and noise, and associated points on the nominal mean multipath pulse. The random noise (Rayleigh amplitude distribution) is such the that SNR (relative to the peak) is 30 dB, which is the minimum SNR used by the WASP system

estimation of the slope of the leading edge, and a third point P_3 (square) which back-traces the pulse to an amplitude at about the level of the noise defined by the threshold, but well below the threshold level. Because of noise (in particular due to multipath signals) in the leading edge (not shown in Fig. 4.7 as it shows mean shapes), the point P_2 must not be too close to P_1 to allow estimating the slope with adequate accuracy. A typical separation (Δ) would be 0.25 to 0.4 times the pulse rise-time (Yu et al. 2009, Chapter 4); the value in the figure is $\Delta = 0.25$. The trace-back process is defined by a parameter β, so that the pulse amplitude P_3 is given by

$$P_3 = \beta P_1 = \beta Th_{noise} \tag{4.13}$$

For example, again using WASP as a guide, the SNR is at least 30 dB, and the threshold amplitude is -20 dB, so that suggested trace-back would be to an amplitude of -10 dB, or $\beta \approx 0.3$. Given this description of the parameters of the enhanced TOA algorithm, the pulse slope estimate is given by

$$s(\tau_1) = \frac{dP_{mp}(\tau_1)}{d\tau} \approx \frac{(\tau_1 + \Delta)^{3/2} - \tau_1^{3/2}}{\Delta} \tag{4.14}$$

Finally, using P_3 as a starting point and the slope given by (4.14), the back-projection time correction using (4.13) and (4.14) is given by

$$\delta(\tau_1) = P_3/s(\tau_1) = \frac{\beta\tau_1^{3/2}\Delta}{(\tau_1+\Delta)^{3/2} - \tau_1^{3/2}} \tag{4.15}$$

and the corresponding TOA error is

$$\varepsilon = (\tau_3 - \delta)\,\tau_{pulse} = \left[\beta\tau_1^{2/3} - \frac{\beta\tau_1^{3/2}\Delta}{(\tau_1+\Delta)^{3/2} - \tau_1^{3/2}}\right]\tau_{pulse} \tag{4.16a}$$

which by expanding as a truncated series can be approximated to

$$\varepsilon \approx \left[\beta^{2/3} - \frac{8\beta}{3(4+\Delta/\tau_1)}\right]\tau_1\tau_{pulse} \tag{4.16b}$$

Applying (4.16a, 4.16b) with the parameters defined in Sect. 4.2.5.1, the resulting bias error is $0.071\tau_{pulse} = 0.9$ ns, or about half that of the simple threshold algorithm. While (4.16a, 4.16b) suggests that the error can be reduced by using a lower value of the back-tracing parameter β, if too small a value is used the back-tracing will enter the noise section of the pulse, and thus will not result in accurate estimates.

4.2.5.3 Statistical Performance of Enhanced Leading-Edge Algorithm

The analysis of the performance of the enhanced leading-edge TOA algorithm in Sect. 4.2.5.2 uses the mean shape of the leading edge, but in practice the leading edge will vary from pulse to pulse in a random manner. It is shown in (Sharp and Yu 2013) that the amplitude at positions along the leading edge exhibit an independent Rayleigh statistical distribution (if the points are not too close together in relation to the pulse rise-time). Thus the projection correction will have random characteristics, as it is the random slope estimate that is used for the projection correction. Note also that this analysis assumes that the SNR is sufficiently high so that the random Gaussian (thermal) noise can be ignored. This assumption is justified, as in the practical situations of interest the SNR will be at least 30 dB, and often 40 dB or greater; in a NLOS environment, the performance is determined by multipath signals, not Gaussian noise.

From the analysis in Sect. 4.2.5.2, the projection correction is given by

$$\delta = \beta\Delta\frac{P_1}{P_2 - P_1} = \beta\Delta\frac{\rho}{1-\rho} \qquad (\rho = P_1/P_2 \le 1) \tag{4.17}$$

Thus the statistical performance of the projection correction depends on the statistical characteristics of the *ratio* of two Rayleigh distributed independent random variables. This statistical analysis (described the Appendix in Sharp and Yu 2014) shows that the PDF of the ratio ρ is given by

$$PDF_{\rho}(\rho) = \frac{2\omega^2 \rho}{(\rho^2 + \omega^2)^2} \tag{4.18a}$$

where the definitions in Sect. 4.2.5.2 and

$$\omega = \sigma_1/\sigma_2 = \left(\frac{\tau_1}{\tau_1 + \Delta}\right)^{3/2} \tag{4.18b}$$

are used and the σ parameters are associated with of the two Rayleigh distributions at P_1 and P_2. The projection correction PDF can be calculated to be

$$PDF_{\delta}(\delta) = 2\alpha\Delta\omega^2\left(1 + \omega^2\right)\frac{\delta(\delta + \alpha\Delta)}{\left(\omega^2(\alpha\Delta + \delta)^2 + \delta^2\right)^2} \tag{4.19}$$

where the relationship $PDF_{\delta}(\delta)\, d\delta = PDF_{\rho}(\rho)\, d\rho$ is used, and that its integral must equal unity. The plot of the PDF of the projection correction is shown in Fig. 4.8. Also shown is the correction according to (4.16a, 4.16b). As can be observed, the projection correction varies considerably.

Based on the time position τ_3 of the point P_3 on the mean pulse shape (4.11), the projection correction can result in either positive or negative measurement errors. If $\delta < \tau_3$ the resulting TOA measurement will have a positive bias; otherwise the bias will be negative. As most measurements will be the former, the enhanced algorithm will almost always result in a small residual (positive) bias, but in a small proportion of cases the algorithm will return negative biases.

Using (4.19) the mean projection correction can be computed to be unbounded due to the "tail" of the PDF distribution. Thus for a practical implementation this tail needs to be truncated at some suitable limit. The enhanced algorithm thus places a lower limit $s_{\min}$ on the estimated slope (s) of the leading edge of the pulse, as given by (4.14). In practice this means that the search for the point P_1 requires that both $P_1 > Th_{noise}$ and $s > s_{\min}$ conditions must be valid. Alternatively, from (4.17), the projection correction can be expressed as

$$\delta \leq \frac{\rho_{\max}}{1 - \rho_{\max}}\beta\Delta \equiv N\beta\Delta \tag{4.20}$$

where the constraint is on the maximum allowable $\rho = \rho_{\max}$, so that as $\delta = P_3/\frac{dP}{d\tau} = \beta Th_{noise}/s_{\min}$ the minimum allowable slope on the leading edge is given by

$$s_{\min} = \frac{Th_{noise}}{N\Delta} \tag{4.21}$$

Thus the PDF is restricted to an upper limit for the projection correction $\delta = N\alpha\Delta$, where N is an appropriate value, such as $N = 3$. Then from (4.21),

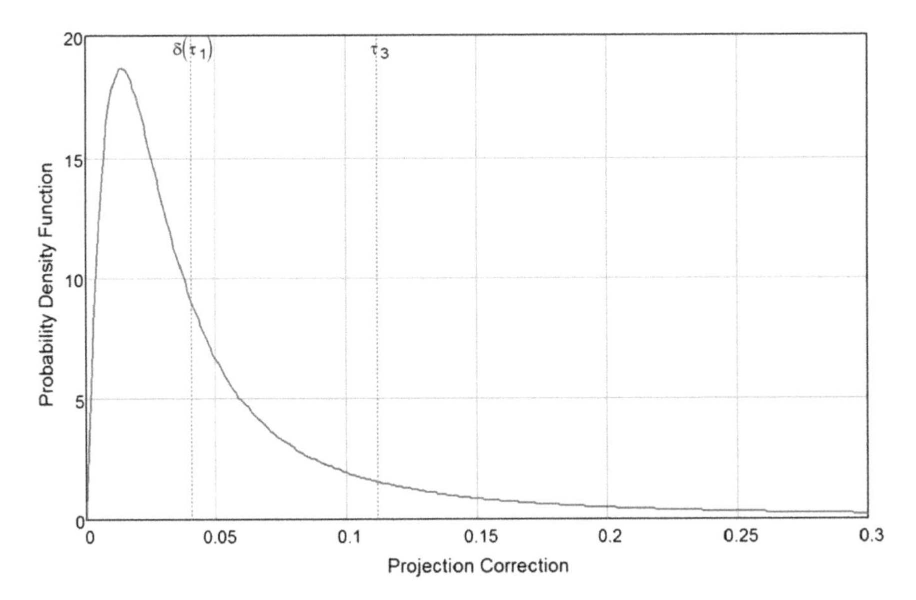

Fig. 4.8 PDF of the projection correction. The parameters are those shown in Fig. 4.7, namely $\Delta = 0.25$ and $\alpha = 0.125$. The correction based on the mean shape of the leading edge is $\delta(\tau_1)$ (see (4.15)). The parameter τ_3 is the time delay of point P_3 based on the mean shape of the leading edge

$s_{min} = 0.167$, or $\rho_{max} = 0.75$. The parameters in (4.20) and (4.21) thus define the upper limit of the PDF distribution, based on characteristics that can be derived from the measurements.

Figure 4.9 shows how the mean and standard deviation of the truncated statistical distribution varies with the parameter N. The suggested value of the $N = 3$ for the parameter requires a tradeoff between the mean error and the probability of a negative bias error.

4.2.6 *Statistical Effect of Signal Bandwidth*

Using the measurement technique described in Sect. 4.2, and the leading-edge TOA algorithm described in Sect. 4.2.5, the TOF can be estimated, and hence range errors can be determined from measured data. These results are based on UWB measurements in a building used for testing positioning systems, as described in Chap. 11; Fig. 11.21 shows a plan of the building. The range errors are calculated as a function of the pulse rise-time from about 0.4 ns (raw UWB data) to 25 ns (appropriate for the bandwidth available in the 2.4 GHz ISM band). The range errors are expected to be a function of the signal bandwidth (or pulse rise-time) and the range.

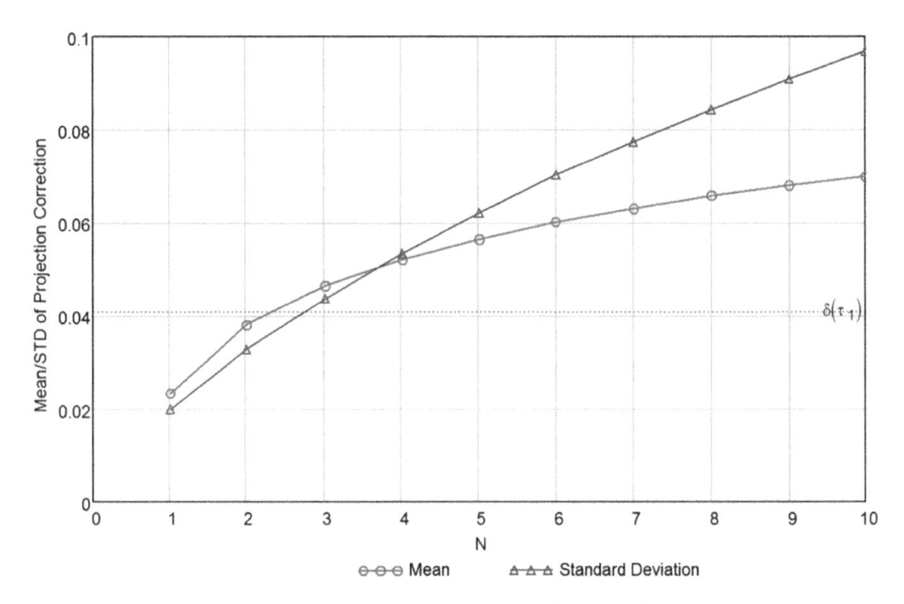

Fig. 4.9 Variation in the mean and standard deviation as a function of the constraint parameter *N*, using the parameters defined in Fig. 4.7

A statistical summary of range error parameters is plotted as a function of the pulse rise-time in Fig. 4.10. As expected, both the mean and standard deviation increase (approximately linearly) with the pulse rise-time. Interestingly the peak errors also have a similar variation with the rise-time as the standard deviation, with the peak errors being about three times the standard deviation. The constraint on peak errors is one of the important features of the Projection algorithm, as constraining peak errors makes position determination more reliable (no large position errors). For rise-times up to about 7.5 ns (or >200 MHz bandwidth), the mean performances are quite similar, which shows there is little added benefit for the increased bandwidth to that available in UWB systems. As the rise-time further increases the mean and particularly the standard deviation increase, but the mean performance with a rise-time of 25 ns (bandwidth 40 MHz) is only about three times that for UW bandwidths. This result indicates that the enhanced leading-edge algorithm is very effective in constraining range errors in low bandwidth signals.

Also shown in Fig. 4.10 are the mean measured range errors, which are the average performance of a number of other studies (Bellusci et al. 2008; Gentile and Kik 2006; Chehri et al. 2007; Alavi and Pahlavan 2006; Wang et al. 2009; Prieto et al. 2009). As can be observed the mean performance of the enhanced algorithm is superior at all bandwidths.

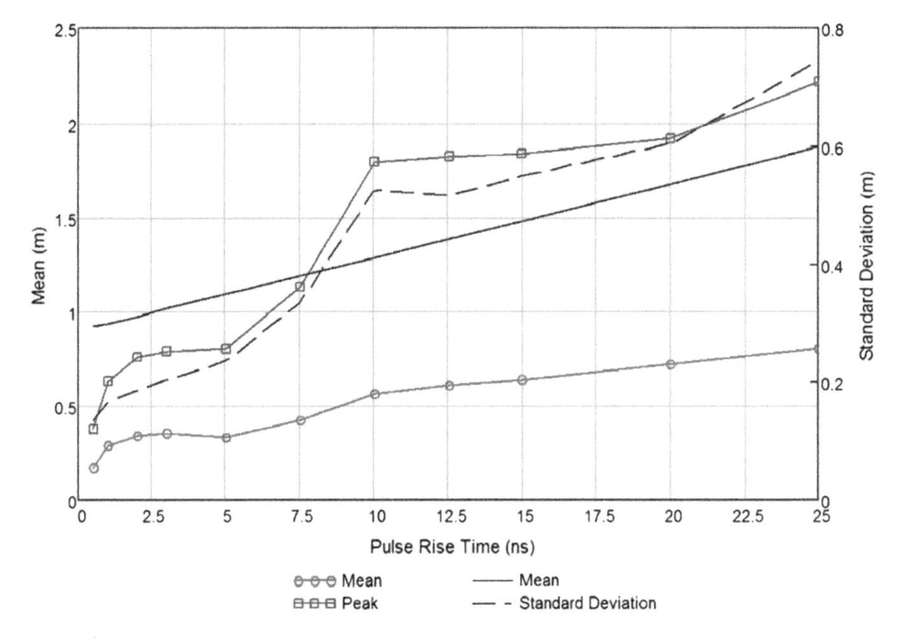

Fig. 4.10 Mean, peak and standard deviation of the range errors plotted as a function of the pulse rise-time. The plot also includes the mean error from other studies

4.2.7 *Measured Indoor TOA Errors*

The scattering of radio signals indoors is complex and depends on the details of the indoor environment, so theoretical calculations are difficult. However, experimental measurements (Alavi and Pahlavan 2006; Bellusci et al. 2008; Gentile and Kik 2006; Alsindi et al. 2009) show that the ranging errors have a bias error which increases approximately linearly with range, and a random component which is largely independent of range. Figure 4.11 shows typical error data using the WASP system (see Sects. 6.4, 11.4.4.3 and 14.5.1 for more details). As shown in Sects. 6.2.2 and 11.3.1.3 the range error model associated with indoor NLOS propagation is

$$\Delta R = (\delta R_0 + \lambda R) + \varepsilon_R \tag{4.22}$$

where the term in brackets is the linear bias model, and ε_R is a random component with zero mean.

For more information on measured range errors, refer to Chap. 11 on system testing. Also refer to Chap. 6 on a position determination method which compensates for bias errors that can be modeled using (4.22). The model (4.22) is based on measurements, but this model is also supported by theoretical calculations in Chap. 13, which provide analytical expressions for the parameters δR_0 and λ in (4.22). These analytical expressions are based on more basic characteristics of

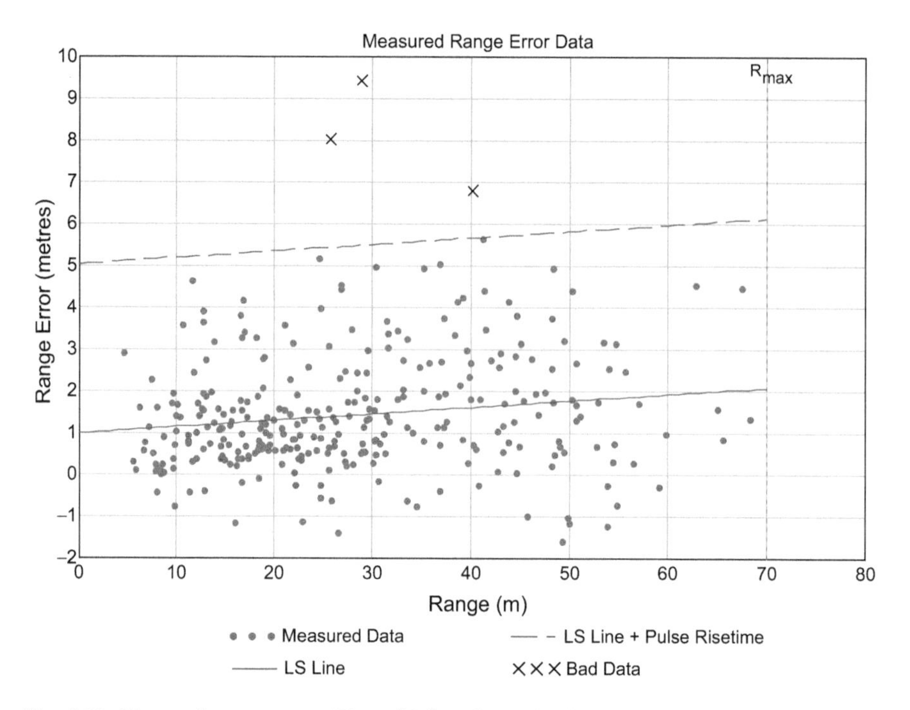

Fig. 4.11 Measured range errors. The solid line shows the linear bias least-squares fit, and the dotted line is one standard deviation of 1.32 m. The zero-range intercept bias is 1.0 m, and the slope is $\lambda = 0.0155$

indoor propagation environments, such as the mean separation of walls, the wall materials, and the size of objects along the propagation path. These theoretical models thus allow estimates of the delay excess to be made in any particular indoor environment, without the need for measurements. As a consequence, a priori calculations can be made that allow a design to be matched to a particular building or building type and the desired positional accuracy.

References

Alavi B, Pahlavan K (2006) Modeling of the distance measure error using UWB indoor radio measurement. IEEE Commun Lett 10(4):275–277

Alsindi N, Alavi B, Pahlavan K (2009) Measurement and modelling of ultra wideband TOA-based ranging in indoor multipath environments. IEEE Trans Veh Technol 58(3):1046–1058

Bellusci G, Janssen GJ, Yan J, Tiberius C (2008) Model of distance and bandwidth dependency of TOA-based UWB ranging error. In: Proceedings of IEEE international conference on ultra-wideband, pp 193–196 (2008)

Chehri A, Fortier P, Tardif P (2007) On the TOA estimation for UWB ranging in complex confined area. In: Proceedings of international symposium on signals, systems and electronics (ISSE), July 2007

Dardari D, Chong CC, Win MZ (2008) Threshold-based time-of-arrival estimators in UWB dense multipath channels. IEEE Trans Commun 56(8):1366–1378

Gentile C, Kik A (2006) An evaluation of ultra wideband technology for indoor ranging. In: Proceedings of IEEE GLOBECOM, pp 1–6 (2006)

Humphrey D, Hedley M (2008) Super-resolution time of arrival for indoor localization. In: Proceedings of the international conference on communications (ICC), Beijing, China, pp 3286–3290, May 2008

Humphrey D, Hedley M (2009) Prior models for super-resolution time of arrival estimation. In: Proceedings of IEEE 69th vehicular technology conference, April 2009

Prieto J, Bahillo A, Mazuelas S, Lorenzo RM, Blas J, Fernandez P (2009) NLOS mitigation based on range estimation error characterization in an RTT-based IEEE 802.11 indoor location system. In: Proceedings of IEEE international symposium on intelligent signal processing, pp 61–66 (2009)

Sathyan T, Humphrey D, Hedley M (2011) WASP: a system and algorithms for accurate radio localization using low-cost hardware. IEEE Trans Syst Man Cybern–Part C 41(2):211–222

Sharp I, Yu K (2013) Improving ranging with analog Wi-Fi radios: methods and analysis. Mob Comput 2(2):43–58

Sharp I, Yu K (2014) Indoor TOA error measurement, modeling and analysis. IEEE Trans Instrum Meas 63(9):2129–2144

Sharp I, Yu K, Guo YJ (2009) Peak and leading edge detection for time-of-arrival estimation in band-limited positioning systems. IET Commun 3(10):1616–1627

Wang W, Jost T, Mensing C (2009) ToA and TDoA error models for NLOS propagation based on outdoor to indoor channel measurement. In: Proceedings of IEEE wireless communications and networking conference (WCNC), Apr 2009

Yu K, Sharp I, Guo YJ (2009) Ground-based wireless positioning. Wiley-IEEE Press

Chapter 5
Time and Frequency Synchronization

5.1 Introduction

The determination of position in a radiolocation system based on time-of-flight (TOF) estimation is closely related to time measurements in the transmitting and receiving nodes. In theory, TOF can be determined from the time of the transmission at the transmitting node and the time of reception at the receiving node. Such a concept implies that the clocks in the nodes are synchronized in time, and thus clearly the accuracy of range measurements is directly related to the accuracy of time synchronization. Such time synchronization can in turn be achieved by a node with a reference clock transmitting a signal which can be received by all the other nodes in the network. As fixed nodes have known positions, the TOF can be inferred from the distance between the nodes, assuming straight-line radio propagation. Thus time synchronisation can be considered as the complement of position determination. At least in theory, if the positions of nodes are known, clocks in the nodes can be synchronized, while if the clocks in the nodes are synchronized then their positions can be determined. While this is true in general, time synchronization in the mobile node cannot be achieved by this method, as the aim is to use the radio transmissions for position determination, rather than synchronisation of the local clock. Thus the direct method of measuring TOF using time-synchronized clocks is not possible in practice, so some other method is required. Two such methods are analysed in detail in this chapter.

The first method applicable to tracking systems is closely related to the above concepts, namely for fixed nodes time is synchronized based on a *timing reference signal* transmitted to all the other nodes from a node with a master clock. By continuously tracking this signal, each fixed node can accurately maintain time (and hence frequency) synchronisation with the master clock, albeit delayed by the propagation time from the reference node. The design and performance of a time synchronization control loop in a node is described in Sect. 5.3. For a LOS situation the propagation delay can be inferred from the known distance between the nodes,

I. Sharp and K. Yu, *Wireless Positioning: Principles and Practice*, Navigation: Science and Technology, https://doi.org/10.1007/978-981-10-8791-2_5

but for the indoor NLOS case, the propagation delay will be somewhat greater by an unknown amount. Accurately determining this delay excess is important, as any errors in time synchronization will be directly reflected in inaccuracies in the computed position. Methods for correcting for these errors are discussed in Sect. 5.2. Another subtler consequence of this method of time synchronization is that the measurement of the time-of-arrival (TOA) of a signal from a mobile node automatically compensates for the internal radio delays in the fixed receiving node. As the TOA measurement of the timing reference signal and the transmission from the mobile node have the same internal delays in the receiving node, using the time-synchronized clock in fixed nodes as the local time reference for TOA determination effectively eliminates these internal delays from the measurement of the propagation delays.

The second method of measuring the propagation delay from a mobile node to a fixed node which does not involve any time synchronization (and hence no timing reference subsystem) is by measuring the round-trip-time (RTT) between two nodes. However, while no time synchronization is required, any time delay in the response cannot be accurately accounted for in the RTT calculation if there is a differential in the clock *frequencies* in the two nodes. For example, if there is a response delay of 1 s and there is a differential frequency error of 1 ppm, then the computed RTT will be in error by 1 μs (or a range error of about 300 m). Thus accurate frequency synchronization is required by this method. Further, the RTT will include the delays through the radio equipment in both nodes, so to determine the propagation delay the internal equipment delays must be accurately known. Thus while the timing reference subsystem is eliminated, another subsystem is required to measure the internal radio delays. The details of the design and performance of the RTT method are described in Sects. 5.4 and 11.3.1.

5.2 Time Synchronization in Tracking Systems

The TOA of a signal at a node in a network must be measured relative to some clock local to the node. Initially, say when first powered on, this clock is unsynchronized to either absolute time or the time local to the network. Before the TOA can be sensibly interpreted, the local clock must either be synchronized to a clock common to the network, or the clock offset relative to some common time reference must be determined. This synchronization has two components, namely the frequency of the oscillations of the clock (usually using a crystal oscillator), and the time offset of some defined timing point (called the *epoch*) in the clock timing sequence, which is assumed to repeat with some system-wide time period. In the case of a direct-sequence spread-spectrum system, this sequence is the pseudo-random code (pn-code) which is defined by the number *chips* in the pn-code.

In this section, a timing synchronization method is discussed based on a system-wide timing signal, which is transmitted from a single timing reference transmitter. This timing reference signal is used by all other nodes in the network to synchronise their clocks. This synchronization can be either determining the time offsets for correction (in the software associated with the position determination) to the measured TOA, or the local clocks can be physically synchronized in the hardware.

5.2.1 Time Synchronisation with Three Base Stations

In this initial example, consider three base stations (at known positions); the task is to synchronize the clocks in each of the three nodes. The time synchronization process should provide an accurate synchronization of the clocks in the base stations after the frequency of the clocks have been synchronized (as described in Sect. 5.3). The time synchronization process essentially involves determining two parameters in each base station, namely the clock offset parameter (ϕ) and the radio delay parameter (Δ), as described in Chap. 4.

The basic process for determining the above-defined parameters is the utilisation of the TOA equation (4.4) for transmissions between nodes. If it is assumed that the propagation delay can be derived from the distance between the nodes (with no measurement errors), then in this case the TOA equation becomes

$$TR_{m,n} = TOA_{m,n} - D_{m,n} = \phi_m + \Delta_n - \phi_n \tag{5.1}$$

For the particular case of three nodes, there will be six such measurements, and as there are two parameters per node, there are a total of six unknown parameters, matching the number of equations. In fact, the problem as outlined above is indeterminate, as the six equations are not independent. However, as the system is independent of any absolute time reference, the time synchronization process cannot be absolute, but must be relative to one of the clocks in the system (the timing reference node). Thus there are in fact only five unknown parameters to be determined, so with six equations the problem is slightly over-defined.

Based on (5.1), and assuming node #1 is the reference $(\phi_1 = 0)$, the system of equations can be defined in matrix form by

$$\begin{bmatrix} TR_{12} \\ TR_{13} \\ TR_{21} \\ TR_{23} \\ TR_{31} \\ TR_{32} \end{bmatrix} = \begin{bmatrix} 0 & -1 & 1 & 0 & 0 \\ 0 & 0 & 0 & -1 & 1 \\ 1 & 1 & 0 & 0 & 0 \\ 0 & 1 & 0 & -1 & 1 \\ 1 & 0 & 0 & 1 & 0 \\ 0 & -1 & 1 & 1 & 0 \end{bmatrix} \begin{bmatrix} \Delta_1 \\ \phi_2 \\ \Delta_2 \\ \varphi_3 \\ \Delta_3 \end{bmatrix} \tag{5.2}$$

$$[\mathbf{TR}] = [\mathbf{A}]\ [\mathbf{X}]$$

The least-squares solution to the linear equations (5.2) is given by

$$[\mathbf{X}] = \left([\mathbf{A}]^T[\mathbf{A}]\right)^{-1}[\mathbf{A}]^T[\mathbf{TR}] \tag{5.3}$$

The matrix associated with transposing, multiplication and inverse is a constant matrix which can be pre-calculated, so that the solution is simply given by the multiplication of this constant matrix by the timing reference (TR) measurement matrix ***TR***.

An alternative (and somewhat simpler) solution can be obtained using (5.1) relating pairs of nodes to first determine the delay parameters (Δ). In this case the measurement equation becomes

$$MD_{m,n} = M_{m,n} - D_{m,n} = (\Delta_m + \Delta_n)/2 = \delta_{m,n} \qquad (n > m) \tag{5.4}$$

This simple set of linear equations has the solution

$$\begin{bmatrix} \delta_1 \\ \delta_2 \\ \delta_3 \end{bmatrix} = \begin{bmatrix} 1 & 1 & -1 \\ 1 & -1 & 1 \\ -1 & 1 & 1 \end{bmatrix} \begin{bmatrix} MD_{12} \\ MD_{13} \\ MD_{23} \end{bmatrix} \tag{5.5}$$

Once the delay parameters are determined, (5.1) can be use to determine the clock phase parameters.

$$\Delta^{tr}_{m,n} = TR_{m,n} - \Delta_n = \phi_m - \phi_n \qquad (n > m) \tag{5.6}$$

However, again it is assumed $\phi_1 = 0$, so the solution for the other two phases is trivial.

As the clock offset parameters have been determined, these offsets can be applied to the time output by the clocks, thus synchronizing time in the nodes, with node #1 the (arbitrary) reference. While these offsets could be applied physically to the clocks, it is usually simpler just to apply this correction in the mathematical processing associated with position determination.

While the above example was limited to the three base station case, it is clear that the above procedure could be extended to the case with more nodes in a mesh network, provided a sufficient number of nodes are mutually within range.

5.2.2 Timing Reference Transmitter Architecture

The procedure described in Sect. 5.2.1 provides timing synchronization to a network based on a mathematical model, but a more direct synchronization of the hardware clocks can be achieved by the use of a timing reference node, to be referred to as the *Timing Reference* subsystem. The timing reference is a special signal transmitted by a special node (or Master Station), which allows other nodes (both fixed nodes/base stations and mobiles) to achieve both frequency and time synchronization. Further, the implementation is typically performed in hardware, whereby the frequency and phase of the local clock is adjusted to match that of the received timing reference signal. Further, if the Master Station is itself synchronized to absolute time using some external timing source such as GPS, then the entire network is synchronized to absolute time. Another advantage of this technique is that it simplifies the conversion of the TOA data to pseudorange data used for position determination.

The architecture of a typical system is as follows. The network timing is based on an oscillator (clock) in the Master Station, which transmits a timing reference signal to the base stations and the mobile nodes. The base stations accurately (typically to better than 1 ns[1]) track the epoch of the timing reference signal, thus providing a local timing reference which is delayed relative to the master clock by a fixed offset due to the propagation and receiver delays. These delays are not known but should be fixed with an accuracy of better than 1 ns (with data smoothing). The mobile nodes can also track the timing reference signal, but the accuracy will be somewhat worse and with unknown relative delay. Additionally, with a moving node, the timing reference signal as received at the moving node will have an offset Doppler frequency, so that accurate frequency synchronization in a moving node is not possible.

The principle of network position determination is that the relative timing of the transmissions from the mobile nodes is unknown (as is the position of the mobile node), and must be determined in the measurement process, and thus the data processing technique should not assume accurate tracking of the timing reference signal by the mobile node. A simple method of eliminating the (epoch phase) of transmissions from the mobile nodes is to use differential TOAs (code phases) as measured at the master/base stations. While the synchronization of the clock in the mobile node may not be accurate enough for accurate position determination, the clock accuracy is more than sufficient for other purposes, such as defining the timing slots in a time division multiple access (TDMA) system typically used in a tracking system.

The standard procedure for determining the pseudorange[2] by base station is described first. The basic components and associated parameters are shown in Fig. 5.1. The Master Station transmits a timing reference signal based on its local clock of phase ϕ_0. This timing reference signal is used by both the base stations and the mobile nodes to synchronize their local clocks. The Master Station also has two additional parameters, namely the delays associated with transmission of the signal to/from its antenna to the receiver/transmitter in a remote node. Similar delays are also associated with base stations. In the case of a mobile node, it is convenient to

[1]This timing accuracy is based on a 10 MHz radio bandwidth, high SNR (>40 dB), and no multipath interference. The accuracy approximately scales inversely with the bandwidth.

[2]A pseudorange is by definition the required range plus an arbitrary constant.

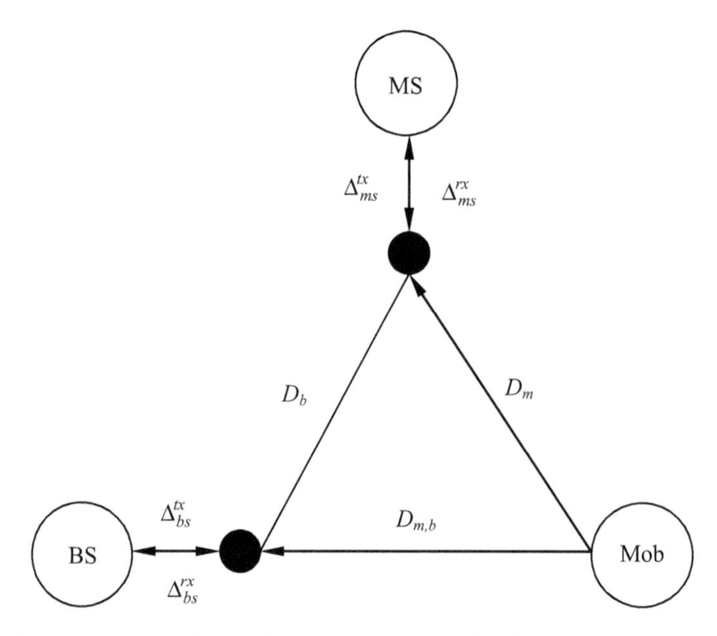

Fig. 5.1 Basic components (Master Station, base station and mobile node) for determination of the pseudorange of the mobile node from the base station

reference the mobile clock at the antenna, as only transmissions from the mobile node are of interest in the analysis.

First consider transmissions from the mobile node to a base station. In the following mathematics, propagation delays with a single subscript are to the Master Station, while propagation delays with two subscripts imply the transmitting and receiving node. The received signal phase from the mobile node at a base station is given by

$$\Phi_b^{mob} = \phi_{mob} + D_{mob,b} + \Delta_{bs}^{rx} - \phi_b \tag{5.7}$$

The received timing reference signal at the mobile from the Master Station is similarly given by

$$\Phi_b^{ms} = \phi_0 + \Delta_{ms}^{tx} + D_b + \Delta_{bs}^{rx} - \phi_b \tag{5.8}$$

The measured signal epoch of the mobile node relative to the timing reference is the difference between these phases, resulting in

$$\Delta\Phi_b^{mob} = \Phi_b^{mob} - \Phi_b^{ms} = \left(\phi_{mob} - \phi_0 - \Delta_{ms}^{tx}\right) + D_{mob,b} - D_b \tag{5.9}$$

Note that the unknown base station clock phase and the base station receiver delay are common, and thus cancel. Thus if the local pn-code epoch is adjusted to zero phase by synchronizing to the pn-code phase from the timing reference signal, the phase of the mobile node signal as measured in the base station is simply the differential phase defined by (5.9). Thus the mobile node signal epoch measured by the hardware is independent of the base station delay and phase parameters.

As the master to base station distance is a known value, the pseudorange of the mobile node as measured at the base station is given by

$$P_{m,b} = \Delta\Phi_b^{mob} + D_b = \left(\phi_{mob} - \phi_0 - \Delta_{ms}^{tx}\right) + D_{mob,b} = \phi_c + D_{mob,b} \tag{5.10}$$

Note that the pseudorange phase term ϕ_c is a function only of the Master Station and the mobile node, and thus is the same for all base stations, and hence complies with the definition of a pseudorange. Equation (5.10) is the basic equation used for position determination. During this process both the mobile node range and the unknown phase term ϕ_c are estimated in the position determination process.

The standard technique of measuring the pseudorange of a mobile node is to use the base stations, as described above. In this case the receiver delay in the base stations is not relevant to the measurement, as the receiver output is a differential measurement between the signal from the mobile node and that from the master station. In the case of the Master Station this differential mode of operation is not applicable, so that the Master Station delay must be determined if the pseudorange to the mobile node is to be measured by the Master Station.

The principle of a system that uses a timing reference for time synchronization is that the base stations measure pseudoranges rather than ranges. Thus the pseudorange is given in (5.10). In the case of the Master Station, consider the tracking of the mobile node signal at the Master Station is similar to that used at the base station, except that there is no timing reference signal as the master clock is local to the Master Station. The TOA of the mobile node signal at the Master Station is thus

$$\Phi_{ms}^{mob} = \phi_{mob} + D_{mob} + \Delta_{ms}^{rx} - \phi_0 \tag{5.11}$$

Equation (5.11) can be arranged so that the pseudorange term of (5.10) appears in the expression, resulting in

$$\Phi_{ms}^{mob} = \left(\phi_{mob} - \phi_0 - \Delta_{ms}^{tx}\right) + D_{mob} + \left(\Delta_{ms}^{rx} + \Delta_{ms}^{tx}\right) = \phi_c + D_{mob} + \Delta_{ms} \tag{5.12}$$

Using Eq. (5.10), the required pseudorange as measured by the Master Station becomes

$$P_{ms} = \Phi_{ms}^{mob} - \Delta_{ms} = \phi_c + D_{mob} \tag{5.13}$$

Note that the pseudorange (5.13) is exactly the same format as the base station pseudorange format in Eq. (5.10). Thus provided the delay term Δ_{ms} can be determined, the Master Station pseudorange can be used in the same manner as the base station pseudorange.

The determination of the Master Station delay parameter Δ_{ms} (a sum of the Master Station receiver and transmitter delays) can be determined by using the base stations as transmitters. With the base station #1 transmitting, the equations for the measured signal epoch at the Master Station and at another base station #2 are as follows

$$\begin{aligned} \Delta\Phi_{bs2}^{bs1} &= (\phi_b - \phi_0 - \Delta_{ms}^{tx}) - D_{bs2} + D_{bs1,bs2} \\ \Phi_{ms}^{bs1} &= \phi_b + D_{bs1} + \Delta_{ms}^{rx} - \phi_0 \end{aligned} \tag{5.14}$$

Thus by subtraction, the resulting equations become

$$\begin{aligned} \Delta\Phi_{bs2}^{bs1} - \Phi_{ms}^{bs1} &= D_{bs1,bs2} - D_{bs1} - D_{bs2} - (\Delta_{ms}^{tx} + \Delta_{ms}^{rx}) \\ \Delta_{ms} &= (\Phi_{ms}^{bs1} - \Delta\Phi_{bs2}^{bs1}) + D_{bs1,bs2} - D_{bs1} - D_{bs2} \end{aligned} \tag{5.15}$$

As all the terms on the right-hand side of (5.15) are either known constants or measured epoch data at the master and base stations, the required Master Station delay can be determined. Further with N base stations, the master station delay can be determined for $(N-1)$ receiving base stations, so that $N(N-1)$ estimates of the Master Station delay parameter can be calculated. Thus by averaging the measurement errors can be minimised. Note that the accuracy of the determination also depends on the ranges between the base stations and between the master/base stations, so that accurate positions of all the fixed nodes are essential for the determination of the Master Station delay parameter.

5.2.3 Performance Measurements Using the Timing Reference System

This subsection defines procedures for estimation the ranging performance, based on the pseudorange measurements at the master and base stations. The aim of the analysis is to provide ranging estimates which are related to known geometric characteristics.

The block diagram of the measurement system is shown in Fig. 5.2. The system consists of a Master Station providing the timing reference, a base station and a mobile node. The mobile node is moved to points in the environment to be tested. The aim of the testing is determining the accuracy of the measurement of the propagation distance from the mobile node to the master/base stations.

The Master Station clock provides the basic timing for the system. This clock reference (ϕ_0) is transmitted (with delay δ_{ms}^{tx}) to both the mobile node and the base station. The reference code-phase is defined at the Master Station antenna (represented by a “dot” in Fig. 5.2). The mobile node transmits a tracking signal, with again the reference defined at the antenna, but with unknown code-phase reference relative to the master clock. The base station receives both these signals, and measures the timing difference between epochs of these two signals. Note that the receiver delay is unknown, but assumed to be the same for both these signals. (The

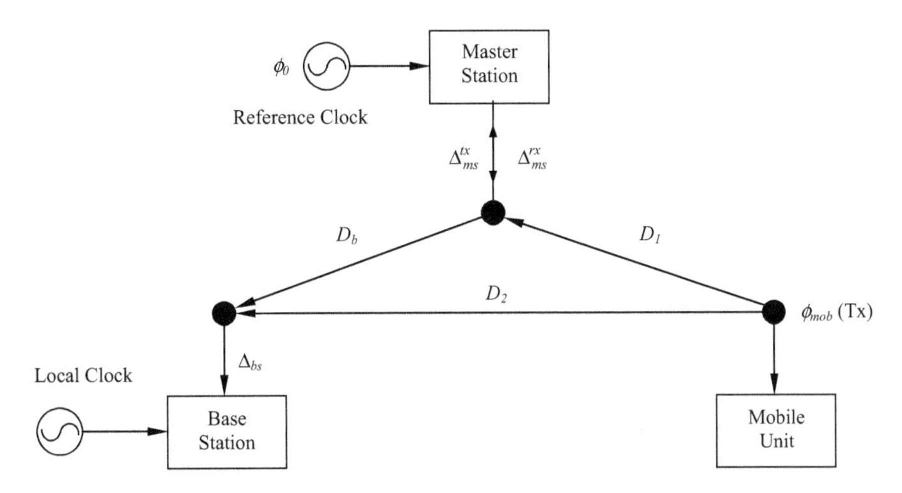

Fig. 5.2 Block diagram of the measurement system, showing all relevant delays

receiver delay is assumed to be independent of the signal strength; this is typically close to being true, but accurate calibration would require a correction factor as a function of signal strength).

Thus the procedure is to compare the measured delays with the predicted performance, and hence determine the measurement errors. The system description (as shown in Fig. 5.2) has many unknown delay parameters, so that the procedure is to develop expressions which eliminate these unknown delay parameters. This is generally achieved by differential measurements, where the unknown parameters are eliminated by the subtraction process. Note in this measurement procedure the mobile node is placed at a known position, and the following analysis is used to estimate the ranging errors based on measurements of the TOA in the master/base stations.

Based on the functional description in Fig. 5.2, the timing reference code-phase at the base station is given by

$$\mu_{tr}^{bs} = \phi_0 + \Delta_{ms}^{tx} + D_b + \Delta_{bs} \tag{5.16}$$

Similarly the mobile node code-phase at the base station is

$$\mu_{mob}^{bs} = \phi_{mob} + D_2 + \Delta_{bs} \tag{5.17}$$

The base station receiver measures the time difference between these two phases, and is given by

$$\Delta\mu_{mob}^{bs} = (\phi_{mob} - \phi_0) + (D_2 - D_b) - \Delta_{ms}^{tx} \tag{5.18}$$

Similarly, the Master Station measurement of the mobile signal code-phase (relative to the master clock) is given by

$$\mu_{mob}^{ms} = (\phi_{mob} - \phi_0) + D_1 + \Delta_{ms}^{rx} \tag{5.19}$$

Because of the unknown phase of the transmitted mobile node signal, the difference between these two measurement is taken, thus cancelling the unknown phase (ϕ_{mob}), resulting in

$$\Delta\mu_{mob} = \left(\Delta\mu^{bs} - \mu^{ms}\right)_{mob} = (D_2 - D_1) - \left(D_b + \Delta_{ms}^{tx} + \Delta_{ms}^{rx}\right) = (D_2 - D_1) - \Delta_{ms}^{b} \tag{5.20}$$

Thus the mobile node differential measurement is equal to the differential distance between the master/base stations, minus a constant related to the delays associated with the Master Station, and master/base stations separation distance. The constant delay offset (Δ_{ms}^{b}) is not known. However, this constant delay is independent of the mobile node, so with two mobile node positions, the difference between the measured differential measurements (double differences) is given by

$$\Delta\mu_{12} = \Delta\mu_{mob1} - \Delta\mu_{mob2} = (D_2 - D_1)_{mob1} - (D_2 - D_1)_{mob2} \tag{5.21}$$

Thus the final differential calculation is the difference between two differential distances to the master/base stations, and is independent of any unknown system constants. However, the above equations assume perfect code-phase measurements with no errors and line-of-sight propagation; actual measurements will include errors, mainly associated with multipath phenomena. Because the theoretical differential measurements are related to geometric distances only, the measurement errors can be estimated, using the known positions of the master/base stations and the mobile nodes to calculate the distances.

For typical measurements the master and base stations are placed close together (a few metres apart), so that it can be assumed there is good signal-to-noise ratio and line-of-sight reception with essentially no tracking errors. The mobile node is moved throughout the environment that is to be tested. The measurement technique requires two measurements and the output data is the difference between these two measurements, so thus individual ranging errors cannot be determined, only their difference which can be compared with the geometric differential distance. While this is a limitation in this test method, usually the requirement is for statistical information on the ranging errors in the environment. Thus if two measurements are made in fairly close proximity (say within a separation of a meter), then it is reasonable to assume the ranging bias errors are the same, but the random errors are uncorrelated with the same standard deviation. Thus if a statistically meaningful number of measurements are made in the immediate vicinity, the differential standard deviation of the measurements will be $\sqrt{2}$ times the ranging standard deviation, and the mean should be zero. Thus bias errors cannot be measured using this technique. The advantage of this technique is that the double differential measurements are (ideally) purely related to geometric factors, and all the unknown equipment parameters are eliminated from the calculation.

However, an extension of the method can measure individual ranging errors. If the mobile node is brought close to the master/base stations so that line-of-sight propagation conditions exist, then from (5.20) the Master Station delay parameter can be determined.

$$\Delta_{ms} = (D_2 - D_1) - D_b - \Delta\mu_{mob} \tag{5.22}$$

Further, if the round-trip time (RTT)—see Chap. 4—between the Master Station and the mobile node is measured then using (4.5) in Chap. 4 the mobile delay parameter can be calculated

$$\Delta_{mob} = 2\left(RTT_{ms,mob} - D_1\right) - \Delta_{ms} \tag{5.23}$$

Assuming these delay parameters remain constant over the period of the measurements, then using these calibrated delays in (4.5) in Chap. 4 the propagation delay can be estimated at other positions by

$$\hat{D}_{ms,mob1} = D_{ms,mob} + \Delta D_{ms,mob} = RTT_{ms,mob} - \frac{\Delta_{ms} + \Delta_{mob}}{2} \tag{5.24}$$

By comparing this delay with the delay associated with the straight line path delay, the ranging delay error (ΔD_1) can be estimated. Note that this measurement technique requires estimation of the radio delay parameters which vary as a function of time (over periods as small as a few minutes). Thus the above calibration technique must be repeated periodically to ensure reliable data. In contrast the double difference method does not require such calibration.

If a large number of such measurements are made, then the nature of the statistics of measurement errors in the environment of interest can be determined. For example, if the measurements are made close to a particular point where it is assumed the error can be described by a bias error plus a random component, then the mean and standard deviation of the measured errors gives the required information for that point. Due to the large number of measurements required, these statistical measurements are quite time consuming if a large area is to be surveyed. Examples of actual measurements in indoor environments are given in Chap. 11.

5.2.4 Calibrating the Timing Reference System

The basic analysis of the performance of the timing reference system in Sect. 5.2.3 assumed that there are line of sight propagation conditions between the Master Station and the base stations, as the theory assumes the propagation delay between the master and a base station is given by the geometric distance propagation delay (D_b). For outdoor systems or in large indoor arenas this assumption works well, but in more typical indoor environments, such as in office buildings, line of sight propagation is typically not present. In this case the measured pseudorange will have an error associated with the excess propagation delay between the Master Station and the base stations. Thus if the delay excess is ΔD_b the pseudorange to a mobile node is given by

$$\begin{aligned} P_{m,b} &= \Delta\Phi_b^{mob} + D_b = \left(\phi_{mob} - \phi_0 - \Delta_{ms}^{tx}\right) + D_{mob,b} - \Delta D_b \\ &= (\phi_c - \Delta D_b) + D_{mob,b} \end{aligned} \tag{5.25}$$

The final expression in (5.25) has a term which varies between base stations, and thus the measured pseudorange is no longer a true pseudorange. If this pseudorange is applied to the position determination algorithm without any correction, the accuracy of the position fix will be degraded. However, the methods described in Sect. 5.2.3 allow the ranging error to be estimated, thus eliminating the errors associated with non-line of sight propagation between the Master Station and the base stations. The details are as follows:

1. The Master Station delay Δ_{ms} and mobile node delay Δ_{tx} parameters are determined using the methods described in Sect. 5.2.3.
2. Take the mobile node to near the base station (bs) whose calibration Δ_{bs} is to be determined. Because the distance is short between the base station and the mobile node, the propagation delay should be close to the straight line delay. Measure the round-trip time between the mobile node and the base station, and hence determine the base station radio delay parameter by

$$\Delta_{bs} = 2\left(RTT_{bs,mob} - D_{bs}\right) - \Delta_{mob} \tag{5.26}$$

3. Measure the round-trip time between the Master Station and the base station, and hence determine the base station delay excess by

$$\Delta D_{bs} = RTT_{ms,bs} - \frac{\Delta_{ms} + \Delta_{bs}}{2} - D_{bs} \tag{5.27}$$

The excess delay from the Master Station to a base station is solely related to the architecture of the building, and thus is invariant,[3] unlike the radio delay parameters. Thus this calibration needs to only be performed once during the setup of the system. Applying this correction to (5.27) allows the measured pseudorange to a mobile node to be independent of any radio propagation delay excesses in the timing reference system. However, delay excesses from mobile nodes to the base stations remain, and are the main source of positional errors. The minimizing of such errors is discussed in detail in Chap. 4.

[3]There will be small transient effects such as people walking in the vicinity of the base station. Installing the base stations high above head level will minimise these transient effects.

5.3 Frequency Control in Tracking Systems

5.3.1 Overview

The timing reference subsystem described in Sect. 5.2 assumes that the base stations and mobile nodes can acquire and track a timing reference signal, thus synchronizing the local clock in each node. This synchronization process includes both the frequency of the clock and alignment of the epoch of the pn-code generator. This section describes a method of achieving both these aims with one feedback control system. For navigation systems such as GPS which transmit continuous signals the control system typically is based on phase-locked loops (PLL). However, for tracking systems, and low powered systems such a wireless sensor networks, the signals are only transmitted intermittently (partly to conserve battery power), so that a PLL is not appropriate; rather the control function is based on a frequency control loop.

The main function of the frequency control loop is to synchronize the frequency of the local oscillator to a close approximation to the local oscillator (master) clock in the timing reference Master Station. For a system based on a direct-sequence spread-spectrum signal the basic input into the loop is the code phase (Φ_{tr}) of the timing reference signal, as determined by the baseband correlator in the receiving node. The node clocks will be synchronized in both frequency and phase if the local pn-code has its signal epoch continuously aligned to the received timing reference signal pn-code epoch, albeit with a propagation delay. For fixed nodes this delay is a known value, particularly after using the calibration processes described in Sect. 5.2. The accuracy of this process is related to the accuracy of the TOA determination. For mobile nodes, the propagation delay is not known (it is what is determined by the position determination process), and the timing reference signal frequency will include a Doppler frequency offset. Thus for a mobile node the synchronization process is much more inaccurate, and clock synchronization cannot be not assumed in the position determination process.

A block diagram of a typical implementation is shown in Fig. 5.3. A summary of the operation of the control loop is as follows:

1. The radio (in a mobile or base station) receives the timing reference radio signal, and converts the RF signal to baseband. The radio is tuned to the RF frequency using a reference frequency derived from a frequency synthesizer module.
2. The radio frequency is generated by a frequency synthesizer which uses the local oscillator output as a reference frequency. Thus if this local oscillator is controlled to match the frequency of the oscillator in the transmitter (timing reference Master Station), then the radio will be accurately tuned to the correct RF frequency.
3. The baseband signal from the radio is processed by a signal decoder (correlator), the output of which is the pn-code epoch (Φ_{tr}) of the timing reference signal relative to local pn-code. The codes are aligned when this offset is zero.

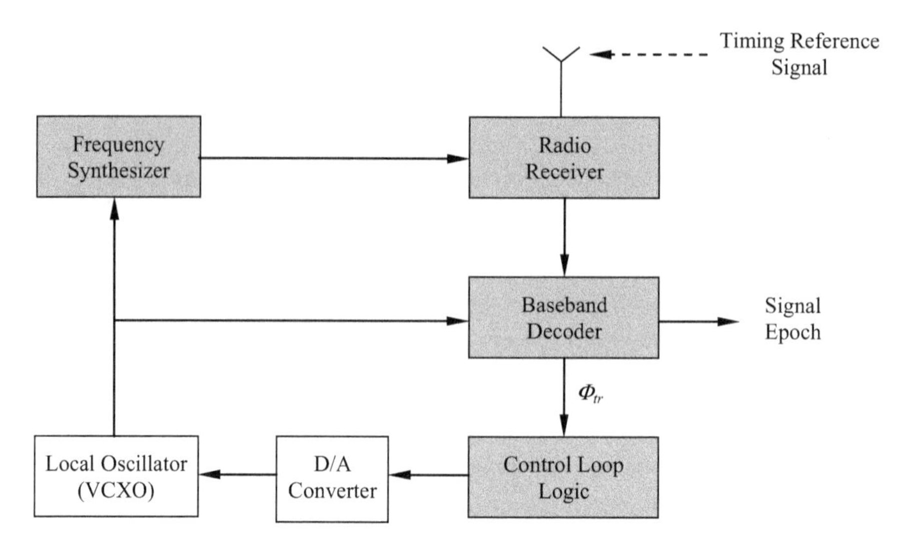

Fig. 5.3 Block diagram of the frequency control and epoch tracking subsystem architecture. The boxes in gray as those components that can be powered down when not required (most of the time), while the white boxes are powered continuously

4. The pn-code epoch signal is input to the control loop logic module. For details of the processing, refer to the following subsection. However, in summary, the purpose of the processing is to output a control signal to a digital-to-analog (D/A) converter for controlling the frequency of the local oscillator.
5. The local oscillator unit is typically based on a voltage controlled crystal oscillator (VCXO). The VCXO is an oscillator whose frequency can be (slightly) altered by the application of a control voltage. Typically, the nominal frequency of the oscillator is 10 or 20 MHz, and the control can change the frequency by up to 50–100 parts per million (ppm). Because of slight variations in the manufacture of VCXO units, and other effects such as frequency variation with temperature and aging of the crystals, the nominal frequency can vary by amount roughly equivalent to the control range. The output from the oscillator is used by both the baseband decoder and the radio, thus locking both these components in frequency and pn-code phase. Thus when the frequency control loop is accurately tracking the timing reference signal (the epoch is tracked to zero), the local oscillator frequency is also accurately matches that in the Master Station.
6. The operation of the frequency control loop can be intermittent, so that when not required some of the components can be powered down. However, the oscillator and its voltage control (D/A converter) must run continuously, as the oscillator defines the continuous flow of time in the node. The components that can be powered down are shown in gray in Fig. 5.3.

Thus in summary, the frequency control module has two main functions, namely:

1. To accurately track the frequency (chip rate) of the pn-code signal in the timing reference signal.
2. To accurately track the epoch (phase) of the pn-code signal, such that the epoch has near zero offset relative the timing reference signal.

The details of how these two functions are simultaneously achieved are described in the next subsection.

5.3.2 *Details of the Frequency Control Loop*

The analysis of the frequency control and epoch tracking loop is based on the block diagram shown in Fig. 5.4, which in turn is based on the architecture shown in Fig. 5.3. However, the block diagram in Fig. 5.4 concentrates on the mathematical processes associated with the operation of the control loop. Although the control loop usually is based on sampled data, the following control loop analysis will be based on classical analog feedback control systems, and in particular the mathematical operations are shown in the Laplace transform domain rather than the time domain. The use of the analog equivalent can be justified if the loop dynamics are much slower than the sampling period.

As explained in the Overview section, the purpose of the control loop is dual purpose, namely to control the frequency of the local oscillator to match the frequency of the timing reference oscillator, and secondly to synchronize the epoch of the pn-code to that of the signal received from the timing reference transmitter. Thus the main outputs the control loop are the differential epoch (which should

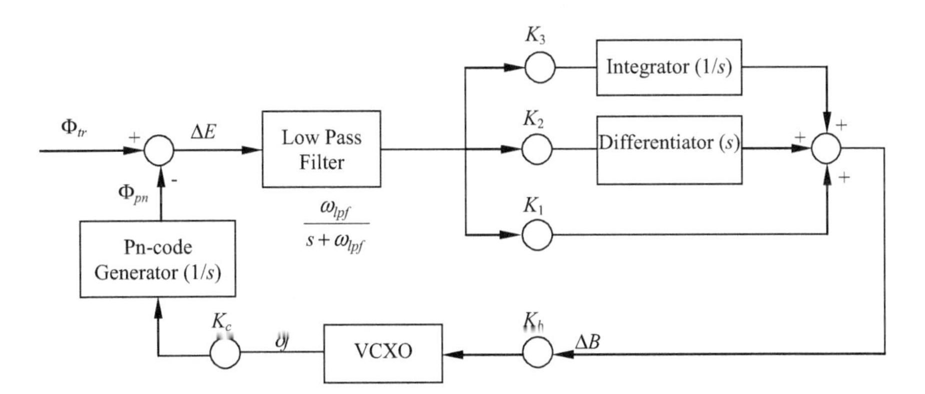

Fig. 5.4 Block diagram of the frequency control and epoch tracking subsystem, based on the analog signal equivalent circuit

track to zero), and the frequency offset of the VCXO (which should nullify any offset frequency in the local oscillator).

A summary of the functional components is as follows:

1. The input signal is the epoch phase of the tracked timing reference signal, as output from the receiver pn-code correlator. The aim of the feedback loop is for the local pn-code generator to track this signal with as little error as possible.
2. The epoch error signal ΔE is generated by subtracting the local pn-code phase from the input timing reference code phase. This error signal will be close to zero when the tracking loop is functioning.
3. As the input signal typically will have some noise associated with the radio and the decoder/correlator, the raw signal is low pass filtered to reduce the noise. As the differential signal will only vary slowly and be close to zero (after the tracking loop has acquired the signal), the bandwidth $(\omega_{lpf} = 2\pi f_{lpf})$ of this filter can be quite low, typically 1–2 Hz.
4. A feedback control signal is now generated from the filtered differential epoch signal. There are three components to the feedback. The first component is simply the error signal multiplied by a weighting constant K_1. This signal tends to drive the loop to zero offset. The second component is the error signal differentiated, with a weighting constant K_2. This feedback signal tends to oppose any rapid change in the error signal, and thus provides damping of any oscillations. The third component is the error signal integrated, with a weighting constant K_3. As the VCXO may have a frequency offset, the feedback signal must include a non-zero component even when the epoch tracking error is zero; this third feedback component supplies this non-zero feedback. This offset signal is supplied by the integrator output.
5. The combined output of three feedback signals represent the binary error signal (ΔB) used to drive the D/A converter. The VCXO input signal (ΔB) to frequency deviation is defined by a scaling factor K_b. The input binary signal ΔB causes the VCXO to generate an output with a frequency deviation δf. This frequency deviation has two functions; one purpose is to null any offset frequency in the local oscillator frequency, and secondly to provide an offset frequency necessary to change the phase of the pn-code generator.
6. The local oscillator output is used to generate the pn-code. As the chip rate of the pn-code may be different from the local oscillator frequency, a conversion constant K_c is applied to the frequency error signal.
7. The VCXO signal is used to generate the pn-code, the phase of which is required to complete the feedback circuit. As phase is the integral of frequency, the transfer function for this block is effectively an integrator. The output signal from the pn-code generator is the local pn-code phase Φ_{pn}, which is subtracted from the input signal phase Φ_{tr} from the timing reference. This completes the feedback loop.

With the above-described feedback loop, the operation is dependent on defining the various weighting constants (K) for correct operation. Some of these are design

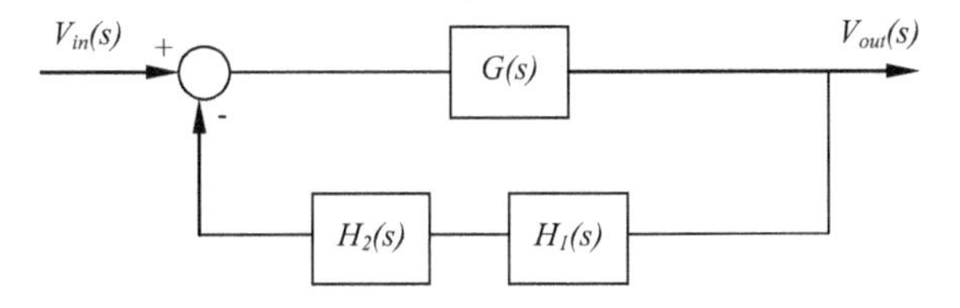

Fig. 5.5 Block diagram of a canonical negative feedback control loop. The blocks in this diagram can be correlated with one (or more) blocks in the frequency control feedback subsystem in Fig. 5.4

decisions, such as the local oscillator to chip rate conversion factor, and the input binary signal to frequency deviation of the VCXO and D/A converter combination. However, the other K factors must be chosen to ensure the feedback loop is stable and has good tracking dynamics. The following analysis derives the closed loop transfer function of the feedback loop, and also describes how these weighting constants are determined for optimum operation.

The above block diagram is based on the analog equivalent circuit of a negative feedback control loop. The canonical equivalent circuit shown in Fig. 5.5 has a transfer function is given by

$$F(s) = \frac{V_{out}(s)}{V_{in}(s)} = \frac{G(s)}{1 + H(s)} \tag{5.28}$$

By comparing the canonical form with the frequency control feedback subsystem, the blocks are as follows:

$$\begin{aligned} G(s) &= \frac{\omega_{lpf}}{s + \omega_{lpf}} \\ H_1(s) &= K_1 + K_2 s + K_s/s = \frac{1}{s}\left[K_2 s^2 + K_1 s + K_3\right] \\ H_2(s) &= K_b \frac{K_c}{s} = \frac{K_{bc}}{s} \end{aligned} \tag{5.29}$$

By substitution into the blocks identified in (5.29), the error transfer function can be calculated to be

$$F(s) = \frac{\delta E(s)}{\Phi_{tr}(s)} = \frac{\omega_{lpf}\, s^2}{s^3 + \omega_{lpf}\left(1 + K_2 K_{bc}\right) s^2 + \omega_{lpf} K_1 K_{bc} s + \omega_{lpf} K_3 K_{bc}} \tag{5.30}$$

Further analysis of the characteristics of this transfer function is difficult due to the cubic function in the denominator. However, the broad characteristics of the loop dynamics can be obtained if the integrator block in Fig. 5.4 is initially eliminated ($K_3 = 0$). In this case the transfer function reduces to a classical second order system, whose characteristics are well known. In particular, if critical second-order system damping is used to minimise the effects of oscillations, it can be shown that (5.30) becomes

$$F(s) = \frac{\omega_{lpf}\, s}{(s+b)^2} \tag{5.31}$$

where

$$b = \sqrt{\omega_{lpf} K_1} = \frac{\omega_{lpf}}{2}(1+K_2) \tag{5.32}$$

Based on the above discussion, an appropriate design procedure is summarized as follows:

1. Based on the second order system with critical damping, the design assumes a transfer function of the form

$$F(s) = \frac{\delta E(s)}{\Phi_{tr}(s)} = \frac{\omega_{lpf}\, s^2}{(s+a)(s+b)^2} \tag{5.33}$$

2. The parameters in the design (ω_{lpf}, a, b) are chosen as follows. The low pass filter bandwidth parameter (ω_{lpf}) is chosen based on the smoothing of the epoch required. The bandwidth is related to the loop update period (δt), and should be chosen such that $\omega_{lpf} \ll 1/\delta t$. The parameters a and b are chosen such that $a \gg b$, so that the loop dynamics is dominated by the time constant $1/b \gg \delta t$. Typically, the "much greater" in the above conditions is at least a factor of 5.
3. By comparing Eq. (5.33) to (5.30), the feedback K parameters are given by

$$\begin{aligned} K_1 &= \frac{b^2 + 2ab}{\omega_{lpf} K_{bc}} \\ K_2 &= \frac{a + 2b - \omega_{lpf}}{\omega_{lpf} K_{bc}} \qquad (K_2 > 0) \\ K_3 &= \frac{ab^2}{\omega_{lpf} K_{bc}} \end{aligned} \tag{5.34}$$

If the design principles described above are followed, a stable loop with adequate loop dynamics can be achieved. An example using these design equations is given in Sect. 5.3.3 following.

If the input phase Φ_{tr} is a unit step function, the output epoch error can be found using inverse Laplace transforms to be

$$\varepsilon_{lpf}(t) = \omega_{lpf}\left[\frac{a}{(a-b)^2} - \frac{bt}{(a-b)}\right]e^{-bt} - \frac{\omega_{lpf}\, a}{(a-b)^2}e^{-at} \tag{5.35}$$

As $a \gg b$, the initial dynamics will be dominated by the a parameter, which has a time constant of $1/a$. For time much greater than $1/a$ the expression reduces to

$$\varepsilon_{lpf}(t) \approx -\frac{\omega_{lpf}\, bt}{(a-b)}\, e^{-bt} \qquad (t > 2/a) \tag{5.36}$$

This expression is negative and has a slow exponential decay with a time constant of about $5/a$, where the "5" is a consequence of condition in subparagraph (2) above. Thus after the fast positive pulse there is a slow small amplitude negative pulse of a period about five times longer than the positive pulse. A numerical example illustrating these characteristics is given in the following subsection.

5.3.3 Example of Control Loop

This subsection describes an example of a control loop, using the design equations given in Sect. 5.3.2. In this example,[4] the sampling period is 80 ms, the chip rate 30 Mchips per second, and the local oscillator frequency is 20 MHz, with a control range of about ± 100 Hz (or ± 5 ppm). In this design a 16-bit D/A converter is used to control the VCXO, which has a control sensitivity factor of 0.003 Hz per bit. The differential epoch (error) is filtered by a low pass filter with a bandwidth of 2.5 rad/ sec (or 0.4 Hz), which with a sampling rate of 12.5 per second results in a loop noise amplitude reduction factor of

$$\rho_{noise} = \sqrt{\frac{\omega_{lpf}\,\delta t}{2 - \omega_{lpf}\,\delta t}} \approx 0.33 \tag{5.37}$$

For this design the a and b parameters are chosen to be respectively 2.5 and 0.5, which results in a loop dynamics time constant of about $1/b = 2$ s for the epoch correction. Using the design Eqs. (5.34) the feedback loop K_1, K_2 and K_3 parameters can be calculated. All these parameters are summarized in Table 5.1.

The performance of the feedback loop is summarised in the Fig. 5.6a–c. The first figure shows the epoch error with a input step function of 1 chip. The feedback loop will track this input signal with an error as shown in Fig. 5.6a. Notice that the main tracking dynamics has a time constant of the order of 2 s in accordance with the above design equations. However, there is also a second slower response with a time constant of about 10 s, as described previously. Both the analog (from (5.35)) and the actual sampled loop dynamics are shown, with only slight difference, as the sampling period is small compared with the loop dynamics time constant.

The second example of the feedback loop performance has simultaneously the above 1 chip step input, plus a 10 Hz offset in the local oscillator frequency.

---

[4]For more details on the WASP system, refer to Sects. 6.4, 11.4.4.3 and 14.5.1.

Table 5.1 Summary of the design parameters for the example design

Parameter	Symbol	Units	Value
Low pass filter bandwidth	ω_{lpf}	rad/sec	2.5
First loop dynamics parameter	a	sec^{-1}	2.5
Second loop dynamics parameter	b	sec^{-1}	0.5
Sampling period	δt	sec	0.08
VCXO control sensitivity parameter	K_b	Hz/bit	0.003
Chip rate conversion factor	K_c	chips/Hz	1.5
Feedback error gain	K_1	bits/chip	242.4
Feedback differentiation gain	K_2	bit-sec/chip	88.9
Feedback integration gain	K_3	bits/sec/chip	55.6

Figure 5.6b shows the epoch error as a function of time, which results in a much larger epoch error than the first case, but the overall loop time constant is again about 10 s. Figure 5.6c shows the corresponding dynamics of the correction of the 10 Hz local oscillator offset error. Thus the loop simultaneously corrects both the epoch error and the frequency error as required, with appropriate damping to ensure there are no oscillations.

5.4 Frequency Control in Mesh Network Systems

Section 5.3 described the frequency control subsystem for a tracking system that uses a timing reference subsystem to synchronize the clocks in the fixed (base stations); such synchronization allows the pseudoranges from the mobile nodes to the base stations to be measured in the base stations. As mesh networks typically do not employ a timing reference subsystem, a different technique is required for frequency synchronization in mesh networks. In particular, as introduced in Chap. 4, a typical design for position determination uses round-trip-time (RTT) measurements, which has the advantage that time synchronization is not required, but very accurate frequency synchronization between the RTT node pairs is required. The following describes in detail how this frequency synchronization is achieved.

In a RTT mesh network, accurate frequency synchronisation is required to ensure that clock drifts between measurements does not result in TOA measurement errors. For example, if the two TOA measurements associated with RTT are 500 ms apart in time, then a drift of 2 ppb represents a timing error in the clock of 1 ns. As typical oscillators have frequency errors of 1–10 ppm, the system must provide a frequency synchronisation correction to reduce this initial error of the order of a factor of 1000:1.

Figure 5.7 shows the measured variation in differential frequency between two oscillators over a long period of time in a temperature-controlled (air-conditioned) office environment. Over short periods of time (up to a few minutes) the differential frequency can be considered constant. However, over longer periods of time the

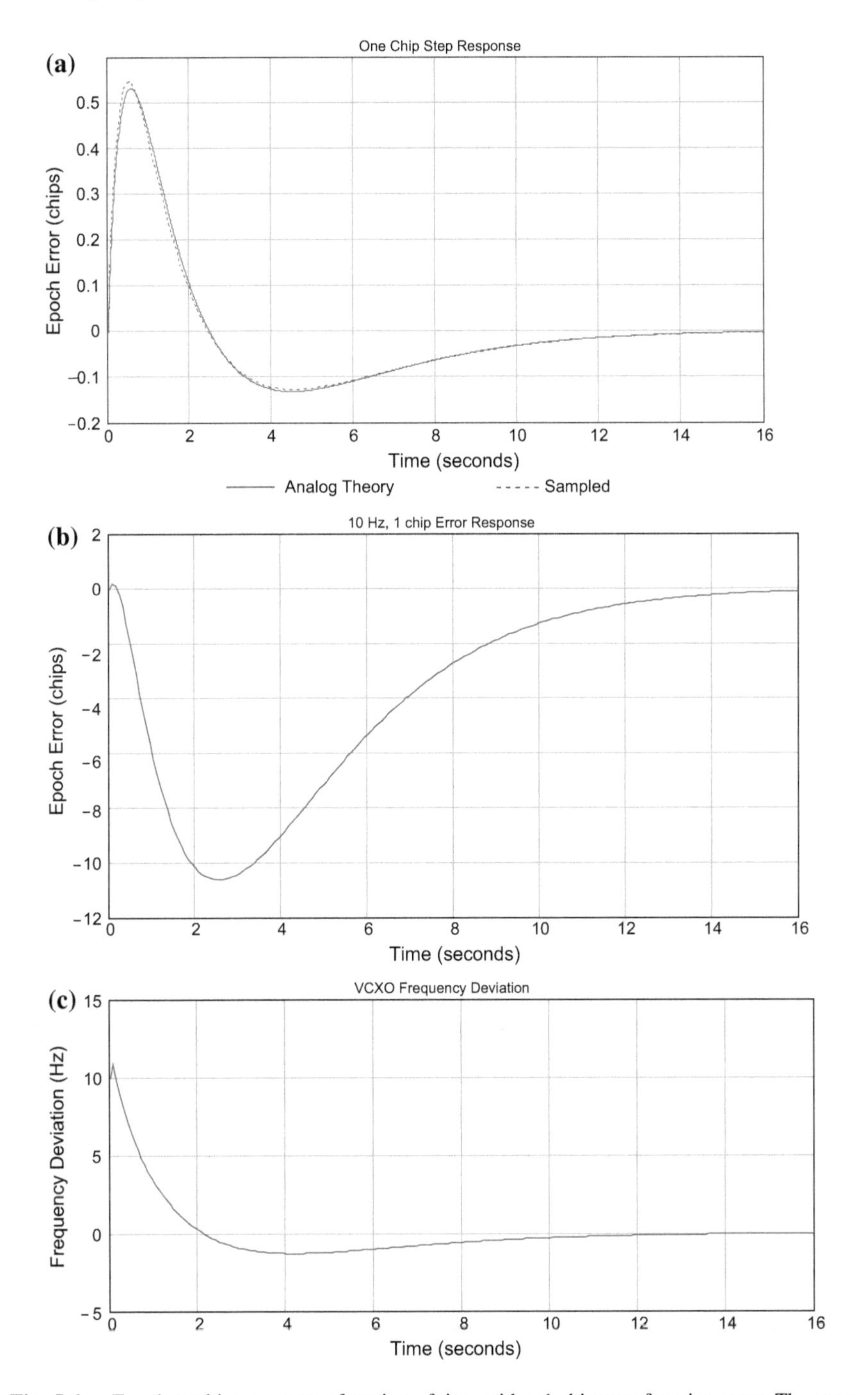

Fig. 5.6 **a** Epoch tracking error as a function of time with a 1 chip step function error. The output error (after the low pass filter) is shown, both the sampled data (dotted) and the analog (solid) versions. **b** Epoch tracking error as a function of time with a 10 Hz local oscillator offset and a 1 chip step function error. **c** Local oscillator tracking error as a function of time with a 10 Hz local oscillator offset and a 1 chip step function error

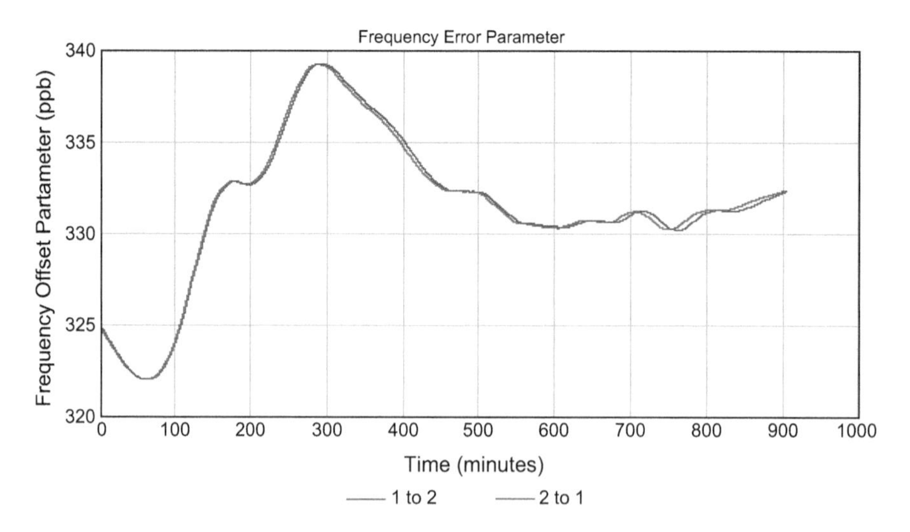

Fig. 5.7 Measured differential frequency parameter between two oscillators of nominal accuracy of 1 ppm. The frequency differential over the 15 h varies by about ±10 ppb. The data in this graph have been smoothed to remove short-term measurement noise

variation is of the order of 20 ppb, so that differential frequency tracking is essential if an overall differential frequency error of not greater than 1 ppb is to be met.

The basic concept for estimation the relative drift between two oscillators (one in the transmitter, and one in the receiver) is to determine the rate of change in the TOA between measurements separated in time of the order of a few seconds. From Chap. 4 it was shown that the TOA measurement with frequency synchronized clocks in nodes m and n can be expressed (see Eq. 4.2) in the form

$$TOA_{m,n} = \phi_m + \Delta_m^{tx} + D_{mn} + \Delta_n^{rx} - \phi_n \tag{5.38a}$$

The TOA estimate from (5.38a) can be modified to account for a *linear* drift in the clock over a short time period (t), resulting in the TOA estimate as a function of time given by

$$TOA_{m,n}(t) = \left(\phi_m^0 + \alpha_m t\right) + D_{mn} + \Delta_n - \left(\phi_n^0 + \alpha_n t\right) \tag{5.38b}$$

where at $t = 0$ the clock phases are the nominal values, but the clock phases slowly change over subsequent time.

Now consider two such measurements at two times t_a and t_b. Then the differential change in TOA is given by

$$\begin{aligned} \delta T_{ab} &= TOA_{m,n}(t_b) - TOA_{m,n}(t_a) = \alpha_m(t_b - t_a) - \alpha_n(t_b - t_a) \qquad (t_b > t_a) \\ \alpha_m - \alpha_n &= \delta T_{ab}/(t_b - t_a) \end{aligned} \tag{5.39}$$

The time differential is determined by the uncorrected clock, but as the measurement error is required to be accurate only to (say) 1:2500, and typical clock errors do not exceed 10 ppm, the resulting drift rate measurements will be of the required accuracy. Note that for this purpose for fixed base stations, accurate TOA data are not necessary, rather the method requires repeatability in measurements for accurate determination of the frequency offsets. Thus for example, TOA errors associated with multipath signal propagation in indoor environments does not affect the procedure, provided such errors are static.

5.4.1 Frequency Synchronization Procedure

The basic concept for estimating the relative drift between two oscillators has been described in (Yu et al. 2009a, b), but the following analysis provides more information on the process and provides analysis of both the stationary case, and also describes a new method for the practically important moving node cases.

The basic method of estimating the differential frequency is to determine the rate of change in the TOA between measurements separated in time. However, as the output from the correlator in the receiver is the signal epoch (not time), the linear clock phase function is defined as

$$\Phi(t, \alpha) = \text{mod}\left(\phi^0 + \alpha t, T_{pn}\right) \tag{5.40a}$$

and the resulting TOA estimate as a function of time is given by

$$TOA_{mn}(t) = \Phi(t, \alpha_m) + (R_{mn} + \delta R_{mn})/c + \Delta_n - \Phi(t, \alpha_n) + T_{noise} \tag{5.40b}$$

The time (t) in (5.40a) for the frequency offset estimation will be over many pn-code periods up to (say) 1 min, and thus it is implied from (5.40a) that ϕ^0 is a constant at the start of transmissions over many pn-code periods if there is no frequency error; in practice for the method to work the transmissions must always commence at the start of a pn-code period, or $\text{mod}(t, T_{pn}) = 0$, where t measured by the local clock. Additionally in (5.40b), a range error has been added due to multipath propagation, plus a random measurement noise T_{noise} (with a zero mean). The mod function in (5.40a) complicates analysis, as the phase function has a sawtooth shape (Yu et al. 2009a, b) with a sudden change of T_{pn} at a period of T_{pn}/α. As α is a small number, this sudden change in the TOA will occur infrequently, and as this change is of known value the measured TOA data as a function of time can be easily corrected by "unwrapping" the raw measurements (either up or down depending on the sign of the drift). With this correction, the mod operation in (5.40a) can be ignored in the analysis of performance, so if the two nodes are stationary and the multipath environment is static, then 5.40b) can be written in the form of a linear equation over time with additive noise, namely

$$TOA_{mn}(t) = \beta_{mn} + \alpha_{mn}\, t + T_{noise} \tag{5.41}$$

where $\alpha_{mn} = \alpha_m - \alpha_n$ and $\beta_{mn} = \left(\phi_m^0 - \phi_n^0\right) + (R_{mn} + \delta R_{mn})/c + \Delta_n$, a constant, at least over the frequency estimation period (T) which is typically of the order of 10 s. If a series of TOA measurements are performed over time (typically with a constant update period of (say) 1 s), then a linear regression analysis can be performed to determine the differential frequency drift parameter α_{mn}. Note that the value of the constant parameter β is not required in determining the differential frequency parameter, so that range measurement errors due to multipath do not affect the estimation of differential frequency. However, the random component will result in errors in the estimate of the differential frequency parameter; this uncertainty can be estimated using the estimated standard deviation in the TOA as given in Chap. 4 by Eq. (4.8). After unwrapping, the regression analysis on (5.41) results in an estimate of the standard deviation in the differential frequency parameter as

$$\sigma(\alpha_{mn}) = \sqrt{\frac{12}{M}}\frac{\sigma_\varepsilon}{T} \tag{5.42}$$

where σ_ε is the standard deviation of the TOA measurement, and is given by (4.8), T is the total period of the sampling, and M is the number of measurements $(M \gg 1)$. Applying (4.8) with a (low) SNR of 20 dB, and 11 measurements with an update period of 1 s (so $T = 10$ s), (5.42) results in an (upper) estimated standard deviation in the differential frequency parameter of 0.15 ns/s (or 0.15 ppb). Thus if the nodes are stationary the frequency offset can be readily estimated to a precision that results in trivial errors in the TOA, even if there are multipath errors in the TOA measurements.

The above procedure is satisfactory for static nodes where the changes in the TOA as a function of time can be interpreted as a consequence of a differential frequency between the two nodes. However, if one node is mobile, the changes in TOA can also be due to movement of the mobile node so that in this case the measurement of the differential frequency by the above method is not possible.

5.4.2 *Estimation of Range Errors Due to Motion*

Before analysing the problem of determining the frequency offset parameter for a moving node, consider the range estimation performance for moving nodes with no frequency offsets. The RTT measurement procedure consists of the mobile node transmitting and stationary nodes in range replying. Each node in the mesh network has an assigned time slot of period T_s for transmissions, which incorporates both the spread-spectrum signals used for TOA measurement and data transmission. The delay in the reply depends on the time slots assigned to the mobile unit and the replying nodes, so a delay of p time slot periods will be in the range $T_s \ldots T_u$, where

$T_u = N_s T_s$ is the update period for position determination. For example, if there are 50 slots of 10 ms, the position update period is 0.5 s. From Eqs. (4.5) and (5.40b) and ignoring the frequency synchronization error, the RTT will be

$$RTT_{m,n} = \frac{1}{2}\left[(\Delta_m + \Delta_n) + \frac{(R_{mn} + \delta R_{mn})_0 + (R_{mn} + \delta R_{nm} + \delta D)_p}{c}\right] + T_{noise} \tag{5.43}$$

where δD is the effective differential distance travelled towards/away from the stationary node by the mobile node in the reply delay period, and is given by $\delta D = vpT_s \cos\theta$, where v is the speed, and θ is the angle between the velocity and range vectors. For the above example the maximum distance travelled is 0.75 m, where it is assumed the mobile unit is attached to a person walking at a speed of 1.5 m/s. As the actual distance will vary depending on the geometry and slot numbers, this distance can be considered as quasi-random, so that from (5.43) the range error due to motion is given by

$$\varepsilon_v = \frac{1}{2}\left[vpT_s \cos\theta + \left(\delta R_0 + \delta R_p\right)\right] + cT_{noise} \tag{5.44}$$

which has three statistically independent components, one associated with the movement of the mobile node, the second associated with multipath propagation range errors, and the third associated with receiver noise. Now consider the statistical properties of the motion-related random process. The slot delay parameter p can be considered as having a statistical uniform distribution $[-N_s/2, N_s/2]$ with a mean delay of $T_u/2$, and the angle between velocity and range vectors can be considered as statistical uniform distribution over $[-\pi, \pi]$, where it is assumed that the replying nodes "surround" the mobile node in a statistically uniform manner. Thus averaging over all anchor nodes and positions in the coverage area the expected value of the range error associated with motion is

$$E[\varepsilon_v] = \frac{vT_s}{2}E[p\cos\theta] = \frac{vT_s}{2}E[p]E[\cos\theta] = 0 \tag{5.45a}$$

as expectations in the last expression in (5.45a) are both zero. Thus on average the motion of the mobile does not result in any range error measured at a point in time delayed by $T_u/2$ relative to the initial transmission by the mobile node. As the unbiased range error reference time will vary with each node pair, for maximum accuracy in position determination a velocity vector correction should be applied. In most cases where the update period is small (as in the above example) this complication is unnecessary. Similarly, the variance of the motion-related error can be calculated to be

$$\mathrm{var}[\varepsilon_v] = \left(\frac{vT_s}{2}\right)^2 E[p^2]E\left[(\cos\theta)^2\right] = \left[\frac{vT_s}{2}\right]^2\left[\frac{N_s^2}{12}\right]\left(\frac{1}{2}\right) = \frac{(vT_u)^2}{96} \tag{5.45b}$$

For example, applying (5.45b) and the numerical values in the above example gives the standard deviation in the range as 0.077 m. As indicated in Chap. 4, the design of an indoor people tracking system should try to limit the ranging standard deviation to about 1 m, so this additional component to the range errors has minimal impact on performance. Further, if the mobile unit moves a distance greater than about a wavelength in a NLOS propagation environment, the two TOA errors associated with the RTT measurement in (5.43) will be largely uncorrelated due to the random scattering of the radio signal. For a system operating in the 5.8 GHz ISM band the wavelength is about 5 cm, and for the above walking case, the distance travelled in one slot period is 1.5 cm, so nearly all of the slot delays will result in uncorrelated measurements. Thus from (5.43b) with two statistically independent random variables δR, the effect of the averaging two TOA measurements reduces the range error standard deviation by about a factor of $\sqrt{2}$. Note that this applies only to the random component, but bias errors typically will be essentially constant over small distances, and thus bias errors do not affect the frequency synchronization process. Note however, it is shown in (Yu et al. 2009a, b, Book Chap. 2, Sect. 2.5) that mounting a mobile unit on the body of a person causes shadowing effects which results in increased geometric dilution of precision (GDOP), so the overall positional accuracy (product of GDOP and ranging error standard deviation) may be worse than that with stationary base stations, despite the improved ranging error standard deviation.

5.4.3 Frequency Synchronization for Mobile Nodes

While the frequency synchronization described in the introduction to Sect. 5.4 and in Sect. 5.4.1 is a simple procedure involving just TOA measurements between two stationary nodes, a modified method is required when one of the nodes is moving. One possible method of determining this motion status is by processing the data from an accelerometer in the mobile node. In particular, if the measured acceleration "noise" is below a threshold, then the node is deemed to be stationary, and the differential frequency parameter is calculated as described in Sect. 5.4.1. When the accelerometer "noise" is above the threshold, a previously measured (when stationary) differential frequency parameter value could be used. However, from Fig. 5.7, even under ideal conditions, the differential frequency parameter will vary slowly. If (say) the maximum allowable drift is 1 ppb, Fig. 5.7 suggests that the period where the previously estimated value can be used is at most several minutes. The length of time will depend on the quality of the oscillator, so if this method is to be used, good quality (and expensive) oscillators are required. As a typical design goal is for cheap hardware, such a strategy is not ideal, because this method would be unreliable if the period of motion without stopping exceeded a few minutes. Thus an alternative method of estimating the frequency offset parameter in nodes which are in motion is highly desirable; such a method is now described.

Before describing the method, it is useful to further refine the method of defining the offset frequency parameter in each node. In the method described in Sect. 5.4.1 only a differential parameter between two nodes is required, but for the moving node case the estimation of the frequency offset parameter for all neighbouring nodes in the mesh network is required, as will be explained below. As only relative frequency synchronization is required, not absolute synchronization, it is appropriate for one node in the mesh network to act as the network frequency reference, and all other nodes are synchronized to this reference node. By definition, this node now is deemed to have a zero offset frequency parameter, namely $\alpha_{ref} = 0$. Further, as all stationary nodes which communicate with the reference node can determine the differential offset frequency relative to the reference node, these nodes also can determine their frequency offset parameter, which can be transmitted in communications data messages with other nodes. Thus by this procedure, with communications throughout the mesh network, all stationary nodes (but as yet not moving nodes) can update in real-time their frequency offset parameter, and transmit this information to other nearby nodes. However, this procedure will result in a slow increase in the frequency offset errors as the number of hops from the reference node increases, so that with N_{hop} hops the standard deviation in the frequency offset parameter will be $\sqrt{N_{hop}}\sigma(\alpha)$, where $\sigma(\alpha)$ is given by (5.42). However, this reduction in the accuracy can be readily compensated for by increasing the measurement time T in (5.42), so the accuracy estimate of 0.15 ppb in Sect. 5.4.1 can be maintained in the multi-hop case. Further, observe that the frequency offset correction process used in correcting raw RTT measurements still only uses the relative offset between adjacent pairs of nodes, so the range error analysis in Sect. 5.4.2 remains valid for multi-hop stationary nodes.

Having established the frequency offset parameter in each stationary node, the next requirement is to establish this parameter for mobile nodes in real-time whilst in motion. The proposed method is similar to position determination using pseudorange data, whereby a range offset common to all measurements is determined as well as the (x, y) position. Because of this similarity with pseudorange least-squares (LS) positioning described elsewhere (Yu et al. 2009a, b, Book Chap. 6), only a summary of the details will be given here. From (5.41), the range measurement between a mobile (m) and stationary node (n) can be expressed as

$$\begin{aligned} R_{mn} = &\left[cRTT_{mn} - \frac{c\alpha_n(t_2 - t_1)}{2} - c\left(\frac{\Delta_m + \Delta_n}{2}\right)\right] + \frac{c\alpha_m(t_2 - t_1)}{2} \\ &- \left[\left(\frac{\delta R_1^{mn} + \delta R_2^{nm}}{2}\right) + \frac{\delta D_{mn}}{2} + cT_{noise}\right] \end{aligned} \tag{5.46}$$

where t_1 is the time of the transmission from the mobile to the stationary node, t_2 is the time of the reply, and δD_{mn} is the effective distance travelled towards/away from an anchor node in this time interval, as described in Sect. 5.4.2. The first term in square brackets in (5.46) is either a measurement or a known value, the second term is only a function of the mobile node (independent of the stationary nodes) with the only unknown the sought-after mobile node frequency offset parameter α_m, and the last term is a random variable, which in Sect. 5.4.2 is shown to be dominated by the range errors associated with the indoor multipath environment.

The scattering of radio signals indoors is complex and depends on the details of the indoor environment, so theoretical calculations are difficult. However, experimental measurements (see Chaps. 4 and 6) show that indoor NLOS ranging errors typically have a bias error which increases approximately linearly with range, and a random component which is largely independent of range, provided a small percentage (typically less than 10%) of large range errors are excluded; this exclusion is justified as they will be rejected in the position determination process (Sathyan et al. 2011; Yu et al. 2009a, b, Chap. 10). Thus the indoors NLOS range error model[5] assumed for the following theoretical analysis is

$$\Delta R = (\delta R_0 + \lambda R) + \varepsilon_R = \beta_R + \varepsilon_R \tag{5.47}$$

where the term in brackets is the linear bias model (β_R), and ε_R is a random component with zero mean. Thus if a LS analysis over all nodes in radio range of the mobile node is performed to determine the four unknowns $(\alpha_m, \beta_R, x, y)$, the last three of which in this case are nuisance variables, then an estimate of the frequency offset parameter can be obtained.

Using model (5.47), (5.45b) and (5.46), an analytical LS analysis (similar to that described in detail in Chap. 6) shows that $E[\alpha_m] = 0$, and the standard deviation in the frequency offset parameter of the mobile node is approximately given by

$$\sigma(\alpha_m) \approx \sqrt{\frac{3}{N_R + 5/9 N_R} \frac{4}{cT_u}} \sigma_{\Delta R} \approx \sqrt{\frac{24}{N_R}} \left(\frac{\sigma_r}{cT_u} \right) \tag{5.48}$$

where $\sigma_{\Delta R}^2 = \frac{\sigma_r^2}{2} + \frac{(\lambda R_{\max})^2}{18} + \frac{(vT_u)^2}{96}$, N_R is the number of stationary nodes in range, and $R_{\max}$ is the maximum radio range. In deriving (5.48) it is assumed that the range measurements to the stationary nodes are statistically independent (because of the random indoor NLOS scattering), and that these nodes "surround" the mobile node in a statistically uniform manner. Notice that the standard deviation is approximately inversely proportional to the square-root of the number of nodes in range, and also inversely proportional to the update period; these relationships are similar to the stationary case given by (5.42).

The simulation results of the frequency offset estimation process are shown in Fig. 5.8. The simulation is based on a mobile unit moving along a circular path with parameters as defined in the caption text for Fig. 5.8. The frequency offset is

[5]The model also is applicable to the LOS case, but with the bias parameter set to zero.

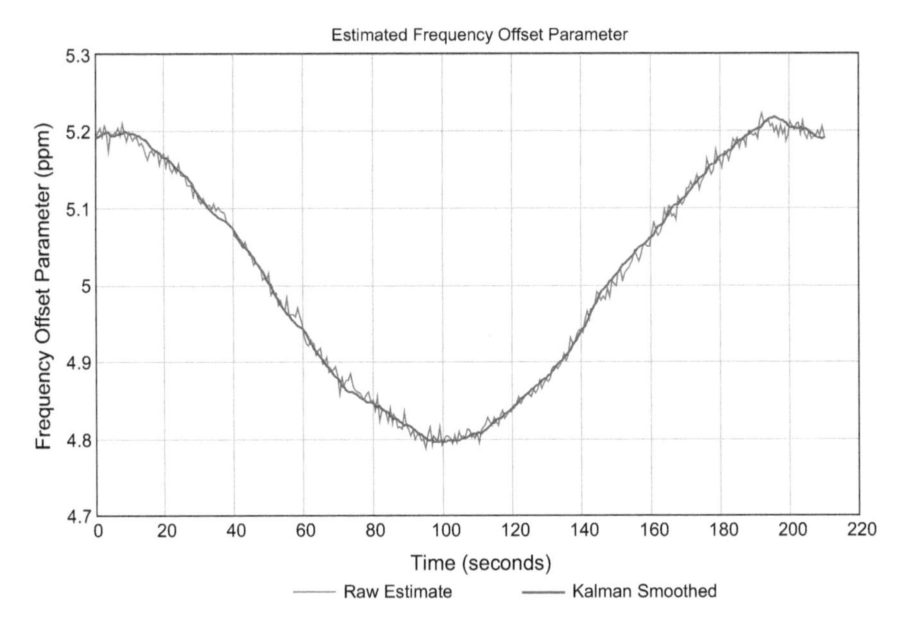

Fig. 5.8 Simulated performance of estimating the frequency offset parameter for a node moving at a speed of 1.5 m/s on a circular path of radius 25 m with an update period of 0.6 s. The coverage area is 100 m square, with 44 nodes uniformly randomly distributed (but with a minimum separation of 7.5 m), of which (on average) 18 are within radio range. The range error model parameters are $\lambda = 0.03$, $\sigma_r = 1$ m, with a maximum radio range of 40 m

represented by a mean of 5 ppm plus a sinusoidal component with amplitude 200 ppb and a period of 200 s, similar to those assumed in the simulation in (Yu et al. 2009a, b). This variation is not intended to model actual frequency variations, but to demonstrate the ability to dynamically track quite rapid variations. The mesh network parameters chosen are typical of the deployment of a mesh positioning system in an indoor office-type environment. Figure 5.8 shows both the raw estimate of the frequency offset parameter, as well as the Kalman[6] filtered smoothed version. The raw data has a mean error in the frequency offset of 1.6 ppb with a standard deviation of 8.1 ppb, compared with the analytical estimate from (5.48) of 7.8 ppb. The filtering is based on a Singer g–h-k Kalman filter (Brookner 1998) for tracking a target whose dynamics have a random acceleration with a exponential autocorrelation function with a time constant of 200 s. The standard deviation of the Kalman filtered data is 4.8 ppb. As described in Sect. 4.4.2, the RTT reply period (T) is a pseudo-random variable (uniform distribution in the range 0–0.6 s in this case), so the associated standard deviation in the range is $c\sigma(\alpha)T/\sqrt{12} = 0.26$ m. Assuming the range bias errors are corrected for, so that only the random range errors ($\sigma_r = 1$ m in this case) remain, then the random effects in the estimation of

[6]See Appendix A for an introduction to Kalman filtering.

the frequency offset parameter will result in the effective standard deviation in the random range error increasing to $\sqrt{1^2+0.26^2}=1.03$ m. Thus the effect of the errors associated with the motion in the determination of the frequency offset parameter for a moving node has only a very minor effect on the ranging accuracy.

References

Brookner E (1998) Tracking and Kalman filtering made easy. Wiley
Sathyan T, Humphrey D, Hedley M (2011) WASP: a system and algorithms for accurate radio localization using low-cost hardware. IEEE Trans Syst Man Cybern–Part C, 41(2):211–222
Yu K, Guo YJ, Hedley M (2009a) TOA-based distributed localization with unknown internal delays and clock frequency offsets in wireless sensor networks. IET Signal Proc 3(3):106–118
Yu K, Sharp I, Guo YJ (2009b) Ground-based wireless positioning. Wiley-IEEE Press

Chapter 6
Enhanced Least Squares Positioning

6.1 Introduction

The design of indoor positioning systems is challenging as the rich indoor multipath radio propagation environment makes accurate TOA measurements difficult. In particular, the scattering of the radio signals results in TOA measurement errors consisting of a biased component as well as a zero-mean random errors typical of LOS outdoor positioning systems, as introduced in Chap. 4, Sect. 4.2.7. In the presence of multipath signals, the classical iterative LS algorithm does not perform well, so that more complex and computationally intensive methods are required to mitigate the effects of multipath bias errors (Yu et al. 2009: Chap. 10; Humphrey and Hedley 2008). These methods include improved TOA estimation such as leading edge detection algorithms, and complex position determination algorithms which attempt to detect large range errors and appropriately weight or eliminate these data in the position determination process. An alternative, simpler approach is analysed in this chapter, based on modeling the bias errors in an indoor environment.

While in an outdoor environment the multipath bias effect cannot be easily modeled due to the great variability in the operating conditions, NLOS indoor measurements show that the TOA bias can be approximately modeled as a linear function of range. See also Sect. 4.2.7 for some measured data of the range bias errors. This effect is based on the quasi-homogeneous structure of many buildings and the consequential signal scattering, so that the zig-zag path from the transmitter to the receiver causes an increase in the measured *range excess* with range, which to a first order is a linear function of range. Additionally, a range-independent delay is associated with the TOA detection of the scattered incident NLOS signals arriving at the receiver. Thus it is suggested that indoor bias errors can be modeled as a constant offset plus a linear range-bias component, namely

I. Sharp and K. Yu, *Wireless Positioning: Principles and Practice*, Navigation: Science and Technology, https://doi.org/10.1007/978-981-10-8791-2_6

$$\Delta r_{bias} = \delta r_0 + \lambda r \tag{6.1}$$

where λ is the range bias error parameter, and δr_0 is the ranging offset error which is common to all the measurement paths. Note that with this ranging error model the δr_0 offset effect turns range measurements into pseudorange measurements. If the two propagation parameters in (6.1) can be estimated, the effect of the ranging bias errors can be significantly reduced, and hence the positional accuracy can be improved to close to the LOS case where only the zero-mean random errors occur. Indeed, it can be shown from (Teunissen 2003) that the LS linearization method described in Sect. 6.2 can approach the Best Linear Unbiased Estimate (BLUE) provided the following conditions apply:

1. The random noise in the range measurements are uncorrelated and Gaussian.
2. The random errors standard deviation is much smaller than the ranges.
3. The standard deviation is the same for each path used for position determination.
4. Range-related bias errors are removed.

Some typical indoor measured range excess data presented in Fig. 6.2 show that the first condition is approximately true, provided the condition 4 is applied. Condition 2 is approximately valid, except for locations near base stations. Condition 3 also seems to be valid, as the error variation appears to be largely independent of range. Further the BLUE solution will approach the minimum geometric dilution of precision (GDOP) if the node geometry is symmetric relative to the location being determined; see Chap. 14 for more details on GDOP. Hence the BLUE solution also approaches the Cramer-Rao Lower Bound (CRLB) due to its equivalence with minimum GDOP (Chaffee and Abel 1994; Levanon 2000). This geometrical condition approximately applies to a mesh network of base stations with approximately (statistical) uniform spatial distribution. Thus in summary, provided the range bias errors are removed (the subject of the following analysis), the LS performance will be close to the CRLB, which is the theoretical lower bound of position errors.

6.2 Enhanced Iterative LS Method

This section summarizes the theory of computing a position based on measured ranges or pseudoranges from fixed nodes in a mesh network. The analysis is essentially the classical theory of radio navigation, but with modifications to include the effects of ranging bias errors.

6.2.1 Position Determination Analysis

The following analysis summarizes the classical least-squares fit approach for the determination of the position of a mobile node using range or pseudorange

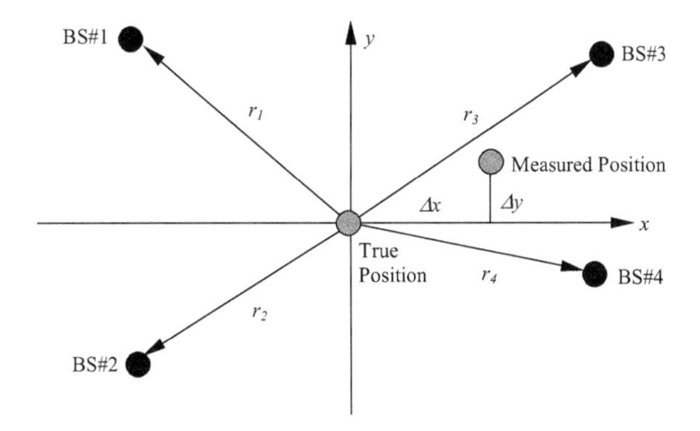

Fig. 6.1 Geometry of the mobile node in relationship to the fixed nodes

measurements to a number of fixed nodes. The measurements will have errors associated with system noise and multipath propagation. The 2D geometry of the problem is illustrated in Fig. 6.1. The true position of the mobile node is located at the origin of the x-y axes. Denoting the mobile and the *i*th base station (BS) positions as (x, y) and (x_i, y_i) respectively, the ranges from the mobile to the base stations are

$$r_i = \sqrt{(x_i - x)^2 + (y_i - y)^2}, \quad i = 1, 2, \ldots, N_R \tag{6.2}$$

where N_R base stations are within the radio range of the mobile unit. Depending on the type of system, the base stations will measure either range or pseudoranges. For range measurements, typical implementation (and in particular the measurements described in Sect. 6.5 and Chap. 4) are based on the round-trip-time (RTT) between two nodes. Note in this particular case the internal delays in the radios and the baseband circuitry needs to be subtracted from the RTT to obtain the propagation delay (and hence the range), so that accurate estimates of internal delays need to be determined in addition to the RTT; see Chap. 4, Sect. 4.4.2 for details. For pseudorange measurements, if the clocks in the base stations are suitably synchronized, the relative transmit timing and the transmitter equipment delays will be common to all the measurements in the base station receivers, and thus form part of the pseudorange offset constant. As a consequence, the delays in the receivers are not required to be determined for pseudorange position determination. Because of these characteristics, pseudorange measurements are often preferred over range measurements, particularly for more accurate positioning. Further, as will be shown pseudoranges have other advantages for indoor positioning when there are measurement bias errors due to the radio propagation characteristics.

For this analysis, it is assumed that the indoor propagation results in both a positive bias which is proportional to the propagation range plus a non-zero offset, and a zero-mean random component which is independent of range—see Eq. (6.1). This assumed error model is supported (at least approximately) by indoor radio propagation measurements (Yu et al. 2009; Alavi and Pahlavan 2006; Alsindi et al.

2009; Alavi and Pahlavan 2003; Gentile and Kik 2006), as well as the measurements described in Sect. 6.5. While measurements show a tendency for the random variation to increase with range, this increase is largely associated with a small fraction of the data which exhibit range errors greater than the time resolution (pulse rise-time) of the TOA measurement. For practical indoor positioning, these large "pathological" errors need to be detected and eliminated (Yu et al. 2009, Chap. 10; Humphrey and Hedley 2008; Sathyan et al. 2011). It will be shown in Sect. 6.5 that when these large errors are eliminated, the residual random errors have a variation which is largely independent of range. Based on these assumptions, the measured pseudorange p_{ε} associated with the ith base station is given by

$$\begin{aligned} p_{\varepsilon_i} &= p_i + \varepsilon(0, \sigma_r) \\ p_i &= f(x, y, \phi; \lambda, \delta r_0) = r_i + c\phi + (\delta r_0 + \lambda r_i) \\ &= (1+\lambda)\, r_i + (c\phi + \delta r_0) \end{aligned} \tag{6.3}$$

where ε is the zero-mean random error with standard deviation σ_r, c is the speed of propagation (0.3 m/ns), and ϕ is the timing-phase of the clock in the mobile node relative to the time-synchronized base stations ($\phi = 0$ for range measurements). Note that in the case $\lambda = 0$, $\delta r_0 = 0$ and (6.3) becomes the classical formulation of the problem, and is typically applied to the indoor case even though it does not accurately describe the problem. In general, the values of λ and δr_0 in a given propagation environment will not be known, but it will be shown later in Sect. 6.4 how these parameters can be estimated. To illustrate these concepts, an example of the measured data from the positioning system described in Sect. 6.5 is shown in Fig. 6.2. Observe the linear trend in the bias, and also that the variation relative to the linear trend appears to be largely independent of range. Also observe that this condition (constant standard deviation) is a requirement for the LS algorithm to provide an optimum solution, as described in the Introduction.

The non-linear equation (6.3) needs to be solved for the position of the mobile node (x, y). Because of the non-linearity an analytical solution is difficult. One standard method (Yu et al. 2009; Sokolnikoff and Redheffer 1966; Sharp et al. 2009, 2012) in such cases is to linearize the equations to find an approximate solution, which can be iterated to obtain the final accurate numerical solution. In particular, (6.3) can be expanded as a first-order Taylor series, resulting in

$$p_{0_i} \approx f(x_0, y_0, b_0; \lambda) + \frac{\partial p_i}{\partial x}\Delta x + \frac{\partial p_i}{\partial y}\Delta y + \frac{\partial p_i}{\partial b}\Delta b \tag{6.4}$$

where the initial (guess) starting point is at (x_0, y_0, b_0), typically chosen at the mean of the surrounding base station coordinates, and the combined bias term is $b = \delta r_0 + c\phi$. This first order (linear) approximation improves as the range error to range ratio approaches zero. Note also that the bias term λ is assumed to be a parameter rather than a variable; the reason for this choice will be explained in Sect. 6.2.2.

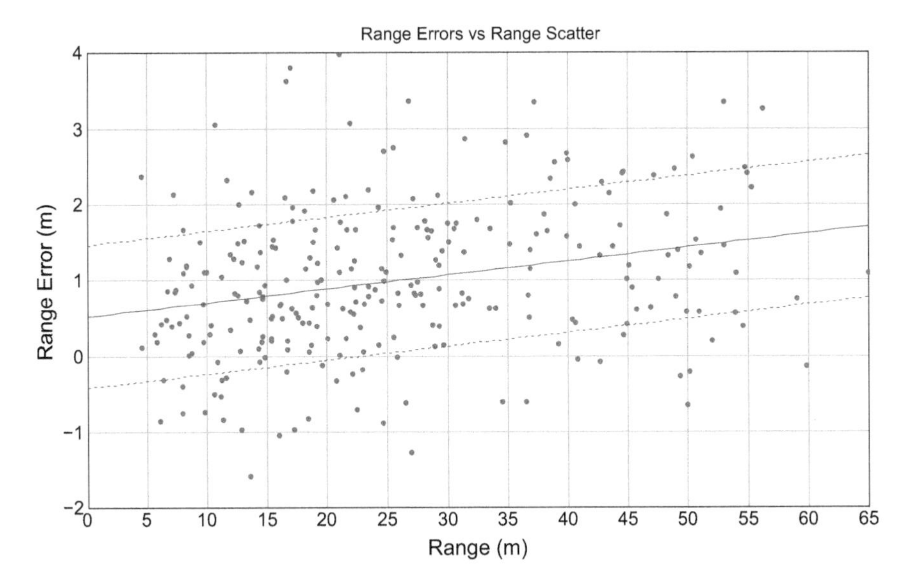

Fig. 6.2 Measured ranging error performance based on RTT measurements between pairs of nodes scattered throughout the building complex shown in Fig. 6.4. The measured parameters are: $\lambda = 0.020$ and $\sigma_r = 0.94$ m. The solid line is the associated linear LS fit model, and the dotted lines are $\pm\sigma_r$ relative to the linear trend line. The zero-range bias error is $\delta r_0 = 0.52$ m

Applying (6.4) with an estimate of the bias parameter defined as $\hat{\lambda}$ results in

$$\begin{aligned} \Delta\hat{p}_i &= \left(1+\hat{\lambda}\right)\left(\frac{x-x_i}{r_i}\right)\Delta x + \left(1+\hat{\lambda}\right)\left(\frac{y-y_i}{r_i}\right)\Delta y + \Delta b \\ &= \left(1+\hat{\lambda}\right)\alpha_i\,\Delta x + \left(1+\hat{\lambda}\right)\beta_i\,\Delta y + \Delta b \end{aligned} \tag{6.5}$$

where $\Delta\hat{p}_i = p_{0_i} - f\left(x_0, y_0, b_0; \hat{\lambda}\right)$, $\alpha_i = \cos\theta_i = \frac{x-x_i}{r_i}$, and $\beta_i = \sin\theta_i = \frac{y-y_i}{r_i}$.

For the classical range case it is assumed that $\Delta b = 0$. Note that (6.5) depends on the angular position θ_i of the mobile node relative to ith base station, and is independent of the associated range.

Equation (6.5) gives the measurements as a function of the positional and bias errors associated with the initial starting point for the iteration. However, the analysis requires estimates of the positional error in terms of the range errors. This is achieved by performing a LS fit on the linear equations represented by (6.5), so that the differences between the measured pseudoranges/ranges and that given by (6.5) are minimised. It is convenient to perform the analysis using matrix algebra, so that (6.5) can be expressed as

$$\mathbf{A\delta} = \Delta\mathbf{p} \tag{6.6a}$$

where for pseudoranges

$$\mathbf{A} = \begin{bmatrix} \left(1+\hat{\lambda}\right)\alpha_1 & \left(1+\hat{\lambda}\right)\beta_1 & 1 \\ \vdots & \vdots & \vdots \\ \left(1+\hat{\lambda}\right)\alpha_N & \left(1+\hat{\lambda}\right)\beta_N & 1 \end{bmatrix}, \ \boldsymbol{\delta} = \begin{bmatrix} \Delta x \\ \Delta y \\ \Delta b \end{bmatrix}, \ \Delta\mathbf{p} = \begin{bmatrix} \Delta\hat{p}_1 \\ \vdots \\ \Delta\hat{p}_N \end{bmatrix} \tag{6.6b}$$

and for ranges

$$\mathbf{A} = \begin{bmatrix} \left(1+\hat{\lambda}\right)\alpha_1 & \left(1+\hat{\lambda}\right)\beta_1 \\ \vdots & \vdots \\ \left(1+\hat{\lambda}\right)\alpha_N & \left(1+\hat{\lambda}\right)\beta_N \end{bmatrix}, \ \boldsymbol{\delta} = \begin{bmatrix} \Delta x \\ \Delta y \end{bmatrix}, \ \Delta\mathbf{p} = \begin{bmatrix} \Delta r_1 \\ \vdots \\ \Delta r_N \end{bmatrix} \tag{6.6c}$$

In typical practical situations the linear equations (6.6a, 6.6b, 6.6c) are overly-defined (more than three base stations for pseudoranges and two base stations for ranges), so that a LS solution can be obtained from

$$\Phi\boldsymbol{\delta} = \mathbf{h} \tag{6.7a}$$

where for pseudoranges

$$\begin{aligned} \boldsymbol{\Phi} &= \mathbf{A}^{\mathrm{T}}\mathbf{A} = \begin{bmatrix} \left(1+\hat{\lambda}\right)^2 \sum_i \alpha_i^2 & \left(1+\hat{\lambda}\right)^2 \sum_i \alpha_i\beta_i & \left(1+\hat{\lambda}\right) \sum_i \alpha_i \\ \left(1+\hat{\lambda}\right)^2 \sum_i \alpha_i\beta_i & \left(1+\hat{\lambda}\right)^2 \sum_i \beta_i^2 & \left(1+\hat{\lambda}\right) \sum_i \beta_i \\ \left(1+\hat{\lambda}\right) \sum_i \alpha_i & \left(1+\hat{\lambda}\right) \sum_i \beta_i & N \end{bmatrix} \\ \mathbf{h} &= \mathbf{A}^{\mathrm{T}}\Delta\mathbf{p} = \begin{bmatrix} \left(1+\hat{\lambda}\right) \sum_i \alpha_i\Delta\hat{p}_i \\ \left(1+\hat{\lambda}\right) \sum_i \beta_i\Delta\hat{p}_i \\ \sum_i \Delta\hat{p}_i \end{bmatrix} \end{aligned} \tag{6.7b}$$

and for ranges

$$\boldsymbol{\Phi} = \begin{bmatrix} \left(1+\hat{\lambda}\right)^2 \sum_i \alpha_i^2 & \left(1+\hat{\lambda}\right)^2 \sum_i \alpha_i\beta_i \\ \left(1+\hat{\lambda}\right)^2 \sum_i \alpha_i\beta_i & \left(1+\hat{\lambda}\right)^2 \sum_i \beta_i^2 \end{bmatrix}, \ \mathbf{h} = \begin{bmatrix} \left(1+\hat{\lambda}\right) \sum_i \alpha_i\Delta r_i \\ \left(1+\hat{\lambda}\right) \sum_i \beta_i\Delta r_i \end{bmatrix} \tag{6.7c}$$

The solution to linear matrix Eq. (6.7a) can be expressed in the form

$$\boldsymbol{\delta} = \left(\mathbf{A}^{\mathrm{T}}\mathbf{A}\right)^{-1}\left(\mathbf{A}^{\mathrm{T}}\Delta\mathbf{p}\right) = \boldsymbol{\Phi}^{-1}\mathbf{h} \tag{6.8}$$

The vector $\boldsymbol{\delta}$ provides an estimate for the correction required to the initial estimate of the three variables Δx, Δy and Δb. Thus better estimates for pseudoranges and ranges are

$$\begin{bmatrix} x_1 \\ y_1 \\ \phi_1 \end{bmatrix} = \begin{bmatrix} x_0 \\ y_0 \\ \phi_0 \end{bmatrix} + \begin{bmatrix} \Delta x \\ \Delta y \\ \Delta b \end{bmatrix} \quad \text{or} \quad \begin{bmatrix} x_1 \\ y_1 \end{bmatrix} = \begin{bmatrix} x_0 \\ y_0 \end{bmatrix} + \begin{bmatrix} \Delta x \\ \Delta y \end{bmatrix} \tag{6.9}$$

where for the range case it is assumed that the constant bias error δr_0 has been independently estimated (see Sect. 6.4) and removed from the measured range data. If this procedure is not performed, then the nominal range data is actually pseudorange data, and must be processed according to the pseudorange method.

Equation (6.8) can be applied iteratively until the increments are sufficiently small. Note that these corrections are not the errors in the position of the mobile node, which are dependent on the measurement errors and the value of the parameter $\hat{\lambda}$, but are increments in the iterative process. As the solution converges, these increments will approach zero in most situations, although with large measurement errors the algorithm may not converge. Also note from a theoretical point of view, if the initial starting point is the true position, the first increment is an estimate of the position error of the LS algorithm, as the algorithm convergence essentially occurs with just one iteration. This procedure is essentially the calculation of GDOP (Levanon 2000; Sharp et al. 2009, 2012), after normalization by the range error standard deviation, and hence directly related to the theoretical limits defined in (Chaffee and Abel 1994; Levanon 2000), as discussed in the Introduction.

6.2.2 Bias Parameter $\boldsymbol{\lambda}$

The solution for the position of the mobile node as given by (6.8) and (6.9) will depend on the bias parameter $\hat{\lambda}$ chosen. The actual radio propagation parameter λ for the operating environment is typically not known, and thus it appears the LS solution should also estimate this parameter. However, the purpose of introducing the parameter λ to the LS solution is to improve the accuracy of the estimated mobile node positions when compared with the standard procedure where this parameter is not used $\left(\hat{\lambda} = 0\right)$. Thus any change from the standard method should not result in a degrading in performance. The following summarizes why it is not appropriate to include $\hat{\lambda}$ as a variable to be determined by the LS solution, but as input parameter determined externally to the positioning calculation:

1. If $\hat{\lambda}$ is included as a variable to be estimated as part of the position determination, there will be a total of four variables to be solved for, and hence requiring at least four pseudorange measurements from four base stations. In contrast, the standard solution only requires measurements from three base station to obtain a position fix. Thus including the parameter $\hat{\lambda}$ will result in fewer positions being determined or less (or no) redundancy, resulting in less reliable positions.
2. The bias model parameter λ is associated with the characteristics of the indoor radio propagation. By definition this parameter can only be estimated by measurements at both short and long range, and thus the $\hat{\lambda}$ parameter represents an average over a considerable area. An estimate based on one set of base station measurements will typically be over a much smaller area, and thus is a less reliable estimate compared with a more global estimate. In particular, averaging over many position estimates will be necessary for a good estimate of the bias model parameter λ.
3. The classical approach of LS position determination effectively decouples the geometric factors (as expressed by the GDOP) and the propagation characteristics as expressed by the ranging error standard deviation. Incorporating the bias model parameter λ into the LS solution will couple these two components of the position accuracy.
4. If $\hat{\lambda}$ is a parameter for the LS positioning calculation rather than a variable, the parameter λ can be estimated by a separate process (as yet undefined) with appropriate smoothing and filtering, such as by applying a Kalman filter (see Appendix A for an introduction) to the raw estimates from individual position fix data.
5. While the optimum solution may seem to occur when $\hat{\lambda} = \lambda$, it will be shown that in fact the optimum which minimizes the mean-square position errors is in fact a different, larger value. Thus $\hat{\lambda} = \lambda_{opt}$ cannot be directly calculated by incorporating $\hat{\lambda}$ into the LS solution.

Thus the algorithm defined in Sect. 6.2.1 assumes that $\hat{\lambda} = \lambda_{opt}$ is determined external to the position determination algorithm; details of this process are described in Sect. 6.4.

6.3 Enhanced LS Algorithm Characteristics

Section 6.2 described the proposed enhancements to the classical LS algorithm to account for bias errors. This section analyses the characteristics of the enhanced algorithm.

6.3.1 Generalized Solution

As shown in Sect. 6.2.1, the performance of the enhanced LS algorithm can be determined by performing an analysis based on the general solution as defined by the matrix Eq. (6.8). However, suppose that a solution to the standard LS algorithm is known (either analytically or numerically), and from this solution it is required to determine the enhanced LS solution without being involved with the details of the standard solution. In particular from (6.8), suppose the standard LS solution with $\hat{\lambda} = 0$ is given by

$$\boldsymbol{\delta}_0 = \boldsymbol{\Phi}_0^{-1}\mathbf{h}_0 \tag{6.10}$$

From (6.7a, 6.7b, 6.7c) and (6.8) it can be observed that the $\boldsymbol{\Phi}$ matrix for an enhanced solution can be expressed as

$$\boldsymbol{\Phi} = \boldsymbol{\Lambda}\boldsymbol{\Phi}_0\boldsymbol{\Lambda} \text{ so that } \boldsymbol{\Phi}^{-1} = \boldsymbol{\Lambda}^{-1}\boldsymbol{\Phi}_0^{-1}\boldsymbol{\Lambda}^{-1} \tag{6.11a}$$

where for pseudoranges or ranges respectively

$$\boldsymbol{\Lambda} = \begin{bmatrix} \left(1+\hat{\lambda}\right) & 0 & 0 \\ 0 & \left(1+\hat{\lambda}\right) & 0 \\ 0 & 0 & 1 \end{bmatrix} \text{ or } \boldsymbol{\Lambda} = \begin{bmatrix} \left(1+\hat{\lambda}\right) & 0 \\ 0 & \left(1+\hat{\lambda}\right) \end{bmatrix} \tag{6.11b}$$

To calculate the enhanced solution $\mathbf{h}$ vector it is convenient to consider the bias and random components of the error separately. Thus considering the bias effect, the first element (1) in the $\mathbf{h}$ vector is given by

$$\begin{aligned} \left(\mathbf{A}^{\mathrm{T}}\Delta\mathbf{p}\right)_1 &= \left(1+\hat{\lambda}\right)\sum_i \alpha_i \Delta p_i = \left(1+\hat{\lambda}\right)\sum_i \alpha_i\left(\lambda-\hat{\lambda}\right) r_i \\ &= \left(\sum_i \alpha_i \lambda r_i\right)\left[\frac{\left(\lambda-\hat{\lambda}\right)\left(1+\hat{\lambda}\right)}{\lambda}\right] \end{aligned} \tag{6.12}$$

The last expression in (6.12) is the product of the standard LS algorithm $\mathbf{h}$ vector component times a constant factor which is related to the actual and estimated propagation bias terms. The second component of the $\mathbf{h}$ vector is similar to the first, with β replacing α. The third component (pseudoranges only) can be similarly calculated as

$$\left(\mathbf{A}^{\mathrm{T}}\Delta\mathbf{p}\right)_3 = \sum_i \Delta p_i = \sum_i \left(\lambda-\hat{\lambda}\right) r_i = \left(\sum_i \lambda r_i\right)\left[\frac{\left(\lambda-\hat{\lambda}\right)}{\lambda}\right] \tag{6.13}$$

Combining these components into a vector, the $\mathbf{h}$ vector related to the bias (b) can be expressed in the form

$$\mathbf{h}^{\mathbf{b}} = \left[\frac{\lambda - \hat{\lambda}}{\lambda}\right] \boldsymbol{\Lambda} \mathbf{h}_0^{\mathbf{b}} \tag{6.14}$$

Thus the enhanced LS bias component of $\boldsymbol{\delta}$ is given by

$$\begin{aligned} \boldsymbol{\delta}^{\mathbf{b}} = \boldsymbol{\Phi}^{-1}\mathbf{h}^{\mathbf{b}} &= \left[\frac{\lambda - \hat{\lambda}}{\lambda}\right] \left(\boldsymbol{\Lambda}^{-1}\boldsymbol{\Phi}_0^{-1}\boldsymbol{\Lambda}^{-1}\right) \boldsymbol{\Lambda}\mathbf{h}_0^{\mathbf{b}} \\ &= \left[\frac{\lambda - \hat{\lambda}}{\lambda}\right] \boldsymbol{\Lambda}^{-1}\left(\boldsymbol{\Phi}_0^{-1}\mathbf{h}_0^{\mathbf{b}}\right) = \boldsymbol{\Omega}_{\mathbf{b}}\boldsymbol{\delta}_0^{\mathbf{b}} \end{aligned} \tag{6.15}$$

where $\boldsymbol{\Omega}_{\mathbf{b}} = \left[\frac{\lambda-\hat{\lambda}}{\lambda}\right]\boldsymbol{\Lambda}^{-1}$, $\mathbf{h}_0^{\mathbf{b}}$ is bias component of the $\mathbf{h}$ vector for the standard LS solution, and $\boldsymbol{\delta}_0^{\mathbf{b}}$ is bias component of the $\boldsymbol{\delta}$ vector for the standard LS solution. Thus if a solution is known for the standard LS algorithm, the corresponding enhanced LS solution can be derived by multiplication by the matrix

$$\boldsymbol{\Omega}_{\mathbf{b}} = diag\left(\frac{\lambda-\hat{\lambda}}{\lambda(1+\hat{\lambda})} \quad \frac{\lambda-\hat{\lambda}}{\lambda(1+\hat{\lambda})} \quad \frac{\lambda-\hat{\lambda}}{1+\hat{\lambda}}\right) \text{ or } \boldsymbol{\Omega}_{\mathbf{b}} = diag\left(\frac{\lambda-\hat{\lambda}}{\lambda(1+\hat{\lambda})} \quad \frac{\lambda-\hat{\lambda}}{\lambda(1+\hat{\lambda})}\right) \tag{6.16}$$

for the pseudorange and range cases respectively. Now consider calculating the random (r) part of the $\mathbf{h}$ vector. The same procedure as described for the bias part results in the expression

$$\boldsymbol{\delta}^{\mathbf{r}} = \boldsymbol{\Phi}^{-1}\mathbf{h}^{\mathbf{r}} = \boldsymbol{\Lambda}^{-1}\boldsymbol{\delta}_0^{\mathbf{r}} = \boldsymbol{\Omega}_{\mathbf{r}}\boldsymbol{\delta}_0^{\mathbf{r}} \tag{6.17}$$

where $\boldsymbol{\Omega}_{\mathbf{r}} = \boldsymbol{\Lambda}^{-1}$ and $\boldsymbol{\delta}_0^{\mathbf{r}}$ is random component of the $\boldsymbol{\delta}$ vector for the standard LS solution. Finally, adding the bias and random components, the total $\boldsymbol{\delta}$ vector is given by

$$\boldsymbol{\delta} = \boldsymbol{\Phi}^{-1}\mathbf{h} = \boldsymbol{\Omega}_{\mathbf{b}}\boldsymbol{\delta}_0^{\mathbf{b}} + \boldsymbol{\Omega}_{\mathbf{r}}\boldsymbol{\delta}_0^{\mathbf{r}} \tag{6.18}$$

Thus if the standard LS solution is known, the enhanced LS solution can be directly calculated using (6.18) without any detailed calculations. This procedure is particularly useful if analytical solutions using the standard method are known, as the enhanced analytical solutions can be obtained with minimal effort; this approach is used in the following section.

6.3.2 Biased Error Solution

Using the generalized theory for the enhanced LS algorithm given in Sect. 6.3.1, this section applies the theory by using the known classical LS algorithm to obtain the enhanced solution; this procedure greatly simplifies the analysis. The distribution of base stations is assumed to have arbitrary geometry, although in practice it often can be approximated by a uniform random distribution (see the simulation in Sect. 6.4.1) surrounding the mobile node. The standard LS solution (Sharp et al. 2009; Teunissen 2003, Chap. 9) can be expressed in terms of the mean and variance of the x-coordinate and y-coordinate position errors. An identical analysis applies to both the x-coordinate and y-coordinate errors, so only equations associated with the x-position errors are given in the following analysis.

The analysis in the references shows that for the bias-error case the x-position error is generally non-zero, and is given by

$$\left(\Delta x^b\right)_0 = \lambda \sum_{i=1}^{N_R} \Gamma_i r_i \tag{6.19a}$$

where the subscript "0" indicates the standard LS algorithm solution, and for pseudorange and range positioning respectively

$$\Gamma_i = \alpha_i \Phi_{1,1}^{-1} + \beta_i \Phi_{1,2}^{-1} + \Phi_{1,3}^{-1} \text{ or } \Gamma_i = \alpha_i \Phi_{1,1}^{-1} + \beta_i \Phi_{1,2}^{-1} \tag{6.19b}$$

and N_R is the number of base stations in range of the mobile node; a similar expression can be derived for the y-position error, except that in this case

$$\Gamma_i = \alpha_i \Phi_{2,1}^{-1} + \beta_i \Phi_{2,2}^{-1} + \Phi_{2,3}^{-1} \quad \text{or} \quad \Gamma_i = \alpha_i \Phi_{2,1}^{-1} + \beta_i \Phi_{2,2}^{-1} \tag{6.19c}$$

respectively for the pseudorange and range cases. For the random case the mean of the x–position error is zero, assuming the mean of the range errors is zero, and the random range errors are uncorrelated. The variance of the x–position error for the random case is given by

$$\left(\sigma_{\Delta x}^r\right)_0^2 = \sigma_r^2 \sum_{i=1}^{N_R} \Gamma_i^2 \tag{6.20}$$

The total error consists of bias and random components. The expected value of the square of the combined bias and random errors is given by

$$\left(\sigma_{\Delta x}\right)_0^2 = \left(\sigma_{\Delta x}^r\right)_0^2 + \left(\Delta x^b\right)_0^2 \tag{6.21}$$

From (6.18) it can be observed that the enhanced LS solution for x-position error involves only multiplication by a constant, so the x-position error for the bias case is given by

$$\Delta x^b = \frac{\lambda - \hat{\lambda}}{1 + \hat{\lambda}} \sum_{i=1}^{N_R} \Gamma_i r_i \tag{6.22}$$

so that with $\lambda = \hat{\lambda}$ the x-position error will be zero using the enhanced LS algorithm. Similarly, the variance will be multiplied by the square of the constant. Thus from (6.17) and (6.20), the variance in the x-position error for the enhanced LS associated with the random range errors is given by

$$\left(\sigma_{\Delta x}^r\right)^2 = \left(\frac{1}{1 + \hat{\lambda}}\right)^2 \sigma_r^2 \sum_{i=1}^{N_R} \Gamma_i^2 \tag{6.23}$$

The total x-position variance is the sum of these two components, similar to (6.21). For a given situation, the total error in the x-position will be a function of both the actual model bias parameter (λ) and the estimated value $\left(\hat{\lambda}\right)$. Observe from (6.22) that if $\lambda = \hat{\lambda}$ the bias component is zero, but interestingly this is not the optimum value of the parameter $\hat{\lambda}$ for obtaining the smallest combined error in Δx. The determination of the optimum value and the consequential performance is considered in the next section.

6.4 Determination of the Bias Parameter

This section describes the analytical method of determining the optimum bias parameter, and how its can be calculated from measured data in a mesh positioning system.

6.4.1 Optimal Bias Parameter

The analysis in Sect. 6.3.2 showed that for the enhanced LS algorithm there are two components of the computed position error, one associated with the random errors and one associated with the bias errors. These errors are a function of the actual range bias error parameter (λ), and also the estimate of this parameter $\left(\hat{\lambda}\right)$. Again the analysis for the y-position error is identical to the x-position analysis, so only x-position error equations are included in the following analysis.

To obtain the optimum solution it is suggested that expected value of the square of the x-position error should be minimized. Thus from (6.22) and (6.23), the expression to be minimized is

$$\begin{aligned}\sigma_{\Delta x}^2 &= (\sigma_{\Delta x}^r)^2 + (\Delta x^b)^2 \\ &= \left(\frac{1}{1+\hat{\lambda}}\right)^2 (\sigma_{\Delta x}^r)_0^2 + \left(\frac{\lambda-\hat{\lambda}}{1+\hat{\lambda}}\right)^2 \left(\frac{(\Delta x^b)_0}{\lambda}\right)^2\end{aligned} \tag{6.24}$$

The optimal position determination performance occurs when (6.24) is minimized by selecting the optimum value of $\hat{\lambda}$. By differentiating (6.24) with respect to $\hat{\lambda}$ and equating to zero, the optimum value can be calculated to be

$$\lambda_{opt} = \lambda + \frac{\rho^2}{1+\lambda} \tag{6.25}$$

where ρ is the dimensionless ratio given by

$$\rho = \frac{(\sigma_{\Delta x}^r)_0}{(\Delta x^b)_0/\lambda} = \sigma_r \left(\sqrt{\sum_{i=1}^{N_R} \Gamma_i^2} \Big/ \sum_{i=1}^{N_R} \Gamma_i r_i \right) \tag{6.26}$$

Note that the term in brackets in (6.26) is solely a function of the geometry of the layout of the nodes in range of the mobile node. From (6.25) it can be observed that the optimum bias parameter to be used in the enhanced LS algorithm is greater than the actual bias parameter associated with the radio propagation. Using (6.26) in (6.25), the mean square of the x-coordinate of the position error for the enhanced LS solution can be calculated to be

$$\left[(\sigma_{\Delta x}^r)^2 + (\Delta x^b)^2\right]_{opt} = (\sigma_{\Delta x})_{opt}^2 = \frac{(\sigma_{\Delta x}^r)_0^2}{(1+\lambda)^2 + \rho^2} \tag{6.27}$$

which is less than the corresponding value based on the standard LS solution, namely

$$\sigma_{\Delta x}^2 = \left(1 + \left(\frac{\lambda}{(1+\lambda)\rho}\right)^2\right) (\sigma_{\Delta x}^r)_0^2 \tag{6.28}$$

An example of the characteristics of the enhanced LS algorithm is shown in Fig. 6.3. The simulated data are for a coverage area of 10,000 m^2 with 46 base stations (similar to the actual system described in Sect. 6.5). The results for measured data are given in Sect. 6.5. The base stations are located within the coverage area with a spatially uniform random distribution. The Fig. 6.3 shows the variation in the standard deviation of the x–coordinate error as a function of the algorithm

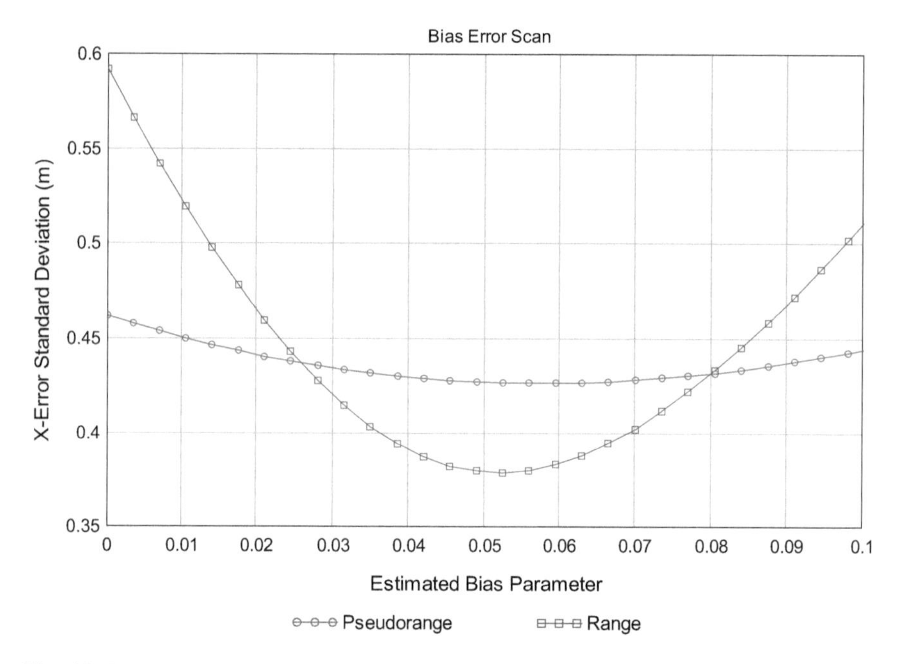

Fig. 6.3 Example of the simulated performance of the enhanced LS range algorithm for pseudorange and range positioning. The parameters are: $\lambda = 0.035$, $\delta r_0 = 0.5$ m $\sigma_r = 1$ m, $R_{\max} = 30$ m and $N = 46$ base stations randomly located in a square area of 10,000 m^2. The data are averaged over 1000 simulated sets of base station locations

estimated bias parameter $\hat{\lambda}$ for both pseudorange and range positioning. As can be observed, using the standard LS algorithm $\left(\hat{\lambda} = 0\right)$ the pseudorange positioning performance is much better than the corresponding range positioning performance, but at the optimum the range performance is superior; in general, pseudorange positioning is less affected by bias errors. This result with bias errors is contrary to the classical solution with only random errors, where the range solution typically has superior accuracy compared with the pseudorange solution.

Thus the recommendation for position determination is as follows. If there is no information on the bias parameter λ, pseudorange position determination should be used, even if the nominal measurements are ranges. However, if there is an estimate of the bias parameter and the measurements are in range, then the enhanced position determination method for range measurements described above should be used.

The performance of a positioning system is often specified in terms of the mean radial error, as this measure can be readily determined from measured positional error data. The above analysis computes the standard deviation in the x-coordinate (and y-coordinate) error only, but this measure can be converted into the equivalent mean radial error. Because the LS algorithm solution involves the summation of random components (see matrix Eq. (6.6)), the statistical distribution of the x and y coordinates will be approximately Gaussian due to the Central Limit Theorem.

Further, the x-coordinate and y-coordinate statistics are statistically independent, with approximately the same standard deviation due to the approximately (statistical) uniform spatial distribution of the nodes in a mesh network. As a consequence the radial distribution will have (approximately) Rayleigh statistics. For a Rayleigh distribution, the mean radial error is related to the 1D Gaussian distribution STD by the multiplication constant $\sqrt{\pi/2}$, so that equations such as (6.27) and (6.28) can be simply converted to the associated mean radial error.

6.4.2 Determining λ_{opt} from Measured Data

The application of the enhanced LS algorithm requires the determination of the optimum value of the bias parameter λ_{opt}. The optimum value is given by (6.25), which shows it is a function of two primary parameters, namely the environmental bias parameter (λ), the random range error STD (σ_r), as well as the geometry of the nodes. As the geometry is known, these two propagation related parameters need to be estimated from measurements (such as in Fig. 6.2) in the actual operating environment using the actual positioning system equipment, as the indoor propagating environment is usually too complex for mathematical modeling. Because the bias parameter requires measurements over a considerable area, the calculated parameters will be an overall average over the propagating environment, typically with one set of values for a whole building; for very large buildings dividing the total area into sub-regions may be appropriate. Assuming that the operating environment is static, the estimation of the parameters could be performed as part of the installation process, although real-time estimation would be more appropriate.

If the propagation parameters are determined as part of the installation process, then a basic survey of a building can be performed as follows. Two nodes of the positioning system are used to measure the round-trip time (RTT), and hence an estimation of the inter-nodal distances. The measurements require careful calibration of the radio equipment delays as well as accurate synchronisation of the frequency in each node; time synchronisation in the nodes is not necessary with the RTT technique (Sathyan et al. 2011; Hedley 2008 and Chap. 5). By accurately surveying the node positions by an independent method (usually using a plan of the building), a set of range errors versus range data can be obtained throughout the building. From this data set the required parameters can be derived, as shown by the LS fit lines in Fig. 6.2. An alternative to the above procedure is to survey the building using only the fixed base stations of the installed system, such as the nodes (base stations) shown in Fig. 6.4. A similar set of data to that described above can be obtained by measuring the range between every pair of base stations in the mesh network, resulting in a set of (at most) $N_{bs}(N_{bs}-1)/2$ range measurements. These measurements can be performed in real-time as a "background" task during the

Fig. 6.4 Layout of site with buildings and the location of the 46 nodes. The area covered is around 10,000 m^2

normal operation of the system, as calibration parameters such as equipment delays will vary over time (Sathyan et al. 2011; Hedley 2008).

Having a set of M (calibrated) radio-measured and surveyed ranges $\left(\hat{R}_m, R_m\right)$ the bias parameter and the range error standard deviation can be determined by a least-squares fitting process. In particular defining the range error as $\Delta R_m = \hat{R}_m - R_m$, the estimated range bias parameters are given by

$$\begin{aligned} \lambda_{est} &= \frac{1}{\Delta}\left[M\sum_m R_m \Delta R_m - \sum_m R_m \sum_m \Delta R_m\right] \\ \delta r_0 &= \frac{1}{\Delta}\left[\sum_m \Delta R_m \sum_m R_m^2 - \sum_m R_m\left(\sum_m R_m \Delta R_m\right)\right] \end{aligned} \tag{6.29}$$

where $\Delta = M\sum_m R_m^2 - \left(\sum_m R_m\right)^2$. The random range error standard deviation can then be calculated by

$$\begin{aligned} \varepsilon_m &= \Delta R_m - (\delta r_0 + \lambda_{est} R_m) \\ \sigma_r &= \sqrt{E[\varepsilon^2]} \end{aligned} \tag{6.30}$$

Using the results from (6.29) and (6.30) in (6.25) and (6.26), an estimate of the optimal parameter λ_{opt} can be calculated, which in turn can be applied to the enhanced LS position determination procedure described in Sect. 6.2.1.

The estimated standard deviations in the bias parameters $(\delta r_0, \lambda_{est})$ and the random range error standard deviation (σ_r) can be obtained from the regression analysis. For σ_r the standard deviation of its estimated value is about $\sigma_r/\sqrt{M}$, and for δr_0 about $\sigma_r\sqrt{3/M}$, where M is the number of measurements. The uncertainty in δr_0 thus adds statistically to the range error standard deviation, resulting in σ_r increasing by the small factor of $\sqrt{1+3/M}$. As the number of such measurements is of the order of $N_{bs}^2/2$, for even moderately large mesh networks the number of measurements will be several hundred; thus the uncertainty in determining σ_r will be $\sqrt{1/M+3/M^2}$, or less than about 8 percent with $M > 200$ (6% for the WARP data in Fig. 6.2). For λ_{est} the estimated value has a standard deviation of about $3\sqrt{2}\sigma_r/MR_{\max}$, which for the WASP data in Fig. 6.2 is 0.0042, or 21% of λ_{est}. The sensitivity of the solution to the value of the bias parameter (λ) is illustrated in Fig. 6.3. For pseudorange positioning, the optimum solution is largely insensitive to the bias parameter, while for range positioning the effect on the bias error is considerably greater. However, even in the later case, due to the flat shape near the optimum, an accuracy of ±20% appears to be adequate. As a consequence uncertainty in determining bias parameter λ_{opt} will not greatly affect the improvement in the accuracy of the position determination using the enhanced LS algorithm.

6.5 Measured Performance

The theory presented in previous sections is tested with actual data from a real indoor positioning system. The layout of the buildings and the nodes is shown in Fig. 6.4. Note that all the nodes (base stations) in this example are indoors, as the analysis is focused on position determination in an indoor environment. However, it is clear from the map of the building layout in Fig. 6.4 that some of the inter-nodal radio paths are partially outdoors; these paths will be less affected by bias errors. The aim of using the experimental data is to confirm the above theoretical analysis, and to determine the improvement factors of the enhanced LS algorithm compared with the standard algorithm for both range and pseudorange data.

The testing of the theoretical results was based on the Wireless Ad hoc System for Positioning (WASP) (Humphrey and Hedley 2003; Sathyan et al. 2011; Hedley 2008), developed by CSIRO. This experimental system operates in the 5.8 GHz ISM band with an effective bandwidth of 125 MHz. Position calculations are based on TOA estimation of a reconstructed wideband spread-spectrum signal with an effective leading-edge rise-time period of 12.5 ns; RTT between node pairs is determined from the TOA measurements. As part of the position determination process large range errors (approximately greater than the rise-time) are eliminated (Sathyan et al. 2011) by a process of comparing the measured range with the range computed from the estimated position. Such a process is possible when there is redundancy in the number of base stations in range. The radio modules also support ad hoc data communications which allows the measured TOA data to be forwarded to a central data logging computer.

First consider the determination of the bias and random parameters from measured data. Figure 6.2 shows a scatter diagram of the measured range errors versus the range using intra-nodal (base station) RTT measurements. Observe that overall the range errors tend to increase with range, but the random component is approximately independent of range. A statistical analysis of the errors as a function of range is summarized in Fig. 6.5. The mean and standard deviation of the error data were estimated in range "bins", so that the variation in these parameters can be estimated as a function of range. As can be observed the mean increases with range (approximately linearly), but the standard deviation is approximately constant with range, as assumed in the theory. At long range the mean errors are somewhat less than the linear trendline; the reason for this is that the long ranges include long obstacle-free paths outside the buildings (between the buildings—see Fig. 6.4), and thus the scattering effect will be less than for an all-indoor path. Note, however, all the nodes are within the buildings. Also note that there is a non-zero offset in the range errors at zero range, as was assumed in the theory. This offset has no effect on

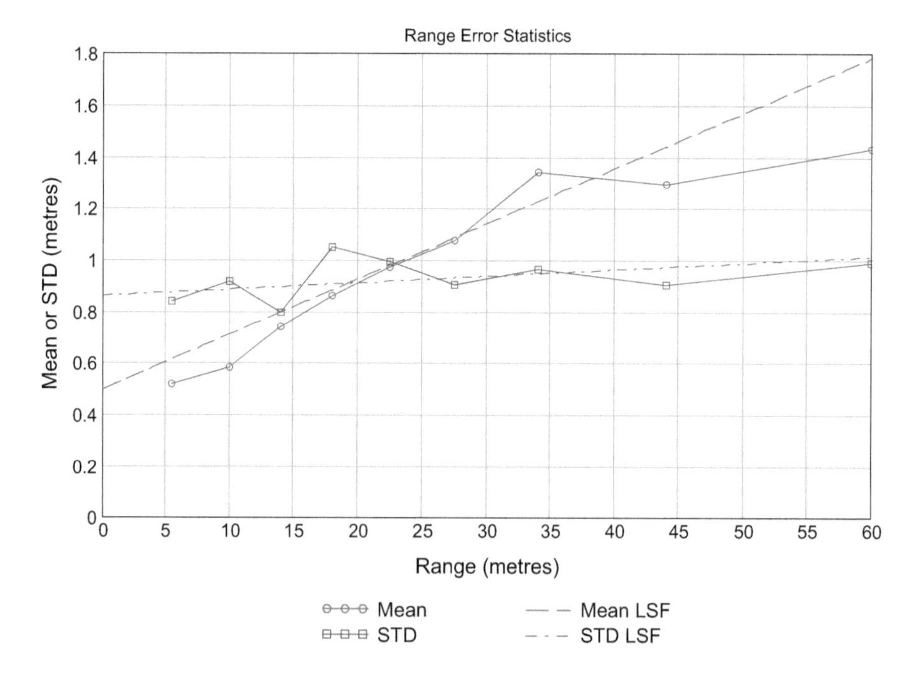

Fig. 6.5 Estimate of the mean and standard deviation of the error data in Fig. 6.2 in "bins" of range. The bins size increases with range so that the number of samples (about 50) in each bin is approximately equal. The plotted points are at the centre of each range bin. The linear trendlines are based on a weighted (using the inverse of variance of the error) least-squares fit (LSF)

the pseudorange positioning performance, but does significantly affect the performance of range positioning.

From this data set, the theory in Sect. 6.3 can be used to obtain the estimated parameters: $\hat{\lambda} = 0.020$ and $\sigma_r = 0.94$ m. Note in this case the bias parameter λ is rather small, so the effects of ranging bias errors will not be very large. The low value of λ is in part due to the sophisticated leading edge algorithm (Humphrey and Hedley 2003) used to determine the TOA, but also the wide effective bandwidth of 125 MHz.

These results can be compared with some other results reported in the literature. The simulations reported in (Alavi and Pahlavan 2003) show that if data with errors greater than the rise-time are eliminated, the range error scatter is approximately independent of range, supporting the WASP results. The measured UWB results in (Gentile and Kik 2006) show a similar characteristic regarding the range scatter, and an approximate linear bias as a function of range. However, the UWB measurements reported in (Alavi and Pahlavan 2006; Alsindi et al. 2009) are only presented in normalized form, the former normalized by $\log(1 + r)$ and the latter by range. As raw (non-normalized) data are not presented, the variation with range cannot be determined with any certainty. For example in the latter case, 1 m error at 10 m range will have the same normalized error as a 10 m error at 100 m range. Further, as there are far more data at short range compared with long range, the normalized statistical distribution is dominated by the short-range data, so that the statistics of the raw data at long range cannot be inferred from the data in (Alsindi et al. 2009). Further, as large range error data are not used for position determination, including such data is not appropriate for realistic performance estimates. Nevertheless, the normalized data (Alsindi et al. 2009) supports the contention that the mean (bias) range errors increase with range, in agreement with the WASP data.

With the above data and the theory described in previous sections, the optimum bias parameter for the enhanced LS algorithm can now be determined. The data in Table 6.1 summarizes the theoretical positioning performance based on the above theory and the measured range error STD, the range bias parameter, and the number of nodes in range. Both the pseudorange and range positioning performance parameters are listed, as well as the parameters summarising the node spatial distribution and the ranging performance, common to both. The random and bias position error parameters listed in Table 6.1 are defined as follows:

$$
\begin{aligned}
(\sigma^r)_0 &= \sqrt{(\sigma^r_{\Delta x})_0^2 + \left(\sigma^r_{\Delta y}\right)_0^2} \\
\left(\Delta^b\right)_0 &= \sqrt{(\Delta x^b)_0^2 + (\Delta y^b)_0^2} \\
(\sigma)_0 &= \sqrt{(\sigma^r)_0^2 + \left(\Delta^b\right)_0^2}
\end{aligned}
\tag{6.31}
$$

which are all invariant to rotation. From the data given in Table 6.1, the following observations can be made:

1. For the standard LS algorithm, the range positioning algorithm has superior performance with random errors (in line with theoretical expectations), but the pseudorange positioning algorithm is superior with bias errors (in line with the above analysis).

Table 6.1 Summary of the theoretical and measured parameters for the indoor test case

Parameter	Pseudorange	Range	Comment
σ_r	0.94 m	0.94 m	Random range error STD
λ	0.020	0.020	Range error bias parameter
$R_{\max}$	55 m	55 m	Average maximum radio range
$\tilde{r}$	33 m	33 m	Average RMS radio range
N_R	11	11	Typical number of nodes in radio range
$(\sigma^r)_0$	0.72 m	0.60 m	Random errors—standard LS algorithm
$(\Delta^b)_0$	0.36 m	0.69 m	Bias errors—standard LS algorithm
ρ	0.0409	0.0178	Random STD to Bias STD ratio
λ_{opt}	0.0220	0.0206	For enhanced LS algorithm
$(\sigma)_0$	0.80 m	0.92 m	Overall STD—standard LS algorithm
σ_{opt}	0.71 m	0.59 m	Overall STD—enhanced LS algorithm
	11%	36%	Improvement factor

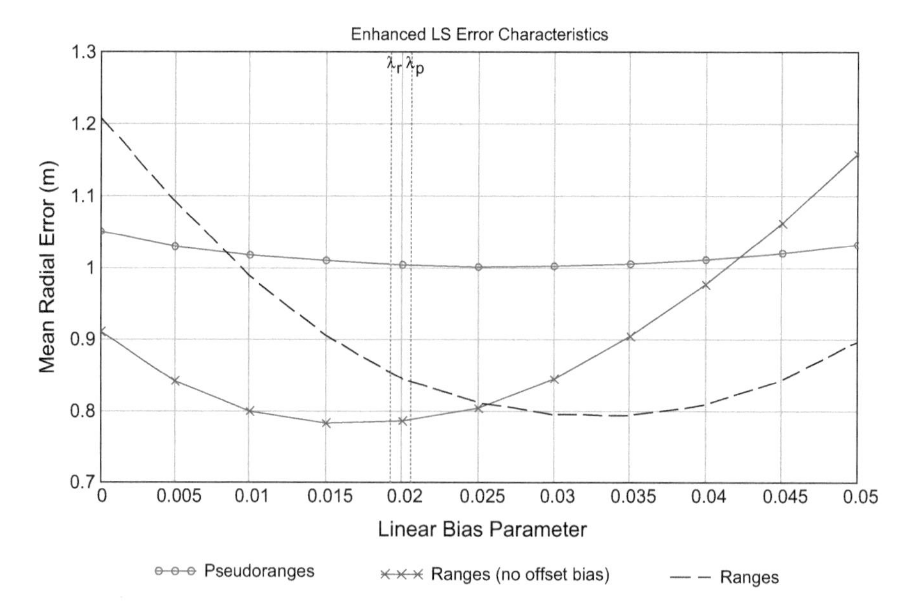

Fig. 6.6 Comparison of the performance of the enhanced pseudorange and range LS positioning algorithm using the measured data. For the range case, the effect of removing the bias offset is also shown

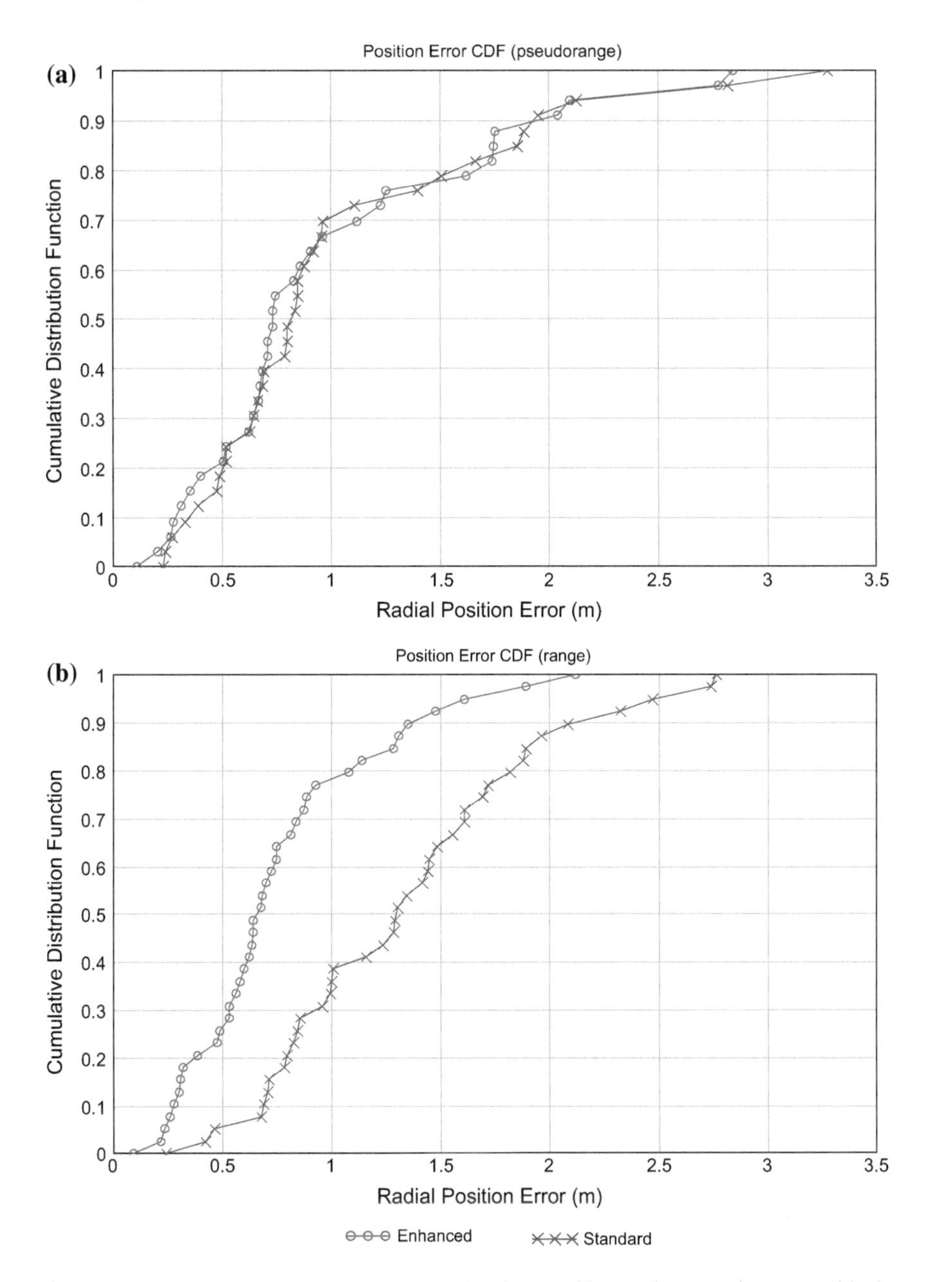

Fig. 6.7 **a** Comparison of the measured positional error CDFs using pseudoranges with the standard and enhanced LS positioning algorithms. **b** Comparison of the measured positional error CDFs using ranges with the standard and enhanced LS positioning algorithms

2. Overall, pseudorange positioning is superior when using the standard LS algorithm, while range positioning is superior when using the enhanced LS algorithm. Note however, that this superior performance of the range-based positioning comes at the cost of the estimation and then the elimination of the constant range error δr_0.
3. The performance gain using the pseudorange-based enhanced LS algorithm is rather modest (11%), while the performance gain using the range-based enhanced LS algorithm is significant (36%).

The pseudorange and range positioning performance as a function of the estimated linear bias parameter is shown in Fig. 6.6. As can be observed, without any correction (standard LS method), the pseudorange performance is superior to the range performance, which is in agreement with the results in Table 6.1; using the enhanced LS algorithm with optimal estimation of the bias parameter $\hat{\lambda} = \lambda_{opt}$ the reverse is true, again in agreement with Table 6.1. Also shown in Fig. 6.6 for the range case is the effect of removing the offset bias (δr_0) from the measured data.

With the standard LS algorithm, the effect of removing the bias offset is significant but not optimum; the optimum using the enhanced algorithm is similar whether the offset bias is removed or not, although the optimum linear bias parameter $\left(\hat{\lambda}\right)$ is different. However, it is also important to note that in actual situations the measurement errors are not known, so that the performance scan shown in Fig. 6.6 cannot be used to find the minimum errors; the actual method of determining an estimate of the optimal value is by applying the method described in Sect. 6.3; these estimates are shown in Fig. 6.6, and can be observed to be close to the true optimum (curve minimum) for both ranges and pseudoranges.

The overall statistical cumulative distribution function (CDF) of the positional errors is shown in Fig. 6.7a (pseudoranges) and Fig. 6.7b (ranges). From Fig. 6.7a it can be observed that for pseudoranges there is little difference between the standard and enhanced algorithms. In contrast, from Fig. 6.7b it can be observed that the performance of the enhanced LS algorithm is considerably better than the corresponding standard algorithm. Thus for example, if the "accuracy" of the positioning system is defined at a CDF of 0.8, the standard LS algorithm has errors of 1.0 m and 1.8 m for respectively pseudorange and range positioning; the corresponding values for the enhanced LS algorithm are 1.1 and 0.8 m. If the median (CDF = 0.5) is used as a reference, the standard and the enhanced range positioning errors are 1.3 m and 0.65 m respectively, resulting in accuracy improvement by a factor of 2.

References

Alavi B, Pahlavan K (2003) Modeling of the distance error for indoor geolocation. In: Proceedings of IEEE wireless communications and networking, pp 668–672, Mar 2003

Alavi B, Pahlavan K (2006) Modeling of the distance measure error using UWB indoor radio measurement. IEEE Commun Lett 10(4):275–277

Alsindi N, Alavi B, Pahlavan K (2009) Measurement and modelling of ultra wideband TOA-based ranging in indoor multipath environments. IEEE Trans Veh Technol 58(3):1046–1058

Chaffee J, Abel J (1994) GDOP and the Cramer-Rao bound. In: Proceedings of position location and navigation symposium, pp 663–668, 11–15 Apr 1994

Gentile C, Kik A (2006) An evaluation of ultra wideband technology for indoor ranging. In: Proceedings of IEEE GLOBECOM, pp 1–6 (2006)

Hedley M, Humphrey D, Ho P (2008) System and algorithms for accurate indoor tracking using low-cost hardware. IEEE/IOA Position, Location and Navigation Symposium, pp 633–640, May 2008

Humphrey D, Hedley M (2003) Super-resolution time of arrival for indoor localization. International Conference on Communications (ICC), Beijing, China, pp 3286-3290, May 2008.

Humphrey D, Hedley M (2008) Super-resolution time of arrival for indoor localization. In: Proceedings of international conference on communications (ICC), Beijing, China, pp 3286–3290, May 2008

Levanon N (2000) Lowest GDOP in 2-D scenarios. IEE Proc Radar Sonar Navig 147(3):149–155

Sathyan T, Humphrey D, Hedley M (2011) WASP: a system and algorithms for accurate radio localization using low-cost hardware. IEEE Trans Syst Man Cybern–Part C, 41(2):211–222

Sokolnikoff IS, Redheffer RM (1966) Mathematics of physics and modern engineering, chapter 10, section 11. McGraw-Hill

Sharp I, Yu K, Hedley M (2012) On the GDOP and accuracy for indoor positioning. IEEE Trans Aerosp Electron Syst 48(3):2032–2051

Sharp I, Yu K, Guo YJ (2009) GDOP analysis for positioning system design. IEEE Trans Veh Technol 58(7):3371–3382

Teunissen P (2003) Adjustment theory—an introduction, VSSD—series on mathematical geodesy and positioning

Yu K, Sharp I, Guo YJ (2009) Ground-based wireless positioning. Wiley-IEEE Press

Chapter 7
Use of Sensors for Position Determination

7.1 Overview of Sensor Position Determination

In Chap. 4 radiolocation time-of-arrival determination was analyzed as a prime method for determining the position of a mobile device due to its high accuracy even in severe multipath conditions indoors. In this chapter an alternative approach to position determination using sensor data is analyzed. In contrast with Chap. 4, the main focus in this chapter is on determining the mobile speed and direction, and the quality of these estimates, rather than determining the position from the data. While the main focus of this book is on radiolocation techniques, the limitations associated with radio propagation, particularly indoors, makes alternative methods an essential adjunct to any practical system. The use of sensors, combined with radio systems in small mobile devices, has become an area of intense research and development in recent years. Because of the availability of small, cheap sensors (and radios), practical systems that are small enough for personal mobile body-worn devices can now be considered practical. Thus the focus of this chapter is on personal mobile devices used for position determination.

The determination of position using estimates of the speed and heading angle has a history of hundreds of years, particularly regarding ship navigation. The method of determination of position using the speed and heading data is known as "dead reckoning". The use of the term "dead reckoning" in this context has been around since at least the seventeenth century.[1] The basis of dead reckoning is simple in principle. The incremental change in position in a given time interval (measured by a clock) can be estimated by integrating the velocity vector (speed and direction). If

[1]See the World Wide Words web site (www.worldwidewords.org) for background on the origin of this terminology. This reference shows that dead reckoning has been in the English language since the early seventeenth century. It had much the same sense then as it does now, that of estimating the position of a vessel from its speed, direction of travel and time elapsed, making use of a log, compass and clock. Note that the term "dead reckoning" (for deduced reckoning) has been erroneously ascribed in the US as the origin of the term in the twentieth century.

I. Sharp and K. Yu, *Wireless Positioning: Principles and Practice*, Navigation: Science and Technology, https://doi.org/10.1007/978-981-10-8791-2_7

the position at the starting point is known (by some other method), then the new position can be estimated. Thus the method requires estimates as a function of time of both speed and the heading angle, both of which can be obtained from appropriate sensors. Note that dead reckoning cannot give absolute positions without some other absolute positioning system (at least at the start of the dead reckoning process, but preferably with continual external reference updates), and thus the method is ideally suited for integration with a radiolocation system. Indeed, it will be shown that dead reckoning and radiolocation are complementary in performance, with the strengths of one method being the weakness of the other, and vice versa. For a more general discussion of the topic of *Data Fusion* of two sets of position data refer to Chap. 17.

The use of a compass in the determination of the heading angle has long history in navigation, and continues to be useful in wireless sensor applications. In typical modern applications the compass is implemented by two orthogonally mounted magnetometers which measure the horizontal component of the Earth's magnetic field. From these two measurement the magnetic heading can be calculated, and assuming the general position on the Earth is known (say within a few hundred kilometers), then the true (geographic) heading can be determined from the magnetic heading by applying the relevant correction call the *magnetic declination.*[2] The main problem with compass heading data (particularly indoors) is that the Earth's magnetic field is disrupted by magnetic anomalies, usually from man-made electrical devices, or structural (steel) components of a building. These anomalies can be relatively short-range in their effect (near the offending electrical equipment or structural element), so that as the mobile device moves through the environment the overall heading in the long-term is satisfactory, but there are often large short-term errors. Thus a method of correcting for these short-term errors is essential for accurate position determination. One method for correcting the compass heading data is by use of a gyroscope. The stabilization of a compass using a gyroscope in marine applications has been common practice for at least a century, but these mechanical devices are not suitable for small mobile devices. Fortunately, small solid-state rate-gyroscopes are now available, although their performance (in terms of stability and accuracy) is far less than the larger mechanical gyroscopes. Nevertheless, as the typical requirement in personal positioning systems is for short-term correction, these solid-state gyros provide significant correction capability when matched with a compass.

The determination of speed of a small mobile device using sensors is a more difficult task than determining the heading angle. The classical approach on aircraft is to use accelerometers, with the integration of the measured acceleration being the required speed. However, this method is not readily applicable to small mobile personal location devices. Firstly, the small solid-state accelerometers commonly

[2]The *magnetic deviation* is another correction factor due to local effects such a materials such as steel in the building structure. Such affects would in general be considered as unknown random heading errors for indoor positioning.

used in mobile devices are typically at least three orders of magnitude less accurate than those used in ships/aircraft. In particular, the sensor outputs have scaling and offset errors which are difficult to accurately compensate for. Further, the mobile device attached to a person is subject to accelerations associated with the walking gait which are much greater than the accelerations associated with the overall movement about the environment. These problems are further exacerbated by the fact that the horizontal acceleration is "corrupted" by components of the Earth's gravity (which the accelerometer also measures), as the mobile device cannot be rigidly attached to the body in a horizontal orientation. As a consequence, other methods are required to measure speed using sensors on a personal mobile device. One such method is described in this chapter, which is easy to implement and provides reasonably accurate estimates of both speed and distance traveled.

Thus with appropriate signal processing, small sensors can provide the speed and heading data necessary for dead reckoning personal navigation. However, practical applications require the integration of the sensor-derived location data with the radiolocation data, so much of the analysis described in this chapter is related to methods of integrating the two data sets. In particular, this chapter describes in detail the use of complementary filtering for improvement in both heading angles and speed, which in turn results in more accurate position determination.

7.2 Sensors for Mobile Location Devices

With the development in recent years of solid-state sensors that are small and relatively cheap, the opportunity has opened up for the incorporation of these sensors into small mobile devices. This section provides an overview of the sensors appropriate for position determination using dead reckoning techniques, or for the integration with radiolocation to improve the positioning accuracy and reliability. The intention of this section is not to describe the characteristics of particular commercially available sensor products, but rather provide an overview of their typical characteristics and how these characteristics are applied to the position determination application.

The sensors appropriate for position determination are magnetometers, rate gyroscopes and accelerometers. In modern mobile devices such as phones and tablets these are small solid-state devices, and are available as packages with three mutually orthogonal sensors. The sensors for heading determination are magnetometers (to form a compass) and rate-gyroscopes which measure the rate of change of an angle (in this case the heading angle). As will be shown in later sections, the combining of the data from these two sensors provide good estimations of the heading angle. The sensor for determining speed is an accelerometer. As the integral of an acceleration gives a speed, two accelerometers in a horizontal plane can in theory provide speed estimates. However, as will be shown in later sections, the limited accuracy of the sensors, difficulties associated with attaching these sensors to the human body, and the complex motion associated with human movement means that this simple approach is not practical. However, it will be shown in Sect. 7.5 that an alternative approach can be used to provide both speed and distance traveled estimates for a person wearing a mobile device with a vertically mounted accelerometer.

7.2.1 Magnetometers

The magnetometers in mobile devices are intended to measure the strength of the magnetic field of the Earth, typically in three orthogonal directions. The magnitude at the Earth's surface ranges from 25 to 65 μT (0.25–0.65 G). The field approximates a magnetic dipole tilted at an angle of about 11° with respect to Earth's rotational axis. The North geomagnetic pole, located near Greenland in the northern hemisphere, is actually the south pole of the Earth's magnetic field, and the South geomagnetic pole is the north pole.

Many smart phones contain miniaturized *micro electromechanical systems* (MEMS) magnetometers which are used to detect magnetic field strength and are used as compasses. The use of a three-axis device means that it is not sensitive to the way it is held in orientation or elevation. Hall effect devices are also popular. Typical characteristics of the magnetometers in mobile devices are summarized as follows:

- Configuration: The components have a Wheatstone bridge configuration that converts magnetic fields into a millivolt output. These Wheatstone bridges are passive components that don't emit any fields or broadband noise. These components are extremely shock and vibration tolerant.
- Resolution: The sensors feature very low noise floors for their size. Typical resolution ranges from 27 to 120 μG.
- Solid-state: The use of semiconductor processes allows the manufacture of very small sensor devices to reduce board assembly costs and improve reliability and ruggedness.
- Set/Reset Straps: On-chip set/reset straps reduce effects of temperature drift, non-linearity errors and loss of signal output due to the presence of high magnetic fields. This feature provides the benefit of an insurance policy against high stray fields. On-chip offset straps may be used to eliminate the effects of hard iron distortion, and to implement a closed loop magnetometer circuit for high performance applications.
- Typical angular accuracy when used for a compass is 0.5° to 2°. Note this is the accuracy of determining the direction of the magnetic field, not the accuracy of the heading direction.
- Size can vary according to the requirements, but indicative package size is small $10 \times 5 \times 2$ mm.

7.2.2 Rate Gyroscopes

Rate-gyroscope measure the rate of change of an angle relative to a specific axis. In typical cases in modern mobile devices there are three axes (roll, pitch and yaw), thus providing a complete measurement of the angular rate of change of the device. While the gyroscope cannot measure absolute angles, relative angular change can be determined by integrating the gyroscope output. By combining these relative data with some independent measure of the absolute angle (such as from a compass) an estimate of the true angle can be determined. However, any measurement offset errors results in a slow drift in the calculated angle, so that gyroscope angle data should be considered as providing only short-term angular information.

The solid state devices in mobile devices are based on inexpensive vibrating structure micro electromechanical systems (MEMS) gyroscopes. These are packaged similarly to other integrated circuits and may provide either analog or digital outputs. In many cases, a single part includes gyroscopic sensors for multiple axes. Some parts incorporate multiple gyroscopes and accelerometers (or multiple-axis gyroscopes and accelerometers), to achieve output that has six full degrees of freedom. These units are called inertial measurement units, or IMUs. Internally, MEMS gyroscopes use lithographically constructed versions of one or more of the mechanisms (tuning forks, vibrating wheels, or resonant solids of various designs). The physical principle of operation is that a vibrating lever tends to want to maintain its plane of vibration, even if the body of the device starts to rotate.

Typical characteristics of the rate-gyroscopes in mobile devices are summarized as follows:

- Typical maximum rate of angle change measurement is the rate ±250 to ±1000 degrees per second.
- The typical angle rate of change sensitivity is 0.03 degree per second for a 1000 degrees per second full scale rotation rate.
- The typical zero rate off 0.04 degrees per second per degree C.
- Non-linearity 0.2% of full scale rate.
- Size can vary according to the requirements, but indicative package size is small $4 \times 4 \times 1$ mm.

7.2.3 Accelerometers

Accelerometers measure the acceleration along the axis of the device. Typically, the solid-state devices used in mobile devices incorporate three orthogonal axes, allowing the magnitude and direction of the acceleration vector to be measured. However, due to Einstein's equivalence principle, the acceleration due to motion and that due to the Earth's gravitational field cannot be distinguished separately by any accelerometer device. As a consequence, the measured data will be a combination of these two effects. Further, as a mobile device will typically be loosely attached to the body of a person, and the orientation can vary considerably, the value of the gravitational acceleration component cannot be readily determined. Thus for example when attached to a person walking the orientation of the accelerometer axis will vary associated with the stride cycle, so that the gravitational component will vary cyclically during each stride, making data interpretation difficult.

The main motivation for the use of accelerometers is that the double integration of acceleration in the horizontal plane (normal to the gravity vector) is the displacement, and thus when combined with the heading angle can be used to dead reckoning position determination. However due the coupling with the gravitational vector and the accelerometer offset error, such simple use of the accelerometer is not possible for personal tracking. However, other techniques, such as described in Sect. 7.7 can be used to estimate the displacement distance associated with walking.

Modern accelerometers are often small micro electro-mechanical systems (MEMS), and are indeed the simplest MEMS devices possible, consisting of little more than a cantilever beam with a proof mass (also known as seismic mass). Damping results from the residual gas sealed in the device. As long as the Q–factor is not too low, damping does not result in a lower sensitivity.

Typical characteristics of the accelerometers in mobile devices are summarized as follows:

- Typical maximum acceleration measurement is ±3 g. (g = 9.81 m/s^2).
- Non-linearity ±0.3%.
- Bandwidth 500–1500 Hz.
- Zero offset error ±0.2 m/s^2.
- The typical zero offset variation is 0.01 percent per degree C.
- Noise 300 μg/$\sqrt{\text{Hz}}$ rms.
- Size can vary according to the requirements, but indicative package size is small 3 × 3 × 2 mm.

7.2.4 *Measurement Errors and Calibration*

7.2.4.1 Overview

The small solid-state sensors used in mobile applications all suffer from the same problem, namely the raw accuracy of the sensor data is limited. While the sensors typically exhibit a repeatability of measurement of one percent or better, the absolute accuracy of the measurements is much less, with errors as much as 5%. Further, some sensors (particularly the rate-gyro) are sensitive to temperature changes,[3] so that the raw measured data can vary over quite short time periods. As a consequence, the raw sensor data must be calibrated to provide data sufficiently accurate for position location applications. Further, this calibration process must be performed in real-time, as the calibration will vary over time. However, in recent years the technology has improved considerably, so that the need for calibration has reduced. Nevertheless, for the best operational performance some sort of calibration process is required.

The raw sensor measured data generally suffer from two calibration errors, one associated with an offset, and one with scaling. Thus in general the raw measured sensor data is related to the "true" value by the relationship

$$m(t) = a(t)\,M(t) + b(t) \tag{7.1}$$

where m is the raw sensor measurement, M is the true value of the measurement, a is the scaling factor, and b is the measurement offset bias. Thus by determining the sensor calibration (as a function of time), the raw measurement can be corrected up to an accuracy of at least 1% and possibly to 0.1% of the full scale measurement. Note that both the measurements and the calibration coefficients are both shown as a function of time, but in practice the calibration coefficients only vary slowly (or practically not at all) as a function of time.

The details of calibrating the various sensor types is beyond the scope of this overview, but some general approaches will be described. In general, the determination of the offset bias is easer than determining the scaling calibration factor.

Typically the calibration process for the scaling factor is only required infrequently, so that the calibration can be performed when the radiolocation system detects good position data,[4] from which heading angles and accelerations can be determined. Further, these values can be stored within the mobile device, so that assuming the calibration only varies slowly over time, the stored values can be used each time the device is turned on. The initial values can be determined during the manufacture of the mobile device.

7.2.4.2 Determination of Offset Bias of Sensors

The determination of the offset bias varies in detail for each type of sensor. However, for the inertial sensors (accelerometer and gyro), a simple method can be

[3]As mobile devices attached to the human body will be affected by the temperature of the body which is typically 10–15 °C above the ambient environmental temperature, the initial attachment of a device to a person will result in a change in temperature of the sensors over a period of several minutes. However, after an initial period, the body temperature ensures the sensors operate at a stable temperature.

[4]For example, in determining a position with redundant range or time-of-arrival data, an estimate of the position error can be made. When this error estimate is small, the sensor calibration process can be initiated.

adopted. If the sensor is stationary, as determined by no significant change in the raw output of the sensor, then the true measured output (acceleration or rate of change of angle) can be assumed to be zero. In this case, application of (7.1) shows that the raw measurement is simply the required offset bias. This procedure can be applied directly to the rate-gyro, but as the accelerometer also measures the Earth's gravity, some additional calibration calculations are required for the accelerometer. In Sect. 7.5 it is shown how the speed and stride length can be determined using a vertically mounted accelerometer, so this case will be used as an illustration of the offset bias calibration process. For an approximately vertically (z) mounted accelerometer, the application of (7.1) results in the equation

$$a_z(t) = a[g + A_z(t)]\cos(\phi(t)) + b(t) \tag{7.2}$$

where $\phi(t)$ is the orientation of the accelerometer relative to the true vertical; this offset is assumed to be small, say less than 20°, but the exact value is assumed to be unknown. With the assumption that the accelerometer output is stationary, the equation becomes

$$a_z = ag\cos(\phi) + b \approx ag + b = b' \tag{7.3}$$

Thus (7.2) can be rewritten in the form

$$a_z(t) \approx aA_z(t)\cos(\phi(t)) + b'(t) \tag{7.4}$$

where the offset bias term includes a component associated with the Earth's gravity. Note that this modified form is essentially the same as the original Eq. (7.1), so with this modified definition the processing is essentially the same as the generic case. While it is assumed in this simple example that the accelerometer is close to vertical, if the vertical angular angle can be estimated (using magnetometers), then the restriction of assuming $\phi(t)$ is small can be removed, and thus a more accurate estimation of the true vertical acceleration can be made.

The determination of the offset bias for the magnetometer can utilize a feature of the magnetometer sensor to simplify the calibration process. Typically, magnetometers can be configured in two alternative modes, where the scaling factor sign can be flipped to either positive or negative polarization. Thus for any orientation of the magnetometer two measurements can be made

$$\begin{aligned} m_1 &= aM + b \\ m_2 &= -aM + b \end{aligned} \tag{7.5}$$

From these two equations the offset bias is simply given by $b = (m_1 + m_2)/2$. In practice, this dual measurement needs be performed only during the calibration phase of setting up the sensor, rather than for every measurement.

7.2.4.3 Determination of Scaling Factor of Sensors

The determination of the scaling factor of sensors is generally more difficult than determining the offset bias. However, it is noted that typically the scaling factor is largely invariant over time, so that one simple strategy is to calibrate the scaling factor at the time of manufacture of the mobile device. As the sensors are often required in two or three orthogonal directions, packages are commercially available with the two or three orthogonal sensors in the one unit. In this case, a strategy that can be used is to assume the scaling factor is the same for each sensor. For example, consider the case of two orthogonal magnetometers used to form a compass. The compass angle can be determined from the equations

$$\begin{aligned} m_x &= aM_x + b_x \\ m_y &= aM_y + b_y \\ \theta &= \tan^{-1}\left[\frac{M_y}{M_x}\right] = \tan^{-1}\left[\frac{m_y - b_y}{m_x - b_x}\right] \end{aligned} \tag{7.6}$$

Thus by assuming the scaling factor is the same for both magnetometers in the package, the scaling factor is not required for the determination of the required heading angle.

This concept can be taken one step further without the need for the assumption that the scaling factors of the two magnetometers are the same. The basis of the method is that the horizontal component of the Earth's magnetic field is nominally a constant independent of the orientation of the two orthogonal magnetometers in a horizontal plane. Thus the equations for the magnetometer calibration are

$$\begin{aligned} m_x - b_x &= a_x M_x \\ m_y - b_y &= a_y M_y \\ M_x^2 + M_y^2 &= M_H^2 = \text{constant} \end{aligned} \tag{7.7}$$

where M_H is the horizontal component of the Earth's magnetic field, and is assumed to be a constant in the operating environment of the mobile device. If these measurements are made as the mobile moves through the environment, and the heading angle changes throughout the 360° range, the data set can be used for the scaling factor calibration. A least squares fit to the data can be performed to determine the scaling factor parameters a_x and a_y. If the value of the Earth's horizontal magnetic field is not known, then this can be assumed to be a constant value (say 1), as absolute calibration is not necessary for the magnetic heading angle to be determined. While the actual horizontal component of the Earth's magnetic field will vary due to local magnetic anomalies (particularly indoors), if these anomalies are assumed to be random with zero mean, then the least-squares fitting process will result in little error in the estimation of the calibration parameters. Note that this calibration procedure can be carried out during the normal operation of the mobile device, as no particular pattern of movement is required for the calibration. Initially the default calibration can be used, which can be progressively updated during normal operation. As more data are used in the least-squares fitting process, the accuracy of the calibration will improve.

The scaling factor for the rate-gyro can be determined by using the compass as a reference. The basic procedure is to compare the incremental change in the heading angle as determined by the compass with the incremental change obtained by integrating the rate-gyro data. As the integration process is sensitive to any offset bias, the offset bias must be determined first as outline above. Even so, the integration time should be restricted to (say) 10 s to minimize the errors due to small residual bias errors. Using this process, the rate-gyro scaling factor can be estimated by the expression

$$a_{gyro} = \frac{\theta_{compass}(t_2) - \theta_{compass}(t_1)}{\int_{t_1}^{t_2} \left(\dot{\theta}_{gyro}(t) - b_{gyro} \right) dt} \tag{7.8}$$

The determination of the scaling factor for the accelerometers and rate-gyros is more complex, and the details will not be given. In the case of the accelerometer, one simple calibration procedure (but which cannot be performed in real-time during normal operation) involves orientating the accelerometer vertically, and then rotating through 180°. The resulting measurements are summarized by the two equations

$$\begin{aligned} a_1 &= ag + b \\ a_2 &= -ag + b \end{aligned} \tag{7.9}$$

Thus the calibration parameters are given by

$$\begin{aligned} a &= (a_1 - a_2)/2g \\ b &= (a_1 + a_2)/2 \end{aligned} \tag{7.10}$$

Thus this calibration method determines both the offset bias and the scaling factor for the accelerometer, but the method is not generally applicable to real-time calibration.

However, one general method that can be used in real time is to match the data from the radiolocation system to that of the sensors, thus determining the sensor scaling factors. Many such methods are possible, but only one will be outlined for the vertical accelerometer scaling factor calibration. It is shown in Sect. 7.5 that the distance traveled by a person walking can be determined by measuring the peak-to-peak acceleration of the vertical accelerometer. The peak-to-peak acceleration measurement will be independent of the accelerometer offset, but is dependent on the scaling factor. Thus if the true vertical acceleration is $\pm A_z$ relative to the

Earth's gravity and the measured acceleration is $\pm a_z$, the distance walked as measured by the radiolocation system is D, and the distance walked according to the accelerometer algorithm[5] is $f(A_z)$, then the accelerometer scaling factor a can be calibrated by solving the equation

$$D = f(A_z) = f(a_z/a) \tag{7.11}$$

7.3 Introduction to Complementary Filters

The determination of position using radiolocation will inevitably involve some positional errors, mainly associated with multipath propagation or even a total loss of data due to low radio signal strength. For a practical system these errors need to be corrected. One possible method of correction is the use of other sensors, such as accelerometers and gyroscopes which can provide distance and heading information by suitable integration of the sensor data. However, the integration process will also introduce positional errors, so that some method of combining radiolocation data with sensor positional data is required. Before this process is examined in detail, an overview the complementary filtering concept provides insights into the basic principles. In later sections, these techniques are expanded to provide practical implementations for the determination of both mobile speed and heading.

7.3.1 *Complementary Filtering—An Overview*

This subsection introduces the technique of complementary filtering, with the particular application of determining the heading angle using angular sensor data and supplementary rate-gyro information. The angular sensor data can be obtained either from magnetometers or from processing the output of the associated radio positioning system. Such systems (particularly the magnetometer data) can be subject to large short-term errors which limit the accuracy of the angular information. However, angular rate data from a rate-gyro can significantly reduce these short-term angular errors by optimally combining both sets of data. In particular, the method described in this section is associated with the complementary filtering of two sets of data, as described in (Pascoal et al. 2000). Much of the introductory theory described is adapted from this paper.

An angular sensor (generically referred to as a "compass") can be subject to short-term transient errors. The characteristics of these errors is that they often

[5]See Sect. 7.7 for details of this function.

exhibit high frequency components[6] when the mobile device is moving, and thus low-pass filtering can reduce the effects. However, the low-pass filter cutoff frequency must be set to provide sufficient bandwidth for the dynamics of the mobile device, and thus the improvements by simple filtering can be limited. A better performance is possible if angular rate information is combined with the angular data. In particular, integrated rate-gyro angular data have relatively small high frequency noise. However, rate-gyros suffer from bias errors; as the angular data are obtained by integration of the rate-gyro data, any bias errors progressively translate into angular errors. Thus the rate-gyro data must be high-pass filtered (with a filter zero at 0 Hz), such that the effects of the bias errors are removed. Thus the compass and rate-gyro data are complementarily filtered, with a low-pass filter for the compass and a high-pass filter for the rate-gyro. When these two sets of filter outputs are suitably combined, the angular errors in the compass are very significantly reduced.

Radiolocation positional errors exhibit a characteristic of good long-term stability, but possibly poor short-term accuracy. Because multipath errors can vary on a spatial basis on the order of a wavelength (5–10 cm for the systems typically used for radiolocation indoors operating in the 2.4 GHz of 5.8 GHz ISM bands), the short-term accuracy for a moving mobile device can be poor. However, on a larger spatial scale the accuracy is stable over time. Now consider positional data based on integrating sensor data. If the initial position is known (say from the radiolocation system), then integrating for a short period of time will result in good positional accuracy. However, as the integration period increases, small offset errors in the sensor data will build up, so that the long-term positional accuracy is poor. From this brief description it can be observed that these two approaches to determining position are complementary, with the strengths of one system the weakness of the other, and vice versa. Thus if a method of combining these two imperfect position determination methods can be devised, the overall positional accuracy should reflect the strengths of both systems, with good short-term and long-term performance. One such method that is particularly appropriate to sensor data is called a Complementary Filter, the characteristics of which are analyzed in detail in the following subsections.

7.3.2 Basic Complementary Filter

This subsection presents the basic theory for complementary filtering, based on data from a compass and a rate-gyroscope. While this is a specific example of

[6]As the transient errors can occur suddenly as a function of position, these errors also occur suddenly as a function of time for a moving device. Sudden changes in the time domain day imply the signal has a wideband spectrum when compared with those associated with the movement of a body.

complementary filtering, the principles are readily adapted to other situations (as described in later sections).

The heading data to be complementary filtered are derived from a compass (based on two magnetometer sensors mounted at right-angles), and a single gyroscope mounted normal to the plane defined by the magnetometers. As the compass angular data are often subject to magnetic anomalies of short spatial duration, correction of these compass errors using the integrated gyroscope rate data is sought by use of the gyroscope data.

The theory of complementary filtering of compass and rate-gyro data can be explained based on the simplest types of low-pass and high-pass filters. Consider the Laplace transform $\psi(s)$ of the compass heading data decomposed as follows

$$\begin{aligned}\psi(s) &= \frac{s+k}{s+k}\psi(s) = \frac{k}{s+k}\psi(s) + \frac{s}{s+k}\psi(s) \\ &= \left[\frac{k}{s+k}\psi(s)\right]_{compass} + \left[\frac{1}{s+k}\dot{\psi}(s)\right]_{gyro}\end{aligned} \tag{7.12}$$

as the s operator is effectively a differentiator, converting the heading angle into angular rate of change as generated by the gyro. In (7.12) the parameter k determines the break frequency of the low-pass and high-pass filters, and is chosen according to the dynamics of the particular application. In practice both the compass and the rate-gyro outputs are subject to errors, so that above equation suggests that a better estimate of the compass angle can be obtained from the measured sensor data as follows

$$\hat{\psi}(s) = \boldsymbol{F_1}(s)\psi_{compass}(s) + \boldsymbol{F_2}(s)\dot{\psi}_{gyro}(s) \tag{7.13}$$

where $\boldsymbol{F_1}(s)$ and $\boldsymbol{F_2}(s)$ are respectively lowpass and highpass filters. For implementation of the complementary filters, (7.13) is rearranged into the form

$$\hat{\psi}(s) = \frac{1}{s}\left[k\left(\psi(s) - \hat{\psi}(s)\right) + \dot{\psi}(s)\right] \tag{7.14}$$

For stability in the calculation, the differential equation associated with the filtering is arranged so that only integration $(1/s)$ is required. The block diagram associated with (7.14) is shown in Fig. 7.1. Notice that the implementation involves one integrator, and that a feedback loop is involved in the calculation. The inputs are the measured compass and rate-gyro data. The estimated compass heading angle $\hat{\psi}$ is output from the filter.

The filter implementation described above assumes analog signals, but the actual implementation is typically based on sampled data. However, the above analysis can be extended to a sampled data system by z-transform analysis rather than Laplace transform analysis. In particular, the Laplace to z–transform conversion (accurate for frequencies up to about half the Nyquist rate) is given by

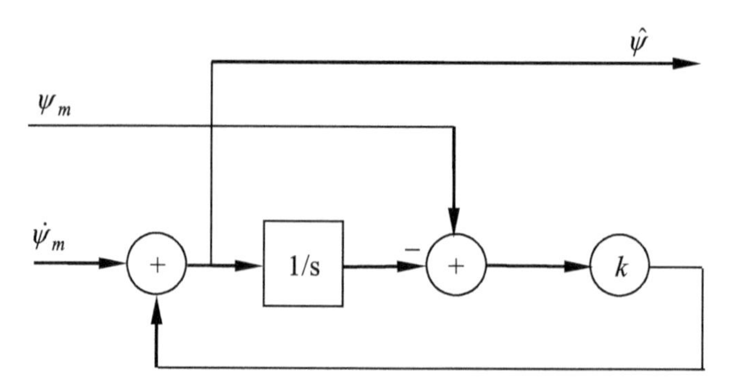

Fig. 7.1 Block diagram of the complementary filter structure defined by (7.14)

$$s \rightarrow \frac{2}{T}\frac{z-1}{z+1} = \frac{2}{T}\frac{1-1/z}{1+1/z} \tag{7.15}$$

Note that in the z-transform domain, multiplication by 1/z is equivalent to a delay by one sample period T. Applying (7.15) to the complementary filter equation (7.14), the resulting sampled data equation is

$$\hat{\psi}_n = \frac{1}{2/T+k}\left[k\left(\psi_n + \psi_{n-1}\right) + \left(\dot{\psi}_n + \dot{\psi}_{n-1}\right) + (2/T - k)\hat{\psi}_{n-1}\right] \tag{7.16}$$

Equation (7.16) represents the digital implementation of the complementary filter. Now consider the performance based on a simulation with the following parameters:

Mobile speed:	15 m/s.
Radius of path:	100 m.
Compass error:	Sinusoid with amplitude 20°, period 1 s.
Compass noise:	Gaussian with zero mean, standard deviation 1°.
Gyro noise:	Gaussian with zero mean and 1 degree per second standard deviation.
Sample period:	T = 80 ms.
Filter k parameter:	0.2 rad/s.

The typical performance of the filter is illustrated in Fig. 7.2. The simulated path consists of three sections, an initial straight line segment, a complete 360° circular path, and then another straight line segment. The raw compass output is simulated with sinusoidal and Gaussian errors, with a peak error of around 25°. After the complementary filter, the errors are as shown in Fig. 7.3, with a standard deviation of 0.6°. This error is minimized with k in the range 0.1–0.2, which is a compromise between minimizing the bandwidth to reduce noise and maximizing the bandwidth to minimize dynamic tracking errors. Thus the complementary filter has reduced the measurement errors by about a factor of 20:1 in this case. The simulation suggests

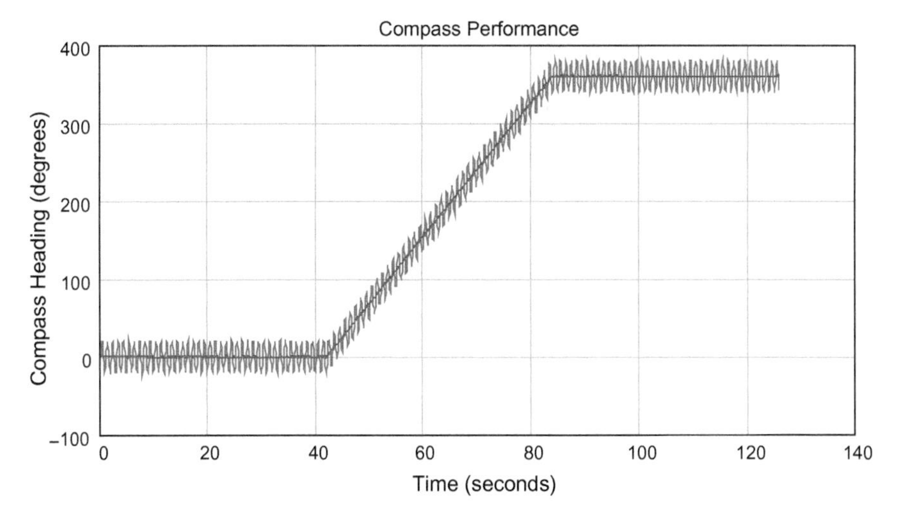

Fig. 7.2 Raw compass output and output from complementary filter

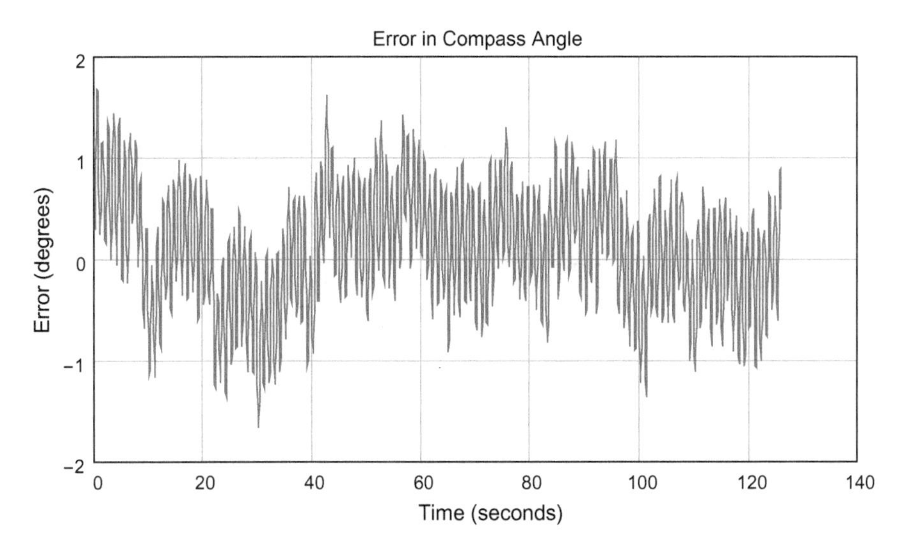

Fig. 7.3 Details of the filtered compass output

that compass errors can be kept to below 1° with respect to short term transient effects (magnetic anomalies). Any anomalies with periods of 10 s or greater will be tracked rather than filtered.

From this illustrative example it is clear that complementary filtering potentially can very much improve the raw sensor data. However, the actual performance very much depends on the characteristics of the sensor errors, and the dynamics of the motion.

7.3.3 Enhanced Filtering to Remove Rate-Gyro Bias Errors

The simple filter defined in Sect. 7.3.2 is shown to effectively filter transient compass errors using a rate-gyro with zero bias errors. In practice, rate-gyros cannot be sufficiently accurately calibrated such that bias errors can be ignored. The simple, cheap rate-gyros used in wireless sensor networks may typically have an output noise standard deviation of at least 1 degree per second. If the (constant) bias is set at 1 degree per second, the error characteristics are as shown in Fig. 7.4. Even with this small bias error the output angle error reaches 6°. Clearly for practical implementation the errors of bias errors in the rate-gyro must be removed.

The effect of the rate-gyro bias error can be removed by dynamically estimating the gyro bias, and subtracting this bias estimate from the measured output before applying the rate data to the filter. In practice this can be achieved by adding an integrator, the output of which is an estimate of the gyro bias error. The input to the integrator is the error between the raw compass output and the complementary filter output, weighted by a suitable factor k_2. The modified equations are

$$\begin{aligned} \hat{\psi} &= \frac{1}{s}\left[k_1\left(\psi_{compass} - \hat{\psi}\right) + \left(\dot{\psi}_{gyro} - \dot{\psi}_{bias}\right)\right] \\ \dot{\psi}_{bias} &= \frac{1}{s}\left[k_2\left(\psi_{compass} - \hat{\psi}\right)\right] \end{aligned} \tag{7.17}$$

By combining the two equations in (7.17), the effective complementary filter transfer function is

$$\hat{\psi}(s) = \frac{k_1 s + k_2}{s^2 + k_1 s + k_2}\psi_{compass}(s) + \frac{s}{s^2 + k_1 s + k_2}\dot{\psi}_{gyro}(s) \tag{7.18}$$

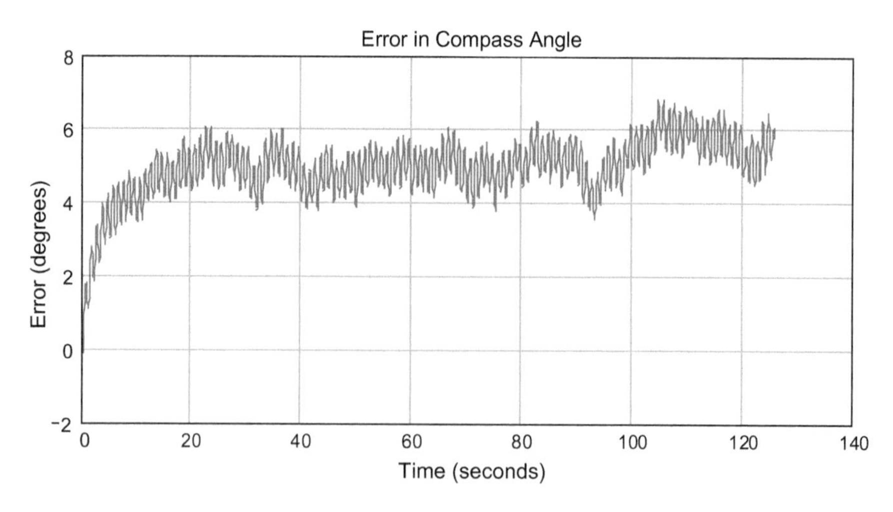

Fig. 7.4 Effect or a 1 degree/second bias error in the error output of the complementary filter. This result should be compared with Fig. 7.3 which has zero rate-gyro error

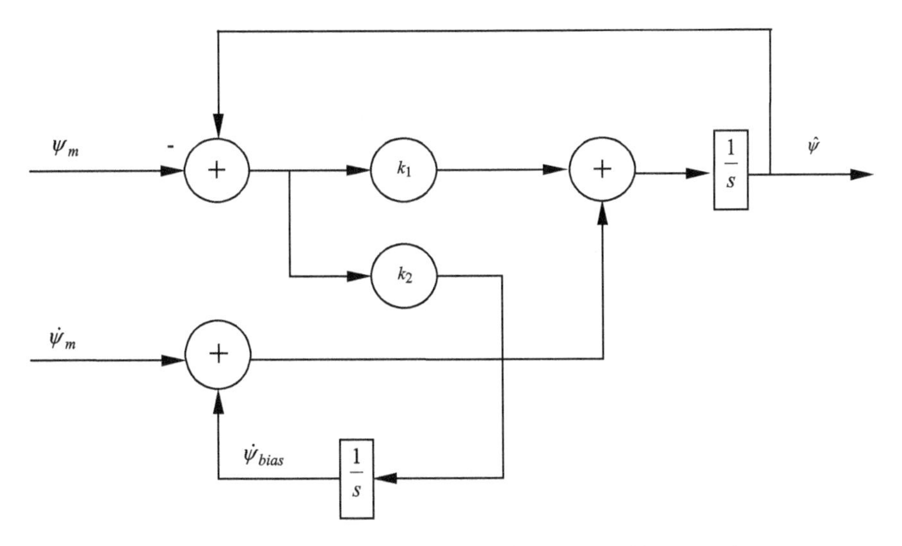

Fig. 7.5 Block diagram of the complementary filter with rate-gyro bias correction

Thus again the complementary filter consists of a low-pass filter of the compass data and a high-pass filter for the rate-gyro. In the case of the gyro bias, the filtered output will be zero as required. The filter parameters are chosen to reflect the dynamics of the particular application. For critical damping, the second parameter k_2 is

$$k_2 = \frac{k_1^2}{4} \tag{7.19}$$

Thus in this case $k_2 = 0.01$. The block diagram of the modified complementary filter is shown in Fig. 7.5.

Using the z-transform conversion formula (Eq. 7.15), the Laplace transform expressions defined by (7.17) can be converted into sampled data form

$$\begin{aligned} \dot{\psi}_{bias_n} &= \dot{\psi}_{bias_{n-1}} + \frac{k_2 T}{2}\left[(\psi_n + \psi_{n-1}) - \left(\hat{\psi}_n + \hat{\psi}_{n-1}\right)\right] \\ \hat{\psi}_n &= \frac{1}{2/T + k}\left[k(\psi_n + \psi_{n-1}) + \left(\dot{\psi}_n + \dot{\psi}_{n-1}\right) + \left(\dot{\psi}_{bias_n} + \dot{\psi}_{bias_{n-1}}\right) + (2/T - k)\,\hat{\psi}_{n-1}\right] \end{aligned} \tag{7.20}$$

where the "compass" and "gyro" tags have been dropped for clarity in the expressions. The two expressions in (7.20) are coupled functions of the nth sample of the estimated angle $\hat{\psi}_n$ and estimate bias $\dot{\psi}_{bias_n}$, so that these equations cannot be directly used in this form. While these two simultaneous equations can be solved for these two variables, the resulting expressions are somewhat complex. A simpler strategy is to note that assuming the middle sample value is approximately the mean of the previous and next samples, so that

$$\hat{\psi}_n \approx 2\hat{\psi}_{n-1} - \hat{\psi}_{n-2} \tag{7.21}$$

Using this expression, the bias equation becomes

$$\dot{\psi}_{bias_n} = \dot{\psi}_{bias_{n-1}} + \frac{k_2 T}{2}\left[(\psi_n + \psi_{n-1}) - \left(3\hat{\psi}_{n-1} - \hat{\psi}_{n-2}\right)\right] \tag{7.22}$$

The modified version of the bias equation only contains references to the complementary filter output from earlier iterations, and thus can be evaluated directly without having to solve the simultaneous equations defined by the equations pairs in (7.20). This result can then be applied directly to the second equation in (7.20) to calculate nth iteration of the filter output.

The performance of the modified complementary filter is shown in Fig. 7.6. In the first segment up to about 40 s the bias integrator is slowly adjusting to null out the bias of 2 degrees per second. After this initial period, the performance is essentially identical to the zero bias case (described in Sect. 7.3.2), with a standard deviation of the compass output of 0.6°. Thus the enhanced complementary filter is insensitive to rate-gyro bias errors, and effectively corrects for transient compass errors.

The effect of the bias integrator in correcting for a gyro bias error is shown in Fig. 7.7, where it can be observed that the initial bias error of 2 degrees per second is nulled in about 40 s. In general, there is a trade-off between the settling time of the filter and the output errors. For example, if the filter parameters are $k_1 = 0.04$ and $k_2 = 0.04$, then the settling time is reduced to 10 s, but the steady-state RMS error increases to 0.67°.

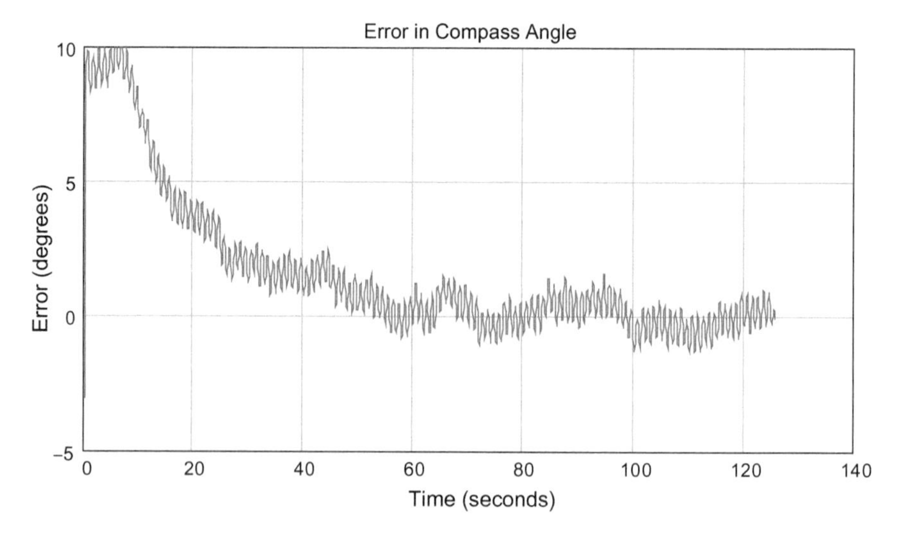

Fig. 7.6 Complementary filter output error with gyro bias correction

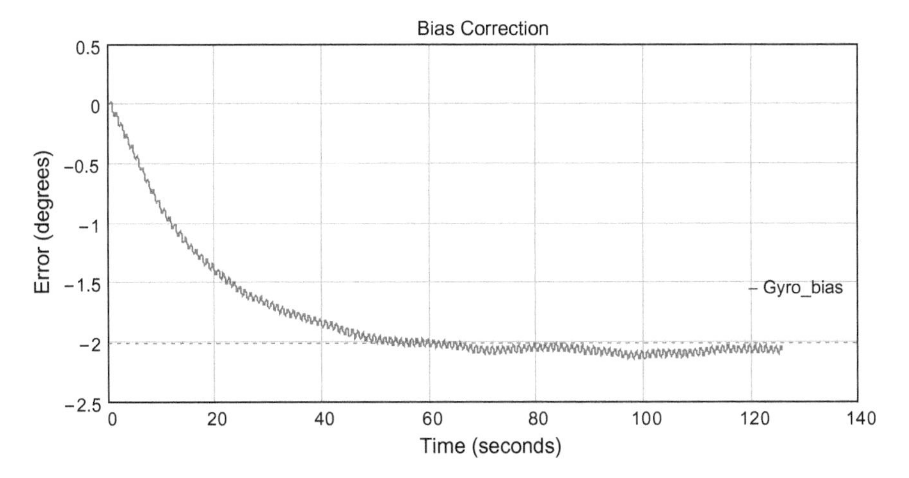

Fig. 7.7 Bias integrator output as a function of time. The bias error is effectively eliminated after the initial settling period, so that the error in the filter output is unaffected by rate-gyro bias errors

7.3.4 Filtering Rate-Gyro Scale Factor Errors

The output of the rate-gyro will suffer errors from both bias and scaling factor calibration errors. As shown in Sect. 7.3.3, the effect of bias errors can be eliminated by the addition of a integrator, which estimates the static bias error. The effect of scaling errors will now be considered.

The measured output of the rate-gyro with bias and scaling calibration errors can be represented by the equation

$$\dot{\psi}_m(t) = a + (1+b)\,\dot{\psi}(t) \tag{7.23}$$

where a is the bias error parameter and b is the scaling factor error parameter. For a perfectly calibrated rate-gyro, both these parameters are zero. Now consider the performance in two special cases appropriate to human walking motion. For this case, the heading angle is either constant when walking in straight lines, or has a (approximately) constant rate of change when walking around corners. In the former case the nominal rate-gyro output is nominally zero, and in the latter case the output is nominally a constant. In particular, from (7.23), the measured output will be for these two cases

$$\begin{aligned} &\dot{\psi}_m = a && (\dot{\psi}(t) = 0) \\ &\dot{\psi}_m = a + (1+b)\,\dot{\psi}_c = a' && (\dot{\psi}(t) = \dot{\psi}_c = \text{constant}) \end{aligned} \tag{7.24}$$

Thus in both these cases the rate-gyro output can be corrected by a bias parameter which is constant during motion either in a straight line or walking around a bend. However, the value of the bias correction will vary according to the particular situation, so that the pseudo-bias parameter (a') will be tracked by the filter. Thus the filter must attempt to track these changes to minimize the effect of the scaling factor error. From (7.18) and (7.24), the compass angle error with bias and scaling errors is given by

$$\hat{\psi} - \psi = \Delta\psi(s) = \frac{s}{s^2 + k_1 s + k_2}\left[a + b\dot{\psi}(s)\right] \tag{7.25}$$

Thus, for example, with the step change in the heading from a straight line path to a constant turn-rate bend, the time-domain error function (using an inverse Laplace transform) is given by

$$\begin{aligned} \Delta\psi(t) &= \frac{Vb}{2\pi\omega R_b} e^{-\alpha t} \sin \omega t \\ \alpha = k_1/2 & \qquad \omega = \sqrt{k_2 - \alpha^2} \end{aligned} \tag{7.26}$$

where V is the speed and R_b is the radius of the bend. For the underdamped case, $k_2 > \alpha^2$. Thus for a period of time somewhat greater than $1/\alpha$ the error will exponentially approach zero. In the above example with $k_1 = 0.2$ the error will decay with a time constant of 20 s.

An example of the tracking of the effective bias error is shown in Fig. 7.8. In this case, the scaling factor is set at 0.1, the other parameters $k_1 = 0.4$ and $k_2 = 0.1$, and the random components are set to zero to allow the dynamic performance of the filter to be more easily observed. Using (7.26) the peak error in the estimated compass heading is 1.0°. To provide reasonably fast pseudo-bias tracking, the filter parameter k_1 should be large. However, a large k_1 will affect the filter bandwidth, decreasing the noise reduction effect due to low pass filtering. Thus there must be a

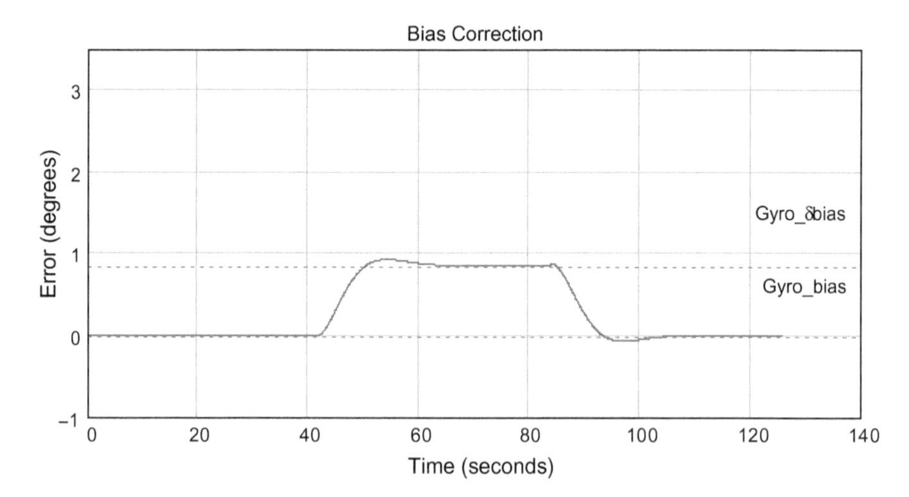

Fig. 7.8 Effect of scaling factor errors, showing the tracking of the pseudo-bias error. The filter parameters are $k_1 = 0.4$ and $k_2 = 0.1$

trade-off between the noise reduction and the bias tracking errors due to scaling factor calibration errors. Optimum performance occurs when the filter bandwidth is reduced and the dynamics underdamped. With such parameters, the error performance is essentially the same as without any rate-gyro scaling factor errors.

7.3.5 *Measured Performance*

The performance of an actual rate-gyro is presented in the subsection. The basic rate-gyro was first calibrated to obtain the bias and the scale factor. The bias is determined simply by measuring the output when the rate-gyro is stationary, with the bias being the mean of the output. The scale factor is obtained by rotating the rate-gyro 180°, and integrating the output. The difference angle (180°) divided by the integrated outputs (from the beginning to the end) is the required scaling factor. The resulting precision of the digitized data is about 0.05 degrees per second per bit.

Figure 7.9 shows the results of the measured rate-gyro noise when stationary. The standard deviation is 0.1 degrees per second, which is equivalent to 2 bits. Because of the bias calibration, the mean value is essentially zero.

The performance of the rate-gyro when rotated through 90° is shown in Fig. 7.10. Note that the complementary filter ensures that the integrated gyro data accurately tracks the compass data. The detailed plot of the difference between the

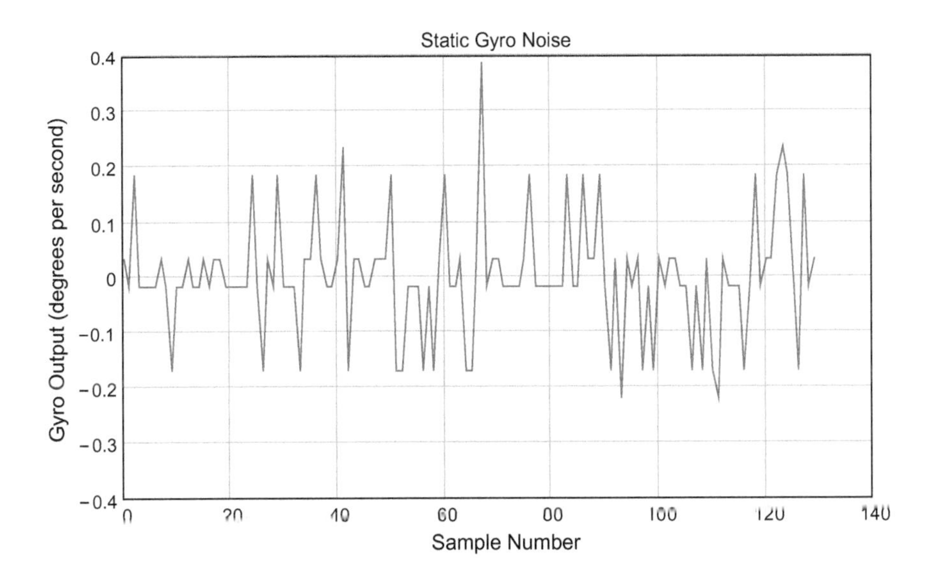

Fig. 7.9 Measured rate-gyro noise when stationary

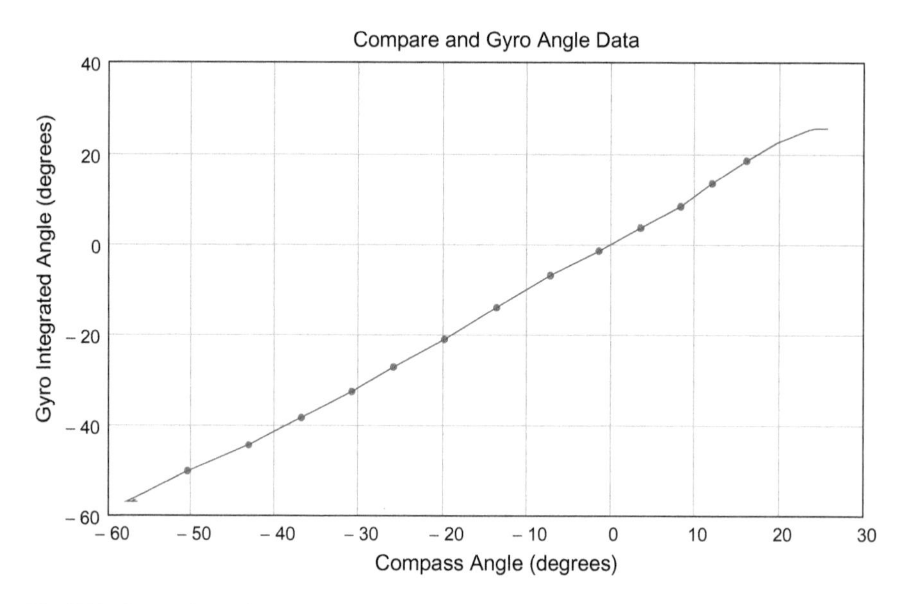

Fig. 7.10 Comparison of rate-gyro integrated output and the magnetometer angle

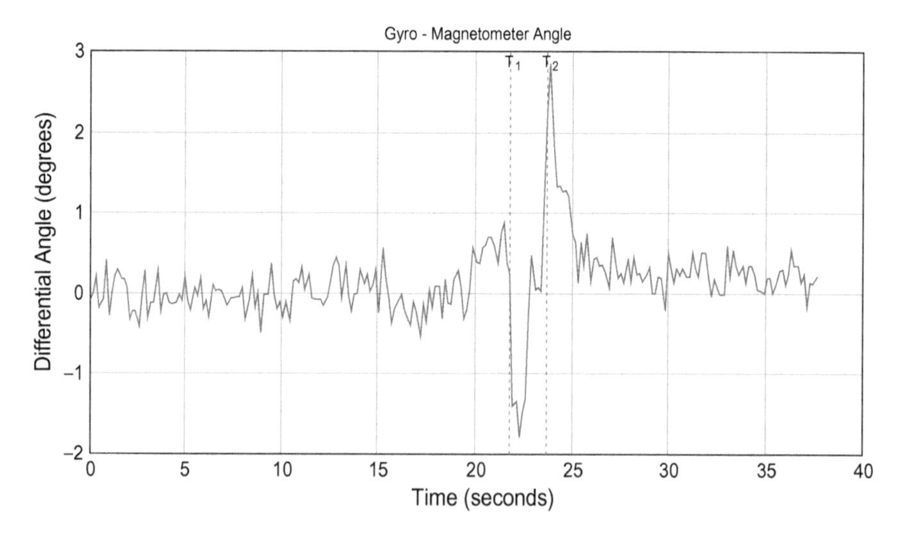

Fig. 7.11 Difference between the magnetometer (compass) angle and the integrated rate-gyro angle. The 90° rotation occurred between times T_1 and T_2

output is shown in Fig. 7.11. When stationary the differential angle is very small (peak errors of the order of ±0.3°), but when rotated the errors increased to about ±2.5° due to the dynamics of the complementary filter.

7.4 Velocity Determination Using Radiolocation and Inertial Data

7.4.1 *Introduction*

Section 7.3 introduced the concept of the complementary filter for improving the quality of heading data obtained from two complementary sensors, one with good short-term accuracy and the other with good long-term stability. In this section we will consider another example where complementary filtering techniques can be applied, namely to improve the accuracy of speed estimates. In particular, this section describes the procedures for accurately determining the velocity (speed and direction) of a mobile device, based both on the radiolocation data and the inertial sensors.

Accurate velocity determination is essential for the dead reckoning positioning, as the projected position is simply the integral of the mobile velocity (speed and heading). In the following analysis the determination of velocity is considered separately in its two components. In both cases the concept is similar, namely the raw speed/heading data are determined from the radiolocation subsystem, but these noisy estimates are corrected using inertial sensor data. The accuracy of the estimates can be improved by the application of the inertial sensor data using complementary filters. The concept of the complementary filter in both cases is as described in the introductory Sect. 7.3.

7.4.2 *Theoretical Analysis*

This subsection provides a theoretical analysis of the performance of processes which optimally combine both the radiolocation and inertial sensor data. The processing is based on the technique of complementary filtering, as introduced in Sect. 7.3. This technique is applied to both the determination of speed (using a longitudinal accelerometer and differential radiolocation data), and the heading angle (using rate-gyro and differential radiolocation data). The basic complementary filter concept is shown in Sect. 7.4.2.1 to be equivalent to *feedforward* processing, which has the advantage over a complementary filter implementation as it requires only one filter; this filter is an arbitrary lowpass filter designed to match the characteristics of the process noises. The approach taken in estimating the speed is similar to that described for the mobile heading angle (described in Sect. 7.3), namely the use of the complementary nature of the radio-derived speed and the speed determined from the integration of the accelerometer data.

The radiolocation data can be used to provide a simple estimate of the speed by subtracting two positions τ seconds apart. Thus the x-coordinate of the speed estimate at time t is given by

$$\hat{v}_x(t) = \frac{x(t) - x(t - \tau)}{\tau} \tag{7.27}$$

A similar expression applies for the y-coordinate of the speed, where the position is given in the orthogonal coordinates (x, y).

Because the estimate involves data with a delay of τ, it is clear that the speed estimate must also be delayed relative to the true speed at time t. From (7.27) it seems reasonable to expect a delay of the order of $\tau/2$; in fact, as will now be shown, the delay is exactly $\tau/2$ regardless of the position function $x(t)$. To obtain this result, consider the Fourier transform of the speed estimate equation. The true spectrum of the speed is given by

$$x'(t) \Rightarrow j\,2\pi f X(f) \tag{7.28}$$

(The dash indicates the time derivative of the position x). The spectrum of speed estimate is obtained by taking the Fourier transform of (7.27), resulting in

$$\frac{x(t) - x(t - \tau)}{\tau} \Rightarrow \frac{X(f) - X(f)e^{-j2\pi f\tau}}{\tau} = [j\,2\pi f X(f)]\,[sinc(f\tau)]\,\left[e^{-j\pi f\tau}\right] \tag{7.29}$$

Thus the spectrum of the speed estimate is the true spectrum multiplied by two modifying functions. The last function can be recognized as a delay by $\tau/2$, while the sinc function can be interpreted as a low pass filter $H(f)$. Thus the delay is half the time separation used in the speed calculation, as suspected. However, the speed estimate is also distorted by the sinc filter, so that the speed estimate will not be just a delayed version of the true speed. Note however, that the delay and lowpass filter function are solely functions of the time separation parameter τ, and independent of the position function $x(t)$ or $y(t)$. Thus the estimated speed derived from both $x(t)$ and $y(t)$ will also have a delay of $\tau/2$ and will be filtered by the same sinc function. However, in the case of the sinc function filter, the speed distortion will be minimal if speed dynamics have frequencies such that $f \ll 1/\tau$. Thus the time separation parameter should be chosen so that the associated filter bandwidth is much greater than the "signal" bandwidth. Figure 7.12 shows a typical spectrum of the speed of a mobile device attached to a walking person. As can be observed, the spectrum has a bandwidth of less than 0.1 Hz, so that if (say) $\tau = 1$ s, the sinc function at 0.1 Hz is 0.9836, which represents minimal distortion.

The above analysis shows that the estimated speed will be distorted by a sinc function lowpass filter. However, the positional data will be distorted by both random noise and systematic errors mainly due to multipath propagation. However, the sinc filter does have a useful purpose, namely in relation to the random positional errors. Figure 7.12 shows that the radio-speed contains wideband noise when compared with the integrated accelerometer data. By expanding the frequency band plotted, Fig. 7.13 shows the radio-speed spectrum has nulls associated with the sinc filter. Thus by filtering the wideband noise associated with the radiolocation data, the noise component of the estimated speed is reduced. Reducing the bandwidth

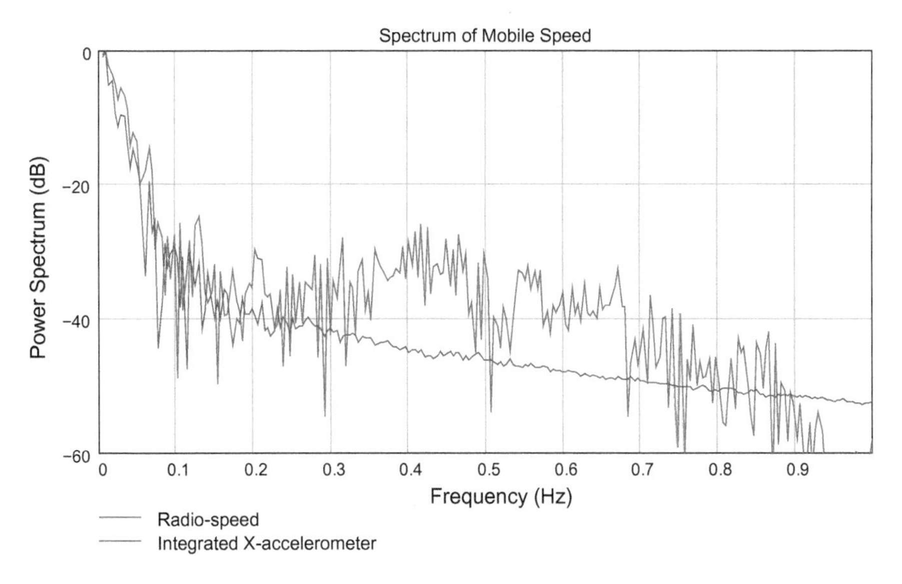

Fig. 7.12 Spectrum of the mobile speed associated with walking around a building. The "noisy" curve is the radio-speed estimate of the spectrum and includes radio positional error noise (about 0.1 to 1 Hz). The other curve is the integrated accelerometer and contains little high frequency noise

reduces this noise, but in this simple implementation the reduced bandwidth results in increased delay in the speed estimate.

Now consider an example with a speed of 15 m/s (appropriate for a motor vehicle in urban areas) and a time separation τ of 1 s; the separation distance will be 15 m which is much greater than the assumed random positional errors of the radiolocation system. Thus the speed standard deviation is given by

$$\sigma_v = \frac{\sqrt{2}\sigma_x}{\tau} = \sqrt{2}\sigma_x f_{bw} \tag{7.30}$$

where f_{bw} is the sinc filter bandwidth, and the $\sqrt{2}$ is due to the use of two positional data points with assumed independently random errors of standard deviation σ_x in calculating the speed. With a quite accurate positional standard deviation of 3 m and a time separation of 1 s the corresponding speed standard deviation is about 5 m/s, which is a large variation in a speed of 15 m/s. This random error in the speed can be reduced by increasing the time separation parameter (to greater than 1 s), but as described above this results in time delay errors. As the spectral bandwidth of the true speed would be of the order of about 0.1 Hz or less due to slow changes of speed in an urban environment, it is clear that the maximum time separation should be less than 10 s (say a maximum of 5 s) to minimize the sinc filter distortion. With this maximum time separation, the random error will be of the order of 0.85 m/s, which represents a speed measurement error of about 5%.

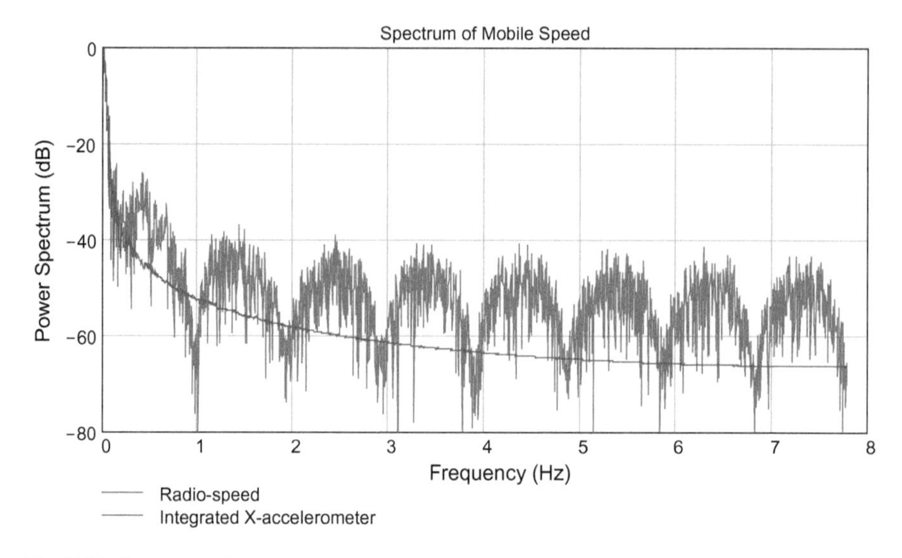

Fig. 7.13 Spectrum of the mobile radio-speed. The time separation parameter is approximately 1 s. Observe that the spectrum has nulls at $1/\tau$ as expected for a sinc filter

However, the associated delay of the speed estimate will be 2.5 s, which will result in considerable systematic errors when accelerations occur. For example, with an acceleration of 1 m/s^2 (or 0.1 g) the corresponding speed error is 2.5 m/s at the second measurement point. A reasonable approach to selecting a suitable separation delay is to have the systematic and random errors equal. From the above equations, the resulting expression for the time separation parameter is

$$\tau = \sqrt{\frac{2\sqrt{2}\sigma_x}{A}} \tag{7.31}$$

where A is the acceleration. Using an acceleration of 1 m/s^2 and a positional standard deviation of 3 m, the corresponding separation is 2.9 s according to (7.31); the corresponding speed error (both the mean and standard deviation) is of the order of 1.5 m/s. The above simple analysis shows that a simple calculation of speed using the radiolocation position data (with some measurement errors that are quite low for GPS in urban areas) results in rather poor estimates of the speed. A more sophisticated approach is to use a Kalman filter (Brookner 1998; see Appendix A for an introduction), however the performance would not be significantly better than the above simple method. Clearly significantly better estimates would be highly desirable; the suggested approach is to use an accelerometer to improve the accuracy of a speed estimate.

7.4.2.1 Equivalence of Complementary Filtering and Feedforward Processing

The classical complementary filter was introduced in Sect. 7.3, which shows that with two filters (one a lowpass and the other a highpass) the output errors can be greatly reduced by appropriately combining two complementary measurements, one with good short-term accuracy and the other with good long-term accuracy. This subsection shows that an alternative technique, namely a feedforward filter, is functionally equivalent but with the advantage that only one filter is required for the implementation. This section shows that these two techniques are mathematically equivalent.

The theory of the complementary filter, as applied to the determination of the mobile speed V, is summarized in Fig. 7.14. The input data are the radio-speed (as determined by (7.27)) and longitudinal accelerometer data. The integration of the accelerometer output is an estimate of the speed to within an arbitrary constant. It is assumed that this constant is determined by using the radio-speed to initialize the integrator.

For a complementary filter the radio-speed data are lowpass filtered and the integrated accelerometer data are highpass filtered before combining to obtain the corrected estimate of the speed.

$$
\begin{aligned}
V_{out}(s) &= H(s)V_{radio}(s) + [1 - H(s)]\ V_{acc}(s) \\
&= H(s)\ [V(s) + N_{radio}] + [1 - H(s)\ ][V(s) + N_{acc}] \\
&= V(s) + H(s)\ [N_{radio} - N_{acc}] + N_{acc}
\end{aligned}
\tag{7.32}
$$

Notice that the lowpass and highpass filters are complementary, this being a fundamental requirement of the method. Thus the output is the required signal, plus

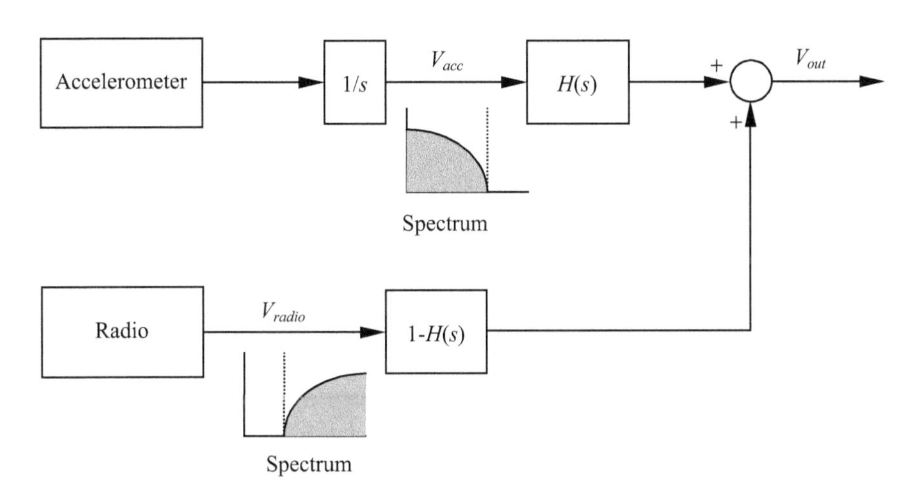

Fig. 7.14 Block diagram of the complementary filter

somewhat complex noise terms. The complementary filter output can be further simplified if the radio-speed noise has only high frequency components (above a cutoff frequency), and the integrated accelerometer noise only has low frequency components (below the cutoff frequency). In this case the radio-noise is effectively eliminated by the filtering process leaving the accelerometer noise unaltered, so that (7.32) becomes

$$V_{out}(s) \approx V(s) + [0 - N_{acc}] + N_{acc} = V(s) \tag{7.33}$$

Thus providing the radio and accelerometer noise have the required spectral characteristics the complementary filter output is approximately the uncorrupted mobile speed. If there is some overlap between the noise components, the output will contain some noise components (but less then the original radio-speed noise).

The filter transfer function $H(s)$ is applied in two places in the above implementation. For practical implementation it is convenient to rearrange the system block diagram so that there is only one filter element. The new block diagram is shown in Fig. 7.15. This rearranged diagram is normally referred to as a "feedforward" circuit. The principle of a feedforward system is that by differencing the outputs from two sources the corrupted error signal can be estimated. This error signal is then subtracted from the original corrupted signal to obtain the uncorrupted output. In this particular implementation, the correction signal is estimated by lowpass filtering $(H(s))$ the difference signal, which because of the particular spectral characteristics of the noise signals results in an estimate of the accelerometer noise. Note however, that the feedforward circuit and the complementary filter circuit are merely different arrangements of the same physical processes.

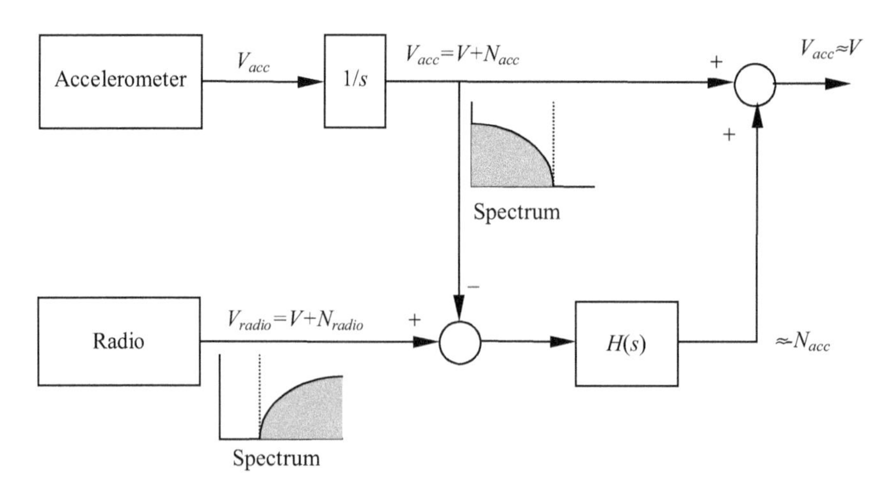

Fig. 7.15 Block diagram of feedforward processing, requiring only one filter. Note the integrated accelerometer speed data has only low frequency components, and the radio speed tends to have only high frequency components

Thus the feedforward circuit effectively combines the integrated accelerometer data and the radio-speed data, such that the noise components of both are eliminated (or at least significantly reduced). This result is dependent on the complementary nature of the noise spectral components of each signal, as illustrated in Fig. 7.15. In practice it is likely that there will be an overlap between the two noise spectra, so that the optimum filter bandwidth needs to be determined.

7.4.2.2 Feedforward Implementation for Estimation Mobile Speed

The feedforward process described in Sect. 7.4.2.1 provides an effective method of combining the radio-speed and integrated accelerometer speed data such that the output speed estimate is superior to both the individual estimates. However, the processing requires the application of a lowpass filter whose bandwidth is a function of the noise characteristics of both data sources. The simplest implementation assumes that the spectral characteristics of both data sources are stationary, so that the optimum filter bandwidth can be determined (say by simple trial and error). However, a more sophisticated approach is for an adaptive filter to be used, which automatically adjusts to the varying nature of the input signals. Observing that in Fig. 7.15 the lowpass filter has an input "signal" which is the integrated accelerometer noise and input "noise" from the radio-speed estimation, the requirement is to adjust the bandwidth so that the "signal" distortion by the filtering process is minimized while at the same time the "noise" is reduced by the maximum extent. Such an optimum filter is a suitably configured Kalman filter. Details of the implementation of the Kalman filter are postponed to Sect. 7.7.4.

Based on these concepts the classical feedforward processing technique has been extended to improve performance in this particular application.

1. In the case of random noise, the optimum filter is an appropriate Kalman filter, which adjusts the filter bandwidth according to the "signal" and "noise" characteristics.
2. The speed determination using the radiolocation data involves a delay and lowpass filtering. The integrated accelerometer data (speed estimate) thus requires an equivalent filter to minimise the errors.
3. The radiolocation data suffers from systematic (multipath) errors as well as random noise-like errors. To cater for these systematic errors, the Kalman filter adapts to adjust the bandwidth to reject systematic errors.

A further refinement to the basic feedforward processing is required in this case, as the radio-speed estimate has errors associated with a time delay and a lowpass filter—see Sect. 7.4.2.1 above for details. As this error does not occur for the integrated accelerometer, a corresponding filter must be applied to the integrated accelerometer data before the error signal is calculated. Equation (7.29) shows that the filter characteristics applied to the radio-speed includes both a time delay and a sinc lowpass filter. As the signal processing is in the time domain, it is appropriate to express the filtering process to the radio-speed (and hence the integrated accelerometer data) as a time domain process. The time domain impulse response is the inverse Fourier Transform of the frequency domain transfer function, and the output speed can be expressed as the convolution of the true speed with the impulse response $h(t)$, namely

$$\begin{aligned} H(f) &= [sinc(f\tau)]\left[e^{-j\pi f\tau}\right] \\ h(t) &= \frac{1}{\tau} rect_{\tau}(t-\tau/2) \\ \hat{v}(t) &= h(t) * v(t) \end{aligned} \tag{7.34}$$

For a discrete implementation, the convolution integral can be approximated by a summation, so that if the time separation parameter τ is m samples of digitized data, the filtered output is given by

$$\hat{v}_{n+1} = \hat{v}_n + \frac{1}{m}[v_{n+1} - v_{n-m}] \tag{7.35}$$

Equation (7.35) is applied to the speed estimate obtained from the integrated accelerometer data. In the ideal case, with error-free accelerometer and radio position data, the feedforward error signal will be zero (except the integrated accelerometer constant of integration). Thus the output speed will be the integrated accelerometer speed, but corrected for the constant of integration. Thus the lag and filtering errors of the radio have been corrected, and the constant of integration correctly compensated for, so that in this ideal case the feedforward output is the correct speed with no errors.

The particular implementation of the feedforward circuit to determine the speed is shown in Fig. 7.16. (See Sect. 7.7.4 and Appendix A for details of the Kalman filter implementation). In addition to the filtered input, the Kalman filter produces an estimate of the accelerometer bias. The Kalman filter implementation has the advantage that the filter bandwidth is automatically adjusted to provide the best estimate of the accelerometer noise. The lowpass filter associated with the integrated accelerometer input to the summer is implemented as per (7.35).

7.4.2.3 An Illustrative Example

To illustrate the concepts of the feedforward processing and the correction to the radio-speed estimation errors, a simple idealized example is given in this subsection. In this one-dimensional example, the speed is initially constant (10 m/s), then the constant acceleration is applied for 5 s to increase the speed to 15 m/s, after which the speed is maintained constant. Initially the position and accelerometer outputs are assumed to be noise free, as shown in Fig. 7.16. In the figure, the time lag effect on the radio speed can be clearly observed, but the feedforward output is essentially identical to the "true" speed.

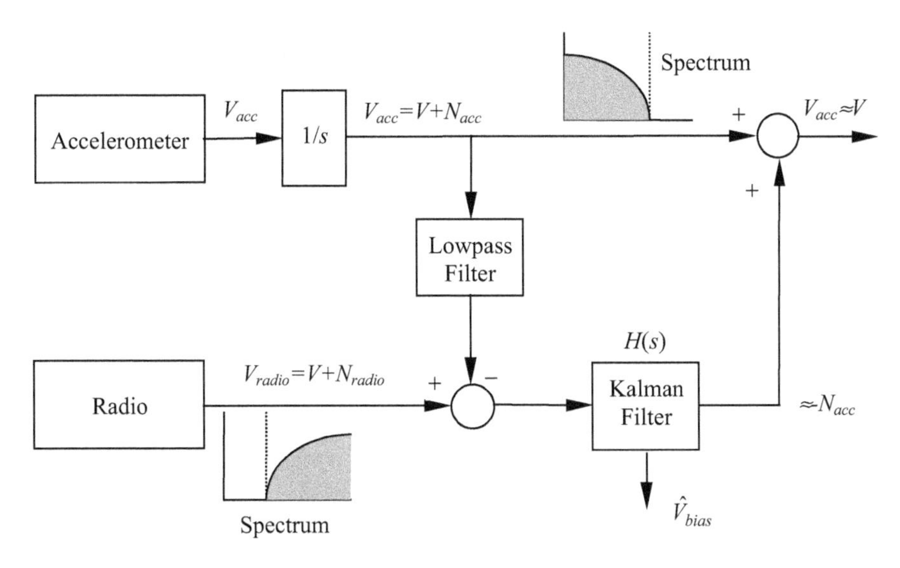

Fig. 7.16 Block diagram of the complementary filter, including the lowpass filtering of the integrated accelerometer data and a Kalman filter for the feedforward lowpass filter H(s)

Thus the feedforward processing of the ideal accelerometer and radio data results in essentially a perfect speed estimation. However, actual data will be corrupted with noise. Figure 7.17 shows the performance with ideal accelerometer data, but with the position data corrupted by Gaussian noise with zero mean and standard deviation of 0.3 m. From Chap. 4, Eq. (4.8) this performance can be achieved with a pulse rise-time of 100 ns (signal bandwidth 10 MHz) and a signal-to-noise ratio of 40 dB, or a signal bandwidth of 60 MHz and signal-to-noise ration of 25 dB. As can be observed, the feedforward output closely follows the "true" speed profile. A crude "figure of merit" can be defined as the sum of the standard deviation of the speed error plus the absolute value of the mean error (capturing both bias errors and noise errors). Using this measure, the raw data has an error of 0.66 m/s, while the feedforward output has an error of 0.11 m/s. Thus in this example with ideal accelerometer data the feedforward processing reduces the speed errors by a factor of 6:1.

Now consider the effect of noise in the accelerometer data. When accelerometers are attached to people, the variation in the orientation of the accelerometer due to the gait results in noisy data. For this demonstration the accelerometer noise is set at a standard deviation of 0.2 m/s^2. In addition to the random noise the accelerometer measures systematic accelerations associated with the stride rate of the gait. However, these systematic accelerations integrate to zero, and contribute little to the speed error. For example, a sinusoidal component with an acceleration of 0.2 m/s^2 with a stride rate of 2 per second would result in a sinusoidal error signal with a standard deviation of 0.107 m/s. Thus the accelerometer noise has only a marginal effect on the speed error. Figure 7.18 shows the speed profiles with accelerometer noise, with essentially the same performance as with no noise shown in Fig. 7.17.

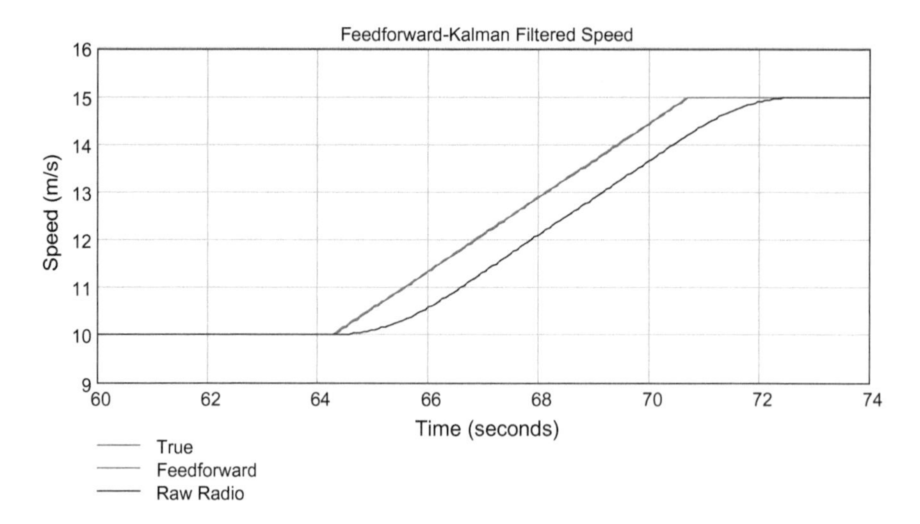

Fig. 7.17 Comparison of the true speed, the radio speed and the feedforward filter output. The time separation parameter for calculating the radio speed is 2 s, and the acceleration is about 0.8 m/s^2. The speed error due to the time lag is 0.8 m/s. Observe that the "true" and feedforward output speeds are essentially identical, thus correcting for the radio-speed time lag effect

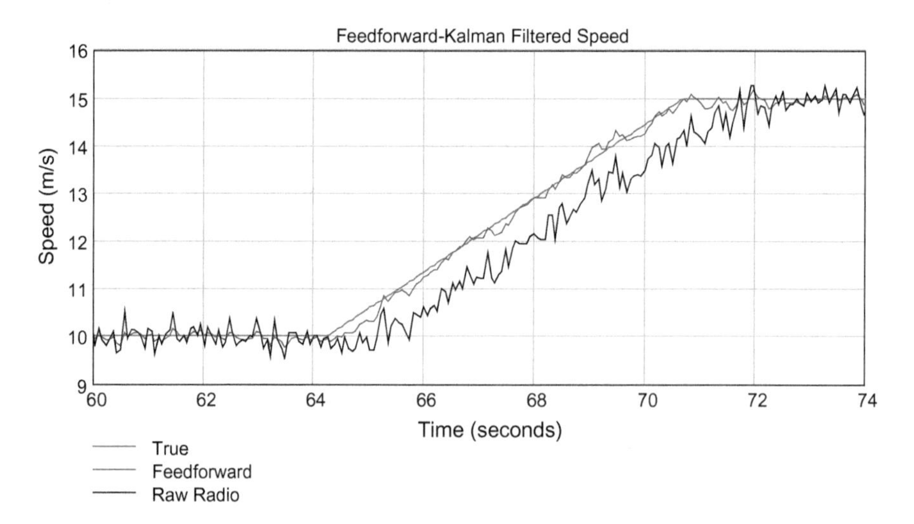

Fig. 7.18 Comparison of the true speed, the radio-speed and the complementary filter output. The time separation parameter for calculating the radio speed is 2 s, and the acceleration 0.8 m/s^2. The speed error due to the time lag is 0.8 m/s. The position data are corrupted by Gaussian noise with a standard deviation of 0.3 m. The standard deviation of the raw radio speed is 0.25 m/s, with a mean error of −0.41 m/s. In contrast, the feedforward output has a mean lag close to zero and a standard deviation of 0.100 m/s

Thus in conclusion for this illustrative example, with random errors in both the positional and accelerometer data, the feedforward process with the appropriate filtering of the integrated accelerometer data has essentially zero mean lag and a very small standard deviation. In contrast, the raw radio-speed has a standard deviation of 0.25 m/s and a lag of about 1 s. Further, unlike the raw radio-speed, the error is insensitive to the time separation parameter used to calculate the raw radio-speed.

7.4.2.4 Performance with Measured Data

The illustrative example in Sect. 7.4.2.3 shows that in the ideal case the feedforward processing of the radio-speed estimate and the integrated accelerometer data results in close to an error-free estimate of the true speed. Further, when typical random measurement errors are included, the feedforward speed estimate has no lag and a measurement error standard deviation of the order of 0.1 m/s. This subsection determines the performance of actual measured data.

Because the true speed is not known in this experiment, the performance of the speed estimates can only be indirectly inferred. For testing the measurement of the speed of a mobile device without directly measuring the speed, two mobile devices were attached on the same vehicle. By independently measuring the position using two tracking units attached to the vehicle, the differential speed (which should be zero) can be determined without any direct knowledge of the true speed. The estimated accuracy of the position for this test system is 0.3 m, as used in the illustrative example. The above illustrative example shows that the actual speed error expected will be of the order of 0.1 m/s (or 0.14 m/s for the differential speed assuming independent random errors). Figure 7.19 shows the raw measured differential speed with a time separation parameter of 3 s. The measured standard deviation is 0.16 m/s, which is broadly in line with the above expectations.

The second illustration of performance is shown in Fig. 7.20. The radio-speed estimate is known to be both delayed and smoothed relative to the true speed. The data in this figure is based on a time separation parameter of 3 s, and hence the delay in the raw speed estimate will be 1.5 s. Reducing the time separation will reduce the time delay, but increase the random errors. As was shown in Sect. 7.4.2.1, the optimum performance will occur when the separation parameter is about 1 s, and the corresponding speed error is 0.5 m/s for both the random and lag components. Observe that the radio-speed estimates are delayed relative to the feedforward estimates, and that the measurement noise is reduced in the feedforward data. In this case the radio-speed estimates are up to 2 m/s in error.

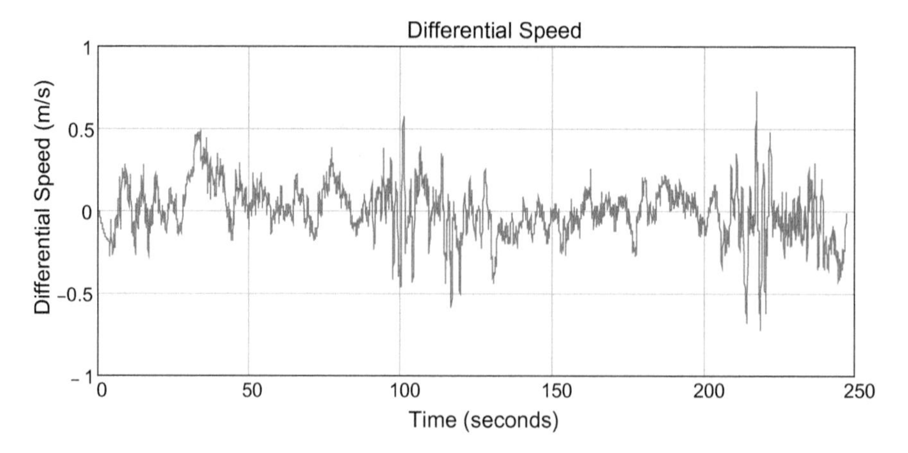

Fig. 7.19 Measured differential speed with two mobiles mounted on the same vehicle. The standard deviation is 0.16 m/s, and the mean error is –0.004 m/s. The largest errors occur near the 100 and 220 s points, where the GDOP (for an explanation of GDOP, see Chap. 14.) is worst

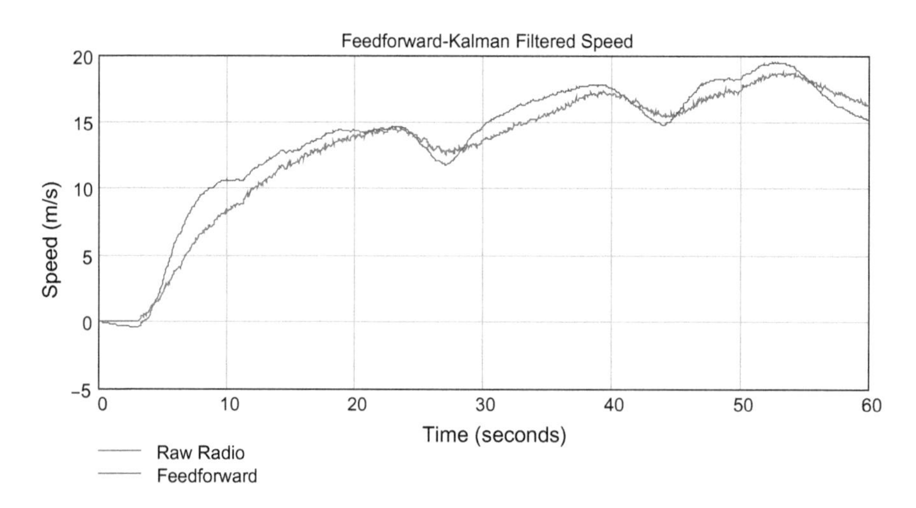

Fig. 7.20 Radio and feedforward speed estimates. The time lag and smoothing of the radio speed estimate can be clearly observed. As a result, the radio speed estimates are in error by up to 2 m/s

7.5 Assessment of the Benefits of Using Gyros and Accelerometers

The previous sections described a system that uses the output from a rate-gyro to effectively reduce the noise from the radio heading data, and accelerometers to improve the accuracy of measuring speed. In Sect. 7.4 a method was described using a feedforward process for estimating the speed, but the same technique can also be applied to improved the estimation of the heading angle using a rate-gyro. The feedforward process described in Sect. 7.4.2.2 and Fig. 7.16 includes a Kalman filter, which acts as a low pass filter.

A possible simpler implementation is to only use the Kalman filter to reduce the radio noise, without the complication of the rate-gyro. While the Kalman filter process is usually described as a feedback loop, it is shown in Appendix A that the filter can be represented by a standard transfer function that is the ratio of a quadratic and a cubic polynomial functions which are defined by just three parameters. Further, with time sampled data such as used in all positioning systems, the Kalman filter transfer function can be implemented as a standard finite impulse response (FIR) filter, which simply uses the current time sample and a few of the past sample data to calculate the filter output. Thus the Kalman filter is simple to implement, and computationally simple.

The question is thus: What is the benefit of including the rate-gyro/accelerometer data in calculating angles/speeds? The Kalman filter at the output of a radio heading module can effectively reduce the radio angle noise. Further, the Kalman filter automatically adjusts the filter bandwidth to provide the optimum compromise between "signal" distortion caused by the filter and the "noise" reduction associated with the filter. The signal bandwidth will have a dominant affect on the Kalman filter bandwidth. With the radio heading data applied directly to the Kalman filter, the track dynamics of the mobile will determine the degree of filtering of heading signal. Highly mobile (angle) dynamics will result in high radio angle noise, as the Kalman filter bandwidth must be large to track the dynamics. However, the input to Kalman filter in the feedforward circuit case is independent of the mobile dynamics, as the heading angles from the radio and gyro as first subtracted, resulting in the "signal" being the integrate bias error signal; this signal *always* has a very low bandwidth regardless of the mobile dynamics, so that the resulting Kalman filter in the feedforward implementation shown in Fig. 7.16 will also have a lower bandwidth than in the radio-only Kalman filter. Thus the feedforward system (with a rate-gyro or accelerometer) should have superior performance. However, if the mobile (angular) dynamics have only low spectral components (for example when the heading is constant along a straight path), then there will be little benefit from including the rate-gyro.

Figure 7.21 shows the spectrum of the track angle dynamics as determined for a typical horse race,[7] together with the integrated rate-gyro bias. The spectral data show that heading angle has only very low frequency components, comparable (but somewhat larger) with the integrated rate-gyro bias. Thus from the above analysis there will be little benefit from measuring the turn rate and applying these data in a complementary filter. However, non-random systematic errors (mainly associated

[7] See Chap. 10 for details of the horse racing application of radiolocation technology.

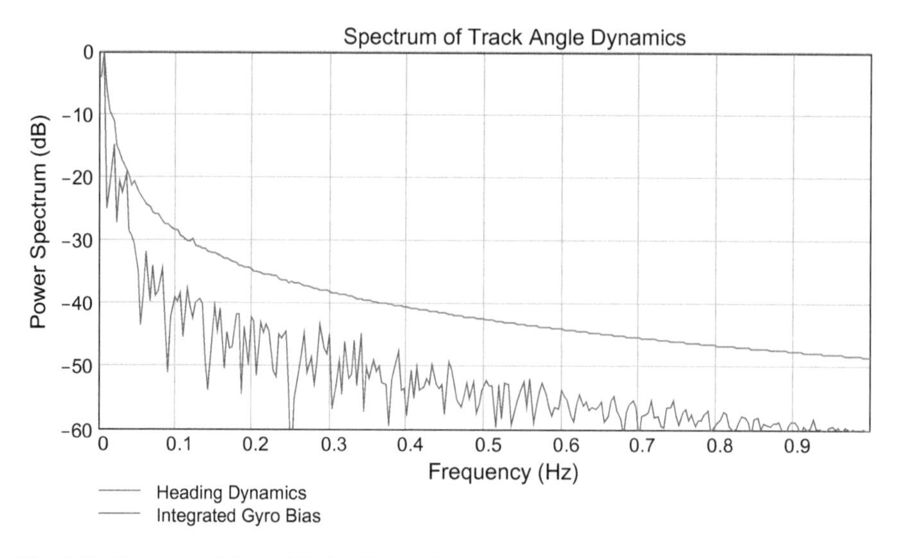

Fig. 7.21 Spectrum of the mobile heading angle dynamics on a typical horse race track, together with the spectrum of the integrated rate-gyro. Both spectra have very low spectral characteristics, with most of the spectral energy below 0.1 Hz

with multipath) can be more effectively overcome with the use of the rate-gyro data. For more details, refer to Sect. 7.6.5 on the reduction of multipath errors using the rate gyro as a reference.

7.6 Heading Angle Using Radiolocation and Gyroscope Data

7.6.1 Overview

The determination of the heading angle from sensor data alone was considered in Sect. 7.3 to illustrate the concept of complementary filtering. In this section we will again consider the determination of the heading angle, but this time the gyroscope data will be used to stabilize the noisy angle data obtained from radiolocation position data. In Sect. 7.3, feedforward techniques were used, together with the use of a Kalman filter. This section provides details of the implementation of the Kalman filter.

The heading angle can be derived from two positional data points, but because the positional data are noisy and must be spaced close in time (say a maximum of 1 s) due to the rapid dynamics of human motion, the resulting angular data are very noisy. Thus some method of smoothing the data is desirable. In this case, a rate-gyroscope will be used.

7.6.2 Basic Theory

The basic heading data can be obtained from the radiolocation position data. If the mobile position is defined in two dimensions by the functions $x(t)$ and $y(t)$, then the heading angle is given by

$$\hat{\psi} = \tan^{-1}\left[\frac{y_n - y_{n-m}}{x_n - x_{n-m}}\right] \tag{7.36}$$

where the data are m sample periods (Δt) of the radiolocation positional data apart. While the above procedure is very simple to implement, it is well known that numerical differentiation results in significant errors due to measurement errors; the random errors associated with the position fix translate into heading errors. These random heading errors can be minimized by increasing the separation time parameter (m), but this process also increases the "lag" error, and fast dynamics of human motion cannot be followed. Thus if the period used in the numerical differentiation is T, and the heading turn rate is $\dot{\psi}$, then the systematic heading error is approximately $\dot{\psi}T/2$. However, as the turn rate is measured by the rate-gyro, this error can be compensated for. In particular, if the path between the two measurement points is curved, then the heading angle given by (7.36) is approximately the true angle with a time lag of $(m/2)\Delta t$. (This lag time is exact for straight line paths and constant radius paths). Thus if the rate-gyro is used to measure the turn rate during the "lag period", the corrected heading angle at sample n is given by

$$\psi_n \approx \hat{\psi}_n + \frac{1}{2}\int_{(n-m)\Delta t}^{n\Delta t} \dot{\psi}_{gyro}(t)\, dt \approx \hat{\psi}_n + \frac{1}{2}\left(\psi_{gyro}(n) - \psi_{gyro}(n-m)\right) \tag{7.37}$$

Thus the heading angle can be estimated without any lag if both the radio and gyro data are used. Note that this correction does not require absolute heading angles from the integration of the rate-gyro data, as only differential gyro angle is used. Note also that the sample rate of the gyroscope needs to be at least 20 per second, which is typically a much greater sample rate than the associated positional sample rate. This high sample rate does not impose any significant processor load; in comparison, radiolocation position determination is very processor intensive, and thus the update rate is typically much lower, say 1–2 times a second.

To estimate the performance of the mathematical procedures outlined above, the statistical error performance must be analyzed. From the analysis of positional errors from the radiolocation sub system it is known that the error distribution can be in general described by an error ellipse "centered" on the true position (Yu et al. 2009, Chap. 9). The size and shape of this ellipse depends both on the ranging measurement errors of the radiolocation system as well as the geometry of the receiving base stations relative to the mobile. Providing the number of base stations is sufficiently large and positioned to approximately "surround" the mobile, the

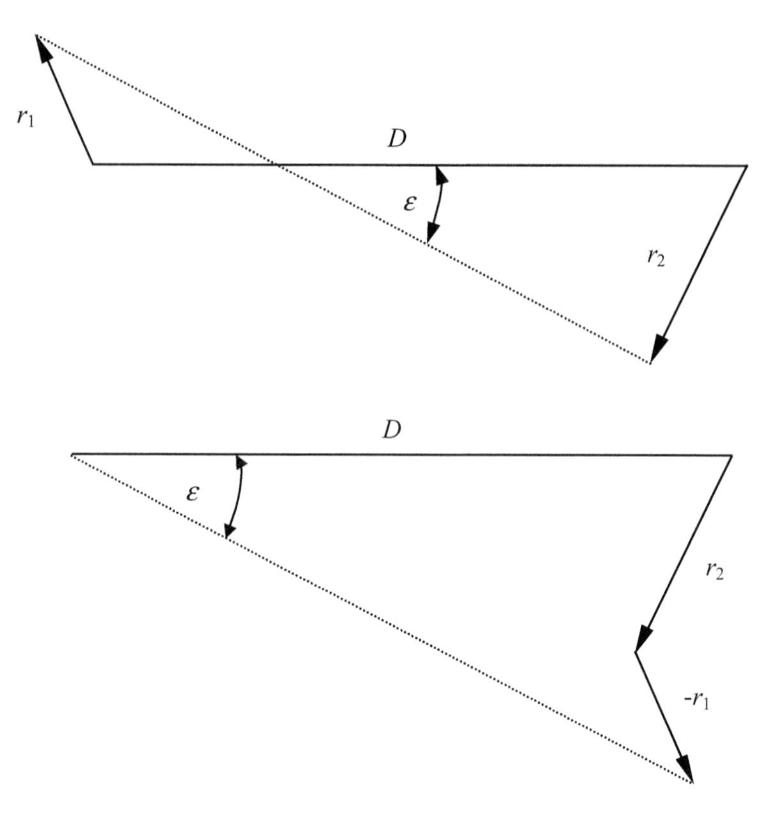

Fig. 7.22 Geometry for the calculation of the heading angle from differential positions

error ellipse will be approximately circular in shape. Thus in the following analysis a circular error probability distribution will be assumed. Further, because the calculation of the position is a least-square process which involves the addition of the ranging measurements, the distribution of errors will be approximately a 2-dimensional Gaussian function due to the Central Limit Theorem.

Consider the geometry for calculating the heading from two positions a distance D apart, as shown in Fig. 7.22. The position errors r_1 and r_2 result in an angular error ε in the heading angle. The second diagram shows that this error is equivalent to a single vector $\mathbf{r}_2 - \mathbf{r}_1$. The amplitude of the vectors $\mathbf{r}_1$ and $\mathbf{r}_2$ have a 2–dimensional Gaussian (Rayleigh) distribution, with uniform angular distribution and a variance parameter σ^2. Because the sum (and difference) of two Gaussian distributions is also a Gaussian distribution, the differential error vector $\mathbf{r}_2 - \mathbf{r}_1$ will also have a Rayleigh distribution, but with a variance parameter $2\sigma^2$. The 2–dimensional probability density function is thus given by

$$q(r,\phi) = \frac{r}{4\pi\sigma^2} \exp\left[-\frac{r^2 - 2Dr\cos\phi + D^2}{4\sigma^2}\right] \tag{7.38}$$

where r is the magnitude of the differential error.

By integrating (7.38) with respect to r (from zero to infinity), the error angle ϕ probability density function becomes

$$q(\phi) = \frac{e^{-D^2/4\sigma^2}}{2\pi} + \frac{D\cos\phi}{\sqrt{4\pi}\sigma} e^{-(D\sin\phi/2\sigma)^2} \left[\frac{1}{2} + \mathbf{\Phi}\left(\frac{D\cos\phi}{\sqrt{2}\sigma}\right)\right] \tag{7.39}$$

The complexity of (7.39) hides its essential characteristics. Two special cases can be considered. When the separation distance D is zero, the angle is indeterminate so that the angular PDF becomes uniform with $q(\phi) = 1/2\pi$. The other case of interest is when the distance D is much greater than the positional error parameter σ. In this case the error angle ϕ is small, so that using small angle approximations in (7.39) the PDF becomes

$$q(\phi) \approx \frac{D}{\sqrt{2\pi}\sigma_\phi} e^{-\frac{\phi^2}{2\sigma_\phi^2}} \qquad \left(\sigma_\phi = \sqrt{2}\sigma \big/ D\right) \tag{7.40}$$

Equation (7.40) can be recognized as a Gaussian distribution with zero mean and a standard deviation of σ_ϕ. Thus the standard deviation in the heading angle is given by the ratio of the differential positional error parameter to the differential distance. This random error decreases with increasing separation D. For intermediate distance separations the full expression given by (7.39) must be used, but roughly it will be a combination of these two limiting cases. With the lag correction process defined in (7.37) the lag errors from the raw radio positions can be minimized but not totally eliminated. In practice, an accuracy of better than 1° can be achieved with integration times of 1 s; the maximum practical differential period is about 2 s.

Consider a numerical example, with a radiolocation sample rate of 2 per second, a mobile speed of 2 m/s, a turn rate of 30 degrees per second, and a positional accuracy of 0.5 m. In this case, if the separation is one sample ($m = 1$ or $D = 1$ m), then $\sigma_\phi = 40.5°$. Further, in this case the time lag effect results in an offset error of 7.5°. Thus the raw heading angle derived from the radiolocation positions is very poor, so that some form of correction and smoothing of the raw radiolocation derived angular data is required.

7.6.3 Heading Determination

In Sect. 7.3 introducing the concept of a complementary filter the primary method of determining the heading angle is by a compass, but a more common approach is to use the radio position data. If the mobile position is defined in two dimensions by the functions $x(t)$ and $y(t)$, then the speed in two dimensions is simply obtained by the differentiation of the functions, so that the heading angle is given by

$$\psi = \tan^{-1}\left[v_y/v_x\right] \tag{7.41}$$

The speed functions can be estimated from the position functions by numerical differentiation. If the position is determined at a constant period Δt, then from (7.41) the heading estimate is given by

$$\begin{aligned} \hat{\psi} &= \tan^{-1}\left[v_y/v_x\right] \approx \tan^{-1}\left[\frac{(y_n - y_{n-m})/\delta t^m}{(x_n - x_{n-m})/\delta t^m}\right] \\ &= \tan^{-1}\left[\frac{y_n - y_{n-m}}{x_n - x_{n-m}}\right] \qquad (\delta t^m = m\Delta t) \end{aligned} \tag{7.42}$$

While the above procedure is very simple to implement, it is well known that numerical differentiation results in significant errors due to measurement errors. The random errors associated with the position fix translate into heading errors. These random heading errors can be minimised by increasing the separation time parameter (m), but this process also increases the "lag" error. Thus if the period used in the numerical differentiation is T, and the heading turn rate is $\dot{\psi}$, then the systematic heading error is approximately $1/2\dot{\psi}T$. However, as the turn rate is measured by the rate-gyro, this error can be compensated for. In particular, if the path between the two measurement points is curved, then the heading angle given by (7.42) is approximately the true angle with a time lag of $(m/2)\Delta t$. (This lag time is exact for straight line paths and constant radius paths). Thus if the rate-gyro is used to measure the turn rate during the "lag period", the corrected heading angle at period time sample n is given by

$$\psi_n = \hat{\psi}_n + \int_{(n-m/2)\Delta t}^{n\Delta t} \dot{\psi}_{gyro}(t)dt = \hat{\psi}_n + \left(\psi_{gyro}(n) - \psi_{gyro}(n - m/2)\right) \tag{7.43}$$

Thus the heading angle can be estimated without any lag if both the radio and gyro data are used.

To estimate the performance of the mathematical procedures outlined above, the statistical error performance must be analysed. From the analysis of position errors from the radio-location sub system it is known that the error distribution can be in general described by an error ellipse "centred" on the true position. The size and shape of this ellipse depends both on the pseudo-ranging measurement errors of the radio-location system as well as the geometry of the receiving Base Stations relative to the mobile. Providing the number of Base Stations is sufficiently large and positioned to approximately "surround" the mobile, the error ellipse will be approximately circular in shape. Thus in the following analysis a circular error

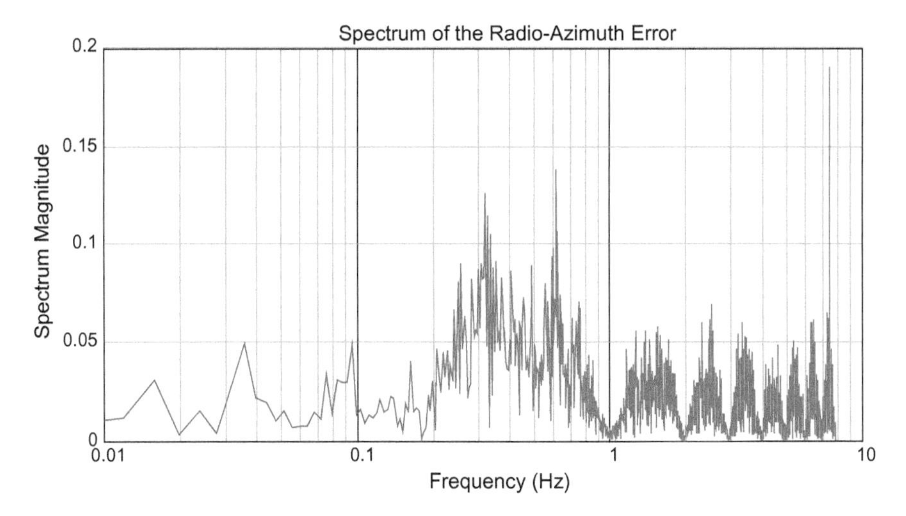

Fig. 7.23 Spectrum of the heading angle errors derived from differential radio positions. Observe that the errors occur mainly above 0.2 Hz. The nulls in the spectrum at above 1 Hz is a consequence of the differential data processing

probability distribution will be assumed. Further, because the calculation of the position is a least-square process which involves the addition of the ranging measurements, the distribution of errors will be a 2-dimensional Gaussian function due to the Central Limit Theorem.

Consider now some data measured on a horse racing track. The error in the heading angle is difficult to estimate exactly, as there is no independent angle measurement. However, the errors can be approximately estimated by assuming that the heading angle is a smooth slowly varying function of time. Using this technique, the angle errors from both the radio and rate-gyro were calculated. With a separation of 16 sample periods (1 s), the radio angle standard deviation was about 1.6°. Applying (7.40) with the horse speed of 60 kph the positional accuracy of the radio location system is 0.33 m, which is in broad agreement with expectations. However, for application with the complementary filter the spectral characteristics of the errors are of interest. Figure 7.23 shows the spectrum of the radiolocation based heading errors. Note that the spectrum has null located at frequencies $n(n\delta t)$ ($n = 0, 1, 2, \ldots$). Additionally, the spectrum near zero frequency is small, as at low frequencies (much less than the sample rate) the error approaches zero. Effectively the spectrum of the error is zero below about 0.1 Hz.

Figure 7.24 shows the corresponding errors for the integrated rate-gyro. The majority of the errors occur at frequencies below 1 Hz. The complementary nature of the spectra is evident in these figures, with the radio-based errors being mainly above 0.1 Hz and the integrated rate-gyro errors below 1 Hz. Thus the errors are somewhat complementary in nature, as there is some overlap between the two spectra. The complementary nature of the data means that improved performance is possible by combining the two angle measurements in a complementary filter.

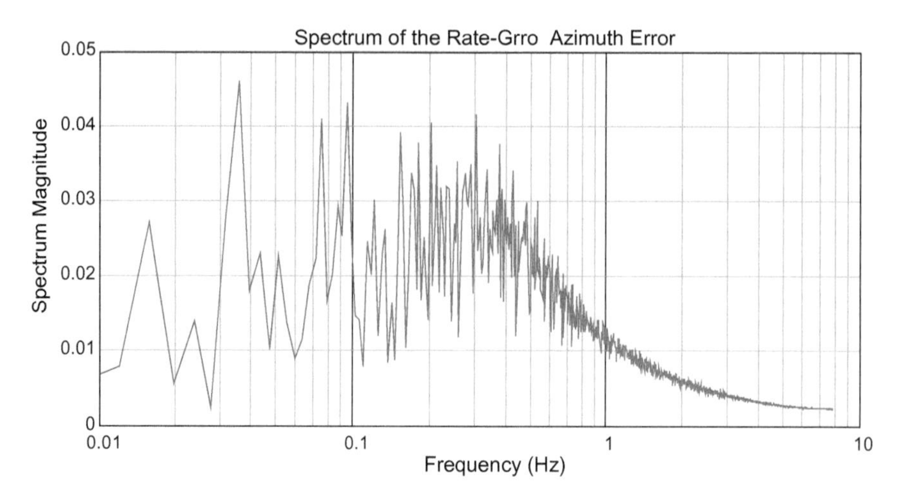

Fig. 7.24 Spectrum of the heading angle errors derived from differential radio positions. Observe that the errors occur mainly below 1 Hz

However, as there is some overlap in the spectra, the filtering processes will result in some residual errors. The optimum filter cutoff frequency is about 0.5 Hz.

7.6.4 Kalman Filter Design

This subsection describes the implemention of the Kalman filter whose broad functions were introduced in Sect. 7.6.1. The theory and implementation of Kalman filters is not presented, but an introduction to the main concepts are presented in Appendix A; rather the design matrices required for a generic Kalman filter implementation are derived for the particular application.

With reference to Fig. 7.25, the Kalman filter is required to perform a number of functions, summarised as follows:

1. Perform the lowpass filtering of the differential heading signal (radio heading minus integrated gyro heading).
2. Estimate the rate-gyro bias, so that a correction can be applied to the raw gyro output.
3. Estimate the difference between the next radio heading data and the projected gyro data, thus allowing heading "glitches" to be detected, and their effects removed.

The design of the Kalman filter is based on the characteristics of the rate-gyro. The rate-gyro nominally measures the heading turn rate, but the measured data are corrupted from two sources. The first error source is simply random (Gaussian)

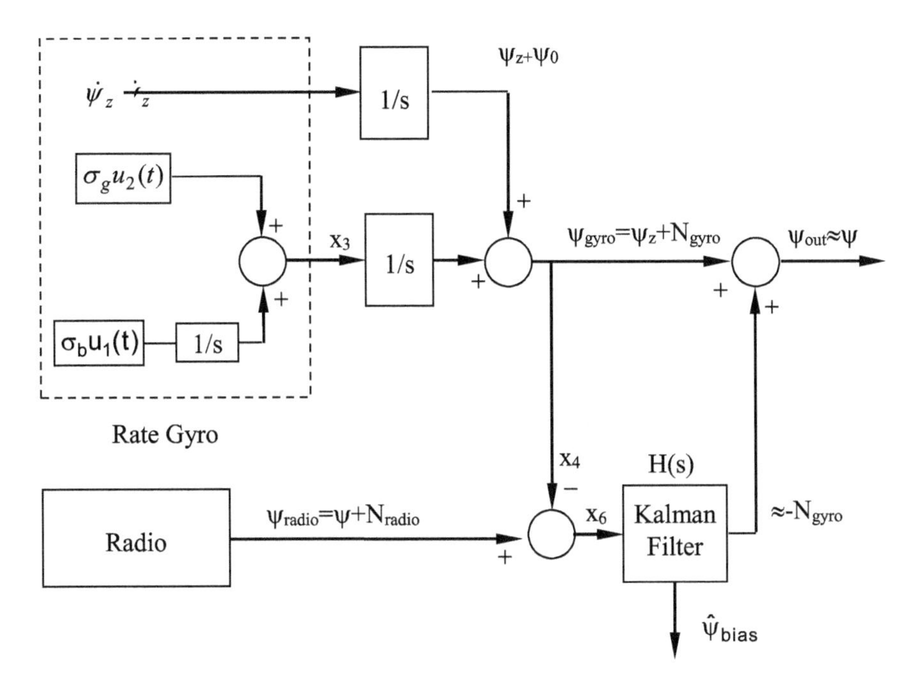

Fig. 7.25 Block diagram of the complementary filter, including bias correction

noise at the output. This noise for a moving mobile is dominated by motion effects rather than sensor electronic noise. For a car mounted mobile the noise standard deviation is about 2 degrees per second, while the stationary noise is about 0.1 degrees per second. The second error source is associated with random variations in the gyro bias. The bias error varies slowly over time, and can be modelled by a random walk process. The Kalman filter is designed to estimate this random bias, so that the gyro output can be corrected. The following description assumes the reader is familiar with the concepts of Kalman filters, as presented in Appendix A. Although the following is based on the angle estimation case using a rate-gyro sensor, the design process for the speed estimation using an accelerometer is essentially the same.

For the Kalman filter in Fig. 7.25, the integrated rate-gyro is the "signal" and the radio angle noise is considered to be measurement noise. For implementation of a Kalman filter, the model dynamics and statistical model must be in the standard form represented by the equation

$$\dot{\mathbf{X}} = \mathbf{FX} + \mathbf{Gu} \tag{7.44}$$

where $\boldsymbol{X}$ is the state vector, $\boldsymbol{F}$, $\boldsymbol{G}$ are matrices defining the dynamical and statistical models, and $\boldsymbol{u}$ is a column vector of unit Gaussian random distributions.

Now consider the system as defined in the block diagram in Fig. 7.25. The random walk process for the gyro is defined by the differential equation

$$\dot{x}_1 = \sigma_b u_2(t) \tag{7.45}$$

where σ_b is the standard deviation of the random bias process. The output from the gyro is given by

$$\begin{aligned} x_4 &= (\psi_z + \psi_0) + \int x_3 dt \\ \dot{x}_4 &= \dot{\psi}_z + x_3 = \dot{\psi}_z + x_1 + x_2 \end{aligned} \tag{7.46}$$

where ψ_0 is the constant of integration. The differential signal input to the Kalman filter is

$$x_6 = x_5 - x_4 = (\psi_z + \sigma_{radio} u_3(t)) - \left((\psi_z + \psi_0) + \int x_3 dt\right) \tag{7.47}$$

The radio angle noise is considered to be measurement noise, and thus is not treated as part of the dynamical model, so that the differential of equation becomes

$$\dot{x}_6 = -x_3 = -x_1 - \sigma_g u_2(t) \tag{7.48}$$

where σ_g is the standard deviation of the rate-gyro noise at its output. Thus writing the above equations in the matrix form results in standard form (compare with Eq. (7.44))

$$\begin{bmatrix} \dot{x}_6 \\ \dot{x}_1 \end{bmatrix} = \begin{bmatrix} 0 & -1 \\ 0 & 0 \end{bmatrix} \begin{bmatrix} \dot{x}_6 \\ \dot{x}_1 \end{bmatrix} + \begin{bmatrix} \sigma_g & 0 \\ 0 & \sigma_b \end{bmatrix} \begin{bmatrix} u_2 \\ u_1 \end{bmatrix} \tag{7.49}$$

The next step in completing the Kalman filter design is to calculate expressions for the dynamics matrix $\mathbf{\Phi}$ and the covariance matrix $\mathbf{Q}$ (see Appendix A, Section A.2.3). The method used is based on the van Loan technique which calculates both $\mathbf{\Phi}$ and $\mathbf{Q}$ using the following partition matrix expressions

$$\begin{aligned} \mathbf{A} &= \begin{bmatrix} -\mathbf{F} & \mathbf{G}\mathbf{G}^T \\ 0 & \mathbf{F}^T \end{bmatrix} \qquad \mathbf{B} = \exp[\mathbf{A}\delta T] \\ \mathbf{B} &= \begin{bmatrix} * & \mathbf{\Phi}^{-1}\mathbf{Q} \\ \mathbf{0} & \mathbf{\Phi}^T \end{bmatrix} \end{aligned} \tag{7.50}$$

where δT is the sample period for the discrete Kalman filter. The exponential expression terminates after three terms, so explicitly calculating the exponential as a series expression results in

$$\mathbf{\Phi} = \begin{bmatrix} 1 & \delta T \\ 0 & 1 \end{bmatrix} \qquad \mathbf{Q} = \begin{bmatrix} \sigma_g^2 \delta T + \frac{\sigma_b^2 \delta T^3}{3} & -\frac{\sigma_b^2 \delta T^2}{2} \\ -\frac{\sigma_b^2 \delta T^2}{2} & \sigma_b^2 \delta T \end{bmatrix} \tag{7.51}$$

The parameter σ_g is the gyro noise standard deviation; a typical value is 2 degrees per second, and the parameter σ_b is the bias random walk; a typical value is 1 degree per second per second.

The final step is to define the measurement accuracy matrix $\boldsymbol{R}$ (see Appendix A, Eq. (A.3)). The nominal accuracy of the radio angle is as defined by (7.40), with the separation distance D determined from the speed $v(t)$ and the times separation τ, so that for random positional errors

$$\sigma_{radio} = \frac{\sqrt{2}\sigma_{position}}{v(t)\tau} \tag{7.52}$$

where $\sigma_{position}$ is the positional accuracy estimate obtained from the least-squares fitting procedure, but also incorporating GDOP. The $\mathbf{R}$ matrix in this case is simply

$$\mathbf{R} = \sigma_{radio}^2 \tag{7.53}$$

However, the radio position is also affected by systematic (multipath) errors, in addition to the random errors, so that an additional term associated with the multipath errors is required. These additional features are considered in the next subsection.

7.6.5 Measurement Errors

One of the basic requirements in the Kalman filter design is to specify the measurement error matrix $\boldsymbol{R}$. For the particular implementation associated with the heading angle measurement this is simply the "radio noise". For a simple model with only random angle errors, (7.53) will suffice. However, with multipath errors, a more complicated approach is required. Because the Kalman filter estimates the gyro bias, the radio noise can be estimated from the difference between the radio angle and the (smooth) integrated gyro (corrected for bias errors), namely

$$\Delta\psi_{radio\_noise} = \psi_{radio} - \int \left(\dot{\psi}_{gyro} - \dot{\psi}_{bias}\right) dt \approx \psi_{radio} - \psi_{out} \tag{7.54}$$

The last approximation follows from the fact that $\psi_{out} \approx \psi$.

An example of the radio noise is shown in Fig. 7.26. These data used the final approximate expression in (7.54). Most of the estimated errors are random, but large multipath "glitches" can be observed near the 60 and 210 s point. These radio angle glitches will appear in the output if appropriate corrective action is not taken.

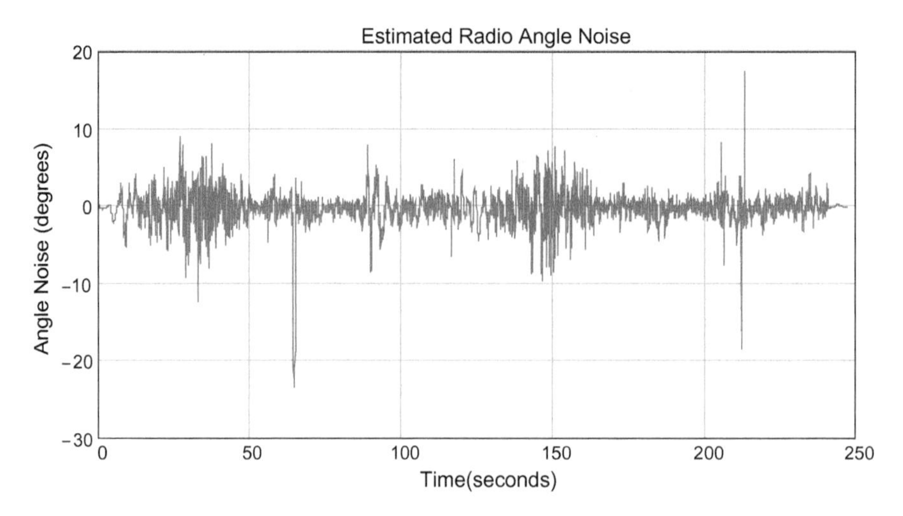

Fig. 7.26 Example of estimated radio noise. Most of the data are random, but larger multipath errors can be observed near the 60 and 210 s point. The 60 s data glitch is artificially added to the radio angle data to test the performance of the feedforward correction system

While Kalman filter theory is based on the assumption of random measurement errors, the general effect of the measurement matrix $\boldsymbol{R}$ is to modulate the bandwidth of the Kalman filter. If the measurement accuracy is low, then the bandwidth is reduced so that the output relies more heavily on the past history. Thus the Kalman filter will tend to filter out any glitches. Thus it is proposed that the Kalman filter is extended with the $\boldsymbol{R}$ matrix defined as follows

$$\begin{aligned} \sigma_{radio}^2 &= \left(c_1 \sigma_{fit\_error}\right)^2 + \left(c_2 \Delta\psi_{radio\_noise}\right)^2 \qquad (\sigma_{radio} > \sigma_{\min}) \\ \sigma_{radio}^2 &= \sigma_{\min}^2 \quad \text{otherwise} \end{aligned} \tag{7.55}$$

where σ_{fit_error} is determined by the least-squares fit procedure associated with the position determination, $\sigma_{\min}$ is the minimum acceptable radio measurement error, and c_1 and c_2 are appropriate weighting constants to "tune" the Kalman filter to provide the required performance. The $\sigma_{\min}$ parameter can be determined using (7.52) where $\sigma_{position}$ is the normal position accuracy (say 30 cm in this example).

The bandwidth modulation functionality can be observed in Fig. 7.27. In this case an "artificial glitch" one-second is added to the data. Observe that the bandwidth of the Kalman filter drops significantly.

The typical performance of the Kalman filter in conjunction with the feedforward system is illustrated in Fig. 7.28. This diagram is based on actual measured positional data. The positional glitch translates into two associated radio angle glitches separated by the time difference used for the angle calculation (1 s in this case). Observe that both the "normal" random errors and the glitch errors are effectively filtered, albeit with a small output delay. The parameters c_1 and c_2 can be

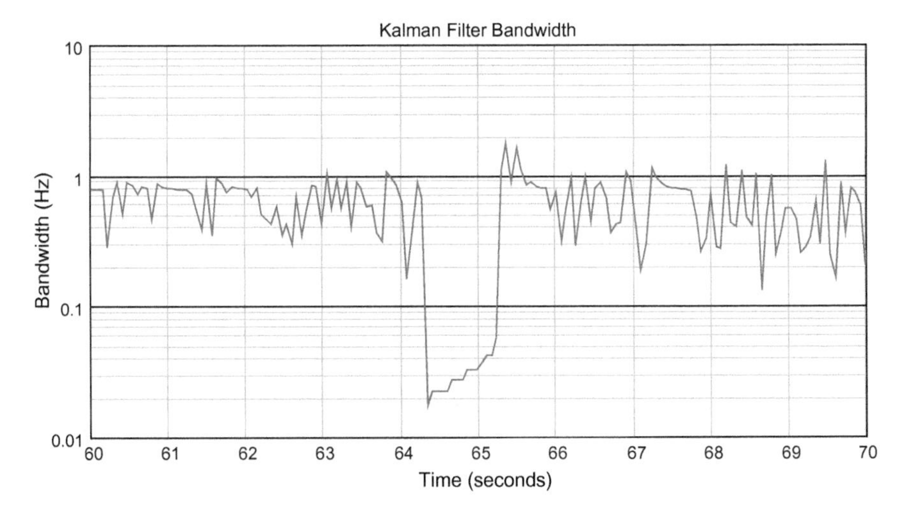

Fig. 7.27 Example of the behaviour of the Kalman Filter in a one-second glitch. The bandwidth of the filter is about an order of magnitude less in the glitch compared with the "normal" radio data. The weighting constants used are $c_1 = c_2 = 1$

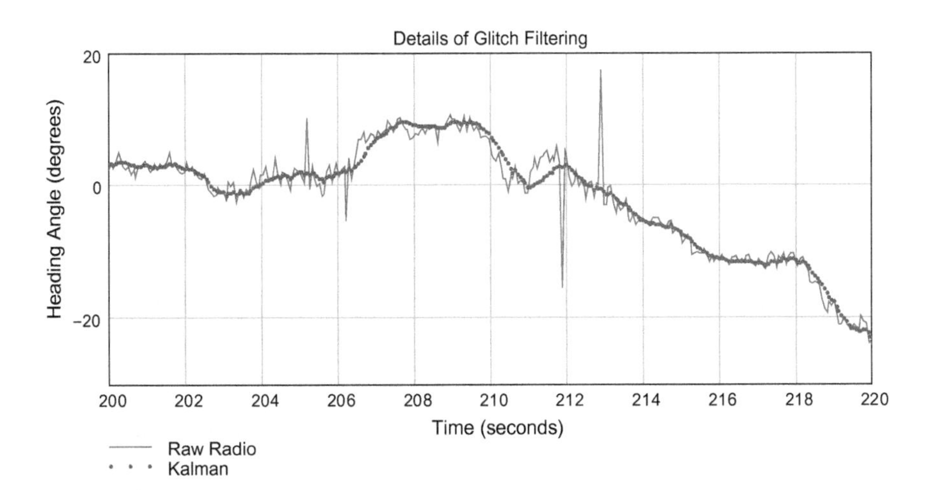

Fig. 7.28 Example of the behaviour of the Kalman Filter and the feedforward correction circuit for typical radio measurement glitches. Observe that the glitches appear in pairs, with a separation equal to that used in the angle determination process

used to determine the length of the glitch removal. With both these parameters set at unity, glitches of up to 1 s can be effectively removed.

Thus in conclusion, the rate-gyro data are compared with the radio data to obtain a suitable estimate of the radio angle noise. As explained in Sect. 7.5 the gyro data has little direct benefit unless the mobile angle dynamics have a bandwidth much greater than the bandwidth of the rate-gyro bias drift. In the horse racing case, the typical angle dynamics have low bandwidth, so the direct benefits are small. However, the integrated rate-gyro data are useful in detecting the multipath errors in the radio, so that multipath induced angle errors can be largely eliminated.

7.7 Determining Stride Length Using Accelerometer

The previous sections considered using sensors to estimate the heading angle and the speed, from which the position can be derived by dead reckoning techniques. In the case of tracking people walking (say in a building), an alternative approach is to count the number of strides and the stride length using an accelerometer, and with the addition the heading data the position can be estimated without the need to estimate speed.

The determination of the number of strides taken is relatively simple, as vertical accelerometer data show a clear cyclical pattern associated with walking. Such devices (called pedometers) are readily available for counting steps, but the determination of the stride length is more difficult. The theoretical approach of double integration of a horizontal accelerometer fails in a practical situation where the accelerometer is attached to a person. As with the heading example in Sect. 7.3, the integrated accelerometer data suffers from both offset and scaling errors, but the accelerometer has an additional source of error due to the Earth's gravity. Any misalignment in a horizontal accelerometer will result in a acceleration component due to gravity. Further, for the case of tracking people with attached devices, the orientation of the sensor and the rigidity of the attachment to the body will be quite variable. As these implementation details are difficult to overcome, an alternative approach is appropriate.

A method of determining stride length from accelerometer data is described in an application note from Analog Devices (Weinberg 2002), a company that manufactures accelerometers. This application note is the starting point for the techniques described in this section. Various experiments are described for measuring the stride length and the associated accelerometer data. These data are analyzed to develop algorithms with improved accuracy. Simple mathematical modeling of the walking process is also described, and the theoretical and empirical models compared. The design aim is to achieve a stride length estimate accurate to 5% for all people (both male and female), all sizes of people, and for all walking speeds. Such variability in the characteristics of people and their walking styles makes the determination of suitable algorithms difficult.

7.7.1 Overview of Theory of Stride Length Determination

The development of mathematical models for the walking process is a difficult task. One approach is to develop a detailed bio-mechanical model of the walking process, and hence determine the stride characteristics. Because of the complexity of the human body and the motion associated with walking, this approach is not adopted, but rather an empirical approach is used. In complex problems such as determining the stride length of walking, one approach is to undertake a dimensional analysis of all the variables and parameters that could affect the stride length. This analysis allows the number of variables to be greatly reduced, so that more simple functional relationships can be observed. By performing curve fitting, simple analytical formulae can be derived from measured data.

Despite the complexity of the mechanics of walking, a simple mathematical model of walking provides useful insight into the empirical curve fitting process. While the mathematical model is too simple for accurate stride length estimates, the formulae do provide insight into the types of empirical formulae that can be fitted to the measured data. Moderately good agreement between the theoretical and empirical models gives added confidence to the validity of the formulae derived from the measured data.

7.7.2 Dimensional Analysis

The determination of a model of stride length as a function of measured data derived from the accelerometer data is difficult due to the complexities of human motion. In such cases, an alternative approach is to perform a dimensional analysis on the data and parameters that could affect the stride length model. While the stride length could be affected by many parameters, the main ones (somewhat restricted by what can be measured) are as follows:

1. The peak-to-peak acceleration amplitude (A_p) measured by a vertically mounted accelerometer.
2. The stride rate (R).
3. The height of the person (H).

The first two variables can be directly measured from the accelerometer data, while the third parameter provides some scaling on the size of the human body, where presumably the walking style is somewhat related to the size of the body. Other parameters that may influence the stride length model include the leg length, sex, mass, age and physical fitness, but these extra parameters would probably over-complicate the model without much benefit. See Sect. 7.7.3 for more information on the relationship of body size with sex. Thus in the proposed model these secondary parameters are all summarized in one parameter, namely the height of the person.

From the above discussion, the basic problem can be summarized as finding a suitable function for the stride length (S)

$$S = f(H, R, A_p) \tag{7.56}$$

For dimensional analysis the function is assumed to be of the form

$$S = KH^a R^b A_p^c \tag{7.57}$$

where a, b, c and K are constants to be determined. In this model, the parameters are only functions of length (L) and time (T). Equating the powers of these fundamental dimensions results in the following equations

$$\begin{aligned} L:&\quad a + c = 1 \\ T:&\quad -b - 2c = 0 \end{aligned} \tag{7.58}$$

Thus there are two equations and three unknowns, so that the solution must be expressed in terms of one of the power constants. While the choice is arbitrary, the choice of using c has some physical merit, as will be shown. Thus (7.58) can be expressed as

$$\begin{aligned} a &= 1 - c \\ b &= -2c \end{aligned} \tag{7.59}$$

Using the results of (7.59), the stride length function becomes

$$\begin{aligned} S &= KH^{1-c} R^{-2c} A_p^c \\ \left[\frac{S}{H}\right] &= \hat{S} = K\left[\frac{A_p}{HR^2}\right]^c \end{aligned} \tag{7.60}$$

The term in brackets on the right side of (7.60) will be referred to as the "Stride Number" N_s, as it can be observed to be a dimensionless number, as is the normalized stride length $\hat{S}$. A more general hypothetical function would then be of the form

$$\hat{S} = g(N_s) \tag{7.61}$$

Notice that N_s can be determined from known or measured data, and thus once the stride length function $g(N_s)$ is known, the stride length can be estimated. The dimensional analysis does not provide any information on the specific function, but rather shows how the data should be organized to achieve a "universal" function coupling the more basic parameters. The hope is that all types of walking for all people can be described by the one function. The actual function is determined by the analysis of actual measured data, where $\hat{S}$ is plotted as a function of N_s.

Table 7.1 Data on the dimensions of the human body, as defined in US MIL STD 1472D. The footwear is assumed to be 3 cm

Parameter	Male	Female
Average height	174 cm	163 cm
Standard deviation of height	7.0 cm	4.7 cm
Average leg length	84 cm	76 cm
Standard deviation of leg	4.6 cm	4.8 cm
Leg-to-height ratio	0.482	0.465
Leg + footwear-to-height ratio	0.500	0.485

7.7.3 Data on the Human Body

This section provides summary data on the human body that is relevant to the determination of the stride length during walking. The stride length will depend on the size and shape of the human body, so that some data are required so that the variations can be assessed. The primary parameter of interest in walking is probably the leg length, but generally people do not know their leg length, but rather their height. Thus as an alternative, the body height could be used plus the leg-to-height ratio. These data are expected to be different for males and females. Table 7.1 shows the average data as defined in the US military standard 1472D (MIL-STD-14720 1989). As can be observed, although the heights and leg lengths are quite different, the ratio of these lengths are approximately the same, although males have a slightly longer leg length in proportion to their height. The data in the table are measured at the crutch rather than the hip, and thus does not accurately represent the leg length which is the desired parameter for the model.

The leg length from the hip (and including any footwear) was measured for the people used in the measurements described in Sect. 7.7.5 following. These data indicated a mean leg-to-height ratio of 0.539 for men and 0.530 for women with very little variation in both cases. However, these ratios are considerably greater than those specified in Table 7.1, but does include footwear; this may suggest that the people participated in the testing were different from the American military personnel, or that there have been some changes in recent years regarding body proportions.

Thus given a person's height a reasonable indication of the leg length can be derived, as required in the model. For the stride length algorithms, the leg-to-height ratio will be assumed to be a constant value of $\alpha = 0.535$. Although this includes footwear, this would be the correct parameter in nearly all practical cases where people being tracked have footwear.

7.7.4 Simple Dynamical Model of Walking

Before considering the measured data for determining the required stride length function, this section develops a simple dynamical model of walking to establish a theoretical framework for the empirical models. The model is based on a simplified

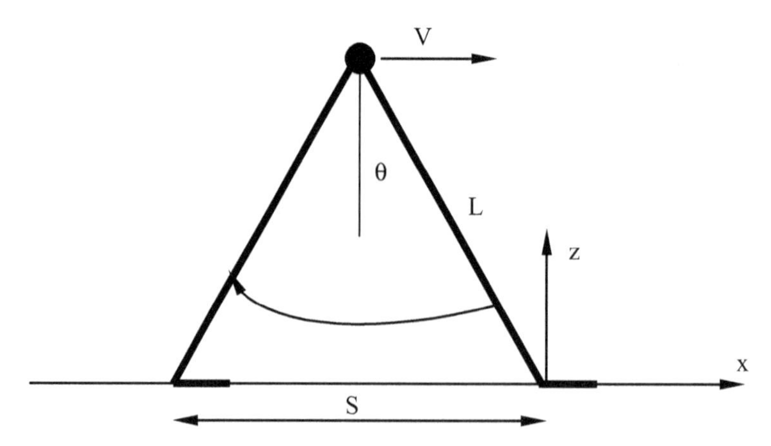

Fig. 7.29 Geometry of walking. The walking motion swings the leg through an angle from +θ to −θ, and as a result of the stiff leg the pivot point moves up and down in the *z*-direction

walking style. During a stride it is assumed that the leg providing thrust remains straight, while the second leg is off the ground in a motion preparing for another stride. As only the leg on the ground can give any (significant) vertical thrust (there is some inertial thrust from the leg off the ground), the vertical acceleration can be based solely on the geometry of the motion of the leg touching the ground. This geometry is shown in Fig. 7.29, where L is the length of the leg and S is the stride length. With the x-z coordinate reference point at the position of the foot at the start of the step, the position of the pivot point (hip) on the body trunk is given by

$$\begin{aligned} x &= -L \sin\theta \\ z &= L\cos\theta \end{aligned} \tag{7.62}$$

The dynamics of the pivot point will be very similar to that of the body trunk where it is assumed the accelerometer is attached. Further, it is assumed that the body moves with a constant speed (V) during the walking process, so that

$$\begin{aligned} V &= \frac{\partial x}{\partial t} = -L\cos\theta\,\dot{\theta} \\ V_z &= \frac{\partial z}{\partial t} = -L\sin\theta\,\dot{\theta} = -L\sin\theta\left[\frac{-V}{L\cos\theta}\right] = V\tan\theta \end{aligned} \tag{7.63}$$

The vertical acceleration can be calculated by a further differentiation, resulting in

$$A_z = \frac{\partial^2 z}{\partial t^2} = V\sec^2\theta\,\dot{\theta} = -\frac{V^2\sec^3\theta}{L} = -\frac{S^2R^2}{L\cos^3\theta} \tag{7.64}$$

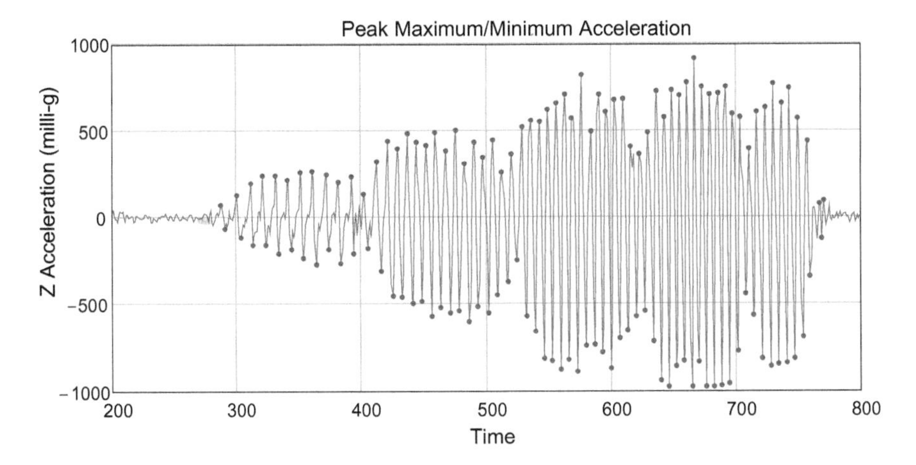

Fig. 7.30 Measured acceleration showing the symmetrical variation of the accelerometer data about the origin. Note that the origin will be zero due to the highpass filtering of the accelerometer data

The acceleration that can be measured most easily is the peak-to-peak acceleration during a stride. The above equations describe the acceleration after the initial impulse when the foot hits the ground. Because of the inertia of the body, this impulse is effectively lowpass filtered by the body in a manner that is difficult to analyze. However, it is known that the average vertical acceleration is zero, so that to the first order of approximation the positive and negative acceleration will be the same. This effect is shown in Fig. 7.30 for actual measured data.

The peak acceleration will occur at the start of the stride when the angle is $\theta = \theta_0$. To a first order of approximation this angle can be set to the mean value (determined below) for men and women, with typical values being 28° for men and 25° for women. The exact value is not too critical as in (7.64) the cosine of this value is used, and thus for small angles there is not much variation.

The differential peak-to-peak acceleration is thus given by

$$A_p = 2A_z(\theta_0) = \frac{2S^2R^2}{L\cos^3\theta_0} \tag{7.65}$$

From the dimensional analysis it is useful to express the result in normalized form, resulting in the expression

$$\begin{aligned} \hat{S} &= K\sqrt{N_s} \\ K &= \sqrt{\frac{\alpha\cos^3\theta_0}{2}} \qquad L = \alpha H \end{aligned} \tag{7.66}$$

Thus as predicted by the dimensional analysis the (normalized) stride length is a function of the Stride Number. The parameter α relating the length of the leg to the height of the person is approximately a constant equal to 0.535 for human bodies, as shown in Sect. 7.7.3. The typical value of K is calculated to be 0.431 for men and 0.455 for women, using the data described previously. Thus to this first order of approximation the algorithm K parameter is nearly constant, independent of men and women. Notice also that the dynamics equation is in the same form as described in Sect. 7.7.2, although the dimensional analysis had no knowledge of the particular dynamics of walking.

Although not of direct interest in determining the stride length, the above model can be rearranged to provide information on the stride angle of the leg. Noting that the stride angle is given by

$$\theta = \sin^{-1}\left(\frac{S/2}{L}\right) \tag{7.67}$$

and using Eq. (7.66) the stride angle is given by

$$\theta = \sin^{-1}\left(\frac{K}{2\alpha}\sqrt{N_s}\right) = \sin^{-1}(0.415\sqrt{N_s}) \tag{7.68}$$

Thus the stride angle is directly related to the stride number (according to this mathematical model). As will be shown in Sect. 7.7.5 following the typical value of the stride number for "normal" walking is 1.3 for men and 1.0 for women, so that the corresponding stride angles are 28° and 25° respectively. As will be shown, the model derived from the measured data (see Sect. 7.7.5) is closely similar to the theoretical model, but the K constant is about 10% smaller, resulting in stride angles also about 10% smaller.

Finally, it is observed that the determination of the stride length is equivalent to determining the walking speed, as the stride length combined with the stride rate gives the walking speed, namely

$$V = SR = K\sqrt{HA_p} \tag{7.69}$$

Thus the (horizontal) walking speed is solely a function of the peak vertical acceleration (with the person's height as a parameter), and is *independent* of the stride rate. The stride rate is difficult to measure accurately on a step-by-step basis, whereas the accelerometer peaks are easy to measure. However, the peak accelerometer readings exhibit considerable noise, as the impulse associated with the foot hitting the ground cannot be measured accurately with relatively low data sampling rates. Thus speed estimates will exhibit considerable noise, and would require some form of filtering, probably using a Kalman filter.

7.7.5 Measurements

7.7.5.1 Overview

To determine the relationship between the measured acceleration, the stride rate and the height of the person, a series of experiments were performed on a variety of people. The basic experiment consisted of the subjects walking a straight line path of 27 m in length at three different speeds, namely "slow", "normal" and "fast"; the definition of these terms was left to each particular person to interpret, as the aim was to obtain test data over a variety of conditions rather than defining specific walking styles. The walkers were instructed to walk at a constant rate (slow, normal or fast) over the length of the track, but in practice some variations in speed and style can be expected during each test. The data measured were the vertical acceleration which were processed to obtain the peak-to-peak acceleration and the stride rate. From the number of steps and the path length the average stride length and stride rate could be determined. Assuming that the subject walked at a constant rate, then the (average) stride length could be expressed as follows

$$\bar{S} = f(H, \bar{R}, \bar{A}_p) \tag{7.70}$$

Thus for each experiment one data point of the function could be determined. However, care must be taken in the averaging process if there is significant variation in the stride rate and the peak acceleration.

The stride rate data for one particular person is shown in Fig. 7.31. As can be observed, the stride rate is difficult to accurately measure on a step-by-step basis, so that averaging is required. As the stride rate was usually fairly consistent, a simple average is adequate for input to Eq. (7.70). Note however that the number of steps in each test was about 35, and the accurate counting of steps is difficult (the first/last step are quite difficult to detect accurately), so that the averaged data has an accuracy of around $\pm 3\%$.

The peak-to-peak acceleration corresponding to the stride data in Fig. 7.31 is shown in Fig. 7.32. The acceleration data show random variations about a constant value for the "slow" and "normal" cases, but for the "fast" case there is a very noticeable drop-off in the acceleration as the testing progresses. This drop off in the acceleration is due to fatigue, and is related to physical fitness. The data showed that most people cannot maintain a "fast" walking gait for more than a few seconds. In such cases a simple averaging of the data is not satisfactory. To average the data correctly it is observed from the theoretical model in Sect. 7.7.4 that the stride length is proportional to the square-root of the acceleration, so that the averaging of the acceleration data should use the following process

$$\overline{A_p} = \left[\frac{1}{N} \sum_{n=1}^{N} \sqrt{A_{p_n}} \right]^2 \tag{7.71}$$

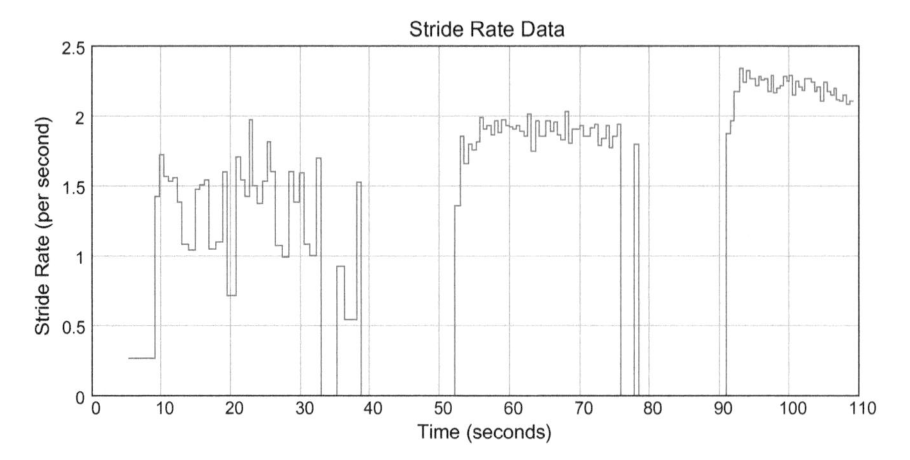

Fig. 7.31 Estimated stride rate showing the three different stride patterns (slow, normal and fast) with a pause in between measurements. The stride rate is difficult to measure accurately on an individual step-by-step basis. Note at the fastest stride rate there is evidence of fatigue, as the high stride rate could not be maintained even to a period as short as 20 s. Also note at the slowest stride rate the rate is quite variable; this result indicates that it is difficult to walk smoothly at a slower rate than the normal gait. The data are from a female of height 1.55 m

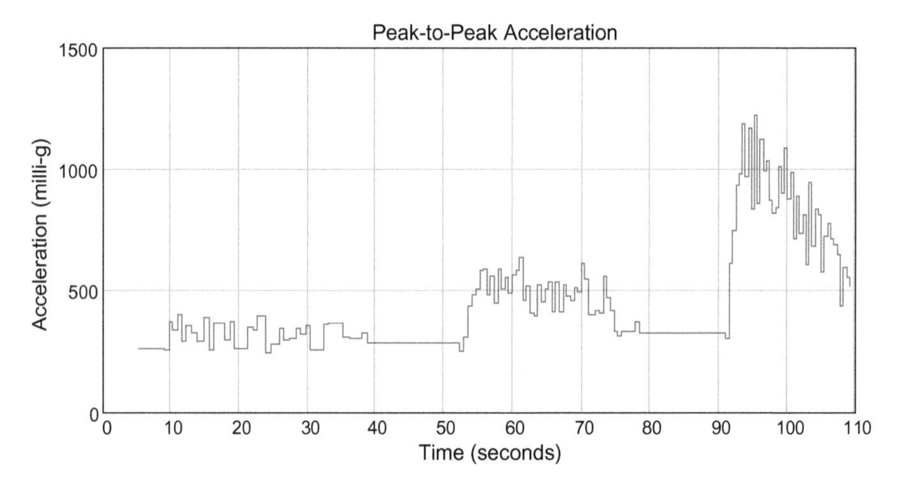

Fig. 7.32 Measured peak-to-peak acceleration for the three walking speeds, plotted for each step. There are no data between each run, so that the acceleration plotted between runs is an artifact of the plotting process. The data are from a female of height 1.55 m

This average acceleration can then be used in (7.70) to determine the required function. Details are given in the following subsections.

Methods of estimating walking speed using an accelerometer were presented in Sect. 7.7.4. The first method is to calculate the speed from the stride length and

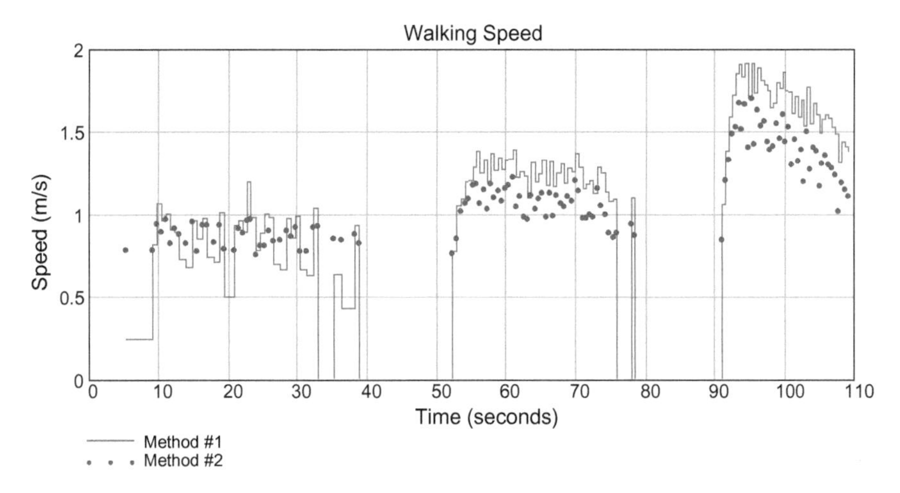

Fig. 7.33 Calculated walking speed for the same data as in Figs. 7.31 and 7.32. The results of the two different methods are shown with reasonably good agreement. The data are for a female of height 1.55 m

stride rate, while the second method calculates the speed from the peak accelerometer data directly (Eq. 7.68). An example of the results are shown in Fig. 7.33. At low walking speeds the second method appears to be superior, while at normal or fast walking speed the two methods appear to have similar performance but slightly different numerical results. The noise in the speed data is of the order of ±0.1 m/s.

7.7.5.2 Determining the Empirical Models

The empirical models of stride length as a function of the stride rate and the peak-to-peak acceleration can be determined by defining an appropriate equation with variables/parameters, namely the height of the person, stride rate and the peak vertical acceleration, together with fixed parameters to be determined by least-squares fitting to the measured data. Before considering several such models, it is appropriate to examine the general characteristics of the measured data.

The first characteristic considered is the stride length as a function of the stride rate; some examples are shown in Fig. 7.34. In general, the stride length increases with the stride rate, but there is some evidence that at very fast walking pace the stride length tends to decrease. The other noticeable observation from the plot is that while each person has a consistent functional relationship between the stride length and the stride rate, each person has a different functional relationship. As a result, stride *rate* is a poor predictor of stride *length*. This means that simple pedometers which can measure average stride rates accurately will result in poor accuracy in determining the distance traveled if the speed of walking varies considerably.

Fig. 7.34 Stride length as a function of the stride rate for three different people. The stride length increases with the stride rate for each person, but each person has a different functional relationship. In particular, male and female data have different characteristics (diamond, triangle: male, square: female). In each case the trendline model fitted to the data is a second-order polynomial

The second characteristic considered is the stride length as a function of the peak-to-peak acceleration. An example from the same experiments in Fig. 7.33 is shown in Fig. 7.35. The data show that the male data agree quite well with one another, but the female data exhibit a quite different characteristic. The general characteristic is that the stride length increases slowly with the peak acceleration. Thus accelerometer peak-to-peak data provides a much better estimate of stride length than does the stride rate, but there is still considerable variation from person to person. Clearly there is not a simple relationship between the stride length and the stride rate. Also plotted are power-law fits to the data, as well as the model proposed by Analog Devices, as explained in Sect. 7.7.5.3 following. Again it is evident that there is variability between individuals, and in particular between male and female. It is also noted that the algorithm proposed by Analog Devices does not fit the data very well.

7.7.5.3 Analog Devices Model

A model for calculating the stride length from accelerometer data was proposed by the Analog Devices company (manufacturer of accelerometers) in the Application Note AN–602 "Using the ADXL202 in Pedometer and Personal Navigation Applications". The algorithm described is an empirical match of the stride length data to the measured peak-to-peak accelerometer data, and is given by

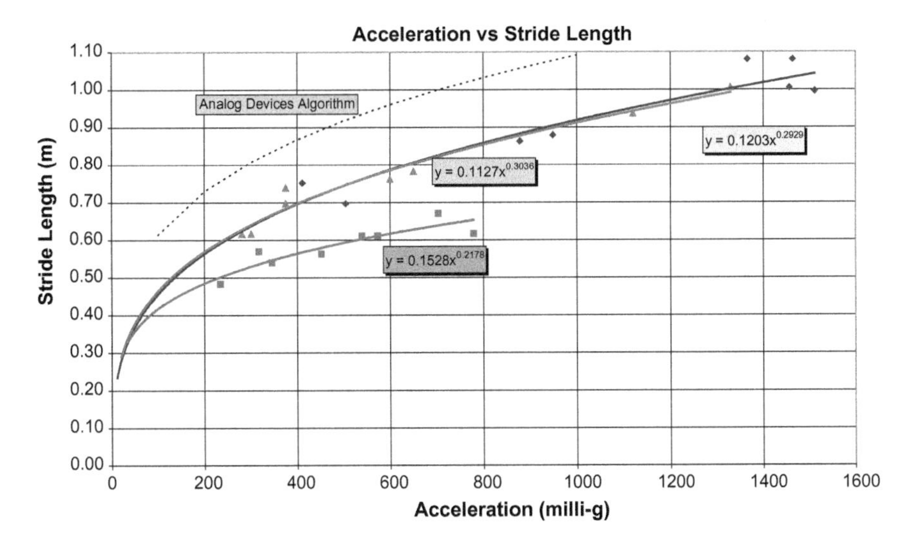

Fig. 7.35 Stride length as a function of the peak-to-peak acceleration. The models are based on a power law as suggested by the Analog Devices application note. The male and female (diamond, triangle: male, square: female) relationships are quite different, but there is good agreement between the male data

$$S = K\sqrt[4]{A_p} \tag{7.72}$$

where A_p is measurement in milli-g, and $K = 0.194$, with the stride length in meters. The claimed accuracy of the stride length is ±8%, presumably the standard deviation in the stride length error relative to the true stride length. Details of the measurements are not given, but it is claimed the measurements were taken over a variety of subjects and leg lengths. The performance of the AD algorithm was checked to confirm the performance. Some typical data are shown plotted in Fig. 7.35. The measured and Analog Devices algorithm data are not in very good agreement. However, the general shape of the data suggests a power law of the form

$$S = KA_p^a \tag{7.73}$$

where K and a are determined from a least-squares fit to the data. As shown in the Fig. 7.35 a power law can fit quite well to individual data sets, but the overall fit is poor between different people. The data for men are in quite close agreement, but the data set for the woman is quite different. When the data were fitted using the Analog Devices algorithm, but with a least-squares fit for the K parameter only over the complete data set (see Sect. 7.7.5.4 for more details), the resulting fit was $K = 0.147$ m, with a RMS stride length error of 8.0 cm or 11.8%. Some further improvement can be made if a least squares fit to (7.73) is used. The results of the least-squares fitting are $K = 0.104$ and $a = 0.302$ (or very nearly a cube-root function). The corresponding RMS errors are 7.8 cm or 11.2%. Thus there is only a marginal improvement over the AD quartic-root results.

From these results it appears that the AD algorithm was based on a limited data set and walking styles, so that the performance is significantly worst than that claimed.

7.7.5.4 Dimensional Analysis Model

Based on the dimensional analysis the normalized stride length is expected to be a function of the stride number N_s. From the analytical analysis, the proposed model is

$$\hat{S} = KN_s^p \tag{7.74}$$

By fitting the measured data to this model the parameters can be determined to be $K = 0.394$ and $p = 0.485$. This model can be compared with the analytical model, where $p = 0.5$, and $K = 0.43$ for women and $K = 0.45$ for men. Thus the measured data agrees rather well with the model parameters.

The data for all the people tested are shown plotted in Fig. 7.36 together with the model curve given by (7.74). The data for each person are individually plotted. As can be observed the data show some scatter about the least-squares fit curve, but the

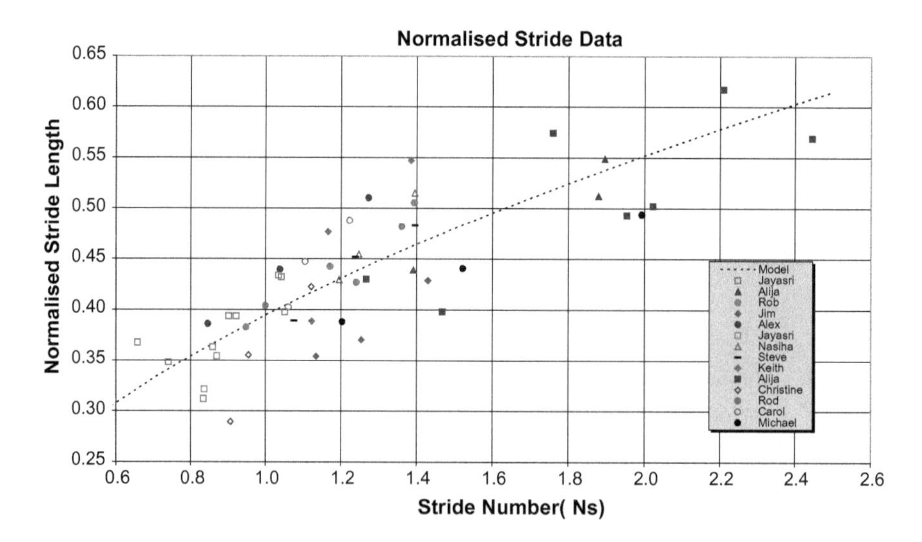

Fig. 7.36 Plot of the measured stride length (normalized by the height of the person) against the Stride Number N_s. Note that the data includes the three different walking rates (slow, normal, fast), and includes both men and women

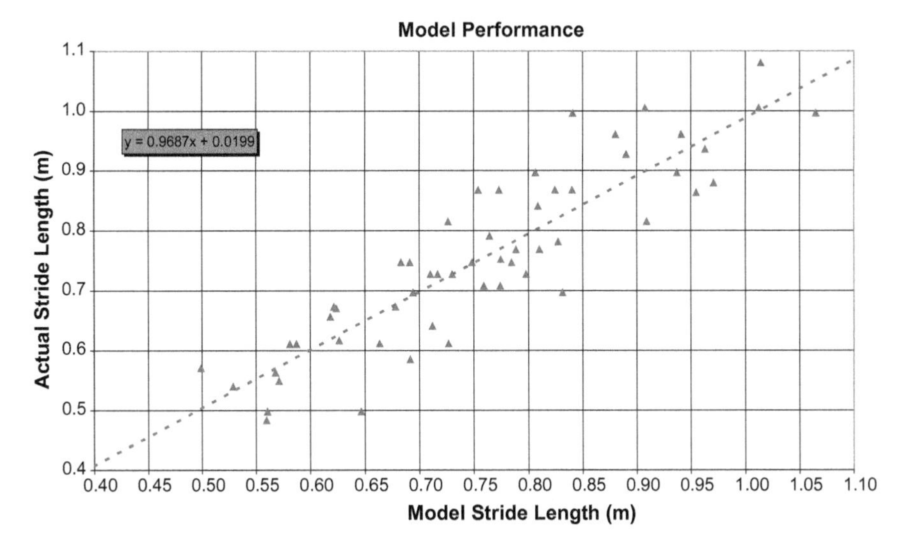

Fig. 7.37 Plot of the model stride length versus the actual stride length. The dotted line is the least-squares fit straight line. Ideally the equation should be $y = x$, so the least-squares fit is quite close to this ideal result

normalized stride length is typically within ±0.05 of the model estimate. The model stride length is plotted against the actual stride length in Fig. 7.37. The standard deviation in the error in the stride model is 7.6 cm.

7.7.5.5 Generalized Model

The dimensional analysis provides a suggested generalized model for the stride length (see Eq. 7.56) of the form

$$S = KH^a R^b A_p^c \tag{7.75}$$

This equation can be used for the least-squares fitting to the data to determine the four parameters of the model. The result is: $K = 0.259$, $a = 1.029$, $b = -0.278$ and $c = 0.340$. While this model has no particular theoretical merit, the resulting performance of the Generalized Model is superior to the Dimensional Analysis model described in Sect. 7.7.2 above (compare Figs. 7.37 and 7.38). The comparison of the model and actual stride lengths is shown in Fig. 7.38, where the RMS error is 4.7 cm (compared with 7.6 cm for the Dimensional Analysis model). The actual parameters of the model are quite sensitive to the measured data, suggesting there is no underlying theoretical model. However, the least-squares fit parameters suggest an approximate model of the form

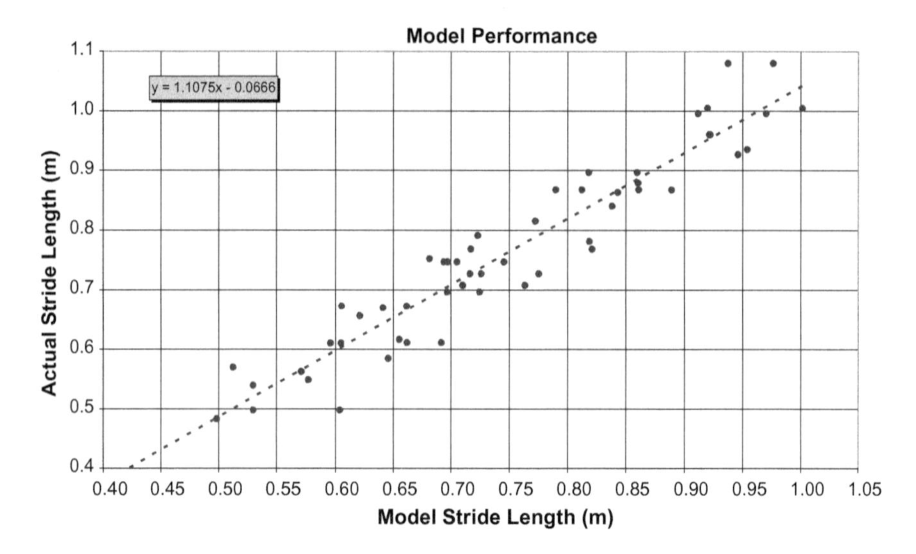

Fig. 7.38 Plot of the model stride length versus the actual stride length. The dotted line is the least-squares fit straight line

$$\hat{S} \approx K \left[\frac{A_p}{R} \right]^p = K_1 V_z^p \tag{7.76}$$

where V_z is the peak vertical speed. Thus this model suggests that the normalized stride length is related directly to the peak vertical speed which can be estimated by the integration of the vertical (z) accelerometer. However, as the estimation of the vertical speed from the vertical acceleration is also subject to measurement errors associated with the integration of the accelerometer data, this form of model is not recommended for practical implementation.

7.7.5.6 Summary of Model Performances

The previous subsections provided several different models for estimating the stride length. This subsection summarizes their performances in Table 7.2. The Analog Devices algorithm uses only the accelerometer data and one scaling parameter, and has the worst performance of all the algorithms. The Power algorithm is similar to the Analog Devices algorithm, but has two parameters, the scaling parameter and the power parameter. However, the performance is only marginally superior to the Analog Devices algorithm. The algorithm based on the dimensional analysis uses the stride rate and the person's height as well as the peak accelerometers data. The algorithm has two parameters to fit to the measured data, and results in a modest improvement in the performance. The Generalized algorithm is based on a general fitting of the data used in the dimensional analysis algorithm, and has four

Table 7.2 Summary of stride length algorithms and their performances

Model	RMS error (cm)	RMS error (%)
Analog devices	8.0	11.8
Power	7.8	11.2
Dimensional	7.6	9.4
General	4.7	7.8

parameters to fit to the measured data. The Generalized algorithm has a significant improvement over the other three algorithms, but is more difficult to implement using the measured accelerometer data.

These results suggest that the vertical accelerometer data can be effectively used in a dead reckoning system for people walking. If we assume an accuracy of 10%, a stride length of 0.8 m, a stride rate of 1.8 per second, then with an error budget of 2 meters the distance traveled before exceeding the error budget is 20 m, or 25 strides or 14 s elapsed time. This analysis suggests that an indoor positioning system based on stride length determination by the methods described in this section will require updates (say from a radiolocation system) typically at least every 20 m.

References

Brookner E (1998) Tracking and Kalman filtering made easy. Wiley

MIL-STD-14720 (1989) Human engineering design criteria for military systems, equipment, and facilities, Mar 1989

Pascoal A, Kaminer I, Oliveira P (2000) Navigation system design using time-varying complementary filters. IEEE Trans Aerosp Electron Syst 35(4)

Weinberg H (2002) Using the ADXL202 in pedometer and personal navigation applications. Application Note AN-602, Analog Devices

Yu K, Sharp I, Guo YJ (2009) Ground-based wireless positioning. Wiley-IEEE Press

Chapter 8
Indoor WiFi Positioning

8.1 Introduction

This chapter studies WiFi based wireless positioning in complex indoor environments. WiFi positioning makes use of the infrastructure (WiFi access points) widely deployed in indoor environments such as office buildings, teaching buildings, hospitals, and shopping centers. It is also a fact that WiFi technology has been adopted in billions of electronic devices such as smartphones. Although WiFi positioning is cost-effective, it suffers the drawback of low positioning accuracy and hence innovative techniques are required to enhance WiFi positioning accuracy. This chapter mainly presents several recently proposed methods for improving WiFi positioning performance.

Wireless positioning is important for a wide range of civilian and military services and applications. In an outdoor environment, the Global Navigation Satellite System (GNSS), including U.S. Global Positioning System (GPS), Russia's GLONASS, China's BeiDou Navigation Satellite System (BDS), and EU's Galileo, can provide reliable positioning, navigation, and timing services. However, in an indoor environment including underground tunnels and mines, the GNSS signals are blocked by building structures and other obstacles, so that a GNSS receiver is typically not able to provide valid position information. Hence, a local positioning system is required to locate and track the targets of interest or provide navigation services for such as pedestrians and mobile robots.

Over the past few decades, many different systems and approaches have been proposed for indoor positioning, including radio based positioning, inertial sensors based positioning, visible light based positioning, and magnetic field based positioning. Among these technologies, radio positioning is probably the most studied technology. This may be related to the fact that the early navigation systems and the current four global navigation satellite constellations all make use of radio signals, as well as the fact that wireless communication is often an integrated part of modern devices such as smart phones. Radio positioning exploits the measurement of signal

I. Sharp and K. Yu, *Wireless Positioning: Principles and Practice*, Navigation: Science and Technology, https://doi.org/10.1007/978-981-10-8791-2_8

parameters and utilizes either parametric or non-parametric approaches for position determination. The parameters include signal strength, time-of-arrival, angle-of-arrival, frequency difference of arrival, and carrier phase.

This chapter focuses on the use of the signals transmitted from access points (APs) and received by devices such as a smart phone carried by a pedestrian for positioning in indoor environments. Such devices incorporate hardware and software components for the measurements of the received signal strength (RSS) or RSS indicator (RSSI), and provides application programmer interfaces (APIs) that allow the development of applications (Apps) for various location services. Thus such a positioning technique does not require any additional infrastructure or hardware modification, only software is needed to collect RSS data and algorithms for position determination.

8.2 Access Point Selection

In a typical office building a smart phone may receive signals from many access points, ten or more being common in modern office buildings. Although it is usually true that the use of RSS measurements from more APs may result in better positioning accuracy, the computational overhead will increase accordingly and hence energy consumption will increase. To reduce computational complexity and energy consumption in user mobile device, it is appropriate to choose a subset of the APs instead of all the available APs for position determination. The purpose is to ignore some access points which have poor signal quality, so that the computational burden is reduced significantly without significantly degrading the positioning accuracy. In the following subsections three AP selection schemes are introduced, namely the maximum RSS mean (MM) method (Youssef et al. 2003), information gain based method (Chen et al. 2006; Deng et al. 2011), and mutual information (MI) based method (Zou et al. 2015). Further methods of RSS positioning are described in Chap. 15, which also provides extensive measurements, both static and mobile, of the performance and signal characteristics in an office-type environment.

8.2.1 *MM Method*

The MM method selects the APs which have the strongest mean RSS values received at mobile devices. It is based on the consideration that the strong RSS is also associated with higher signal-to-noise ratios, and hence the reduced measurement noise, and thus more accurate position estimates can be produced. The implementation of this selection method is simple and it can be readily carried out online. This method has been implemented in the clustering and RSS distribution based location system developed by Youssef et al. (2003).

8.2.2 Information Gain Based Method

Suppose that the location coverage area is divided into n grids and there are totally m detectable APs. The centers of the grids may be treated as the reference or calibration points. In this case, the calibration points will be uniformly distributed. In some cases, the calibration points are not exactly selected in such a grid-based way, and they may not be uniformly distributed for simplicity and efficiency. The ith AP is denoted AP_i and the jth calibration point is denoted G_j. In the offline phase where signal data at calibration points are collected and stored in a database, the signal strength from AP_i and received at G_j is averaged, and taken as the value of the ith feature of G_j. In the case where an AP is not detected at a calibration point, then the feature of the missing AP takes on a default value which is the minimum signal strength that can be received in the environment. The *information gain* criterion for AP selection proposed in (Chen et al. 2006) is to choose those APs which have the highest discriminative power. The discriminative power of an AP_i is measured by information gain (InfoGain for short) which is defined as

$$InfoGain(AP_i) = H(G) - H(G|AP_i) \tag{8.1}$$

where $H(G)$ is the entropy of the grids when the feature value of AP_i is not known, and is defined as

$$H(G) = -\sum_{j=1}^{n} \Pr(G_j) \log \Pr(G_j) \tag{8.2}$$

where $\Pr(G_j)$ is the prior probability of G_j, which is uniformly distributed if a user is equally likely anywhere in the grid. $H(G|AP_i)$ is the conditional entropy of grids given the feature value of AP_i, defined as

$$H(G|AP_i) = -\sum_{v} \sum_{j=1}^{n} \Pr(G_j, AP_i = v) \log \Pr(G_j|AP_i = v) \tag{8.3}$$

where the conditional probability $\Pr(G_j|AP_i = v)$ can be calculated by

$$\Pr(G_j|AP_i = v) = \frac{\Pr(AP_i = v|G_j)\Pr(G_j)}{\Pr(AP_i = v)} \tag{8.4}$$

Here v is one possible value of signal strength received from AP_i and the summation is taken over all possible values of signal strength of AP_i. After the information gain is calculated for all APs, the AP data are sorted in a descending order based on their InfoGain values and the top APs with the highest InforGain values are selected for online positioning.

The InfoGain criterion deals with the discriminative ability of each AP independently, while (Deng et al. 2011) considers the discriminative ability of a group of APs jointly, using a method termed the *joint information gain* (JIG) method. The feature values of APs can be correlated so that a joint treatment could produce a performance gain. Suppose that a subset of k APs (AP_1, AP_2, …, AP_k) are used for positioning, and hence there are totally C_m^k combinations of subsets of APs. The joint location information gain of the jth combination is defined as

$$\begin{aligned} InfoGain(AP_1, AP_2, \ldots, AP_k) &= H(G) - H(G|AP_1, AP_2, \ldots, AP_k) \\ &= H(G) + \sum_{v_1} \cdots \sum_{v_k} \sum_{j=1}^{n} P_b P_c \end{aligned} \tag{8.5}$$

where

$$\begin{aligned} P_b &= \Pr(G_j, AP_1 = v_1, \ldots, AP_k = v_k) \\ P_c &= \log \Pr(G_j | AP_1 = v_1, \ldots, AP_k = v_k) \end{aligned} \tag{8.6}$$

The subset of APs with the highest joint information gain values is selected for online positioning. Clearly, the computational complexity is increased significantly by the above procedure, but this computation is carried out and the AP selection is completed during offline phase, so that mobile device does not have any energy consumption increase during online real-time position determination phase.

8.2.3 Mutual Information Based Method

Suppose that m APs have been deployed in an indoor environment. At some locations, the mobile device may detect all the APs, while at other locations, the mobile may not be able to detect some of the APs. In this case, the minimum detectable RSS value is assigned to those undetected APs as mentioned earlier. Thus an online RSS vector is formed as

$$\mathbf{r}_{ol} = \begin{bmatrix} RSS_{ol}^1 & RSS_{ol}^2 & \cdots & RSS_{ol}^m \end{bmatrix} \tag{8.7}$$

The objective is to select q APs with the best discriminative ability from all the m APs to reduce computational complexity and probably improve positioning accuracy. The first step of the *mutual information* (MI) based method (Zou et al. 2015) is to compute the mutual information between each pair of available APs (AP_a and AP_b) by

$$MI(AP_a, AP_b) = H(AP_a) + H(AP_b) - H(AP_a, AP_b) \tag{8.8}$$

where $H(AP_a)$ and $H(AP_b)$ are the entropies of AP_a and AP_b, respectively. $H(AP_a, AP_b)$ is the joint entropy of AP_a and AP_b, which is defined as

$$H(AP_a, AP_b) = \sum_{v_2} \sum_{v_1} \Pr(RSS_{ol}^a = v_1, RSS_{ol}^b = v_2) \times \log \Pr(RSS_{ol}^a = v_1, RSS_{ol}^b = v_2) \tag{8.9}$$

where v_1 and v_2 are the possible RSS values received from AP_a and AP_b, respectively, and $\Pr(RSS_{ol}^a = v_1, RSS_{ol}^b = v_2)$ is the joint probability. After computing the mutual information for all the pairs of APs, the pair of APs with the minimum mutual information is selected. Such a selection criterion is based on the observation that a smaller mutual information indicates less dependence between the two APs. The next steps are to select one more AP at each step so that the mutual information among the selected APs always remains the minimum. In the ℓth step, the AP selection is performed by

$$s_{\ell+1} = \arg \min_{z \notin \{s_1, \ldots, s_\ell\}} MI(AP_{s_1}, \ldots, AP_{s_\ell}, AP_z \tag{8.10}$$

where

$$MI(AP_{s_1}, \ldots, AP_{s_\ell}, AP_z) = H(AP_{s_1}, \ldots, AP_{s_\ell}) + H(AP_z) - H(AP_{s_1}, \ldots, AP_{s_\ell}, AP_z) \tag{8.11}$$

There are $m - \ell$ choices for AP_z, producing $m - \ell$ mutual information values. Since $H(AP_{s_1}, \ldots, AP_{s_\ell})$ was already obtained in the previous step, one only needs to calculate $H(AP_z)$ and $H(AP_{s_1}, \ldots, AP_{s_\ell}, AP_z)$ which is determined by

$$\begin{aligned} H(AP_{s_1}, \ldots, AP_{s_\ell}, AP_z) = \sum_{v_1, \ldots, v_\ell, v_z} & \Pr(RSS_{ol}^{s_1} = v_1, \ldots, RSS_{ol}^{s_\ell} = v_\ell, RSS_{ol}^{z} = v_z) \\ & \times \log \Pr(RSS_{ol}^{s_1} = v_1, \ldots, RSS_{ol}^{s_\ell} = v_\ell, RSS_{ol}^{z} = v_z) \end{aligned} \tag{8.12}$$

where $v_1, \ldots, v_\ell$ and v_z are the possible RSS values received from $AP_{s_1}, \ldots, AP_{s_\ell}$ and AP_z, respectively, and $\Pr(RSS_{ol}^{s_1} = v_1, \ldots, RSS_{ol}^{s_\ell} = v_\ell, RSS_{ol}^{z} = v_z)$ is the joint probability. Since this is an online selection scheme, the issue of mismatch between offline and online measurements due to environmental variation can be avoided.

8.2.4 Performance Comparison Among Different Selection Methods

In this section experimental results are used to evaluate the performance of the different AP selection methods (Hua et al. 2017). Figure 8.1 shows a rectangular positioning area with dimensions of 8 m × 8 m in an office building, where six

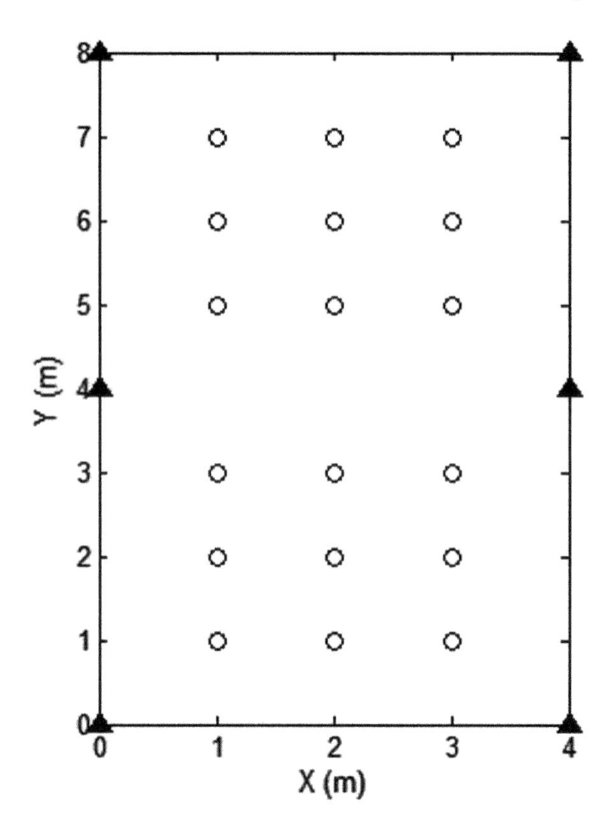

Fig. 8.1 Deployment of calibration points (triangles) and target points (circles) in the positioning area used for testing

calibration points and 18 target points are selected. The weighted k–nearest neighbor (WKNN) algorithm, which is discussed in Sect. 8.4, is used for target position estimation. In the experiment, a smart phone (Millet) is used for receiving the AP signals, and all detectable APs in the building are considered. Four different performance indexes are used to evaluate the radial position error (distance between the estimated and true mobile phone positions): maximum error (MAXE), root mean square (RMS) error, mean error (ME), and error standard deviation (STD). The 50 RSS samples are used to produce 50 position estimates. Table 8.1 shows the comparison of the error statistics between three different selection methods (JIG, MI, and MM) when the number of APs ranges between four and eight. It can be observed that MI and MM methods outperform JIG considerably in terms of both mean and RMS errors, indicating that online real-time selection of APs can achieve better performance. The MI selection method only performs slightly better than MM selection method, so that the simpler MM method is a good practical choice. Except for the case of four APs, the performance is not very sensitive to increased numbers of APs, so that five or six APs is sufficient for position determination.

The RMS errors in Table 8.1 show that the results for the three AP selection methods converges to values of 2 ± 0.1 m using WKNN positioning algorithm. This positioning method is based on a square grid of calibration points with a side

Table 8.1 Positional performance comparison between three different AP selection methods

Selection method	Number of APs	MAXE (m)	ME (m)	RMS (m)	STD (m)
JIG	4	3.90	2.49	2.61	0.81
	5	3.59	2.17	2.27	0.71
	6	3.57	2.11	2.27	0.85
	7	4.36	2.13	2.36	1.05
	8	3.28	1.96	2.11	0.82
MI	4	3.99	1.74	1.99	1.00
	5	3.44	1.78	1.99	0.90
	6	3.07	1.74	1.92	0.84
	7	3.11	1.74	1.91	0.81
	8	3.18	1.75	1.91	0.79
MM	4	2.97	1.93	2.05	0.69
	5	2.90	1.84	1.98	0.74
	6	2.91	1.81	1.94	0.73
	7	2.87	1.84	1.97	0.75
	8	2.89	1.83	1.96	0.75

length $S = 4$ m. Thus these results show that the RMS error is about $0.5S$. It is shown in Chap. 15, Eq. (15.13) that assuming a mobile device is randomly located within the square grid the position error based on RMS average of the ranges to the four corners of the square is given by $S/\sqrt{6} = 0.41S$. Thus the results in Table 8.1 are generally consistent with the theoretical expectation. However, these results were produced in one particular environmental area, so further experimentation is necessary in different indoor areas to confirm these conclusions.

8.3 RSSI Measurement Accuracy Improvement

In an indoor environment with walls and other structures, multipath propagation conditions usually prevail, causing significant fluctuation in the RSSI observed by a mobile device, which is also called fading. See Chap. 9, Sect. 9.5 for more details on the signal variability in a non-line-of-sight propagation environment. To deal with multipath fading to improve RSSI measurement accuracy, a number of different techniques have been proposed in the literature and two of these are studied in this section.

8.3.1 Mean Algorithm

The simplest way to mitigate the effects of multipath fading is to collect RSSI samples over a period of time, and the average of the samples is used as the RSSI value when the mobile device is either static or moving. In fact, such an averaging technique (or mean algorithm) has been widely used in wireless communications and wireless positioning. When selecting the time interval for averaging, the device speed should be taken into account. For instance, the distance traveled by the device during the time interval should be smaller than twice the required position accuracy.

The mean algorithm is effective in reducing multipath fading effect on RSSI measurement. However, RSSI is also affected by other factors. For instance, shadowing such as due to the presence of a pedestrian between transmitter and receiver especially close to either of them will considerably reduce the RSSI. In addition, the Wi-Fi radio channel is generally shared by different systems or devices such as various Bluetooth devices and microwave ovens, which use the same frequency band. These devices may produce co-channel interference if they are nearby and in operation. Such interference may also decrease RSSI considerably. That is, averaging may not be an effective way for mitigating the effect of shadowing and co-channel interference on RSSI.

8.3.2 *Mean Maximum Method*

As mentioned above, the RSSI is significantly affected by different factors, including multipath fading, shadowing, and interference. A strong signal would be mainly affected by fading, while the weak signal may be affected by one or more of these three factors. To reduce the effect of shadowing and co-channel interference, the weakest RSSI values should be excluded. Therefore, the maximum RSSI values should be used for localization purpose as proposed by Xue et al. (2017), which is termed *mean maximum* method for convenience. Since the maximum RSSI values are also typically affected by fading (or possibly constructive interference which actually increases the signal above the local mean), averaging can be used to reduce the effects of fading to achieve better localization performance. Averaging also produces fractional RSSI values instead of integer values provided by device, enabling smaller differential distance, which will be discussed later. The flowchart of the mean maximum method is shown in Fig. 8.2.

8.3.2.1 Path-Loss and Signal Strength Spatial Resolution

In free space the radio signal power attenuation or path-loss follows the fundamental formula

$$\mathrm{FSPL} = \left(\frac{4\pi d}{\lambda}\right)^2 = \left(\frac{4\pi f d}{c}\right)^2 \tag{8.13}$$

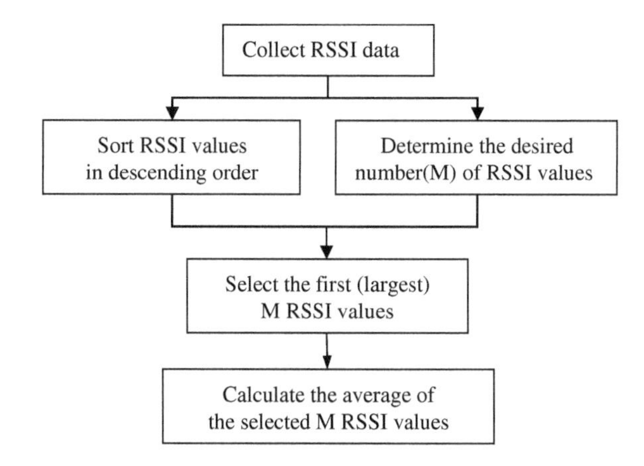

Fig. 8.2 Flowchart of the mean maximum method for the improved RSSI measurement

where d is the distance between the transmitter and receiver, f is the signal frequency, λ is the signal wavelength, and c is the speed of propagation. In a non-free space indoor environment, the signal path loss does not comply with (8.13), so the following empirical log-distance path-loss model is widely employed:

$$PL = P_t - P_r == PL_0 + 10\eta \log_{10}\left(\frac{d}{d_0}\right) + \varepsilon \tag{8.14}$$

where PL is the total path-loss in dB, P_t is the transmitted signal power, P_r is the received signal power at distance d, PL_0 is path-loss over the reference distance d_0, η is the path-loss exponent, and ε is the measurement noise. The noise ε can be either simply modeled as a Gaussian random variable or as a Laplace variable based on experimental results.

The parameter d_0 is set to be a specific value such as one meter for convenience. By ignoring the measurement noise, there are two unknown parameters PL_0 and η in (8.14) which can be determined using recorded signal strength data in experiments conducted in the positioning area of interest. For instance, the mobile device is moved from one position to another with known distances to the APs, and the RSSI recorded in a database. After calculating the path loss with a known effective transmitting power, a group of simultaneous linear path loss measurement are obtained. A simple but often effective linear least-squares technique is typically used to determine the parameters of PL_0 and η in (8.14). See Chap. 15, Sect. 15.3.2 for more information about this technique. This procedure is completed in the offline phase. In the online real-time or positioning phase, the unknown distance d is calculated by

$$d = d_0 10^{\frac{PL - PL_0}{10\eta}} \tag{8.15}$$

where the measurement noise is ignored, PL_0 and η are the known two model parameters, and PL is replaced with the real-time path loss measurement derived from the RSSI measurements in the mobile device.

Typically, the AP transmitter power can be treated as constant, and the AP antenna gain and the antenna gain in the mobile device are roughly uniformly distributed. Then, this calculation does not require information about these parameters. Otherwise, these data need to be extracted from an appropriate database that is generated from information which includes the locations of the APs in the building or buildings in a large coverage area.

The exponent η depends on both signal frequency and propagation environment. As reported in (Rappaport 2002), the exponent η ranges basically between 1.7 and 3.0 for nine different building types and signal frequency between 900 MHz and 60 GHz. Note that the value of η also needs to be included in the online data base, which can be accessed via the data capability of the WiFi network of APs in the building.

Now consider two position points which have distances d_i and d_j to a transmitter, respectively. From (8.15) the differential distance between the two points can be readily obtained as

$$\begin{aligned}\Delta d_{ij} &= d_i - d_j \\ &= d_0 10^{\frac{PL(d_i) - PL_0}{10\eta}} - d_0 10^{\frac{PL(d_j) - PL_0}{10\eta}}\end{aligned} \tag{8.16}$$

In a typical office building with rooms separated by concrete walls and corridors, the exponent η can be selected to be 3. Also it is convenient that d_0 is chosen to be 1 m and for this example calculation PL_0 is set at a typical value of 20 dB (Xue et al. 2017). Then Table 8.2 displays the relationship between a range of RSSI

Table 8.2 Relationship between RSSI, distance and differential distance

RSSI (dBm)	D (m)	Δd (m)	RSSI (dBm)	D (m)	Δd (m)
−100	464.159		−100	464.159	
−99	429.866	34.293	−99.9	460.610	3.549
−98	398.107	31.759	−99.8	457.088	3.522
…	…	…	…	…	…
−70	46.416	3.703	−71	50.119	0.386
−69	42.987	3.429	−70.9	49.736	0.383
…	…	…	…	…	…
−40	4.642	0.370	−44.5	6.556	0.051
−39	4.299	0.343	−44.4	6.506	0.050
…	…	…	…	…	…
−22	1.166	0.093	−20.2	1.015	0.008
−21	1.080	0.086	−20.1	1.008	0.008

values and the corresponding propagation distances. The differential distance between each pair of adjacent RSSI values is also listed. The left panel shows the integer RSSI from −100 to −21 dB with a 1 dB interval, while the right panel uses a RSSI interval of 0.1 dB. All the distances and the differential distances are calculated using (8.15) and (8.16) as well as the given RSSI values. Note that the RSSI provided by the system is an integer number, while the average of a number of RSSI observations can be a fraction.

As shown in Table 8.2, given the same RSSI difference, larger RSSI produces smaller differential distance, providing better spatial resolution. That is, to achieve better positioning accuracy, higher RSSI should be employed. Another observation is that differential RSSI of 0.1 dB produces much smaller differential distance (and hence much better spatial resolution) than differential RSSI of 1 dB. Therefore, it is an advantage to use average of a number of RSSI observations instead of individual RSSI observations for position determination.

8.3.2.2 Selection of Number of Maximum RSSI Measurements

As indicated in Fig. 8.2, one needs to determine how many maximum RSSI measurements should be used to achieve the best possible RSSI estimate associated with each access point. Intuitively, the RSSI quality would be better if the selected maximum RSSI measurements vary more smoothly from the largest to the smallest. That is, the curve smoothness index (Borel 1998) can be used as a measure to evaluate the quality of the RSSI measurements, which is defined as

$$S = \sum_{k=2}^{\ell-1} \sqrt{\left(RSSI_k - \frac{RSSI_{k-1} + RSSI_k + RSSI_{k+1}}{3} \right)^2} \tag{8.17}$$

where ℓ is the number of sample points on the curve, $RSSI_k$ is the mean of the M selected RSSI values at the kth position point. The smaller the value of S is, the smoother the signal RSSI curve is, and the better the quality of the RSSI data. Figure 8.3 shows an example of the curve smoothness index versus M associated with six APs when the RSSI data were collected as the mobile moved along a corridor in an office building (Xue et al. 2017) shown in Fig. 8.4, where the dotted line is the trajectory. As M goes from 5 to 20 m, the smoothness index of each access point does vary much, and the minimum value of the smoothness index can be easily determined. However, the smoothness index variation patterns of the six APs are significantly different from each other and do not have a consistent variation pattern. For instance, the minimum smoothness index of AP1 occurs at $M = 5$, while that of AP4 happens at $M = 20$. Thus it is not feasible to choose the best M based on the smoothness indices of the individual access points. To handle the problem, we may use the sum of the individual results as a measure, that is,

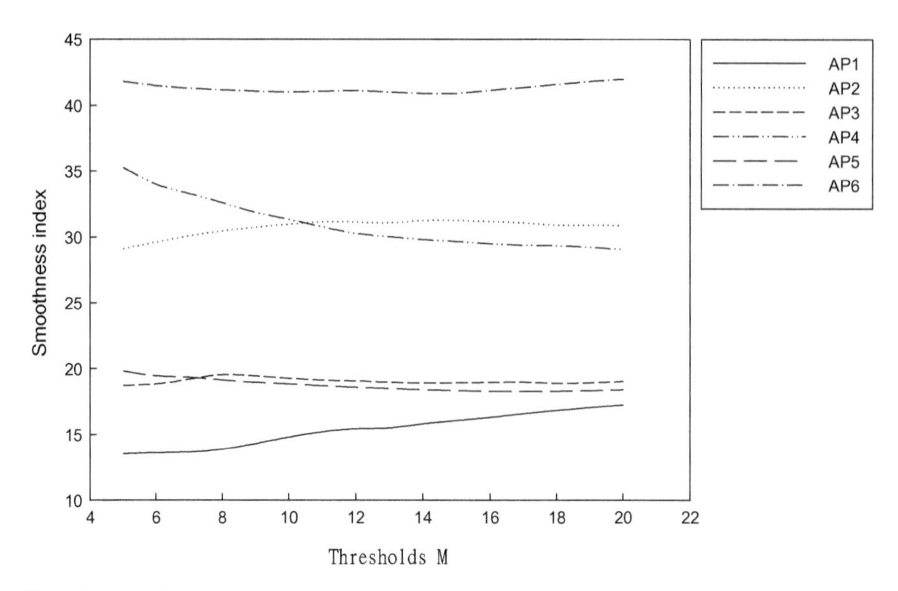

Fig. 8.3 The effect of M on smoothness index. The blue grids are just used as a background

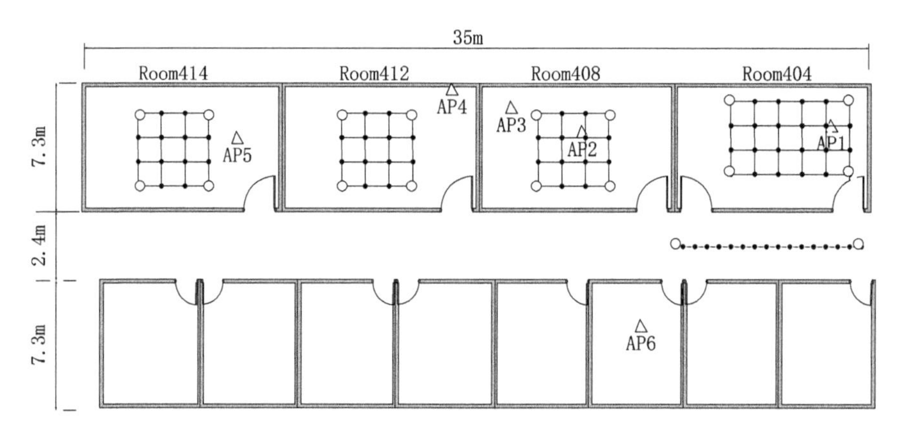

Fig. 8.4 The diagram of experimental site which is on the fourth floor of an office building at Wuhan University

$$S_{sum} = \sum_{i=1}^{N} S^{(i)} \tag{8.18}$$

where N is the number of access points and $S^{(i)}$ is the smoothness index of the ith AP calculated by (8.17). Figure 8.5 displays the summed smoothness index for 16 different M values. It can be seen from the figure that the minimum summed smoothness index occurs when $M = 13$, although the variation is not significant.

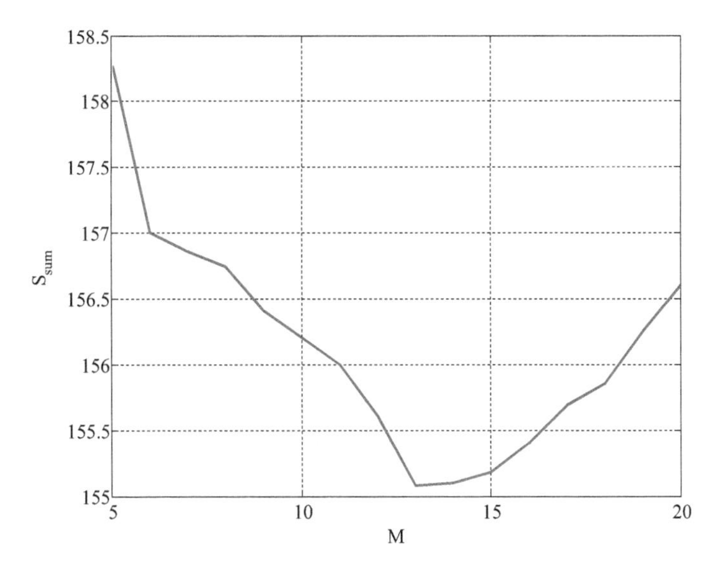

Fig. 8.5 Summed smoothness index

This selected M value is particularly suited for positioning in the corridor area, but it would be also suited for positioning in other areas such as rooms on the same floor, where the best M value might be slightly different.

In the literature there are a number of different techniques for handling the issue of RSSI variation caused by multipath fading and other factors. They include the widely used averaging technique mentioned earlier (Wang et al. 2012), Kalman filtering (Mansour 2014), and particle filtering (Chao et al. 2008). It is interesting to compare the performance of the different algorithms. Figure 8.6 shows the RSSI measurements in the corridor processed by the four algorithms using RSSI data associated with the six access points. The basic RSSI variation trend related to each access point is consistent with the rule that the signal attenuation is greater at longer propagation distance. The relatively small-scale variation is mainly caused by multipath fading and possibly body orientation variation. On average the variation associated with the method proposed in (Xue et al. 2017) has the smallest variation among the four algorithms. This is further verified by the calculated smoothness indexes as displayed in Table 8.3. It is difficult to see from the data related to individual access points which method has the smallest smoothness index. However, in terms of the summed smoothness index, the mean maximum method achieves the smallest variation.

Alternatively, as indicated in (Xue et al. 2017), the two-dimensional RSSI spatial distribution can be used to evaluate the quality of RSSI. Figure 8.7 shows four contour maps of the WiFi signal strength, respectively produced by the four algorithms. The signal strength data were collected in room 404 (shown in Fig. 8.4), and the signal was transmitted from AP1. The number of *saltation points* may be used to evaluate the quality of the RSSI data. A saltation point in the

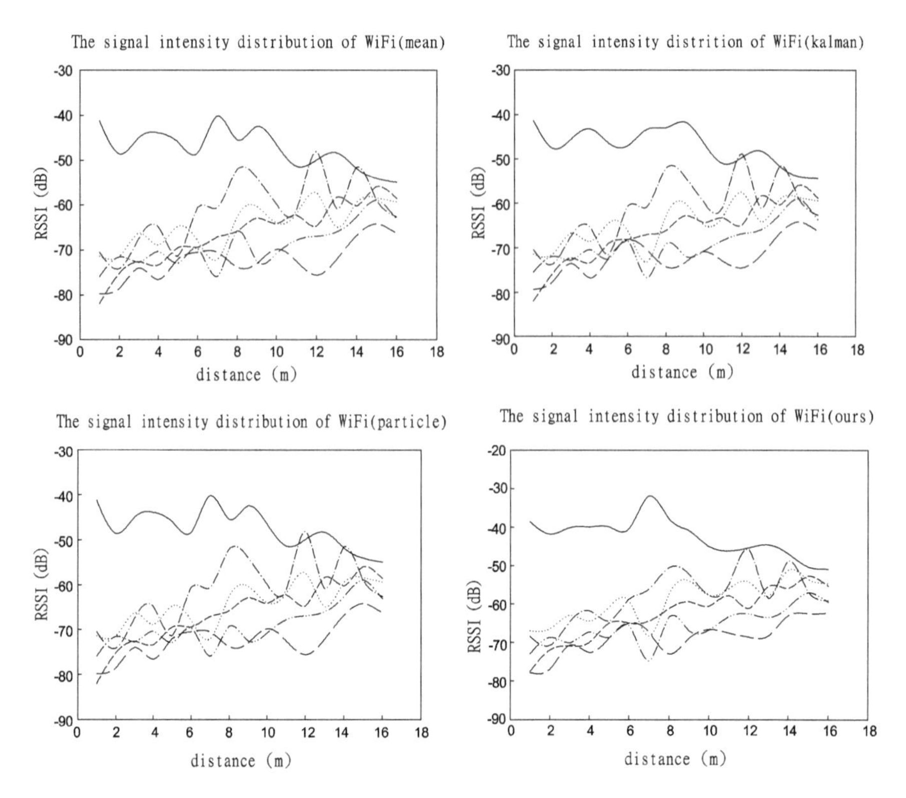

Fig. 8.6 The distribution of the RSSI observations processed by four different algorithms. Each RSSI curve corresponds to a specific access point

Table 8.3 Comparison of the smoothness indexes of four algorithms

	AP1	AP2	AP3	AP4	AP5	AP6	All APs
Mean	23.81	31.71	21.03	28.49	17.97	53.83	176.84
Kalman filter	15.16	33.99	20.03	24.84	17.13	52.51	163.66
Particle filter	23.81	31.75	21.07	23.90	17.13	53.85	171.51
Mean maximum	15.51	31.08	18.98	30.03	18.49	41.00	155.08

contour map is a point where the RSSI is significantly different from those of the surrounding points. More saltation points in a contour map means worse quality of RSSI data. Table 8.4 shows the number of saltation points produced by the four algorithms associated with four rooms and six APs. The total number of saltation points are 50, 46, 46, and 29 respectively for the mean algorithm, Kalman filter, particle filter, and the mean maximum algorithm. Clearly, the mean maximum algorithm produces the minimum number of saltation points.

Smoothness index and number of saltation points are measures for evaluating the quality of RSSI data. However, it is more important to use positional accuracy to

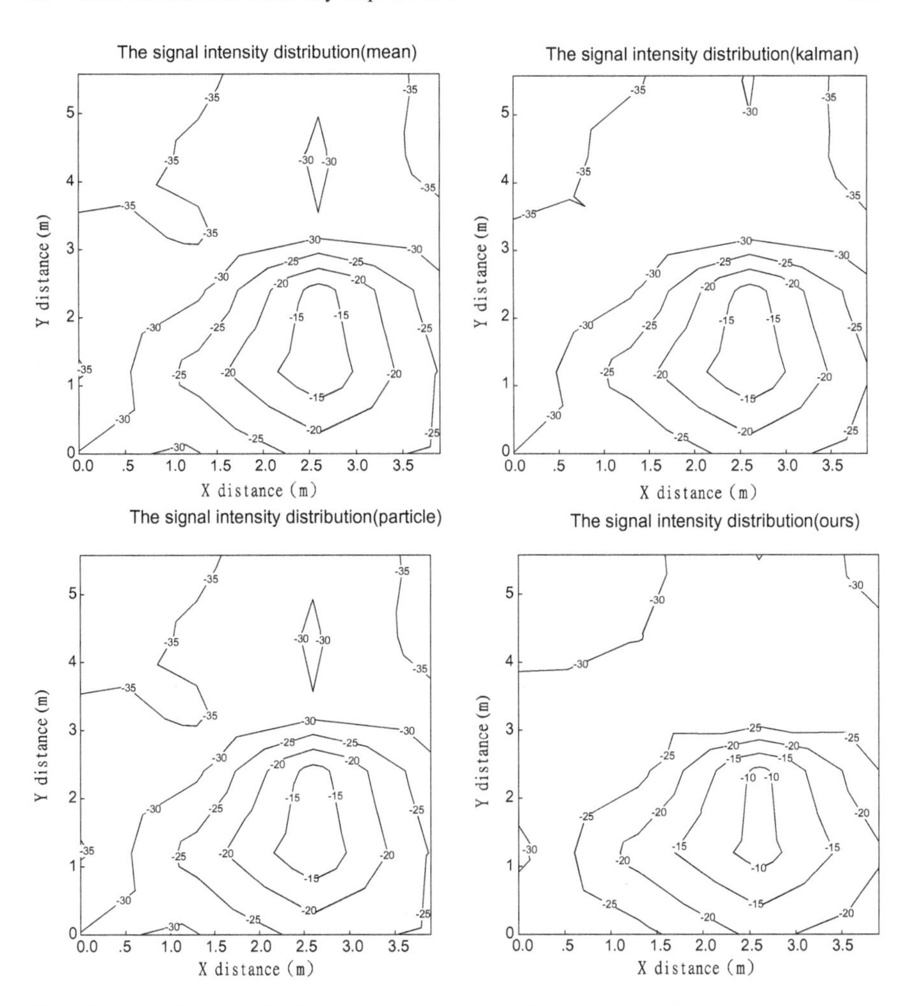

Fig. 8.7 Signal intensity distribution (contour map) of AP1 in room 404

Table 8.4 Comparison of number of saltation points of four algorithms in four rooms with six access points

	AP1	AP2	AP3	AP4	AP5	AP6
Mean	{3, 2, 0, 2}	{2, 0, 2, 2}	{1, 1, 3, 3}	{2, 3, 2, 2}	{3, 2, 2, 3}	{4, 1, 2, 3}
Kalman filter	{2, 1, 1, 2}	{2, 0, 1, 2}	{1, 2, 2, 3}	{2, 3, 2, 2}	{3, 2, 2, 2}	{4, 1, 1, 3}
Particle filter	{3, 2, 0, 2}	{2, 0, 2, 2}	{1, 1, 3, 3}	{2, 2, 2, 1}	{3, 2, 2, 2}	{3, 1, 2, 3}
Proposed	{0, 2, 0, 1}	{2, 0, 1, 0}	{1, 0, 2, 2}	{1, 1, 1, 1}	{3, 2, 2, 1}	{2, 1, 0, 3}

evaluate the RSSI data quality processed by the different algorithms. Table 8.5 shows the positional accuracy of the k-nearest neighbor position determination method which is widely used for WiFi positioning. The cumulative distribution

Table 8.5 Comparison of location accuracy of four algorithms in terms of CDF

Accuracy (m)	0.1 (%)	0.2 (%)	0.3 (%)	0.5 (%)	1 (%)	1.3 (%)
Mean algorithm	9.17	23.33	27.50	36.67	71.67	83.33
Kalman filter	10.00	25.00	29.17	40.00	77.50	91.67
Particle filter	9.17	24.17	28.33	38.33	73.33	85.83
Mean maximum method	13.30	33.33	37.50	45.83	85.83	100

function (CDF) is used as the accuracy index and the RSSI data are the same as those used for the calculation of smoothness index and the contour map. CDFs of up to six positional errors are listed. The results are in good agreement with those in Tables 8.3 and 8.4, indicating smoothness index and number of saltation point are useful indexes for the evaluation of the quality of RSSI data.

8.4 Weighted K-Nearest Neighbors Algorithm

One of the algorithms often considered for WiFi based position determination is the weighted k-nearest neighbors algorithm (Jekabsons et al. 2011). As shown in Fig. 8.8 the positioning procedure involves two phases, the offline calibration phase and the real-time positioning phase. In the offline phase a group of n points in the positioning coverage area are chosen as calibration points, whose two-dimensional position coordinates are known, and denoted as $\mathbf{p}_i = \begin{bmatrix} x_i & y_i \end{bmatrix}^T$, $i = 1, 2, \ldots, n$. At the ith calibration point the measurements of the received signal strength associated with the m access points are recorded to form a RSS vector $\mathbf{r}_i = \begin{bmatrix} r_{i,1} & r_{i,2} & \cdots & r_{i,m} \end{bmatrix}^T$. Since indoor multipath propagation results in signal fading, the received signal can vary significantly with time, both in the short term due to people moving, and in the long term due to changes in the structure and furnishings of the building. To reduce these multipath effects, the average over a period of time of the raw measured RSS vectors is saved in a database for each calibration point. A radio map or a database can be constructed as

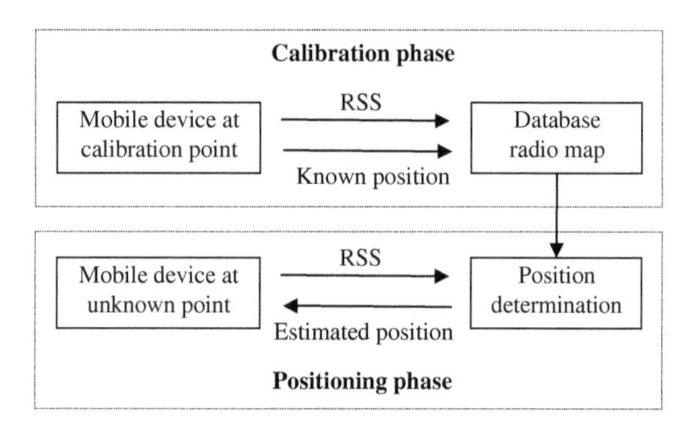

Fig. 8.8 Calibration and positioning phases of fingerprint WiFi positioning

$$DB = \begin{bmatrix} \mathbf{p}_1 & \mathbf{p}_2 & \cdots & \mathbf{p}_n \\ \mathbf{r}_1 & \mathbf{r}_2 & \cdots & \mathbf{r}_n \end{bmatrix}^T \tag{8.19}$$

In addition to multipath signal fading, RSS measurements are also affected by orientation of the pedestrian or vehicle. At specific orientations the RSS of some access points can be significantly reduced due to body blockage. If the antenna is not omni-directional, the RSS can also be affected by antenna orientation. Thus, taking orientation into account will significantly improve positioning accuracy. At each calibration point, typically RSS measurements at four different orientations (say facing north, south, west and east) are recorded individually in the database. That is, four RSS vectors of length m, instead of only one m-vector, are produced and recorded, resulting in the expanded database

$$DB = \begin{bmatrix} \mathbf{p}_1 & \cdots & \mathbf{p}_1 & \mathbf{p}_2 & \cdots & \mathbf{p}_2 & \cdots & \mathbf{p}_n & \cdots & \mathbf{p}_n \\ \mathbf{r}_1^{(1)} & \cdots & \mathbf{r}_1^{(4)} & \mathbf{r}_2^{(1)} & \cdots & \mathbf{r}_2^{(4)} & \cdots & \mathbf{r}_n^{(1)} & \cdots & \mathbf{r}_n^{(4)} \end{bmatrix} \tag{8.20}$$

where $\mathbf{r}_i^{(j)} = \begin{bmatrix} r_{i,1}^{(j)} & r_{i,2}^{(j)} & \cdots & r_{i,m}^{(j)} \end{bmatrix}^T$ and $r_{i,\ell}^{(j)}$ is the RSS received at the ith calibration point under the jth orientation and associated with the ℓth access point.

In the real-time positioning phase, at every sampling instant a RSS m-vector is produced in the mobile device from AP signals in radio range. Each vector component corresponds to a specific access point and averaging is used to generate each RSS vector component. The number of measurements and the measurement duration used for averaging depends on the position update rate and the mobile velocity. The mobile position denoted as $\mathbf{p} = \begin{bmatrix} x & y \end{bmatrix}^T$ is treated as at the center point in the time averaging period. The real-time RSS vector is produced in the mobile device and transmitted to the central database using the WiFi data transmission capability. The central processing associated with the database then calculates the Euclidean distance in the RSS signal space between the database calibration vectors and the real-time RSS vector, and is defined as $\mathbf{r} = \begin{bmatrix} r_1 & r_2 & \cdots & r_m \end{bmatrix}^T$. The k calibration points with the shortest RSS Euclidean distances are selected. Since orientation is considered, the same calibration point may be counted multiple times (up to the number of orientations). For convenience, the corresponding k RSS m-vector is defined as

$$\tilde{\mathbf{r}}_i = \begin{bmatrix} \tilde{r}_{i,1} & \tilde{r}_{i,2} & \cdots & \tilde{r}_{i,m} \end{bmatrix}^T, \ i = 1, 2, \ldots, k \tag{8.21}$$

The reason behind such a selection is based on the assumption that if a calibration point is closer to the mobile, the RSS distance is shorter. The mobile position estimate $\hat{\mathbf{p}} = \begin{bmatrix} \hat{x} & \hat{y} \end{bmatrix}^T$ is then calculated according to

$$\hat{\mathbf{p}} = \sum_{i=1}^{k} \frac{w_i \mathbf{p}_i}{\sum_{j=1}^{k} w_j} \tag{8.22}$$

where w_j is the weight which can be defined as the inverse of Euclidean distance between the database RSS vector and the jth selected RSS vector defined by (8.21):

$$w_j = 1/d(\tilde{\mathbf{r}}_j, \mathbf{r}) = \left(\sum_{i=1}^{m} (r_i - \tilde{r}_{j,i})^2 \right)^{-\frac{1}{2}} \tag{8.23}$$

8.5 Radius Based Domain Clustering Method

The radius based domain clustering (RDC) method proposed in (Zhang et al. 2017) makes use of all available APs data to estimate the position of mobile device. The method relies on the consideration that each AP provides at least some location information of the target position. The distribution pattern of the intermediate position estimates would contain some information about the location of the mobile device. That is, the target is more likely located in the densest area on the distribution. Figure 8.9 shows the flow chart of the RDC method.

For clarity, a few definitions are made as follows. A *regional* visible AP is an AP whose signal can be detected at *all* the calibration points within a room-level region. If the signal from a regional visible AP can also be detected at the mobile device's location, the regional visible AP is termed as a *common AP* of the calibration points and target point of the region. The candidate region is the region which has the largest number of common APs associated with a mobile device AP list. It is most likely that the mobile device is located in the candidate region which is thus selected for further position determination.

Once the room-level region is determined, suitable calibration points in the database can be selected. This selection is based on the RSS Euclidean distance between the calibration point and real-time RSS data measured in the mobile device, and is defined as

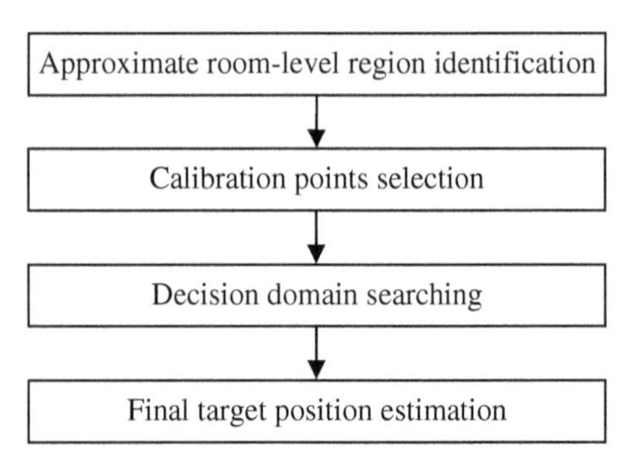

Fig. 8.9 Flow chart of RDC method

$$d_{c_j} = \sqrt{\sum_{i=1}^{m} (RSS_{c_j}^i - RSS_t^i)^2} \tag{8.24}$$

where m is the number of common APs, $RSS_{c_j}^i$ is the RSS received from the ith AP at the c_jth calibration point, and RSS_t^i is the RSS received at the mobile device. The calibration points with the smallest RSS Euclidean distances are chosen, which are the nearest neighboring calibration points in the m-dimensional signal space. Each common AP is used to produce an intermediate mobile device position estimate using (8.22) where k is typically set to be three. A larger k may degrade positioning accuracy since more distant calibration points will be used.

Then, the concept of circular error probable (CEP) can be applied to provide a statistical estimate of the accuracy of the computed position. The decision domain associated with an intermediate position estimate is defined as a circle whose center is at the intermediate position estimate, and which exactly covers 50% of the m intermediate position estimates, that is $m_1 = m/2$, or $m_1 = (m+1)/2$ if m is an odd number. There are a total m decision domains and the size of each decision domain may be determined as follows. The distance from the center to each of the $m_1 - 1$ intermediate position estimates is calculated. Then the distances are sorted in ascending order and the $(m_1 - 1)$th distance in the sorted distance sequence is equal to the radius of the decision domain.

Finally, one finds the smallest radius among all the m radii and the corresponding decision domain is considered as the best and used for mobile position estimation, namely

$$\hat{x}_t = \frac{1}{m_1} \sum_{j=1}^{m_1} \hat{x}_j^{(u)}, \ \hat{y}_t = \frac{1}{m_1} \sum_{j=1}^{m_1} \hat{y}_j^{(u)} \tag{8.25}$$

where $(\hat{x}_j^{(u)}, \hat{y}_j^{(u)})$ is jth intermediate position estimate in the selected decision domain. That is, the final target position estimate is the average of the m_1 intermediate position estimates in the selected decision domain.

The performance of the RDC method has been evaluated by conducting an experiment in the office building of the School of Geodesy and Geomatics, Wuhan University. The experimental site consists of three rooms of the same dimensions of $10\,\mathrm{m} \times 7.2\,\mathrm{m}$ as shown in Fig. 8.10. In each room, six calibration points and ten target points are selected. 180 RSS samples were collected at each target point for position determination. Since the RSS sampling interval is 1 s, it took 3 min to collect the RSS data at each target point. Such a long period is too long for real-time positioning, but this method will tell the best achievable performance of the algorithm.

Table 8.6 shows the performance of the RDC method in terms of mean error, error STD, RMS error, and maximum error. For comparison, the performance of the popular WKNN method is also shown. Except for the maximum error, the RDC method outperforms WKNN method significantly, with the RDC error about 20%

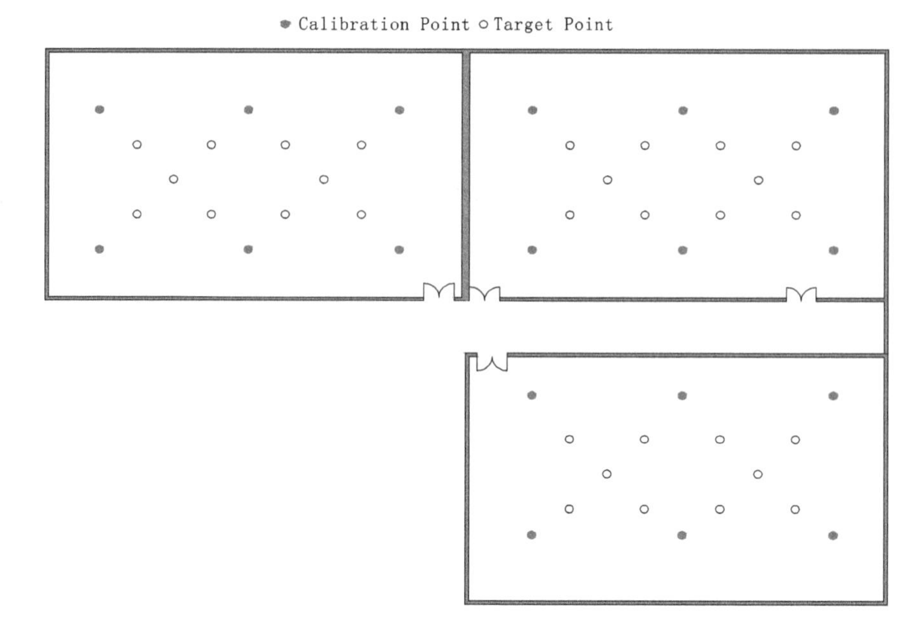

Fig. 8.10 Distribution of calibration points and target points in the experimental site. Each of the three rooms has dimensions of 10 m by 7.2 m. The calibration points are on the corners of a 4 m square grid

Table 8.6 Performance comparison between the WKNN approach and RDC method

Method	Number of APs	Mean error (m)	STD of error (m)	RMS of error (m)	Maximum error (m)
WKNN	4	1.41	0.77	1.69	2.86
	5	1.37	0.79	1.66	2.93
	6	1.47	0.78	1.75	2.92
	7	1.51	0.73	1.76	2.78
	8	1.50	0.75	1.76	2.80
RDC	–	1.18	0.62	1.40	2.82

less than the WKNN method. The RMS error for the RDC method is 1.4 m, which is equivalent to $0.35S$, where $S = 4$ m is the size of the square calibration grid. This compares with the theoretical value of 0.41S given by Eq. (15.13) in Chap. 15. While there is reasonable agreement, the small sample size means that these results should be taken with some caution. Nevertheless, the difference in performance between the WKKN and RDC methods seems to be statistically significant.

Figure 8.11 shows the cumulative distribution function (CDF) of the two methods. WKNN method uses five different numbers of APs, from four to eight. For position errors less than about 1 m both methods appear to be statistically quite similar, with no apparent benefit in using more than four APs in the WKKN

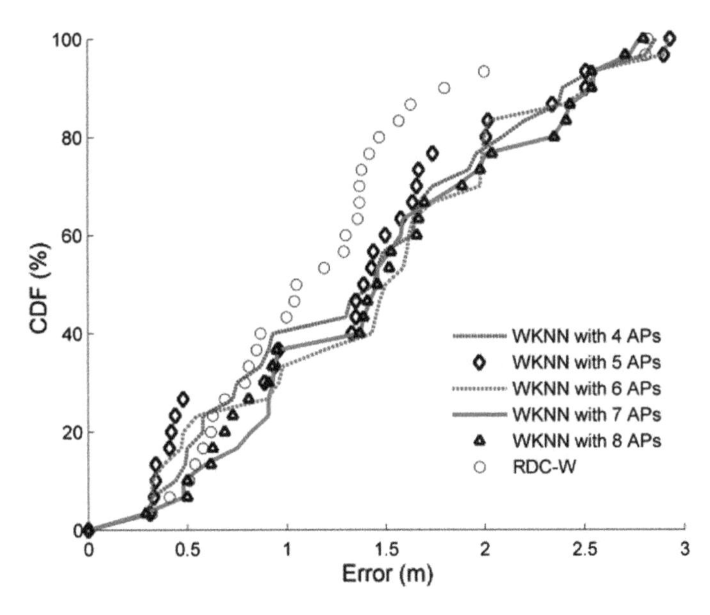

Fig. 8.11 Cumulative distribution function of positional error of WKNN method and RDC method

method. For errors greater than 1 m the RDC method is clearly superior, with the error being about 0.2 m less for the same CDF greater than 0.5.

Acknowledgements Kegen Yu would like to thank two Ph.D. students, Weixing Xue and Wei Zhang, for providing the experimental results, and Professor Xianghong Hua for useful discussions.

References

Borel CC (1998) Surface emissivity and temperature retrieval for a hyperspectral sensor. In: Proceedings of IEEE international geoscience and remote sensing symposium, Seattle, WA, USA, pp 546–549

Chao CH, Chu CY, Wu AY (2008) Location-constrained particle filter for RSSI-based indoor human positioning and tracking system. In: Proceedings of IEEE workshop on signal processing systems, Washington, D.C., USA, pp 73–76

Chen Y, Yang Q, Yin J, Chai X (2006) Power-efficient access-point selection for indoor location estimation. IEEE Trans Knowl Data Eng 18(7):877–888

Deng Z, Ma L, Xu Y (2011) Intelligent AP selection for indoor positioning in wireless local area network. In: Proceedings of international ICST conference on communications and networking in China, Harbin, China, pp 257–261

Hua X, Zhang W, Yu K, Qiu W, Zhang S, Chang X (2017) Performance analysis for AP selection strategy. In: Proceedings of China satellite navigation conference, Shanghai, China, pp 325–333

Jekabsons G, Kairish V, Zuravlyov V (2011) An analysis of Wi-Fi based indoor positioning accuracy. Sci J Riga Tech Univ 47:131–137

Mansour MF (2014) Kalman filter for indoor positioning. Patent US20140368386

Rappaport TS (2002) Wireless communications principles and practices. Prentice-Hall

Wang X, Song SZ, Li M (2012) Design of personnel position system of mine based on the average of RSSI. In: Proceedings of IEEE international conference on automation & logistics, Zhengzhou, China, pp 239–242

Xue W, Qiu W, Hua X, Yu K (2017) Improved Wi-Fi RSSI measurement for Indoor localization. IEEE Sens J 17(7):2224–2230

Youssef M, Agrawala A, Shankar AU (2003) WLAN location determination via clustering and probability distributions. In: Proceedings of IEEE international conference on pervasive computing and communications, Fort Worth, Texas, USA, pp 143–150

Zhang W, Hua X, Yu K, Qiu W, Chang X, Chen X (2017) Radius based domain clustering for WiFi indoor positioning. Sens Rev 37(1):54–60

Zou H, Luo Y, Lu X, Jiang H, Xie L (2015) A mutual information based online access point selection strategy for WiFi indoor localization. In: Proceedings of IEEE international conference on automation science and engineering (CASE), Gothenburg, Sweden, pp 180–185

Chapter 9
Link Budgets and System Design for Positioning Systems

9.1 Introduction

The design of terrestrial radio navigation systems tends to focus on the details of the method of determining the position of a mobile device, based on the properties of the radio signal, and the details of detection and subsequent signal processing. However, the starting point in any design will be in response to a system specification which defines the requirements of each application. These high level requirements are often rather limited in scope, and can vary considerably from application to application. While the widespread adoption of GPS in mobile devices means that the default solution is typically based on applying the capabilities of GPS to a particular application, specific radiolocation applications are such that GPS is not able to meet the requirements; it is these types of applications that are the focus of this chapter. In particular, these will include indoor applications where GPS performs poorly or not at all, applications where there are much higher accuracy requirements than can be achieved by GPS even in outdoor applications, and even security applications where vulnerability to jamming of the signal makes GPS unviable.

The design process for a radiolocation system with specific needs means that a "one solution fits all" cannot be used. The requirements specification will dictate the overall design approach for the system, from which the detailed designs of the subsystems are derived. Thus while the principles of radiolocation systems are generic in character, the specific requirements of each application can result in very different solutions. Thus the approach in this chapter is to first outline some general principles used in an overall system design, and then specific details are given in some detail for specific cases which shows the diversity of particular solutions. These will vary from short-range, high accuracy indoor systems, to systems that are designed to cover a whole city.

I. Sharp and K. Yu, *Wireless Positioning: Principles and Practice*, Navigation: Science and Technology, https://doi.org/10.1007/978-981-10-8791-2_9

9.2 Overview of Design Process

The overall design of a new terrestrial-based (radio) positioning system would typically commence with a set of requirements, which would define the main operational aspects of the system. Because of the universal nature of GPS, these requirements would include aspects which preclude the use of a GPS-based system. These requirements might include the required accuracy, location of operation such a indoors, or high update rates for the positional information. Given the high level specification, the basic design can be based on the methods described in this chapter.

The intention of a high level design is to establish the basic parameters of the system, such as the operating frequency and signal bandwidth, method of position determination, operating details such as the use of fixed base stations or more mobile nodes, radio link parameters for both position determination and data communications, size, weight and battery life for the mobile devices, and the user interface to the system. In the initial design process, typically several concept designs are investigated, so that tradeoffs in the designs can be investigated. For example, the radio link range will determine the number of nodes required to cover the required operational area, and to minimise the costs and logistics of setting up the system the operational range should be a large as possible. However, the range depends on parameters such as the operational frequency, signal protocols, transmitter power and receiver sensitivity, so that many possible solutions need to be investigated. Further, the operational range may have impact on the positional accuracy requirements. For example, to maximise the range a low operating frequency is required, but such relatively low frequencies would also have low signal bandwidths. If the system requires high accuracy (say better than 1 m), then simple solutions using signal strength methods are not viable, so that more sophisticated methods which use wide bandwidths would be required, which are only available at higher radio frequencies. As the transmission losses increase with the radio frequency, the operating range is reduced. Thus typically a tradeoff between system accuracy and system transmission range is necessary.

To provide a high-level framework for these initial tradeoff studies, some broad principles can be used to determine the possible design that meets the functional requirements. Much of the initial design analysis can be based on simple design formulas, which can be implemented conveniently in the form of a spreadsheet. This approach allows the broad characteristics of a system design to be established, before more detailed design activities are commenced. The process may be iterative, whereby more detailed analysis requires modification to the high-level design, or some performance or cost compromises may be necessary. The following subsections in this chapter describe the main design methods that can be applied to most new radio-based positioning systems.

9.3 Link Budget Parameters

The initial step in the design of any radiolocation system is to determine the overall nature of the radio links in the system. For a terrestrial system the components are usually divided into "base station" and mobile devices. The base stations are normally fixed in location, but potentially they can also be mobile devices whose positions are known. The mobile devices are the components that move within a coverage area, and whose position is to be determined by radio communications between the components. These communications can be classified into four types of radio transmission links (from a transmitting node to a receiving node), namely radiolocation transmissions to and from the mobile devices, and data transmissions to and from the mobile device. In general these transmission are all different in character, and thus need to be designed individually. However, in some systems not all four link types are present. For example, in some systems the "mobile" devices can function as "base stations", in which case there are only two types of links, one to radiolocation and one for data. Further, the data and radiolocation functions can often be combined into the one overall signal protocol, but even in this case the signal-to-noise ratio (SNR) required for satisfactory operation will usually be different. Further the effective SNR is not only a function of the received RF signal level, but depends on the signal processing within the receiver. Broadly all these concepts can be grouped under the term "link budget", with one budget for each type.

The link budget is broadly defined by the parameters associate with the path from the transmitter to the receiver. The main components are associated with the radio transmission from the transmitter to the receiver, but also includes some baseband parameters associated with the signal processing in the receiver. The effective transmitter power (including the antenna gain) will be attenuated by the transmissions to the radio receiver, so that the effective received power (including antenna gain) will be a function of the propagation distance as well as the characteristics of the radio propagation environment. What is typically specified in any design is the minimum distance that the position system requires for satisfactory operation, so that a link budget will specify the allowable propagation loss for satisfactory operation, based on a minimum allowable RF signal input to the receiver. The receiver RF front-end will convert the signal to baseband where usually there is further signal processing before the final decoding of the signal for both radiolocation and data functions. This signal processing often will result in "process gain", whereby the output SNR is enhanced relative to the input SNR, namely the process gain is defined as

$$G_p = SNR_{out}/SNR_{in} \tag{9.1}$$

The input SNR is related to the effective received radio signal power, and the noise is associated with the RF receiver noise characteristics (and possibly man-made interference noise), which is typically defined as a function of the RF signal bandwidth. While the methods of receiver process gain can vary considerably, the general principle is the SNR is improved by integrating the input signal over a period $T \gg 1/BW$, where BW is the RF signal bandwidth. The process gain can often be quite large (>20 dB) for the radiolocation function, but in the case of the data transmissions improving the SNR at the input to the data decoder is at the expense of a corresponding reduction in the output data rate (bits per second). However, the data requirements for a radiolocation system are usually quite modest, so that it may be possible to use the same process gain for both functions. The effect of process gain on the system design is to increase the range for the same transmitter power, or to reduce the transmitter power for the same range. In the case of low power transmissions from mobile devices this is very important, as the low transmitter powers significantly increases the times between battery recharging.

The radio transmission process can be summarised in terms of a Link Budget, whereby all the major hardware components and associated parameters are defined, so that the received signal power can be estimated. For calculations it is convenient to define the link budget in decibels, so that various elements in the budget can be added rather than multiplied. Thus the link budget can be expressed in the form

$$P_{rx} = P_{tx} + G_{tx} - L(R) + \Delta L + G_{rx} \tag{9.2}$$

where

P_{rx} received power (in dBm, or dB relative to a milliwatt)
P_{tx} transmit power (dBm)
G_{tx} transmitter antenna gain relative to an hypothetical omnidirectional antenna (in dBi)
$L(R)$ propagation loss (in dB) as a function of range R
ΔL loss excess (in dB), in addition to the basic loss model $L(R)$
G_{rx} receiver antenna gain (dBi). May be the same as the transmitting antenna

The basic concept in any Link Budget calculations is that received signal power is estimated, given the transmitter power, the antenna gains, and the propagation loss which will be a function of range. For the receiver to function correctly there will be a lower minimum signal required, and thus the maximum operating range can be estimated if a suitable range-loss model is applied; see Sect. 9.4 for more details.

The details of link budgets will be considered in following subsections, but Table 9.1 provides a summary of the most importance parameters in the design of a link budget. The details associated with these parameters are presented in following subsections. It is convenient to present the information in the form of a generic spreadsheet, so that the overall performance characteristics of a system can be determined by simply supplying the parameter values appropriate to that system. From these data the performance of the various links can be easily determined. Note also it is conventual to measure power in dBm, that is power relative to a milliwatt.

Table 9.1 Link budget parameters. Not all parameters are always required

Parameter	Units	Comment
Base station power amp output	dBm	RF output power
Base station antenna gain	dBi	Relative to omni-directional transmissions
Effective radiated power	dBm	Maximum constrained by licensing requirements
Maximum range	m	Maximum range for required coverage area
Minimum range to base station	m	For inter-BS data communications
Minimum range to mobile	m	High received signals can overload sensitive receivers
LOS loss at maximum range	dB	Line-of-sight propagation loss (minimum possible) to BS from mobile
Minimum loss to BS	dB	Minimum possible loss based on LOS propagation
Minimum loss to mobile	dB	Minimum possible loss based on LOS propagation
Fading margin (BS)	dB	Due to multipath propagation and attenuation due to obstacles (such as walls) along the path
Antenna loss due to body	dB	From body-worn devices diffraction losses around a body can be considerable (5–20 dB)
Fading margin (mobile)	dB	Similar to BS case, but with additional component due to blockage by body
Base station antenna gain	dBi	Relative to omni-directional transmissions
Mobile antenna gain	dBi	Internal antenna may result in negative gain
Rx power: max range (BS)	dBm	Effective BS Tx power − LOS loss − Fade margin
Rx power—max range (Mobile)	dBm	Effective Mobile Tx power − LOS loss − Fade margin
LNA gain	dB	LNA is used to improve the receiver noise figure. Mainly used in BS to compensate for lower Tx power from the mobile
Min BS Rx (LNA) input power	dBm	With maximum propagation losses, including fade margin
Max BS Rx (LNA) input power	dBm	With LOS propagation and minimum range. May overload the amplifier
Min Rx input power (mobile)	dBm	With maximum propagation losses, including fade margin
Max Rx input power (mobile)	dBm	With LOS propagation and minimum range
Receiver IF bandwidth	MHz	RF signal bandwidth plus margin for filter characteristics (both front-end RF and Rx IF filters)
Noise figure of receiver (BS)	dB	LNA will improve (reduce) the noise figure
Noise figure of receiver (mobile)	dB	May or may not have LNA to improve noise figure
Baseband bandwidth (BS)	MHz	Baseband filter bandwidth at receiver output
Baseband bandwidth (mobile)	MHz	Baseband filter bandwidth at receiver output

(continued)

Table 9.1 (continued)

Parameter	Units	Comment
Process gain (BS)	dB	Depends on signal protocol and any signal integration in receiver prior to final detection
Process gain (mobile)	dB	Depends on signal protocol and any signal integration in receiver prior to final detection
Receiver noise (BS)	dBm	$10\log(kTBF)$ (effective baseband bandwidth B after process gain)
Receiver noise (mobile)	dBm	$10\log(kTBF)$ (effective baseband bandwidth B after process gain)
Receiver output SNR (BS)	dB	Received mobile signal power − Receiver noise
Receiver output SNR (mobile)	dB	Received BS signal power − Receiver noise
Required minimum SNR (BS, radiolocation)	dB	SNR for reliable measurement of position using signals from mobiles
Required minimum SNR (BS, data)	dB	SNR for reliably receiving and decoding mobile data transmissions at the base stations
Required minimum SNR (mobile, radiolocation)	dB	SNR for reliable measurement of position using signals from base stations
Required minimum SNR (mobile, data)	dB	SNR for reliably receiving and decoding base station data transmissions at the mobiles
Base station receiver margin	dB	Computed minimum SNR − Required SNR
Mobile receiver margin	dB	Computed minimum SNR − Required SNR

Thus transmitter powers are typically of the order of 30 dBm and minimum receiver powers of the order of −100 dBm, equivalent to a power attenuation factor of 10^{13}.

9.4 Correlation and Process Gain

From the overview of the process associated with the link budget, one key aspect is the concept of "process gain" which is associated with improving the SNR at the input to the data decoder or the processing associated with determining the position from properties of the received RF signal. These properties may include the measurement of the time-of-arrival (TOA), the signal strength or the signal phase. By improving the SNR by the application of process gain, the accuracy of the position determination will be improved, as well as the range of detection. The details of processing associated with process gain is beyond the scope of this chapter (see Chap. 3 for more details on spread-spectrum signals), but a broad outline of the concepts are summarised below. However, while the details may vary, the basic concept of process gain is the integration of the raw input signal. If this integration is arranged to be coherent (constant relative phase between the input signal and a reference signal in the receiver), then the output power of the integrator will increase proportional to the square of time of integration. In contrast the incoherent noise will be random relative to the receiver reference signal, and thus the noise output (variance) will increase proportional to the integration time. As a consequence, the output SNR will increase proportional to the integration time.

While the particular methods of implementation of process gain may vary, two important implementations will be briefly described. The first implementation is a simple summer. If the signal is repeated (ideally exactly) N times in a coherent fashion, then the summation process will produce a single version of the signal, but with the SNR improved by the process gain. In a typical implementation, the baseband signal is first digitised, and the summer is implemented digitally. This enhanced digital signal is then processed to determine the TOA, signal strength or signal phase for the position determination.

The second common method which results in process gain is the application of a correlator. A correlator integrates the received signal with a coherent version generated in the receiver. This signal may be the output of the summer described above. One common method for detection of the TOA of the signal is to transmit a direct-sequence spread-spectrum signal, which is used to modulate (multiply) the RF carrier. In this case the received baseband signal is despread by the correlation process, resulting in a triangular-shaped pulse of width of the order of the pseudo-random code "chip" period. This "pulse" can then be used to estimate the TOA. In this case, the process gain associated with the correlation process is equal to the length of the code in binary digits (chips).

A simplified block diagram of a generic receiver architecture is shown in Fig. 9.1. The radio receiver amplifies the RF signal and down converts to the baseband, which is filtered to a bandlimited signal. Because this signal is bandlimited, it can be digitised (sample rate at least twice the signal bandwidth) by an analog-to digital converter. This digital signal is (optionally) summed in a digital accumulator. Finally, this accumulated signal is integrated by a correlator to generate a correlation pulse whose TOA is determined.

The radio system receives the RF signal and translates the signal to baseband. In a typical implementation, the RF signal is first amplified and converted to an Intermediate Frequency (IF). As the local oscillator phase is not synchronized to the phase of the RF signal, the baseband output is typically in the form of in-phase and quadrature components (effectively a complex-number signal), so that the output has two components to be filtered and then converted to digital signals. To ensure no aliasing in the conversion process, the baseband signal is filtered to a bandwidth

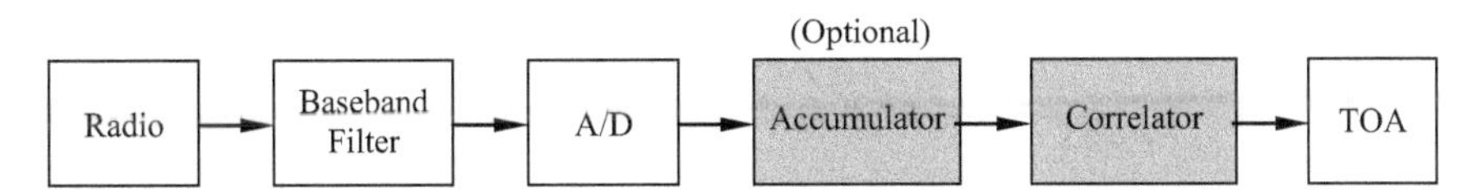

Fig. 9.1 Generic block diagram in a modern radio receiver, with analog components at the front end, and digital components at the back-end

less than twice the sampling rate. While is filtering causes some distortion to the shape of the outputted correlation pulse, this process results in minor errors in estimating the TOA, while significantly reducing the required sampling rate and the consequential signal processing requirements. This reduction in processing power is particularly important for battery-powered mobile devices.

9.5 Radio Propagation

The characteristics of the radio signal propagation are important in designing a positioning system. While in free space a simple inverse-square law can be used to determine the signal strength as a function of range, in a NLOS environment the complexity of the environment makes calculations of the large-scale signal attenuation with distance difficult. Further, while the average variation with distance is important, the medium-scale variation of the signal strength is also an important parameter in the design, particularly for indoor positioning systems. These medium-scale effects are typically on a scale of 5–50 wavelengths, and are associated with major obstacles along the propagation path, such as structural elements of a building. Small-scale signal variations (of the order of a wavelength) also can be considerable, further complicating matters. In particular, constructive and more importantly destructive interference (Rayleigh and Rician statistical fading) occurs on the scale of the order of a wavelength. These dividing lines are only rough guidelines which are useful for engineering analysis of the complex propagation environment of indoor radio propagation at microwave frequencies.

These multipath effects not only vary the signal strength, and thus are important in Link Budget calculations, but also impact on the signal phase (important in angle-of-arrival systems), and on the signal time-of-arrival, which has impact on the estimation of range in pulse-based systems. Thus radio propagation effects have a significant impact on the design of the signal protocols used to mitigate the effects of multipath signals. However, for the initial RF design, the only parameters that need to be considered are the RF frequency and bandwidth, with the latter being largely determined by the positional accuracy requirements. If the reciprocal of the signal bandwidth is defined as $T = 1/B$, then a rough estimate of the positional accuracy is αT where the positional accuracy is measured in time using the speed of radio propagation (30 cm/ns). The parameter α depends on the propagation environment and the SNR, but typically is in the range $0.01 \le \alpha \le 0.1$. This topic is considered in more detail in the case studies in Sect. 9.7.

9.5.1 *Free-Space Propagation*

The basic starting point for signal loss determination is that associated with free-space propagation (also referred to as the line-of-sight (LOS)), as this represents the minimum possible loss from a transmitter to a receiver. From Maxwell's equations it is known that the RF energy reduces as the square of the distance from a point radiating source, so it can be shown that the free space path loss for a range R and a wavelength λ is given by[1]

$$\begin{aligned} l_{free\_space}(R) &= \left(4\pi \frac{R}{\lambda}\right)^2 \\ L_{free\_space}(R) &= 20\log\left(4\pi \frac{R}{\lambda}\right) = \alpha_0 + \beta_0 \log\left(\frac{R}{\lambda}\right) \quad \text{(dB)} \end{aligned} \tag{9.3}$$

For design calculations the propagation loss is specified in decibels, so the second equation in (9.3) is the most appropriate. Note in this case the loss can be expressed as a linear function of the log of the propagation distance measured in wavelengths. Thus the operating frequency has a direct effect on the Link Budget calculations, as the LOS loss increases with frequency. Further, as the positional accuracy is often directly related to the signal bandwidth (particularly TOA systems), and greater bandwidths are available at higher RF frequencies, higher accuracy will be associated with higher propagation losses, and this usually means shorter ranges.

This loss estimate can be applied in any Link Budget calculations, but normally an additional loss excess component is required, which depends on the operating conditions. These are considered in more detail in the following subsections.

9.5.2 Two-Ray Model

While the free-space model provides a basic estimation of propagation losses, in practice the positioning system will operate in an environment where as a minimum there will be reflections from the ground. For positioning systems considered here, the operating frequencies will be at least 200 MHz, but more typically at microwave frequencies to up 10 GHz. For these frequencies, particularly at longer ranges, the ground reflections can cause a significant reduction in the received signal strength relative to free-space propagation. With such shallow reflection angles, the reflection coefficient for radio waves of all polarizations will be close to –1. In this case it can be shown (Bertoni 2000; Jordan and Balmain 1968) that the extra loss over the free-space loss in signal strength is given by

$$\Delta L = 20\log\left(\frac{\lambda R}{4\pi h_1 h_2}\right) \quad \text{dB} \tag{9.4}$$

[1]The convention used is that lower case text (l) is used for the loss in the standard form, and upper case (L) when specified in decibels.

where h_1 and h_2 are the heights of the base station and mobile antennas, R is the range. Note that the loss increases as the height decreases, so that in applications with the mobile device close the ground the extra loss will be large. The total loss when the free space loss is included is given by

$$L = 20 \log \left(\frac{r^2}{h_1 h_2} \right) \quad \text{dB} \tag{9.5}$$

From (9.5) it can be observed that the loss increases by 40 dB per decade in range, compared with 20 dB per decade for free space propagation. Thus the effect of the ground reflections is to greatly reduce the effective range of the propagation from mobile nodes, even though the radiated power remains unaffected by the nearby ground reflections.

To illustrate this effect, consider an example of tracking rowing boats. For more details, refer to the case study in Sect. 9.7.1. With typical deployment the base station heights are about 5 m above water level, and the antenna of the mobile device attached to the rowing scull is about 25 cm above the water. Applying (9.4) at a range of 2000 m[2] and a frequency of 2.4 GHz shows that the extra loss is 24 dB. The free-space loss at that range is 106 dB, so that the total loss is 130 dB. The measured loss from a mobile node to a base station is shown in Fig. 9.2 during a 2000-m race, with the base station being at about the 1000 m point (or midway). The simple two-ray model above predicts a loss of 118 dB at a range of 1000 m, while the free-space loss is 100 dB. The data in Fig. 9.2 shows that the measured loss at 1000 m is about 115 dB, so the two-ray model provides a good estimation of the actual losses. This suggests that applications such as outdoor sports tracking systems should use the two-ray model for the link budget calculations, as the flat, open sports fields closely approximate the simple image reflection used in the two-ray model.

9.5.3 Generalisation of Free-Space Model

The operation of real-world positioning systems will usually be in a complex multipath environment where a simple two-ray model is not an adequate model for predicting propagation losses. As the propagation characteristics both indoors and outdoors can be complex, the typical approach taken is for empirical generic equations to represent the propagation loss, with the parameters determined by appropriate fitting these equations to measured data. One simple approach is to model the losses in a manner similar to the analytical free-space model described in

[2]The standard course length for Olympic-type rowing is 2000 m.

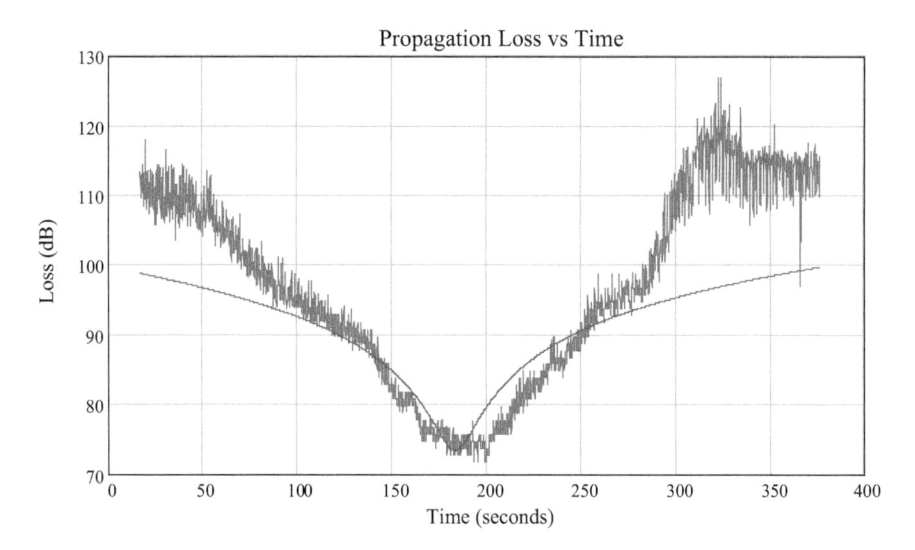

Fig. 9.2 Propagation loss to base station as a function of time during a 2000-m rowing race. The smooth curve is the free-space loss. The high-frequency loss noise is associated with multipath scattering from the rowers, which is maximum at the start/end of the race, and minimal at the halfway point where there is no blockage from the rowers

Sect. 9.5.1. One very common indoor propagation model based on an extension of (9.3) is of the form

$$l = l_0 \left(\frac{R}{\lambda}\right)^{\gamma} \tag{9.6}$$

In this power-law model, the attenuation as a function of distance is controlled by the parameter γ with $\gamma > 2$ as the loss is greater than the free-space case, and l_0 the loss at one wavelength. The loss is assumed to be a function of the propagation distance measured in wavelengths, as with the free-space case. However, as in practice loss is measured in decibels, a more appropriate form for indoor NLOS large-scale propagation loss is

$$\begin{aligned} &L(\mathrm{dB}) = \alpha + \beta \log(R/\lambda) \qquad (L \geq 0 \quad \text{or} \quad R \geq R_{\min}) \\ &\alpha = 10\log(l_0) \qquad \beta = 10\gamma \qquad R_{\min} = e^{-\alpha/\beta} \end{aligned} \tag{9.7}$$

Figure 9.3 also shows the least-squares fit based on model (9.7) for data for a Zigbee (IEEE 15.4) system (2.4 GHz) in an office-type building, as well as the free-space loss for comparison. As can be observed, the measured loss is considerably greater than the free-space loss, and also exhibits random (medium-scale) variation. From the least-squares fit to these data the power law parameter γ is 5.8 (compared with 2 for free-space). As the power law exponent $\gamma = 5.8$ is much

Fig. 9.3 Propagation loss as a function of range based on the raw RSSI signal strength data in an indoor environment. The dashed line is a least-squares fit to a model of the form $L(R) = \alpha + \beta \log(R/\lambda)$ where $\alpha = -35.7$ dB and $\beta = 25.3$ so $\gamma = 5.8$. Also shown is the LOS propagation path loss. The model RMS fit error is 8.5 dB, which is typical for indoor applications

greater than the free space exponent, the losses increase much more rapidly with propagation distance. While simplistic in nature, this simple model is adequate for system design purposes. The main problem is that the model parameters cannot be determined without measurements, but a simple strategy is to apply a power-law exponent in the range $4 \leq \gamma \leq 6$, selected on the empirical basis of how "cluttered" is the environment.

9.5.4 Hata Model (City)

For radiolocation on the scale of a city, a path loss model from a base station to a mobile device (typically on a person or vehicle) is required. The two-ray path loss model described above is not really appropriate, as the loss excess is also associated with medium-scale scattering, shadowing and diffraction effects from obstacles along the propagation path. This type of loss model is of much interest due to cellular phone applications, but it is equally applicable for radiolocation applications (see Sect. 9.7.3). The intention here is not to cover this topic in detail, but to provide details of one typical model, namely the Hata model (Hata 1980), which was originally developed from propagation studies in Tokyo in Japan. The model is mainly intended for city environments, but there are also loss corrections to the city (urban) model for suburban and rural environments. The model is applicable for radio frequencies from 150 to 1500 MHz, base station heights 30–200 m, and mobile heights 1–10 m. In urban areas the loss (in dB) is given by

$$L = 69.55 + 26.16\log(f_M) - 13.82\log(h_{BS}) - \alpha(h_m) + (44.9 - 6.55\log(h_{BS}))\log(R_{km}) \tag{9.8a}$$

where f_M is the frequency in MHz, h_m the mobile height in meters, h_{BS} the height of the base station in meters, and R_{km} the range in kilometers. The term $\alpha(h_m)$ is a correction related to the mobile height

$$\begin{aligned} &\alpha(h_m) = 8.29(\log(1.54h_m))^2 - 1.1 \quad (f_M < 200) \\ &\alpha(h_m) = 3.2(\log(11.75h_m))^2 - 4.97 \quad (f_M > 400) \\ &\alpha(h_m) = 4.145(\log(11.75h_m))^2 + 1.6(\log(1.54h_m))^2 - 3.035 \text{ otherwise} \end{aligned} \tag{9.8b}$$

The model above is for an "urban culture". For "suburban culture" there is a correction to (9.8b), namely

$$-2(\log(f_M/28))^2 - 5.4 \tag{9.8c}$$

and for "rural cultures"

$$-4.78(\log(f_M))^2 - 18.33\log(f_M) - 40.94 \tag{9.8d}$$

While the Hata equations define a fairly complex relationship of the loss as a function of the antenna heights, frequency and range, for a given application where the base station heights, mobile heights and frequency are all constant values, the urban Hata propagation loss can be expressed in the form

$$\begin{aligned} L &= L_0 + \log(R^a/\lambda^b) \\ L_0 &= 134.4 - 13.82\log(h_{BS}) - \alpha(h_m) + 3b \\ a &= 26.6 \\ b &= 44.9 - 6.55\log(h_{BS}) \end{aligned} \tag{9.9}$$

which is in a form similar to (9.7). Indeed if $a = b$ then (9.9) is exactly the same form as the generic model, but the two parameters (α, β) defined in (9.7) have explicit values which are functions of the system characteristics, rather than being determined from the measured losses directly. For typical base station heights used for systems that operate in cities, the base station antenna heights above the terrain level is in the range 15–30 m, or $b = 37 - 35$, so in practice $a \approx b$. Thus the Hata

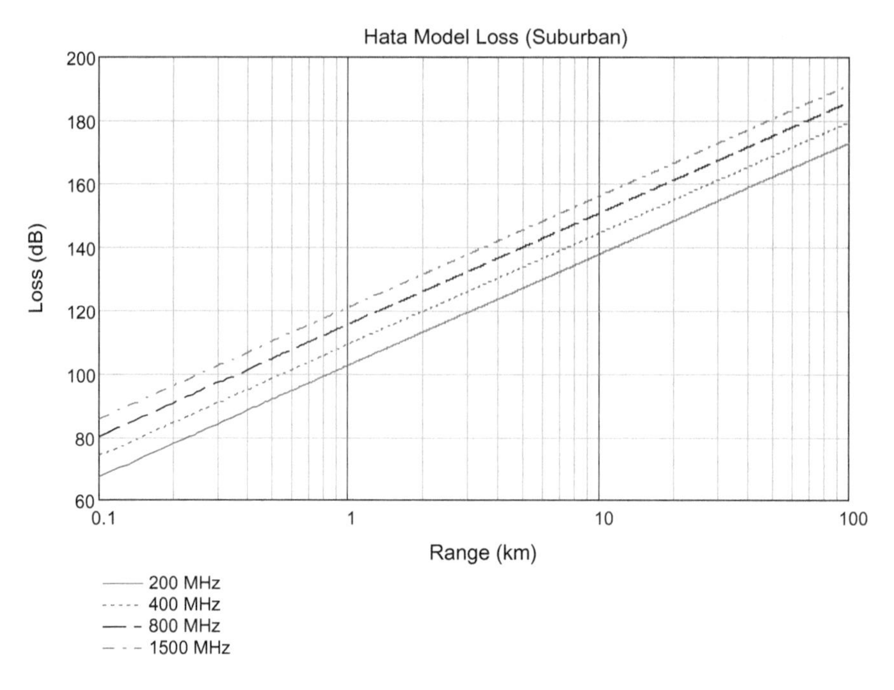

Fig. 9.4 Propagation loss as a function of range based on the Hata model. Base station height is 15 m, mobile height 1.5 m

model is actually close the generic model, with similar characteristics of loss versus range.

The (mean) propagation loss based on the Hata (suburban) model is shown in Fig. 9.4 for three different frequencies in an urban environment. When the range is plotted using a logarithmic scale, the resulting plot is a straight line, as expected from (9.9).

9.6 Statistical Link Budget Performance

The link budget calculations based on the models defined above are deterministic in terms of the basic system parameters, but variations in the propagation losses can only be described in terms of random processes. Thus the propagation loss will be a random variable which is best described by a Probability Density Function (PDF), with the above deterministic estimates the mean loss, with actual losses varying above and below the mean value. The main concern with performance is when the loss exceeds the mean value, and in particular how deep and how often signal fades occur. In practical design considerations no absolute performance can be guaranteed, but rather some temporal and/or spatial specification. For example, the system could be designed to operate satisfactorily x percent of the time at y percent of locations. Given these specifications and an estimate of the signal PDF, a *link margin* can be determined, which is an additional component allocated in the link budget to meet the required statistical performance. The following subsections provides an overview on how the link margin can be estimated in practical situations.

9.6.1 Log-Normal Variation

From the introductory comments on radio propagation, medium-scale variation is defined as being associated with energy scattering and absorption from large objects along the propagation path from the transmitter to the receiver. To measure the medium-scale energy received, the small-scale signal variation must first be averaged over an area of the order of 10 × 10 wavelengths. The deviation between the measured loss (corrected for small-scale fading) and the large-scale model loss can be explained by the medium-scale random variations. Because of the complex nature of the propagation environment, direct deterministic calculation of the signal path attenuation variation on a medium scale is not feasible, so that statistical methods must be adopted. Consider the simplified propagation diagram shown in Fig. 9.5, which shows that the signal amplitude scattering/reflection coefficient (ρ) for each leg of a path from a transmitter to a receiver. Although only one such path is shown, in practice the received signal will be composed of many such paths (p), but initially only the dominant (smallest loss) path will be considered. From the transmitter power and the receiver power, the excess propagation loss[3] is given by

$$\frac{P_{tx}}{P_{rx}} = l_{free\_space}(R) l_p = l_{free\_space}(R) \prod_{n=1}^{N} \rho_n \tag{9.10}$$

where N is the number of significant obstacles along the path. The nature of the path attenuation and scattering losses is undefined, but are assumed to be random with undefined statistics. As the loss measurements will be defined in decibels, it is more useful to consider the log of l_p, so that

$$L_p = 20 \log(l_p) = 20 \sum_{n=1}^{N} \log(\rho_n) \qquad \text{(dB)} \tag{9.11}$$

Thus when expressed in decibels the loss excess is the sum of the scattering coefficients expressed in decibels. As the scattering coefficients are presumably

[3]Here by definition the *path loss* is a number greater than unity, or in decibels the loss is a positive number. Alternatively, the *path gain* is the reciprocal of the loss, and will be less than unity and a negative number in decibels.

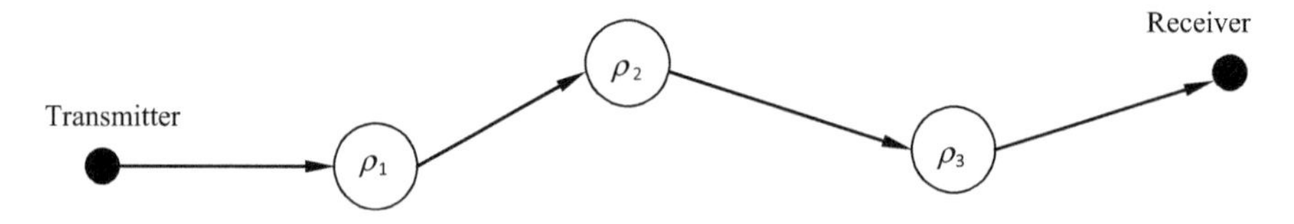

Fig. 9.5 Simplified diagram showing the scattering of RF energy along a path from the transmitter to the receiver

independent random variables, their logarithms also will be random, so the path loss in decibels also will be a random variable. However, as the path loss in decibels is expressed as the sum of a moderately large number independent random variables, invoking the Central Limit Theorem implies as the number of scattering paths becomes large, the statistical distribution will approach the Normal distribution. Thus when the loss is measured in decibels, the expected distribution will be log-normal, at least for a simple dominant path. The smoothing of the medium-scale fluctuations over a larger area, say 100×100 wavelengths, results in the large-scale variation in the propagation loss with distance, such as in the Hata model described in Sect. 9.5.4.

The log-normal distribution will have a large-scale mean which is a function of range (large-scale model), and a standard deviation relative to the mean. The standard deviation depends somewhat on the environment, but typical values lie in the range of 4–8 dB (Parsons and David 2002), including both indoor and outdoor environments. For a link budget, what is required is that the signal exceeds the minimum value a certain percent of the time. Figure 9.6 shows the required fade margin as a function of the probability of a fade occurring due to medium-scale lognormal fading statistics. For example, assuming a 6 dB standard deviation, and a 2% probability of a fade, then from Fig. 9.5 the required fade margin is $6 \times 2.05 = 12.3$ dB. This means that in 98% of locations the signal strength would be adequate for the operation of the link, assuming the large-scale model adequately defines the mean variation of the signal with range. Note also that in some environments small-scale fading may also be important.

9.6.2 *Small-Scale Signal Statistics*

The analysis of the signal variation for an outdoor system is usually limited to the large and medium scale effects, but in indoor applications where multipath and non-LOS propagation conditions predominate, small-scale signal variations can also be important. This is further exacerbated by the higher accuracy and higher update rates typically required in indoor positioning systems.

To consider these effects, it is useful to observe the propagation losses associated with actual measurements indoors. The data shown in Fig. 9.3 show considerable

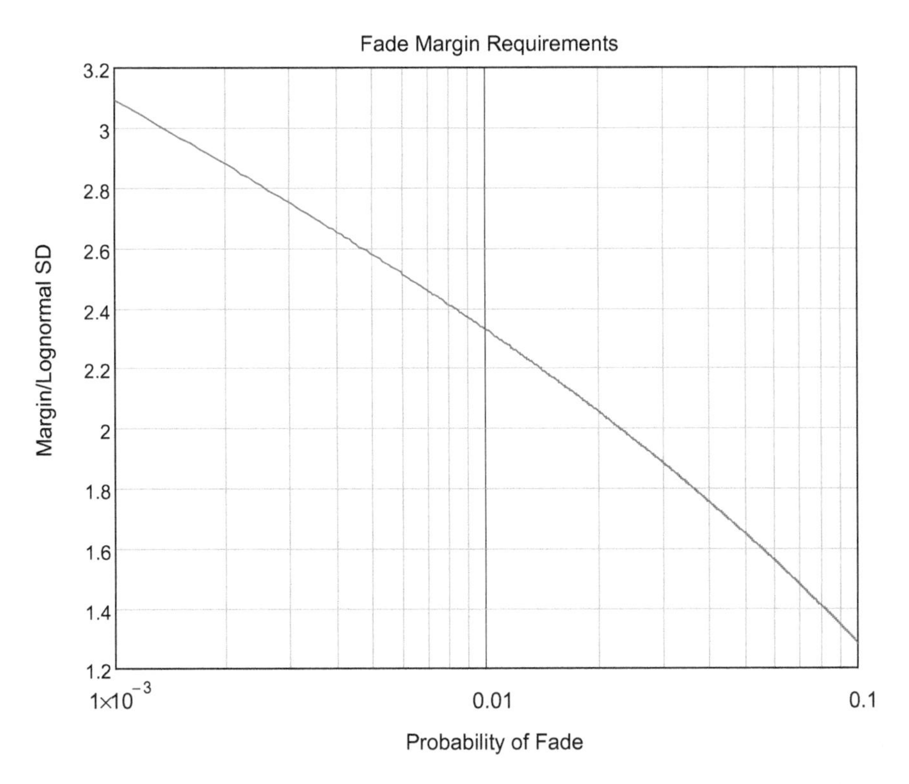

Fig. 9.6 Log-normal fade margin requirements (relative to the log-normal standard deviation) as a function of the probability of a fade

variation relative to the large-scale model defined by (9.7). The data in Fig. 9.3 show variations of 30–35 dB at the same range. In a NLOS indoor environment there are many possible paths between the transmitter and the receiver. At the receiver, the measured signal strength is the (complex) sum of all these signals, that is both the amplitude and the phase of the signal are important. The phase of the signal is closely related to the wavelength (360° phase variation), so phase effects can be expected on distance scales of the order of a wavelength. With no direct signal to dominate, the signal magnitude will be the complex sum of many signals scattered by the building walls and objects in rooms. These signals can combine either constructively or destructively, so that the measured signal can vary significantly from the local large-scale mean used for position determination. A detailed statistical analysis (Yu et al. 2009; Chap. 2) shows that in severe NLOS conditions the small-scale signal amplitude (s) exhibits approximately Rayleigh distribution statistics, which has the following mathematical form for the probability density function (PDF) and the cumulative distribution function (CDF)

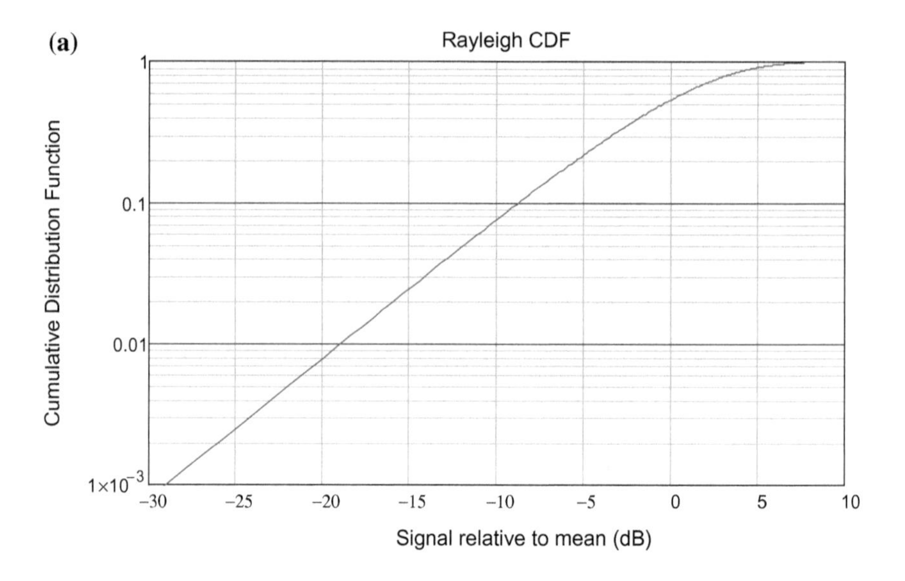

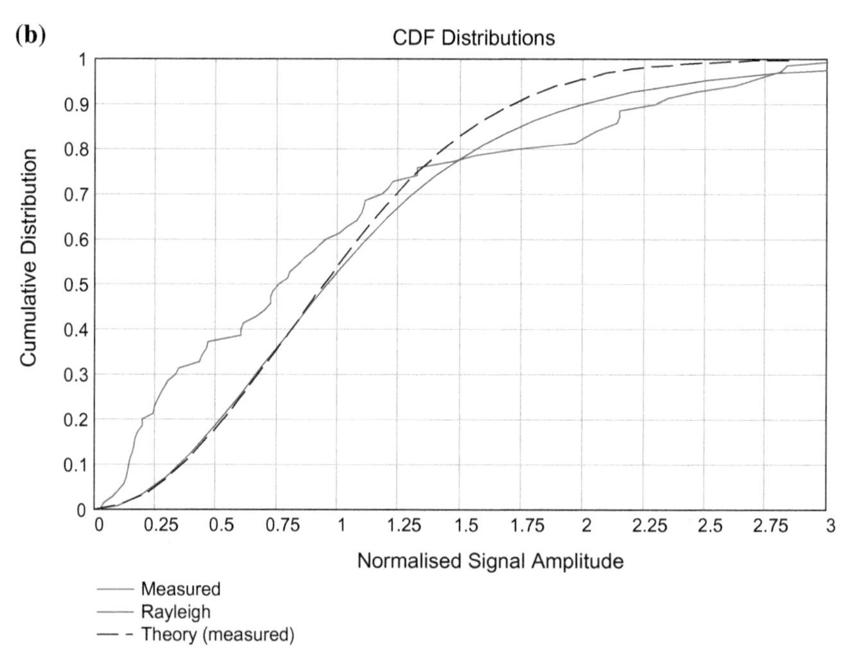

Fig. 9.7 **a** Normalised Rayleigh cumulative distribution function plotted with the normalised signal level in dB. **b** Cumulative distribution function derived from the measured raw data in Fig. 9.3, and the theoretical expected distribution based on normalisation with the mean signal. Also shown is the Rayleigh distribution

$$\begin{aligned} p_R(s) &= \frac{s}{\sigma^2}\exp\left[-s^2/2\sigma^2\right] \\ cdf_R(s) &= 1 - \exp\left[-s^2/2\sigma^2\right] \end{aligned} \quad (9.12)$$

However, the RSSI measurements are in decibels (dBm), so small changes in small signals can result in large changes in the measured RSSI signal; this effect explains some of the large variation in the computed losses shown in Fig. 9.3.

The normalised Rayleigh CDF is shown in Fig. 9.7a, where the signal amplitude is normalised by the signal mean and expressed in decibels, and the CDF is plotted on a log scale so that small probabilities can be observed. For example, the probability that a measured signal strength has a 10 dB fade relative to the mean signal is about 0.075. Note that the large scale mean signal strength is what is used to estimate the link budget.

The data in Fig. 9.3 can be tested to determine approximately if the signal strength data exhibits Rayleigh statistics. Rayleigh statistics are associated with signal fading relative to the local mean signal, which is also affected by unknown medium and large scale propagation attenuation effects. However, while the local mean signal at each measurement point is not known, an estimate can be made based on the multiple measurements near each position. This variation of the signal can be used to estimate the small-scale statistical variation, that is the normalised signal is the RSSI (in non-dB form) divided by the mean signal. In the case of the Zigbee measurements analysed there were four such measurements per location. The results are shown in Fig. 9.7b. The comparison between the measured and theoretical distributions is moderately accurate, but the measured data has a higher probability of occurrence at low signal amplitudes than the theoretical estimate.

Thus for the design of an indoor positioning system, it is recommended that the signal variations are modelled using small-scale Rayleigh fading statistics, rather than the medium-scale log-normal statistics which is recommended for outdoor positioning systems. A consequence of this is that the fade margin required is greater for indoor systems than for outdoor systems. For example, assuming the seme 2% failure rate as given in the medium-scale log-normal example, from Fig. 9.7a the required fade margin is 17 dB, compared with 12 dB in the log-normal case.

9.7 Case Studies of Positioning Systems

This section provides a number of case studies of positioning systems which illustrate the design principles described in previous sections of this chapter. The examples are chosen to illustrate that the same basic design strategy can lead to very different systems, due to the varying operational requirements. In each case a brief function description is given, then the design is largely summarised by tables (spreadsheets) which summarise the design. All the examples are based on actual systems which have been implemented using the methods described in this chapter.

The following high-level design examples shows how the link budget can be used to provide an initial estimate of the signal protocols and hardware design to meet the requirements of a specific application. Using the link budget the next step would typically be to define the design specifications for the various subsystems, such as the RF transmitters and receivers, baseband analog circuits, digital signal processing hardware, and the software, as well as the details of the signal protocols. Such detailed design may result in some refinements in the design, such as reducing the transmitter power (to increase battery life or battery size) in exchange for increasing the process gain or receiver sensitivity. In any case, it is the link budget analysis which ensures the subsystems collectively perform to the requirements of the particular application.

9.7.1 Rowing (2.4 GHz, 100 mW Transmitter, 2000 m)

The first example is a positioning system designed for tracking people or other objects on an outdoor sporting field. The original requirement was for tracking of race horses on a horse racing track, which is an open field of size up to 2 km, but the particular case studied is a rowing application based on an Olympic-size rowing course (2000 m in length). The intention is that the system equipment can be quickly set up at any sports field, similar to the setting up of TV cameras before an event. After the event the equipment can be removed.

The required positional accuracy is 0.5 m, with a position update rate of 10 per second. The portable base stations are located around the outside of the sporting field, and the mobile device is located on an athlete, on a rowing boat, or for horse racing in a saddle bag which is normally used for carrying handicap weights. The base stations and mobile devices are battery powered, with a battery life of at least 12 h. The high update rate and the high accuracy precludes using GPS.

Because of the mobile nature of the system, it is desirable that the system operates in an unlicensed frequency band, such as the ISM bands. Because of the combination of comparatively long range requirements and high accuracy requirements, the only suitable operating band is the 2.4 GHz ISM band (bandwidth typically 80–100 MHz). While the 900 MHz ISM band has lower propagation losses, the lower available channel width (typically 8 MHz) means the it is not possible to achieve a 0.5 m accuracy. Further as GPS (civilian mode) has an accuracy of the order of 5 m, and a channel width of 20 MHz (and an effective signal bandwidth somewhat less), it is clear the system operating with a channel width of only 8 MHz does not meet the required 0.5-m accuracy. Conversely, the 5.8 GHz ISM band has more bandwidth (better potential accuracy), but the higher

Table 9.2 Link budget data for sports tracking system (2.4 GHz)

Parameter	Value	Comment
Mobile power amp output	20 dBm	RF output power from mobile device. Low power to save battery life
Mobile antenna gain	−5 dBi	The internal antenna can be inside a saddle bag
Loss due to body blockage	5 dB	Diffraction loss around a body (such as a person)
Effective radiated power	10 dBm	TX power + Antenna gain – Diffraction loss
Maximum range	1000 m	Maximum range for required coverage area (not all base stations are required to be in range)
LOS loss at maximum range	100 dB	LOS loss (minimum possible) to BS from Mobile
Two-ray loss to BS	115 dB	Rowing example. See Fig. 9.2
Fading margin (BS)	5 dB	Due to multipath propagation and attenuation due to obstacles. See Fig. 9.2
Base station antenna gain	8 dBi	Relative to omni-directional transmissions
Rx power: max range (BS)	−102 dBm	Effective BS Tx power – Propagation loss – Fade margin + Rx antenna gain
Receiver IF bandwidth (B)	56 MHz	RF signal bandwidth using 30 Mchips/s spread-spectrum. Channel width 80 MHz
Noise figure (F) of receiver (BS)	3 dB	LNA at base station will improve the noise figure
Receiver noise (BS)	−94 dBm	$10\log(kTBF)$ (before despreading of signal)
Rx output SNR before despreading of signal	−8 dB	Rx power – Rx noise
Process gain (BS)	35 dB	Based on 255 chip code, accumulation of 15 codes
Receiver noise (BS) at output of correlator/accumulator	−129 dBm	$10\log(kTBF)$ (effective baseband bandwidth $B =$ 15 kHz after process gain)
Receiver output SNR (BS)	27 dB	Radio Rx SNR + Process gain
Required minimum SNR (BS, radiolocation)	20 dB	SNR for reliable measurement of position using signals from mobiles
Base station receiver margin	7 dB	Computed minimum SNR – Required SNR

frequency results in high propagation losses, which typically would mean higher transmitter powers, and hence larger batteries.

Table 9.2 provides the calculations of the link budget for a sports tracking system that meets the above specifications. The table only includes the link from the mobile device to the base stations used for determining the time-of-arrival for position determination. The detailed design will include four link budgets, two for data and two for position determination, but only the path from the mobile to the base station is considered here. The associated data transmission link will have a similar but slightly different link budget, as does the links to the mobile device from the base stations. The details of the implementation are not necessary for the link budget, except in the broadest terms. However, one design limitation is that a signal protocol in the ISM band must be some form of spread-spectrum, so only direct-sequence spread-spectrum protocols are considered n the link budget.

One key design aspect obtained from the link budget in Table 9.2 is associated with the spread-spectrum signal required for accurate position determination (defined as a position error standard deviation of 0.5 m). The 2.4 GHz ISM band has a bandwidth of 80 MHz, which dictates a direct-sequence chip rate of about 30 Mchips/s, after allowing for filtering to ensure the nominal binary signal spectral bandwidth fits within the RF bandwidth. This chip rate will define the accuracy of the TOA detection process, which is a function of the output SNR after any process gain. While the accuracy of TOA detection depends on the details of the detection process, the TOA standard deviation can be estimated (Yu et al. 2009; Chap. 4) by

$$\sigma_{TOA} \approx KT_{chip}/\sqrt{SNR} \tag{9.13}$$

where $K \approx 1$. In this case the chip period T_{chip} is 33 ns, and the required accuracy is 0.5 m (equivalent to about 1.5 ns propagation time), so applying (9.13) requires that the receiver baseband output SNR is about 500 (27 dB). Before the process gain Table 9.2 shows that the SNR is −8 dB, so the process gain required is 35 dB to meet the required tracking accuracy. For a direct-sequence spread-spectrum signal the associated process gain is simply the length of the code in chips (Yu et al. 2009; Sect. 3.3.2.1), so the integration length required is about $10^{35/10} \approx 3150$ chips. A spread-spectrum code with good auto-correlation properties is typically chosen such as a *maximal length sequence* binary code (see Chap. 3, Sect. 3.3.2.1), which has a length of $L = 2^N - 1$, so that given the requirement that the number of chips is at least 3150, one possible design is that the code length is 4095 ($N = 12$). For efficient signal processing of the required correlation process a Fast Fourier Transform (FFT) can be exploited, in which case the processing time required for the correlation is proportional to $L\log_2 L \approx LN = 49152$. However, based on the processing architecture in Fig. 9.1, an alternative is to have a shorter code length, plus an accumulator. The advantage of this design is the processing time associated with the correlation process can be greatly reduced, and with a processing pipeline the accumulation and correlation processes can be performed simultaneously. Thus the suggested design has $L = 255$, $N = 8$ and 15 of these codes are accumulated, so that the total length is 3825 chips (process gain 36 dB), and the correlation computation time is proportion to $L\log_2 L \approx LN = 2048$, or about 25 times less than using a 4095 chip code. The 3825 chips will take 127 μs to transmit at 30 Mchips/s. In this case the time available is sufficient so that the FFT can be performed in software within a general purpose Digital Signal Processor (DSP), rather than in dedicated hardware.

9.7.2 Indoor Position Determination (5.8 GHz, 10 mW Transmitter, 50 m)

The second example of a link budget for a TOA positioning system has a similar positioning accuracy of 0.5 m, but operates indoors. However, while an outdoor system will largely have LOS propagation conditions, an indoor system will largely have non-LOS propagation conditions due to internal walls with a building. These walls and other obstacles will mean the multipath conditions are severe making measurement of the TOA more difficult. Further, the propagation losses will be much greater for a given propagation distance, but the ranges will be much less which partly compensates for the additional attenuation with range. Because of the severe multipath conditions, a smaller TOA time resolution is required if the same accuracy is to be achieved, so that the system described in Sect. 9.7.1 will not meet the accuracy requirement. In a severe multipath environment range errors cannot be estimated using (9.13), as the range errors are related to the pulse distortion due to the multiple delayed signals. While estimates for these multipath errors are possible, for initial design calculations in indoor NLOS environments the suggested standard deviation of the error is 0.1 times the pulse rise-time.

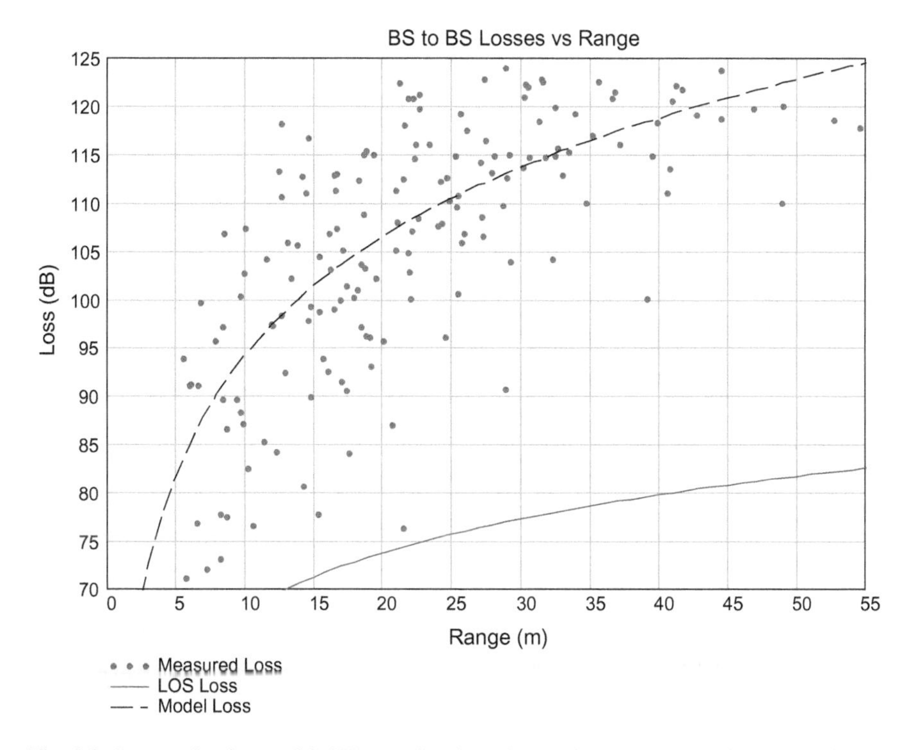

Fig. 9.8 Propagation loss at 5.8 GHz as a function of range between base stations in a building. The model (9.7) parameters are $\alpha = 0.7$ dB and $\beta = 40.9$, so that $\gamma = 4.1$ in (9.6)

Table 9.3 Link budget data for indoor tracking system (5.8 GHz)

Parameter	Value	Comment
Mobile power amp output	20 dBm	RF output power from mobile device. Low power to save battery life
Mobile antenna gain	−3 dBi	Internal antenna
Loss due to body blockage	10 dB	Diffraction loss around body (person)
Effective radiated power	5 dBm	Tx power + Antenna gain – Diffraction loss
Maximum range	40 m	Maximum range for required coverage area (not all base stations are required to be in range)
LOS loss at maximum range	80 dB	Line-of-sight loss (minimum possible) to BS from mobile. See Fig. 9.8
Mean loss excess at 40 m	38 dB	See Fig. 9.8
Fade margin	10 dB	Rayleigh fading margin probability 0.08 (see Fig. 9.7a)
Base station antenna gain	8 dBi	Relative to omni-directional transmissions
Rx power: max range (BS)	−113 dBm	Effective BS Tx power – Propagation loss – Fade margin + Rx antenna gain
Receiver IF bandwidth (B)	80 MHz	RF signal bandwidth.
Noise figure (F) of receiver (BS)	3 dB	LNA at base station will improve the noise figure
Receiver noise (BS)	−92 dBm	$10\log(kTBF)$ (before despreading of signal)
Rx output SNR before despreading of signal	−21 dB	Rx power – Rx noise
Process gain (correlator)	33 dB	Based on 2047 chip code, 80 Mchips/s
Process gain (accumulator)	9 dB	Accumulate 8 symbols
Symbol period	25.6 μs	
Receiver output SNR (BS)	21 dB	Radio Rx SNR + Process gain
Required minimum SNR (BS, radiolocation)	20 dB	SNR for reliable measurement of position using signals from mobiles
Base station receiver margin	1 dB	Computed minimum SNR – Required SNR

As a consequence, this system design is based on the 5.8 GHz ISM band, which can have up to 150 MHz channel width, or a signal bandwidth of around 80 MHz. The associated pulse rise-time is about 15 ns, so that using the above suggestion the estimated standard deviation of the TOA is 1.5 ns, or 0.45 m; this estimate is compatible with the desired accuracy of 0.5 m.

At 5.8 GHz the propagation losses (both in LOS conditions and through walls) will be significantly greater than at 2.4 GHz, and diffraction losses (around corners in buildings, and around a body) will also be significantly greater. Figure 9.8 shows some measured propagation losses between base stations operating at 5.8 GHz, together with the LOS and the model (9.7) losses. Notice the large scatter in the losses relative to the model mean loss, which is a combination of large and small scale effects described in Sect. 9.5. These effects need to be included in the link budget calculations.

The main link budget parameters are shown in Table 9.3. The design is aimed at achieving a TOA tracking accuracy of 0.5 m at a range of 40 m in the presence of a loss excess of 38 dB (see Fig. 9.8), and maximum fades of 10 dB due to body diffraction losses, and 10 dB loss due to multipath signal fades. The symbol length is chosen to be 2047 chips (25.6 μs), so that there are about 40,000 symbols per seconds. These symbols are shared between the mobile devices (say 50), or 800 symbols a second per mobile device, or 80 symbols per transmission assuming a position update rate of 10 per second for all 50 mobiles. The 80 symbols are used for the signal acquisition preamble, TOA estimation and data transmissions (such as the mobile identifier and device status data). Details of the actual protocol is beyond the scope of the link budget preliminary design. However, the link budget assumes 8 symbols are accumulated to improve the output SNR for determining the TOA. This design results in link budget margin of 1 dB, even under the worst case propagation conditions.

Finally, in summary, observe that although link budget methodology and the accuracy requirements of the indoor and outdoor (Sect. 9.7.1) tracking systems are the same, the link budget calculations result in very different designs in terms of the RF signals and their modulating protocols, and as a consequence the hardware is quite different.

9.7.3 *Covert Vehicle Tracking (400 MHz, 100 mW Transmitter, 15 km)*

The first two link budget cases are examples of high accuracy tracking systems that operate in a relatively small coverage area, and are mainly intended for tracking people. This case study is associated with a system for tracking vehicles over an area of a large city typically covering a few thousand square kilometers. While GPS is typically used for such applications, this application is intended for covert operations and to be essentially immune to jamming. Thus while GPS receives very weak signals from satellites (which can easily be jammed), a covert tracking system transmits a weak signal to base stations located throughout a city. To jam such a system all the base stations throughout the city would need to be jammed simultaneously, which is logistically improbable. To minimise the battery power, the transmitter power is limited to just 100 mW (same as the short range tracking systems examined previously), but the required range is 20 km. To achieve this performance a radical design is required, but the performance can be defined by the same principles using a link budget table.

The key to achieving a long range with a low transmitter power is having a very small effective signal bandwidth, and hence a very low system noise. In particular, the suggested method is the multiple access method described in (Yu et al. 2009; Sect. 3.3.2.1). Basically the concept is to exploit the spectral characteristics of the repeated transmission of a direct-sequence spread-spectrum signal, which results in a line spectrum at the code repetition rate. In this example, the code is 1023 chips in length, and the chip rate is 1 Mchips/s, resulting in a (filtered) spectrum which has a channel width of 2 MHz. As the code period is about 1 ms, the spectrum has lines separated by 1 kHz. Further, if the receiver employs an accumulator to sum 10 codes, the SNR in these spectral lines is enhanced (process gain 10 dB), but the effective accumulator filter also has nine intervening nulls at a separation of 1000/10 = 100 Hz (sinc function filter). Thus if other transmissions (carrier frequency) are at relative frequency offsets of 100, 200 … 900 Hz, these transmissions will not interfere with the original transmissions; this effect is mutual between each of the 10 simultaneous transmission, which all occupy the same 2 MHz channel. Thus in this particular example, there are 10 frequency domain multiple access (FDMA) channels, and as each transmission is 10 $\times$ 1 ms for the cumulation process, there can be 100 time domain multiple access (TDMA) transmissions per second per FDMA channel, or effectively 1000 time slots per second for transmission from mobile devices. For example, 1000 different mobile devices could be tracked every second, or 10,000 every 10 s, with 10 FDMA channels all within the same 2 MHz RF channel and all having an effective channel width of also 2 MHz.

The above scheme provides effectively (after despreading) channels of width 100 Hz, and hence very low associated noise. However, to achieve this the transmissions from all mobile devices must be very tightly controlled in RF frequency. For example, if frequency of transmissions is within $\pm$10 Hz and the RF frequency is 400 MHz, the frequency must be controlled to an accuracy of 0.025 ppm; this accuracy is well beyond the accuracy of commonly used crystal oscillators, which typically would have frequency stability of 1–10 ppm. Thus for this system to function correctly a timing reference (spread-spectrum) signal is transmitted from a central location, and the mobile devices control their local oscillators using this timing signal as a reference in a feedback loop. This frequency control concept is described in Chap. 5, and can readily achieve 0.025 ppm accuracy. However, there is a further complication when the mobile device is in a moving vehicle, due to the Doppler effect. The Doppler frequency offset is given by $\Delta f = f_{Rf} v/c$, where f_{RF} is the RF frequency (400 MHz in this case), v is the effective vehicle speed moving to or from the timing reference transmitter, and c speed of RF propagation (300 m/μs). For example, if the maximum vehicle speed is (say) 25 m/s, then the corresponding Doppler offset is $\Delta f = \pm 33$ Hz. As the above-described channel structure has a channel of $\pm$50 Hz, this worst case scenario has frequency offsets that remain within the channel width. Although the accumulator frequency nulls are not as

Table 9.4 Link budget data for vehicle tracking system (400 MHz)

Parameter	Value	Comment
Mobile power amp output	20 dBm	RF output power from mobile device. Low power to save battery life
Mobile antenna gain	0 dBi	Internal antenna
Loss due to signal blockage	8 dB	Device hidden inside of vehicle
Effective radiated power	12 dBm	TX power + Antenna gain – Diffraction loss
Maximum range	15 km	Maximum range for required coverage area (not all base stations are required to be in range)
LOS loss at maximum range	108 dB	Line-of-sight loss (minimum possible) to BS from Mobile.
Mean loss excess at 15 km	36 dB	See Fig. 9.4 (Hata model, 400 MHz)
Fade margin	8 dB	Probability of fade 0.1, with log-normal STD of 6 dB (see Fig. 9.6)
Base station antenna gain	10 dBi	Relative to omni-directional transmissions
Rx power: max range (BS)	−122 dBm	Effective BS Tx power – Propagation loss – Fade margin + Rx antenna gain
Receiver IF bandwidth (B)	1.0 MHz	RF signal bandwidth
Noise figure (F) of receiver (BS)	2 dB	LNA at base station will improve the noise figure
Receiver noise (BS)	−112 dBm	$10\log(kTBF)$ B = 1 MHz, F = 2 dB
Man-made RF noise excess	10 dB	Typical city suburban environment at 400 MHz
Receiver noise (BS)	−102 dBm	
Radio baseband output SNR	−20 dB	Before the receiver signal processing. Signal buried in noise cannot be detected by spectrum analyser—good for covert operations
Rx output SNR	20 dB	Rx power – Rx noise. Accumulator + Correlator process gain 40 dB
Symbol period	10 ms	10 codes accumulated. Sinc filter bandwidth 100 Hz
Required minimum SNR (BS, radiolocation)	20 dB	SNR for reliable measurement of position using signals from mobiles
Base Station receiver margin	0 dB	Computed minimum SNR – Required SNR

effective in this case, the vast majority of vehicles will have much slower speeds (and consequentially small Doppler offset frequencies), and thus overall the degrading effect from Doppler frequency offsets will be minimal.

Given the above description of the covert vehicle tracking system, Table 9.4 provides the details of the link budget. For this application the mobile device is hidden within the vehicle (such as under the vehicle instrument panel, or under the seats), so there is a considerable transmitted signal loss (budgeted value 8 dB) from the device to the outside of the vehicle—effectively the mobile antenna is an exciter for the metallic body of the vehicle to become the effective transmitting element. The (mean) propagation loss is estimated using the Hata suburban model described in Sect. 9.5.4, based on a transmitter height of 30 m and a receiver height of 1.5 m. The signal level (including a 8 dB fade margin) of –122 dBm at the input to the receiver is very low compared with more typical terrestrial radio system (such a cellular phones), which typically have a lower operating signal level of about –100 dBm. The receiver noise is dominated by the man-made environmental noise, which for a suburban environment at 400 MHz is about 10 dB (Sams and Co Engineers 1976) above the basic thermal noise kTB. The resulting noise is –112 dBm for the 1 MHz signal bandwidth. The SNR associated with the output of the radio receiver is –20 dB, so the signal will be buried in noise, and is not detectable on a spectrum analyser. This condition is important in covert operations. However, after the accumulator and correlator process gains (respectively 10 dB and 30 dB) the output SNR is 20 dB, which is close to the minimum suggested requirement for TOA tracking. Applying (9.13) in this case gives a TOA standard deviation of 33 m, so the worst case estimate has an accuracy close to the desired 30 m accuracy. Note also that the multipath accuracy estimate for a 1 Mchip/s code is 0.1 times 300 m, or 30 m. Thus the preliminary performance estimates obtained from the link budget predict that the required performance can be met.

9.7.4 UWB Testbed (6 GHz, 0.1 mW Transmitter, 50 m)

The final case study is not a link budget for a positioning system, but a tool used for measuring the RF propagation characteristics, particularly in an indoor environment. Unlike the other case studies which were based on spread-spectrum transmissions in either ISM bands or dedicated licensed channels, the basis for this link budget is the ultra-wide band (UWB) spectral allocation for appropriate applications, including high-speed data transmissions and highly accurate indoor positioning.

The UWB band is defined as 3.1–10.6 GHz, and the spectral bandwidth of transmissions must be at least 500 MHz. As the operating frequencies overlap other defined bands, both licensed and unlicensed, there is potential for interference with other radio systems, so that US FCC Part 15 rules limit the total transmission power to around 1 milliwatt, and a low power spectral density of −40 dBm/MHz. Thus to obtain any significant range the link budget process gain must be quite large, to compensate for the low transmitter power.

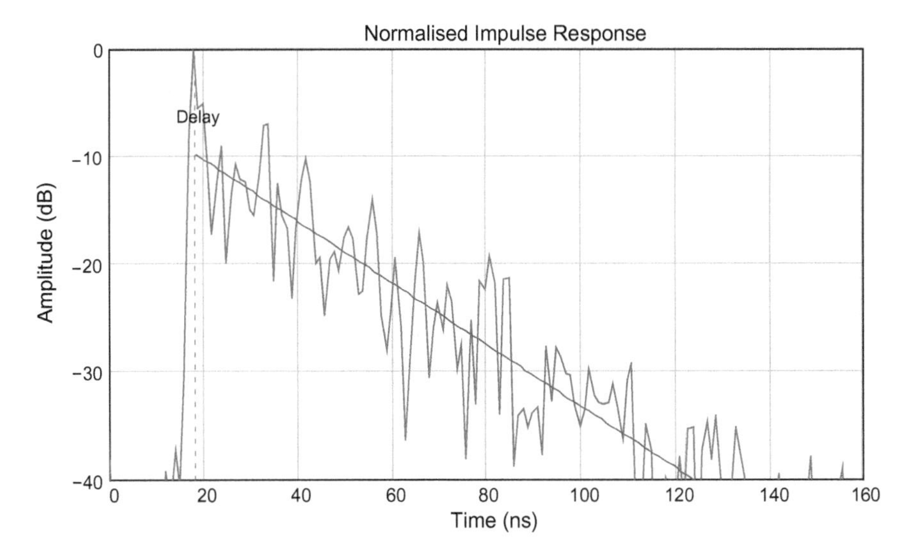

Fig. 9.9 Measured impulse response using a 1 GHz bandwidth at a RF frequency of 5 GHz. The propagation distance is 5.8 m (equivalent to 18 ns), corresponding to 62 dB LOS loss, and the mean measured loss in the frequency domain is 70 dB. The "Delay" parameter is the line-of-sight propagation delay, which is in close agreement with the measured delay. The linear decay slope of the impulse response is −0.27 dB/ns, or the normalised exponential delay parameter is −0.496

The UWB Testbed equipment is intended to test various systems signal protocols, particularly in indoor environments. Additionally, the equipment can be used as a Channel Sounder, which essentially measures the impulse response of the propagation environment. If the bandwidth of the signal is sufficiently large, the time of arrival of all the scattered signals of the transmitted "impulse" can be individually detected. The Testbed equipment described in this link budget analysis has a RF bandwidth of 1 GHz, and can operate in the bands 2–3 and 6–9 GHz. The baseband data are sampled at 2 Gsps. Typically for sounding a 1 Gchip/s direct-sequence spread-spectrum signal is transmitted, which results in a pulse rise-time of 1 ns. The measured impulse response is shown in Fig. 9.9. Notice that the time resolution of about 1 ns allows the straight line path delay to be measured, despite the extensive multipath signals which can have delay excesses of at least 100 ns. Also note that the amplitudes of these scattered signals decay approximately exponentially with delay excess.

Now consider the link budget for the above example, as shown in Table 9.5. The worst case (lowest) signal level required with an SNR of 20 dB is –150 dBm, an extremely low signal level for a radio receiver. As the RF signal bandwidth is 1 GHz to obtain the required time resolution, the raw receiver output noise is comparatively very large at –82 dBm. To be able to detect the weakest signals in the impulse response, the process gain must be very high, in this case 87 dB. To achieve such a high process gain, an integration time 0.5 s in required for a

Table 9.5 Link budget data for UWB Testbed (2–4 GHz, 6–9 GHz)

Parameter	Value	Comment
Transmitter power output	1 dBm	RF output power limited by the FCC rules
Antenna gain	2 dBi	UWB antenna
Effective radiated power	3 dBm	TX power + Antenna gain
Maximum range	50 m	Desired range for channel sounder measurements
LOS loss at maximum range	82 dB	Line-of-sight loss (minimum possible) at 6 GHz
NLOS mean loss excess at 50 m	40 dB	See Fig. 9.8
Weak signals margin	30 dB	Margin for weak signals in impulse response
Receiver antenna gain	2 dBi	UWB antenna
Rx power: max range	−147 dBm	Effective BS Tx power – Propagation loss – Fade margin + Rx antenna gain Minimum desired Rx power: −150 dBm
Receiver IF bandwidth (B)	1.0 GHz	RF signal bandwidth
Noise figure (F) of receiver (Rx)	2 dB	LNA will improve the noise figure
Receiver noise	−82 dBm	$10\log(kTBF)$ (before despreading of signal)
Rx output SNR before despreading of signal	−65 dB	Rx power – Rx noise. Signal deeply buried in noise
Tx/Rx local oscillator stability	0.1 ppb	Uses rubidium reference oscillators
Tx/Rx frequency uncertainty	±0.5 Hz	RF frequency 5 GHz
Maximum frequency error	1 Hz	
Maximum integration time	0.5 s	Maximum time for coherent integration
Process gain (correlator)	87 dB	Correlate 0.5×10^9 chips
Receiver output SNR	22 dB	Radio Rx SNR + Process gain
Required minimum SNR	20 dB	SNR for reliable measurement of impulse response with large delay excess
Receiver margin	2 dB	Computed minimum SNR – Required SNR

1 Gchips/s code. Note that this requires the signal to be sampled at 2 Gsps, and the 1 billion samples of in-phase and quadrature are each stored in the equipment RAM; for 12-bit data 3 GB of RAM is required. This integration time requires that the transmitter and receiver frequencies must be within 1 Hz of the RF signal frequency. This is achieved by the use of Rubidium local oscillators in both the transmitter and receiver units, which have a frequency stability of at least 0.1 ppb. The resulting output SNR is 21 dB.

This example of the link budget calculations shows how the very stringent specification can be met, both in terms of time resolution (1 ns) and very low signal levels (−150 dBm). Normally a small time resolution requires a wide signal bandwidth, which in turn implies high noise levels (low receiver sensitivity). The solution requires very stable oscillators for the transmitter/receiver RF equipment, very fast D/A and A/D converters, large and very fast RAM memory, and the associated digital equipment to store and process large amounts of data very quickly. The resulting equipment allows very detailed environmental impulse responses to be measured at ranges of at least 50 m in an indoor operation.

References

Bertoni H (2000) Radio propagation for modern wireless systems. Prentice Hall PTR

Hata M (1980) Empirical formula for propagation loss in land mobile radio services. IEEE Trans Veh Technol 29(3):317–325

Jordan EC, Balmain KG (1968) Electromagnetic waves and radiating systems. Prentice Hall

Parsons J, David J (2002) The mobile radio propagation channel. Wiley

Sams HW, Co Engineers (1976) Reference data for radio engineers. Chapter: Radio noise and interference. Sams

Yu K, Sharp I, Guo YJ (2009) Ground-based wireless positioning. Wiley-IEEE Press

Chapter 10
Applications for Non-GNSS Positioning Systems

10.1 Introduction

The widespread adoption of GPS in mobile devices has meant that there has been an explosion in applications (or commonly called "Apps" on mobile devices such as cellular phones and Tablets) that use location information to provide useful services. Apart from direct navigation and tracking functions for vehicles and to a lesser extent people, *location assistance services* can be a component of more general applications, such as providing "position stamping" of photos taken on a phone, or providing a list of local providers of services (such as shops, restaurants, etc.) based on the current GPS location. However, this chapter concentrates on a few applications for non-GPS tracking systems, and their associated applications. These applications typically require positional accuracy much greater than can be obtained from GPS (say in the range 0.1–1 m), but not the very high precision that can be obtained (for example) with visual positioning systems for tracking the motion of the human body. While several tracking technologies may be appropriate (including ultrasonics and inertial dead-reckoning), this chapter will be limited to the use of radio-based technologies. Thus this chapter will assume an appropriate radiolocation technology can provide position information at update rates of 1–10 per second with an accuracy of 0.1–1 m, and these data can be used in applications to provide other useful derived information. Thus much of the analysis in this chapter will focus on the methods used to create useful derived data from the raw positional data, and how positional errors affect the accuracy of the derived application output data.

I. Sharp and K. Yu, *Wireless Positioning: Principles and Practice*, Navigation: Science and Technology, https://doi.org/10.1007/978-981-10-8791-2_10

10.2 Overview of Applications

The applications studied in the following sections are associated with the combining of both positional and sensor (mainly accelerometer) data to provide useful information relating to a particular activity, mostly sports related. For such applications the requirements are typically beyond the capabilities of GPS, both in the required accuracy and the update rate. While visual tracking systems can achieve very high accuracy (of the order of a few centimeters) and can track fast moving objects, the application of video technology is limited by requiring many cameras to cover an area the size of a sporting field. Some other specialist equipment is also used in sporting applications, such as the very accurate timing (nominally 0.1 ms) required in motor sports. In this case, loop antennas are required to be embedded into the track to detect when a vehicle passes over the antenna. While such technology can achieve centimeter accuracy, the logistics of installing such equipment means that position information is only available at a few positions around the track. Thus radio-based technology is a good compromise, as it can provide good coverage of sporting arenas with a modest amount of equipment, and has appropriate positional accuracy and update rates required for a wide variety of sports.

The applications considered in this chapter are not those typically available on modern mobile devices, which can have country-wide coverage, but the positional accuracy is in the range of 5–10 m for GPS, or 20–50 m for basic wide-area WiFi positioning. For GPS, indoor positioning is usually not possible, while WiFi positioning is dependent on the distribution of WiFi Access Points that are defined in an appropriate database. These positional accuracies are adequate for typical location services (such as finding to local restaurant), but not for applications requiring precision location information.

The particular case studies considered in this chapter are as follows:

1. Sports Training System (Catapult 2018 visited). This application provides a generic system that can be used for training elite athletes in sports such as football (many different codes), athletics (particularly running related), basketball, rowing and cycling. The system is mainly a training tool, but also can be used during actual sporting events.
2. Motor Sports. This application is an example of tracking racing cars around a loop track to an accuracy of better than 50 cm. This allows race order, speed and partial and full lap times to measured to an accuracy of 1 ms.
3. Horse Racing. The horse racing application provides similar functions to the cars racing application, but also uses sensors to allow other useful data to be determined, such as the stride rate and length, and reaction times at the start of a race.
4. Animal Tracking (animal behaviour research). This application allows the behaviour of animals to be closely monitored for research purposes. The detailed movement of animals allows grazing patterns and herd instinct characteristics to be observed in detail. Sensors also allows other behavioural pattern to be observed and recorded over extended periods, both day and night.

5. Accurate Timing. While a positioning system is mainly used for determining the location of an object, the position data can also be processed to obtain very accurate timing information in activities which involve racing. These include areas such as motor sports, horse racing, athletes, cycling and rowing. By suitable processing of the positional data during a race, timing information to an accuracy of at least 1 ms for speeds up to 250 kph are possible. This performance compares favourably with existing (but more limited) timing methods.

10.3 Sports Training

In modern professional sport achieving the best performance from an athlete is becoming increasingly important. While training for sporting activities has long been performed, getting meaningful data regarding performance was at best rudimentary, particularly as a function of time over long periods of months and even years. Further, real-time data, both during training and during actual sporting events was not technically feasible. However, with the advent of radio-based tracking and data communications using small light-weight mobile devices, coupled with small but powerful computers with considerable data storage capability, a more scientific approach can be adopted in sports, particularly sports such as football (many different types), basketball, athletics, cycling (indoor velodrome), and rowing (2000 m Olympic course). In addition, the availability of small solid-state sensors, such as accelerometers, gyroscopes (angle turn rate measurements) and small mobile heart rate monitors allows further information to be gathered regarding the movement of the human body, and the physiological responses to physical exertion to be recorded. The following provides a description of a modern Sports Training System (STS), which is based on radiolocation technology with additional sensors and data communications. Such systems (Sports Training System) also include the data analysis and data display functions, as well physiological interpretation of the data, all of which are beyond the scope of the overview in this chapter. However, specific details on some of the measurements (particularly relating to positional data) are given in following sections of this chapter.

The concept of a STS is illustrated in Fig. 10.1. A STS is designed to be installed at indoor or outdoor arenas or training areas without any special infrastructure, and is totally portable. The system uses radio data links for communications with a mobile device attached to the athlete, between the STS components, and to a central data base. Access to the data base can be via the Internet, or in real-time to the athlete and/or the coach.

The concept is built around a positioning system which consists of base stations located around the periphery of the sports area, and mobile devices attached to the athletes. The system also includes data communications (typically using wireless LAN—WiFi), and Internet connections to a central Control and Processing Center and an associated Data Base where the real-time data are stored. The data also can

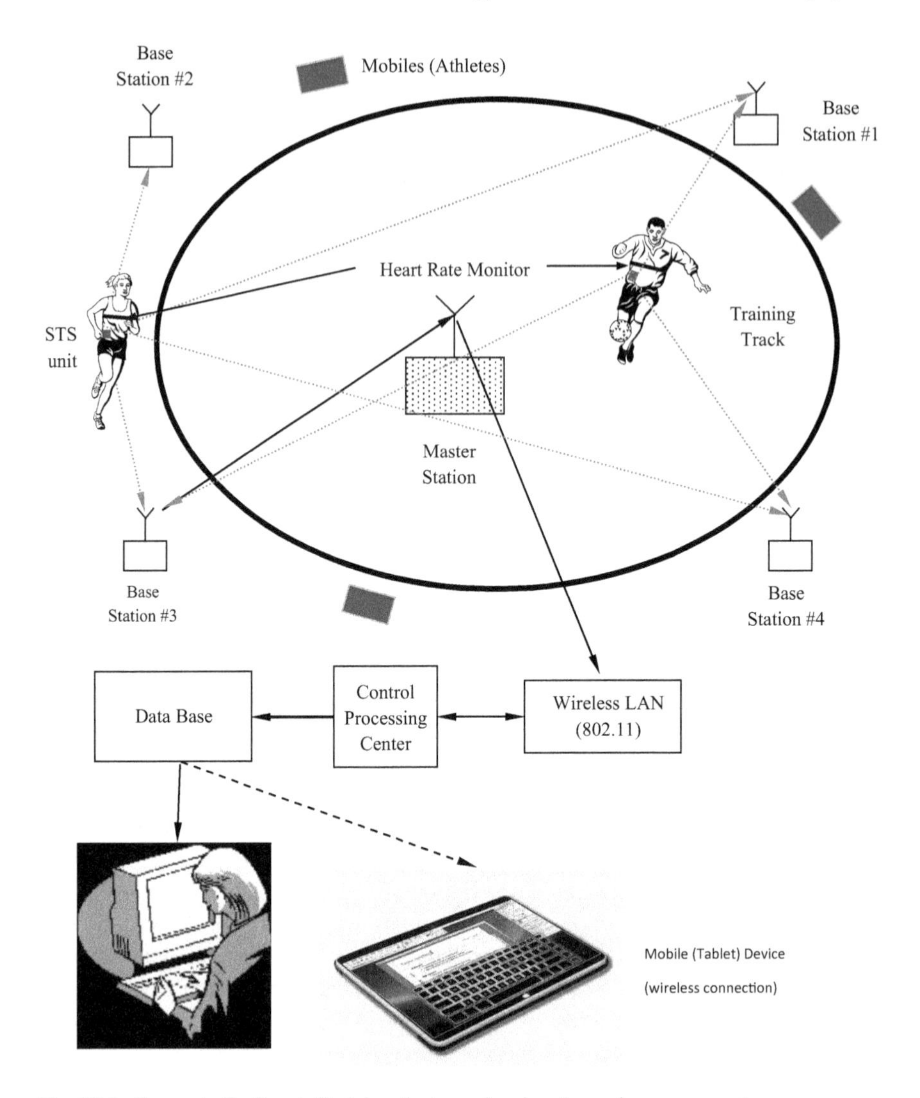

Fig. 10.1 Concept of a Sports Training System, showing the major components

be analysed off-line using appropriate software, or viewed in real-time at the training track by coaches and sports-medicine experts. A typical system could monitor a few tens of athletes with an update rate of at least 10 per second.

The main components of the STS and their typical specifications are summarized as follows. These specifications are typical of the major subsystems required to implement both the hardware and software components of a STS.

10.3.1 Base Station

The Base Stations, which are typically located around the periphery of a sporting field, receive the radiolocation and data transmissions from the mobile devices, and retransmits the information to the Master Station, which also provides the master system time synchronisation.

- Size: 170 mm × 140 mm × 80 mm
- Power: 12 V, 1 A, battery
- Up to 10 Base Stations. Typically requires only four to cover a typical sports field.
- Communicates with up to 24 mobiles
- Radio range: At least 1 km
- Radio transmissions time-of-arrival measurement accuracy (outdoors) 1 ns
- Communicates with Master Station

10.3.2 Master Station

The Master Station, which is typically located at a central location in relation to the sporting field, receives the radiolocation and data transmissions from the Base Stations, and then forwards the information to the Central Data Base via the Internet. The physical communications could be a combination of wireless (WiFi) and wired communications to an Internet Access Point. The Master Station would also provide the accurate local timing reference for determining the time-of-arrival of signals from the mobile devices (on athletes), and for the allocation of time slots for the time division multiple access (TDMA) method of sharing the radio channel.

- Size: 170 mm × 140 mm × 80 mm
- Power: 12 V, 1 A
- Provides communications hub and time synchronisation
- Wireless data link to central data base
- Radio range: At least 2 km
- Provides control signals to other components via the radio

10.3.3 Mobile Device

The mobile device worn by the athlete contains radio transmitters and receivers, and other various sensors to measure athlete motion and other important physiological parameters. The device is typically located in a pocket in a vest, which is typically located between the shoulder blades to ensure minimal interference to the athletic activities. The exact configuration of sensors would depend on the particular sport, but as a minimum would include inertial sensors which can measure the parameters such as the step rate and angular orientation of the body. Commonly a receiver from a heart rate monitor (magnetically coupled to a monitor belt worn around the chest) would also be included. Other possible sensors could measure the breathing rate, or ECG monitor for more detailed heart monitoring. The device would also have a connector (such as USB) to allow other (external) devices to be connected.

- Low Power RF transmissions (100 mW)
- Radio to receive transmissions from the Master Station, providing time synchronisation signals
- Inertial sensors (three-axis accelerometers and gyroscopes), compass
- Bio-sensors: heart rate (magnetic coupling to heart rate monitor), breathing rate. Optional 3-lead ECG
- Other devices: Microphone, speaker
- Size: 100 mm × 60 mm × 10 mm
- Weight: 100 g
- Battery life under continuous operation: 8 h
- Connector to external devices (typically a mini USB connector)
- Position accuracy: 50 cm. Range 2,000 m

10.3.4 Software

The STS software would consist of both real-time control and monitoring functions, as well as processing and displaying data received from the mobile devices. The real-time software would also send the data to a central data base for off-line data analysis. Such sport-medicine data analysis is becoming increasingly sophisticated, and this topic is beyond the scope of this high-level overview.

The basic functions of the real-time software, typically hosted in a Notebook PC, are summarised as follows:

- PC software, Windows operating system. All components designed for networking
- STS Management software for system setup, monitoring and fault diagnostics
- Technical software to display and monitor performance of system
- All data archived to SQL-compatible data base. Data archived for each athlete
- Local or remote real-time access to data. Web-based data base access
- Comprehensive graphical display of all data, user selectable
- Bio-medical software for displaying heart rate, breathing rate and accelerometer gait data
- Derived data include speed, stride rate, stride length, acceleration, turning rate, heading angle

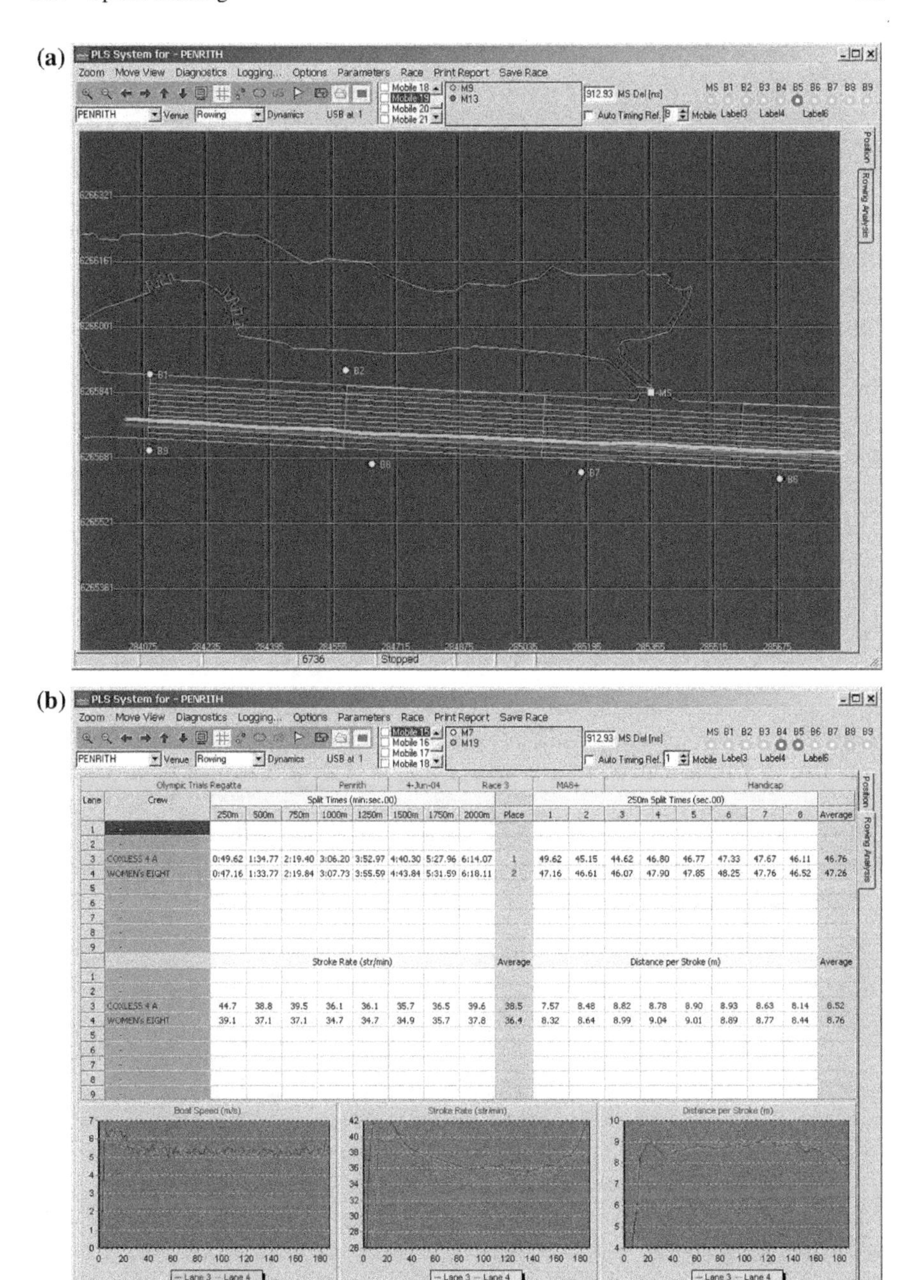

Fig. 10.2 **a** Sample screen showing the positions of the boats at an Olympic rowing course (Sydney International Rowing Course). The thin green lines are the lane markers, and the thick coloured lines the trails of the boats. **b** Sample data showing the real-time timing data, plus other derived information such as the speed and the stroke rate for each boat

- Analysis software to determine athlete fitness data
- Real-time setup for each individual athlete by the coach.

While the core software would be common to all sporting applications, the raw data would be processed differently for each type of sporting application. Such specific processing is beyond the scope of this overview. However, Fig. 10.2 shows two typical displays for a rowing application. The first figure shows a map of the rowing venue, with the path of the boats overlayed. It also shows the location of the base stations surrounding the rowing course. The user interface provides many options related to the STS itself, as well as data on each individual boat. The second figure shows some specific data relating to each boat, including a table of the split times along the 2000 m course, stroke rate and stroke length every 250 m, and graphical information on the boat speed, stroke rate, and stroke length. These data are derived from the positional data from the radiolocation component, and accelerometer data to determine the stroke rate. Although not shown, the data on individual rowers in the boat (such as the heart rate) can also be displayed.

10.4 Motor Sports (NASCAR)

The second application of a tracking system overviewed is associated with motor sports, whereby the position and other data from a racing car can be measured and displayed, often for television. This application is somewhat similar to the tracking of athletes, but the details required are somewhat different. While the technology and application software would be common to many particular motor sports, the following will describe an example for the National Association for Stock Car Auto Racing (NASCAR) in the US.

10.4.1 Deployment

An example of a NASCAR setup at an actual track (Atlanta) is shown in Fig. 10.3. In this case eight base stations are used around the track, as well as a Master Station (MS). For car racing the size of the field is several times greater than typically required for applications such as athletics or football; in this case the track is about 1 km in length. Note also in this application a base station can be placed on the inside of the oval track, as the cars are only on the oval track; a base station on the inside assists in the position determination. To ensure that the base station geometry gives accurate positioning of mobiles, the operator runs the GDOP (Geometric Dilution of Precision) utility using the planned base station locations. A GDOP of 1.5 or better is required for accurate positioning. Figure 10.4 shows actual GDOP at Atlanta speedway. It is noticeable that GDOP figure degrades to about 2 at the

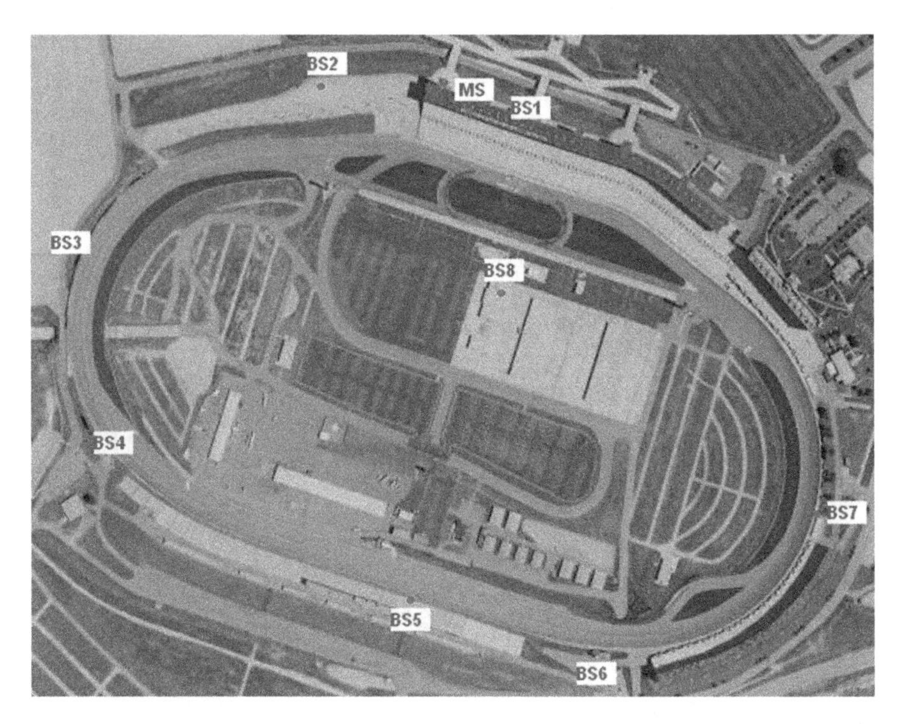

Fig. 10.3 Base station layout at Atlanta speedway; the largest distance is between BS #3 and BS #7 (just under 1 km). Note that BS #2 is located on an extension of the main grandstand which was constructed after the aerial photograph was taken

beginning of pit lane. Additional base station deployment in this area is required to improve accuracy.

The location of the mobile unit for car racing applications presents somewhat of a problem compared with mounting on a person. While ideally the antenna should be mounted on the roof (for the NASCAR case) which has 360° clear view of the base stations, the high speeds and resulting aerodynamics means that internal mounting is necessary, as shown in Fig. 10.5. Also shown is the mobile unit with a cable attached to the antenna. This location of the antenna is far from ideal, as the signal is blocked in directions where the base station direction is in front or to the left of the vehicle. For more details on the effects of internal verses external antenna locations on racing cars, refer to Sect. 10.6.

10.4.2 Functional and Performance Requirements

The requirements for a tracking system for sports car racing covers a number of areas. The first is simply to keep track of the location of each vehicle, so that the race order is known. This is particularly important in NASCAR racing, as the large

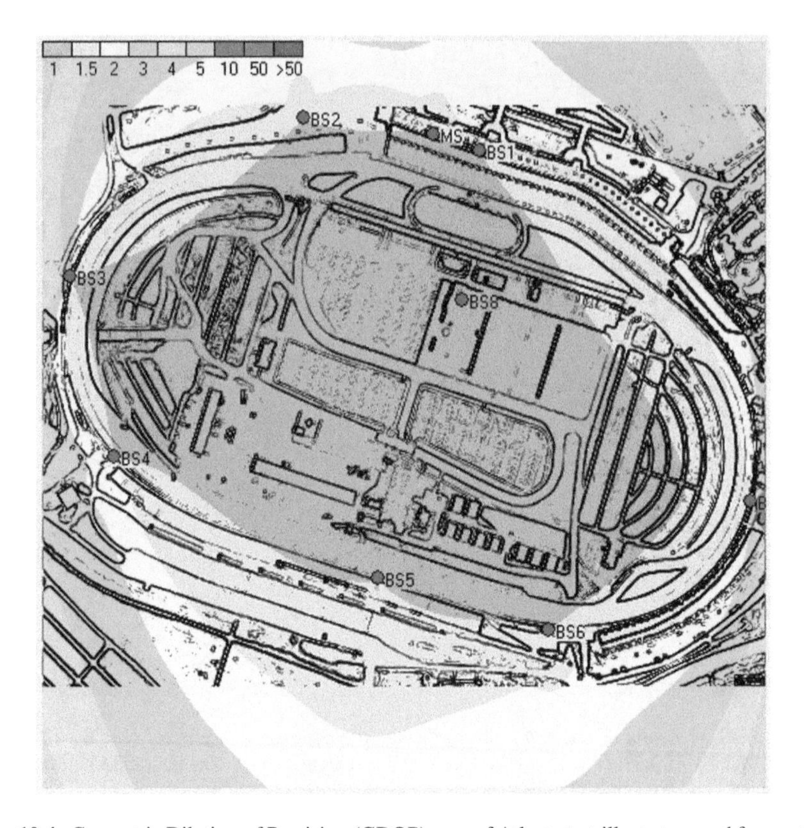

Fig. 10.4 Geometric Dilution of Precision (GDOP) map of Atlanta test illustrates need for another base station in the vicinity of pit lane start (GDOP < 2 is preferable)

number of vehicles and the small track size means it is not possible to determine race order by the position on the track, as cars may have completed different numbers of laps. This tracking task does not require high positional accuracy, so that a GPS based system could be used for this task, as an accuracy of the order of 3–5 m is adequate for this task.

A second more indirect task is to provide timing information such as lap times. Due to the high speeds timing in car racing is recorded to a precision of 1 ms, usually by a special unit located under the vehicle, and loop antennas buried within the track. See Annex A for more details on this technology. Typically, only a few such loops located around the track are used, so that there is little information available on car performance as a function of position along the track. If accurate tracking is used, together with special processing (see Sect. 10.6), then the position information can be converted to very accurate timing information between any two points around the track. For example, to achieve a timing accuracy of 1 ms for a vehicle moving at 250 kph, the corresponding position accuracy is 7 cm.

Fig. 10.5 Photograph showing the location of the antenna inside the vehicle adjacent to the rear windscreen

This capability is beyond the capability of GPS (for fast moving vehicles), but is possible for a tracking system of the type described in Sect. 10.3.

A third capability that is useful in a positioning system that includes sensors is that vehicle dynamics can be measured in real time, along with the corresponding positional information. While professional racing cars will include such sensors and the corresponding radio systems, for other motor sports it is useful to use the sensors contained within the radiolocation mobile device. With the use of accelerometers and gyroscopes, and with appropriate analytic software, detailed data on the acceleration and turning capabilities of the vehicle can be derived.

The performance of a tracking system which uses the 2.4 GHz ISM band with a signal bandwidth of 80 MHz is illustrated in Fig. 10.6a. In this figure the measured vehicle track is overlayed onto a map of the track. The positions are shown as individual dots with an update period of 80 ms. Based on this signal bandwidth the nominal time resolution is the reciprocal of the bandwidth, or 12.5 ns (equivalent to 3.75 m), but the actual positional accuracy will be much better than this figure. In practical implementation, with a signal-to-noise ratio (SNR) of 40 dB, the ranging measurement error (standard deviation) can approach $1/\sqrt{SNR}$ times the nominal resolution, or equivalent to about 4 cm in this case. This figure will be degraded by

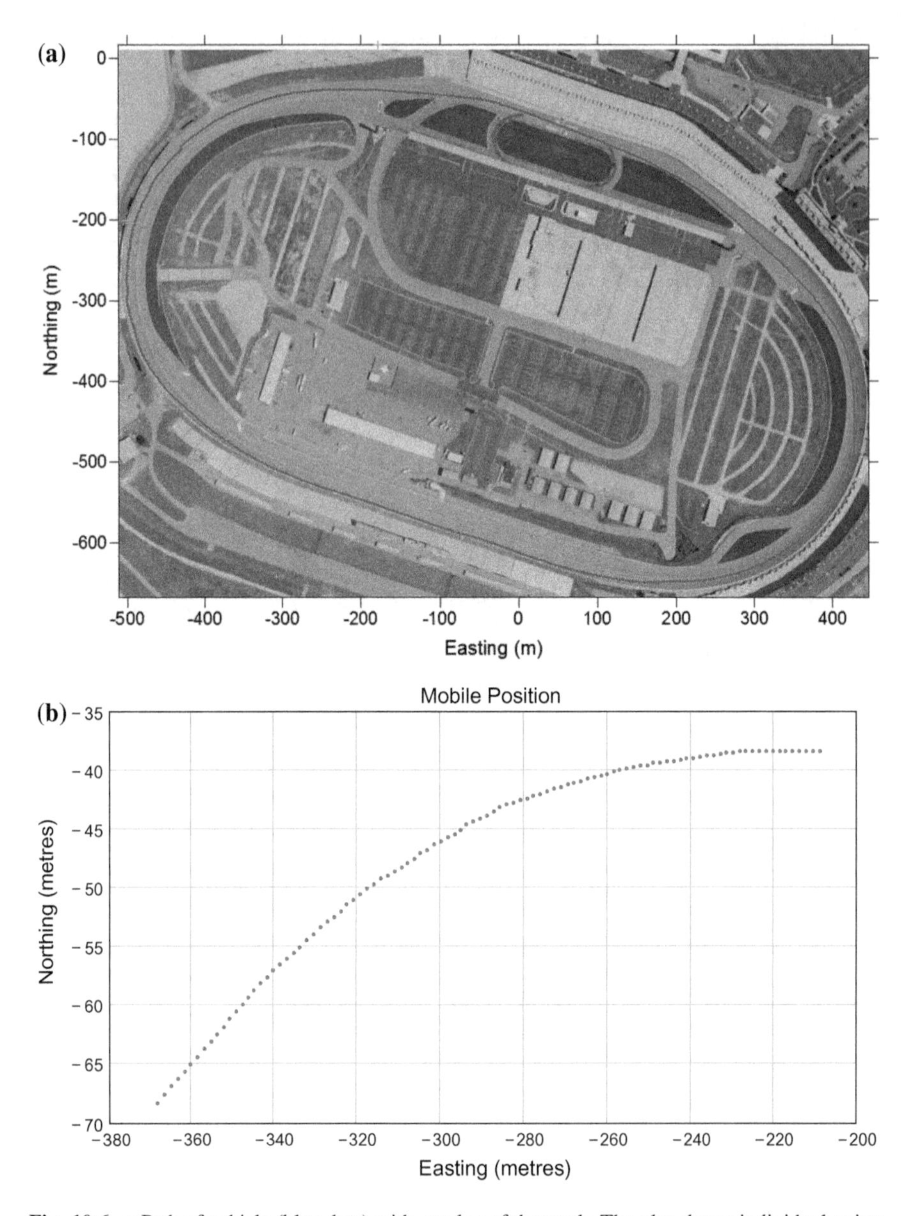

Fig. 10.6 **a** Path of vehicle (blue dots) with one lap of the track. The plot shows individual points of the path, but resolution of the plot limits observing the details. **b** Zoomed in section of the path of vehicle in Fig. 10.6a. The plot shows individual points of the path separated in time by 80 ms

multipath signals and other factors associated with the locations of the base stations and the mobile device on the vehicle. The actual positional accuracy can be estimated by performing a number of tests.

The first test is based on tracking the vehicle around the track. In this case absolute positional accuracy cannot be determined, but relative position accuracy can be estimated from differential positional measurements. With the speed in this test of about 40 m/s, the distance between position fixes is about 3 m, which is difficult to resolve on a map of the size shown in Fig. 10.6a. As the detailed overall results presented in this figure are limited by the resolution of the plotting (and not by the system), a zoomed-in version of part of the track is presented in Fig. 10.6b. Observe that the path is consistent with sub-metre accuracy, with differential measurement noise only a few centimeters.

The absolute accuracy cannot be determined from this path, but an estimate of the absolute accuracy can be derived by comparing the measured ranges to the base stations with the ranges to the surveyed positions. The average error due to the fitting process is shown in Fig. 10.7a, which has a mean error of 1.0 m, and the standard deviation error is 0.45 m. Note however that these errors also include surveying errors associated with K*inematic GPS* (El-Rabbany 2002), which can be a few decimetres. Alternatively, the relative accuracy between adjacent position fixes of the moving vehicle can provide further estimates of the positional accuracy of the system. Figure 10.7b shows the difference between adjacent position fixes (and hence is related to the speed of the vehicle). As can be observed in this case the differential between adjacent positions is about 1.5 m (corresponding to a speed of 67.5 kph), but the differential positional noise is about ± 0.1 m. This good relative positional accuracy allows accurate timing estimates to be made; see Sects. 10.4.3 and 10.6 for more details.

10.4.3 Timing Performance

The primary function of the vehicle tracking system is to determine the location of the vehicle anywhere on the track, but knowledge of position can also be used to estimate the time the vehicle passes a particular line (such as a start/finish line). Note that unlike the use of loop detectors, the "line" is defined only in the computer software, and thus many such lines can be defined. Further, by recording the time on successive laps, lap times can be calculated, as well as recording the number of completed laps. A detailed general analysis of this application of a tracking system is given in Sect. 10.5; the following is some specific details which relate to the NASCAR application.

The method of determining the time the vehicle crosses a line is based on performing a least-squares fit to the position data near the line, assuming the speed is constant for a small period of time, say ± 1 s relative to the crossing time. The analysis provides estimates of the time of crossing the line, the associated speed, and their standard deviations. The standard deviations are inversely proportional to the square-root of the number of samples in the analysis. In the case of the NASCAR, 25 samples were used, so the reduction factor is five. Based on the worst case positional accuracy of 20 cm, and applying the equations in Sect. 10.5, the

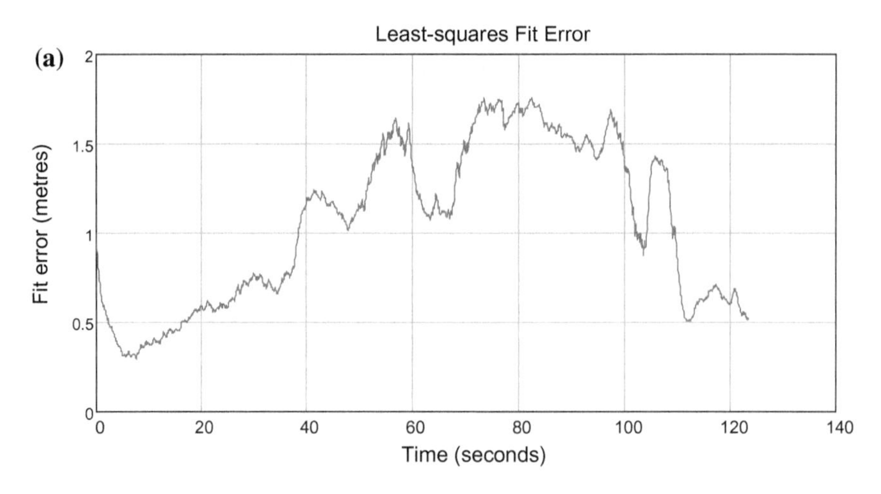

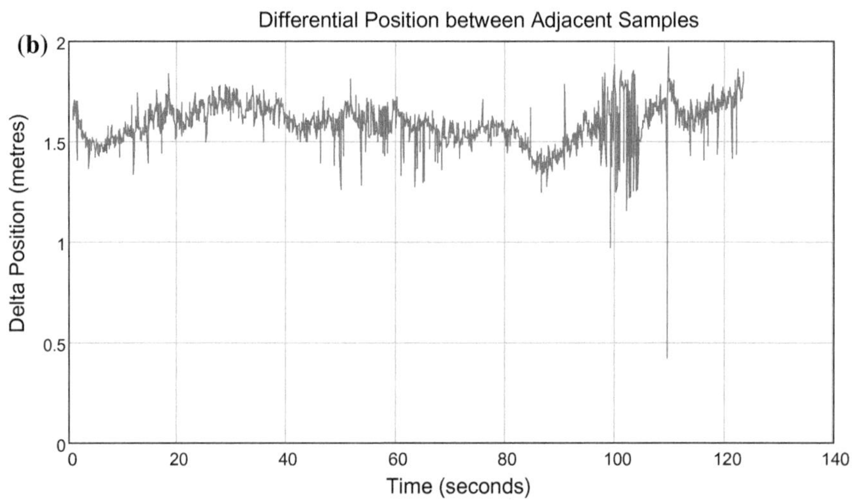

Fig. 10.7 **a** Least-squares fit error of the ranging data to the base stations. This fit error is an indication of the mean statistical error in the absolute positions. The standard deviation of 0.45 m is a better estimate of the absolute positional accuracy. **b** Change in position between adjacent position fixes for the tracking data shown in Fig. 10.6a. Position fixes are calculated every 80 ms. As can be observed, the sample-to-sample position noise is of the order of ±10 cm, but occasionally the differential error is much greater

measurement error in crossing the line is 4 cm, and the corresponding error in the speed estimate is 0.25 kph. Further, based on a maximum vehicle speed of 250 kph, the corresponding worst case timing error is about 0.6 ms. Note that while these estimates are based on relative accuracy (rather than absolute accuracy), the errors in absolute position will be essentially the same for all vehicles, because the measurements are all made at essentially the same position. Thus the errors associated with surveying and multipath propagation are the same for each vehicle, and thus there is very little effect on the relative timing errors between cars. This concept also means that lap times will be accurate, even if there are some absolute positional errors near the measurement line.

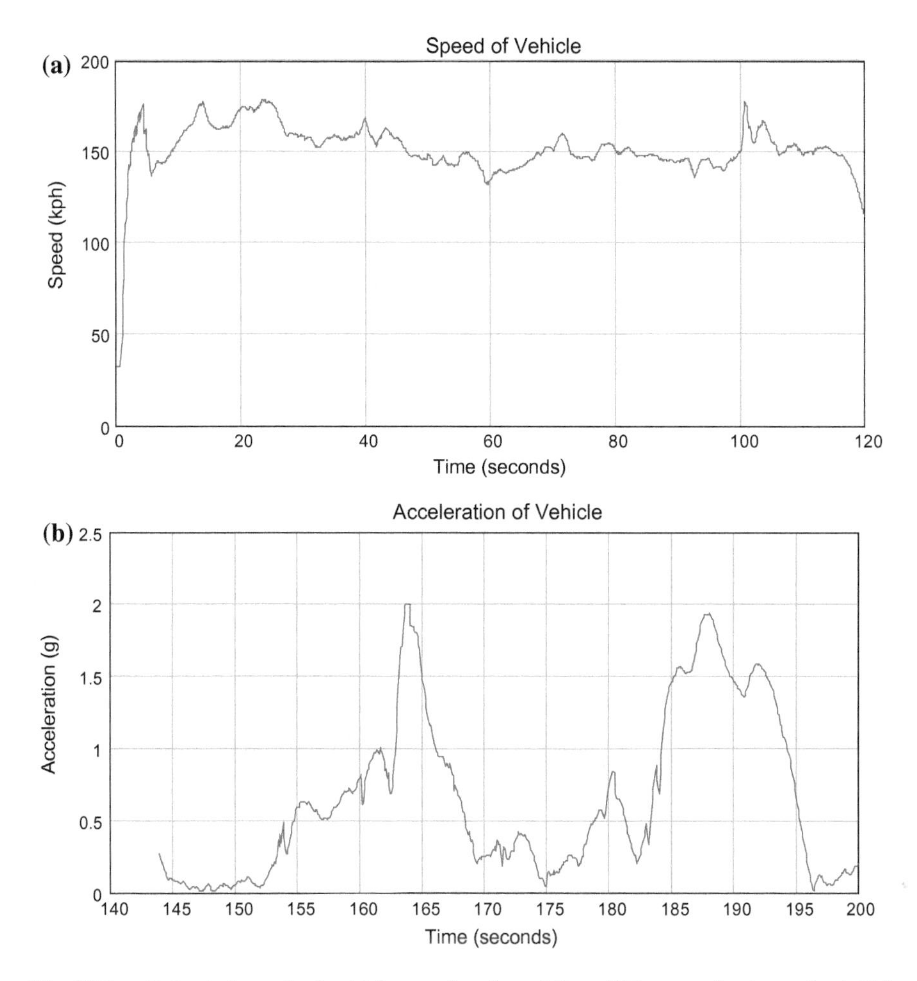

Fig. 10.8 **a** Estimated speed of vehicle as a function of time. This example shows the initial acceleration and then the approximate constant speed of 150 kph for a run with two laps of the circuit, and then slowing down after the second lap. The speed is estimated by a Kalman filter. **b** Acceleration of vehicle expressed in "g". In this example the vehicle accelerated from about 150 kph to about 250 kph. The peak in the acceleration g forces are in the turns due to centripetal acceleration. The acceleration is estimated by a Kalman filter

10.4.4 Other Data

The vehicle tracking system based on position determination allows other useful information to be derived from the positional data. This section briefly presents some of these derived data.

The raw positional data are processed using a Kalman filter, which not only provides improved positional accuracy, but provides estimates of the vehicle speed and the vehicle acceleration. Examples of these outputs are shown in Fig. 10.8a and b. Note however these estimates are generally not very accurate, and thus provide indicative information only.

In addition, the mobile device has inertial sensors to directly measure acceleration and turn rates, and a compass for the heading. If these sensor data are incorporated with the radiolocation data, better estimates of speed, acceleration, heading angle, and position are possible.

10.5 Horse Racing

In Sect. 10.4 the car racing application using radiolocation technology was described. In this section a closely related application is described, namely the tracking of race horses. In particular, this application mainly requires the positions of each horse relative to the leading horse continuously during the race, so that the information can be presented in real-time during the televising of the race.

While the concept is essentially identical to the car racing case, some of the details are peculiar to horse racing. Thus this section describes requirements which are peculiar to horse racing, and utilises the positioning capability together with the sensors on the mobile to obtain other useful information, such as the stride length and stride rate. Such data is useful in assessing the capabilities of each individual horse during a race, and can be useful in training as well as providing information on race tactics based on the capabilities of individual horses.

10.5.1 Deployment

The example described in this example case has a total of five Base Stations and a Master Station, which is the minimum recommended configuration for the horse racing application. The location of the Base Stations and the track configuration are indicated in Fig. 10.10. The base stations are located on towers around the track, which are also used for the cameras for televising the racing. Typical ranges required are up to 1 km.

The horses are tracked using mobile devices which are attached to the horses using pockets in the specially-made saddle cloths. Additionally, a standard cloth

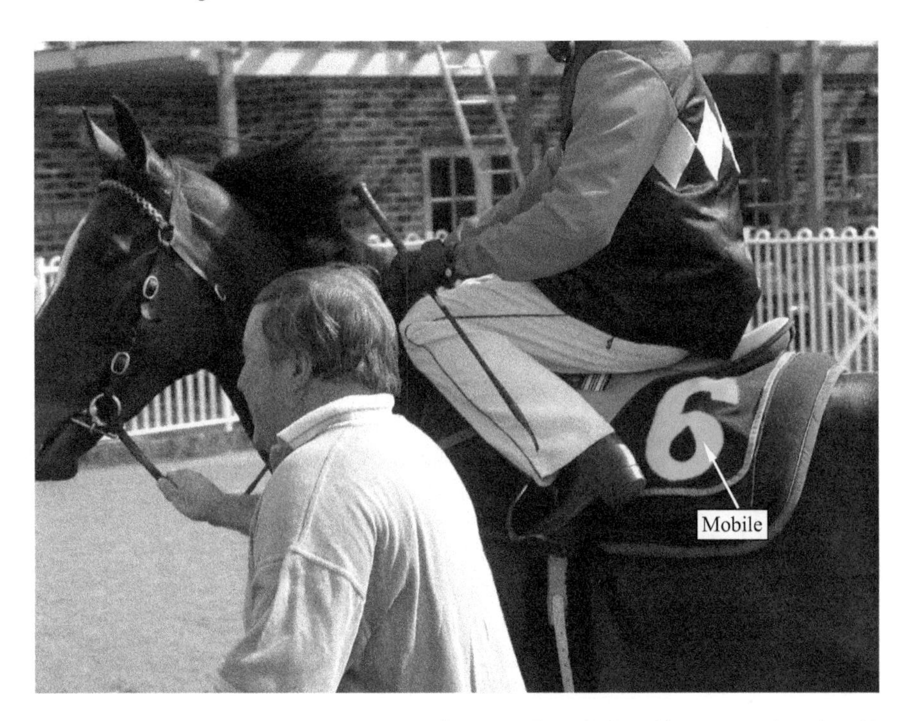

Fig. 10.9 Example of mobile under outer saddle cloth. The actual position was on the other side of the horse, or on the inside when traveling around the track in a clockwise direction. The mobile size used in this example is 110 × 60 × 30 mm, and the weight is 160 g

(with the horse number) covers the mobile unit in the inner saddle cloth. Figure 10.9 shows the typical location of the mobile on the horse.

10.5.2 Overall Tracking Performance

The prime output of the positioning system is the positions of the horses at the rate of 12.5 per second.[1] These positions are computed in Map Grid coordinates (Eastings and Northings), but for convenience the positional data are referenced to mean of the Base Station locations, so that the origin is at a point near the centre of the track. Figure 10.10 shows the track of a horse before, during and after a race. The positions of the Base Stations and the finish line are also shown. At the scale of the plot, the path is smooth, but the presentation is limited by the plotting resolution. A more detailed estimate of the positional accuracy is considered below.

[1]This rate is half the frame rate for the PAL television signal, so that the positional data can be synchronised with the television video.

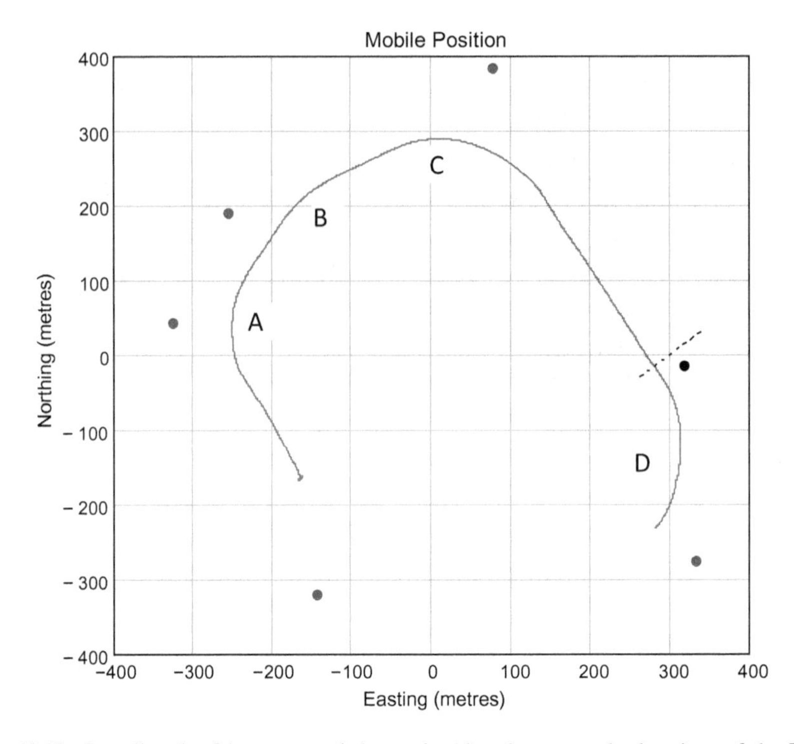

Fig. 10.10 Overall path of horse around the track. Also shown are the locations of the Base Stations and Master Station (black dot) and the finish line (dashed). The bends are labelled A, B, C and D. Details of the track shown in this figure are limited by the plotting resolution rather than the position system itself

A more detailed plot of a section of the path of the horse is shown in Fig. 10.11 near the finish line. In this case the Northing coordinate is plotted as a function of time, with the circles representing individual samples. If the speed is constant, the curve will be a straight line. As can be observed, the data are very close to a straight line, showing that the positional errors are rather small (see below for more details on the accuracy). It is further evident from the diagram that the time of crossing the finish line can be determined accurately, better than 0.01 s, if data interpolation is used, as described in Sect. 10.6.

However, note that the measured time is that of the mobile crossing the line, rather than the traditional measured time related to when the nose of the horse crosses the finish line. The distance of the nose of the horse relative to the mobile can be used to approximately correct the finishing time to the standard time based on the position of the nose of the horse. However, as the head of the horse moves relative to the body, the measured time cannot replace accurate finishing times generated from photo-finish equipment. Refer to Sect. 10.6 for more details of the timing accuracy.

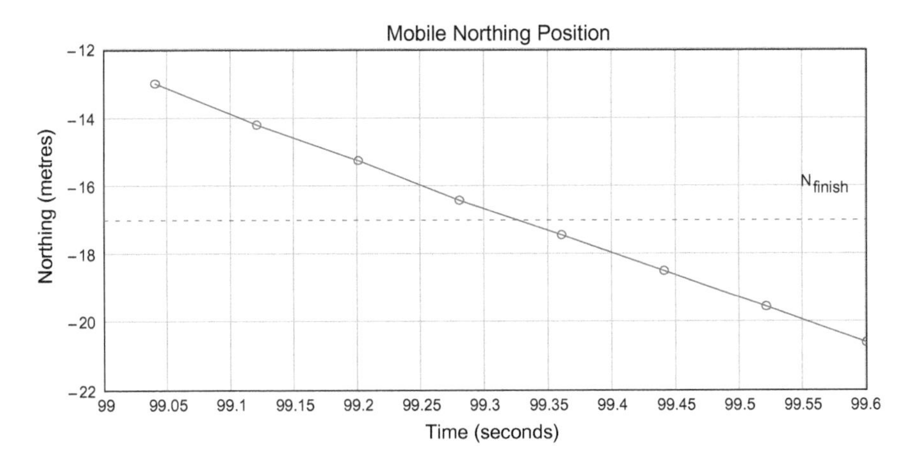

Fig. 10.11 Details of the northing data near the finish line. The individual circles are the positions calculated every 80 ms. By appropriate interpolation the finishing time can be determined to better than 0.01 s

The absolute accuracy of the positioning system cannot be determined from the raw positional data, so that indirect methods must be used to estimate the positional accuracy. One method is based on the least-squares fitting technique to the time-of-arrival measurements at each Base Station, used to determine the position. The derived range measurements at each Base Station can be compared with the range computed from the calculated horse position and the known positions of each Base Station. The root-mean squared (RMS) of the differential ranges can be shown to be directly related to the positional accuracy. Indeed the mean radial error in position (σ_r) is related by the ranging accuracy (σ_d) by the equation

$$\sigma_r = \sqrt{\sigma_x^2 + \sigma_y^2} = (GDOP)\,\sigma_d \tag{10.1}$$

The geometric dilution of precision (GDOP) factor is related to the position of the Base Stations relative to position of the horse, and for good placement of Base Stations around the track (as in this case) the GDOP is close to unity. The plot of GDOP for the data in Fig. 10.10 is shown in Fig. 10.12, which shows the GDOP is close to unity. The spikes in the data are due to the rejection of "bad" data from one Base Station, which reduces the number of base station data used to calculate the position, which slightly increases GDOP. Theoretical analysis shows that GDOP is inversely proportional to the square-root of the number of Base Stations.

As part of the position determination process, a least-squares fit to the measured time-of-arrival data is performed, and the RMS fit error to the noisy data calculated; this fit error data are plotted in Fig. 10.13, together with a smoothed estimate of the errors. The smoothing of the data is associated with the application of a Kalman filter to smooth the raw positional data, based on the known dynamics of the motion of horses during a race. The results show that the accuracy when the horse is in the

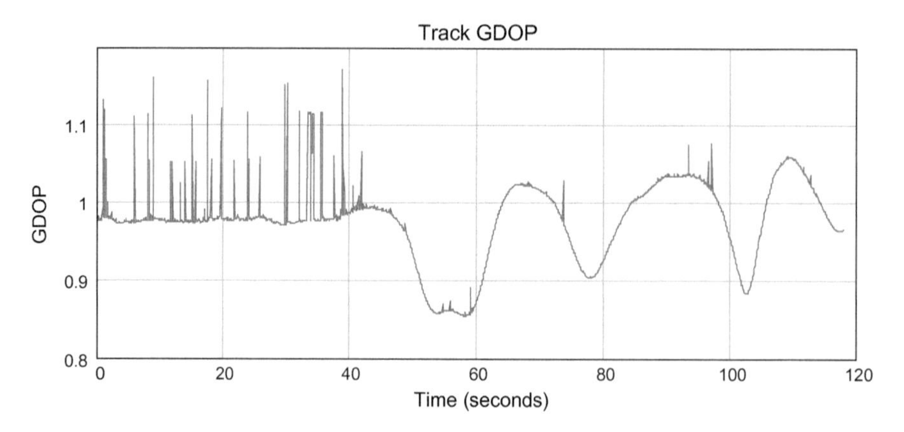

Fig. 10.12 Calculated GDOP for the horse path and the positions of the base stations shown in Fig. 10.10

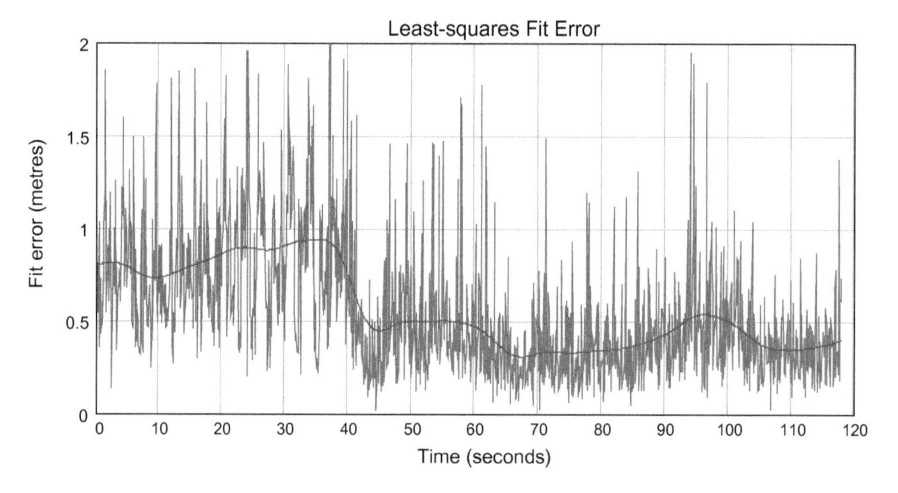

Fig. 10.13 Measured RMS fit error. The red plot is the data on a frame-by-frame basis, and the blue data are a smoothed version. The first 40 s relate to the horse in the starting gate, while the data after the 40 s point are the race data

starting gate (first 40 s) is around 1 m, while during the race the accuracy is better than 0.5 m. The reduction in accuracy in the starting gates is due to the extra signal scattering from the metal starting gate structure.

An alternate approach to the estimation the positional error is based on the assumption that the path of the horse is a smooth function of time. Thus by smoothing the data the error in position can be estimated. In general, this method will underestimate the error, as slowly varying errors will be followed by the smoothing process, and will thus does not appear in the differential position data. The results of the analysis are shown in Fig. 10.14a. The mean radial error is of the

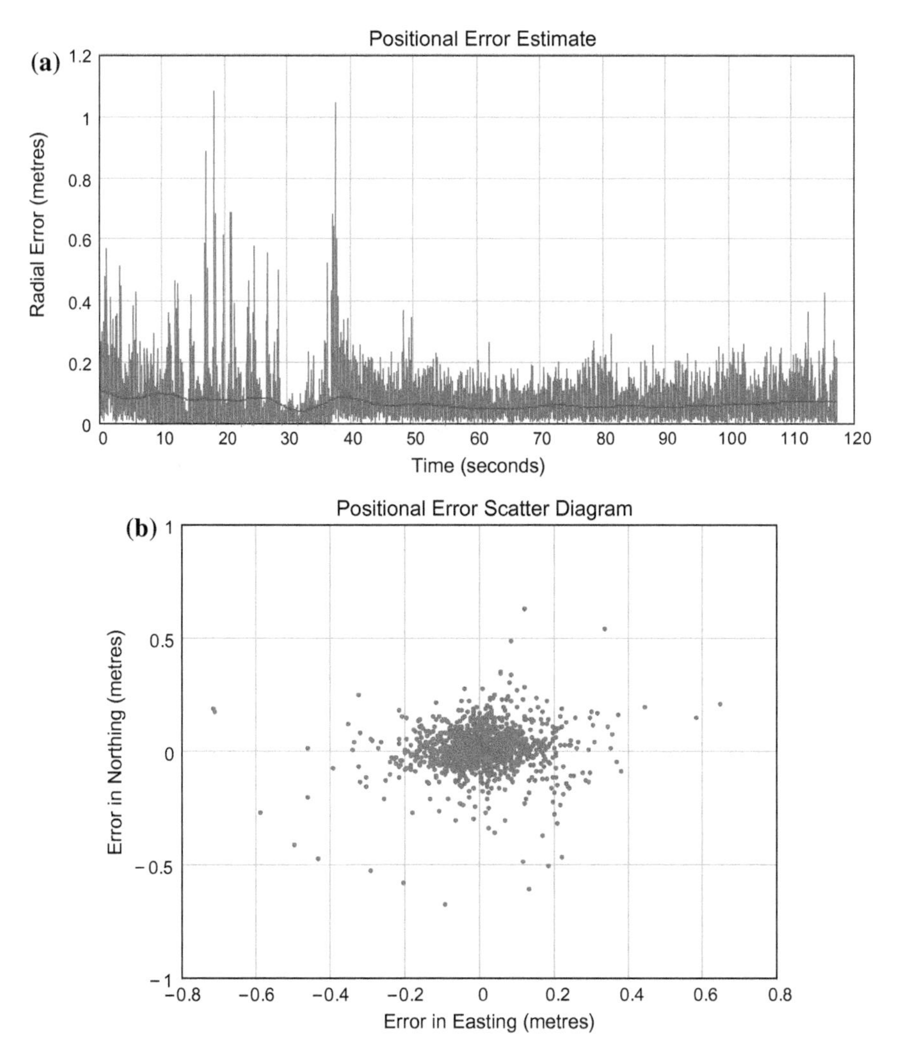

Fig. 10.14 **a** Estimated radial positional error based on the differential between the measured position and a smoothed track. The blue line is the smoothed average of the individual frame data. **b** Scatter diagram of positional errors, based on deviations from the smoothed track

order of 0.15–0.2 m, but peak errors are up to about 1 m. As previously observed the errors are largest when the horse is in the starting gate (first 40 s). When compared with the error estimates derived from the RMS fit, the errors based on path smoothing are about a factor of two smaller.

The 2D scatter of positional errors (based on track smoothing) are shown in Fig. 10.14b. The scatter of points is broadly circularly symmetrical, with most points clustered close to the centre. All the positional errors lie within a circle of radius 1 m.

In conclusion, the positional accuracy of the horse racing tracking system during a race has been estimated to be in the range of 0.15–0.3 m, with peak errors of up to 1 m. These figures more than match the accuracy requirement of 0.5 m deemed to be satisfactory for determining the relative position of horses during a race, but is insufficient to accurately determine the race finishing order.

10.5.3 Relative Tracking Performance

The data presented in Sect. 10.5.3 related to the absolute positional accuracy of individual mobile devices (horses), but the main requirement for horse racing is relative positions between horses during a race. This section presents data on relative positions.

The simplest analysis of relative positions is between two horses. Figure 10.15 shows the distance between two horses before and during the race described previously. For the first 40 s the horses are in the starting gate, so the distance between the horses is constant. The actual data show variations (positional measurement noise) of up to ± 1 m, which is consistent with the estimated accuracy in the starting gate in Sect. 10.5.2. During the race the distance between the horses is not known, but it is expected that the distance will be a slow function of time. The measured distance data in Fig. 10.15 generally show such a slow variation in separation with time, with very small noise. Based on the statistical data in Sect. 10.5.2, the estimated accuracy in the separation distance is in the range of 0.15–0.3 m.

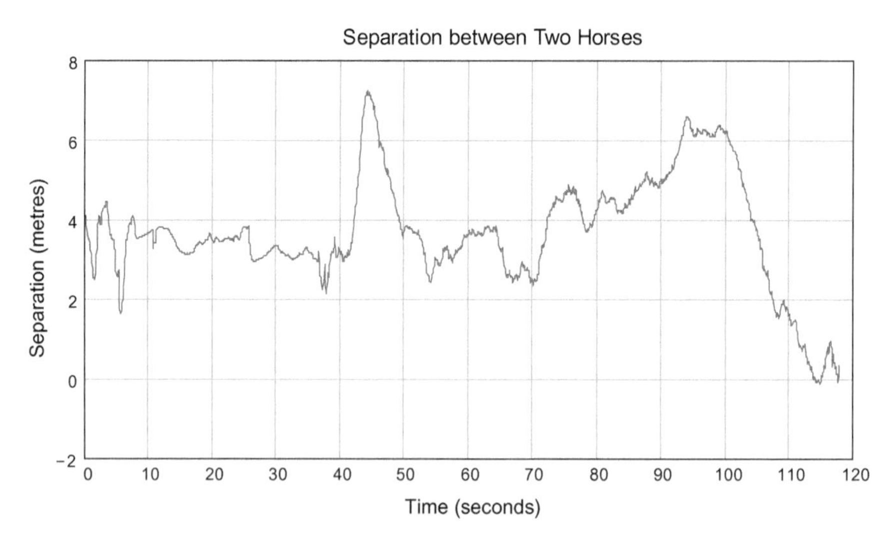

Fig. 10.15 Example of the separation distance between two horses. While the true separation is not known, the accuracy can be inferred by the distance "noise" which is much less than 1 m

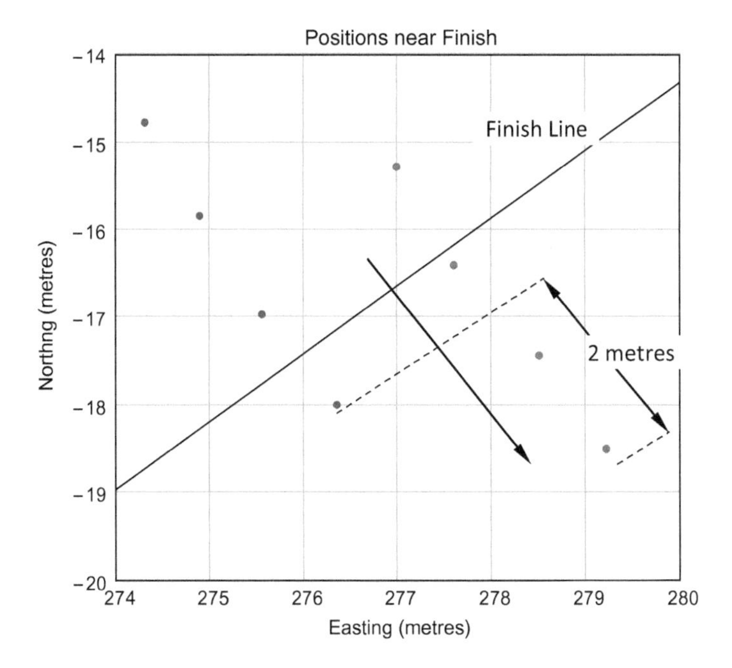

Fig. 10.16 The positions of the first and second horse near the finish line. The separation of the two horses (2 m) at the finish line can be estimated geometrically as shown

The separation between two horses can be considered in more detail, as shown in Fig. 10.16. This example shows the position of the first and second horses near the finish line. Simple geometry allows the separation distance based on the individual position determinations at 80 ms spacing in time. At the typical speed of horses, the position separation is around 1.4 m between position updates. More accurate separation estimates can be determined by performing a least squares fit to the data near the finish line. For more details on the analysis associated with the crossing of a finishing line, refer to Sect. 10.6 following. Because of the averaging associated with the fitting process, the relative position (of the mobile devices) can be determined to an accuracy of an estimated 5 cm.

10.5.4 Sensor Measurements and Performance

The mobile device includes sensors for measuring the acceleration, rotation about the vertical axis, and two magnetometers which can be used as a compass. One purpose of these sensors is to provide positional data (using "dead reckoning" techniques) when the radiolocation system is inaccurate. However, the results reported in this section relate to additional useful information associated with horse racing. These include:

- Determining whether the mobile is moving or stationary, so that the Kalman filtering dynamics can be altered accordingly.
- Determine the start time of the race.
- Determining the reaction time of the horse/jockey at the start of the race.
- Determining the stride rate (and hence the stride length) of the horse.

These auxiliary functions are described in this a subsequent sub-sections.

10.5.4.1 Raw Sensor Data

The raw accelerometer and rate gyro data are shown in Fig. 10.17a and 10.17b respectively. The acceleration is measured in two directions, namely in the longitudinal direction (direction of motion) and vertically (or the z direction). The z-accelerometer also measures the earth's gravity, so the nominal quiescent reading is 1 g. The rate-gyro measures the turn rate (degrees per second) about the vertical (z) direction.

As can be observed from the figures, the raw sensor data have dominant components associated with the movement of the horse (the stride pattern). Thus it is clear that the sensor data can be used to detect when the horse is galloping or is stationary. For more details on the determination of the start time refer to Sect. 10.5.5.1 following.

The raw accelerometer data in Fig. 10.17a show quite large variations of about 20 m/s^2, or 2 g. These variations are a consequence of the oscillatory motion associated with the stride pattern. The relationship between the amplitude (a) of the oscillations, the stride period T_{stride} and the acceleration (A) is given approximately by the formula

$$A = \left(\frac{2\pi}{T_{stride}}\right)^2 a \tag{10.2}$$

For typical accelerations of 2 g and a stride period of 0.4 s (see Sect. 10.5.4.2 following), the amplitude of the oscillations is 8 cm. As the stride length is typically 8 m, the oscillations represent only 1% of the stride length, and thus are not easily observable.

10.5.4.2 Determination of the Stride Pattern

The sensor data show a regular pattern associated with the stride pattern of the horse. A typical spectrum of the accelerometer is shown in Fig. 10.18. In this case the main spectral component is 2.3 ± 0.2 Hz, but a smaller amplitude second order

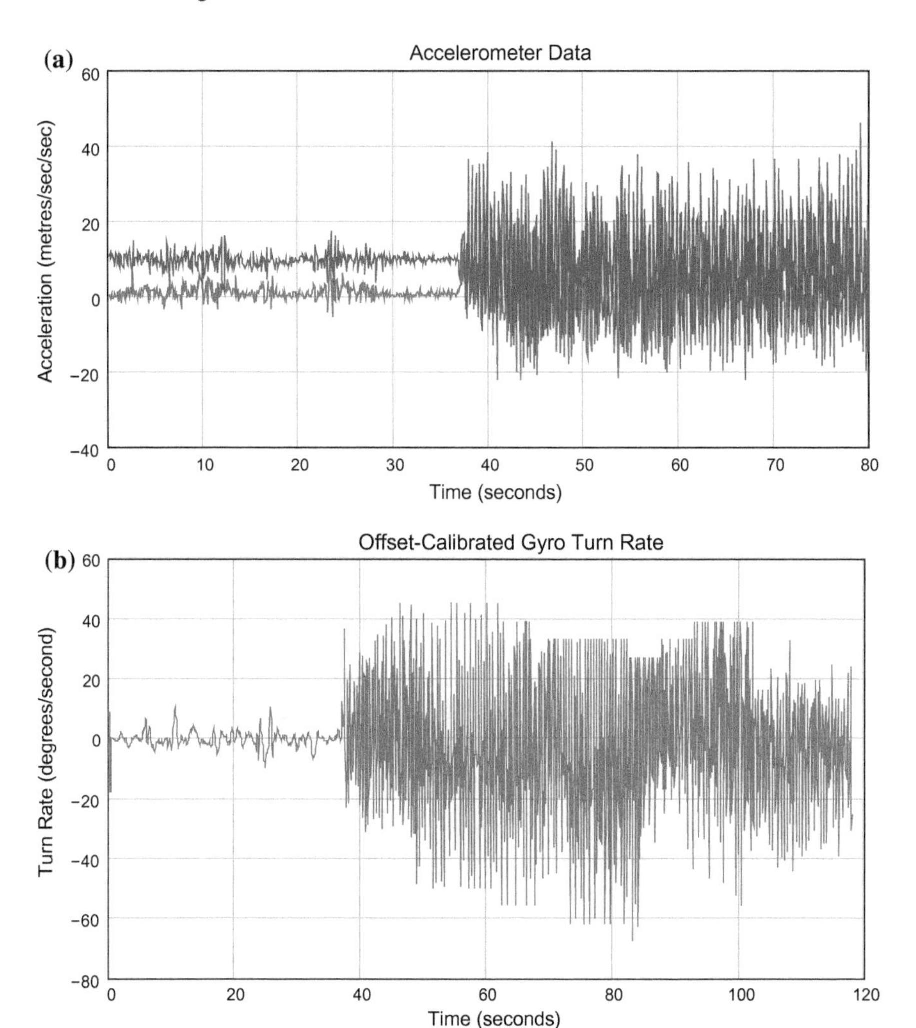

Fig. 10.17 **a** Raw accelerometer data. The red data are the longitudinal acceleration, and the blue data are the vertical acceleration relative to 1 g. For the first 37 s the horse is stationary, before the start of the race. **b** Raw rate-gyro data for the vertical axis. For the first 37 s the horse is stationary, before the start of the race

harmonic can also be observed. By bandpass filtering the accelerometer data, the time of each stride can be determined, and hence the stride rate. Figure 10.19a shows the stride rate of two horses as a function of time throughout the race. The stride data show a surprisingly large difference in the stride rate between horses, with horse #1 (the winner) having a stride rate 0.2 strides per second less than horse #2. Also notice that the stride rate tends to decrease throughout the race (after the initial acceleration at the start of the race).

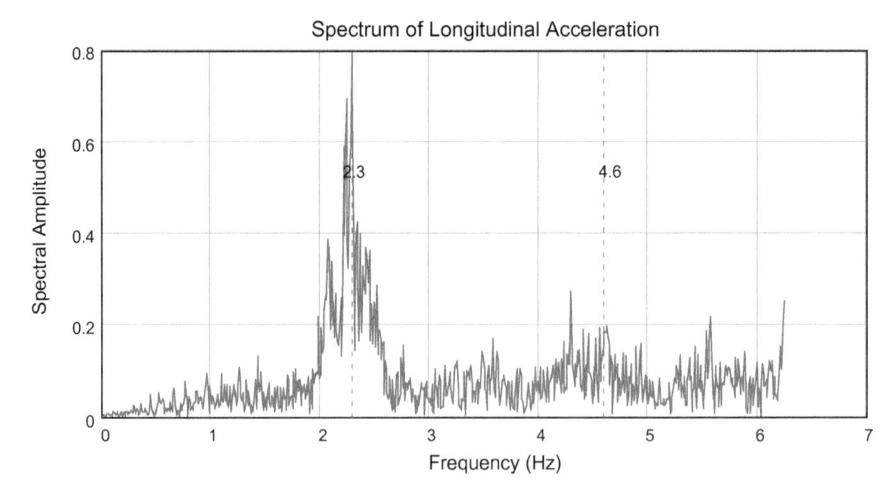

Fig. 10.18 Spectrum of the longitudinal accelerometer. Observe the dominant spectral component at around 2.3 Hz and a weak second harmonic at 4.6 Hz

The measured stride rate data also can be presented in another useful form. Because the speed of the horse is measured by the positioning system (see Sect. 10.5.4.3 following), the stride length of the horse can be estimated. The results of processing the stride rate data in Fig. 10.19a are shown in Fig. 10.19b. Typical stride lengths are in the range of 7–8 m, with little variation throughout the race. It can be observed that the horse with the lower stride rate has a compensating longer stride length. The superior speed of the winning horse was in part due to the longer stride length, rather than a faster stride rate. Such data are useful in assessing the characteristics of each horse, and also the fitness level of a horse. Such information is difficult to obtain by traditional visual observations of the motion of horses.

10.5.4.3 Other Derived Data

This section shows examples of a variety of additional data derived from the positional data. These include:

- Speed
- Longitudinal and lateral accelerations
- Heading angle
- Race order
- Leading or lagging margin
- Distance travelled.

The speed of the horse (associated with the positional data presented in previous sections) is shown in Fig. 10.20. The horse is stationary until about the 37 s point,

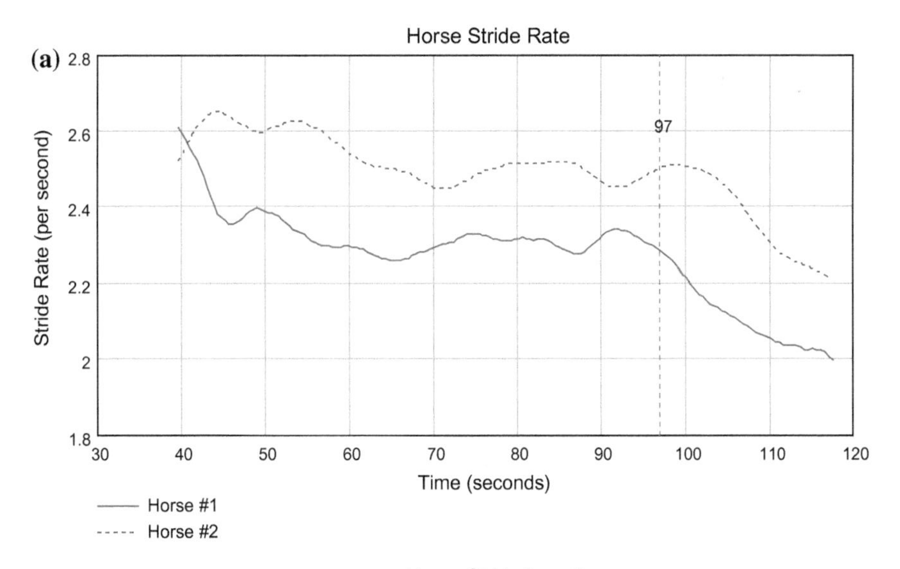

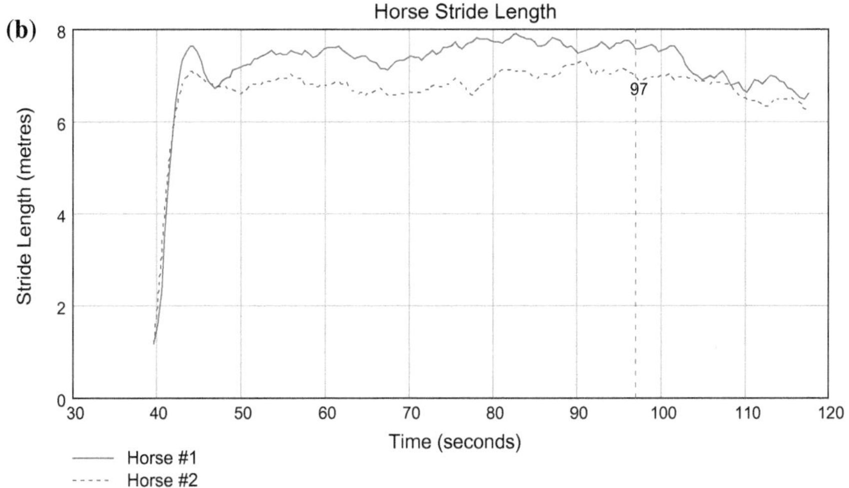

Fig. 10.19 **a** Stride rate derived from the longitudinal accelerometer data. The race starts at the 37 s point and ends at the 97 s point. **b** Computed horse stride length from the stride rate data in Fig. 10.20a. In this case horse #1 won the race by passing horse #2 near the finish

and then accelerates rapidly to reach a peak speed of about 66 kph at the 42 s point. The speed then varies slightly, averaging slightly more than 60 kph until the race end at the 97 s point. The remainder of the plot relates to the post-race slow-down. Notice that the pre race speed is not zero; this is due to the small variations in the position when the horse is in the starting gate; this variation is interpreted (incorrectly) as motion by the filtering process. This effect can be significantly reduced by changing the Kalman filter dynamics when the sensors show that the horse is essentially stationary.

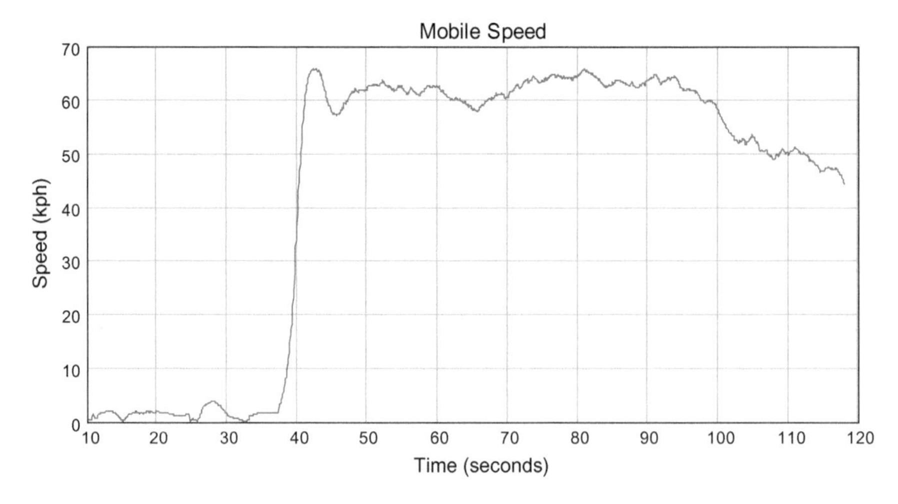

Fig. 10.20 Estimated speed of the horse as a function of time. The race starts at about the 37 s point, and finishes at the 97 s point, for a race time of about 60 s. The official race time was 60.65 s

The Kalman filter also can produce estimates of the acceleration in terms of the map grid coordinates directions, Easting and Northing. These coordinates allow horse positions to be overlayed onto a map of the course. By combining these positional data with the heading data, the longitudinal and lateral acceleration can be computed. The longitudinal acceleration is associated with changes in speed (the time differential of the data in Fig. 10.20), while the lateral acceleration is associated with the centripetal acceleration when negotiating a bend. The acceleration data are shown in Fig. 10.21. For the longitudinal data, the key feature is the initial acceleration (peaking at 0.4 g) at the start of the race; there after the longitudinal acceleration is close to zero. The lateral acceleration has peaks associated with each bend in the track. This acceleration occurs even if the speed (V) is constant, and is given (in g) by the equation

$$A_{lat} = \frac{V^2}{gR} \tag{10.3}$$

For example, at a speed of 60 kph and a radius (R) of 100 m, the acceleration is 0.28 g. The radius can be determined from a map of the course, such as shown in Fig. 10.10. These centripetal accelerations are shown in Fig. 10.22, and can be cross referenced to the map in Fig. 10.10.

The position data also can be used to estimate the heading angle. The heading data associated with the above example are shown in Fig. 10.22. Note that the heading angle as a function of time will be (approximately) straight-line segments, both when on a straight segment of the track and when negotiating a fixed radius

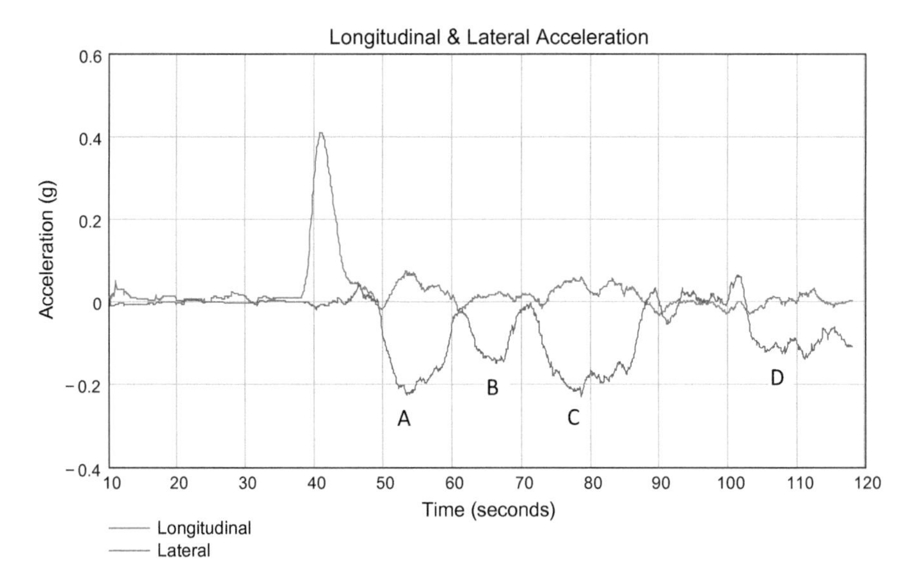

Fig. 10.21 Measured longitudinal and lateral acceleration. The race starts at about the 37 s point, and finishes at the 97 s point. The letters are associated with the lateral accelerations identify the corresponding bends of the track. See Fig. 10.11 for details

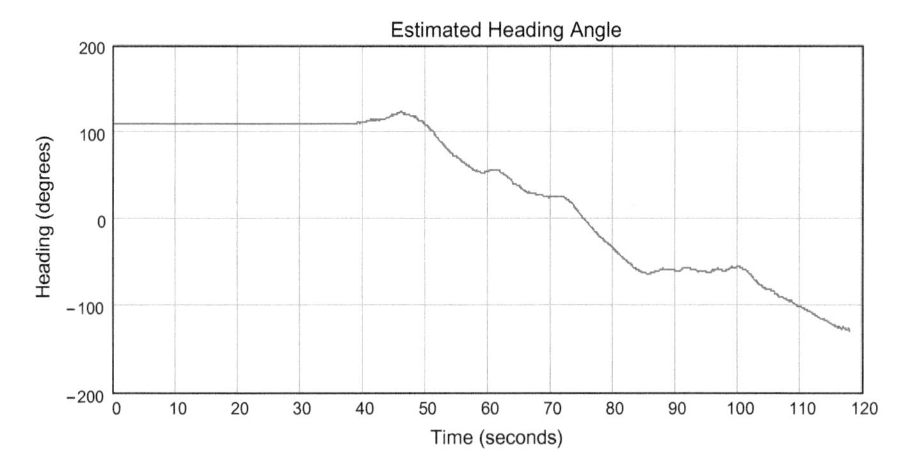

Fig. 10.22 Heading angle as a function of time. When the speed is low (up to the 40 s point) the heading is held fixed. The low-speed heading can be determined from the magnetometer data (compass)

bend at constant speed. The data in the diagram closely follow the expectations. The accuracy of the heading angle is about $\pm 2°$.

The horse positions also can be used to derive the running order. The positions in themselves cannot determine the running order, as the shape of the track is

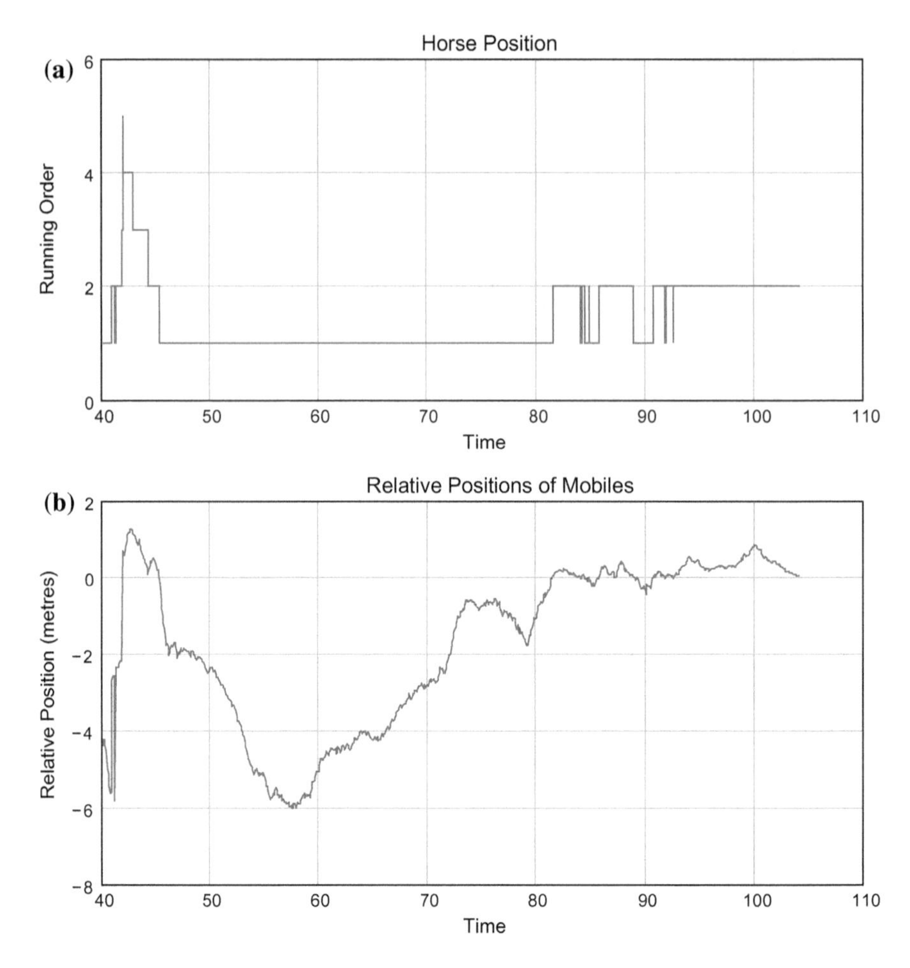

Fig. 10.23 **a** Running order of a particular horse as a function of time. The horse leads (position #1) for most of the race, has a close tussle with another horse in the closing stages (80–93 s), and finally comes a close second at the end of the race (at the 97 s point). **b** The position of a horse relative to the leading horse, or the leading distance if the first horse. This example shows the horse winning (at the 97 s point) by about 1 m

required also. The shape of the track can be determined from a map, and with the horse positions defined in grid coordinates the running order can be determined by simple geometric calculations. Alternatively, with a "pack of horses" spread over a considerable distance, the approximate local shape of the track can be determined, without the need for a map of the race course. Knowing the shape of the track and the individual horse positions, the running order can be determined by projecting the positions onto the track contour. An example is shown in Fig. 10.23a.

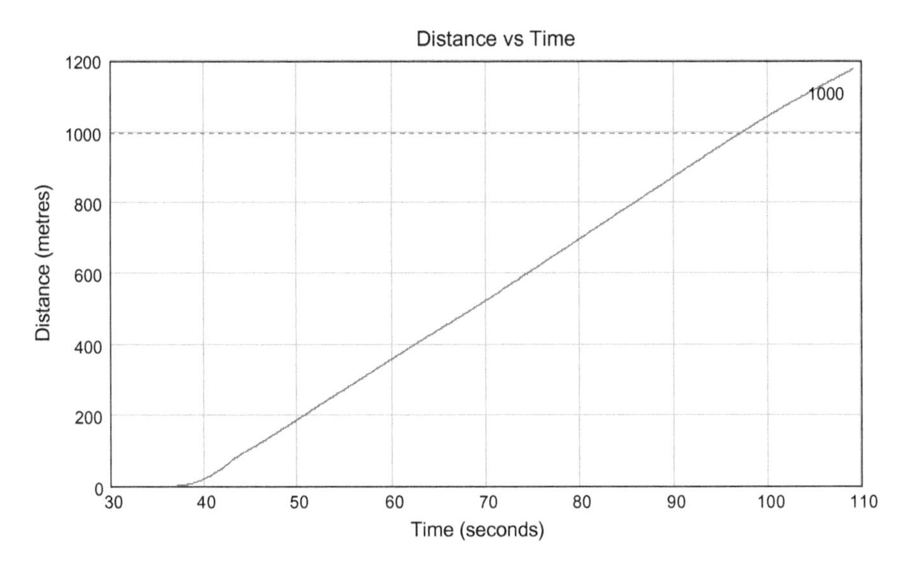

Fig. 10.24 Distance traveled as a function of time. This example shows the winning horse crosses the 1000 m point at a time of 97 s

An extension of the above technique is to calculate the positions of the horses relative to the leading horse (positive relative position), or the distance to the leading horse (negative relative distance). An example is shown in Fig. 10.23b, in this case for the horse that wins the race. The data show that after leading early in the race the horse drops to up to 6 m behind the leading horse, then steadily makes up ground, has a close tussle for the lead in the 80–92 s section, and finally slightly pulls away to win by less than 1 m.

The final example in Fig. 10.24 shows the distance travelled as a function of time. Alternatively the distance to the finish could also be computed, knowing the race distance of 1000 m.

10.5.5 Measured Timing Events During a Race

In any form of racing measuring the time of events during the race, and particularly at the end of the race are very important. Although a positioning system does not directly measure such timing events, the positional data can be used to derive timing information. Further, as will be shown the sensor data can be used to derive additional data associated with events (such as the start time) where the positioning system does not have sufficient accuracy.

10.5.5.1 Start Time

A basic positioning system is not designed to accurately measure the start time of a race using position information only. As the positional accuracy is of the order of 0.5 m (and may be worse due to signal blockage when the horses are close together in the starting gates), the detection of the first movement of the horse is difficult to detect when buried in the position measurement noise. However, a more accurate start time can be inferred from the inertial sensor data. An example is shown in Fig. 10.25. As can be observed, the sensor data are largely quiescent before the start, but there are large oscillations in the data when the horse starts to gallop. In this system the sensor samples the data with a 80 ms period, so that the start time can be inferred to an accuracy of about 80 ms. Note that the inertial data also can be used to estimate the stride period, which in this case is about 0.38 s.

The actual start time of the race will be somewhat earlier than that indicated by the inertial sensors due to the reaction time of the horse and jockey. The reaction time for humans is around 0.25 s, so that the actual race start time will be the inertial start time minus the (unknown) reaction time. Alternatively, the positioning system could be enhanced to input the actual start time. For example, a special mobile unit could be attached to the starting gates, with an input to the mobile unit from the starter's button or lever. As the mobile unit has very accurate time synchronisation, an electrical input could be timed very accurately (essentially with no timing error). In this case accurate time differentials relative to the start time could be determined. If the start time is input directly by such a method, the reaction time

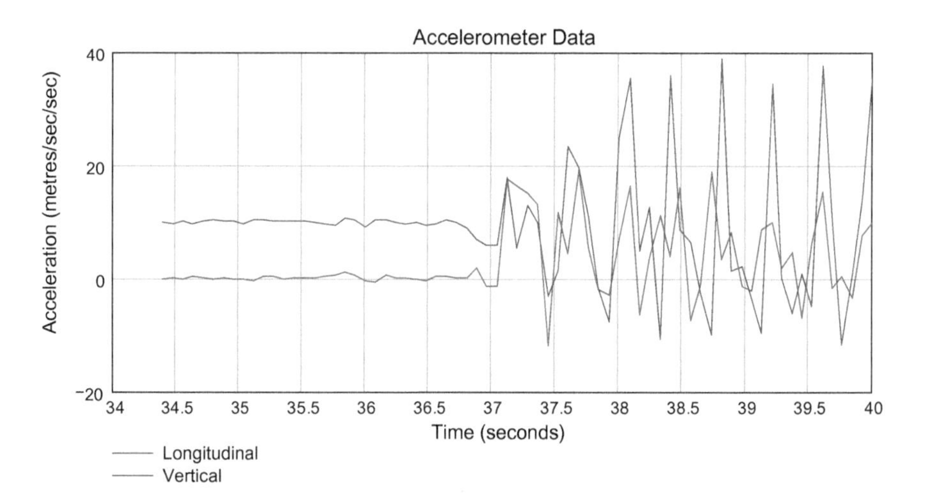

Fig. 10.25 Accelerometer data near the start time. The vertical accelerometer shows that the start of movement occurs at about the 36.8 s point. The longitudinal accelerometer shows a slightly later time of 37 s. The vertical accelerometer shows that the initial stride period is 0.38 s, or a rate of 2.63 strides per second

of the horse could be estimated, rather than being used to estimate the start time. Thus in a practical implementation, the reaction time would be an output from the system.

10.5.5.2 Finish Time

The main time of interest in a horse race is the finish time. Thus if the start time of the race is known or measured (see Sect. 10.5.4.1 above), the elapsed time of the race can be determined. Clearly the method can be extended to any point other than the finish line, such as the marker distances every 100 m.

The measurement of the finish time can be estimated from the positional data near the finish line. Figure 10.26a shows an example with two horses in close proximity, with the dotted time representing the finish line. The problem is to determine when the mobile device crosses the finish line. Note that this line is defined into processing software using a map of the race course, and unlike current video technology there is no additional equipment required. Because of the high update rate of the positioning system (80 ms in this case), the distance travelled between updates is relatively small, about 1.3 m, as can be observed in the figure. To obtain an estimate of the time of crossing the line a simple linear interpolation could be performed to estimate the crossing time. However, superior accuracy can be obtained by adopting a more sophisticated least-squares fitting approach. For a detailed analysis of determining accurate times of crossing any line defined across the track at any point during a race, refer to Sect. 10.6.

By an appropriate interpolation of the positions on either side of the line, an accurate estimate of the finish time of each horse can be determined. The difference between the measured distance data and the least-squares fit polynomial is shown in Fig. 10.26b. These differential data represent the measurement noise, and in this case is of the order of 2–3 cm. In theory (see Sect. 10.6) the (relative) error in the estimated crossing distance is this standard deviation divided by the square-root of the number of samples (25 in this case), so that the theoretical accuracy is of the order of 5 mm; this improvement is due the averaging effect of the least-squares fitting process. The absolute accuracy (as opposed to the relative accuracy) is much larger, typically around 5 cm. At a typical speed of 60 kph, the absolute distance accuracy translates into a timing measurement accuracy of 3 ms.

A further piece of useful information can be determined from the polynomial fit data. By differentiating the polynomial, the speed can be determined as a (linear) function of time. Thus the speed at the time of crossing the reference line can be estimated. A further differentiation can determine the acceleration, but the acceleration is usually of little interest. Thus the position data can be processed to obtain an accurate time of when the mobile crosses a reference line. However, the official race timing is based on when the nose of the horse crosses the finish line (not the mobile device located in the saddle bag) so that such measurements cannot be used to determine the winner of a race. While a correction time can be calculated, based estimated distance from the mobile unit to the nose of the horse (about 1.5 m).

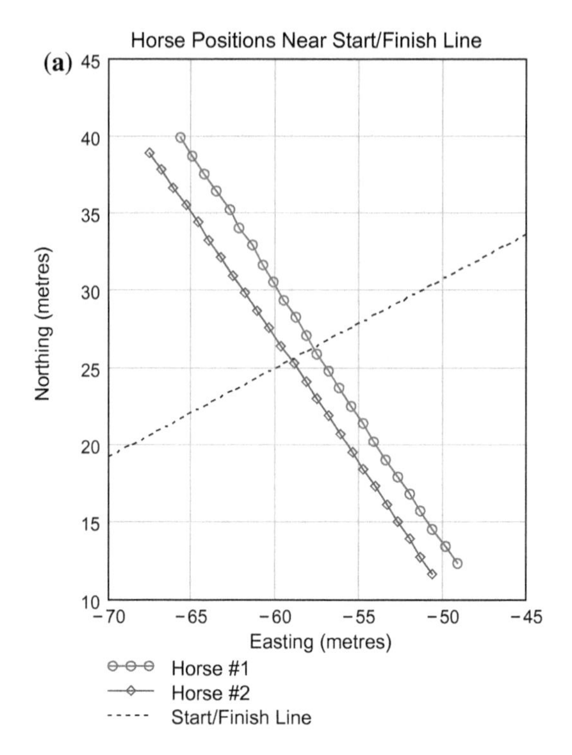

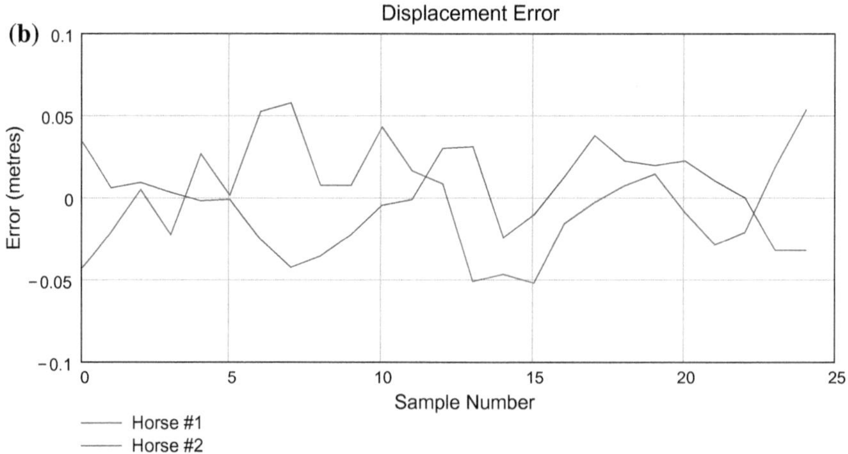

Fig. 10.26 **a** Path of two horses near the finish line, showing the individual positions at 80 ms spacing. The circles and diamonds are the positions of two horses. **b** The calculated displacement noise (difference between the measured displacement and the least-squares fit model). The RMS noise is 31 mm for horse #1 and 23 mm for horse #2. The absolute accuracy will be worse than the RMS noise

Table 10.1 Summary of the race time data (in seconds) for each horse

Parameter	Horse #1	Horse #2	Horse #3	Horse #4	Horse #5	Horse #6
Movement start	36.80	36.88	37.12	37.04	36.96	36.96
Reaction time	0.25[a]	0.33	0.57	0.49	0.41	0.41
Estimated race start	36.55	36.55	36.55	36.55	36.55	36.55
Estimated mobile finish	99.252	97.256	97.347	98.744	97.686	98.307
Correction to nose	−0.092	−0.084	−0.084	−0.090	−0.084	−0.084
Estimated race time	62.61	60.62	60.71	62.10	61.05	61.67
Margin	1.99	0.00	0.09	1.48	0.43	1.05

[a]Assumed value of the fastest reaction time

However, this is only an approximation. as this distance varies due to the relative motion of the head of the horse relative to its body trunk. At a typical speed of 60 kph, this correction would be of the order of 90 ms.

Applying the above-described methodology to the race data used as the example throughout this section, the timing data can be calculated as summarised in Table 10.1. The estimated winning time is 60.62 s compared with the "official" race time of 60.65 s. (The method and accuracy of determining the official race time is not known, but the timing resolution is 0.05 s). Thus there is good agreement between the positioning system times and the official race times. Further, if an accurate start time is available to the positioning system (as explained previously), the timing precision will be of the order of 5 ms, or 10 times better than the current "official" timing system.

10.5.5.3 Time to Distance

The measured position as a function of time can be used to estimate the time to a specific distance. Note that the "distance" is not the distance traveled by the horse, but is the "distance-made-good". Thus the distance calculation also must have knowledge of the shape of the race track as well as the positions of the horses. If a detailed map of the race course is not known, then the shape of the course can be estimated based in the positions of the horses.

Figure 10.27a shows the concept of determining the "distance-made-good" where the horse positions are projected onto the running rail. The alternative approach based on the positions of the horses is shown in Fig. 10.27b, where the track shape is estimated by the fitting a least-squares curve to the positions on all the horses in the race. In both cases, once the projected positions are known, the integration of the projected positional data provides the time-to-distance data for each horse. An example of the time-to-distance data (based on method 2 above) is shown in Fig. 10.27c. By appropriate interpolation the time to any distance can be estimated.

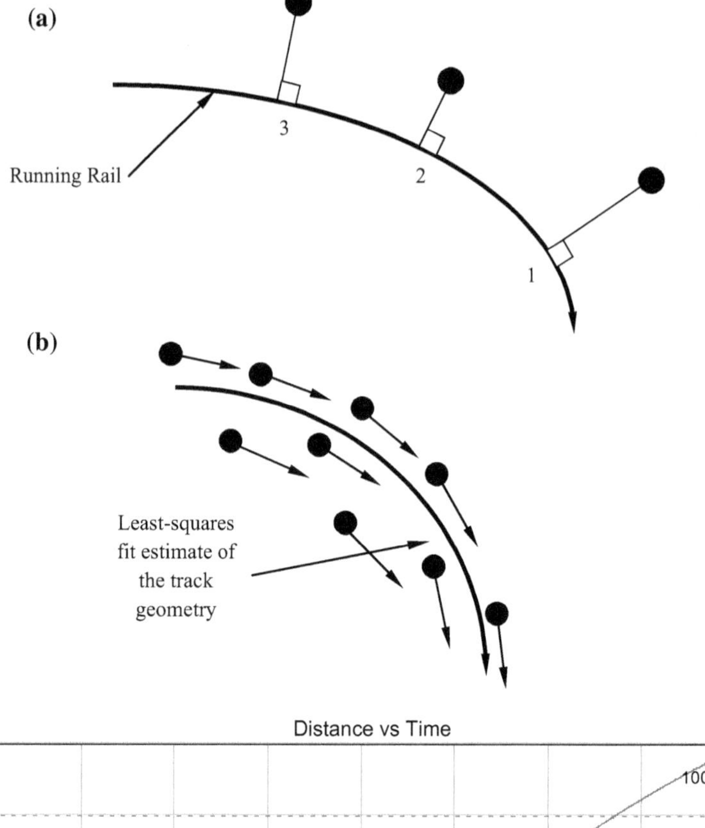

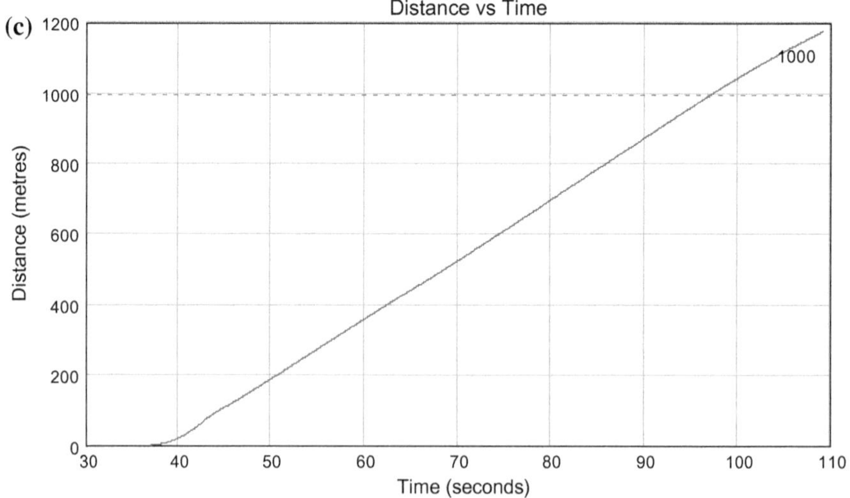

Fig. 10.27 **a** Geometry for determining distance-made-good from running-rail geometry. **b** Geometry for determining the local shape of the track. **c** Plot of distance travelled as a function of time. By appropriate interpolation, the time to any distance can be calculated. The race distance of 1000 m is shown as an example

10.6 Accurate Timing of Crossing a Line Across a Track

A common function associated with tracking in sports applications is providing "split" or lap times during a race, and finishing times at the end of a race. Typical sports where these functions are required include car and horse racing, athletics, rowing and velodrome bicycle events. While cameras can be used for measuring the finish times and the race order, split times are difficult unless multiple cameras are deployed around the track. This section describes how a tracking system can be used to provide very accurate (better than 1 ms) timing at any point in the coverage area. In practice, the timing will be associated with the mobile device (attached to a car, horse, bicycle or person) crossing a virtual line defined perpendicular to the track. This concept is showing in Fig. 10.28a, with data in this case associated with the car racing case study described in Sect. 10.4.

A wide-area tracking system can provide positional data at all positions around a track, and thus these positional data can be processed to provide timing information. The main problem with this approach is that the accuracy of the wide-area system, based on radiolocation operating typically in the shared ISM bands, is typically of the order of 0.2–0.5 m, and thus the timing data accuracy will be limited by these positional errors. For example, consider tracking a person in athletics, where the running speed is of the order of 10 m/s. If the positional accuracy is 0.2 m, the corresponding timing accuracy is 20 ms. Clearly an accuracy of 0.2 m is not sufficient to determine race finishing order, as in comparison video accuracy is of the order of 1 cm. However, it will be shown by processing multiple positions of the mobile device in the vicinity of the line across the track, the distance errors corresponding time errors can be greatly reduced. Also, as a bonus the speed can also be estimated to great accuracy.

The basic idea of determining the time of crossing a line across the track is to estimate the perpendicular distance to the line based on the radiolocation positions. Figure 10.28a shows an example of the positions measured near the start/finish line for the car racing case study. As expected, the vehicle track is close to a straight line, so that an estimate of the tracking errors can be made. If the positional errors have a statistical circularly-symmetric 2D distribution, then the one-dimensional error in any direction will be approximately Gaussian, while the radial error distribution will exhibit approximately Rayleigh statistics. An example of the measured one-dimensional distributional errors relative to the least-squares fit curve for the NASCAR case is shown in Fig. 10.28c, which shows the raw tracking errors have a standard deviation σ_d of 0.2 m. The data are taken ± 1 s relative to the reference time.

Now consider the processing of the raw positional data. The start/finish line divides the plane into two regions, so that the distances can have an associated sign. The time of crossing the start/finish line is when the distance is zero. The time axis is based on detecting the sample near where the sign of the distance changes sign,

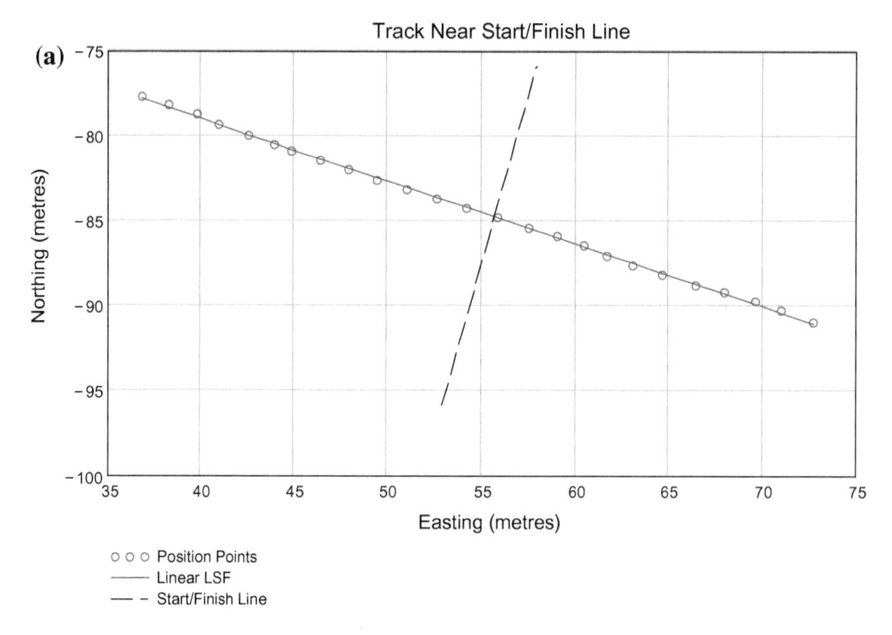

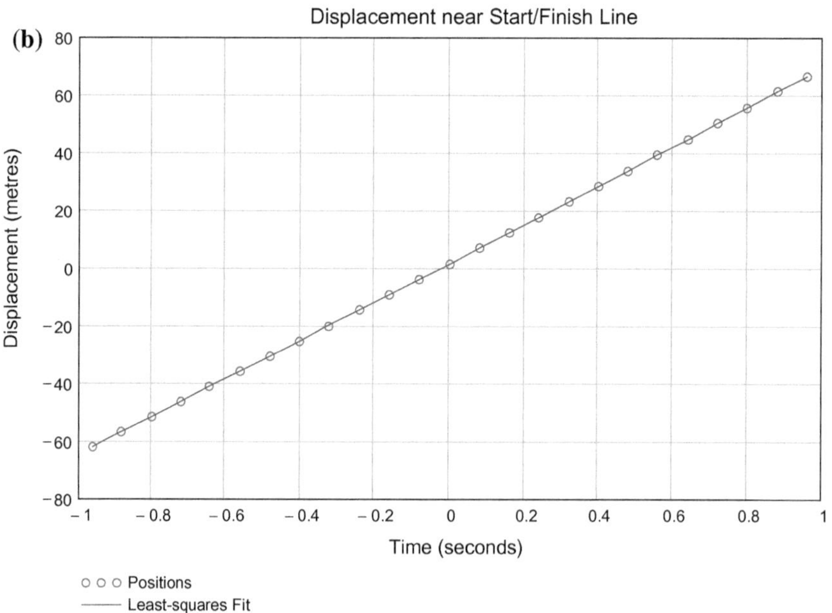

Fig. 10.28 **a** Plot of the radiolocation position near the start/finish line. The red dots are the positions determined by the positioning system, and the blue line is the fitted second order polynomial line. The 25 data points are for positions 1 s before and after the dashed start/finish line. **b** Plot of the displacement near the start/finish line. The red dots are the positions determined by the computed displacement from the finish line every 80 ms, and the blue line is the least-squares second-order polynomial. The finishing time t_0 is when the displacement is zero. **c** Displacement error from the fitted curve in Fig. 10.9b. The standard deviation is $\sigma_d = 0.068$ m

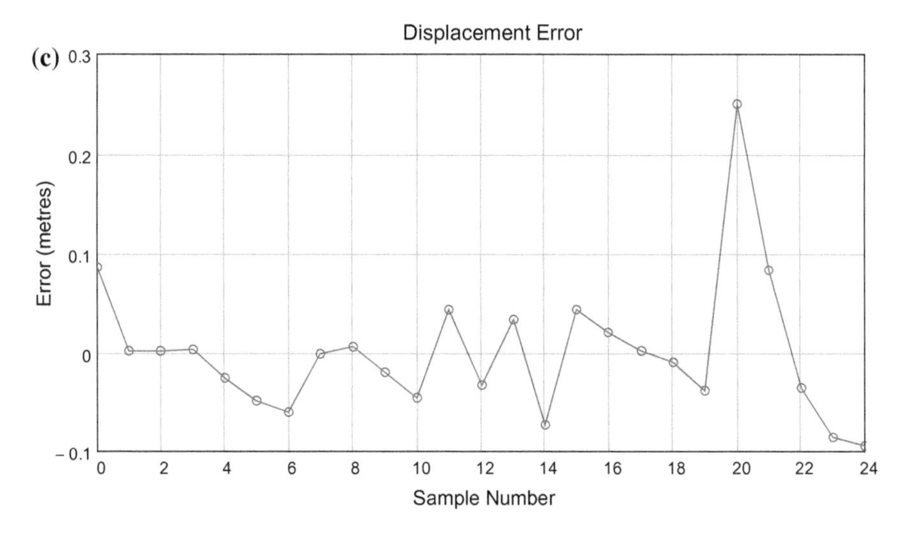

Fig. 10.28 (continued)

and normalising time relative to this reference time—see Fig. 10.28b. These distance-time data can be used to estimate the time of crossing the start/finish line. If the speed is constant, the expected relationship is a straight line; the least-squares fit to the data are also shown in the figure, which confirms that the speed was indeed close to constant. By determining the line slope (mobile device speed) and the intercept distance of the least-squares fit line, the time when the distance is zero can be estimated. Further, because of the averaging effect of the least-squares fitting process, the timing accuracy is considerably better than that inferred from the basic positional accuracy of the positioning system.

The timing accuracy can be inferred from the statistics associated with linear regression. From the least-squares fit analysis, the displacement (s) is modelled by the linear equation

$$s(t) = s_0 + v_0 t \qquad (s_0 < 0) \tag{10.4}$$

See the example shown in Fig. 10.28b. The time of crossing the line is when the distance is zero, so that the required time estimate for zero distance is given by

$$t_0 = -\frac{s_0}{v_0} \tag{10.5}$$

The accuracy of this time estimate is related directly to the accuracy of the estimate of the speed (or the slope of the line) and the intercept point (s_0). With the data sampled symmetrically about the zero distance point, it can be shown that the speed and the intercept distance estimates are independent random variables with standard deviations given by

$$\begin{aligned} \sigma_v &\approx \sqrt{\frac{12}{N}}\left(\frac{\sigma_d}{T}\right) \qquad T = (N-1)\delta t \\ \sigma_s &= \frac{\sigma_d}{\sqrt{N}} \end{aligned} \tag{10.6}$$

where N is the number of samples, and δt the sampling period. The standard deviation decreases as N increases, but a linear approximation of displacement over time is only valid if the variation in speed over the time interval is small. Thus a compromise on the number of samples used is necessary; the above example uses $N = 25$ samples, or ± 1 s.

The least-squares fitting procedure provides an estimate of the vehicle speed as a by-product of the timing estimate. The accuracy (see (10.6)) is related to the one-dimensional displacement accuracy σ_d, the number of samples, and the length of the timing window. For example, for the above example with $\sigma_d = 6.8$ cm as in Fig. 10.28c, $N = 25$, and $T = 2$ s, the estimated speed accuracy is 0.024 m/s, and the displacement reference origin accuracy is 1.4 cm.

The statistics of the line-crossing time can now be estimated. From Eq. (10.5) with small (random) errors in both the speed and intercept distance, the time for zero distance is given by

$$t_0 = -\frac{\bar{s}_0(1+\delta_s)}{\bar{v}_0(1+\delta_v)} \approx -\frac{\bar{s}_0}{\bar{v}_0}(1+\delta_s)(1-\delta_v) \approx \bar{t}_0(1+\delta_s-\delta_v) \tag{10.7}$$

The deltas in Eq. (10.7) are random variables with zero means and standard variations related to those given in Eq. (10.6), namely

$$\begin{aligned} \sigma_v &= \bar{v}_0\sqrt{\overline{\delta_v^2}} \\ \sigma_s &= \bar{s}_0\sqrt{\overline{\delta_s^2}} \end{aligned} \tag{10.8}$$

From Eq. (10.7) and using the fact that the two random variables have zero mean and are statistically independent, it can be observed that the mean $\bar{t}_0$ is an unbiased estimate of the time of zero displacement, and the standard deviation in the time estimate is given by

$$\sigma_t = \bar{t}_0\sqrt{\left(\frac{\sigma_v}{\bar{v}_0}\right)^2 + \left(\frac{\sigma_s}{\bar{s}_0}\right)^2} \approx \frac{2\sigma_d}{\sqrt{N}v_0} \tag{10.9}$$

to a first order of approximation. Alternatively, assuming the system spatial accuracy is independent of the vehicle speed, the spatial accuracy σ_x is given by

$$\sigma_x = v_0 \sigma_t = \frac{2\sigma_d}{\sqrt{N}} = \sqrt{\frac{8}{\pi N}} \sigma_{pos} \tag{10.10}$$

where σ_{pos} is the 2D positional accuracy of the positioning system, assuming circular error distribution of the positional errors. Thus the spatial accuracy is directly related to the displacement accuracy and the (square-root) of the number of samples. For example, with $\sigma_d = 20$ cm and $N = 25$ samples the spatial accuracy is 4 cm. At a speed of 300 kph, the corresponding timing accuracy is 0.5 ms.

The current timing technology used in motor sport is based on loops placed across the track at a few key locations (particularly the start/finish line), and a small tag located under the vehicle. This technology is based on short-range signal strength measurements to estimate the time of crossing the loop. For more details, refer to Annex A. To check the accuracy of the positioning system lap time estimates, a limited number of tests were performed by comparing the times calculated using the above method with the official track timing system.[2] The tests were performed at the Eastern Creek raceway in Sydney Australia, using a V8 Supercar. The tests used two units, one internal to the car, and one with an external roof-mounted antenna. The tracking system used six base stations located around the track, as shown in Fig. 10.29. The track is located between the surrounding small hills with elevations of 10–20 m relative to the track, with the base stations located on the hills. Because of the hilly terrain, line-of-sight paths to the track is not possible from all base stations. Further, a building adjacent to the start/finish line blocks the direct paths to the three southern base stations. Thus this test was performed with difficult operating conditions due to the radio propagation, so that test results are likely to be at the lower end of performance. For example, the radio reception at the southwest and southeast portions of the track are located in valleys between hills, and the associated track path shown in Fig. 10.29 shows significant tracking errors. However, the path adjacent to the start/finish line does not exhibit any observable errors at the scale of the map.

The results of the tests are summarised in Table 10.2. The testing was performed using several mobile units with antennas both internal to the car and externally mounted on the roof of the car. The computed lap times based on the positioning system measurements are shown to a precision of 0.1 ms, and to 1 ms for the track loop measurement (Tag) system (normal reported racing precision). Note that while the claimed accuracy of the Tag system is 0.1 ms, the actual output data are recorded only to a millisecond. Also shown is the estimated vehicle speed at the finish line, and the estimated distance and time accuracy based on the above theoretical analysis.

---

[2]The track timing used in the test is based on a loop system from Dorian Industries. The claimed accuracy is 0.1 ms.

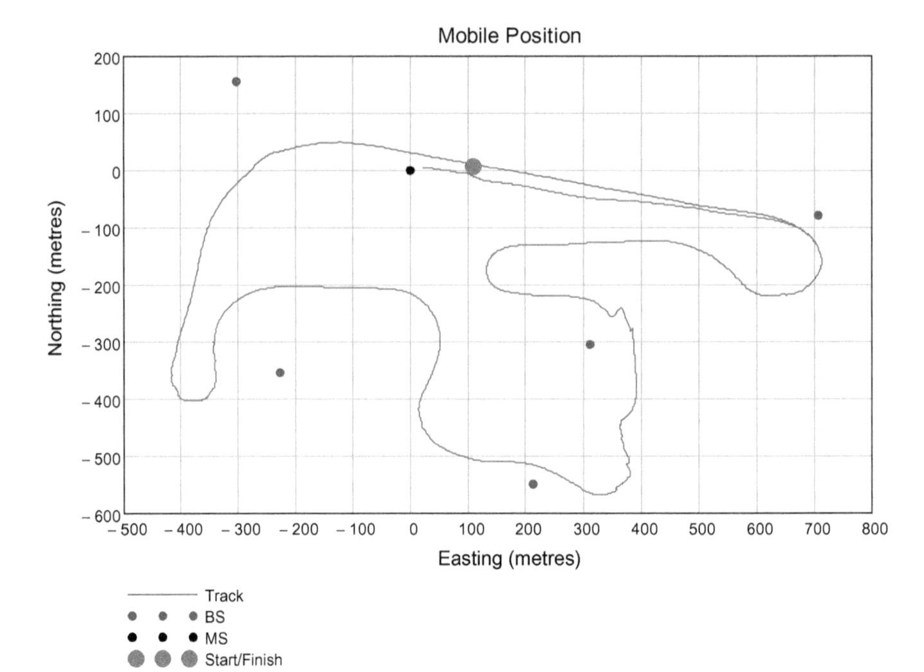

Fig. 10.29 Map of the track showing the location of the base stations, an example of the tracked path of the car, and the location of the start/finish line. The tracked path shows errors in the southeast portion where the hilly terrain limits the coverage by the base stations. Also observe the car entering the pit lane at the end of the tracked path

Table 10.2 Summary of the results of lap times using both the track loop system and the positioning system

Run	Lap	Mobile	Measured (s)	Tag (s)	Delta (ms)	Speed (kph)	Distance SD (m)	Estimated accuracy (ms)
1	1	10 (External)	94.1813	94.184	−2.7	241.6	0.043	0.26
	2	10 (External)	94.2433	94.240	3.3	241.6	0.040	0.24
2	1	06 (Internal)	98.3063	98.297	9.3	233.2	0.254	1.57
	2	06 (Internal)	96.7646	96.767	−2.4	234.5	0.474	2.96
	3	06 (Internal)	96.7929	96.781	11.9	234.5	0.436	2.68
3	1	10 (External)	93.8478	93.847	0.8	243.3	0.050	0.30
		08 (Internal)	93.8544	93.847	7.4	232.3	0.412	2.54

(a)

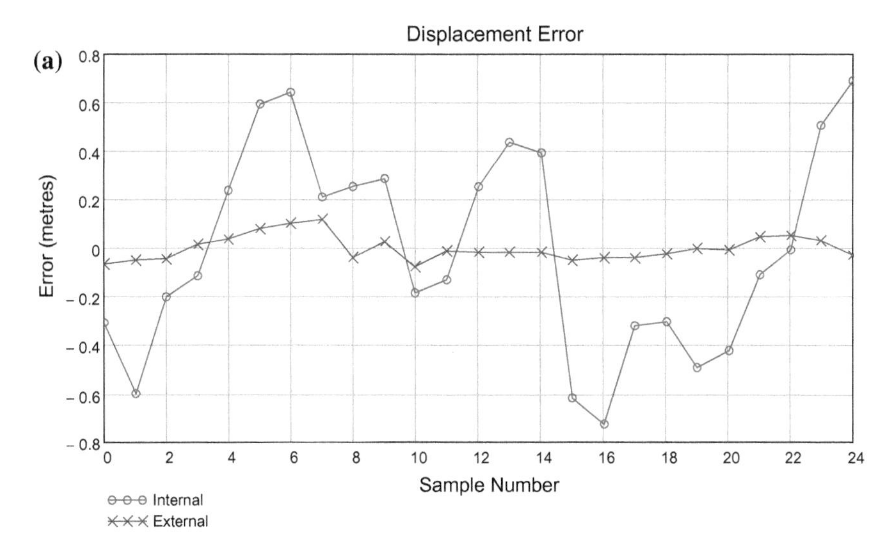

(b)

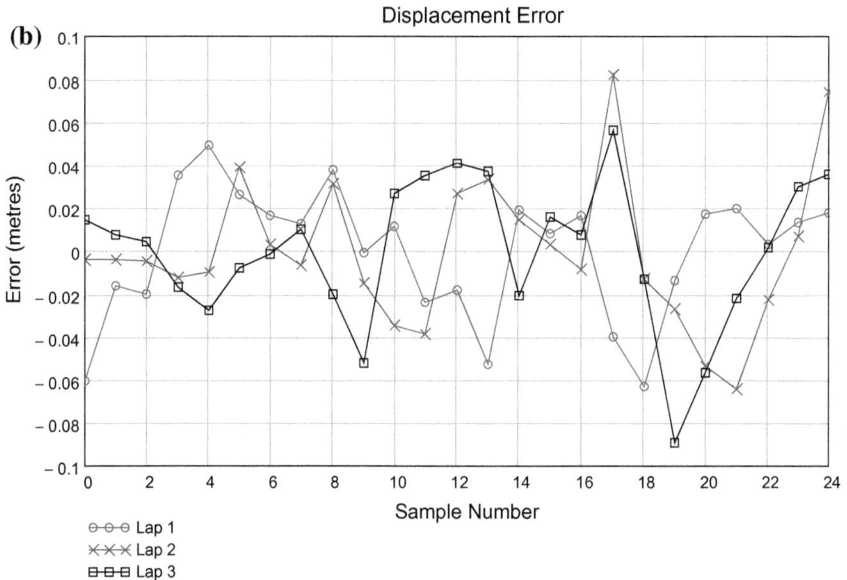

Fig. 10.30 **a** Comparison of the internal and external antenna estimated displacement errors for the Run 3 case, where both internal and external antennas were used during the same test run, so that the relative tracking performance could be compared. The standard deviation for the external antenna is 5, and 41 cm for the internal antenna. **b** Comparison of the estimated displacement errors for three laps (Run 1) using an external antenna. The standard deviations are (3.4, 3.3, 7.8) cm

The first important observation is the difference in performance between the results with internal and externally mounted antennas, as shown in the table and in Fig. 10.30a. As can be observed the errors when using an internal antenna (standard deviation of about 0.4 m) are much greater than those achieved with an external antenna (about 5 cm). The reason for this big difference is not fully understood, but a possible explanation is that the internal antenna effectively excites currents in the body of the vehicle, which in turn act as the effective transmitting antenna. These emissions are thus not a single point source, but are related to the size of the upper body of the vehicle, say a variation of ±0.5 m.

The results for an external antenna over three laps are shown in Fig. 10.30b. The error patterns are generally random both between laps and within each run of 25 samples, although errors in the 16–24 section appear to be more correlated between laps, perhaps indicative of more systematic (multipath) errors. These results show that the random errors with an external antenna have a standard deviation σ_d of typically 5 cm. In this case applying (10.9) and a speed of 240 kph the standard deviation of the time measurement is 0.3 ms.

In Table 10.2 there are three lap times using both an external antenna and where independent track timing system are available; Run 1 (laps 1, 2) and Run 3 (lap 1). In the first case (raw data shown in Fig. 10.30b) the positioning system errors are small, and the estimated timing errors standard deviations are about 0.25 ms. However, the difference between the positioning system lap times and the official track timing system lap times (see Table 10.2) are respectively −2.7 and +3.3 ms. As the positioning system raw data appears to be highly consistent it is difficult to believe these errors are not associated with the official track timing system. For example, a 3 ms timing error would correspond to a position error of 0.2 m at 240 kph, which is much greater than the error data shown in Fig. 10.30b. The third comparative data is from Run 3. In this case the agreement between the two measurement systems is much better, with a differential of only 0.8 ms. However, the estimated measurement error in the positioning system is only 0.3 ms, so the 0.8 ms error is again more likely to be due to the official track measurement system. While the available data are limited, the conclusion is that the time measurements using the positioning system with external antennas is better than 1 ms in accuracy, and probably of the order of 0.3 ms. Although the official track timing system has a claimed accuracy of 0.1 ms, these test measurements do not support this claim, with the actual accuracy probably of the order of 2–3 ms.

10.7 Animal Tracking

The tracking of animals for research has been long been performed, but the context of such research has usually been with wild animals, whose roaming range or migration patterns are of interest; in such situations low accuracy and infrequent position updates are sufficient. In this section the focus will be on domesticated livestock on farms, where the areas of interest are much smaller (down to a paddock of a few hectares), and the movements need to be tracked at high update rates, and with high accuracy to observe animal patterns of behaviour.

While the applications considered in previous sections are associated with sporting applications, this section considers how the combination of position location technology and sensors could be used in the management of livestock,

Fig. 10.31 **a** Sheep with mobile mounted on a collar strap. **b** Steer with mobile mounted on a collar strap

such as sheep and cattle. Such a system in the future could take advantage of GPS positioning and cellular mobile phone technology to monitor the status of livestock. However, the monitoring described here is based on a trial at CSIRO Livestock Industries at Armidale Australia. For this trial, GPS was not used, but a positioning system similar to that described for the sporting applications was used. This system has much better accuracy than GPS, and allows animal behaviour to be observed in considerable detail. The tracking of the animals was based on placing the mobile devices on three sheep and three cattle, as shown in Fig. 10.31a and b. The animal behaviour seemed to be unaffected by the attached units.

While the information in this section is related to a very limited trial for research purposes, the type of information that could be obtained from such a tracking and sensor system may be applicable in the future to more commercial situations. In the examples given below, the grazing pattern of three sheep and three cattle (in separate paddocks) were analysed.

10.7.1 Cattle Grazing Pattern

The grazing pattern of cattle can be analysed by observing the movement over a period of time. When grazing cattle will not move very far over short periods (few minutes) of time. A sample is shown in Fig. 10.32, which shows the movement of a

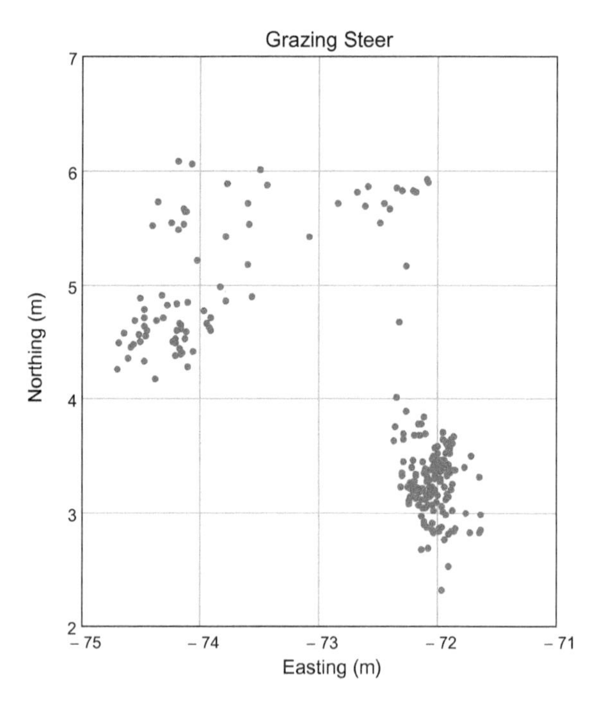

Fig. 10.32 Position of grazing steer over a period of 100 s. The update period is 0.5 s. Notice how after about a minute the steer moved a few meters to a new area before continuing to graze

single steer over about 100 s. The movement is just a few meters; note however that at this fine time and spatial resolution some of the apparent motion is measurement noise. This pattern is similar to other steers, which tend to graze in close proximity.

This grazing pattern can be analysed further by determining the circle of minimum diameter that encloses the positions of the three cattle. Further, the centre of the circle can be defined as the grazing point, which can be plotted as a function of time. Figure 10.33a shows the grazing point over a 100 s period, whereby the displacement can be observed to be about 5 m. The corresponding radius of the circle is shown as a function of time in Fig. 10.33b. Most of the time the three cattle are within 3 m of one another. If this distance increases to 6 m, the most distant steer quickly moves back to a closer distance to the other steers.

Thus the grazing pattern shows a collective movement of about 2 m a minute with a radius of typically 3–4 m for this case of 3 steers, or a spatial density of about 1 steer per 10 m^2. However, as the cattle are moving at a collective rate of about 2 m/min, there is little grazing pressure within any given area.

10.7.2 Herding Cattle

Another topic of interest is the management of the movement of livestock. In Fig. 10.34 three cattle are initially grazing alone (top left in figure) for about 1 min, and then are herded as a group for another minute. As can be observed the herded cattle generally moved as a tight group, similar to when grazing, as shown also in Sect. 10.7.1. This behaviour herding behaviour contrasts significantly with sheep (see Sect. 10.7.5 following), which tend to move in a more chaotic fashion when being herded. At one point during the herding an attempt was made to cut out the "black" steer from the other two, but this was resisted and it soon returned to the other two. This result shows the herding is a natural instinct of domesticated cattle, which make the movement of a herd of cattle relatively easy.

10.7.3 Feeding Hay to Cattle

Another example with cattle is the movement pattern while grazing over a period of about 10 min, as shown in Fig. 10.35. At the start the three steers were fairly closely grouped (within about 15 m) at the lower right. After about 4 min, some hay was delivered about 40 m away to the northwest. The cattle immediately moved to the hay. Two of the cattle remained eating the hay for the remainder of the 10 min, but after about 2 min one steer moved away and started grazing again by itself. This behaviour pattern shows while there is a herding instinct, cattle can also act as individuals. This pattern of behaviour is different than in sheep, as will be shown in later sections.

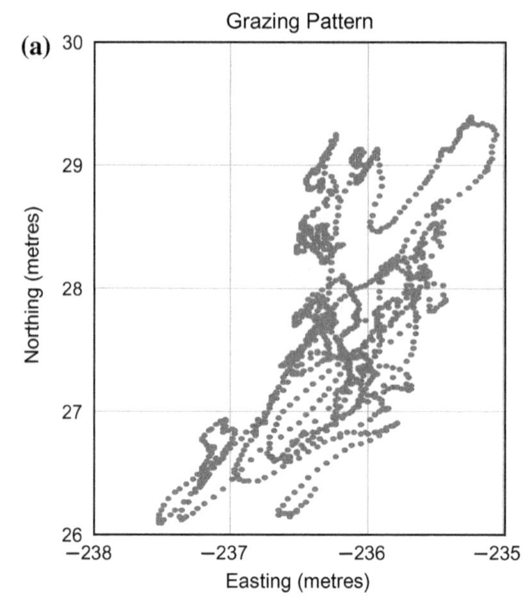

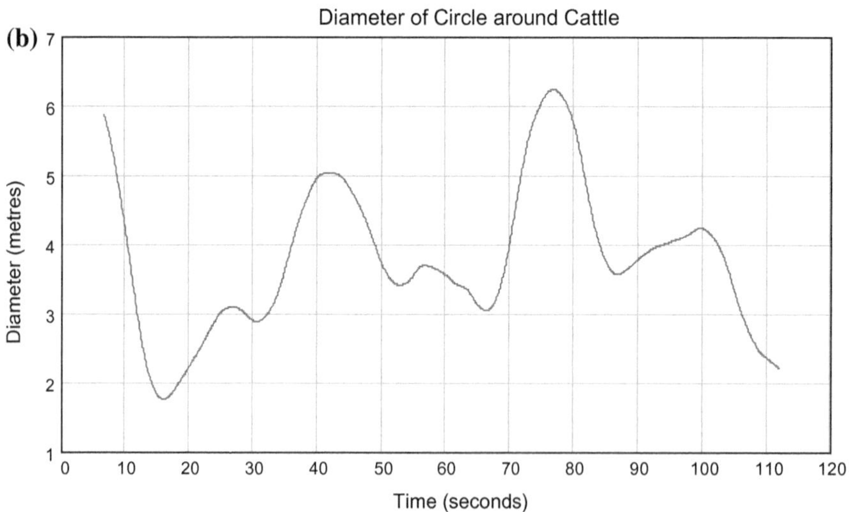

Fig. 10.33 **a** Collective cattle grazing pattern over a 100 s period. The movement over this period is about 5 m. **b** Grazing diameter of three cattle as a function of time. The steers maintain a formation as they slowly move. The mean diameter is 3.9 m, and the standard deviation 1.2 m

10.7.4 Sheep Grazing in Flock

An example of sheep grazing is shown in Fig. 10.36. This figure tracks three sheep in the flock (other sheep in flock not shown) for about 350 s. During this time the

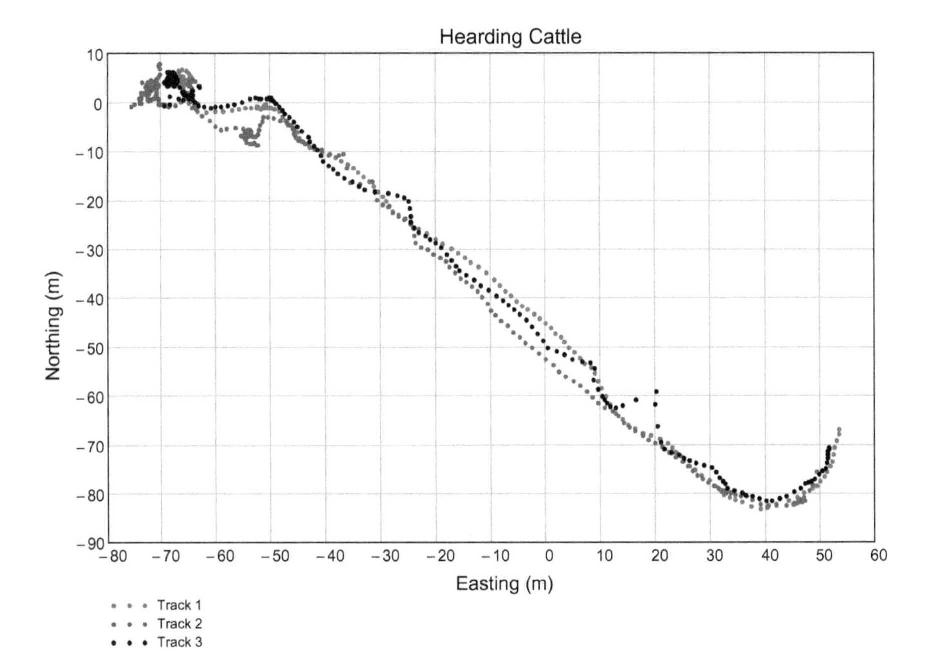

Fig. 10.34 Herding three cattle, from top left to bottom right. Observe the attempt to "cut" one steer from the other two at the point (20, −60). The update period is 320 ms. The total elapsed time is about 2 min

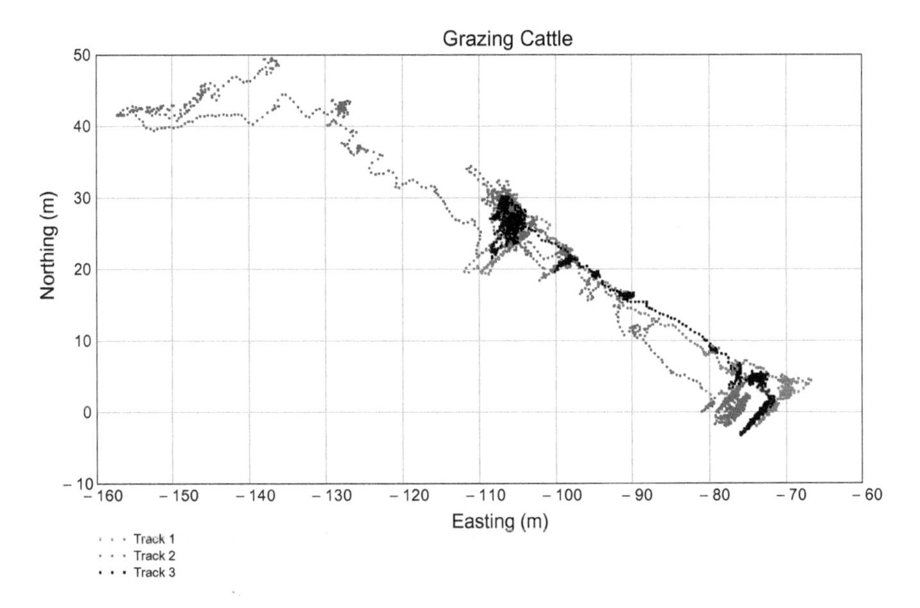

Fig. 10.35 Plot of cattle grazing over a 10 min period. The update period is 0.4 s

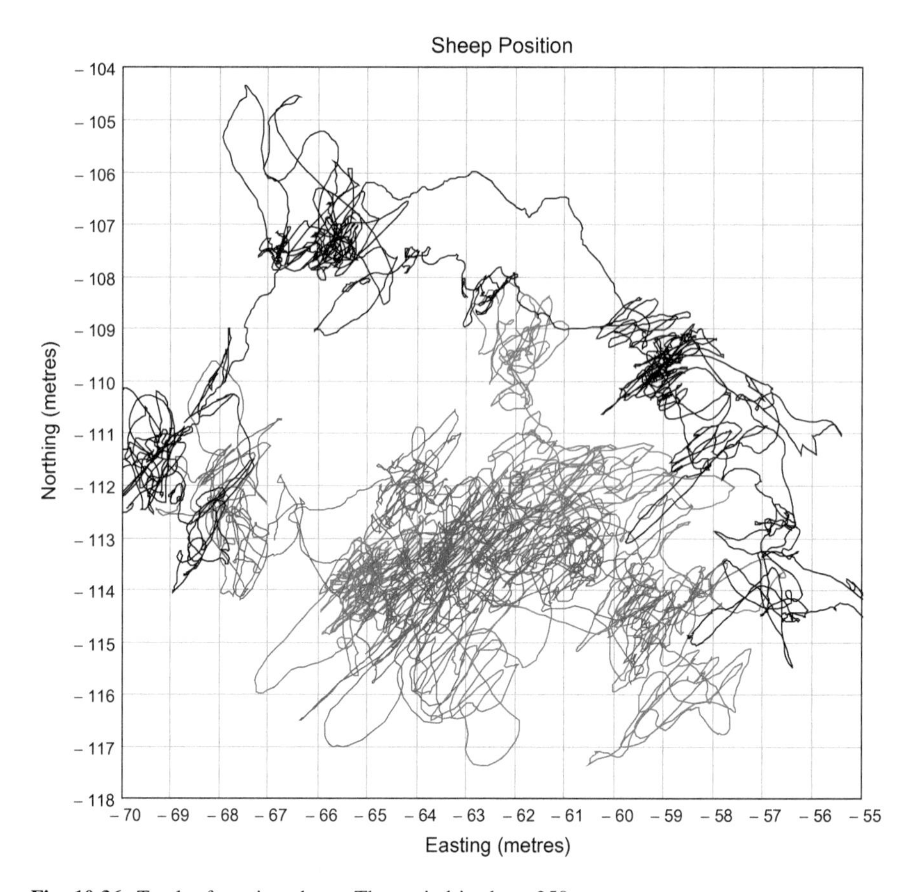

Fig. 10.36 Track of grazing sheep. The period is about 350 s

three sheep remained within an area of about 15 m by 15 m. Two sheep kept very close together and moved very little distance. The third sheep moved more, but kept within about 10 m of the other two sheep. The overall pattern appears random, although the sub-50 cm movement is associated with position measurement noise. Overall the grazing pattern of sheep appears to be more static then for cattle, and the separation between animals is less than for cattle.

10.7.5 Herding Sheep in Flock

An example of herding the flock of sheep, which includes the original three sheep is shown in Fig. 10.37; the other sheep in the flock are not shown in the figure. The herding starts in the southwest, with the intended movement towards a pen at the top of the figure. The chaotic nature of the path shows the difficulty in herding

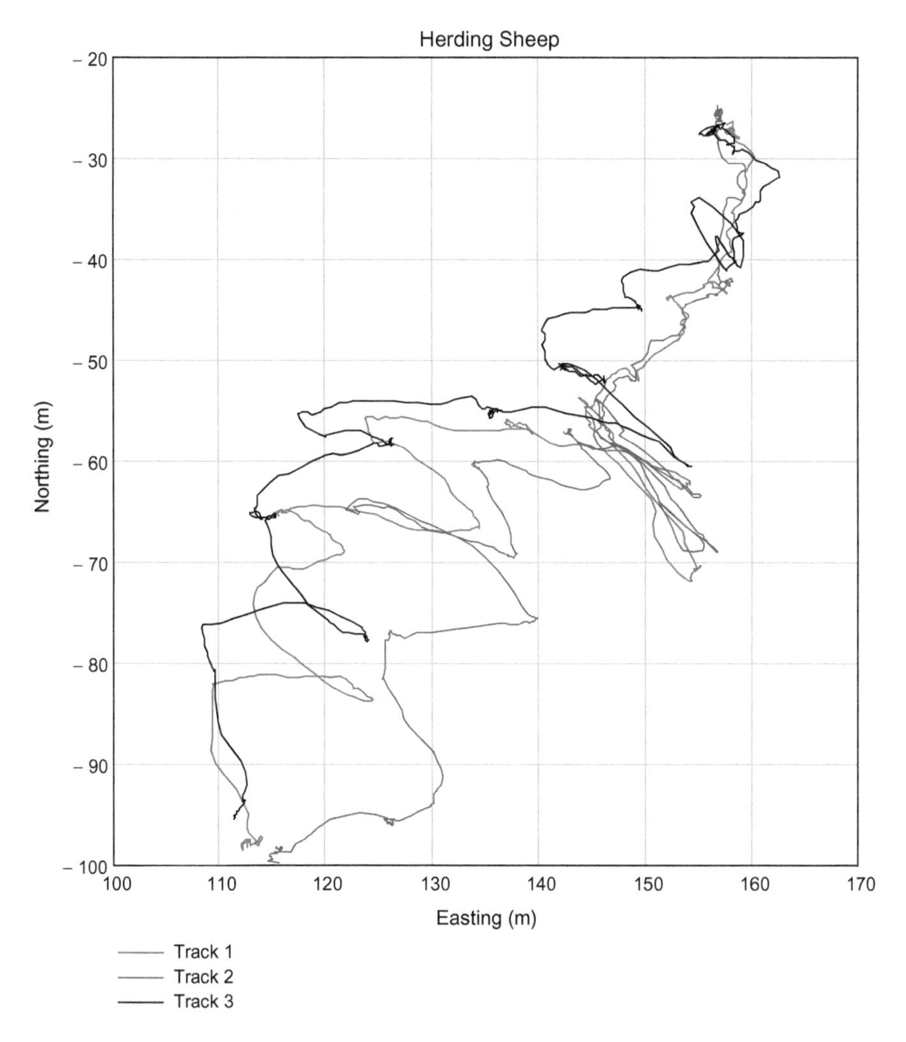

Fig. 10.37 Herding of sheep into pen located near the point (157, −27) at the top right. The chaotic paths of the sheep are evident. Notice that the "red" sheep tends to seek either of the other two sheep it was grouped with originally, and separated from the rest of the flock

sheep in the required direction, which contrasts with the herding pattern of cattle shown in Fig. 10.34. The paths in the figure show the sheep tend to want to move at right-angles to the overall desired direction of travel, so that path has a zig-zag characteristic This pattern should be contrasted with the relatively straight path associated with the herding of cattle in Fig. 10.34. The other noticeable pattern is that the "red" sheep seeks one of the original three sheep (which were originally grazing separated from the rest of the flock) during the flock herding process.

Annex A—Estimation of the Performance of a Loop Timing Tag System

Introduction

This Annex describes the timing technology used in motor sports, but with a particular reference to Formula 1 racing requirements. The Formula 1 cars incorporate a Timing Tag which provides information on the lap times, number of laps and speed of the car at certain positions around the track. The details of the operation of the Tag are not described here, but rather the general operational and performance characteristics are analysed in some detail. These operational characteristics, particularly the timing accuracy are of interest when compared with the timing accuracy that can be obtained from a tracking system which provides position data around the whole of the track. In contrast, the loop system can only provide timing information at a few points around the track, as it is not logistically feasible to deploy a large number of loops around the whole track.

The basic physical layout of the Timing Tag system in a car and the associated antennas buried in the track is shown in Fig. 10.38. The Tag/transponder is mounted near the underneath side of the car, and transmits a signal downwards to antennas buried in the track. Two antennas are used for the estimation of the time the car passes over the antennas; Sect. "Analysis" provides a brief summary of the mode of operation. From this analysis, it is possible to obtain the required signal-to-noise ratio necessary for the performance stated for the Timing Tag system. The stated accuracy of the system is 0.1 ms, so that for a car travelling at 100 m/s, the positional accuracy is about 1 cm. However, there is some doubt that this is achieved, based on comparison tests described in Sect. 10.6.

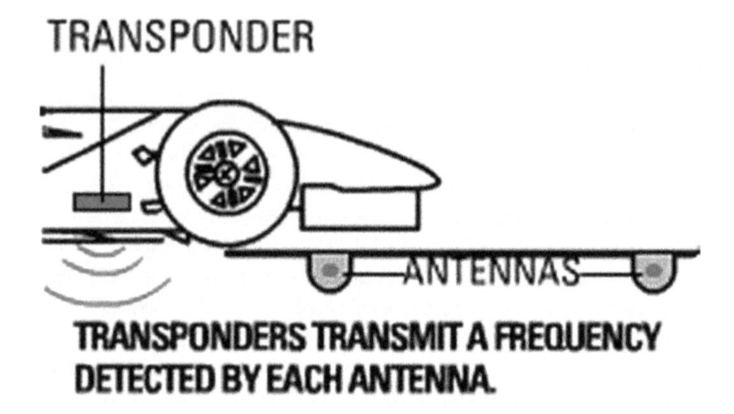

Fig. 10.38 Diagram of the geometry of the Tag on the car, and the ground antennas

Analysis

Overview

The details of operation of the proprietary Timing Tag used in F1 and other motor sports have not been published, but reasonable estimations can be made based on experience in RF tracking systems. What is currently known is that the each Tag operates at a separate frequency; as there are a maximum of 22 cars in a F1 race, there must be at least 22 frequency channels.

The simplest possible implementation is to measure the short range signal strength as the car passes over an antenna buried in the track, so that the time of the received "pulse" can be used for the time estimate. The antenna in the car is typically around 15 cm above the track, and assuming that the signal strength is dominated by the inverse range (near field) effect, the 3 dB pulse width is about twice the height H, or (say) 30 cm. If the car is travelling at 100 m/s, the pulse width (3 dB points) is about 3 ms. As the stated accuracy is 0.1 ms, the time resolution implies that the position of the pulse must be determined to an accuracy of about 1/30 of the pulse period. This accuracy is possible but difficult, as the absolute magnitude and shape of the pulse is not known, and would vary from car to car. Thus some method of normalisation is desirable. As can be observed in Fig. 10.38, the actual system utilises two antennas; the following method provides a solution which is simple to implement and can meet the timing accuracy requirements.

Analysis of Timing Detection

The timing detection system is assumed to be based on measuring the received signal at the two antennas. These signals are then both summed and subtracted, and finally the ratio computed. Thus this method calculates a function which is independent of the absolute signal strength, and is only weakly dependent on the shape of the received pulse. The Tag antenna is assumed to be a patch antenna with a radiation pattern given by $\cos(\theta)$. Thus if the height of the antenna above the ground is H and the range R, the received signal at the antenna in the track is given by

$$E(\theta) = \frac{A\cos(\theta)}{R} = \frac{A}{H}\cos^2(\theta) = E_0\cos^2(\theta) \tag{10.11}$$

where A is some (unknown) constant. Now consider two antenna D apart, with the origin of the x-axis at the mid-point. Clearly the signals will be symmetric about this point, so that this point also represents the position for which the time is to be measured. Thus setting $t = 0$ at this reference point, the received signal strength of the two antennas is given by

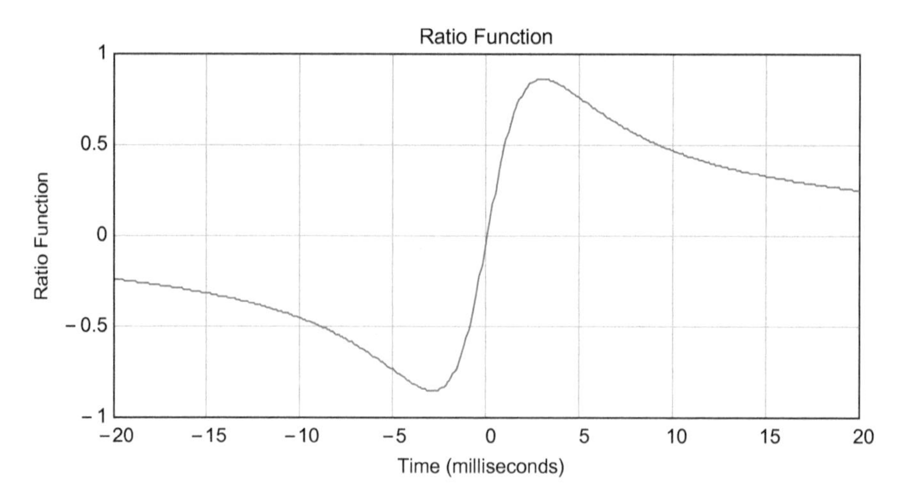

Fig. 10.39 Ratio function with parameters $H = 15$ cm, and $D = 50$ cm

$$E_1(\theta) = E_0 \frac{H^2}{H^2 + \left(vt + \frac{D}{2}\right)^2} \qquad E_2(\theta) = E_0 \frac{H^2}{H^2 + \left(vt - \frac{D}{2}\right)^2} \tag{10.12}$$

where v is the car speed, and E_0 is a constant reference signal strength. The sum and difference of these two signals is thus given by

$$E_{sum}(\theta) = E_2(\theta) + E_1(\theta) \qquad E_{dif}(\theta) = E_2(\theta) - E_1(\theta) \tag{10.13}$$

so the ratio of the difference signal to the summed signal is given by

$$\rho(t) = \frac{E_{dif}(\theta)}{E_{sum}(\theta)} = \frac{E_2(\theta) - E_1(\theta)}{E_2(\theta) + E_1(\theta)} = \frac{Dvt}{H^2 + \left(\frac{D}{2}\right)^2 + (vt)^2} = \frac{Dvt}{R_1^2 + (vt)^2} \tag{10.14}$$

Note that the ratio function is independent of the signal strength, and is zero at $t = 0$. Thus by detecting the position where $\rho(t) = 0$ the required timing data has been found. Further, near the origin $(vt \ll R_1)$, the function is approximately linear. Alternatively when $(vt \gg R_1)$ the ratio approaches D/vt which approaches zero when the Tag is far from the receiving antennas. The function is shown plotted in Fig. 10.39. In the ideal case, the time where $\rho(t) = 0$ can be determined exactly, but measurement errors will result in some computed timing errors. These errors are considered in the next section.

Effect of Noise

The measurement technique described above provides the necessary timing by detecting the time when $\rho(t) = 0$. However, the actual measurements will have random noise, so that the determination of the time will also have some random errors. The required design accuracy of the Timing Tag system is 0.1 ms, so that this figure can be used to estimate the required signal-to-noise ratio.

The following analysis assumes that the receiver output is corrupted by Gaussian noise with the same power in both receivers, but uncorrelated. In this case the ratio function can be expressed as

$$\rho(t) = \frac{E_{dif}(t) + n_2(t) - n_1(t)}{E_{sum}(t) + n_1(t) + n_2(t)} \tag{10.15}$$

The main interest is when the signal is much greater than the noise, so that Eq. (10.15) can be approximated by

$$\begin{aligned} \rho_n(t) &= \frac{\frac{E_{dif}}{E_{sum}} + \frac{n_2 - n_1}{E_{sum}}}{1 + \frac{n_1 + n_2}{E_{sum}}} \approx \left[\rho + \frac{n_2 - n_1}{E_{sum}}\right]\left(1 - \frac{n_1 + n_2}{E_{sum}}\right) \\ &\approx \rho + \frac{n_2 - n_1}{E_{sum}} - \rho\frac{n_1 + n_2}{E_{sum}} \end{aligned} \tag{10.16}$$

As the assumed noise has zero mean, the expected (mean) value of the ratio (10.16) is an unbiased estimate of the ratio. Now consider computing the variance of the ratio function. As the noises are statistically independent, the variance of the ratio is given by

$$\begin{aligned} \mathrm{var}[\Delta\rho_n(t)] &= \frac{2\rho^2\sigma_n^2}{E_{sum}^2} + \frac{2\sigma_n^2}{E_{sum}^2} = 2\left[\frac{\sigma_n^2}{E_{sum}^2}\right](1 + \rho^2) \\ &= \frac{2(1+\rho^2)}{\gamma_0} \qquad \gamma_0 = \left.\frac{E_1^2}{\sigma_n^2}\right|_{t=0} \end{aligned} \tag{10.17}$$

where γ_0 is the receivers signal-to-noise ratio (SNR) at the origin of the measurements ($t = 0$). As from Eq. (10.16) the mean error in the ratio function is zero (assuming the noise has zero mean), the standard deviation of the error in the ratio function is given by

$$\sigma_{\Delta\rho} = \sqrt{\frac{2(1+\rho^2)}{\gamma_0}} \approx \sqrt{\frac{2}{\gamma_0}} \tag{10.18}$$

Further, from Eq. (10.14) the standard deviation of the ratio function near the origin can be expressed as

$$\sigma_{\Delta\rho} = \left(\frac{Dv}{R_1^2}\right)\sigma_t \tag{10.19}$$

where σ_t is the standard deviation in the timing measurement, and it is assumed the timing measurement errors δt are small so that $v\delta t \ll R_1$. Thus by equating the two estimates of the error in the ratio function the standard deviation in the timing estimate is given by

$$\sigma_t = \left(\frac{R_1^2}{Dv}\right)\sqrt{\frac{2(1+\rho^2)}{\gamma_0}} \approx \left(\frac{\sqrt{2}R_1^2}{Dv}\right)\frac{1}{\sqrt{\gamma_0}} \qquad \left(R_1 = \sqrt{H^2 + (D/2)^2}\right) \tag{10.20}$$

Thus the timing accuracy is a function of the geometry (H, D) of the Tag and receiver, the speed (v), and the SNR at the receiver at the measurement point a distance R_1 from the Tag to $t = 0$. However, constraints on the geometry and the speed means that the timing accuracy is mainly controlled by the receiver signal-to-noise ratio. The timing accuracy is related to the speed, but the position accuracy is independent of the speed, and is given by

$$\sigma_x = v\sigma_t = \sqrt{\frac{2}{\gamma_0}}\frac{\left(\frac{D}{2}\right)^2 + H^2}{D} \tag{10.21}$$

Consider a numerical example, with $H = 15$ cm, $D = 50$ cm, and peak SNR is $\gamma_0 = 30$ dB. The positional accuracy according to Eq. (10.21) is 7.6 mm. Alternatively the SNR required for a timing accuracy of 0.1 ms at 80 m/s is 29.5 dB. Thus the required accuracy can be achieved, but with a fairly high SNR. In practice such a higher SNR should be readily achieved due to the very short range.

The conclusion is that theoretical performance estimates of a simple signal strength measurement system is capable of very accurate timing measurements, meeting the 0.1 ms requirement of F1. However, the actual accuracy based on the testing described in Sect. 10.6 casts some doubts on the actual performance.

References

Catapult (2018, visited) Sports Training System. www.catapultsports.com

El-Rabbany A (2002) Introduction to GPS: the global positioning system. Artech House, Norwood, MA

Chapter 11
System Testing

11.1 Introduction

One important aspect in the design and development of a positioning system is the performance testing to ensure the actual equipment meets the design requirements. While this testing would include the testing of all the subsystems (including the RF receiver and transmitter, analog baseband components, digital signal processing, and processor software), this chapter concentrates on the overall integrated performance testing.

Consider the general problem of testing the accuracy of a positioning system in general (not specific to any particular technology). In particular, for a positioning system, determining the accuracy and reliability of the position determination in a variety of operating conditions is the most fundamental requirement in any system testing. While comparing the "true" and measured positions may seem a simple, if tedious task, in practice this is difficult to accomplish in many situations of practical interest.

Some positioning systems may be able to be tested using a high-accuracy reference system. For example, one possibility is the use of a very expensive (military grade) inertial system as a reference, but this is not appropriate to general (civilian) applications (Schubert et al. 2008; Elkaim et al. 2008). The problem with inertial reference systems is that the errors integrate over time, so that frequent recalibration at known reference points is necessary to maintain adequate accuracy. Kinematic GPS (El-Rabbany 2002) may be a possible alternative for outdoor situations, but typically they can only be used in static surveying applications (Teunissen and Kleusberg 1998; Annex A) with good line-of-sight conditions, as multipath propagation conditions with a moving receiver destroys the RF signal phase that is used by GPS receiver for accurate position determination. Another possible accurate reference system could be based on video tracking technology (centimeter accuracy), but the use of such systems would be limited to line-of-sight conditions over a rather limited area. For example, indoor testing would be limited

I. Sharp and K. Yu, *Wireless Positioning: Principles and Practice*, Navigation: Science and Technology, https://doi.org/10.1007/978-981-10-8791-2_11

to a single room within the building. In summary, there currently appears to be no suitable technology for dynamic non-line-of-sight situations, particularly indoors, but even outdoor accuracy testing can be challenging.

Most practical location-based applications involve the tracking of a mobile device as it moves throughout its operating environment. Further, it is often the case that the system being dynamically tested has an accuracy which exceeds other (reference) systems with which positional data is be compared. To obtain meaningful accuracy estimates, ideally the reference positions should have an accuracy at least ten times greater than the accuracy of the system being tested. A typical approach in many situations has been to abandon dynamic accuracy testing for static testing, whereby the mobile device is stationary, and the static positions are determined by some classical surveying method, such as using a theodolite. While this approach can provide some positional accuracy data, these results can not provide reliable estimates when the mobile unit is moving, particularly at high speeds, or if the operating environment is dominated by multipath propagation for a radio-based system. As the types of positioning systems can vary greatly, and alternative more accurate reference systems are usually not available, more universal testing methods are required; such methods are described in detail in this chapter.

From the above discussion it is clear that it is highly desirable to measure positional errors in the normal operating environment, including both stationary and moving mobile devices. The results of such measurements are usually not required on an individual point-by-point basis, but rather the overall positional statistics are required. Thus there is an opportunity to develop a technique which only estimates the overall statistics, typically in the form of a cumulative distribution function (CDF) of the position error. Using the CDF, the "accuracy" may be specified (say) as the error which is only exceeded 50% (median) or 10% (conservative estimate) of the time, depending on the requirement.

While the focus of the testing described in this chapter concentrates on determining the positional accuracy of a system, for radio-based systems testing will also include measurements of the nature of radio propagation, particularly in indoor environments. Radio propagation in a multipath non line-of-sight environment is complex, which challenges the design of both data transmissions as well as accurate position determination. The topic of obtaining high data rates in indoor situation is important for many modern radio technologies, but is beyond the scope of this chapter. While data communications will form part of any radiolocation system, the required data rate are rather low, and typically data transmission is obtained by low bit-rate modulation of the signal used for position determination. However, the nature of the signal propagation does play a significant part in the performance of a radio positioning system, so this chapter also provides some details on the measurement of the signal characteristics, particularly as a function of range. As range information will be available from any positioning test scheme, the signal strength measurements can be easily incorporated into and testing scheme for positioning. Such data provides useful statistical data at varying scales, from centimeters to tens of meters.

11.2 Types of Measurements for System Performance Determination

Receiver performance requires measuring the particular parameter used for position determination. For example, the signal amplitude, signal phase, or the of arrival of the pulse.

11.2.1 Signal Strength Measurements

The measurement of signal strength at a base station or at the mobile unit is usually rather simple, as most receivers incorporate a receiver signal strength indicator (RSSI). The implementation of the RSSI is usually in conjunction with the receiver automatic gain control, whereby the feedback control voltage is a measure of the signal strength. Thus the RSSI is an indirect method of measuring the signal power, and feedback voltage needs to be calibrated with a known range of input RF signal levels to obtain reasonably accurate signal levels. Further, the feedback mechanism will typically have a rather long time constant, so that fast changing signal levels cannot be measured. Also the calibration is typically with an unmodulated RF signal, so the measured value with signal modulation will not be a true indication of the signal power. Further, the calibration in typical commercially available radios (such as chip radios used in mobile devices) would varying from radio to radio, so that the supplied calibration curve is only accurate to a few decibels. Testing of a wide range of devices (Lui et al. 2011) has shown that the nominal RSSI reading can be quite inaccurate, so that any testing should be based on a calibration for each individual unit.

Finally, accurate signal strength measurements depend on accurate knowledge of the gain of the receiving antenna. When the antenna is internal to the mobile unit, this can be difficult to determine, so that correction factor associate with the antenna adds further uncertainty to the signal strength measurements. Even if the mobile unit has a connector for an external antenna, when the mobile unit is close to a body during measurements the effective gain of the antenna is not known. In such circumstances this uncertainty should be considered as part of the multipath operating environment, and thus the antenna gain variability will add to the overall signal variability. However, such variability can be assessed as part of any comprehensive testing program.

Thus overall the RSSI is not an accurate measure, but can be useful in providing information of the variability of the signal strength within the coverage area of a radiolocation system.

11.2.2 Environmental Impulse Response Measurements

The signal strength measurements described in Sect. 11.2.1 provide some information relating to the effects of multipath propagation. However, to obtain a more fundamental understanding of the effects of multipath propagation, and its affect on time-of-arrival positioning systems, more complex measurement methods are required. As the focus of this chapter is on testing actual positioning systems rather than the physics of radio propagation, this section only provides a brief summary of techniques that allow the environmental impulse function to be measured. The impulse response in a multipath environment is the (theoretical) received signal as a result of transmitting an ideal impulse (delta function). The received signal will be composed of a sequence of delta functions delayed by their respective propagation times. The first signal to arrive will be any line-of-sight path, and subsequent multipath signals will be delayed relative to this direct path. The shape of this environmental impulse response has a direct effect on the performance of any time-of-arrival positioning system.

There are basically two methods of measuring the impulse response. In all cases ultra-wideband measurements are required; UWB is usually defined as transmissions with a bandwidth of at least 500 MHz. The first method uses a Network Analyser to measure the spectrum of the signal received at one antenna (simulating a Base Station) from the signal transmitted from another antenna (simulating the transmissions from a "mobile" device). Network Analysers are typically used to measure the transmission of signals through a hardware circuit by scanning the input signal across the frequency band of interest, and recording the received signal. The Network Analyser effectively measures the scattering parameter S_{21} across the band by performing an inverse Fourier transform of the received signal across the band. In the case of propagation experiments, the environment (plus antennas) are the "network" whose properties are measured.

While the first technique is a frequency domain method, the second method is based on time domain measurements. The general principle is that a wideband pulse is transmitted, and the environmental impulse is recorded, again in the time domain. In practice transmitting high powered pulses is impractical, so that an alternative method is used—namely by the use of a spread-spectrum signal. As such signals are commonly used in actual positioning systems (such as GPS), this method is closely aligned with the techniques used in real systems, but the UWB spread-spectrum signals are used to allow the details of the impulse response to be observed. The usual method is to use a direct-sequence spread-spectrum signal, which transmits a RF signal phase modulated by a binary sequence with special auto-correlative properties. In particular, the cross-correlation when displaced by one of more binary "chips" is essentially zero, so the transmitted signal (when cross-correlated with its own code) is essentially a narrow pulse in the time domain. By performing a similar correlation at the receiver the impulse response can be determined, to a time resolution equal to the chip period. Thus by using a high chip rate (typically hundreds of megahertz), a detailed measurement of the environmental impulse response can be determined.

Measurements of the wideband impulse response (by either method) can be used in the design and testing of actual positioning systems. For TOA-based systems, the performance at various signal bandwidths can be estimated by convolving the measured impulse response with the actual (nominal) time response of the actual system.

This allows various TOA detection algorithms to be tried using real-world measured impulse data, before the actual equipment is constructed. For example, given a particular (indoor) operating environment and an accuracy requirement, the signal protocol and detection algorithms can be determined to meet these requirements. Further, when the actual system is being tested, the results can be compared with those predicted by the analysis based on the measured impulse response.

11.2.3 TOA and Time Synchronisation Measurements

A common method used in radiolocation systems is based on the time-of-arrival (TOA) of the signal from the mobile device at a receiver such as a base station. While these measurements are nominally based on the detection of a signal pulse (such as in UWB systems), actual implementation typically (such as with GPS) involves the reconstruction of a pulse from a spread-spectrum signal using signal processing techniques. In particular, correlation techniques allow the reconstruction of a pulse using the particular code used by the transmitter, and with some restrictions simultaneous transmissions from multiple transmitters (code division multiple access—CDMA) can be decoded due to the (almost) orthogonal relationship of the correlation processes. For a navigation system (such as GPS) the transmitter is located in a "fixed" component and the receiver in the mobile component, while for a tracking system the roles are reversed. The advantage of the navigation mode is that an unlimited number of mobile devices can receive the transmitted signals simultaneously. This is possible as the transmitting satellites are remote from the receiver, so that all the simultaneous signal are of approximately the same signal level. In contrast, for terrestrial positioning system, particularly those operating indoors, the signal strengths from the base stations could vary greatly, so that strong signals swamp weak signals (near-far effect). Thus in such circumstances the signals have to be transmitted sequentially in allocated time slots (time division multiple access). Thus the system must include some method of defining time to all the components (base stations and mobile devices), so that each component can transmitter in its allocated time slot. Thus for tracking systems the testing needs to include the performance of both the time synchronisation and the TOA accuracy. Note that time synchronisation also implies that the clock *frequencies* will also be synchronised, which also provides very accurate radio frequencies.

Thus in summary, in a TOA-based tracking system accurate TOA measurement is required for a number of critical functions, namely:

1. The prime function is to allow accurate position to be determined using pseudorange measurements. The pseudorange method also requires time synchronisation in all base stations, but importantly not in the mobile units.

2. The time slot TDMA structure can be determined in each component. One element, referred to as the Master Station, is used as the reference clock, and other components then synchronize to this clock.
3. By measuring the change in the TOA over time (many seconds) accurate clock frequency synchronisation can be performed.
4. The above timing structure allows ranges (rather than pseudoranges) to be determined using a round-trip time measurement technique (see Sect. 11.2.4), whereby the reply can occur in the appropriate transmission time slot rather than immediately replying. This method greatly improves the efficiency of the TDMA system.

11.2.4 Round Trip Time Method

While a common technique in radiolocation systems based on TOA measurements of a signal allows position determination to be performed without any time synchronization in the mobile device, such a system requires very accurate frequency and time synchronization in the "fixed" components such as base stations. However, if ranges are directly measured instead of pseudoranges, then such synchronisation is not required, although low accuracy TDMA slot synchronization is still necessary. In such a system, each module (fixed and mobile) simply transmits in its time slot an "acknowledge" reply timed relative to the measured TOA of the original signal, so that the round-trip time can be measured in the original transmitting node from all replying nodes. As the original transmission and the reply are all timed relative to the clock in the originating node (and not the receiving node), the measured RTT are independent of the clock phases. In contrast, the one-way transmissions TOA will depend on phase of the transmitting (mobile) device, but not on the receiving clock phase due to clock synchronization—hence the measurement of pseudoranges. However, the RTT measurement will include the combined transmit and receiving delays of both the nodes involved with the RTT measurement. Thus for the RTT to function as a range measurement both then delay parameters must be known. For the base stations these delays can be estimated by using the RTT measurements between base stations and the known fixed inter base station distances. While a nominal value of the mobile delay will be known, its value will vary over time, so that this would translate into computed position measurement errors. However, the RTT measurement errors associated with range measurements to base stations will have a common delay calibration error, so in fact the RTT data for ranges from mobile units to base stations should in fact be treated as pseudoranges. Thus the RTT method results in data of a similar nature to the more traditional method involving clock synchronization on the base stations, but can be achieved with simpler more ad hoc positioning methods.

Thus in summary, testing of the RTT method for accuracy depends on the underlying accuracy of the TOA measurements, plus the accuracy in determining the delay parameters associated with the fixed base stations.

11.3 Positioning Measurements—Accuracy Determination

In the design and testing of a radio positioning system, accuracy evaluation is important since positional accuracy is one of the key requirements for a variety of applications and services. The performance of a positioning algorithm implemented in a real positioning system can be examined through computer simulation based on a model which describes the characteristics of the received radio signal (Yu et al. 2009; Qi et al. 2006; Yu 2007; Miao et al. 2007). However, in general simulations cannot accurately represent the performance of real positioning systems as the modeling is imperfect due to environmental complexity and system hardware effects. Thus in practice the accuracy of a positioning system must be examined through field trials (Fontana et al. 2003; Fontana 2004; Hedley et al. 2008; Lewandowski and Wietfeld 2010; Priebe et al. 2011).

The measurement of positional errors of a moving object can be difficult, as by definition an error is determined by comparing the estimated position of the mobile unit with some more accurate reference position. As it is often the case that the positioning system under test is the most accurate positioning system available in its operating environment, typically the procedure is to measure the positional errors at fixed stationary points in the coverage area, with the position of the fixed points determined by some independent surveying method. Alternatively, reference points can be surveyed throughout the coverage area, and to determine a positional error the experimenter selects a reference point and indicates (say by clicking a button) when the reference point is reached (Modsching et al. 2006). Such techniques, while simple in concept, are very time consuming, and do not actually measure the true performance of a moving mobile device.

11.3.1 Static Measurements

The simplest testing of the performance of a positioning system is to perform static measurements associated with position determination. These measurements are based on comparing the results from a positioning system (such as range or position) with the expected (true) results based on simple geometric measurements using classical surveying results. However, such measurements can be extended to provide other vital information associated with position system performance. In particular, by comparing the true and measured data other data such as the calibration of the positioning system equipment and determining the characteristics of the operating environment (particularly important for indoor positioning) can be determined.

Thus this section describes various testing that can be performed with positioning system equipment which is largely aimed at equipment calibration and environmental propagation characteristics.

11.3.1.1 Radio Equipment Delay Calibration Measurements

The determination of the propagation delay used for position estimation is based on the measurement of the time-of-arrival (TOA) of a radio signal at a receiver. However, the TOA measurement encompasses the total transmission delays from the transmitter to the receiver as measured relative to the clocks in the (mobile) transmitter and the receiver (stationary node or anchor node/base station), and includes the delays in the baseband hardware, the RF and IF sections of the radios, as well as the propagation delay. This concept is described in Chap. 4, Sect. 4.2.2, which gives a detailed description of the transmit and receive delays in the radio equipment between node m and node n. A block diagram of the transmission path and the associated parameters is shown in Fig. 4.2 in Chap. 4. Note that the transmission can be from node m to node n or in the reverse direction. For typical positioning systems, nodes both transmit and receive, so that it useful to consider the TOA measurements between a pair of nodes. Thus applying (4.5) twice, the resulting two TOA measurements are

$$\begin{aligned} TOA_{mn} &= \phi_m + R_{mn}/c + \Delta_n - \phi_n \\ TOA_{nm} &= \phi_n + R_{nm}/c + \Delta_m - \phi_m \\ RTT_{mn} &= (TOA_{mn} + TOA_{nm})/2 = R_{mn}/c + (\Delta_m + \Delta_n)/2 \end{aligned} \tag{11.1}$$

where reciprocity in the propagation delay has been used (Jordan and Balmain 1968). Thus by combining two TOA measurements into a single round trip time (RTT) measurement, the clock phase parameters can be eliminated. Thus in this case, the measured data only involves the sought-after range measurement and the delay parameter associated with each node. If the radio delay parameters are known (preset) values, then the two TOA measurements combined can be used to estimate the range. Alternatively for calibration, if the RTT is measured and the range is a known value (or a known cable delay if the two nodes are connected by a cable for calibration purposes), then the combined delays $\Delta_m + \Delta_n$ can be estimated. Note that this measurement cannot estimate the individual node delays. However, by combining such measurements with its neighbors, redundancy in the equations allows the radio delay of each node to be determined.

Now consider using (11.1) to determine the delay parameter for a particular node. This can be achieved if three units are used. In particular, a cable is used to connect two nodes at a time; thus there are three pairs of nodes. The delay through the cable can be determined from its length and propagation constant, so (11.1)

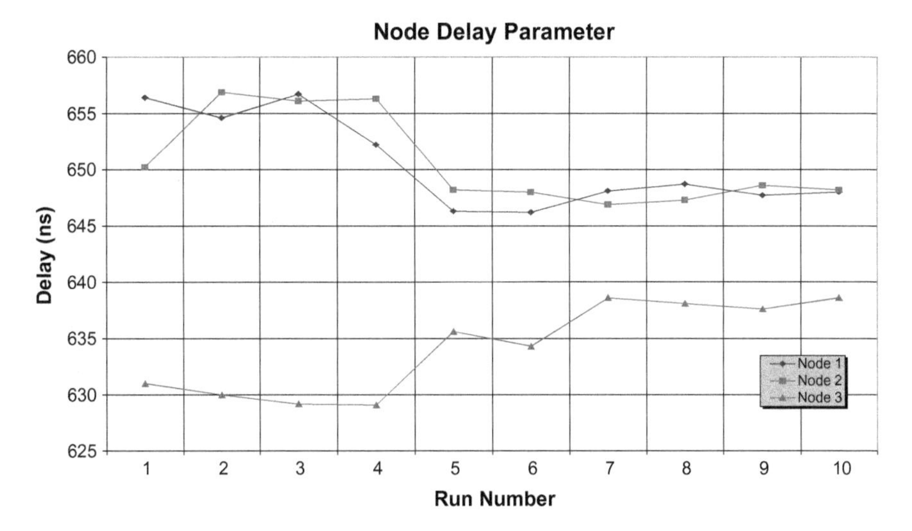

Fig. 11.1 Variation in the node delay parameter for three different nodes as a function of time. The data are the average delay over about a period of 3 min (run number period). The particular chip radio is a Maxim MAX2828

represents three linear equations which can be solved for the three delay parameters. Then subsequent nodes can be calibrated using one of the original three nodes, and again using the cable and applying (11.1). This calibration method should be very accurate, as the SNR will be high and the cable delay can be accurately measured.

This calibration method was used to measure the delay associated with a chip radio; the results are shown in Fig. 11.1. Notice the measured delays are around 650 ns, which is equivalent to a propagation range of around 200 m. While the radios are nominally identical, observe that one has a delay characteristic somewhat different from the other two. This shows that for accurate range measurements it is not sufficient to simply use the nominal delay in determining the range using (11.1). Also observe that the delay varies over time. The measurements were made from a "cold start", so that the temperature increased until it reached an equilibrium value in about 15 min. This change in delay is about 10 ns, which is equivalent to about 3 meters of propagation distance. Thus changing ambient temperatures will have a significant affect on the measured range, so that some form of temperature compensation is necessary. Alternatively, some form of real-time calibration is necessary. Such a method is described next section.

11.3.1.2 Real Time Delay Calibration Measurements

While it is possible to perform calibration measurements by connecting the modules by cables, such measurements are typically impractical in real-world situations, so that measurements should normally be performed in real-time in situ "over the air".

However, such a procedure will also incur some additional measurement errors, so that real-time calibration will not be as accurate as the manual calibration procedure using a cable. However, real-time calibration will track effects such a delay variation with temperature and ageing effects, so overall the accuracy will improve over simply using nominal calibration values.

The basis for estimating the radio delays is a model of the measurements errors, plus the redundancy in the inter-node measurements. The range measurements will include the following errors:

1. Errors in the internal radio delay parameters, as described above.
2. For indoor situations with multipath and NLOS propagation a bias model is required, as the measured propagation delays will be greater than the geometric range between nodes. Typical measurements show that the bias can be adequately modeled as a linear function of range.
3. The multipath propagation also results in an additional random component with (by definition) zero mean and unknown variance. In general, these errors will not be Gaussian, but may be approximately so.
4. Due to receiver noise, the RTT will have random errors which are a function of the receiver signal-to-noise ratio (SNR) in both nodes. This ranging noise can be modelled as a Gaussian process. This component can be minimized if the SNR is high, but an actual installation may include long propagation paths, and hence low to moderate receiver SNR.

As accurate RTT measurements rely on accurate frequency synchronization between the two nodes, any frequency errors translate into ranging errors. Such frequency offset measurements are insensitive to the errors described in subparagraphs (1–3) above, but are sensitive to the receiver noise described in subparagraph (4). As a consequence, these frequency synchronization errors can be modelled as a random Gaussian process with zero mean.

The calibration process is based on applying (11.1) with a known inter-nodal distance (from a survey of the location of the nodes), and estimates of the radio delay parameters (default equipment delay parameters). Then the measurement errors (δM) between two nodes can be represented by the equation

$$\delta M_{mn} = M_{mn} - R_{mn} - c\left(\hat{\Delta}_m + \hat{\Delta}_n\right)/2 = (\delta R_0 + \lambda R_{mn}) + c\left(\frac{\delta\Delta_m + \delta\Delta_n}{2}\right) + Noise \tag{11.2}$$

where δR_0 is the (unknown) bias model error at zero range and λ is the (unknown) linear model range bias parameter associated with multipath propagation, $\delta\Delta$ is the calibration error (to be determined) in a node internal radio delay parameter, and *Noise* is a random variable with zero mean and with approximately Gaussian statistics. As the measurement error model is linear and the various error mechanisms are uncorrelated with one another, a linear least-squares (LS) analysis can be used to obtain an unbiased estimate of the model parameters. However, a direct LS analysis using (11.2) is not possible. In particular, the δR_0 parameter cannot be determined independently from the delay calibration parameters. However, if it is assumed that the delay calibration errors are random with zero mean, these random errors initially can be incorporated into the noise, so that the LS equation becomes

$$\delta M_{mn} = (\delta R_0 + \lambda R_{mn}) + Noise \tag{11.3}$$

which can be solved for estimates of the δR_0 and λ parameters. Having established an estimate of δR_0 and λ, the equation for a measurement associated with nodes m and n then becomes

$$M_{mn} - R_{mn} - c\left(\hat{\Delta}_m + \hat{\Delta}_n\right)/2 - \delta R_0 - \lambda R_{mn} = \varepsilon_{mn} = \frac{c}{2}\left(\delta\Delta_m + \delta\Delta_n\right) + Noise \tag{11.4}$$

If the number of nodes in the mesh network is N_{nodes}, then the number of such measurements will be up to $N_{nodes}(N_{nodes} - 1)/2$, but because of radio range limitations the typical number of neighboring nodes with radio communications for satisfactory measurements per node will be of somewhat smaller. In any case, the total number of measurements N_{meas} is far in excess of the number of nodes (and hence their delay calibration parameters), and thus the considerable redundancy allows a LS procedure to be used to estimate the error in the delay calibration. The (11.4) model can be expressed in matrix form

$$\mathbf{\Gamma d} = \frac{2}{c}E \tag{11.5a}$$

where $\mathbf{d}$ is the vector of N_{nodes} elements of calibration delay corrections, $\mathbf{E}$ is an error vector with N_{meas} elements of ε_{mn}, and $\mathbf{\Gamma}$ is a $N_{meas} \times N_{nodes}$ matrix of 0 and 1 elements associated with the pattern of communications between nodes m and n. The LS solution for the delay calibration errors is

$$\mathbf{d} = \frac{2}{c}\left[\mathbf{\Gamma}\,\mathbf{\Gamma}^{\mathrm{T}}\right]^{-1}\left[\mathbf{\Gamma}^{\mathrm{T}}E\right] \tag{11.5b}$$

However, for this method to be practical the accuracy in estimating the node calibration parameters should ideally be within the desired specification of a standard deviation (STD), say not greater than 1 ns for indoor positioning. The source of this uncertainty is noise term in (11.4), and is largely associated with the range measurement noise due to multipath propagation, provided the receiver SNR is sufficiently high. To estimate the STD of the radio delay calibration, an analytical LS statistical analysis using (11.5b) is required. Analysis of the symmetrical square matrix $\mathbf{\Gamma}\,\mathbf{\Gamma}^{\mathrm{T}}$ shows that it consists of a diagonal with elements of value equal to the number of neighboring nodes, while the associated row/column consists mainly of zeros, but also with some elements (in number equal to the number of neighbouring nodes) of value one; the pattern of ones is associated with the pattern of radio communications between nodes. For analytical statistical calculations (but not for the numerical calculations involving (11.5b)) it is appropriate to set all the off-diagonal elements to zeros, which greatly simplifies the analytical calculations with minimal error, and does not require detailed knowledge of the radio communications pattern to estimate statistical performance. With this simplification, the solution (11.5b) for a particular node p can be expressed in the form

$$\delta\hat{\Delta}_p = \frac{2}{c}\left[\frac{1}{N_{neighbors_p}}\right]\left[\sum_{n=1}^{N_{nodes}} \mathrm{F}(p,n)\varepsilon_{p,n}\right] \tag{11.6}$$

where $\mathrm{F}(p,n)$ is 1 if nodes (p,n) communicate, otherwise the function is zero. Using (11.6) it is clear that the expected value $E\left[\delta\hat{\Delta}_p\right] = 0$ as the errors have a zero expectation after the biases are removed in (11.4); thus the delay calibration estimate using (11.4) is unbiased. Similarly, noting that the measurement errors are uncorrelated, the variance of the estimated delay calibration is given by

$$\begin{aligned} E\left[\delta\hat{\Delta}_p^2\right] &= \left[\frac{2/c}{N_{neighbors_p}}\right]^2 E\left[\left(\sum_{n=1}^{N_{nodes}} \mathrm{F}(p,n)\varepsilon_{p,n}\right)^2\right] \\ &= \left[\frac{2/c}{N_{neighbors_p}}\right]^2 N_{neighbors_p} E\left[\varepsilon_{p,n}^2\right] \end{aligned} \tag{11.7}$$

which results in the expression for the STD

$$\sigma\left(\delta\hat{\Delta}_p\right) \approx \frac{2\sigma_\varepsilon/c}{\sqrt{N_{neighbors_p}}} \tag{11.8}$$

where σ_ε is the STD of the range measurement noise in (11.2) after the bias errors are removed, and is assumed to be the same for each node in the mesh network. The actual number of neighboring nodes will vary throughout the mesh network, and the individual node's errors are statistically independent. Summing up the variances, an "average" number of nodes applicable for calculating an overall calibration accuracy using (11.8) is

$$\bar{N}_{neighbors} = \frac{N_{nodes}}{\sum_{n=1}^{N_{nodes}} \left(N_{neighbors_n}\right)^{-1}} \tag{11.9}$$

Thus given a specified ranging accuracy, (11.8) can be used to determine the average number of nodes required to meet the desired calibration accuracy. For example, for the WASP system (see Sect. 11.4.4.3) indoors $\sigma_\varepsilon \approx 0.8\,\mathrm{m}$ (see Fig. 11.2), so applying (11.9) a 1 ns calibration accuracy requires on average 28 neighboring nodes; such a high number is impractical, so the desired calibration accuracy of 1 ns cannot be met in this case. As the geometric dilution of precision

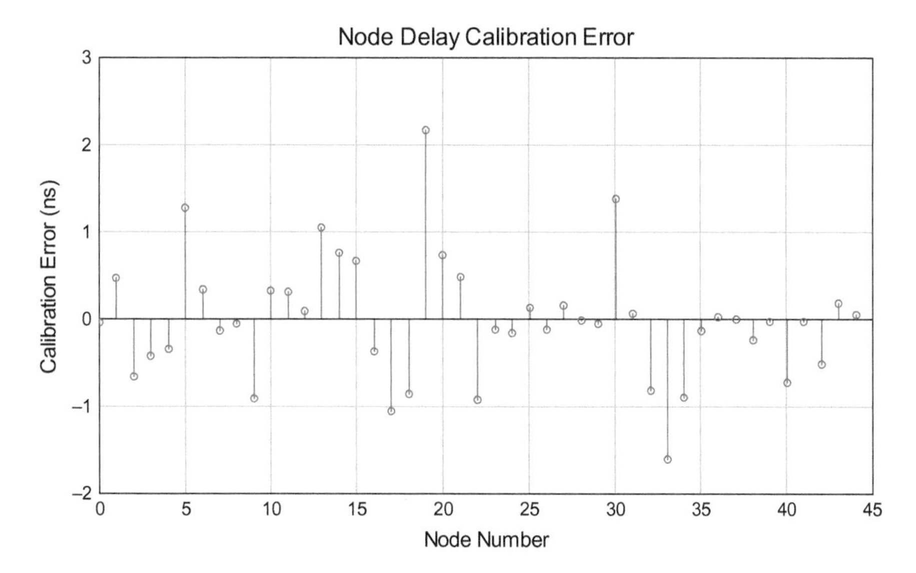

Fig. 11.2 Residual radio delay errors after applying the real-time calibration. The mean error ($\mu_{\delta\Delta}$) is −0.01 ns and the STD ($\sigma_{\delta\Delta}$) is 0.7 ns

(GDOP) is about $2/\sqrt{N_{neighbors}}$ for a mesh network (Sharp et al. 2009), for positioning to meet a 0.5 m accuracy requirement (and with $\sigma_{\varepsilon} = 0.8\,\text{m}$) the number of neighbouring nodes required is about 10; applying this number to (11.11) gives the calibration accuracy as 1.7 ns, or equivalently 0.5 m. This calibration error increases the overall range error to $\sqrt{0.8^2 + 0.5^2} = 0.94$ m for position determination using the above real-time delay calibration procedure.

The above method was applied using the 46 nodes deployed throughout a building complex, and RTT data between nodes recorded using the cable-calibrated radio delay data as the nominal delay value. The test was performed a few weeks after the cable calibration, so there may be some ageing effects. Both measurements were made indoors in a temperature controlled environment, so the operating temperatures should be similar in both cases.

While the number of nodes in the network was large (46), the mean number of neighboring nodes in this mesh network is nine, as the range between nodes limited the number of nodes in range with sufficiently high SNR. Only high SNR (>28 dB) data were used, with 100 samples averaged to largely remove temporal noise from the measurements. These RTT data were then used to estimate the node calibration errors using the above technique, which ideally should result in near-zero calibration errors if there were no errors in the real-time calibration procedure. Applying the above calibration process showed that the STD was 2.6 ns, greater than the expected result of 1.8 ns using (11.4) and (11.5a, 11.5b). This shows that there is probably a drift in the internal radio delays since the original manual calibration. Applying the estimated corrections to the radio delay parameters and recalculating gives the residual errors STD of 0.7 ns, as shown in Fig. 11.2. These results are within the desired calibration accuracy of 1 ns, and better than the theoretical value; this example emphasises the need for real-time calibration, even if a pre-installation calibration is performed.

The above calibration procedure was based on measurements using anchor nodes in a mesh network, but the procedure also could be applied to mobile nodes when they are stationary and have had their position previously determined. In this case, there will be some errors in their computed position, which would flow through to the estimation of the node delay parameter. However, over time these errors would be random with zero mean, and the delay parameters would only vary slowly over time. Thus more accurate node delay parameters could be obtained using a Kalman filter.

11.3.1.3 Indoor Propagation Modeling Measurements

The indoor propagation environment is complex, and usually the non line-of-sight conditions apply. Under such circumstances the assumption of straight line propagation paths used in position determination is not valid. In particular, when methods such as the RTT described previously for radio delay calibration are used, the measured range exceeds the true range in most cases. Similarly, for receiver signal strength positioning the propagation loss will be in excess of the free-space loss. For improved positioning performance some method of modeling the range bias or loss excess is required. This section describes tests that enable a model to be determined for the particular operating environment where the positioning system is installed. Note that as this operating environment will vary from building to building, there can be no one set of model parameters that can be universally applied to all buildings.

To determine the parameters of a model for a particular indoor location some measurements of the range bias or loss excess is required. Such measurements could be performed by moving a mobile unit to locations throughout the coverage area and measuring range or signal strengths at the base stations installed in the building. The location of the mobile unit needs to be determined by some independent surveying method, so that the positioning system data can be compared with the "true" ranges. Such a manual procedure is a simple but time consuming process, and would typically be performed as part of the positioning system installation. However, this approach is not particularly satisfactory, as it only captures the data at the time of installation, and does not cater for changes over time. Thus the recommended approach is to use the base stations as test units, whereby each in turn acts as the transmitter and the measurements are made at the other base stations. As the base stations are at known locations the (straight-line) distance of propagation is known, so the range bias and loss excess can be determined. Such a procedure will by definition encompass the whole of the coverage area, and the number of measurements will quite large, of the order of the square of the number of base stations. Further, this process can be performed as a background task of the positioning system, so that any changes over time can be automatically tracked, and the associated models updated.

Based on these concepts, the details of measurement procedures and associated analysis are provided for both range bias and signal loss excess in the following sub-sections.

Range Bias Measurements and Modeling

For this analysis, it is assumed that the indoor propagation results in both a positive bias which is proportional to the propagation range plus a non-zero offset, and a zero-mean random component which is independent of range. This assumed error model is supported (at least approximately) by indoor radio propagation measurements (Alavi and Pahlavan 2006; Alsindi et al. 2009; Alavi and Pahlavan 2003; Gentile and Kik 2006). Some typical measured data are shown in Fig. 11.3. While there is considerable scatter in range error data, there is a clear linear trend of the errors increasing with range. In contrast, the random variation relative to the linear trend line appears to be independent of range.

Based on these assumptions, the measured pseudorange p_ε associated with the ith base station is given by

$$\begin{aligned} p_{\varepsilon_i} &= p_i + \varepsilon(0, \sigma_r) \\ p_i &= f(x, y, \phi; \lambda, \delta r_0) = r_i + c\phi + (\delta r_0 + \lambda r_i) \\ &= (1 + \lambda)\, r_i + (c\phi + \delta r_0) \end{aligned} \tag{11.10}$$

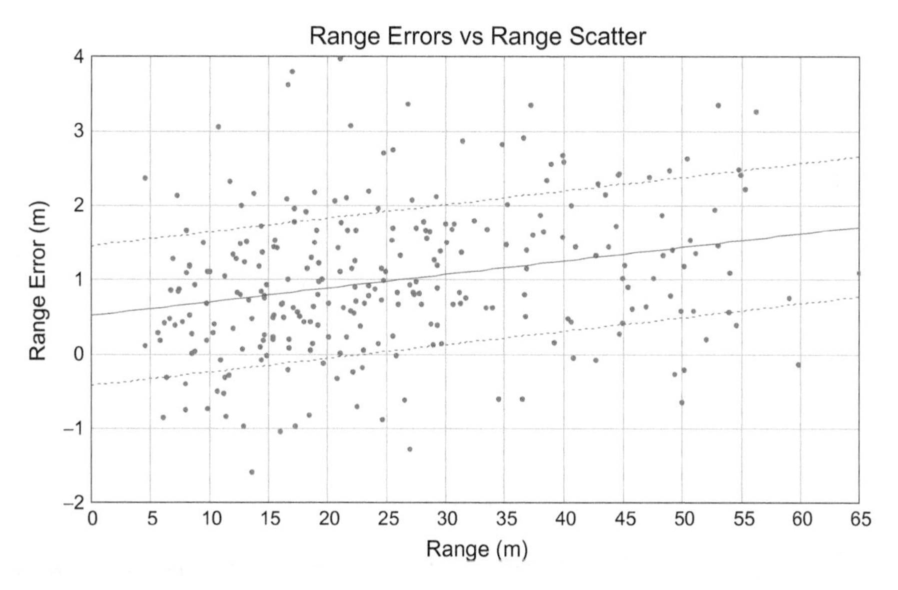

Fig. 11.3 Measured ranging error performance based on RTT measurements between pairs of nodes scattered throughout the building complex Shown in Fig. 11.20. The measured parameters are: $\lambda = 0.020$ and $\sigma_r = 0.94$ m. The solid line is the associated linear LS fit model, and the dotted lines are $\pm\sigma_r$ relative to the linear trend line. The zero-range bias error is $\delta r_0 = 0.52$ m. The radio frequency is 5.8 GHz and the signal bandwidth 125 MHz

where ε is the zero-mean random error with standard deviation σ_r, c is the speed of propagation (0.2998 m/ns), and ϕ is the timing-phase of the clock in the mobile node relative to the time-synchronized base stations ($\phi = 0$ for range measurements). Note that in the case $\lambda = 0\,, \delta r_0 = 0$ and Eq. (11.10) becomes the classical formulation of the problem, and is typically applied to the indoor case even though it does not accurately describe the problem. In general, the values of λ and δr_0 in a given propagation environment will not be known, but using the above measurement procedure data such as in Fig. 11.3 can be analysed using linear regression techniques. The range model (11.10) can be used for position determination, and effectively removes range bias errors from degrading the accuracy of the system. Note also that the random range data associated with the measurements shown in Fig. 11.3 allows the positional accuracy to be estimated.

Propagation Loss Excess Measurements and Modeling

The propagation loss is important in the operation of a radiolocation system, both for the data links between nodes and for signal strength positioning for estimating the ranges to mobile units. The signal strength is measured in a base station and mobile unit by the receiver signal strength indicator (RSSI), which typically measures the received power in the receiver IF in association with a automatic gain control (AGC) feedback system. The control voltage in the feedback loop can be calibrated to measure the received power, usually the dBm. Note however that such measurements are not a particularly accurate estimate of the received RF signal, and in particular are affected by the effective gain of the receiving antenna. This is particularly relevant for the mobile case, where the mobile unit is typically close to the body which can significantly block the signal from a base station. Thus using the nominal mobile unit antenna gain is not particularly accurate in any received signal calculations.

The RSSI reading can be related to the propagation loss by the *Link Budget* (see Chap. 9), so that the propagation loss can be determined from the effective transmitter power and the receiving antenna gain. This propagation loss can be considered as the sum of the free-space loss and a *loss excess* component, which is a consequence of additional signal losses due to processes such as signal scattering, diffracting and losses through walls and other intervening obstacles. However, in some uncommon situations the loss can be negative, which is a consequence of constructive signal interference. For example, as base stations will be mounted on walls the path between two base stations which have line-of-sight propagation (say along a corridor) will have reflections from the wall that can effectively double the received signal amplitude (6 dB signal enhancement).

An example of the propagation loss between pairs of base stations are shown in Fig. 11.4, plotted as a function of range. The data are based on the RSSI and the known transmitter power and the base station antenna gains. The frequency is 5.8 GHz. Also shown are the LOS loss and a loss excess model. Almost universally modeling of propagation losses indoors assumes a power law with $p > 2$, similar in nature to the free space loss law (with $p = 2$). Such a model has the advantage that when the loss in decibels is plotted as a function of the log of the range the result is

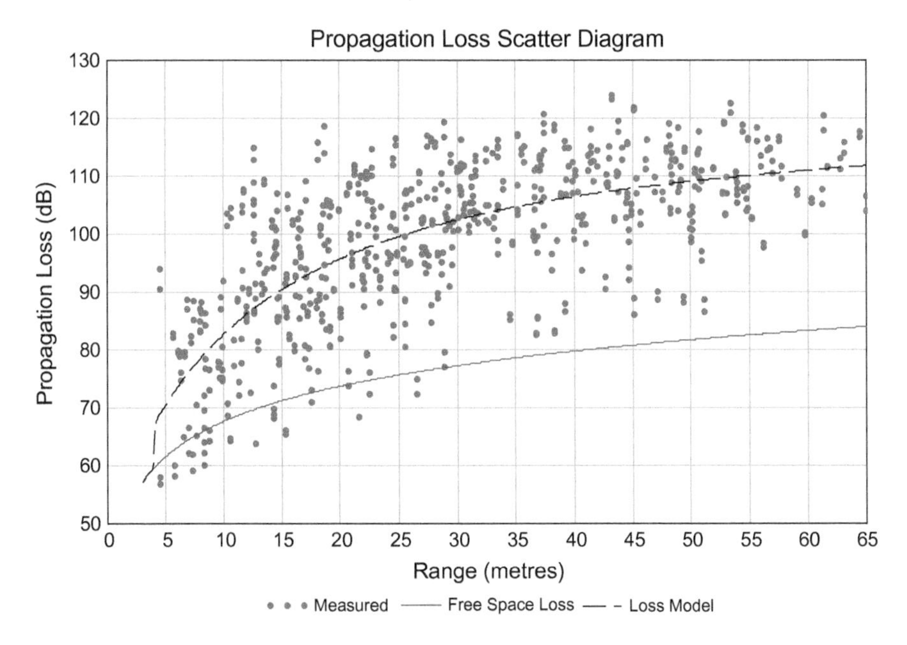

Fig. 11.4 Propagation loss scatter diagram for pair of base stations (46 in total) in a building complex. Also shown is the associated free-space loss. The loss excess model parameters are $\Delta L_{max} = 28$ dB, $R_0 = 4$ m, and $R_1 = 13$ m. The standard deviation between the model and the measured data is 10 dB. The transmission frequency is 5.8 GHz

a straight line, so that linear regression analysis is possible. However, this nonlinear range plot distorts the least-squares fitting process, and thus it is not a true least-squares fit to the data, and is in fact sub-optimum. With the data in Fig. 11.4 a better fit to the measured data as a function of range is obtain with an exponential loss excess model of the form

$$\begin{aligned} \Delta L(R) &= \Delta L_{max}\left(1 - \exp\left(-\frac{R}{R_1}\right)\right) \qquad (R > R_0) \\ &= 0 \quad \text{otherwise} \end{aligned} \tag{11.11}$$

where ΔL_{max} is the maximum loss excess at large range, R_0 is the maximum LOS range where there are no intervening walls, and R_1 is a parameter which is related to the scattering losses associated with the building architecture, and will vary from building to building. The model is limited to the maximum range where the signal strength is adequate for communications between nodes. Notice that the data scatter is quite large, of the order of 40 dB at a given range. This scatter is largely independent for ranges greater than 10 m. The large scatter is a consequence of some long paths are totally or in part along corridor (with minimal loss excesses), while other paths are through multiple intervening walls. Also note that in some cases the loss is less than the free-space loss, implying some form of constructive interference. However, such cases are limited to shorter ranges where there is LOS propagation such as along corridors. It can be concluded that the large scatter of losses as a function of range makes range determination from signal strength measurement difficult. For data communications the large scatter suggests the reliable communications range can be quite short, but in some cases ranges of 50 m or more are possible.

11.3.2 Dual Measurements at Fixed Separation

While static error measurements are simple to implement, there is a severe limitation to its use in that the position accuracy data cannot be determined for a moving mobile device, which is typically the normal operating environment for many applications. The design of the tracking algorithms typically includes some type of data smoothing, which recognises the dynamics of the object being tracked. Clearly the dynamics of a person walking is very different from a fast-moving racing car, so the optimum tracking performance is very dependent on the particular application; thus optimum tracking performance will utilise some form of specific data filtering to reduce measurement noise. Further for radiolocation, multipath induced signal fading can result in large but spatially specific position errors, which can be partially removed by suitable filtering when moving, but not when stationary. As a consequence, static measurements tend to *overestimate* the errors when compared with a moving mobile unit. Thus a method that *estimates* the accuracy of a positioning system tracking moving objects is important in testing a positioning system. However, if an independent but very accurate reference system is required then such testing may be difficult to perform. Thus some method that can estimate the positional accuracy while tracking a moving object without the use of a reference system is highly desirable. It is important to note that what is required from the testing is the overall statistical performance of the positioning system in its normal operating environment, rather than a collection of individual position measurement errors. While such statistical information can be obtained from a large collection of (static) individual measurements, such individual measurements are not necessary for obtaining statistical information.

The following subsections describe methods based on placing two mobile units on the object being tracked, and using differential measurements to estimate the accuracy of the positioning system. The concept is that while the absolute position errors may not be able to be measured, differential errors can be measured as the two mobile units are attached at fixed locations on the object being tracked. Thus statistics on the differential errors can be determined, and from these data the absolute position error statistics can be inferred.

While the above procedure is simple to implement, the limitations of the method should be noted. The fundamental assumption in the method is that the measured position errors of the two devices are random and statistically independent. Clearly any common mode of position errors will not be captured in any differential measurements. This restriction may limit its use in applications where the motion of the object being tracked (such as a walking person) affects the accuracy of the position determination. Thus for example inertial-based positioning may have common-mode errors due to the common accelerations associated with a person walking, so differential measurements may provide little useful information. In the case of radiolocation, the signal will be affected by a multipath propagation environment which will be (at least partially) common to both attached devices. However, if the devices are separated by (say) at least a few wavelengths, then the Rayleigh fading components will be uncorrelated in the two devices, so that differential measurements provide useful positional accuracy information. However, medium scale (10–100 wavelength) log-normal signal variations would be highly correlated for two devices (say) 1 m apart, so that positioning based on signal strength could not be accurately tested by this method. However, positioning systems based on time-of-arrival measurements will be mainly affected by Rayleigh fading signals, so that the differential method would be applicable in this case.

11.3.2.1 Separation Distance Method

The first method described is based on analysing the statistics of the separation distance between the two mobile units, as determined from the positions simultaneously reported for the two units. As the true differential separation distance is a known constant, the difference between the measured and true separation distances will be related to the measurement accuracy of the positioning system. This procedure will work at any mobile speed and any path taken by the mobile, and can be used both indoors and outdoors. However, a mathematical procedure and error modeling is required to convert the differential distance measurement data to an estimate of the single device position errors statistics.

To obtain useful numerical results using the differential measurement method, a statistical error model is required. As the computed position errors will be a consequence of many sources of errors, such as radio propagation related measurement errors, measurement errors within the equipment itself, and possibly combining data from many sources (such as multiple base stations), the computed position will be a consequence of summing many random processes. While these processes will vary between positioning systems, and the statistics of these processes are not known, application of the Central Limit Theorem means the position errors will tend to have Gaussian statistics. For the following analysis only 2D position determination is considered, which is the most common requirement. Further, it is assumed in the following analysis that the measurements of position are in two orthogonal directions (x, y), and that the errors in these two directions are statistically uncorrelated with a Gaussian probability distribution, each with zero mean and the same standard deviation. These assumptions may not always be appropriate, but would be applicable in many practical situations.

Now consider that two devices are fixed to the object being tracked, separated by a fixed distance D. The positions are determined by the system as (x_1, y_1) and

(x_2, y_2). However, as only differential positions are required in this method, the "true" positions can be normalised to $(0, 0)$ and $(D, 0)$ by suitable translation and rotation of the axes; this procedure can be performed for each measurement as a function of time. However, as there will be measurement errors, this normalization process will result in the second position as $(D, +\delta x_2 - \delta x_1, \delta y_2 - \delta y_1)$, where the delta terms are the measurement errors in the rotated/translated (x, y) coordinate system used by the analysis. Now based on the above discussion these errors are assumed to have Gaussian statistics, with zero mean and standard deviation of σ_0 for both devices, as they are assumed to have (nominally) identical hardware and performance, and over time are operation in an identical environment. Further, assuming that the errors are uncorrelated, the joint position error probability statistics can be represented as

$$p(x, y) = \frac{1}{2\pi\sigma^2} \exp\left[-\frac{(x - D)^2 + y^2}{2\sigma^2}\right] \tag{11.12}$$

where $\sigma = \sqrt{2}\sigma_0$. Equation (11.12) can be recognized as representing Rician statistics (Yu et al. 2009, Sect. 2.1.5), which can expressed in polar coordinates $R = \sqrt{x^2 + y^2}$ as

$$p(R, \sigma) = \frac{R}{\sigma^2} \exp\left[-\frac{R^2 + D^2}{2\sigma^2}\right] I_0\left(\frac{RD}{\sigma^2}\right) \tag{11.13}$$

The Rician statistics can be conveniently the normalised by the parameter σ so that the normalised separation is $r = R/\sigma$, and a Rician factor defined as $K = D^2/2\sigma^2 = D^2/4\sigma_0^2$, which is related to the fixed separation distance of the two devices and the Gaussian 1D position errors. While the true separation is D, the mean (normalized) measured separation is given by

$$\mu_r(K) = \sqrt{\frac{\pi}{2}}\, e^{-K/2}[(1 + K)I_0(K/2) + KI_1(K/2)] \tag{11.14a}$$

and the corresponding variance is

$$\mathrm{var}_r(K) = 2(K + 1) - \mu_r(K)^2 \tag{11.14b}$$

For example, if the two devices are separated by 1 m, and the 1D positional accuracy standard deviation is 0.5 m, then $K = 1$. Note in this case the normalise mean is $\mu_r = 1.81$, and the corresponding mean separation is 1.28 m, which is somewhat different from the true separation of 1 m. From (11.14b) the standard deviation of the measured separation is 0.6 m. Thus note that this model shows that the mean measured separation distance will not be an unbiased estimate of the true separation distance, so that a simple calculation based on averaging the measured differential distance between the true distance D and the measured distance will give incorrect estimates of the positioning system positional error statistics. The correct analysis is as follows. The 1D measurement standard deviation σ_0 can be estimated from the measured separation standard deviation by solving the (11.15) for $\sigma = \sqrt{2}\sigma_0$

$$\sigma_{meas} = \sigma\sqrt{\mathrm{var}_r(D^2/2\sigma^2)} \tag{11.15}$$

This solution can then be used to define the statistical performance of the positioning system, such as described in (Yu et al. 2009, Sect. 2.1.5).

11.3.2.2 Differential Range Method

The method described in Sect. 11.3.2.1 uses the differential positions determined from two mobile devices to estimate the system positional accuracy. This can be applied to any type of positioning system in any operating environment, but is based on modeling the statistical errors of the positioning system. Another approach described in this section is based on differential ranges to a single base station, and does not require any modeling of the position errors. However, the method is restricted to radiolocation systems which use a network of base stations for position determination, although the test method only requires one base station. The method is based on measured range or pseudorange data to the two mobile units from a single base station. Pseudoranges are ranges plus an unknown range offset which is common to the two ranges to the two mobile units. The use of pseudoranges is common in radiolocation technology (such as GPS), and thus the method could be widely applied in many practical situations.

The measurement of range (or pseudorange) based on a property of a radio signal can be of several different types, including based on signal strength or on time-of-arrival (TOA) measurements. However, in all cases, one device transmits a signal, and a second device receives the signal. To determine the range, properties of the transmitting and receiving devices are required. For example, for the signal strength method, the transmitter effective power, the receiving device antenna gain and the calibration of the receiver signal strength indicator (RSSI) are required for determining the range. For the TOA method, the transmitter and receiver equipment delays are required. Such information must be determined as part of the testing process. However, for the TOA method some additional information is required associated with the clocks in the devices, which are used for the time measurements. For example, some method of time synchronization is required between the transmitter and the receiver. An alternative method is to use round-trip time (RTT) measurements, where unit 1 transmits the signal which is received by unit 2, which then transmits a signal back to unit 1. As the transmit and receive functions in unit 1 use the same clock, no time synchronization is required. The test method described following is based in this RTT method. Thus using this method, the RTT can be expressed as

$$RTT_{12} = \Delta_1 + \Delta_2 + 2/cR_{12} \quad (11.16)$$

where the delay parameters Δ are the combined transmitter and receiver equipment delays, which in general are not known accurately, and R_{12} the range between the two units. For static measurements the range R_{12} can be initially determined by surveying (for example with Differential GPS—see Annex A) or by use of a map, so the combined delays $\Delta_1 + \Delta_2$ can be estimated by using (11.16), and a surveyed inter-unit distance and the measured RTT_{12}. Subsequently this calibrated delay data can be used in calculating the range (again using (11.16)) when one of the units (number 2) is attached to a moving object. However, without an accurate reference positioning system, this measured range cannot be used to determine the range error. However, if a second mobile unit (number 3) is used also attached to the moving object along with unit 2, then the range measurement error characteristics can be inferred by the equation

$$RTT_{12} - RTT_{13} = (\Delta_1 + \Delta_2) - (\Delta_1 + \Delta_3) + 2/c(R_{12} - R_{13}) \quad (11.17)$$

The differential ranges can be calculated using (11.17), which will be small as the two mobile units will be close together, so their ranges to unit 1 will be very similar as the ranges (R_{12}, R_{13}) are much greater than their separation distance. Further as will be explained, an accurate estimate of the true differential range can also be calculated from simple geometry, so that the differential range errors of the moving test object can be calculated as a function of time.

The first performance check is to determine that the equipment noise is within specifications. The receiving equipment noise in the RTT measurements has two main sources. The first is associated with the radio receiver characteristics, such as the receiver effective bandwidth and the radio noise figure. As the main testing aim is to estimate the range errors due to effects such as multipath propagation, the testing should be conducted with high signal levels (and hence high signal-to-noise ratios (SNR)) to minimise range errors associated with radio noise. If the receiver output SNR is at least 40 dB, these range errors can be kept relatively small; see below for some estimates. The second source of errors is associated with the baseband and digital processing required to determine the TOA. For example, the analog baseband signals must be digitised to a limited number of bits, typically 8 or 10 bits. Assuming that the signal is in the range $[-1, 1]$ and is digitised to N_{bits}, and has a signal power of unity, then the associated SNR with this digitisation is given by (Carlson 1968)

$$SNR_{DtoA} = 10\log(3) + 20N_{bits}\log(2) \approx 5 + 6N_{bits} \quad \text{(dB)} \quad (11.18)$$

For 8-bit conversion, the associated SNR is 53 dB, which is much greater than the typical output SNR from the radio. The data processing associated with determining the TOA will further reduce the digital SNR, but the main source of TOA noise is likely to be from the radio. If the TOA is determined from measuring the arrival time of a (noisy) radio signal pulse of rise-time τ_{rise}, then the standard deviation in the TOA will be given approximately by (Yu et al. 2009, Chap. 4)

$$\sigma_{TOA} \approx \tau_{rise}/\sqrt{SNR} \tag{11.19}$$

For example, with the suggested minimum SNR of 40 dB the standard deviation in the TOA will be $\sigma_{TOA} = 0.01\tau_{rise}$. For the test data described below the pulse rise-time is 100 ns, so the associate measurement noise standard deviation will be 1 ns, or equivalent to 30 cm in range. This expectation can be compared with the measured performance data.

To estimate the noise from the radio and the baseband analog and digital components, the raw differential range data is derived using (11.17). The noise is then estimated by subtracting a smoothed version from the raw differential range data. An example of this process is shown in Fig. 11.5. These test data are from measurements made on a race course (size about 700 m by 500 m), with the master station (unit 1) placed in the approximate center of the course, and the two mobile units located (separation 1.5 m) on a vehicle being driven at a speed of about 15 m/s. The fast changing range noise from each unit will be statistically uncorrelated, so the noise standard deviation of a unit (assuming the same performance for each unit) will be $1/\sqrt{2}$ times the standard deviation of the raw data minus the smoothed data.

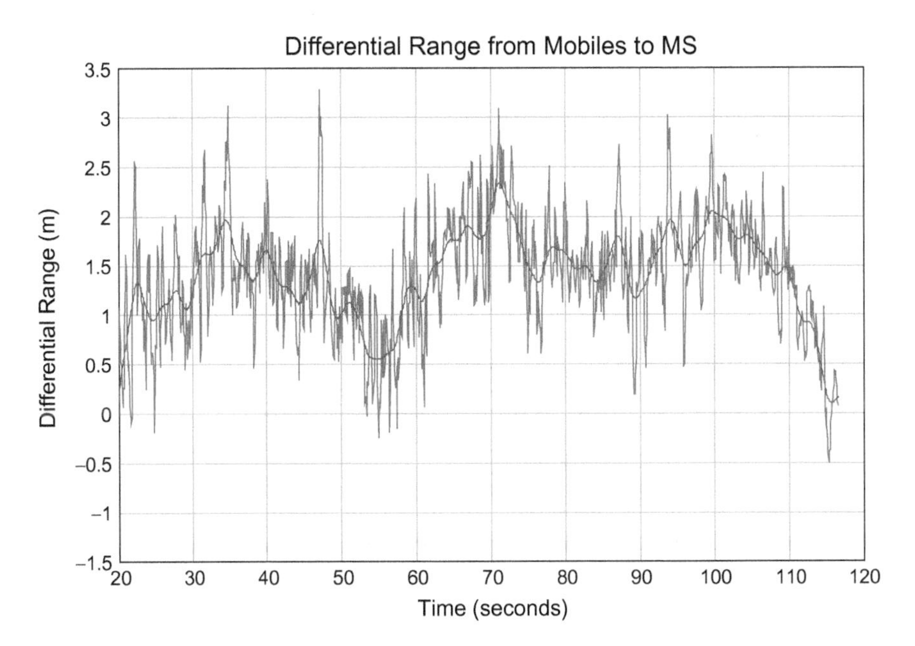

Fig. 11.5 Measures differential range to mobiles from the master station (MS). Also shown is the smoothed signal which is used to estimate the measurement noise

The first measurements were made at a stationary starting point at a range of about 400 m. This resulted in the unit noise standard deviation of 7 cm, which applying (11.19) implies a SNR of 53 dB. The estimated SNR based on the measured signal levels are 47 dB for unit 1 and 54 dB for unit 2, so the measured noise standard deviation is close to expectations, showing the equipment is working correctly.

The second measurements were made when the vehicle was moving around the track, with range varying from 150 to 400 m. Figure 11.5 shows the measured differential range data, and the smoothed curve which is used to estimate the noise as explained previously. The inferred unit measurement noise is 25 cm, which is considerably larger than the corresponding static measurement of 7 cm. The reason for this difference is that there is extra noise associated with small-scale multipath signals, which vary as the vehicle moves around the track. Note that this noise is in addition to the range errors associated with larger-scale effects which are considered separately below.

The corresponding probability density function of the noise in Fig. 11.5 is shown in Fig. 11.6. The interesting characteristic of this result is that the errors match a Laplace distribution (double-sided exponential function), rather than a Gaussian distribution. The reason for this is that errors are associated with small-scale multipath effects, which result in the received radio impulse response being distorted, and consequently results in TOA measurement errors. These effects

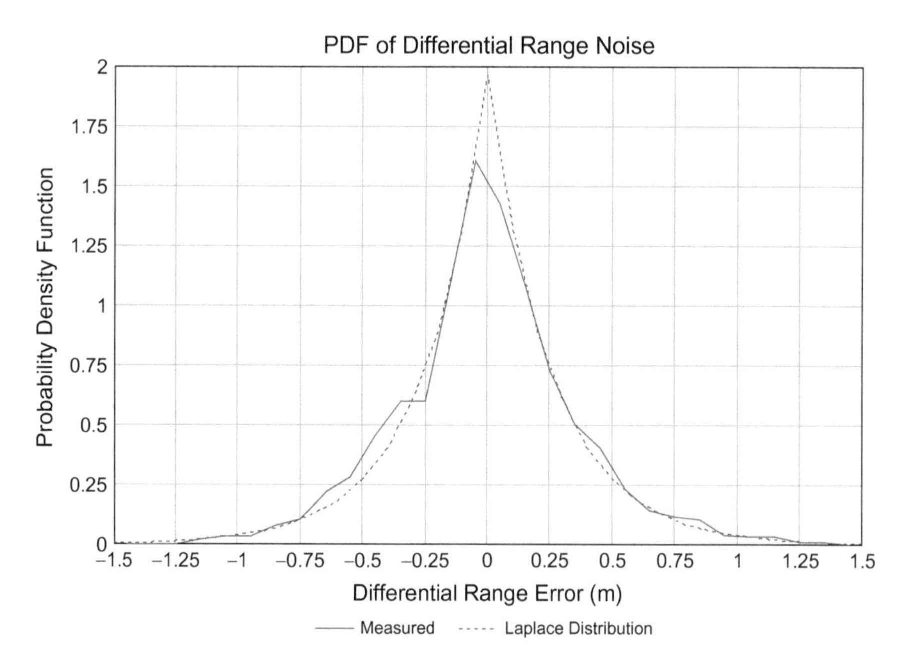

Fig. 11.6 Measured error probability density function for the noise shown in Fig. 11.5. Also shown is the Laplace distribution with the same standard deviation

are known to approximate exponential decay with delay excess, and as a result the TOA errors also exhibit exponential effects.

The third data analysis is aimed at determining the main range errors, which are associated the medium to large scale radio propagation effects. The measured data is the same as shown in Fig. 11.5, but the comparison data is the computed differential range based on geometric calculations. For this test a GPS was also used to track the moving vehicle. The GPS antenna was located midway between the two test units, which had a separation of 1.5 m. From the GPS position data, the range to the vehicle from the master station can be calculated, as well as an estimate of the vehicle heading angle. Based on these data the differential range to the two mobile units can be calculated by simple geometry. The results of comparing the measured differential range data in Fig. 11.7 with the geometric differential range results in a mean radial error of 0.9 m.

Now the important point to note with this analysis is that as the range from the master station to the vehicle is hundreds of meters, the computed geometric differential range is insensitive to GPS positional errors. Even if the GPS position has errors of (say) 10 m the corresponding errors in the geometric differential range data will be small, only a few centimetres. The reason for this is the range to the vehicle from the master station is very much greater than the separation of the two test units mounted on the vehicle. Thus the GPS is only used to provide low quality positional data, and is not an accurate reference for range error measurements. Indeed, it would be possible to use the position system itself to provide the course positional

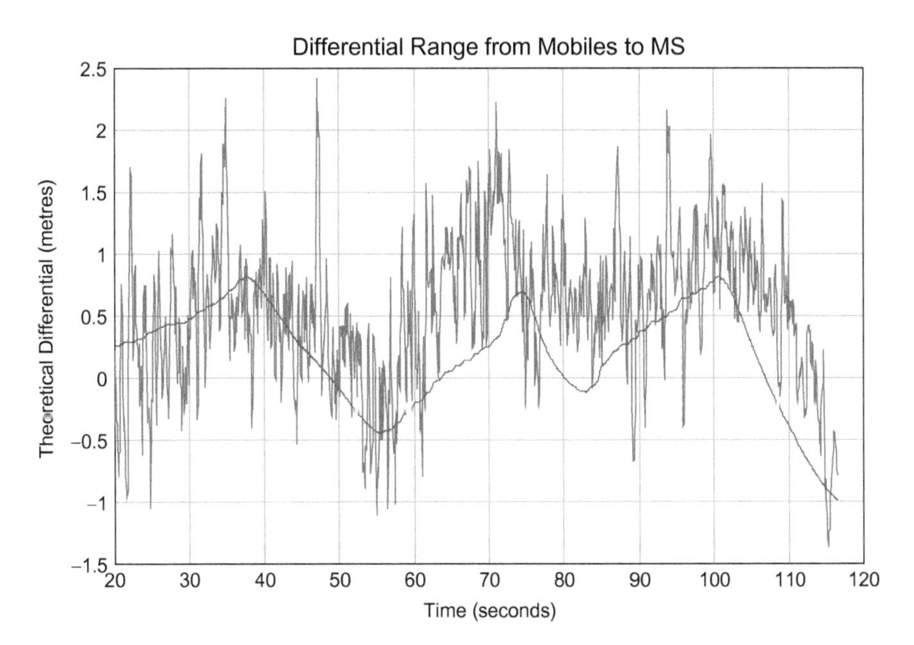

Fig. 11.7 Comparison of the measured (noisy data) and geometric differential range data as a function of time while driving around the track at 60 kph

data, so in practice the position system itself can be used to obtain accurate estimates of the range errors (to centimeter precision), even if the raw positioning accuracy is much worse (say of the order of a few meters). This strange situation means that no additional positioning equipment is necessary to obtain accurate estimates of range errors. The only constrain is that the ranges to the moving object must be much larger than the separation of the two units used for differential range measurements. Note the actual path, speed and direction can be arbitrarily varied, making this test method very flexible.

11.4 Cross-Track Based Approach

The differential measurement methods described in Sect. 11.3 allows estimates to the made of the accuracy of a positioning system. While these methods allow estimates to be made over a wide range of operating conditions, one requirement is that two units must be place on the tracked object, and that only differential data can be measured. While differential data can be useful in estimating overall performance, it would be desirable to estimate true position errors from a moving object without the need of a very accurate separate positioning system. This section describes such a test procedure that can be widely applied to any 2D tracking system in any environment with minimal constraints on the path, speed (including stationary) and the direction of travel.

From the above discussion it is clear that it is highly desirable to measure positional accuracy in the normal operating environment, including both stationary and moving mobile devices. The results of such measurements are usually not required on an individual point-by-point basis, but rather the overall positional statistics are required. Thus there is an opportunity to develop a technique which only estimates the overall statistics, typically in the form of a cumulative distribution function (CDF) of the position error. Thus the "accuracy" may be specified (say) as the error which occurs 50% (median) or 90% of the time, depending on the requirement. The technique to be described provides such information with no additional equipment, and only simple data processing requirements.

The problem of determining positional errors of a radiolocation system is illustrated in Fig. 11.8. A mobile device is moving relative to fixed base stations which "surround" the mobile location. Signals emanating from the mobile device are received by base stations located within radio range, and data related to the characteristics of the received signal are used to estimate the position, which will have errors. The mobile device is assumed to be moving along a locally straight-line segment of the path, so that the 2D positional error can be defined by two coordinate errors, Δx and Δy. The required positional error is then given by $\Delta r = \sqrt{\Delta x^2 + \Delta y^2}$. Now assume that the path of the mobile device is known, either as part of the testing procedure, or because the path is restricted. For example, vehicles will drive on roads for which maps are available, or people in buildings

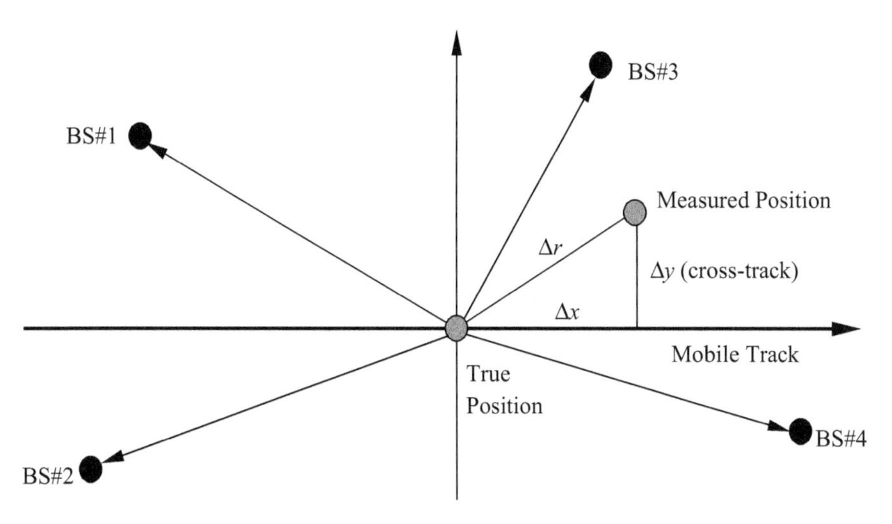

Fig. 11.8 Geometry of a mobile on the nominal track and surrounding base stations, showing the cross-track error measurement

walk along specific paths such as corridors. It is further assumed that the accuracy of these maps is better (say by a factor of 10 or more) than the positional accuracy of the positioning system. For outdoor systems, such as GPS, the positional accuracy maybe typically around 5–20 m, depending on the local multipath environment, whereas typical maps are more accurate. For indoor positioning, the positional accuracy may be around 1–2 m, but the accuracy of building maps would be at least 10 times better. Thus reference maps are usually available to allow the relative *cross-track* positions to be measured by simply plotting the positions on a map, and dropping a perpendicular onto the known track, as shown in Fig. 11.8. Note that while the cross-track error (Δy) can be estimated, the along-track error (Δx) cannot be simply determined, as the true position of the moving object is unknown.

From the above description, provided there is an accurate map, and the mobile moves along a defined path the cross-track errors can be determined, either by a manual plotting procedure, or by an appropriate computer program. This latter case means that large quantities of positional data can be logged and processed with little manual effort, thus obtaining good statistical information due to the large number of positional data samples. While cross-track statistics provide some useful information on positional accuracy, the actual requirement is to obtain statistics on the positional error (Δr). Thus a key part of the method is devising a method to convert cross-track error statistics into overall positional error statistics.

The key idea in performing this conversion is recognising that overall the statistics in the cross-track and along-track directions should be quite similar. For the measurements the path should include many changes in direction, so the local along-track reference used for an individual measurement is not in any way related to a fixed geographical direction. Indeed, for good sampling of the coverage area, many track path directions should be incorporated into the test track. In the case of radiolocation where the current location will be "surrounded" by base stations, it is reasonable to assume that the error distribution about the nominal position will be approximately circular. More specifically, geometric dilution of precision (GDOP) analysis (Yu et al. 2009, Chap. 9; El-Rabbany 2002) shows that in general the error contours of equal probability are in the form of ellipses, but with base stations deployed symmetrically about the mobile position the contour becomes a circle, with identical cross-track and along-track error statistics. Further, in such cases it is shown in GDOP analysis that the cross-track and along-track errors are statistically independent, and this is true regardless of the particular direction of orthogonal cross-track and along-track directions chosen. In practice, this equality will not be true at each location, but would be expected to be approximately true when averaged over a large number of points on a path that has frequent changes of direction. For example, local geometric effects may result in the main errors being orientated in the along-track direction, which will not be detected by the cross-track method. However, if the direction is changed frequently, this underestimation in one direction is compensated by overestimation in an orthogonal direction.

Another situation that can result in estimation errors is where there is a rapid change in velocity (speed and/or direction). In such cases where raw data filtering (such as by a Kalman filter) is employed, the filtering can result in along-track over or under shooting, which is not detected by the proposed cross-track method. However, during system testing the changes in speed should be within the specification of the tracking system appropriate for the application; in practical situations people tend to walk at an approximately constant speed, and vehicles on roads tend to move with relatively low accelerations/decelerations, well within the tracking dynamics of systems such as GPS. Errors also can occur with sudden changes in direction, such as at corners. This situation is specifically considered in Sect. 11.4.3 and in particular in Fig. 11.11; the described technique minimizes the error estimation in this case.

Another case of interest is associated with a mobile device located on the body of a person, which can result local bias position errors due to signal blockage by the body. By traversing the path twice, once in each direction, such biases can be averaged out.

A practical issue associated with implementation is the sign attached to the cross-track errors. From the above discussion, circular probability contours would imply that the plus and minus data should be statistically equivalent. The definition of the sign of the error is arbitrary; for example, errors to the right of track could be defined as positive, and those to the left negative. However, if this symmetry is assumed, then it is simpler to just record the magnitude of the cross-track errors, and to then assume symmetrical statistics in the data processing. Indeed, geometric symmetry can be guaranteed, if the track chosen is a round-trip "there and back", with the sign of the cross-track errors on the outbound track being the opposite for the return track. For such a case there will be minimal potential bias in the measured data.

11.4.1 Statistical Analysis

Based on the above discussion, the statistical analysis of the cross-track data is now described. Consider that the cross-track error data has been gathered as shown in Fig. 11.8, and its probability density function (PDF) has been estimated by performing a histogram analysis on the data. This one-sided (magnitude only) cross-track error PDF is defined as $\hat{p}_{xt}(x)$ where x denotes the cross-track error,[1] so the (assumed) symmetrical two-sided error PDF is given by

$$p_{xt}(x) = \frac{1}{2}\hat{p}_{xt}(|x|) \qquad (-\infty \leq x \leq \infty) \tag{11.20}$$

The along-track errors are assumed to have the same statistics, and also the cross-track and along-track errors are assumed to be statistically independent, so the joint 2D probability is

$$p(x,y)\,dx\,dy = p_{xt}(x)\,dx\,p_{xt}(y)\,dy \tag{11.21}$$

This 2D probability distribution in Cartesian coordinates is now converted to polar coordinates in the form

$$\begin{aligned} q(r,\theta)\,rd\theta\,dr &\equiv p(x,y)\,dx\,dy \\ &= p_{xt}(r\cos\theta)\,p_{xt}(r\sin\theta)\,rd\theta\,dr \end{aligned} \tag{11.22}$$

Now consider the properties of the 2D distribution. From (11.20) and (11.21), it can be seen that the function is symmetrical about both the x-axis and the y-axis. Further, from (11.21) one can see that the equation $p(x,y) = p(y,x)$ always holds. Therefore, there is 8–way symmetry in the assumed 2D probability function. Using this property and integrating out the θ variable, the required distribution can be described as a function of the radial positional error (r), and is given by

$$q(r) = rf(r) \tag{11.23}$$

where

$$f(r) = 8\int_0^{\pi/4} p_{xt}(r\cos\theta)\,p_{xt}(r\sin\theta)\,d\theta \tag{11.24}$$

Thus given the PDF of the measured cross-track data, the function $f(r)$ can be computed numerically, and then the required radial error PDF $q(r)$ is produced by (11.23). Alternatively, if there are analytical expressions for the cross-track errors,

[1] In the following analysis Δ has been dropped for simplicity.

analytical expressions of the PDF can be derived. Two important analytical cases are considered in the following sub-section.

11.4.2 Statistical Error Models

In this section analytical models are used to test the reconstruction algorithm described in Sect. 11.4.1. The method assumes a statistical model for the PDF of the radial (r) errors, then uses this model to generate some cross-track errors, and finally the method in Sect. 11.4.1 is used to reconstruct the radial distribution. This reconstructed distribution can then be compared with the original distribution, which ideally should be the same. However, the reconstruction process involves two important assumptions, namely:

1. The cross-track and along-track errors are statistically independent.
2. The cross-track and along-track errors have the same statistical distribution, which is an even function.

These assumptions may not be exactly true in either measured or simulated data. In the case of the simulated data the model/reconstruction results can be checked, but this is not possible with the measured data. However, if measured data approximates an analytical model which performs satisfactorily in simulation, then it is reasonable to assume that the algorithm described in Sect. 11.4.1 provides a satisfactory method of determining approximately the positional error statistics of the measured data.

11.4.2.1 Gaussian Statistical Error Analysis

The positional error statistics of a radio tracking system are directly related to the ranging error statistics. If the range or pseudorange measurement errors are due to receiver noise, then the noise statistics will be Gaussian, which translates into a Rayleigh distribution for 2D positional errors. In the line-of-sight (LOS) case, ranging errors due to multipath propagation also can be approximated by a Gaussian model (Alavi and Pahlavan 2006; Alsindi et al. 2009). In the case where the measurement errors are dominated by multipath effects in a non-line-of-sight (NLOS) environment, the measurement errors will not be Gaussian. However, in the case where the number of base stations exceeds the minimum of three, a least-squares (LS) position determination process involves summation of the measurements. If the number of such base stations is moderately large (say 5 or more), the Central Limit Theorem (CLT) can be invoked, and the positional error statistics can again be approximated by a Gaussian model—the higher the number of base stations the better the approximation becomes (Yu et al. 2009, Sect. 2.1.4). Thus there are strong theoretical reasons for suggesting a Gaussian statistical model. If it is further assumed that the cross-track and along-track statistics are the same, both Gaussian and statistically independent, then applying (11.22) in this case results in the joint probability

$$p_g(r,\theta)\, r\, d\theta\, dr = \frac{r}{2\pi\sigma^2}\exp\left(-\frac{r^2}{2\sigma^2}\right) d\theta\, dr \tag{11.25}$$

where σ is a parameter that can be used to match the model to measured data. Integrating out the (absent) θ component gives

$$p_{rayleigh}(r) = \frac{r}{\sigma^2}\exp\left(-\frac{r^2}{2\sigma^2}\right) \tag{11.26}$$

which is recognized as a Rayleigh distribution, and is due to the assumed symmetry and the properties of the Gaussian distribution. In following analysis, the positional error statistics are presented in terms of the Complementary CDF (or CCDF), as the CCDF allows low-probability, large errors to be readily observable on a logarithmic plot, such as in Fig. 11.9. From (11.26), the corresponding CCDF for the radial error is given by

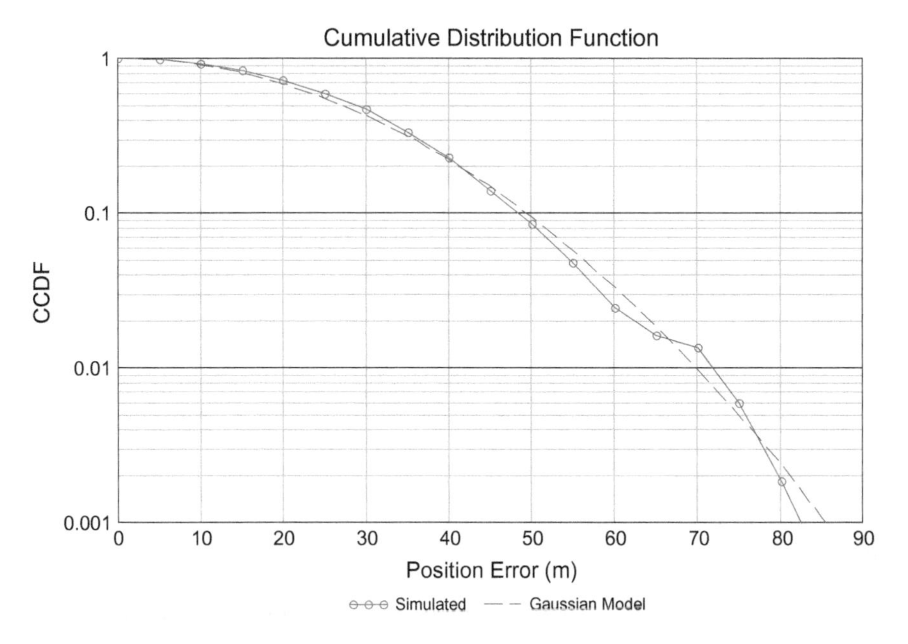

Fig. 11.9 Simulated and model CCDF for the Gaussian error

$$C_g(r) = 1 - \int_0^r \frac{z}{\sigma^2} \exp\left(-\frac{z^2}{2\sigma^2}\right) dz = \exp\left(-\frac{r^2}{2\sigma^2}\right) \tag{11.27}$$

If the CCDF is plotted using a log-scale, the corresponding curve will be parabolic.

It is instructive to use this model to generate some simulated cross-track data, which then can be used to test the reconstruction algorithm. In this case two random variables (radial error (r) and azimuth angle (θ)) are required, and the simulated cross-track error data can be generated by

$$x = \left|\sigma\sqrt{-2\ln(U(1))}\sin(U(2\pi))\right| \tag{11.28}$$

where $U(a)$ is a uniform random variable in the range [0, a]. With data generated using (11.28), the algorithm described in Sect. 11.4.1 can be applied to estimate the required radial CDF statistics which should be similar to those based on analytical Eq. (11.23). Note that in this case the two conditions described in the introduction of this section apply, so the reconstruction should be essentially perfect, with the only discrepancies being introduced by the numerical process in generating the histogram of the cross-track data. The results of a simulation are shown in Fig. 11.9. The data are based on 500 simulated cross-track samples, with the number of samples chosen to be similar to practical situations. To generate the histogram a bin size of 1 m was used. The Gaussian model parameter σ is 24 m, typical of the measured data for the outdoor system as described in Sect. 11.4.4.1. As can be observed, the reconstruction algorithm essentially generates the required radial CDF, bearing in mind the logarithmic nature of the plot. As the number of points increases, the simulation results become increasingly close to the analytical performance. Thus if the measured cross-track and along-track data approximate the same Gaussian distribution, the reconstruction algorithm should generate accurate estimates of the error CDF, provided the number of measured points is sufficiently large. As a minimum, at least several hundred points should be used.

11.4.2.2 Exponential Statistical Error Analysis

The assumption that cross-track errors are Gaussian is appropriate for scenarios where the range measurements are corrupted with Gaussian noise, or where multipath propagation exists, but the number of base stations within radio range is large. However, a common case is that there are multipath propagation and the number of base stations is rather limited, say 4–6. Further, under NLOS propagation conditions, some large ranging measurement errors will occur (Alavi and Pahlavan 2003), resulting in large positional errors greater than that expected from a Gaussian distribution. In this case the error distribution is non-Gaussian, with typically the “tail” of the distribution much larger than the Gaussian case. In such cases an alternative model for the cross-track errors is a Laplace distribution with a PDF given by

$$p_{xt}(x) = \frac{1}{2a} e^{-|x|/a} \qquad (-\infty \le x \le \infty) \tag{11.29}$$

where a is a parameter which defines the rate of exponential decay of the cross-track errors. Note that the assumed distribution is symmetrical (even function) about the x-axis, as required by the reconstruction algorithm. Thus from (11.22) the probability distribution in polar coordinates is given by

$$q(r,\theta)\, d\theta\, r\, dr = \frac{r}{2a} \exp\left[-\frac{r}{a}(|\cos\theta| + |\sin\theta|)\right] d\theta\, dr \tag{11.30}$$

Applying the 8-way symmetry property as explained in Sect. 11.4.1, the function $f(r)$ in (11.24) is given by

$$\begin{aligned} f(r) &= \frac{8}{2a}\int_0^{\pi/4} \exp\left[-\frac{r}{a}(\cos\theta + \sin\theta)\right] d\theta \\ &= \frac{4}{a}\int_{\pi/4}^{\pi/2} \exp\left[-\frac{\sqrt{2}r}{a}\sin\phi\right] d\phi \end{aligned} \tag{11.31}$$

There is no closed form for the integral in (11.31). However, observe that the sine function over the range of the angles in the integral in (11.31) varies slowly. One possibility is to approximate the sine function by a quadratic function, but the simplest approach is to replace the sine function by its average over the range of the integral; such an approximation was found to have minimal effect on the accuracy of the result. Applying this simplification and carrying out the integration result in

$$\hat{q}(r) = \frac{\pi}{a} r \exp\left[-\frac{4r}{\pi a}\right] \tag{11.32}$$

However, for this expression $\hat{q}(r)$ to be a PDF its integral over all position errors must be unity. Thus from (11.32) the normalization constant is

$$\int_0^{\infty} \hat{q}(r)\, dr = \frac{\pi^3 a}{16} \tag{11.33}$$

so normalizing (11.32) by the result in (11.33) yields the exponential model radial error PDF as

$$
\begin{aligned}
q_e(r) &= \left[\frac{4}{\pi a}\right]^2 r \exp\left[-\frac{4r}{\pi a}\right] = \frac{r}{\beta^2}\exp\left[-\frac{r}{\beta}\right] \\
\beta &= \frac{\pi}{4} a
\end{aligned}
\tag{11.34}
$$

which can be recognized as the Gamma probability distribution $\Gamma(k, \beta)$ with shape factor $k = 2$. The mean of the distribution is $\mu = 2\beta = (\pi/2)\,a$, and the variance of the distribution is $\text{var} = 2\beta^2 = \pi^2 a^2/8$. From (11.34) the associated CCDF is

$$
C_e(r) = 1 - \int_0^r q_e(z)\,dz = \left(1 + \frac{r}{\beta}\right)\exp\left[-\frac{r}{\beta}\right] \tag{11.35}
$$

When the CCDF is plotted using a logarithmic scale, the asymptotic curve is linear with the position error, in contrast to the quadratic function for the Gaussian model. The results of a simulation are shown in Fig. 11.10. The Exponential model parameter a is 22.5 m, again typical of the measured data for the outdoor system described in Sect. 11.4.4.1.

The processing method is the same as described for the Gaussian model case. The data are based on 500 simulated cross-track measurements. As can be observed, while the match is generally good, the match is not quite as good as in the

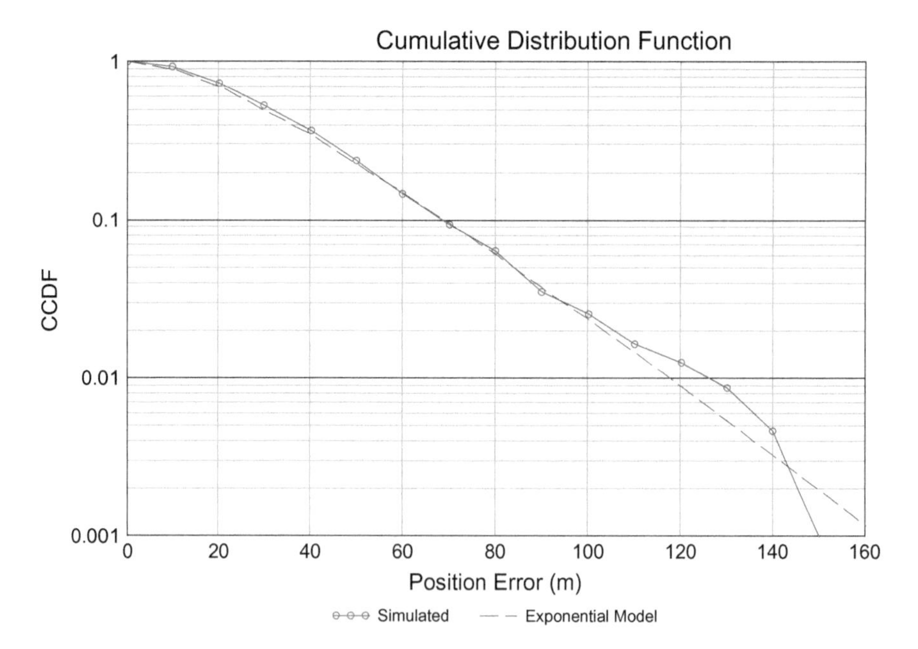

Fig. 11.10 Example of simulated and modeled radial error CCDF for the exponential cross-track error model

Gaussian case. This result is expected; the development of the Rayleigh radial error model requires no mathematical simplifications, whereas the development of the analytical Gamma radial error model requires some simplifications. Furthermore, increasing the number of data samples does not significantly improve the match, so measured cross-track errors that have approximately exponential characteristics will result in some imprecision in CDF estimation, particularly for large positional errors.

11.4.3 Measured Data Processing

This section describes the general procedures for gathering raw cross-track data and subsequent data processing for real positioning systems. The particular data processing for two such systems, one outdoors and one indoors, is considered in Sect. 11.4.4.

The procedure for determining the cross-track data is for the mobile device to follow a predefined track, and the associated tracking data of the positioning system are recorded into a computer file for later replay and data processing. The accurate implementation of the cross-track data processing method requires that the errors in movement of the mobile device along the predefined track are much smaller than the errors in the tracking system. This requirement usually can be met for both indoor and outdoor systems. For an indoor system the tracking errors may be 1–2 meters, which implies the errors in moving along the required track must be of the order of 10 cm. Such accuracy is possible by simply accurately walking along the nominal track. For an outdoor system, say tracking a vehicle on a road, the positioning system accuracy is typically 10–30 m, which implies that the vehicle must be driven along the predefined track to an accuracy of (say) 2 m. This means that the vehicle should stay in the centre of a lane, typically the one next to the curb, to achieve the required accuracy. As the lane width is typically about 3.5 m wide, achieving 2 m accuracy is not difficult. Note that continuous constant-speed progress along the track is not required by the method, so the vehicle can be driven at any speed, and stop and start as required by the traffic flow, although fast acceleration or deceleration should be avoided. Thus in both the indoor and outdoor cases the basic motion along a defined track with the desired accuracy can be readily achieved.

The second step in the data processing is to compare the measured track positions with some type of map. For an outdoor system a road map is typically used, or photographic maps from sources such as Google Earth. For indoor systems, an architectural drawing of the building is typically used as the reference. In both cases it is important that the accuracy of the map is much better than the accuracy of the tracking system. For indoor systems this is not a problem, as building layout drawings typically have accuracy of better than a few centimeters. For outdoor systems, the accuracy of the maps may be more of a problem, especially maintaining the accuracy over a large geographical area. However, the absolute accuracy of even the most basic road maps is typically 10 m and often much better. Information on the accuracy of the grid-based street directory maps used in the generation of the cross-track data in this analysis is not publicly available. However, the publisher states that the maps are based on Differential GPS and aerial mapping, so it is probable that the accuracy is better than 5 m. The raster scanned digital maps have a pixel size of about 2 m. For Google Earth maps only the pixel resolution is specified rather than the accuracy. The resolution in major cities in US, Europe and Australia can be as small as 15 cm, but the more general resolution is 2.5 m. Thus while not ideal, the general availability of road maps with the required accuracy means that determining the cross-track data with an accuracy of at least 2–5 m should be possible, and greater accuracy is possible with more accurate maps.

The third step in the data processing is to plot the measured tracking data as an overlay onto the reference map. While this could be performed manually on paper maps, more typically this is performed by a computer using digital maps and the recorded track data. However, even if the map with the track overlayed is generated digitally, it may be appropriate to generate the cross-track error data manually from a printed version of the computer-generated map. The reason for this is that computer recognition of the nominal track is a non-trivial task, whereas human recognition is usually simple. If a computer program is used, the unreliability of map matching data processing may result in errors in the estimated cross-track data. Further, the map may have a relatively coarse pixel resolution, again limiting the accuracy. With manual processing the cross-track errors can be measured simply with a ruler. In this case the measurement accuracy on the paper may be (say) 1 mm, which will limit the accuracy of the estimated cross-track data. Thus regardless of the method chosen, typically the cross-track estimates may be limited in accuracy and incur quantization errors.

For practical implementation of computer-based cross-track error determination, the track is divided into straight-line segments; the shortest perpendicular distance to one of these segments is assumed to be the required cross-track error data. This procedure can lead to errors in some circumstances, for example as illustrated in Fig. 11.11; the common example of a right-angled corner is shown, but the concept can be extended to other angles associated with adjacent line segments. For point 1 the cross-track distance is correctly measured to a horizontal track segment, and similarly point 2 to a vertical segment. For point 3 a perpendicular to the extension of the horizontal segment would be a false cross-track estimate, and would underestimate the true positional error. The nominal path is shown as two segments meeting at right-angles, but the actual path would be a curve of finite radius. The cross-track error in such a case could be estimated by the line projected onto a circular path; however, in practice the circle is shrunk to a point, and the cross-track is defined by the distance from the measurement position to the intersection point between the two segments. Using the above-described technique, a computer program can be derived to determine the best estimate of the cross-track errors for any track consisting of straight-line segments.

The fourth step in the processing is the generation of a histogram from the measured data, and hence the PDF of the cross-track data. This PDF is the starting

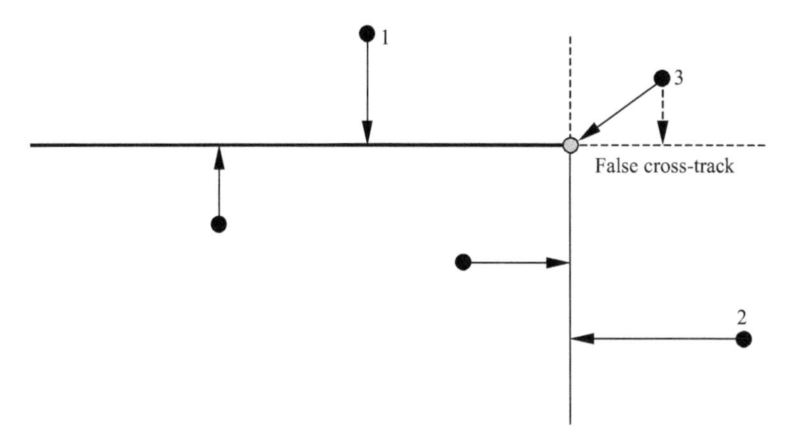

Fig. 11.11 Examples of determining cross-track errors near a corner

point for the algorithm described in Sect. 11.4.1. The PDF is related to the histogram (H) by the relationship

$$PDF_b = \frac{H_b}{N_x B} \tag{11.36}$$

where b is the bin index, B is the bin size (in meters) and N_x is the number of cross-track data samples. While the generation of the PDF from a histogram is simple, the estimated PDF depends on the bin size chosen. An example PDF from actual measured data is shown in Fig. 11.12 for bin sizes of 5 and 10 m; this example is considered in more detail in next section. As can be observed the estimated PDF varies considerably with the choice of bin size. While small bin sizes result in good resolution in the PDF, the number of samples in each bin may be too small for an accurate estimate of the bin PDF magnitude. Conversely, increasing the bin size to increase the number of samples in each bin produces poorer resolution in the computed PDF. This problem can be resolved if the number of samples is large. However, obtaining a large number of data points can be very time consuming, particularly if the generation of the cross-track data is performed manually, as in the example in Fig. 11.12.

To at least partially overcome these difficulties in estimating the PDF, an alternative data processing technique is suggested. Instead of using the histogram method, the raw cross-track data are sorted in ascending order. These data are plotted against n/N_x, $n = 0 \ldots N_x$ which can be recognized as the CDF of the cross-track data. The data from the above example is shown in Fig. 11.13. The "staircase" curve is the raw CDF data generated by this procedure. The reason for this shape is the coarse quantization in the cross-track data, due to the quantization errors from using the paper-and-ruler method described above.

However, what ideally is required is a smooth CDF which would occur with no quantization errors in the measured cross-track data. One method that this result can

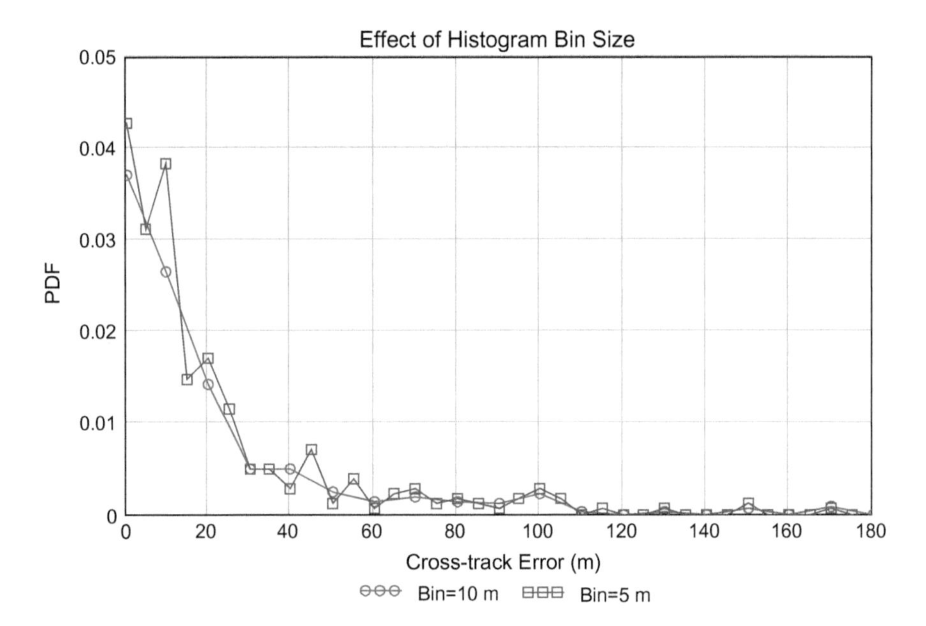

Fig. 11.12 Example of computed cross-track PDF for two bin sizes. The number of samples used to generate these PDFs is 365

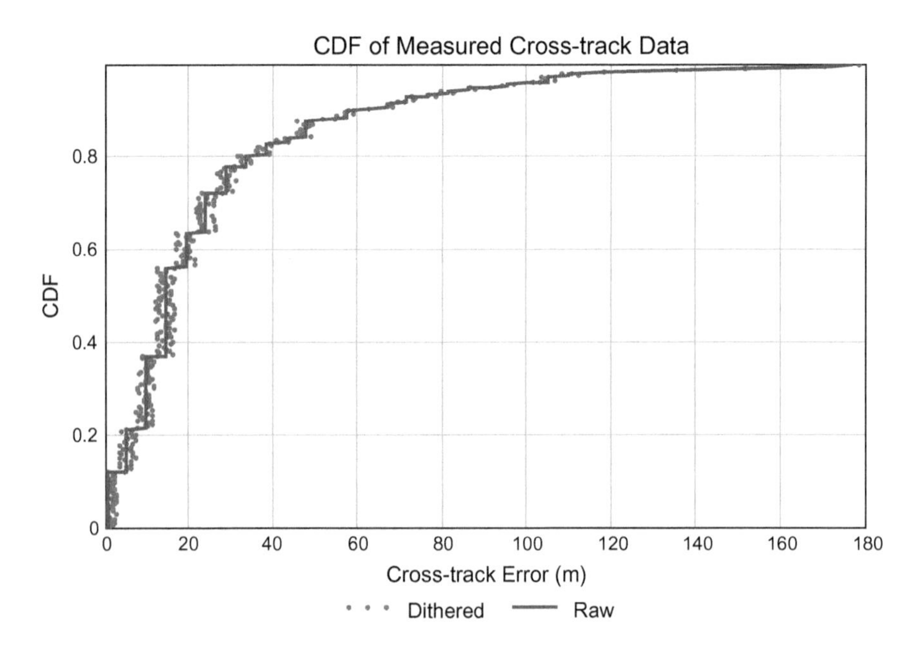

Fig. 11.13 Example of computed cross-track data plotted in sorted order against the CDF (solid curve). The "staircase" plot is due to cross-track quantization, in this case 4.8 m

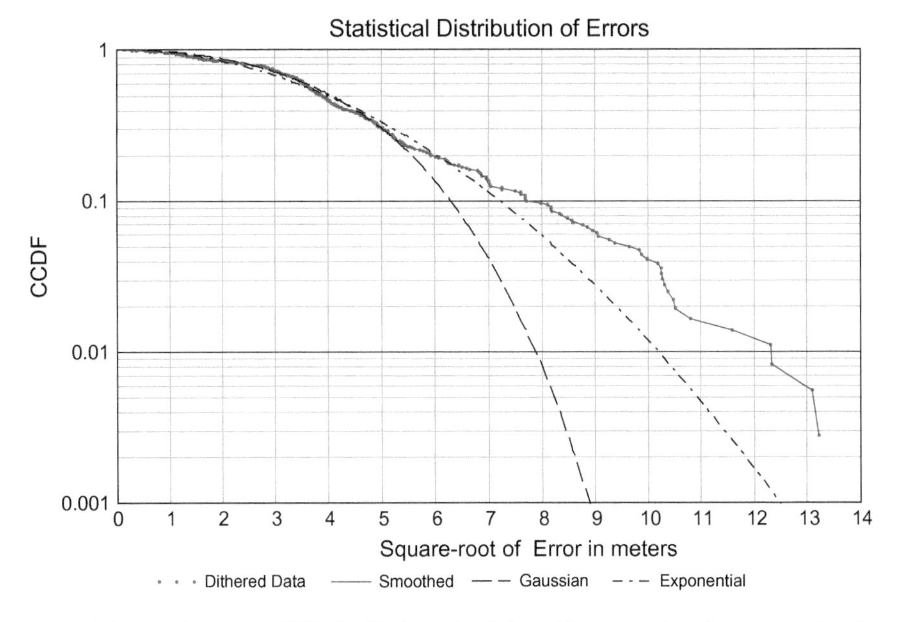

Fig. 11.14 Complementary CDF distribution of radial position errors based on processing the quantized measured cross-track data with dithering. The error is plotted as a square-root to compress the x–axis, making the characteristics of the curve clearer. Also plotted are the smoothed version of the data, and the Gaussian (σ = 24 m) and Exponential (a = 23 m) LS models

be achieved, at least to a reasonable approximation, is by dithering the data, as shown by the dots in Fig. 11.13. The dithering process can be described by the equation

$$\begin{aligned} X_n &= U(\delta/2) \text{ if } Xsort_n = 0 \\ &= Xsort_n + U(\delta) - \delta/2 \text{ otherwise} \end{aligned} \tag{11.37}$$

where $U(x)$ is the uniform distribution described previously, and δ is the quantization interval of the data. The dithered data are resorted to obtain an estimate of the CDF with minimal quantization errors. The result is shown in Fig. 11.14 (dots), as well as a smoothed version plotted as a continuous curve. Also plotted are the LS fit[2] to the Gaussian and Laplace (Exponential) models described in Sect. 11.4.2. From this plot it is evident that the Exponential model more accurately describes the data over the full range of cross-track errors, but the Gaussian model is more accurate for small cross-track errors. Clearly neither model provides an adequate description of the measured cross-track data.

[2]Note that the logarithmic plot in Fig. 11.14 exaggerates fit errors for small values of the CCDF. The least-squares fit does not use the log scale.

The above procedure allows a good estimate of the radial error CDF to be determined, even if there are large quantization errors and a limited number of samples. However, the procedure described in Sect. 11.4.2 requires the PDF rather than the CDF. One method would be to differentiate the CDF data to obtain the required PDF, but because of problems with numerical differentiation of noisy data an alternative approach is suggested. In particular, simulated cross-track data with the same estimated statistics can be generated using the above calculated CDF in the formula

$$x_n = CDF_{sm}[U(1)] \tag{11.38}$$

where CDF_{sm} is the smoothed version of the estimated CDF function shown in Fig. 11.14. The number of such simulated cross-track data can be large, say 10,000, and this data set can then be used for generating the histogram and PDF as described previously, but without the problems associated with data quantization and limited measured data. This computed PDF can then be used to estimate the radial position errors, as described in Sect. 11.4.1.

11.4.4 Two Practical Examples

The theory and techniques described in the previous sections can be applied to particular cases of actual measured data. Two cases are considered, one outdoor case with a coverage area of a city, and another indoor case where the coverage area is within a building.

11.4.4.1 Quiktrak Performance

The first practical case considered is data from a tracking system called Quiktrak (Hurst 1989), which provides vehicle tracking and data services within a city, typically covering thousands of square kilometers. The Quiktrak system used for the measurements described below is located in Sydney Australia, with 14 base stations covering a geographical area of approximately 40 $\times$ 40 km^2. The system operates at about 400 MHz with a bandwidth of 2 MHz. The nominal accuracy of the system is 30 m. For a typical location the number of base stations in radio range of the vehicle is 4–6. One of the main applications of the Quiktrak system is in stolen-vehicle recovery, where its ability to resist radio jamming gives it an advantage over GPS-based technology. The tracking accuracy of Quiktrak is similar but somewhat worse than GPS, but the radio spectrum is only one-tenth that used by GPS.

The measured data were obtained by driving a vehicle on roads, and recording the track positional data. The vehicle path is about 20 km in length, and the sample period is 5 s. These data are then plotted on a road map of the city to obtain the raw cross-track errors. The method described in Sect. 11.4.3 is then used to manually determine the cross-track data from printed maps; the associated measured PDF is shown in Fig. 11.15.

Note that the PDF is plotted using a logarithmic scale so the low probability errors can be observed in the plot. Also plotted in the figure are the LS fit to the measured data for both the Gaussian and Exponential models described in Sect. 11.4.2. As can be observed in this case, the measured cross-track errors do not well match across the entire error range for either the Gaussian or Exponential models of error distributions. For errors up to 50 m the Gaussian model is the closest match, but for larger errors the Exponential model is better. Note in particular that the Gaussian model severely underestimates the probability of tracking errors exceeding about 60 m. While these errors may occur rather infrequently, large errors can have an important impact on practical applications of the Quiktrak technology. For the particular track analysed the large positional errors are associated with the path being located either in a deep valley or near high-rise buildings where both signal blockage and bad GDOP conditions occur.

To provide a better match to the measured data, an alternative model is suggested, namely a weighted combination of both the Gaussian and Exponential models (Alavi and Pahlavan 2003). Such a hybrid model describes the ranging error

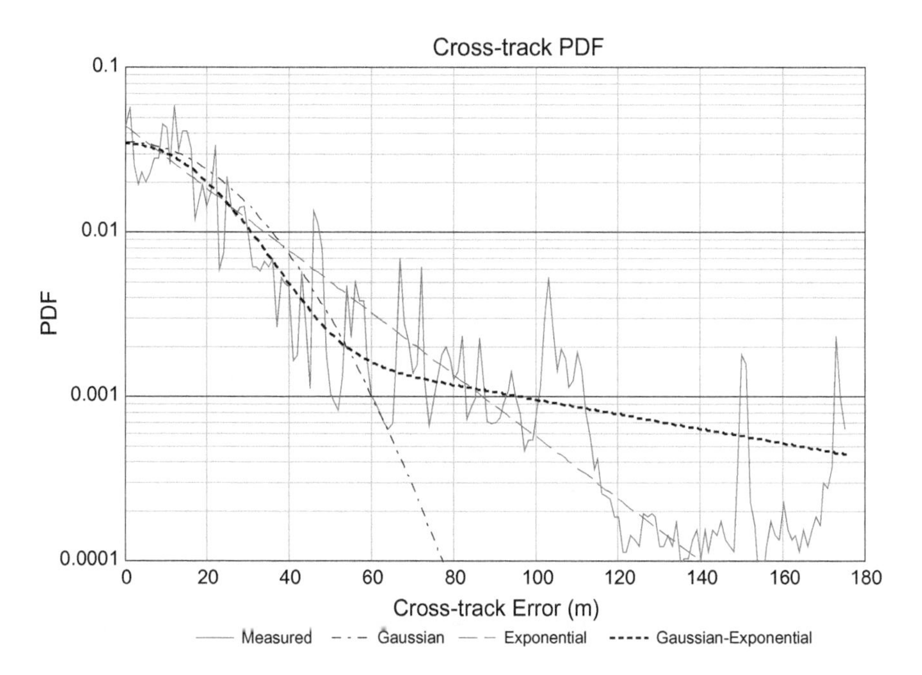

Fig. 11.15 Comparison of the PDF of measured cross-track errors, and the PDF based on the Gaussian, Exponential and Gaussian-Exponential models. The Gaussian model parameter is $\sigma = 23$ m, and the Exponential model parameter is $a = 23$ m, and the Gaussian-Exponential model parameters are $\sigma = 19$ m, $a = 98$ m, and $\alpha = 0.74$

distribution well in an indoor multipath propagation environment, as evidenced by the good match between the model and the experimental data. This combined Gaussian-Exponential model allows a better match as the Gaussian component can describe the low to medium sized errors, and the exponential component the large errors. Thus the proposed model for the cross-track PDF is of the form

$$\begin{aligned} p_{ge}(x,\sigma,a,\alpha) &= \alpha\left(\frac{2}{\sqrt{2\pi}\sigma}\exp\left(-\frac{x^2}{2\sigma^2}\right)\right) \\ &\quad + (1-\alpha)\left(\frac{1}{a}\exp\left(-\frac{x}{a}\right)\right) \qquad (x \geq 0) \end{aligned} \tag{11.39}$$

where α is a weighting parameter for the two one-sided distributions. The three model parameters are again determined by a LS fit to the measured data. The comparison of this model with the measured cross-track data is shown in Fig. 11.15, which shows generally good matching across the entire cross-track error range. Because the Gaussian-Exponential model is simply the weighted sum of the analytical models described in Sect. 11.4.2, the analytical model for the radial error distribution is the same weighted summation, resulting in the CCDF

$$\begin{aligned} C_{g-e}(r,\sigma,\alpha,\beta) &= \alpha\ \exp\left(-\frac{r^2}{2\sigma^2}\right) \\ &\quad + (1-\alpha)\left(1+\frac{r}{\beta}\right)\exp\left[-\frac{r}{\beta}\right] \end{aligned} \tag{11.40}$$

The radial error CCDF model can be compared with the CCDF generated from the measured cross-track data using the method described in Sect. 11.4.2. The results are shown in Fig. 11.16. As can be observed, the match is good across all position errors, although the model slightly overestimates large position errors. The parameters for the model were again determined using a LS fit; observe that the parameters are somewhat different from those obtained by performing a LS fit to the cross-track data in Fig. 11.15. In a similar manner to the CCDF model in (11.40) the mean radial error is computed by the weighted sum of the means of the Rayleigh and Gamma distributions, and is given by

$$\mu_r(\sigma,a,\alpha) = \sqrt{\frac{\pi}{2}}\alpha\sigma + \frac{\pi}{2}(1-\alpha)\,a \tag{11.41}$$

For the parameters defined in Fig. 11.16 the mean radial error is 41 m, about 75% of which is due to the exponential component. Alternatively, the performance can be defined as the radial error which occurs with less than a specified probability. For example, in this case 50% of position estimates have errors less

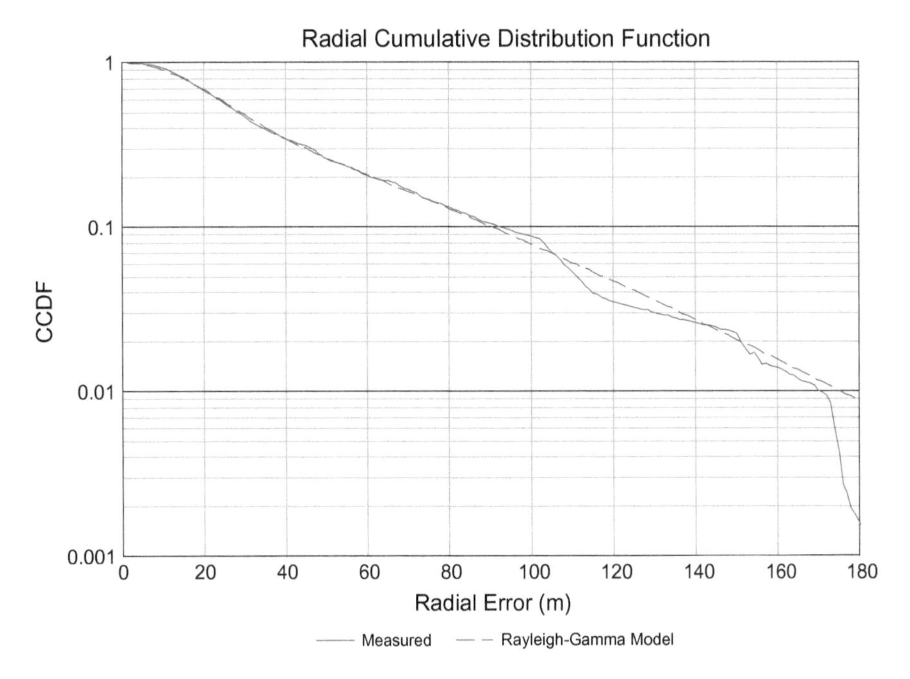

Fig. 11.16 Comparison of estimated CCDF of the measured radial error data and the CCDF based on the Gaussian-Exponential cross-track error model. The Gaussian parameter is $\sigma = 17$ m, the exponential parameter is $a = 38$ m, and the weighting parameter $\alpha = 0.50$

than 28 m (median error). However, because of the long error tail, if the 90% point is taken as a performance measure, the accuracy is about 92 m. This result clearly shows the distribution of errors is important in defining the "accuracy" of a tracking system, as one number cannot provide an overall measure of performance.

11.4.4.2 Indoor Tracking System Performance—Simulated

To further test the method, two more examples associated with indoor tracking are analysed: one example through simulation and the other with data obtained from a real system. The simulated data are used to test the method, as the simulation can generate both cross-track data and the position errors, so the error statistics reconstructed from the cross-track data can be compared with the true positional statistics.

The simulation is based on an indoor system operating in the 5.8 GHz ISM band with a bandwidth of 125 MHz, and approximately represents the real system described in Sect. 11.4.4.3. The maximum radio range is assumed to be 35 m. The simulated office-type building has a size of 40 × 40 m, with nine base stations located approximately uniformly throughout the building, as shown in Fig. 11.17, and can be compared with the real system shown in Fig. 11.20. Also shown is the

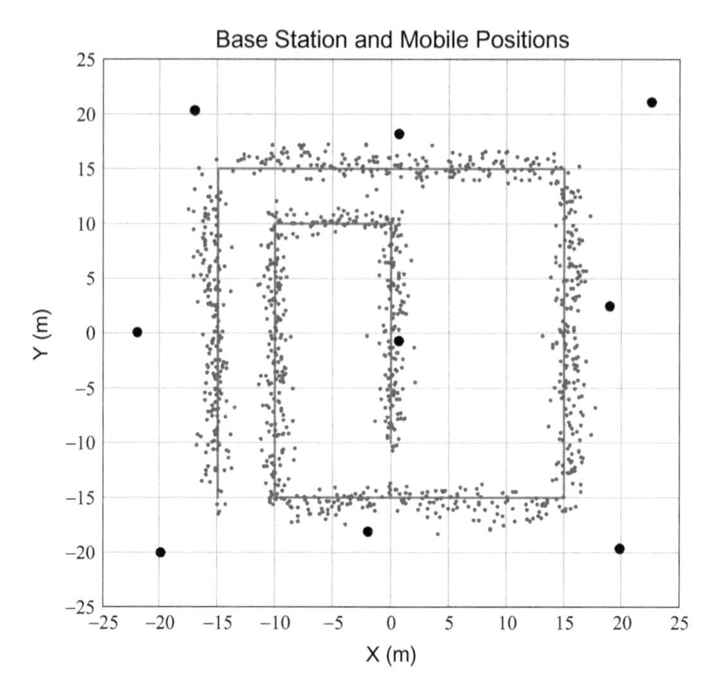

Fig. 11.17 Geometrical details of a "building" with 9 base stations, a path in the building, and the corresponding positions "measured" by the simulated positioning system

path on which the performance is measured, and the estimated positions based on a simulation of the positioning system.

The indoor situation results in ranging errors, which are modelled by Gaussian noise with a standard deviation of 1 m and a bias error of 10 cm/m of propagation range. Note that while each individual measurement error has Gaussian statistics with a constant standard deviation (independent of range), the bias (mean) error varies with range, so that overall a simple Gaussian model of the ranging errors is not possible. The "measurements" are associated with walking along the indicated path at a speed of 1.5 m/s, with an update period of 0.1 s. The position errors and the cross-track errors are recorded from the simulation, for processing by the same cross-track computer program as used for the real system described in the next sub section.

The results from the processing of the simulated cross-track data are shown in Fig. 11.18, including the cross-track reconstruction as well as the CCDF of the simulated measurement position errors. Also shown is the LS fit Rayleigh-Gamma model distribution. As can be observed, the reconstructed distribution generally matches well the actual distribution, except at very low probabilities where accurate reconstruction is difficult due to the limited number of samples in the data set. The Rayleigh-Gamma radial error model matches the data quite well, although it somewhat overestimates the low-probability events—a conservative estimation result.

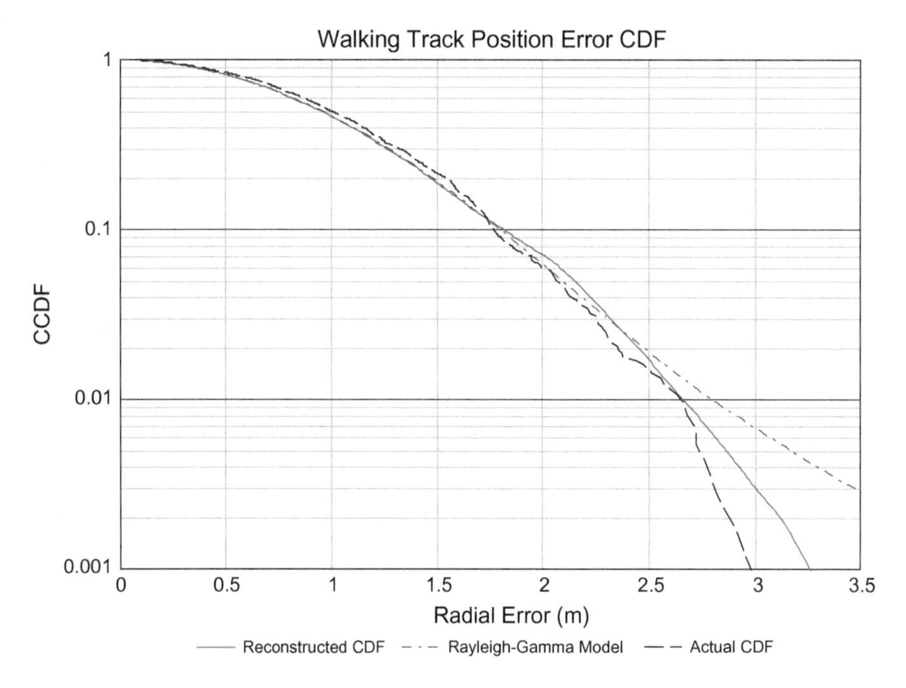

Fig. 11.18 Position error CCDF comparison of the actual (simulated measured) errors, the reconstructed errors from the cross-track data, and the LS fit Rayleigh-Gamma model

Because of the generally good agreement between the actual and reconstructed CDF data, the general reconstruction method has been validated in a simulation mode which mimics actual indoor situations, and thus the method could be expected to perform well in real indoor situations with actual measured data, as described in the next sub section.

11.4.4.3 Actual Indoor System Performance

The cross-track processing method was tested using data from a tracking system called Wireless Ad hoc System for Positioning (WASP) (Hedley et al. 2008; Humphrey and Hedley 2008; Sathyan et al. 2011), developed by CSIRO. This experimental system operates in the 5.8 GHz ISM band with an effective bandwidth of 125 MHz. Position calculations are based on time-of-arrival (TOA) estimation of a reconstructed wideband spread-spectrum signal with an effective leading-edge rise-time period of 12.5 ns; round-trip-time (RTT) between node pairs is determined from the TOA measurements. Each TOA measurement is performed within 200 μs, so for mobile measurements at walking speed of (say) 1 m/s the distance traveled during the measurement is only 0.2 mm, so each measurement is effectively at one point in time. The measurements using the WASP system were performed within an office-type building with 29 nodes covering an area of approximately 2000 m^2. The layout of the nodes and the building is shown in Fig. 11.20. The building construction is varied, including concrete with some plasterboard, steel and wood.

The test performed consisted of strapping a mobile unit onto the hip of a person (see Fig. 11.19), and walking around a pre-defined figure-8 path, as indicated in Fig. 11.20.

The path was traversed twice, with the walking direction reversed for the second traverse. Because of the blocking effect of the person's body the base stations that communicate with the mobile unit will depend to some extent on the orientation of the person. Thus while the nominal path is the same for both passes around the track, the radio conditions will be different for each pass, resulting in different measured paths; this effect can be observed in Fig. 11.20. The testing of the accuracy of the system of a person walking through a building is a key requirement in validating the design aim of an accuracy of better than 1 m.

The positions of the mobile were estimated using a least-squares (LS) positioning algorithm based on pseudoranges. Although the measured data are nominally ranges, the performance of the pseudorange LS algorithm is superior in an indoor environment due to range bias errors associated with indoor radio propagation (Yu et al. 2009, Sect. 5.4.3). The measured and nominal paths (as shown in Fig. 11.20) were computer-processed to determine the cross-track error. The cross-track error is determined by calculating the minimum perpendicular distance from a point to a straight-line segment of the nominal walking track, as described in Sect. 11.4.3. The errors of walking along the nominal path is estimated at ±15 cm, which is much smaller than the positional errors of the WASP; thus the measured cross-track data will be largely associated with the positional errors of the WASP.

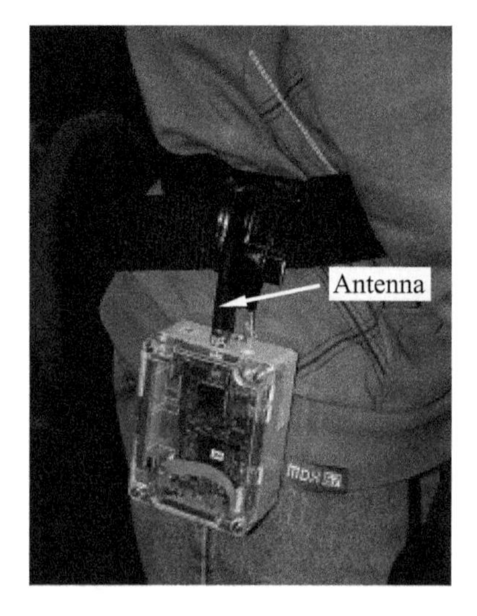

Fig. 11.19 WASP system mobile unit on the hip of a person. The vertically-orientated antenna is located on the top of the unit. Note that because of the close proximity to the body considerable signal blockage will occur

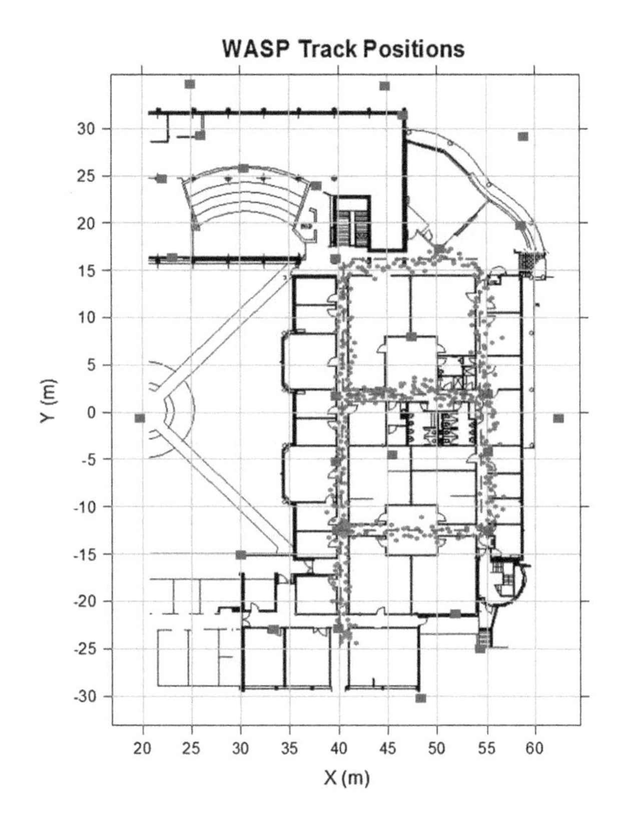

Fig. 11.20 Measured track (without any track smoothing) inside a building using the WASP system. The dots are the measured locations, and the dashed line is the nominal track. The mobile unit was attached to a person walking around the building. The square dots are the locations of 29 base stations. The calculated mean positional error is 0.76 m

The measured cross-track statistical data are summarized by the PDF in Fig. 11.21. As can be observed, the peak of the PDF occurs at near zero error, and then the PDF decays for greater positional errors, with most of the errors confined to a maximum of about 1.5 m, but with a noticeable "tail" to the distribution. These raw cross-track PDF data were processed using the algorithm described in Sect. 11.4.3. The resulting CDF is shown in Fig. 11.22. From this CDF the mean error can be calculated by

$$\mu_r = r_{\max} - \int_0^{r_{\max}} CDF(r)dr = \int_0^{r_{\max}} CCDF(r)dr \qquad (11.42)$$

where $r_{\max}$ is the maximum range error in the CDF defined when $CDF = 1$. The mean error for the walking track using this formula is 0.76 m. Also shown is the LS Rayleigh-Gamma model with parameters: $\sigma = 0.58$ m, $a = 0.55$ m, and $\alpha = 0.75$, so the cross-track distribution is mainly Rayleigh, but with a significant exponential component associated with the distribution tail noted previously.

The practically important effect of the degrading of performance by mounting a mobile device on a person walking around a building can be estimated from the

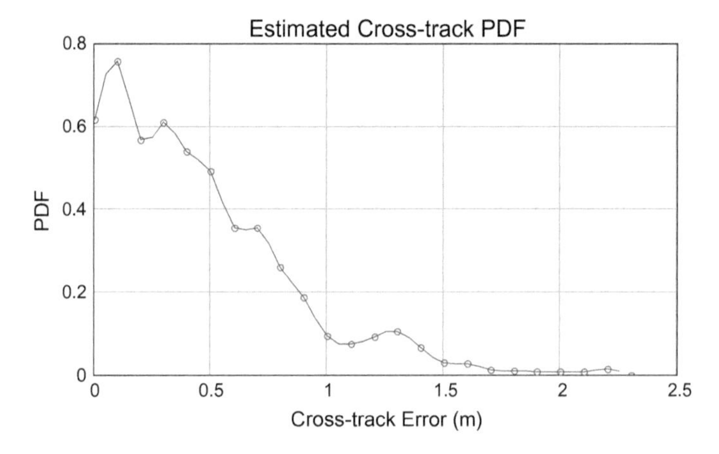

Fig. 11.21 Measured cross-track error PDF for the track shown in Fig. 11.20. The PDF of the measured data are denoted by dots, and the estimated continuous PDF function is denoted by the line connecting the dots

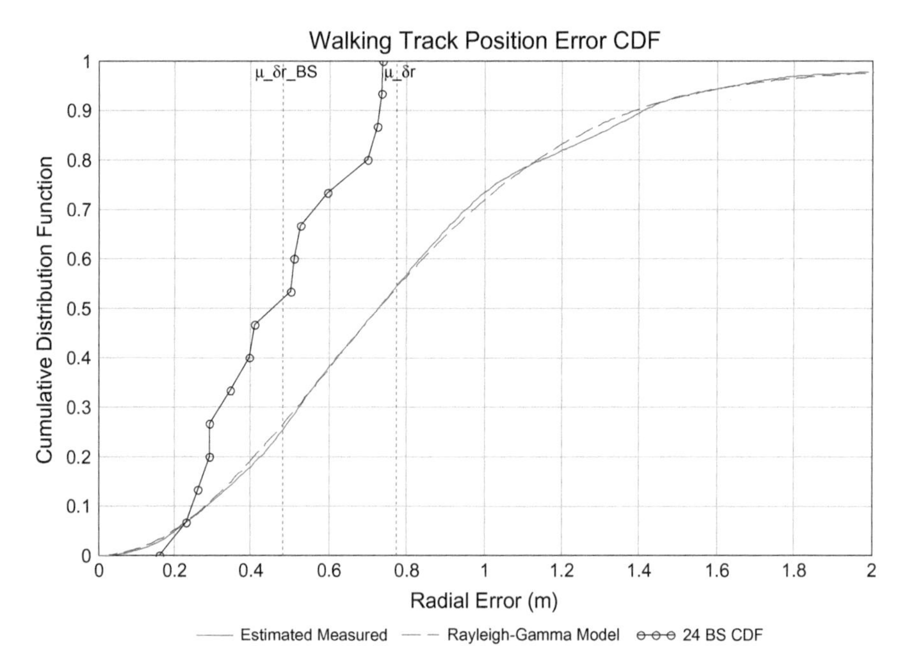

Fig. 11.22 Estimated radial position error CDF computed from the cross-track error distribution shown in Fig. 11.21. Also shown is the Rayleigh-Gamma model fit to the data, and the error distribution based on 24 static measurements in the same general area distributed approximately uniformly throughout the building, as shown in Fig. 11.20

data shown in Fig. 11.20. Typically, measurements within a building are performed using static positions (such as at base stations—see Sect. 11.3.1) without the blocking effect of a body. For this reference case, the CDF of the position error was also determined using static measurements at 24 points located throughout an area approximately defined by the walking track. In this case there is no body-blocking effect, and the temporal effects were largely removed by averaging 100 positional estimates at each of the 24 locations. However, as the signal-to-noise ratio for the measurements was set at a high threshold value of 28 dB, the effect of averaging to remove receiver thermal noise was small, effectively reducing the mean error due to thermal noise in Fig. 11.21 to less than 1 cm. These static measurements provide a lower bound for the positional errors of the WASP system in the building, so that the degrading effects of walking around the building can be assessed. As no filtering of the dynamic data shown in Fig. 11.20 was performed, and the overall radio propagation effects (Rayleigh fading) throughout the area of the test are similar for both the static and dynamic tests, the main difference in performance can be attributed to mounting the mobile device on the body.

The resulting CDF for the 24 static points is also shown in Fig. 11.22, with an associated mean error of 0.48 m; if no temporal averaging was performed, this would increase to 0.49 m, as explained above. As the only effective difference between the static and dynamic measurements after accounting for the temporal effects is due to the mounting of the mobile unit on the body, it can be concluded that the effect of placing the mobile device on a person results in a degrading of mean positional accuracy by 0.76 m−0.49 m = 0.27 m, or about a degrading factor of 1.6. Thus placing a mobile unit on the body of a person significantly reduces the positional accuracy.

11.5 Signal Strength Measurements—Mobile Case

Signal strength measurements associated with a mobile device are important for both position determination and data transmission. Typically, such testing measurements are restricted to the static case (see Sect. 11.3.1.3) as these measurements are easier to perform. However, for a positioning system the movement and the location of the mobile device in relationship to the body can significantly affect the signal strength of the signal from a base station. This section provides data and modeling of measured mobile signal strengths within buildings. The following analysis is based on measurements made while walking along the corridors of a building (same as described in Sect. 11.4.4.3), and does not include locations in rooms within the building. Some measurements were also performed in an adjacent building, so that ranges at times exceeded 50 m. The frequency of the measurements is 2.4 GHz (ISM band).

Propagation loss is determined from the known transmitter power, antenna gains, and the measured receiver signal strength indicator (RSSI). However, if the positioning system uses spread-spectrum signals, the receiver sensitivity can be extended by utilising the receiver process gain, so that the total link budget is of the order of 145 dB (Yu et al. 2009, Appendix B). This total loss is considerably greater than that expected from measurements using RSSI data only, which typically has a link budget of about 100–110 dB. Thus these measurements based on spread-spectrum signals can provide data with a considerable additional margin, which will allow the measurement of signal fades associated with the indoor multipath environment.

The test system is capable of measuring the link loss in two directions, both to and from the mobile device to a base station. The exact transmitter powers and antenna gains, as well as the RSSI calibration are not known, so that the measurements are calibrated using a starting point with known line-of-sight loss. The measurements are made with a 80 ms update period, so that quite fine detail of the signal strength as a function of position (person walking) can be logged. However, at the typical walking speed of 1.5 ms/s, the measurements are taken at about one wavelength separation, and thus the sub-wavelength fine structure of the signal strength cannot be accurately measured. In particular, the multipath environment can result in very deep signal fades over small sub-wavelength distances, but these fades cannot be accurately measured. An example of the signal strength as a function of time is shown in Fig. 11.23. Three scales of signal fluctuations can be observed. The fast variation is associated with Rayleigh fading mechanisms at the scale of a wavelength. The medium scale effects are of the order of 10–100 wavelength (1–10 m), and are related to the local signal scattering from the building structure and objects such as furnishings within building. The largest scale is

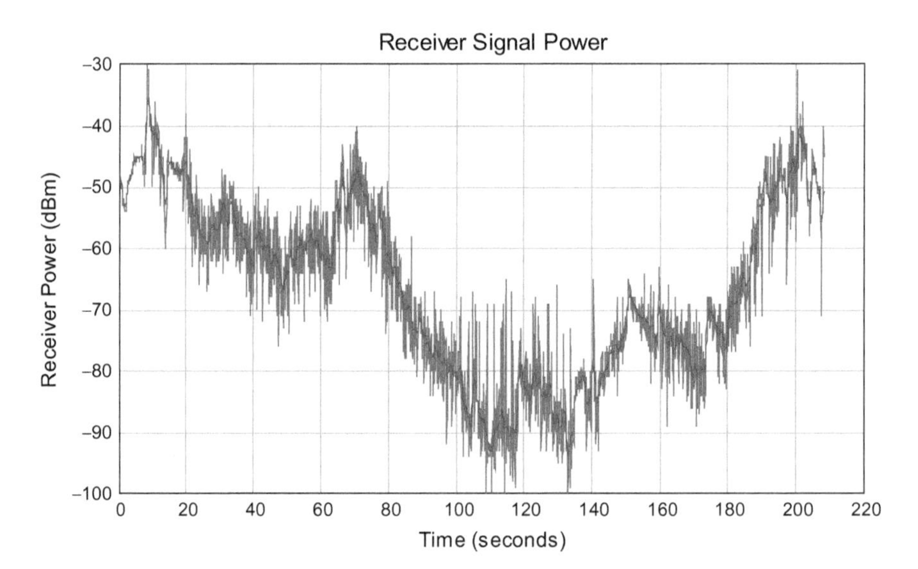

Fig. 11.23 Measured receiver signal strength as a function of time. Also shown is a smoothed estimate at allows the fast-changing signal noise to be estimated. The noise standard deviation is 4 dB

associated with the variation in signal strength as a broad function of range. The characteristics of this variation are considered in more detail below.

The measured signal strength needs to be recorded as a function of range. As the range measurement ideally needs to be sub-metre and the radio ranging is typically not this accurate, an alternative measurement technique is required. The method used is a combination of map matching and inertial sensor detection of movement. In particular, a rate-gyroscope is used to detect when the walker turns a corner associated with a corridor. Figure 11.24 shows the rate-gyroscope z-axis data during the walking path. As can be observed the time of each turn can be accurately measured, as can be the walking start and stop times. The approximately ±5 degrees per second data between turns are associated with hip swings which occur when walking, and thus the timing of each step can also be estimated, if required. These times are matched to the associated positions on the map of the building. As the position on the map can be matched to the time in the logged data file, the intermediate positions between the corners can be estimated to an accuracy of better than 0.5 m by interpolation. These positions as a function of time are then used to calculate the range between the mobile and the base station.

One significant effect on the measured signal loss for the mobile case is the effect of the body in blocking or enhancing the received signal. This effect is illustrated in

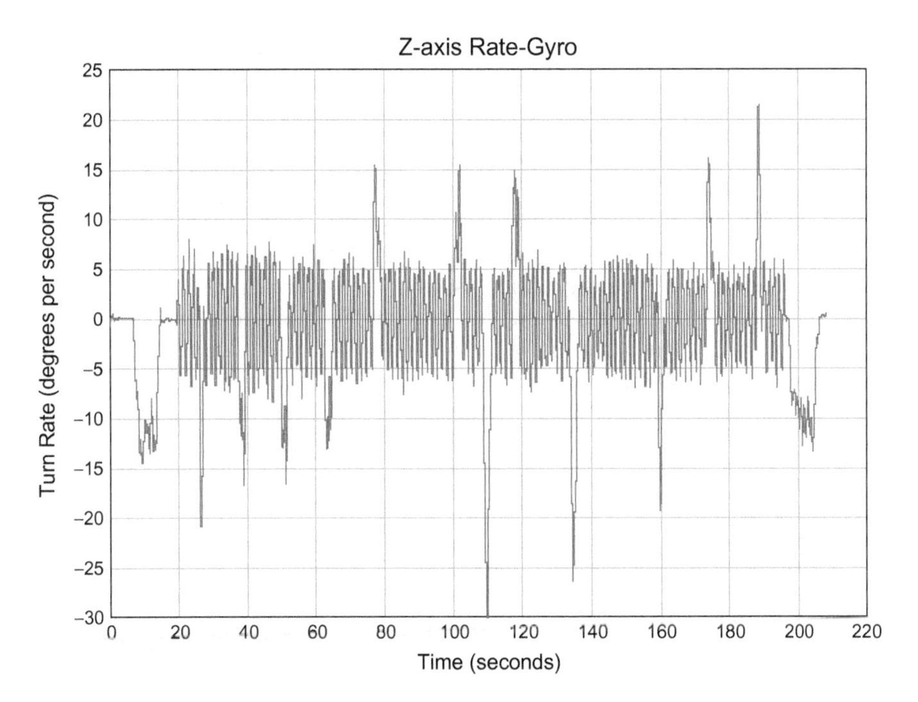

Fig. 11.24 Measured rate-gyroscope data. The large peaks indicate the time of a path turns (clockwise or anticlockwise) at the turn in the walked path. The data starts and ends with a 360° rotation, which can be used to calibrate the rate-gyroscope data

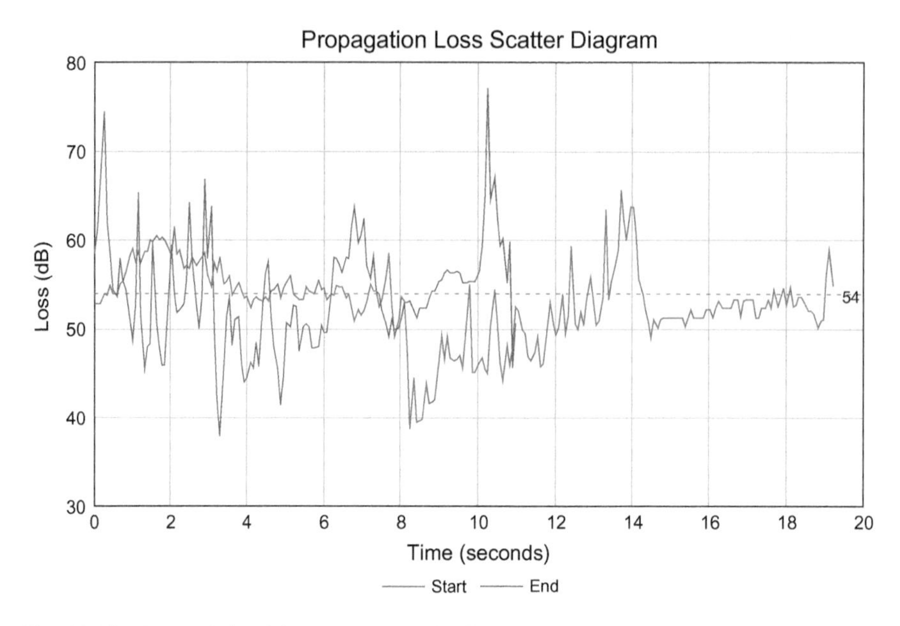

Fig. 11.25 Measured signal loss associated with 360° turns at the start and end of the walk

Fig. 11.25 (and previously analysed in Sect. 11.4.4.3 at a different frequency of 5.8 GHz). The walk started and ended close (5 m) to the base station with LOS propagation, so the expected free-space loss is 54 dB. Although there was LOS propagation, the was also considerable multipath signals from reflections from adjacent walls. The start and end point were nominally the same, but not identical, so the multipath effects would be expected to differ for each measurement. As can be observed from Fig. 11.25 the mean loss in both cases are close to the expected value, but these is considerable variation as the body turned, with both positive and negative variations. The peak extra loss is about 20 dB, and the peak loss reduction is about 15 dB. These results show that with mobile measurements there will be considerable extra signal loss "noise" associated with a mobile device attached to a person (see Fig. 11.19) in addition to that associated with propagation effects in static measurements. In both cases for the data in Fig. 11.25 the standard deviation of the loss noise associated with body rotation is 4.4 dB.

For the design and performance estimation the characteristics of the signal variation needs to be modeled. This modeling involves an estimation of the large scale loss as a function of range, and the statistical variation relative to this local mean. For the large scale the propagation loss is modeled as the nominal free-space loss as a function of range, plus the loss excess also as a function of range.

The measured loss excess for the data in Fig. 11.23 together with the loss excess model (11.11) is shown in Fig. 11.26. At short range (less than 5 m) LOS propagation conditions exist (no recorded data), while at longer ranges there is no LOS. The apparent anomalies at about 20 and 30 m is actually due to the path being normal to the direct range vector, so that the distance was approximately constant

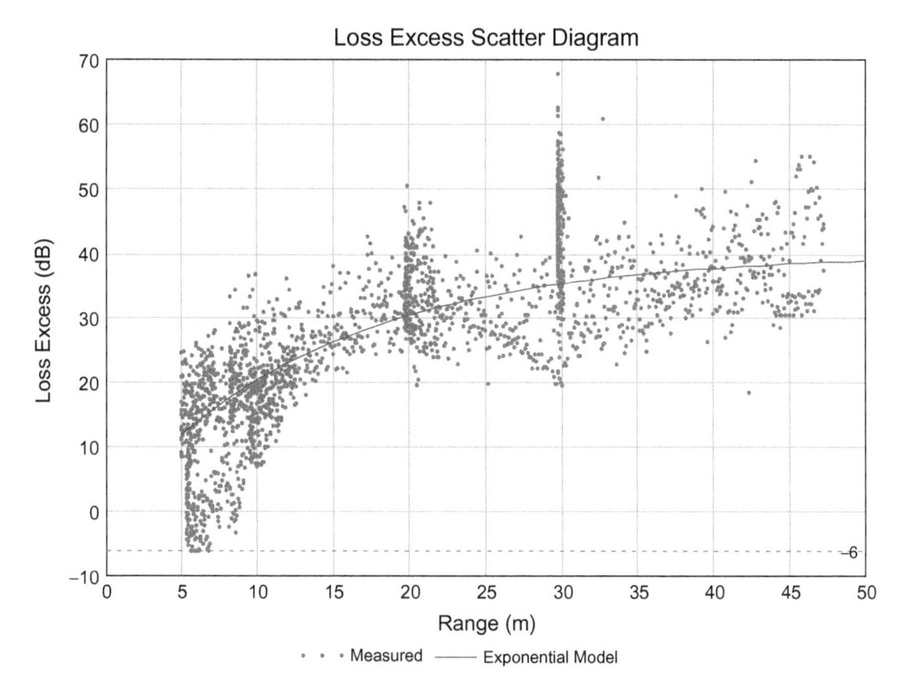

Fig. 11.26 Scatter diagram for the path loss excess above free-space for the data in Fig. 11.23. The model parameters are $\Delta L_{max} = 40$ dB, $R_0 = 4$ m, and $R_1 = 14$ m. The standard deviation between the model and the measured data is 8 dB. The transmission frequency is 2.4 GHz

for about 10 meters in each case; these NLOS measurements show that there can be a large variation in signal (30 dB) over a relatively short distance travelled, even though the range is approximately constant.

The corresponding probability density function is shown in Fig. 11.27. These statistical variations include both the small and medium scale signal variations measured in decibels. Typically, the statistics of the medium scale variations are assumed to be log-normal, and the small scale variations in a NLOS environment are considered to exhibit Rayleigh statistics. However, it is evident that the statistics for this data set are far from showing Gaussian characteristics.

An alternative statistical analysis is based on converting the data from decibels to a linear format for the signal amplitude, again relative to the large-scale model (11.11). In this case the signal magnitude is a positive random variable greater than zero. The signal amplitude statistics of the difference between the measured loss data and the model loss are shown plotted in Fig. 11.28. In this representation in a NLOS environment the nominal distribution will exhibit Rayleigh statistics, but again the match with the measured data is not particular good. One alternative that can be used in a fading signal environment such as in these measurements is the more general Nakagami statistics (Cheng 2001; Zhang 2002). The Nakagami distribution is given by

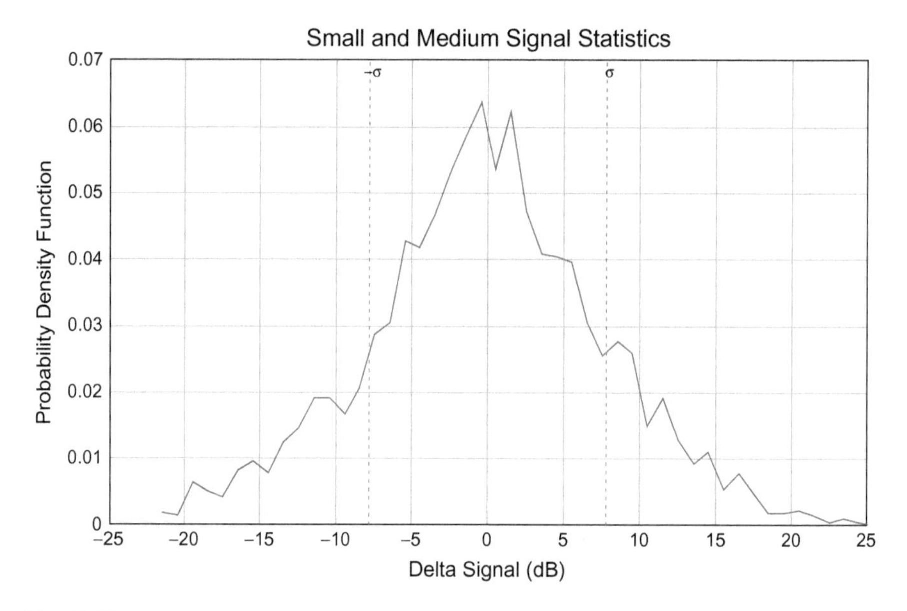

Fig. 11.27 Statistical variation between the measured loss and the large-scale model. The standard deviation is 8 dB

$$p(x) = \frac{2}{\Gamma(m)}\left(\frac{m}{2\sigma^2}\right)^m x^{2m-1}\exp\left(-\frac{mx^2}{2\sigma^2}\right) \tag{11.43}$$

The Nakagami-m distribution covers a wide range of fading conditions. When $m = 1/2$ it is a one-sided Gaussian distribution. For $m = 1$ the Nakagami distribution reduces to the Rayleigh distribution, which is the theoretical distribution when there are a large number of scattered signals with no direct signal, and random phase with uniform distribution. Alternatively for Nakagami statistics, the signal power $(y = x^2)$ has a Gamma distribution given by

$$\begin{aligned} q(y) &= \frac{y^{m-1}e^{-y/\beta}}{\Gamma(m)\beta^m} \qquad (y \geq 0) \\ \beta &= \frac{2\sigma^2}{m} \end{aligned} \tag{11.44}$$

For the case when $m = 1$, the distribution of power is exponential.

The fitting of the model to the measured data depends on the model. For the Rayleigh distribution the only parameter (σ) can be estimated from the sample mean (μ) by the expression

$$\sigma = \sqrt{\frac{2}{\pi}}\mu \tag{11.45}$$

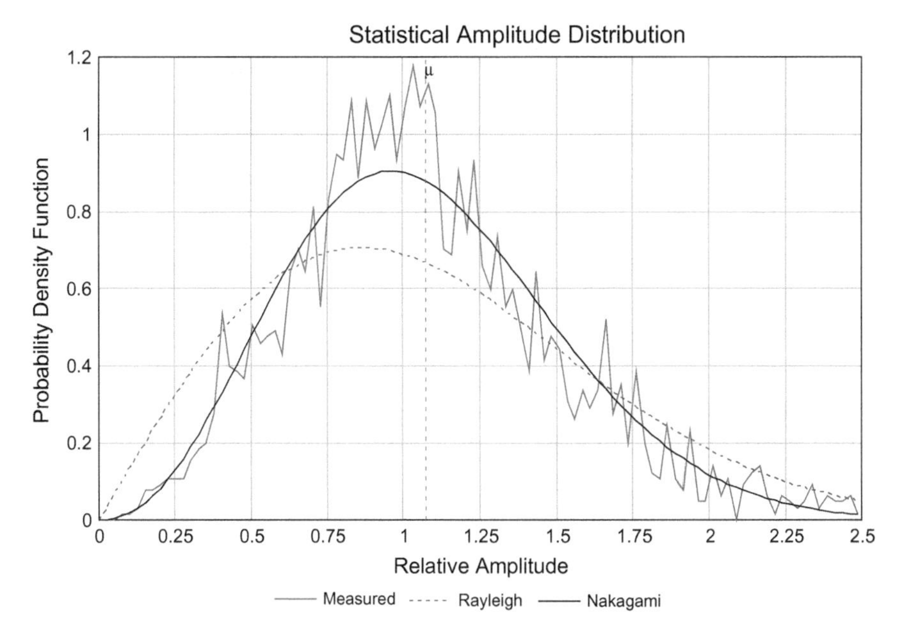

Fig. 11.28 Statistics of the measured differential data, representing the small-scale signal amplitude variation. The data are compared with the Rayleigh and Nakagami fitted to the measured data. For the Nakagami distribution $m = 1.6$

For the differential data in Fig. 11.28, the Rayleigh parameter is $\sigma = 0.86$.

For the Nakagami distribution there are two parameters to be determined. Using the method of moments on the Nakagami distribution the estimate of the m is given by (Julian and Norman 2001; Zhang 2002)

$$\begin{aligned} \hat{m} &= \frac{\hat{\mu}_2^2}{\hat{\mu}_4 - \hat{\mu}_2^2} \\ \hat{\mu}_k &= \frac{1}{N}\sum_{k=1}^{N} x_n^k \end{aligned} \tag{11.46}$$

where x is the signal amplitude given by $x = 10^{-\Delta\bar{L}/20}$. Using this method the value of m for the differential loss data in Fig. 11.28 is $m = 1.60$.

Annex A—Accuracy Evaluation of Differential GPS

Introduction

This annex provides an accuracy evaluation of Differential GPS (DGPS) system with one Base and one Rover unit. The motivation for this objective is associated with the determining the accurate positions of Base Stations for outside positioning systems.

Ideally, a system would need very accurate positions of Base Stations and this could be obtained by standard surveying methods (millimeters accuracy), such as using a Theodolite (angle measurements) or a Total Station which incorporates both a angle and distance measurements. This is a very time-consuming and expensive method, so a more elegant approach is to use GPS for the surveying task. Due to the accuracy limitation (3—15 m) an ordinary GPS system (Selective Availability inactive) would be inadequate. Therefore, Differential or Kinematic versions of GPS are to be considered. DGPS improves the accuracy of the non-differential version by 5—10 times. This is achieved by applying differential corrections to the measured pseudoranges of the rover GPS unit. The following provides information on the expected accuracy of DGPS based on test measurements.

For the simplest version of Differential GPS (shown in Fig. 11.29) a Base unit and a Rover unit is required. The Base unit is positioned at a known (surveyed) location. Differential corrections are generated as the difference between the currently calculated Base position and the fixed (surveyed) position. Corrections are then transmitted (by cable or radio) to the Rover unit to correct currently measured Rover position.

Accurate Base positions can be obtained either by classic surveying methods (Theodolite/Total Station equipment) or by long-term averaging using the GPS receiver itself. As mentioned above, the first method is expensive requiring specialist surveying equipment. The second method on the other hand takes a long time, but can use relatively cheap GPS units. To get a reasonably accurate positions the Base GPS unit has to be in "self-surveying" mode for about 24 h.

Measurements

Several tests were carried out with an initial setup with the units only a few meters apart, and thus could be connected by a cable. The measurements were performed on the flat roof of a building, but it is surrounded by other low-rise buildings and large trees which could partially obscure the path to GPS satellites. The initial test was aimed at determining how long the DGPS measurements need to be averaged to obtain a sufficiently accurate measurement of the Rover position. The initial tests used 15 min of averaging. However, to obtain a good initial reference for the Base

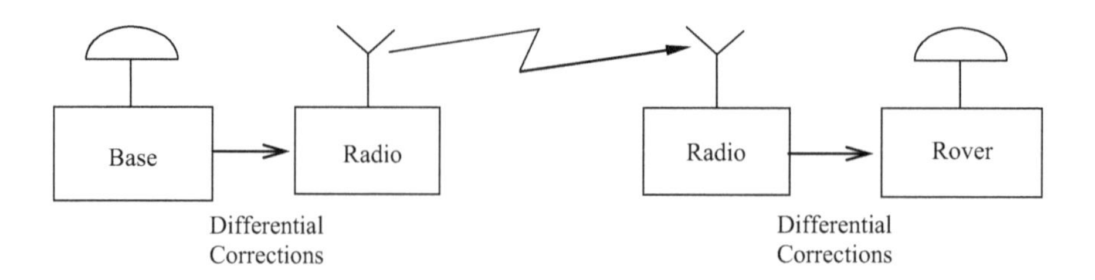

Fig. 11.29 Setup for Differential GPS

unit, 24 h of positional data were averaged. In three 15 min measurements (taken at different times of day) the obtained accuracies were 0.2, 0.23 and 0.35 m. Figure 11.30 shows an example plot of the variation in measured positions over the 15 min period. As can be observed there is a generally chaotic variation over time, with no real pattern or correlation between the two units, and with considerable error from the 24 h average.

Quite clearly the 15 min averaging interval is not long enough to get reasonable accuracy but it improves the instantaneous readings significantly. However, changing the averaging time will not necessarily improve the accuracy of the position. In one particular measurement the estimate of accuracy was getting worse with longer averaging. This is shown in table below.

Averaging time (min)	Distance (m)	Error (m)
5	84.12305	−0.80695
10	84.17194	−0.75806
15	83.94985	−0.98015
20	83.52374	−1.40626
25	83.32507	−1.60493
30	83.26382	−1.66618

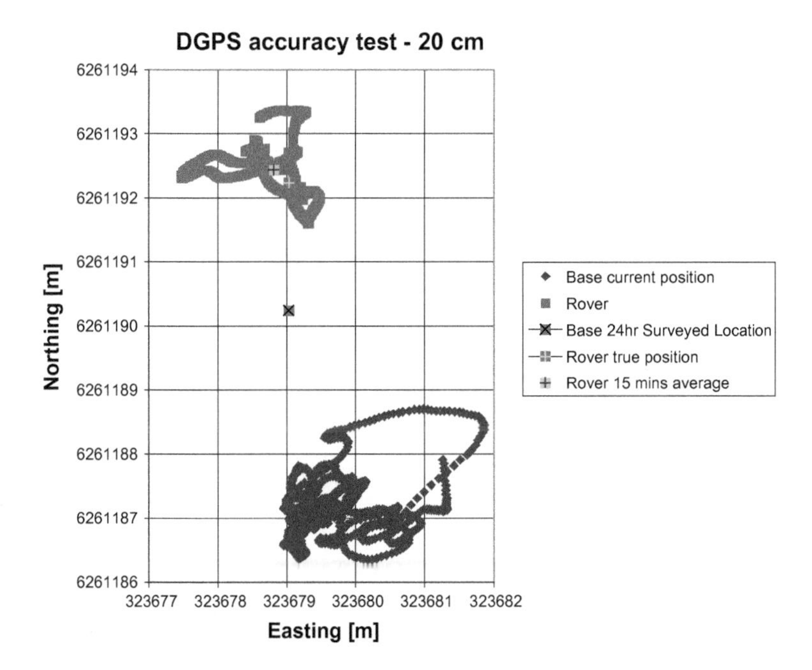

Fig. 11.30 Base and Rover units' positions recorded over 15 min; averaged Rover position has a 0.2 m error in distance from the Base unit

In the second part of the testing, a similar test-setup was used but this time with a pair of radios so that differential corrections could be sent over a longer distance. Position measurements of Rover unit were taken at 2, 5, 10, 20, 40 and 80 m from the base station. The results were shown in Fig. 11.31. It is somewhat inconclusive that accuracy decreases with the bigger separation between Base and Rover units.

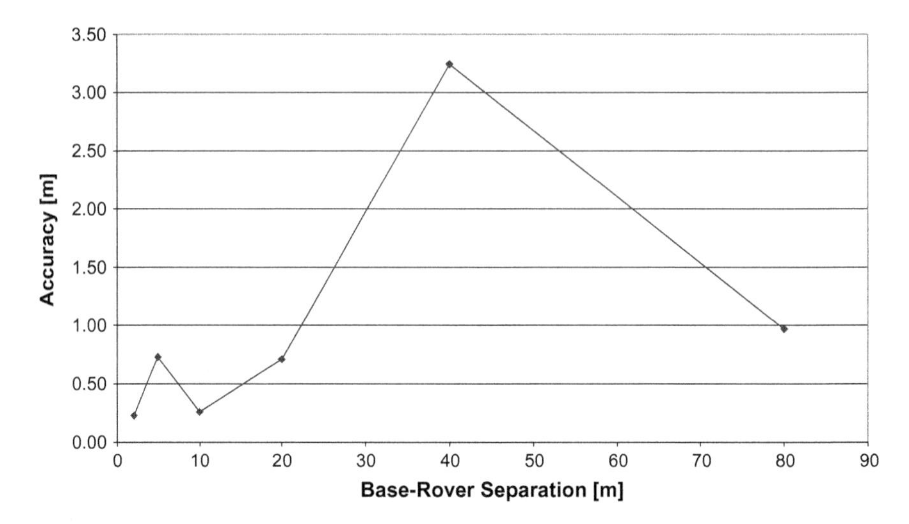

Fig. 11.31 Average error with 15 min of averaging as a function of the separations distance. Observe the errors are not particularly correlated with separation distance

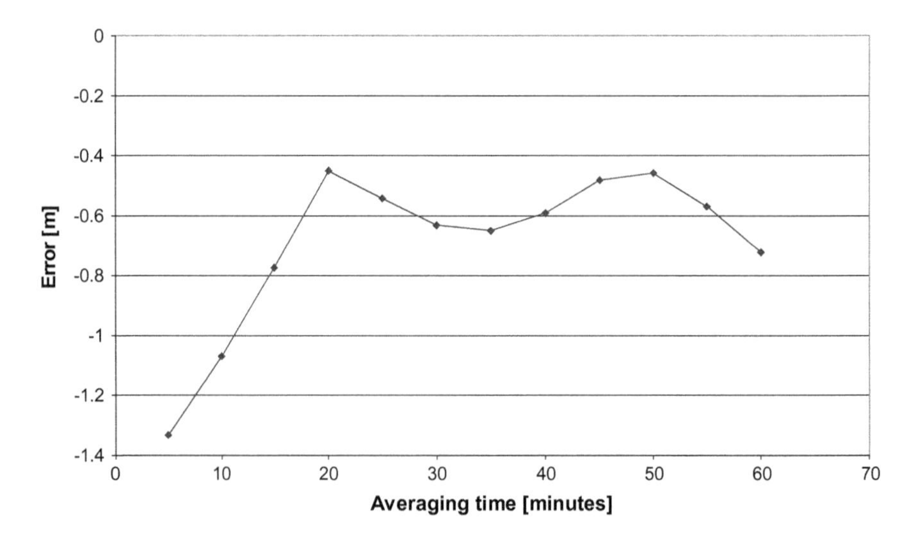

Fig. 11.32 Differential error between DGPS base and rover unit; error is calculated after 5, 10, 15…60 min of averaged positions

The final test was carried out at a race course, which is largely free of obstacles. The separation distance between Base and Rover unit was 671.9 m. The true distance between Base and Rover antennas was measured with a Total Station to millimeters accuracy. The Base GPS unit was given roughly the correct position coordinates and differential corrections were transmitted to the Rover unit using a radio link. Position data were logged over 1 h. The differential distance between the Base (given coordinates) and Rover unit (recorded data) was calculated with 5, 10, 15…60 min averaging times. The measurement error is then calculated as difference between true (Total Station) distance and differential distances calculated from position averages. The results are shown in Fig. 11.32. This measurement suggests that after 15 min, differential accuracy is better than 0.8 m in an open field situation.

References

Alavi B, Pahlavan K (2003) Modeling of the distance error for indoor geolocation. In: Proceedings of IEEE wireless communications and networking conference, pp 668–672, Mar 2003

Alavi B, Pahlavan K (2006) Modeling of the TOA-based distance measurements error using UWB indoor radio measurements. IEEE Commun Lett 10(4):275–277

Alsindi N, Alavi B, Pahlavan K (2009) Measurement and modelling of ultra wideband TOA-based ranging in indoor multipath environments. IEEE Trans Veh Technol 58(3):1046–1058

Carlson A (1968) Communications systems. Section 7.4, McGraw Hill

Elkaim G, Lizarraga M, Pedersen L (2008) Comparison of low-cost GPS/INS sensors for autonomous vehicle applications. In: Proceedings of IEEE/ION position, location, and navigation symposium, pp 1133–1144

El-Rabbany A (2002) Introduction to GPS: the global positioning system. Artech House, Norwood, MA

Fontana RJ (2004) Recent system applications of short-pulse ultra-wideband (UWB) technology. IEEE Trans Microw Theory Technol 52(9):2087–2104

Fontana RJ, Richley E, Barney J (2003) Commercialization of an ultra wideband precision asset location system. In: Proceedings of IEEE conference on UWB systems and technologies, pp 369–373

Gentile C, Kik A (2006) An evaluation of ultra wideband technology for indoor ranging. In: Proceedings of IEEE Globecom, pp 1–6

Hedley M, Humphrey D, Ho P (2008) System and algorithms for accurate indoor tracking using low-cost hardware. In: Proceedings of IEEE/IOA position, location and navigation symposium, pp 633–640, May 2008

Humphrey D, Hedley M (2008) Super-resolution time of arrival for indoor localization. In: Proceedings of IEEE international conference on communications (ICC), Beijing, China, pp 3286–3290, May 2008

Hurst GC (1989) QUIKTRAK: a unique new AVL system. In: Proceedings of vehicle navigation and information systems conference, 1989, pp A60–A62, Sep 1989

Jordan E, Balmain K (1968) Electromagnetic waves and radiating systems, 2nd edn. Prentice-Hall

Julian C, Norman BC (2001) Maximum-likelihood based estimation of the Nakagami m parameter. IEEE Commun Lett 5(3):101–103

Lewandowski A, Wietfeld C (2010) A comprehensive approach for optimizing ToA-localization in harsh industrial environments. In: Proceedings IEEE/ION position location and navigation symposium, pp 516–525

Lui G, Gallagher T, Li B, Dempster AG, Rizos C (2011) Differences in RSSI readings made by different Wi-Fi chipsets: a limitation of WLAN localization. In: Proceedings of international conference on localization and GNSS, Tampere, Finland, pp 53–57

Miao H, Yu K, Juntti M (2007) Positioning for NLOS propagation: algorithm derivations and Cramer-Rao bounds. IEEE Trans Veh Technol 56(5):2568–2580

Modsching M, Kramer R, Hagen KT (2006) Field trial on GPS accuracy in a medium size city: the influence of built-up. In: Proceedings of workshop on positioning, navigation and communication (WPNC), pp 209–218

Priebe S, Nacob M, Kurner T (2011) AoA, AoD and ToA characteristics of scattered multipath clusters for THz indoor channel modelling. In: Proceedings of European wireless, pp 188–196, Apr. 2011

Qi Y, Kobayashi H, Suda H (2006) On time-of-arrival positioning in a multipath environment. IEEE Trans Veh Technol 55(5):1516–1526

Sathyan T, Humphrey D, Hedley M (2011) WASP—A system and algorithms for accurate localization using low-cost hardware. IEEE Trans Syst Man Cybern: Part—C, 41(2):211–222

Schubert R, Richter E, Wanielik G (2008) Comparison and evaluation of advanced motion models for vehicle tracking. In: Proceedings of international conference on information fusion, pp 1–6

Sharp I, Yu K, Guo Y (2009) GDOP analysis for positioning system design. IEEE Trans Veh Technol 58(7):3371–3382

Teunissen P, Kleusberg A (1998) GPS for geodesy. Springer

Yu K (2007) 3-D localization error analysis in wireless networks. IEEE Trans Wirel Commun 6 (10):3473–3481 (2007)

Yu K, Sharp I, Guo Y (2009) Ground-based wireless positioning. Wiley-IEEE Press

Zhang QT (2002) A note on the estimation of Nakagami-m fading parameter. IEEE Commun Lett 6(6) (2002)

Chapter 12
Sub-meter Indoor Ranging Performance Using Wi-Fi Chip Radio

12.1 Introduction

The development of mobile radio systems has been dramatic, this being in part due to the development of chip radio systems, including the integration of both the analog radio transmitter and receiver, as well as the associated digital signal processing. While the main focus of the developments has been digital radio communications, in more recent times the availability of integrated single chip Global Positioning System (GPS) receivers (El-Rabbany 2002) has allowed the rapid development of position-based services in devices such as "smart" phones. Simultaneously, radio technology has been extended to indoor environments for wireless data communications, particularly the widespread adoption of IEEE 802.11 (Wi-Fi) as the defacto standard. With the availability of integrated chips, Wi-Fi has become standard in both Notebook Personal Computers and smart phones, with Wi-Fi base stations for data communications common in both the workplace and the home.

However, indoor radiolocation is not commonly available, although applications such as tracking people and assets (Bahl and Padmanabhan 2000; Fontana et al. 2003; Fontana 2004; Hedley et al. 2008) are commonly cited as future developments. In such indoor environments GPS does not work well because the satellite signals are blocked by the buildings, and typically non-line-of-sight (NLOS) multipath conditions predominate. Further, the accuracy requirements of indoor positioning are typically much greater than the corresponding requirements for outdoor positioning. For example, GPS accuracy outdoors is typically around 5–10 meters (El-Rabbany 2002), while for the indoor case with applications such as tracking people, an accuracy of 1 m or better is desirable to allow location determination to a particular room inside a building. As GPS cannot meet these operational requirements indoor, some other technology is required.

Because of the complexity of the indoor propagation environment, achieving 1 meter radio ranging accuracy is difficult. While UWB systems can have good

I. Sharp and K. Yu, *Wireless Positioning: Principles and Practice*, Navigation: Science and Technology, https://doi.org/10.1007/978-981-10-8791-2_12

indoor ranging error performance of the order of 10–20 cm, the severe limitation on the available transmitter power limits the range to typically 10 m (Roy et al. 2004; Ghassemzadeh et al. 2005; Alavi and Pahlavan 2006; Bellusci et al. 2008; Gentile and Kik 2006; Alsindi et al. 2009). Thus to cover large indoor spaces a large number of nodes is required, which is logistically difficult and expensive. An alternative is to use the industrial, scientific and medical (ISM) bands (particularly the 2.4 and 5.8 GHz bands) as used by Wi-Fi devices, where the much higher allowable transmitter power (FCC Part 15) results in indoor ranges of typically 40–80 m, and hence far fewer nodes are required to cover the same area. However, with maximum bandwidths of respectively 80 MHz and 150 MHz the time resolution of ISM-band positioning systems is much less then UWB systems, and the positional accuracy is considerably worse (Yu et al. 2009b). With a bandwidth of up to 150 MHz, recent experimental ISM-band positioning systems have demonstrated that indoor positioning accuracy of 0.5 m is possible (Hedley et al. 2008; Sathyan et al. 2011; Yu et al. 2009a), and thus meets the above-defined requirement for practical applications. While the development of specific new technologies (such as IEEE 802.11az) may meet future needs, a commercially attractive alternative is to adapt the current Wi-Fi data communications technology for indoor radiolocation. In particular, this chapter shows how current generation Wi-Fi chip radios (with bandwidths of 20 MHz) can be used to achieve a performance equivalent to a 150 MHz radio by use of appropriate signal processing. Because these chip radios were not designed for radiolocation, some of their characteristics make the design challenging. The chapter thus describes various techniques to overcome the limitations of current Wi-Fi radios, and provides theoretical estimations of the performance.

12.2 Overview of Related Work

Before considering the specific details of the proposed use of current-generation Wi-Fi chip radios, this section briefly overviews some related work in the general area of Wi-Fi positioning.

1. Intel research has developed a Wi-Fi positioning system based on TOA ranging, which requires some hardware modifications to implement (Golden and Bateman 2007). The range is computed through performing a number of different exchanges of messages between an access point and a stationary mobile. Although some advanced signal processing techniques are used, the range root-mean-square error (RMSE) associated with 5 access points is 2.66 m.
2. Typically, in currently proposed systems, two-way TOA (or single round-trip-time (RTT)) measurements are employed for range measurements, such as using the packet pairs DATA and ACK in the IEEE 802.11 standards. In (Llombart et al. 2008) using the existing Wi-Fi hardware a pure-software based ranging and positioning approach is proposed and implemented. Instead of

using packet pairs, packet sequences (RTS, CTS, DATA and ACK) are exploited to generate multiple RTT measurements, resulting in improved measurement accuracy, although still only 4 m.
3. The development of a WLAN channel sounder for IEEE 802.11b is reported in (Ciurana et al. 2009). The sounder directly uses the physical layer characteristics of Wi-Fi systems to estimate the channel impulse response. Using a multipath propagation model and conducting a Monte-Carlo simulation, the resolution of the system is found to be nine nanoseconds, equivalent to 2.7 m.
4. A strategy is proposed in (Jemai and Kurner 2008) to reduce the position determination latency without degrading positioning accuracy. In particular, a method is described to optimize the tradeoff between a short RTT response time and the position update period. (Hoene and Willmann 2008) investigates the scalability issue associated with the Wi-Fi network based positioning.

In summary, it is observed that although the above-mentioned Wi-Fi based ranging and positioning techniques can achieve an accuracy better than the corresponding received signal strength based systems, the reported indoor accuracy (2.5–4 m) is still not considered to be satisfactory for many practical applications. Also, current literature has no comprehensive theoretical performance analysis of the different techniques, nor do they address practical issues associated with implementation using chip radios.

12.3 Range and Pseudorange Measurements

This section summarizes a general theory of determining ranges or pseudoranges between nodes in a network using TOA measurements; see Chap. 4 for a general introduction into the techniques and performance for TOA measurements. Although some related work has been described elsewhere (Hedley et al. 2008; Sathyan et al. 2011; Yu et al. 2009a), the analysis in this chapter provides more in-depth information relating to chip radios, and the associated performance analysis which forms the underlying theoretical framework in subsequent sections. The measured data in this and subsequent sections relates to an experimental positioning system called the Wireless Ad hoc System for Positioning (WASP) (Sathyan et al. 2011), developed by CSIRO. This experimental system operates in the 5.8 GHz ISM band with an effective bandwidth of 125 MHz, and is designed to achieve an indoor positioning accuracy of 0.5–1 m accuracy based on a mesh network of nodes.

12.3.1 Transmitter to Receiver Timing Measurements

The determination of the propagation delay used for position determination is based on the measurement of the TOA of a radio signal at a receiver. However, the TOA measurement cannot be used directly to measure the propagation delay, as the total delay from a transmitter to a receiver includes many other delays associated with the radio equipment—see Chap. 4 for more details. Thus before TOA can be used to determine range, it is important to understand the details of the delays through the radio equipment. In this initial analysis it will be assumed that the clocks in the transmitter and receiver have the same frequency, but later this restriction will be removed. However, it is important to note that the clocks are not synchronized in time, so that the clocks will have a random offset relative to an absolute time reference.

The TOA measurement encompasses the total transmission delays from the transmitter to the receiver as measured relative to the clocks in the transmitter (mobile) and the receiver (stationary node or anchor node/base station), and includes the delays in the baseband hardware, the RF and IF sections of the radios, and the propagation delays. A block diagram of the transmission path and the associated parameters is shown in Fig. 4.2 in Chap. 4. Note that transmission can be from node m to node n or in the reverse direction.

The transmitter baseband electronics has a clock which has a (unknown) offset relative to absolute time. Further, it is assumed that the transmissions are based on a pseudo-random noise code (or pn-code) of finite length, which repeats at the code period; it will be assumed that the code length is greater than the total propagation delay, so ambiguity does not occur. As the time measurement has a repetition period, it is appropriate for the (absolute) clock offset to be referred as a phase offset (ϕ) rather than an (unknown) absolute time offset. The baseband pn-code is used to modulate the RF signal in the radio, which outputs the radio signal via an antenna. This process involves delays in the radio, filters, and cables, all of which are of unknown value. The total delay in the transmitter process is defined as Δ_m^{tx} for node m. Note that the delays in the radio transmitter, particularly the filters can be considerably greater than the propagation delay between the nodes, and further these delays will vary (slowly) over time. The reason for the variability is mostly associated with filter delays varying due to temperature and ageing effects. Further the length of the cable from the radio to the antenna can vary from node to node.

Now consider the TOA measurement in more detail. If the clocks are assumed to be accurately synchronized in frequency (see Sect. 12.5 for details of the frequency synchronization), then the local time in both the transmitter (tx) and the receiver (rx) can be defined by a phase offset determined at local clock time $t = 0$. With reference to Fig. 4.2 and Eq. (4.4) in Chap. 4, the TOA measurement from node m to n (ignoring measurement noise) can be represented by the expression

$$TOA_{m,n} = \phi_m + R_{mn}/c + \Delta_n - \phi_n \tag{12.1}$$

where ϕ is the clock offset phase and c is the propagation speed of 0.2998 m/ns. Thus the TOA measurement can be expressed in terms of the clock phases in the transmitter and the receiver, the propagation delay, and the combined transmitter and receiver delays in the *receiving* unit only. As a consequence, only two parameters in each node need to be defined, namely the local clock phase and the combined radio delays. Thus (12.1) is the basic starting point in position determination.

For typical indoor positioning systems, nodes both transmit and receive, so that it useful to consider the TOA measurements between a pair of nodes. Thus applying (12.1) twice, the resulting two TOA measurements are

$$\begin{aligned} TOA_{mn} &= \phi_m + R_{mn}/c + \Delta_n - \phi_n \\ TOA_{nm} &= \phi_n + R_{nm}/c + \Delta_m - \phi_m \\ RTT_{mn} &= (TOA_{mn} + TOA_{nm})/2 = R_{mn}/c + (\Delta_m + \Delta_n)/2 \end{aligned} \tag{12.2}$$

where reciprocity in the propagation delay has been used (Jordan and Balmain 1968). Thus by combining two TOA measurements into a single round-trip time (RTT) measurement, the clock phase parameters are eliminated. In this case, the measured data only involves the sought-after range measurement and the delay parameter associated with each node. If the radio delay parameters are known (preset) values, then the two TOA measurements combined can be used to estimate the range. However, this typically results in low accuracy measurements due to the variability in the radio equipment delays over time.

12.3.2 Timing Measurement Errors

The analysis in Sect. 12.3.1 assumed ideal TOA measurements without any errors, but in an actual system there will be measurement errors associated with multipath propagation and receiver noise. To minimise multipath errors, TOA measurements need to be based on the leading edge of the despread received wideband spread-spectrum signal. For the WASP system, a leading edge threshold algorithm (Sathyan et al. 2011) is used, so the following discussion will be limited to that particular implementation; other leading edge algorithms (Sharp et al. 2009b; Yu et al. 2009b, Chap. 4) have similar characteristics. For frequency synchronization, as described in Sect. 12.5, only the Gaussian noise performance is important, while for radio delay calibration, as described in Sect. 12.6, the multipath performance is the most important. The following subsections summarize these respective performances.

12.3.2.1 Theoretical Gaussian Noise Performance

The theoretical performance of a pulse *threshold* algorithm in the presence of Gaussian noise can be calculated by assuming the nominal triangular pulse shape (autocorrelation function of the pn-code) is corrupted with random noise with zero mean and a specified standard deviation σ_n. In such a case the signal-to-noise ratio (γ) with the amplitude (A) of the pulse without noise normalized to unity, is $\gamma = 1/\sigma_n^2$. The threshold for the detection of the leading edge has a nominal amplitude of $a_{th}A$, or simply a_{th} in the normalized form. The rise-time of the leading edge is τ_{pulse}. While this triangular shape is only an approximation for the bandlimited signal case, provided the threshold level is not too small $(a_{th} > 0.05)$ typical shape distortions are sufficiently small (Sharp et al. 2009b and Chap. 4, Sect. 4.2.3) that they have a minor effect on the estimation of the TOA error due to Gaussian noise. Using techniques similar to those described in (Sharp et al. 2009b) it can be shown that the standard deviation in the TOA error measured by the threshold algorithm is given by

$$\sigma_{\varepsilon} = \tau_{pulse}\sqrt{\frac{1 + a_{th}^2}{\gamma}} \approx \frac{\tau_{pulse}}{\sqrt{\gamma}} \tag{12.3}$$

where the last approximation is because the threshold amplitude is small to reduce the effects of multipath signals. It is informative to compare the expression (12.3) with the Cramer-Rao Lower Bound (CRLB) (Urkowitz 1983). For a pulse with a signal bandwidth B, the standard deviation of the measurement error is constrained by the lower bound

$$\sigma_{CRLB} = \frac{1}{\sqrt{8}\pi B\sqrt{\gamma}} \tag{12.4}$$

Note in this case the SNR is measured after the correlation process, which enhances the baseband SNR, and hence reduces the errors. In particular, the correlation process results in a process gain given by

$$G_p = \gamma_{out} - \gamma_{in} \tag{12.5}$$

where the output and input SNR values are measured in decibels. For the WASP system the process gain is about 31 dB, so applying (12.3) the error standard deviation σ_{ε} is reduced by the considerable factor of 15.5 dB. Thus the error given by (12.3) relative to the lower limit given by (12.5) is given by

$$\rho = \sigma_{\varepsilon}/\sigma_{CRLB} = \sqrt{8}\pi\left(\tau_{pulse}B\right) \tag{12.6}$$

Note that B is the despread signal bandwidth (not the RF channel bandwidth), and also that the ratio is independent of the SNR. For the WASP system $\tau_{pulse} = 14.5$ ns, and $B \approx 60$ MHz, so that the dimensionless rise-time bandwidth product is $\tau_{pulse}B = 0.87$ which is a typical value (near unity) for direct-sequence spread-spectrum positioning systems. Finally, the ratio for the WASP is $\rho = 7.73 = 17.8$ dB. Thus the leading edge algorithm, which is optimized to reduce the effects of multipath signals, is far from the lower bound with Gaussian noise corruption. However, from (12.3), with a typical operational SNR greater than 30 dB, the theoretical standard deviation in the TOA error is 0.46 ns = 0.14 m, and in the limit with SNR around 50 dB (see Fig. 12.2) the TOA error is 0.014 m. This result shows that Gaussian noise will have a minor effect on the potential positional accuracy of a practical system using the 5.8 GHz ISM band.

12.3.2.2 Multipath TOA Errors

The scattering of radio signals indoors is complex and depends on the details of the indoor environment, so theoretical calculations are difficult. However, experimental measurements (Alavi and Pahlavan 2006; Bellusci et al. 2008; Gentile and Kik 2006; Alsindi et al. 2009) show that the ranging errors have a bias error which increases approximately linearly with range, and a random component which is largely independent of range. Figure 12.1 shows typical error data using the WASP system. Thus for the remainder of this chapter the range error model associated with indoor NLOS propagation is assumed to be given by

$$\Delta R = (\delta R_0 + \lambda R) + \varepsilon_R \tag{12.7}$$

where the term in brackets is the linear bias model, and ε_R is a random component with zero mean. The range error constant bias error is δR_0, and is largely associated with the leading-edge detection process rather than directly with radio propagation. The bias error is assumed to increase linearly with range (λ m/m) due to the multipath scattering effects which cumulatively increase with range.

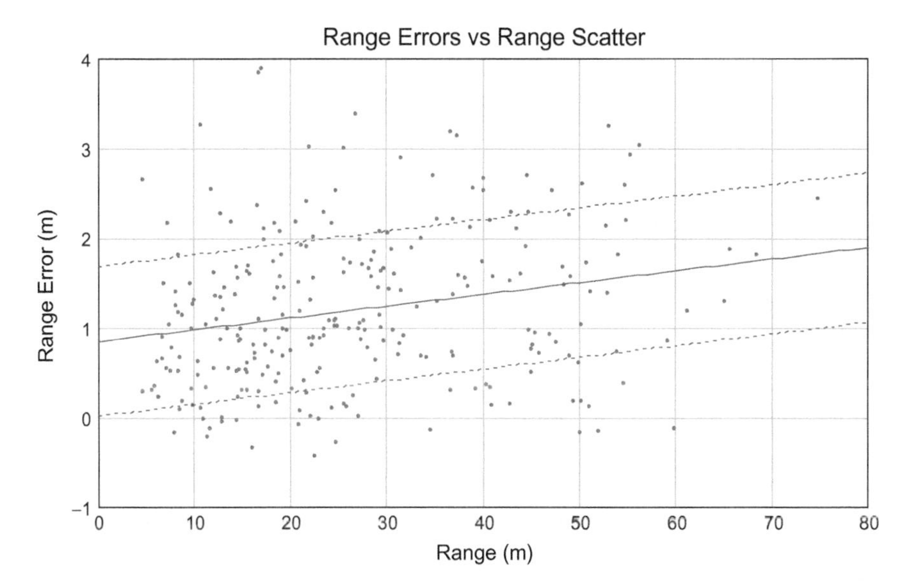

Fig. 12.1 Measured range errors. The solid line shows the linear bias least-squares fit, and the dotted lines are ± one standard deviation of 0.83 m

12.3.2.3 Measured Noise Performance Data

The noise performance of the chip radio and the noise reduction associated with the despreading of the spread-spectrum signal can be used to estimate the output SNR, and hence the TOA noise from (12.3). The chip radio includes a receiver signal strength indicator (RSSI), which is typically measured by a square-law detector monitoring the power in the receiver IF. Note that the RSSI measurement is actually the combined receiver signal plus noise power, and is unreliable as the signal power approaches the noise power. A more sophisticated method which takes into account the receiver process gain (G_p) to extend the range to well below the noise level. Using the method described in (Yu 2009, Appendix B), the signal power is given by

$$S = \frac{RSSI}{1 + G_p/\gamma_{out}} \tag{12.8}$$

The output SNR (γ_{out}) has two noise contributions, one from the receiver thermal noise, and a second associated with digital noise. The digital noise is a consequence of limitations in the data processing, including quantization noise from the A/D conversion, limited precision arithmetic and multi-channel reconstruction noise (see Sect. 12.4). This digital noise limits the maximum output SNR, even if the input signal is very large. Assuming that this digital noise is proportional to the signal power, the output SNR is given by

$$\gamma_{out} = \frac{G_p S}{N_0 + \eta S} \xrightarrow{S \gg N_0} G_p \eta \tag{12.9}$$

where the receiver thermal noise is $N_0 = kTBF$, F is the receiver noise factor, and η is the digital noise parameter. For the WASP system design, the receiver parameters are:

Spread spectrum channel bandwidth (B): 125 MHz
IF bandwidth: 18 MHz
Noise figure (F): 4 dB
Noise (N_0): −97 dBm
Process gain (G_p): 31 dB

The measured performance of the WASP receiver in an indoor environment meters is shown in Fig. 12.2, together with the least-squares fit to the model defined by (12.8) and (12.9). Because of the variations associated with the indoor multipath environment, there is some scatter about least-squares model, but overall there is a good fit between the model and the measured data. Note also that using (12.8) signal levels down to about –107 dBm can be measured, 10 dB below the receiver noise level.

Using the lower limit of 28 dB for the SNR and (12.3), the upper limit of the Gaussian noise-related error in the TOA is 0.6 ns, or equivalently 0.18 m, while at

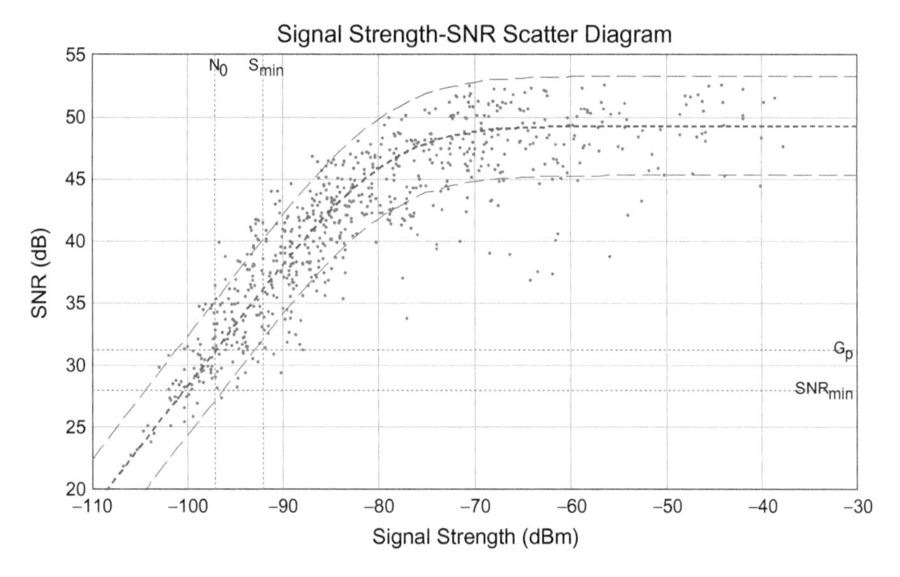

Fig. 12.2 Measured SNR characteristics for the WASP system deployed inside a building. The signal strength is estimated from the receiver RSSI, but extended using the receiver process gain. The receiver noise power (N_0) and the process gain (G_p) are shown, as well as the minimum limits for the SNR and signal strength used for position determination. The dotted model curve is based on (12.9), while the dashed lines represent variations of ±4 dB relative to the least-squares fit. Also shown is the very conservative lower limit $S_{\min}$ of −92 dBm used for position determination

the limiting SNR the value is 1.4 cm. Thus the errors in TOA measurement errors in an indoor environment are dominated by multipath effects.

12.4 Multi-segment Concatenation Performance

To meet the design goals outlined in Sect. 12.1 the effective time resolution of the TOA measurements must be of the order of a 10 ns, or alternatively a bandwidth of at least 100 MHz. While such a bandwidth is available in the 5.8 MHz ISM band, Wi-Fi chip radios designed for IEEE 802.11a operation have a bandwidth of 20 MHz, and thus cannot meet this design requirement. The transmission approach adopted in the WASP system is somewhat similar to that proposed in (Saberinia and Tewfik 2004, 2008) for wireless personal networks using multiband OFDM, namely to sequentially scan in time across the ISM band, and to reconstruct the wideband signal by appropriate signal processing of the logged channel data. However, as the receiver must be retuned for each channel, the data from each channel are not globally coherent (although coherent within each channel), even after an appropriate settling period for the receiver local oscillator; a coherent wideband signal is essential for performing the despreading of the spread-spectrum signal. This important complication in the reconstruction is not considered in (Saberinia and Tewfik 2004, 2008). One possible method to ensure a coherent total signal is to overlap the bands in frequency, so that the overlapping frequency sections can be used to determine the differential phase between adjacent channels. In the WASP system, eight overlapping bands of bandwidth 18 MHz (total 144 MHz) are concatenated into a 125 MHz wideband signal, with about 2.5 MHz of overlap between adjacent channels.

12.4.1 Procedure for Transmitting/Reconstructing the Wideband Signal

As this chapter focuses on the complications associated with using a chip radio rather than the details of signal processing, the following is only a summary of the procedure for generating the multi-segment signals for transmission, and the subsequent reconstruction in the receiver.

1. The spectrum of the wideband time-domain spread-spectrum transmitter baseband signal $s_{tx}(t)$ is subdivided into overlapping segments; typically the frequency overlap is 10–15%.
2. The power in each segment is normalized for equality. This table of normalization constants is also known by the receiver to allow renormalization in the reconstruction process. This procedure ensures constant SNR for each received segment. The inverse Fourier transform of the normalized segment spectral data is the time-domain segment data transmitted.
3. Each time-domain segment is transmitted sequentially with the appropriate center frequency. Additionally, there is a period between each transmission to allow the receiver to retune and stabilize (in phase) the receiver local oscillator for the next transmission.
4. The receiver calculates the spectrum of each time-domain signal segment, and uses the normalization table to recover the spectrum of each segment of the wideband signal.
5. The relative phase between adjacent segments is estimated using the procedure described in detail in Sect. 12.4.2 following.
6. The differential phase data are used to rotate the phase of each segment to reconstruct the (coherent) wideband spread-spectrum baseband receiver signal $s_{rx}(f)$.
7. The despreading of the wideband signal is performed using the computationally efficient Fourier transform method. In particular the correlation diagram (correlogram) is generated by the operation

$$C(\tau) = \mathrm{FFT}^{-1}(\mathrm{FFT}(s_{rx}(t))\mathrm{FFT}(s_{tx}(t))^{*}) \tag{12.10}$$

Note that the Fourier transforms of the received segment signals required in step (5) are used in the calculation of the correlogram, so little additional computations are required to calculate the differential phases. Also note that the Fourier transform of the transmitted signal is invariant, and are simply stored as a data table in the receiver. The receiver uses the wideband correlogram to estimate the TOA of the signal.

12.4.2 Recovering the Differential Phase

This subsection describes the method of estimating the differential phase between two adjacent segments, and determines the statistical accuracy of the estimate. Consider two segments of overlapping complex spectral data. The receiver output can be represented as

$$\begin{aligned} &\text{Segment 1}: \ A_1 S(f) e^{j\phi_1} \\ &\text{Segment 2}: \ A_2 S(f) e^{j\phi_2} \end{aligned} \tag{12.11}$$

where the (random) phases are unknown, but the relative amplitudes (A_1, A_2) are known by the receiver from the normalization table described previously. To recover the differential phase while reducing the effects of signal noise, the following *frequency-domain* correlation is performed

$$\begin{aligned} C &= \int_0^{\Delta f} \left(A_1 S(f) e^{j\phi_1}\right) \left(A_2 S(f) e^{j\phi_2}\right)^* df \\ &= A_1 A_2 e^{j(\phi_1 - \phi_2)} \int_0^{\Delta f} |S(f)|^2 df = A_1 A_2 e^{j(\phi_1 - \phi_2)} E_{seg} \end{aligned} \tag{12.12a}$$

where E_{seg} is the energy in the overlapping segment, and receiver noise has been ignored. As the segment energy is a real number, the differential phase can be recovered by

$$\phi_1 - \phi_2 = \Delta\phi_{12} = \arg(C) \tag{12.12b}$$

While the above analog correlation process can recover the differential phase, in practice the receiver output will include noise, and the signals are in a sampled digital form. Thus the receiver correlator output from (12.12a) will be with digital processing

$$C = A_1 A_2 e^{j(\phi_1 - \phi_2)} \sum_{i=1}^{N_{ovr}} \left(\left(\hat{S}_1^i \right) \left(\hat{S}_2^i \right)^* \right) = A e^{j(\phi_1 - \phi_2)} \sum_{i=1}^{N_{ovr}} \hat{S}_{12}^i = A e^{j(\phi_1 - \phi_2)} \hat{C} \quad (12.13a)$$

where the ith sample of the digital noisy complex spectra is given by

$$\begin{aligned} \text{Segment 1}: \quad \hat{S}_1^i &= \left(S_R^i + n_{11}^i \right) + j\left(S_I^i + n_{12}^i \right) \\ \text{Segment 2}: \quad \hat{S}_2^i &= \left(S_R^i + n_{21}^i \right) + j\left(S_I^i + n_{22}^i \right) \end{aligned} \quad (12.13b)$$

and where the random noise n has the statistical properties $E[n] = 0$ and $\text{var}[n] = \sigma_n^2$, and the four noise components in (12.13b) are statistically independent. Thus unlike the integration in (12.11), the summation in (12.13a) is complex, and thus performing (12.12b) using (12.13a) will result in an error in the differential phase estimate. Now consider properties of the real and imaginary components of $\hat{S}^i$. The expected value of the real component of ith sample is $E\left[\text{Re}\left(\hat{C}_i\right)\right] = \left(S_R^i\right)^2 + \left(S_I^i\right)^2 = P_s^i$, so the summation over the N_{ovr} overlapping samples in the overlapping region is again the energy in the overlapping segment as in (12.12a). The corresponding variance is

$$\text{var}\left[\text{Re}\left(\hat{C}_i\right)\right] = E\left[\text{Re}\left(\hat{C}_i^2\right)\right] - E\left[\text{Re}\hat{C}_i\right]^2 = \sigma_n^2 \left[\left(S_R^i + S_I^i\right)^2 + 2\left(S_R^i\right)^2 + 2\sigma_n^2 \right] \quad (12.14)$$

Now as transmitted signals are pseudo-random the real and imaginary spectral components are quasi-random with a mean given by $E[S_R] = E[S_I] = 0$ and a variance $\text{var}[S_R] = \text{var}[S_I] = \sigma_s^2$, and the real and imaginary components are statistically independent. Thus applying these conditions to the N_{ovr} overlapping samples in the segment gives

$$\begin{aligned} E\left[\text{Re}\left(\hat{C}\right)\right] &= 2N_{ovr}\sigma_s^2 \\ \text{var}\left[\text{Re}\left(\hat{C}\right)\right] &= 4N_{ovr}\sigma_s^2\sigma_n^2(1 + 1/2\gamma_s) \end{aligned} \quad (12.15)$$

where the pre-correlation segment spectrum SNR is $\gamma_s = \sigma_s^2/\sigma_n^2$, as the (normalized) segment power is $2\sigma_s^2$ and the noise power is $2\sigma_n^2$. From (12.15) the output SNR (γ) for the real component of $\hat{C}$ is

$$\gamma = \frac{\left(2N_{ovr}\sigma_s^2\right)^2}{4N_{ovr}\sigma_s^2\sigma_n^2(1 + 1/2\gamma_s)} = \frac{\sigma_s^2 N_{ovr}}{\sigma_n^2(1 + 1/2\gamma_s)} = \frac{N_{ovr}}{1 + 1/2\gamma_s}\gamma_s \quad (12.16)$$

Thus this correlation has a process gain of $N_{ovr}/(1 + 1/2\gamma_s) \approx N_{ovr}$, where the last approximation is a consequence of the segment pre-correlation SNR typically being much greater than unity. A similar analysis for the imaginary component of $\hat{C}$ gives

$$\begin{aligned} E\left[\mathrm{Im}\left(\hat{C}\right)\right] &= 0 \\ \mathrm{var}\left[\mathrm{Im}\left(\hat{C}\right)\right] &= 3N_{ovr}\sigma_s^2\sigma_n^2(1+1/3\gamma_s) \approx 3N_{ovr}\sigma_s^2\sigma_n^2 \end{aligned} \tag{12.17}$$

The imaginary component of $\hat{C}$ is responsible for errors in estimating the differential phase as defined by (12.12b). In (12.15), the mean of the real part of $\hat{C}$ is much greater than its standard deviation, so for the following estimate of the standard deviation of the differential phase error only the mean will be used. Combining this approximation with (12.17) gives

$$\sigma_{\Delta\phi} \approx \frac{\sqrt{3N_{ovr}\sigma_s^2\sigma_n^2(1+1/3\gamma_s)}}{2N_{ovr}\sigma_s^2} = \sqrt{\frac{3(1+1/3\gamma_s)}{4\gamma_s}}\frac{1}{\sqrt{N_{ovr}}} \quad \text{(radians)} \tag{12.18}$$

Thus given the segment SNR, the standard deviation of the differential phase error decreases inversely as the square root of the length of the overlapping segment. However, increasing the overlapped part of the segment reduces the transmission efficiency, so in practice this is limited to a small fraction of the total segment length.

Consider a typical numerical example, with parameters similar to the WASP system. To ensure adequate performance the signal SNR is $\gamma_s \geq 0$ dB, and the length of the overlap is $N_{ovr} = 64$ samples out of a total segment length of 512 samples. Applying (12.18) gives $\sigma_{\Delta\phi} \leq 0.125$ radians. While this phase error is quite small, the variance of the differential phase error (relative to the first segment) increases proportional to the segment number, so the relative phase error in the last segment can be quite large. The cumulative effect of these errors is considered in the next subsection.

12.4.3 Reconstructed Signal Correlation Performance

Having determined the statistics of the differential phase error between adjoining segments, the overall effect on the wideband *time-domain* correlation of the spread-spectrum signal is now estimated. Although the wideband spectrum power varies across the band (maximum in the center and approaching zero at the edges), the power in each transmitted segment is normalized so that the transmitted power in each segment is the same, and hence the SNR in each segment is approximately the same. As a consequence, from (12.18) the standard deviation of the phase error in each segment will be similar. Further, the contribution of each segment to the overall time-domain correlation (without phase errors) will also be approximately similar. Thus using the first segment as a phase reference, the relative effect of the differential phase errors on the correlator output is given by

$$\Gamma_c = 1 + \sum_{s=1}^{N_{seg-1}} e^{j\Phi_s} \tag{12.19a}$$

where

$$\Phi_s = \sum_{n=1}^{s} \Delta\phi_n \tag{12.19b}$$

is the summation of phase errors accumulating in a random walk fashion across the reconstructed received time-domain signal. What is of interest is the magnitude of (12.19a), which would involve a square-root operation, making statistical calculations difficult. Thus as an alternative, the statistics of the magnitude-squared (power) is analysed, so that with N_{seg} segments the relative power is

$$\begin{aligned}\Gamma_{cc} &= \Gamma_c\Gamma_c^* = \left(1 + \sum_{s=1}^{N_{seg-1}} e^{j\Phi_s}\right)\left(1 + \sum_{s=1}^{N_{seg-1}} e^{j\Phi_s}\right)^* \\ &= N_{seg} + 2\sum_{s=1}^{N_{seg-1}} \cos(\Phi_s) + 2\sum_{s=2}^{N_{seg-2}} \sum_{t=s+1}^{N_{seg-1}} \cos(\Phi_s - \Phi_t)\end{aligned} \tag{12.20}$$

To further simplify the statistical calculations associated with (12.20), as the phase terms are generally relatively small (see numerical calculations below) the cosines in (12.20) are approximated by the truncated Taylor series $\cos x \approx 1 - x^2/2$. Thus the expectation (mean) of (12.20) can be approximated by

$$E[\Gamma_{cc}] = N_{seg}^2 - \sum_{s=1}^{N_{seg-1}} E[\Phi_s^2] - \sum_{s=2}^{N_{seg-2}} \sum_{t=s+1}^{N_{seg-1}} E\left[(\Phi_s - \Phi_t)^2\right] \tag{12.21}$$

By applying (12.19b) and assuming the differential phase errors of each segment pairs have the same variance and are statistically independent (due to the random phase associated with the frequency scanning process by the receiver) the expectations in (12.21) become

$$\begin{aligned}\sum_{s=1}^{N_{seg-1}} E[\Phi_s^2] &= \sum_{s=1}^{N_{seg-1}} E\left[\left(\sum_{n=1}^{s} \Delta\phi_n\right)^2\right] = \sum_{s=1}^{N_{seg-1}} \sum_{n=1}^{s} E[\Delta\phi_n^2] \\ &= \left[\frac{N_{seg}(N_{seg}-1)}{2}\right]\sigma_{\Delta\phi}^2 \quad \sum_{s=2}^{N_{seg-2}} \sum_{t=s+1}^{N_{seg-1}} E\left[(\Phi_s - \Phi_t)^2\right] \\ &= \sum_{s=2}^{N_{seg-2}} \sum_{t=s+1}^{N_{seg-1}} E\left[\left(\sum_{n=1}^{s} \Delta\phi_n\right)^2\right] = \left[\frac{N_{seg}(N_{seg}-1)(N_{seg}-2)}{6}\right]\sigma_{\Delta\phi}^2\end{aligned} \tag{12.22}$$

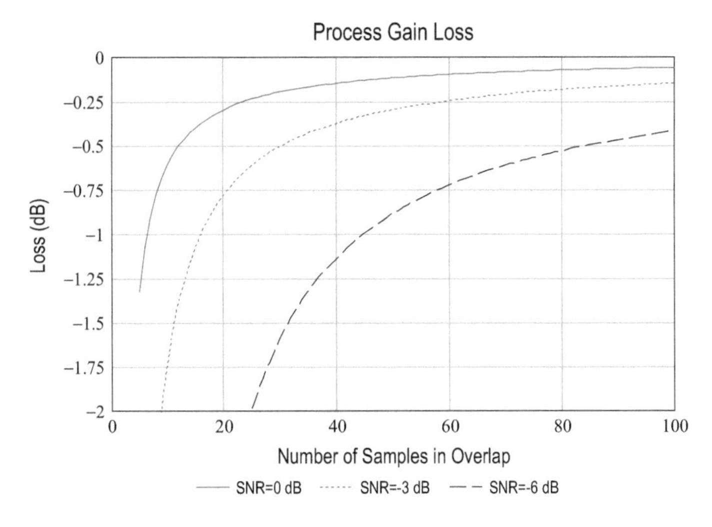

Fig. 12.3 Variation in the process gain loss as a function of the number of samples in the overlap and the input SNR (γ_s)

where $\sigma_{\Delta\varphi}$ is given by (12.18). Substituting these results into (12.21) and normalizing by the zero phase error value N_{seg}^2 gives the expected relative correlator output power as

$$E\left[\hat{\Gamma}_{cc}\right]\Big/N_{seg}^2 = 1 - \left[\frac{\left(N_{seg}^2 - 1\right)}{6N_{seg}}\right]\sigma_{\Delta\phi}^2 \approx 1 - \frac{N_{seg}\sigma_{\Delta\phi}^2}{6} \qquad (12.23)$$

Equation (12.23) gives the loss in the process gain due to the phase errors. For example, from the numerical example in Sect. 12.4.2 with $\sigma_{\Delta\phi} = 0.125$, applying (12.23) gives a trivial loss in the process gain of 0.1 dB. Figure 12.3 shows the loss in Process Gain as a function of the number of samples in the overlap and the input SNR. For minimal impact on the performance the number of samples should be greater than 50 and the input SNR greater than –3 dB; these requirements are met by the WASP system.

12.5 Frequency Offset Estimation

The analysis in Sect. 12.3.1 assumed perfect frequency synchronization throughout the mesh network. This section briefly summarizes a frequency synchronization technique; the details of the method of frequency control in a mesh network are described in Chap. 5, Sect. 5.4.

Accurate frequency synchronization in the mesh network is required to ensure that clock drifts between the transmit signal and receiving a reply does not result in RTT measurement errors. For example, if the two TOA measurements associated with the RTT measurement are one second apart in time, then a differential drift of 1 ppb in the clocks in the two nodes represents a RTT measurement error of 1 ns. As typical low-cost oscillators have frequency errors of the order of 1–10 ppm, the system must provide a frequency synchronisation correction to reduce this initial error by at least a factor of 1000:1.

The details in Sect. 5.4 show for both static and mobile nodes the frequency can be monitored and synchronized throughout a mesh network to an accuracy of better than 1 ppb, by using a technique of measuring the change of round-trip-time between a pair of nodes over time. This method is not affected by multipath ranging errors as only differential time measurements are used. Thus with this accurate frequency synchronization, the ranging errors in the RTT method due to frequency errors is minimal.

12.6 Radio Delay Calibration

As both TOA and RTT measurements involve propagation delays, including those through the radio equipment, accurate determination of range requires accurate calibration of the internal radio delays, both in the transmitter and the receiver. As chip radios are designed for data communications where these delays are not important, the performance of Wi-Fi chip radios is not ideal for accurate positioning applications. Some data associated with the delay parameter for three different units as a function of time after power-on is shown in Fig. 12.4. As can be observed, there is considerable variation (10 ns) as a function of time due to temperature warming effects in the hardware. Also observe that the delay parameter varies considerably between units. Thus in general it is not possible to a priori define the radio transmitter delay with any precision, as the delay can vary by many nanoseconds in periods as short as a few minutes. Thus unless this variability is tracked, the estimation of the radio propagation delay will have errors relating to the variability of the internal radio delays.

In theory, the calibration of the radio delay is simple for anchor nodes whose inter-nodal distance is known. The RTT as given by (12.2) can be used in reverse to determine the combined radio delay of the two nodes in the RTT process, given the known inter-nodal distance. By combining such measurements with its neighbors, redundancy in the equations allows the radio delay of each node to be determined. However, the RTT measurement will also include measurement errors associated with the indoor propagation environment (which typically are much greater than the desired 1 ns calibration accuracy), and thus the estimated radio delays may not have the required accuracy.

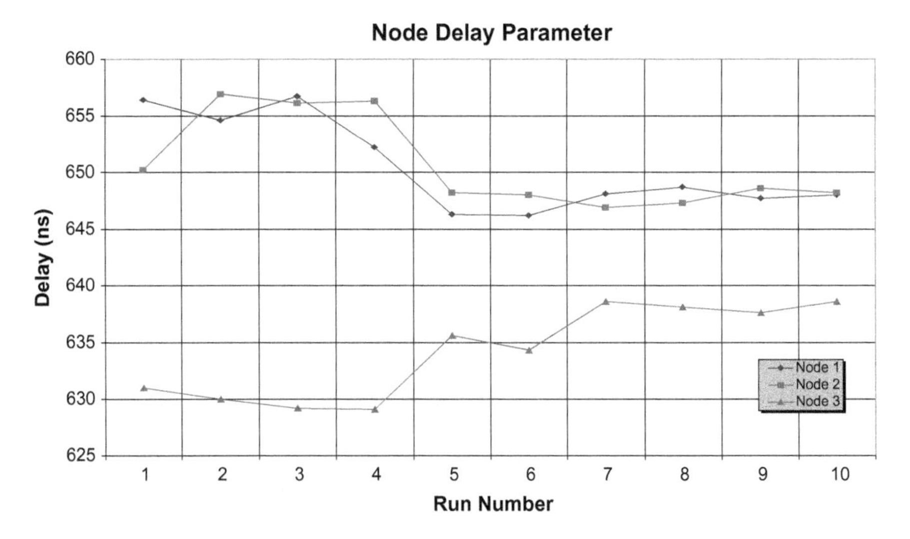

Fig. 12.4 Variation in the node delay parameter (mainly associated with filters) for three different nodes as a function of time. The data are the average delay over about a period of 3 min (run number period). The particular chip radio is a Maxim MAX2828

The basis for accurately estimating the radio delays is a model of the measurements errors, plus redundancy in the inter-node measurements. The range measurements will include the following errors:

1. Errors in the radio delay parameters, as described in the introduction to this section.
2. For indoor situations with multipath and NLOS propagation the model is defined by (12.7).
3. In addition to the deterministic bias ranging errors defined by (12.7), there will be a random component with (by definition) zero mean and unknown variance. In general, these errors will not be Gaussian, but may be approximately so.
4. Due to the receiver noise, RTT measurements will have random errors which are a function of the receiver SNR in both nodes. This ranging noise can be modelled as a Gaussian process.

As accurate RTT measurements rely on accurate frequency synchronization between the two nodes, any frequency errors translate into ranging errors. Such frequency offset measurements are insensitive to the errors described in subparagraphs (1–3) above, but are sensitive to the receiver noise described in subparagraph (4). As a consequence, these frequency synchronization errors can be modelled as a random Gaussian process with zero mean. The errors associated with frequency synchronization are summarized in Sect. 12.5, with details in Chap. 5, Sect. 5.4.

Based on the above description, the range measurement (M) between two nodes can be represented by the equation

$$M_{m,n} = R_{m,n} + \left(\delta R_0 + \lambda R_{m,n}\right) + c\left(\frac{\delta\Delta_m + \delta\Delta_n}{2}\right) + Noise \tag{12.24}$$

where $\delta\Delta$ is the error in the node internal radio delay parameter, and *Noise* is a random variable with zero mean and approximately Gaussian statistics. As the measurement error model is linear and the various error mechanisms are uncorrelated with one another, a linear least-squares analysis can be used to obtain an unbiased estimate of the model parameters. However, a direct least-squares analysis using (12.24) is not possible. In particular, the δR_0 parameter (range error at zero range) cannot be determined independently from the delay calibration parameters. However, if it is assumed that the delay calibration errors are random with zero mean, these random errors *initially* can be incorporated into the *Noise*, so that the least-squares equation initially becomes

$$M_{m,n} - R_{m,n} = \left(\delta R_0 + \lambda R_{m,n}\right) + Noise \tag{12.25}$$

which can be solved for estimates of the δR_0 and λ parameters. Having established an estimate of δR_0 and λ, the equation for a measurement associated with nodes m and n then becomes

$$M_{m,n} - R_{m,n} - \delta R_0 - \lambda R_{m,n} = \varepsilon_{m,n} = \frac{c}{2}\left(\delta\Delta_m + \delta\Delta_n\right) + Noise \tag{12.26}$$

If the number of nodes in the mesh network is N_{nodes}, then the number of such measurements will be $N_{nodes}(N_{nodes} - 1)/2$, but because of radio range limitations the typical number of neighboring nodes with satisfactory radio communications will be less, typically of the order of 5–10 in practical situations. In any case, the total number of measurements is far in excess of the number of nodes (and hence their delay calibration parameters), and thus a least-squares procedure can be used to estimate the delay calibration. Writing the (12.26) model in the matrix form

$$\mathbf{\Gamma} d = \frac{2}{c} E \tag{12.27a}$$

the least-squares solution for the delay calibration parameters **d** is given by

$$\mathbf{d} = \frac{2}{c}\left[\mathbf{\Gamma}\,\mathbf{\Gamma}^{\mathrm{T}}\right]^{-1}\left[\mathbf{\Gamma}^{\mathrm{T}} E\right] \tag{12.27b}$$

However, for this method to be practical the accuracy in estimating the node calibration parameters should ideally be within the desired specification of a standard deviation not greater than 1 ns. The source of this uncertainty is noise term in (12.26), this largely being associated with the range measurement noise due to multipath propagation, provided the receiver SNR is sufficiently high. To determine the standard deviation of the radio delay calibration, a statistical analysis using

(12.27b) is required. Analysis of the symmetrical square matrix $\boldsymbol{\Gamma}\boldsymbol{\Gamma}^{\mathrm{T}}$ shows that it consists of a diagonal with elements of value equal to the number of neighboring nodes, while the associated row/column consists mainly of zeros, but also with some elements (in number equal to the number of neighbouring nodes) of value one; the pattern of ones is associated with the pattern of radio communications between nodes. For analytical statistical calculations (but not for the numerical calculations involving (12.27b)) it is appropriate to set all the off-diagonal elements to zeros, which greatly simplifies the analytical calculations with minimal error, and does not require detailed knowledge of the radio communications pattern to make statistical performance estimates. With this simplification, the solution (12.27b) for a particular node p can be expressed in the form

$$\delta\hat{\Delta}_p = \frac{2}{c}\left[\frac{1}{N_{neighbors_p}}\right]\left[\sum_{n=1}^{N_{nodes}} \mathrm{F}(p,n)\varepsilon_{p,n}\right] \tag{12.28}$$

where $\mathrm{F}(p,n)$ is 1 if nodes (p,n) communicate, otherwise the function is zero. Using (12.28) it is clear that the expected value $E\left[\delta\hat{\Delta}_p\right] = 0$ as the errors have a zero expectation after the biases are removed in (12.26); thus the delay calibration estimate using (12.26) is unbiased. Similarly, noting that the measurement errors are uncorrelated, the variance of the estimated delay calibration is given by

$$\begin{aligned} E\left[\delta\hat{\Delta}_p^2\right] &= \left[\frac{2/c}{N_{neighbors_p}}\right]^2 E\left[\left(\sum_{n=1}^{N_{nodes}} \mathrm{F}(p,n)\varepsilon_{p,n}\right)^2\right] \\ &= \left[\frac{2/c}{N_{neighbors_p}}\right]^2 N_{neighbors_p} E\left[\varepsilon_{p,n}^2\right] \end{aligned} \tag{12.29}$$

which results in the expression for the standard deviation

$$\sigma_{\delta\hat{\Delta}_p} \approx \frac{2\sigma_\varepsilon/c}{\sqrt{N_{neighbors_p}}} \tag{12.30}$$

where σ_ε is the standard deviation of the range measurement noise in (12.26) after the bias errors are removed, and is assumed to be the same for each node in the mesh network. The actual number of neighboring nodes will vary throughout the mesh network, so as the individual node's errors are statistically independent and summing up the variances, an "average" number of nodes applicable for calculating an overall calibration accuracy using (12.30) is

$$\bar{N}_{neighbors} = \frac{N_{nodes}}{\sum_{n=1}^{N_{nodes}} N_{neighbors_n}^{-1}} \tag{12.31}$$

Thus given a specified ranging accuracy, using (12.31) in (12.30) the average number of nodes required to meet the desired calibration accuracy can be estimated. For example, for the WASP system indoors $\sigma_\varepsilon \approx 0.8$ m (see Fig. 12.1), so applying (12.30) a 1 ns calibration accuracy requires on average 28 neighboring nodes; such a high number of nodes is impractical, so the desired calibration accuracy of 1 ns cannot be met in this case. As the geometric dilution of precision (GDOP) is about $2/\sqrt{N_{neighbors}}$ for a mesh network (Sharp et al. 2009a), for positioning to meet a 0.5 m accuracy $(GDOP\sigma_\varepsilon)$ requirement the number of neighbouring nodes required is about 10; applying this number to (12.30) gives the calibration accuracy as 1.7 ns, or equivalently 0.5 m. This calibration error increases the overall range error to $\sqrt{0.8^2 + 0.5^2} = 0.94$ m for position determination using the above real-time delay calibration procedure. Thus there is only a modest degrading in the positional accuracy due to the less than ideal accuracy of calibrating the node delay parameters.

The above technique was used to process WASP mesh nodal RTT data in an indoor environment. In this case the radio delay parameters were first determined manually using a cable of known electrical length to interconnect two nodes at a time; the measured calibration have errors estimated to be less than 0.5 ns. This procedure is both labor-intensive and time consuming, and would not be considered practical in a commercial environment, but is appropriate for experimental investigations.

The 46 nodes were deployed throughout a building complex, and RTT data between nodes recorded using the cable-calibrated radio delay data. The mean number of neighboring nodes in this mesh network is 9. Only high SNR (>28 dB) data were used, with 100 samples averaged to largely remove temporal noise from the measurements. These RTT data were then used to estimate the node calibration errors using the above technique, which ideally should result in near-zero calibration errors if there were no errors in the real-time calibration procedure. Applying the above calibration process showed that the standard deviation is 2.6 ns, greater than the expected result of 1.8 ns using (12.30) and (12.31). This shows that there may be a drift in the delays since the original manual calibration. Applying the estimated corrections to the radio delay parameters and recalculating gives the residual errors standard deviation of 0.7 ns, as shown in Fig. 12.5. These results are within the desired calibration accuracy of 1 ns, and better than the theoretical value. This test result emphasizes the need for real-time calibration, even if pre-installation calibration is performed.

The above calibration procedure was based on measurement using anchor nodes in a mesh network, but the procedure also could be applied to mobile nodes when they are stationary and have had their position determined previously. In this case, there will be some error in their computed position, which would flow through to the estimation of the node delay parameter. However, over time these errors would be random with zero mean, and the delay parameters would only vary slowly over time. Thus more accurate node delay parameters could be obtained using a Kalman filter in a similar manner to that described Sect. 12.5 for estimating the frequency offset parameter.

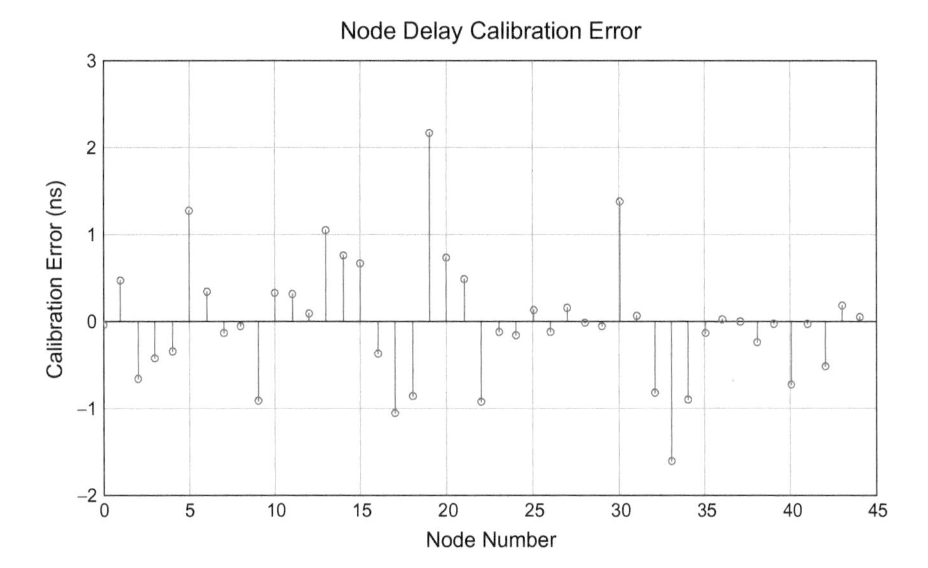

Fig. 12.5 Residual radio delay errors after applying the real-time calibration. The mean error $(\mu_{\delta\Delta})$ is −0.01 ns and the STD $(\sigma_{\delta\Delta})$ is 0.7 ns

12.7 Concluding Remarks

This chapter investigated the application of IEEE 802.11/Wi-Fi chip radios for use in indoor positioning systems. Because of the desire for cheap implementation in commercial applications, it is suggested that Wi-Fi chip radios could be used for radiolocation in addition to data communications. Achieving the desired sub-meter positional accuracy indoors with this type of radio is difficult, as the design of these chip radios is aimed at data communications rather than radiolocation. To achieve this goal, three main areas where innovative design is required have been identified, namely achieving an effective wide bandwidth, frequency synchronization in a mesh network with mobile nodes, and calibration of internal radio delays, all in a radio-hostile indoor multipath propagation environment.

To achieve effective wide bandwidth transmissions in the ISM bands used by Wi-Fi, it was shown that concatenation of 20 MHz channel spread-spectrum data is possible by suitable phase alignment of the individual channel data to achieve a coherent wideband signal with a loss in process gain of less than 1 dB. To avoid the need for *time* synchronization in a mesh network, a difficult task in an indoor environment, a RTT technique is described which only requires accurate *frequency* synchronization of the nodes in the network rather than time synchronization. Analysis shows that frequency synchronization to better than 1 ppb is achievable for both anchor and mobile nodes. Finally, although the Wi-Fi chip radios have large temporally-variable internal delays, it is shown that in a mesh network with sufficient inter-nodal radio communications that in situ calibration of these internal delays can be achieved to an accuracy of about 1.5 ns, despite the indoor NLOS multipath propagation conditions.

The overall goal is to achieve sub-meter ranging accuracy. Measurements using a mesh network with 46 nodes showed that by applying the above-described techniques the residual random range error standard deviation is 0.8 m; the associated positional accuracy is about 0.5 m. Such radiolocation capability can be integrated into existing IEEE 802.11 data communications using the same chip radio with only the addition of some signal processing hardware, typically based on a digital signal processor. The positioning performance of such a system, while somewhat worse than a UWB-based system, should be satisfactory for most indoor applications at a fraction of the cost.

References

Alavi B, Pahlavan K (2006) Modeling of the distance measure error using UWB indoor radio measurement. IEEE Commun Lett 10(4):275–277

Alsindi N, Alavi B, Pahlavan K (2009) Measurement and modelling of ultra wideband TOA-based ranging in indoor multipath environments. IEEE Trans Veh Technol 58:1046–1058

Bahl P, Padmanabhan V (2000) RADAR: an in-building RF-based user location and tracking system. In: Proceedings of IEEE conference on computer communications (INFOCOM), pp 775–784

Bellusci G, Janssen G, Yan J, Tiberius C (2008) Model of distance and bandwidth dependency of TOA-based UWB ranging error. In: Proceedings of IEEE international conference on ultra-wideband, pp 193–196

Ciurana M, Barcelo-Arroyo F, Llombart M (2009) Improving the performance of TOA over wireless systems to track mobile targets. In: Proceedings of IEEE international conference on communications workshops, pp 1–6, June 2009

El-Rabbany A (2002) Introduction to GPS: the global positioning system. Artech House, Norwood, MA

FCC Part 15 (Title 47 of the Code of Federal Regulations transmission rules, Part 15—Radio frequency devices, subpart C—Intentional radiators)

Fontana R, Richley E, Barney J (2003) Commercialization of an ultra wideband precision asset location system. In: Proceedings of IEEE conference on UWB systems and technologies, pp 369–373 (2003)

Fontana R (2004) Recent system applications of short-pulse ultra-wideband (UWB) technology. IEEE Trans Microw Theory Technol 52(9):2087–2104

Gentile C, Kik A (2006) An evaluation of ultra wideband technology for indoor ranging. In: Proceedings of IEEE Globecom, pp 1–6

Ghassemzadeh S, Greenstein L, Kavcic A, Sveinsson T, Tarokh V (2005) UWB indoor delay profile model for residential and commercial environments. IEEE Trans Veh Technol 54 (4):1235–1244

Golden S, Bateman S (2007) Sensor measurements for Wi-Fi location with emphasis on time-of-arrival ranging. IEEE Trans Mob Comput 6(10):1185–1198

Hedley M, Humphrey D, Ho P (2008) System and algorithms for accurate indoor tracking using low-cost hardware. In: Proceedings of IEEE/IOA position, location and navigation symposium, pp 633–640, May 2008

Hoene C, Willmann J (2008) Four-way TOA and software-based trilateration of IEEE 802.11 devices. In: Proceedings of IEEE personal, indoor and mobile radio communications (PIMRC), Cannes, pp 1–6, Sept 2008

Jemai J, Kurner T (2008) Broadband WLAN channel sounder for IEEE 802.11b. IEEE Trans Veh Technol 57(6):3381–3392

Jordan E, Balmain K (1968) Electromagnetic waves and radiating systems, 2nd edn. Prentice-Hall

Llombart M, Ciurana M, Barcelo-Arroyo F (2008) On the scalability of a novel WLAN positioning system based on time of arrival measurements. In: Proceedings of workshop on positioning, navigation and communication (WPNC), pp 15–21, Mar 2008

Roy S, Foerster J, Somayazulu V, Leeper D (2004) Ultrawideband radio design: the promise of high-speed, short-range wireless connectivity. Proc IEEE 92(2):295–311

Saberinia E, Tewfik A (2004) Enhanced localization in wireless personal area networks. In: Proceedings of IEEE Globecom, pp 2429–2934

Saberinia E, Tewfik A (2008) Ranging in multiband ultrawideband communication systems. IEEE Trans Veh Technol 57(4):2523–2530

Sathyan T, Humphrey D, Hedley M (2011) WASP: a system and algorithms for accurate radio localization using low-cost hardware. IEEE Trans Soc Man Cybern—Part C 41(2):211–2221

Sharp I, Yu K, Guo Y (2009a) GDOP analysis for positioning system design. IEEE Trans Veh Technol 58(7):3371–3382

Sharp I, Yu K, Guo Y (2009b) Peak and leading edge detection for time-of-arrival estimation in band-limited positioning systems. IET Commun 3(10):1616–1627

Urkowitz H (1983) Signal theory and random processes. Artech House

Yu K, Guo Y, Hedley M (2009a) TOA-based distributed localization with unknown internal delays and clock frequency offsets in wireless sensor networks. IET Signal Proc 3(3):106–118

Yu K, Sharp I, Guo YJ (2009b) Ground-based wireless positioning. Wiley

Chapter 13
TOA Error Modeling and Analysis

13.1 Introduction

The design and performance estimation of indoor positioning systems is challenging as the rich multipath indoor radio propagation environment makes accurate range measurements difficult. In particular, the scattering of the radio signals results in ranging measurement errors consisting of a biased component as well as the random error typical of line-of-sight (LOS) outdoor positioning systems. Such systems typically calculate positions either from the determination of range based on time-of-flight measurements, or more commonly by the application of pseudoranges estimated from time-of-arrival (TOA) measurements (Dardari et al. 2008; Ibraheem and Schoebel 2007; Chehri et al. 2007; Wang et al. 2009; Sharp et al. 2009). In both cases, the characteristics of the range errors directly affect the computed position accuracy. These concepts have been described in Chap. 4 associated with the design of TOA systems, but the underlying mechanism of the associated biases (as summarised by Eq. (4.22) in Chap. 4) has not been established. This chapter proposes a analytical model for explaining indoor NLOS ranging errors based on the shape of the leading edge of the received pulse, and the delay excess associated with walls and other objects in indoor environments. The model is based on the fundamentals of the causes of ranging errors indoors, and thus is expected to explain these errors over a wide range of operating parameters in positioning systems. To check this assertion, the chapter compares the model estimates with measured data.

13.2 Indoor Radio Propagation Overview

Indoor radio propagation is complex, with multiple scattering, diffraction and signal attenuation associated with walls and other obstacles along the path from the transmitter to the receiver. Because of this complexity, analysis requires considerable

I. Sharp and K. Yu, *Wireless Positioning: Principles and Practice*, Navigation: Science and Technology, https://doi.org/10.1007/978-981-10-8791-2_13

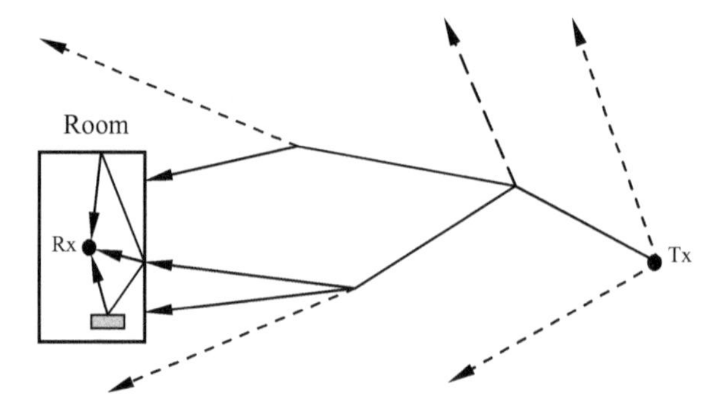

Fig. 13.1 Simplified diagram of indoor signal transmission, with multiple paths due to signal scattering. The dashed lines indicate paths that do not reach the receiver (Rx)

simplifications. One possible approach is to adopt the *Uniform Geometrical Theory of Diffraction* (McNamara et al. 1990), so that the radio signals can be represented as rays, incorporating both reflected and diffracted paths. With this interpretation of the scattered signal, the received signal can be expected to be composed of multiple individual components with varying amplitudes and delays depending on the path from the transmitter to the receiver.

Consider the transmission of a signal from a transmitter (Tx) to a receiver (Rx) in a (simplified) representation of an indoor environment, as shown in Fig. 13.1. Because of the scattering of the signal, there will be multiple paths between the transmitter and the receiver. However, only a small number (N_p) of these paths need to be considered in practice, as most of the paths will be so highly scattered and attenuated that their received signal strengths will be negligible.

Now consider a typical example of an ultra-wideband (UWB) measurement in an indoor environment, as shown in Fig. 13.2. These UWB results are similar to those referenced in Chap. 4, Sect. 4.2.3.4. The impulse response shows that the signal scattering results in a large number of received signals, with a general trend of decreasing amplitude as a function of the delay excess. With these high bandwidth measurements, individual scattered signals down to the measurement resolution (in this case 1 ns) can be observed. Clearly to accurately measure the TOA, the system should use the first detectable signal, namely using the leading edge of the impulse.

Based on the above discussion and Fig. 13.1, indoor ranging errors have two main components, namely the errors that accumulate along the path associated with scattering and passing through walls, and errors associated with the determination of the TOA in the receiver. Thus NLOS ranging errors can be expressed as the sum of two types of errors

$$\varepsilon_{NLOS} = \varepsilon_{TOA}(\tau_{pulse}) + \varepsilon_{path}(R, f_c, BW, N_{walls}) \tag{13.1}$$

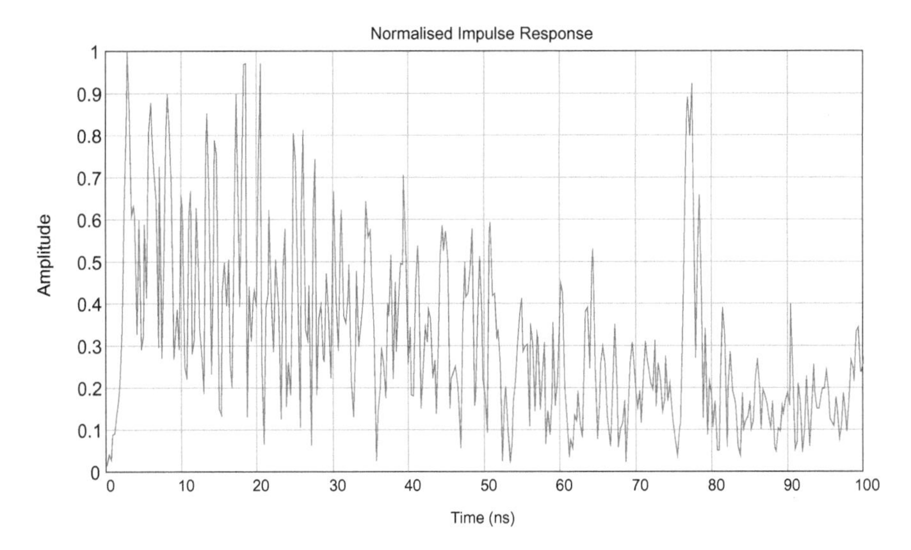

Fig. 13.2 An example of the impulse response of a 2–6 GHz UWB signal in a building (with time relative to the straight-line delay). The NLOS path length is 21 m. The nominal resolution of the measurements of individual signals is about 1 ns. Notice the curved leading edge

The first component of the error in TOA measurements will be a function of the TOA detection algorithm and the rise-time of the leading edge of the receiver baseband pulse (τ_{pulse}). The second component of the path error due to NLOS propagation will be related to the path length (R), the RF signal carrier frequency (f_c) and its bandwidth (BW), and the number of walls through which the path passes (N_{walls}). The propagation of the straight-line path through multiple walls can result in severe attenuation of this signal, such that it may be below the detection level of the TOA algorithm. In this case the first detectable signal path will be longer than the straight-line path, resulting in a bias error. Intuitively this bias error would be expected to increase as the path range increases and/or the number of walls along the path increases.

The two errors defined in (13.1) will be random variables, the statistics of which will be investigated later in subsequent sections. The TOA error can be analyzed based on the detection algorithm and the statistics of the shape of the leading edge of the pulse used for TOA measurement. Refer to Chap. 4 for details of some TOA detection algorithms. In a NLOS environment the received signal consists of many multipath components. At longer ranges the first signals arriving at the receiver through the shortest path are often too weak to be detected (see Fig. 13.3). However, as far as the TOA algorithm is concerned the delay of the first detectable signal can be treated as a pure delay excess (in addition to the Euclidean propagation path delay). Further, observe that relative delay excesses of greater than the rise-time of the first detectable pulse cannot affect the measurement of the TOA

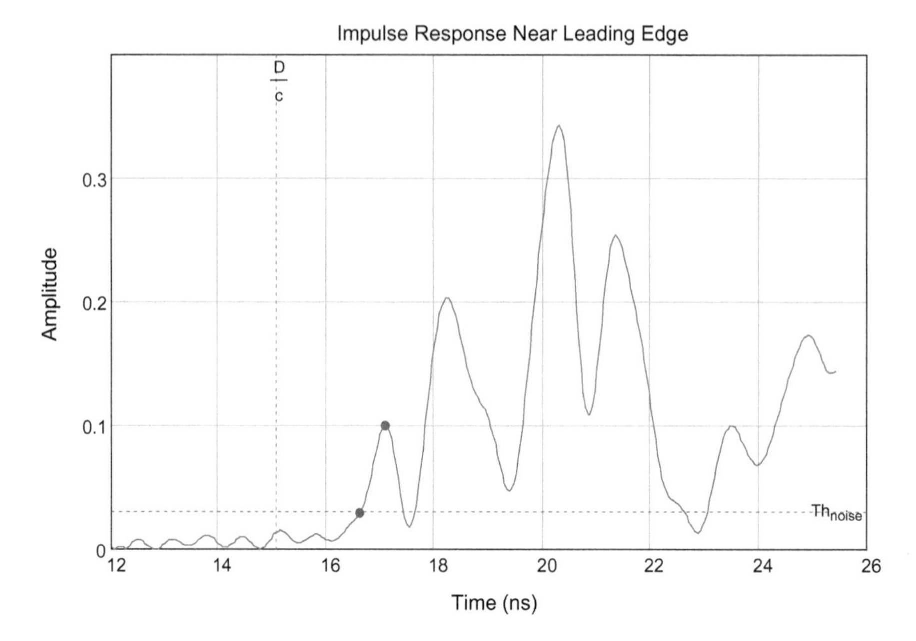

Fig. 13.3 Example of a TOA measurement using a Threshold algorithm. The signal is UWB 6–9 GHz, with a range of $D = 4.5$ m (15 ns) through a wall. Note the concave curved leading edge. The signal *Epoch* is measured where the signal exceeds the noise threshold level, and is shown by a dot. The local peak on the leading edge is also shown by another dot. Note that this local peak has a much smaller amplitude than the signal peak. Also observe the additional delay before the first detectable signal is not an error attributable to the TOA detection algorithm

when using a leading edge detection algorithm. Thus when evaluating the statistical performance of the TOA algorithm, one only needs to consider the shape of the leading edge of the pulse in the range $0 \leq \varepsilon_{TOA} \leq \tau_{pulse}$. The details of this topic are considered in Sect. 13.6.

The second random component of the range error is related to the characteristics of the multipath scattering along the path. This random error can be considered as having a mean (bias) error and a random component. Details of this topic are considered in Sect. 13.6.

13.3 Measurements

The investigations into the Time-of-Flight (TOF) performance in an indoor environment in this section are based on a set on UWB spectral measurements. The data set consists of a total of 12 NLOS paths (as summarized in Fig. 13.4 and Table 13.1), with ranges from 6 to 21 m. The spectral data were measured using a Network Analyzer (at two locations), with two UWB antennas (height 1.5 m)

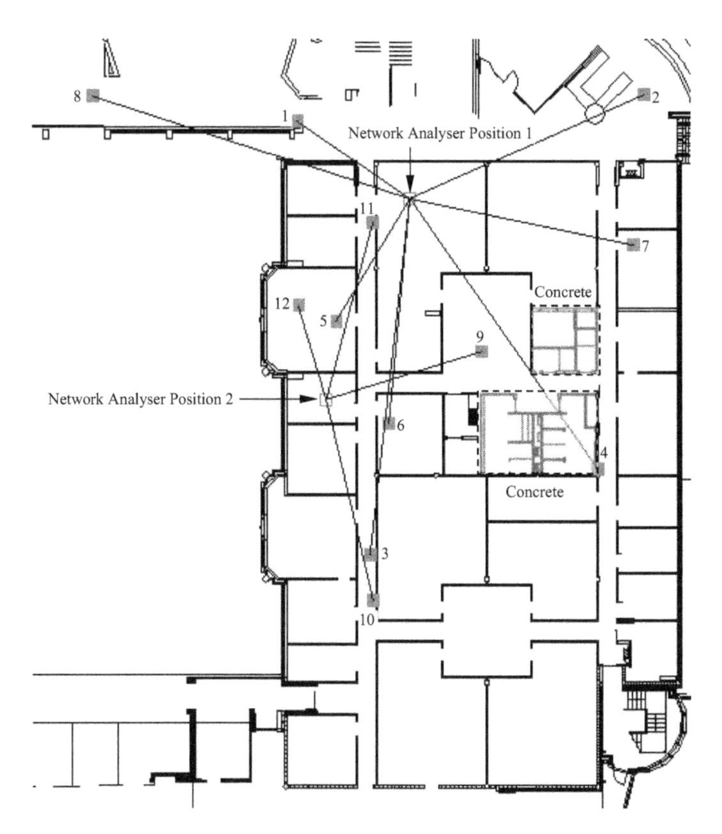

Fig. 13.4 Indoor measurements showing the map of the building and the 12 paths. The Network Analyzer (2 locations) is shown as a square, and the remote antenna location (12 locations) is shown as a square dot. The dashed boxes marked "Concrete" are of steel-reinforced concrete construction which is more opaque to radiowaves than other internal parts of the building. The mean distance between walls is 4.3 m

located at either end of each path, and covering the bandwidth of 2–6 GHz. The number of walls along a path is determined directly from the path lines shown in Fig. 13.4.

For the measurements summarized in Fig. 13.4, the Network Analyzer was located at two different locations (squares in the figure), and the remote antenna (at the end of the long cable) was located at 12 positions (square dots). The Network Analyzer measures the 4 GHz bandwidth spectrum of the transfer function at 1600 points (or 2.5 MHz separation). This logged spectrum is processed to generated the equivalent time-domain impulse response by an inverse Fourier transform. This impulse response is further processed by a TOA algorithm to provide an estimate of the propagation range.

Table 13.1 Summary of the characteristics of the paths shown in Fig. 13.4

ID	Range (m)	Walls	Error (m)	Path description
1	8.18	2	0.21	Laboratory, corridor, vestibule
2	15.33	2	0.14	Laboratory, vestibule
3	20.99	4	0.27	Laboratory, corridor, office, corridor
4	19.48	4	0.37	Laboratory, atrium, concrete construction block
5	8.47	2	0.27	Laboratory, corridor, office
6	13.22	2	0.02	Laboratory, corridor, office (no objects along path)
7	13.77	2	0.21	Laboratory, office, corridor, office
8	20.00	4	0.38	Laboratory, corridor, office, external brick wall, library
9	9.80	2	0.03	Office, corridor, office (corner), atrium (no objects along path)
10	12.10	3	0.36	Office, office, office, corridor
11	10.72	2	0.07	Office, office, corridor
12	5.73	1	0.08	Office, office

13.4 Effects of Walls on Range Errors

One of the key characteristics of indoor NLOS range-errors is that errors tend to increase with increasing range. The delay excesses associated with the path propagation clearly will be related to the architecture of each building and its contents, so that any analysis must utilize such information. Measured range error data (Yu et al. 2009; Bellusci et al. 2008; Alavi and Pahlavan 2003, 2006; Alsindi et al. 2009; Prieto et al. 2009; Sharp and Yu 2013) show that the bias errors tend to increase with range in a linear fashion, namely

$$\varepsilon_{bias} = \varepsilon_0 + \lambda R \tag{13.2}$$

where the constant term ε_0 is associated with the effects of multipath propagation on the *shape* of the received pulse leading edge and the TOA algorithm details, and the λ parameter is associated with the delay excesses along the propagation path. Using the measurement technique described in Sect. 13.3 the TOF can be estimated, and hence range errors can be determined from the measured data. For example, Fig. 13.5 shows range error data for two different bandwidths (or pulse rise-times), namely 0.5 ns (raw UWB data) and 25 ns (appropriate for the bandwidth available in the 2.4 GHz ISM band). It was shown in Sharp and Yu (2014) that the number of walls along a propagation path can be used as a *proxy* for a more detailed analysis of signal scattering. For example, in Fig. 13.5 the range error data are plotted as a function of the number of walls along the path for two different pulse rise-times. Although there is considerable data scatter, in both cases an approximate linear relationship between the range errors and the number of walls is evident.

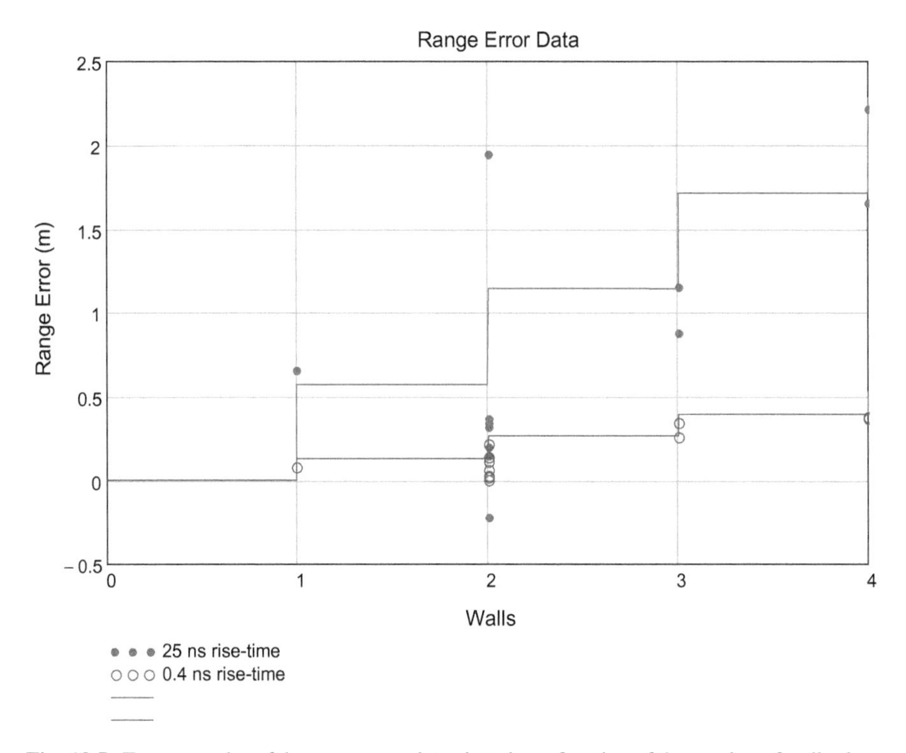

Fig. 13.5 Two examples of the range error data plotted as a function of the number of walls along the propagation path. Also shown is the associate least-squares fit estimate of the delay excess through walls, plotted as a stepped function at each wall. The delays are 10 cm per wall for 0.5 ns rise-times and 56 cm per wall for 25 ns rise-times. The data scatter is greater for the narrowband signal

From the data presented in Fig. 13.5, there is clearly a correlation between range errors and the number of walls along the path. However, the nature of this relationship is not clear, as the delay excess is also a function of the signal bandwidth, which is not related to signal propagation. Thus the analysis is split into two components, the path propagation characteristics, and the effects of signal bandwidth.

13.4.1 *Effect of Wall Material on Delay Excess*

The most obvious direct effect walls have on the measured delay excesses is the increase in the propagation delay associated with the material of walls. Most internal walls in buildings are made of dielectric materials such as wood, fiberboard, concrete, brick or glass, which can be penetrated by radiowaves, albeit with a loss in the signal strength and a delay excess over free-space propagation. While both

Table 13.2 Delay excesses of various materials through walls of various thicknesses

Material	Relative dielectric constant	Typical thickness (cm)	Delay excess (cm)
Wood	2	2	0.8
Fiberboard	1.2	2	0.2
Brick	3.7–4.5	10	9–11
Concrete blocks	2.3	5	2.6
Glass	5	1	1.2
Paper (books)	2.7	15	10

effects are relevant to indoor positioning systems, range errors are associated with the delay excess, while signal attenuation through walls will limit the range of measurement due to a reduction in the receiver signal-to-noise ratio. Note however that as metallic walls are impenetrable to radiowaves, such walls are not considered, as the large associated path excesses (Gentile and Kik 2006) around these walls make such environments unsuitable for radiolocation based on TOF measurements.

First consider the normal incident penetration of a wall whose material thickness is W (often the wall will include hollows which are not included in this measure) with a relative dielectric constant ε_r, then the associated delay excess (expressed as an incremental distance in free-space) is given by

$$\delta\Delta_{wall} = W(\sqrt{\varepsilon_r} - 1) \tag{13.3}$$

The relative dielectric constant of typical wall materials, and the associated normal-incident delay excesses are given in Table 13.2. Table 13.2 also shows the typical effective thickness of walls of different types, and the associated delay excess based on (13.3). From Table 13.2 and Fig. 13.5 it can be observed that the delay excesses (with the exception of the delay through brick/concrete blocks and through books in a bookshelf) are much smaller than most measured delay excesses even with UWB signals, so that the large measured delay excesses cannot be explained by the delay excess associated with the material of walls.

13.4.2 Statistical Analysis of Wall Delay Excess

If the RF signal incident angle is not normal to the wall the delay excesses will be greater than that given by (13.3). The propagation excess associated with walls is complex requiring detail EM analysis. However, some basic properties can be established without such analysis. The propagation excess at an incident angle θ through the wall results in the effective wall thickness (W_{eff}) given by

$$W_{eff} = W/\cos\phi = W\bigg/\sqrt{1 - \frac{\sin^2\theta}{\varepsilon_r}} \quad (0 \le \theta \le \pi/2) \tag{13.4}$$

where ϕ is the refracted angle given by Snell's law.

In general the direction of propagation through walls is not known, but can be assumed to have the uniform random distribution function $U(\pi/2)$. Using the transform expression $U(\theta)\,d\theta = \text{PDF}(W)\,dW$ and (13.4), the normalized effective wall thickness probability density function (PDF) can be shown to be

$$\text{PDF}(w, \varepsilon_r) = \frac{2\sqrt{\varepsilon_r}}{\pi} \frac{1}{w\sqrt{w^2(w^2-1) - \varepsilon_r(w^2-1)^2}} \tag{13.5}$$

where $w = W_{eff}/W$ and $1 \le w \le w_{\max} = \sqrt{\varepsilon_r/(\varepsilon_r - 1)}$.

From (13.5) it can be observed that the PDF is infinite at either end of the distribution, that is at normal incidence and grazing incidence. Thus the effective width of the wall can be considered a random variable whose value will be somewhat greater than the physical wall width. Using (13.5) the mean and the standard deviation of the effective width can be calculated. For example, the mean is given by

$$\begin{aligned} \mu &= W \int_1^{w_{\max}} w\,\text{PDF}(w, \varepsilon_r)\,dw \\ &= \frac{2W\sqrt{\varepsilon_r}}{\pi} \int_1^{\sqrt{(\varepsilon_r-1)/\varepsilon_r}} \frac{1}{\sqrt{w^2(w^2-1) - \varepsilon_r(w^2-1)^2}}\,dw \end{aligned} \tag{13.6}$$

The integral in (13.6) is in the form of an elliptic integral of the first kind, and it can be shown that the analytical result of the integral is

$$\mu = \frac{2W}{\pi} K\left[\frac{1}{\varepsilon_r}\right] \tag{13.7}$$

where $K(m)$ is the complete elliptic integral of the first kind. A similar analysis shows that the standard deviation of the effective wall width is given by

$$\sigma = W\sqrt{\sqrt{\frac{\varepsilon_r}{\varepsilon_r - 1}} - \frac{4}{\pi^2}\left(K\left[\frac{1}{\varepsilon_r}\right]\right)^2} \tag{13.8}$$

An example of the PDF for the case $\varepsilon_r = 2$ is shown in Fig. 13.6. Also shown is the associated mean (μ) as given by (13.7). Observe that the PDF is infinite at both normal incident and grazing angles. However, even in the latter case the effective width due to the refraction of the incident ray is not significantly greater than the wall width. Further, as the refraction increases with the refractive index, higher values of ε_r result in an even smaller spread of effective wall widths. As a result, the delay through walls is not significantly greater than that associated with the normal incident ray, regardless of the angle of incidence. Note also that the above effects are only a function of the radio frequency (not the bandwidth), although the variations with frequency in the refraction index at microwave frequencies is minimal.

13.5 Effect of Path Propagation on Delay Excess

The analysis in Sect. 13.4 shows that the delay excess through walls results in a slope parameter (λ in (13.2)) much less than the measured range error slope parameter. For example, from Table 13.2 with wooden walls with the mean wall separation for the building shown in Fig. 13.4 of 5 m, the slope is about $\lambda_{wall} \approx 0.002$, whereas the measured value (see Fig. 13.5) is about 0.02 for UWB signals, based on 5 m between walls. Thus the measured data show that the main contribution to the bias errors increasing with range is not caused directly by the propagation delay through walls.

13.5.1 Preliminary Discussion on Path Delay Excess

The path-related delay excess is clearly related in some manner to the scattering of the signal in the indoor environment. This environment is dominated by the architecture of the building, in particular the walls which separate the building space into rooms. The objects within rooms will vary according to the type of building and room, but for the office environment (Fig. 13.4) where the data were measured, typical objects included semi-opaque objects such as desks, chairs/people, bookshelves/books, and opaque objects such as computers and computer-monitors, metal cabinets, and metal-backed whiteboards. These objects have horizontal widths of around 0.5–1.5 m, and are often located next to the walls. Thus walls are suspected to be indirectly related to delay excesses associated with the diffraction around the objects if they (approximately) lie along the straightline path between the transmitter and the receiver. However, some "rooms" normally have no objects at all, such as corridors and large open areas such as atriums (path 9) and vestibules (paths 1, 2) in Table 13.1. If the path is associated with these types of rooms, then the delay excess would be expected to be small, as observed in the data. Even if there are objects in the room, there is a possibility that the direct path will not pass through any (significant) object, so again the delay excess would be small. Thus if a large number of measurements are made, a statistical distribution of range errors can be expected.

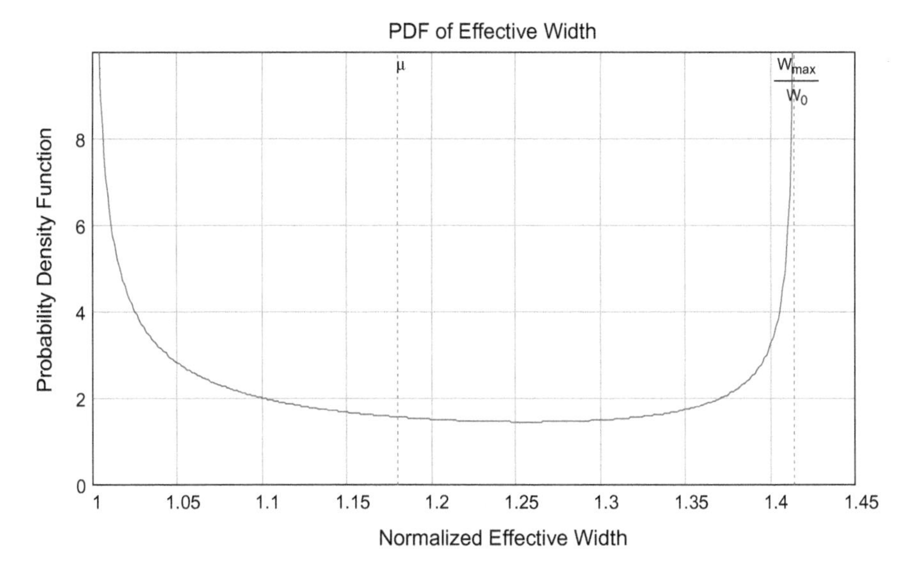

Fig. 13.6 Example of the normalized PDF with $\varepsilon_r = 2$. Also shown is the mean $\mu = 1.18$ normalized effective width. The corresponding standard deviation is $\sigma = 0.145$

The above discussion strongly suggests that the path delay excess is related to the propagation characteristics within rooms. With reference to Fig. 13.7, three possible causes of the observed measured delay excess characteristics are considered.

1. The first type of propagation considered is scattering from the walls and possibly large objects within the room. With typical room sizes of about 5 m, the scattering delays would be expected to be of the same order, as illustrated in

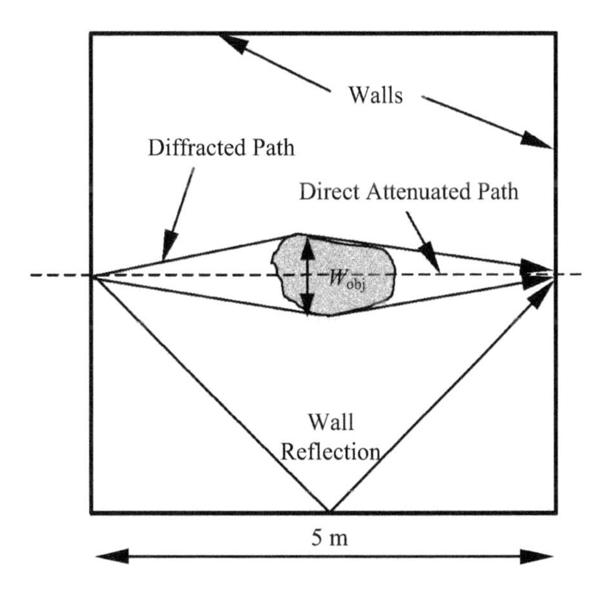

Fig. 13.7 Geometry of signal reflections and diffraction around an object within a room. Also shown is a typical wall reflection

Fig. 13.7. Scattering delays from large objects may be somewhat smaller but still of the order of a few meters. However, the measured delay excesses (Table 13.1) are of the order of 0.05–0.4 m. Thus this type of scattering, while present, is not responsible for the measured delay excesses, which are based on the first detectible signal to arrive at the receiver.

2. The second type of propagation effect considered is the delay excess associated with propagation through large obstacles along the propagation path. Many of the typical objects listed above are opaque to radiowaves, so this mechanism does not apply to these objects. Other objects, such as books in bookcases would have the appropriate delay excesses (see Table 13.2), but such cases would typically be rather rare. Further, these types of semi-opaque obstacles would also incur considerable signal attenuation, particularly when compared with diffraction losses associated with the signal passing around the obstacle where the diffraction angles are rather shallow. Thus while the delays associated with passing through large obstacles may occur, these would often result in such low signal strengths at the receiver as to be mostly undetectable.
3. The third possible source of the delay excesses is associated with the diffraction around objects. If for simplicity the analysis is restricted to two dimensions, there will be two main diffraction paths around each obstacle. (A three-dimensional analysis does not alter the general thrust of the argument). The diffracted signal losses will be generally rather modest (much less than passing through the object) as the amount of diffraction of the signal around an object (see Fig. 13.7) is typically rather modest. In such cases the Geometrical Theory of Diffraction (McNamara et al. 1990) can be used to estimate the delay excesses associated with diffraction around objects. Each obstacle will result in two diffracted signals which then will be incident on the next object, and so on, so the number of paths will grow exponentially, namely 2, 4, 8, …, and thus the scattered signal at the receiver will be very complex. However, of these many received signal paths, only one (the shortest delay) is of interest in determining the measured delay excess.

Based on the above discussion, the third type of phenomenon is considered the most likely cause of the measured delay excesses, and thus will be analyzed in more detail in the next subsections. Note however, even though there are many other types of signal scattering, they are not relevant to determining the delay excess associated with UWB signals.

13.5.2 Basic Assumptions and Constraints for Model

This subsection provides a description of the assumptions and constraints required to develop a suitable model for the delay excesses in a NLOS indoor environment. Using these assumptions, an analytical mathematical model will be developed which hopefully is in agreement with the empirical model (13.3). The development of the model is based on the following conditions and assumptions:

1. The diffracted path can be adequately described by diffracted rays, based on the Geometrical Theory of Diffraction. Annex A gives an example of measured delays around an object; the results compare favorably with the expectations from the ray theory. Implicitly this theory assumes that the signal wavelength is much smaller than the size of objects, which will be true for frequencies in the 2.4 and 5.8 GHz ISM bands, and all UWB signals which are defined as being between 3.1 and 10.6 GHz.
2. There are two diffracted rays around each obstacle. The delay excesses will depend on the position of the obstacle relative to the incident ray. This relative position can be considered a random variable, but over a large number of such cases there is overall an equal probability of either the right or left diffracted path being on the path of minimum overall path delay.
3. The effective diffraction width of an obstacle associated with a path is in the range $\left[0, W_{obj}\right]$, where W_{obj} is defined in Fig. 13.7. This effective diffraction width $\left(W_{dif}\right)$ is measured relative to the nominal straightline path from the object to the receiver. If one of the diffracted paths has W_{dif}, then the other diffracted path effective width is $W_{obj} - W_{dif}$. The shortest delay excess at an object will be associated with the smallest of these two effective widths. However, the ray associated shortest delay excess at an object may not be associated with the shortest overall path from the transmitter to the receiver.
4. In the analysis, the various angles of interest are all assumed to be small, so that small angle approximations can be used, such as $\theta \approx \sin\theta \approx \tan\theta$. This assumption is based on the fact that the size of objects is much smaller than the size of rooms.
5. The distance between obstacles (d) is assumed to be much greater than the diffraction width. As obstacles are often located along walls, the separation is often of the order of the wall separation distance; from Fig. 13.4 this is about 5 m on average. Thus the angular size of the object is defined as $\theta_{obj} = W_{dif}/d \ll 1$.
6. Although the separations between obstacles will vary, to simplify the analysis it will be assumed that the separation is a constant distance $(\bar{d} = D)$, namely the mean distance between obstacles along the propagation path. Although this assumption is used in the development of the analytical model, it will be shown that the final result only involves the angular sizes of objects, so this constant distance assumption is not particularly relevant to the application of the final analytical model.
7. In the development of the model, all the measurements (angles, distances) are relative to the straightline path between the transmitter and the receiver. By convention but without constraining generalization, the path direction is defined as the x axis, and the object width is measured in the y direction. By this definition, the

transmitter and the receiver lie on the x-axis, with a transmitter-receiver separation (range) of R.

8. The total propagation path length can be determined from the summation of the individual path segment vectors length. From constraint (7) the sum of the y-component of these vectors must be zero, so $\sum_{n=1}^{N} \theta_n = 0$ for the path vector angles, where there are N obstacles along the path. Further, from assumption (2), the probability density function of these angles is symmetrical about the origin, and thus $E[\theta] = 0$. From assumption (3), the PDF will have the angular spread $[-\theta_{\max}, \theta_{\max}]$, where $\theta_{\max}$ is the *maximum* object angular width of an obstacle along the path.
9. The transmitter and receiver are in rooms at either end of the path. If the average object separation distance is D, it is assumed that the transmitter/receiver start/end points are a distance $D/2$ from the adjacent object on the propagation path. In general, there is no information about the location of the transmitters and receivers, but this assumption appears reasonable given devices will be typically located on people in the middle of the room. The exact location of the transmitters and receivers has very little influence on the overall delay excesses averaged over many paths.

13.5.3 Development of the Analytical Model

Based on the assumptions and constraints defined in Sect. 13.5.2, a model of the delay excess can be developed. The signal at the receiver will be a very complex mix of many components of the scattered signals which makes analytical analysis very difficult. In such cases a simulation of the scattering process could be performed, based on the assumptions and constraints defined previously. Typically, this approach would use ray-tracing techniques for the many scattered paths, and then determine the shortest delay to estimate the TOA measurement error. To obtain statistical information, a Monte-Carlo simulation could be performed. Note however that the ultimate goal of the analysis is to determine an analytical formula for the λ parameter in (13.2), so that this considerable effort and detail is ultimately lost in the final mean value of λ. As the λ parameter is based on averaging the range errors over various ranges, the model will be based on average range error performance, and not the performance of each simulated path. Thus the complexity of the propagation environment is not explicitly expressed in this one number, suggesting that a simpler statistical-based approach is more appropriate. Further, such simulation methods do not lead to an analytical result, so the information from a simulation is essentially the same (but less reliable) as that obtained by the analysis of the measured range error scatter. Thus for a tractable analytical analysis, an alternative approach is required. Such an approach is the basis for the model described below.

First, based on the above assumptions, observe that range excesses are solely related to the statistical properties of the path segment vector angles, provided the path segments are assumed of equal length (assumption (6)). Further, individual vector angles are not required, rather the statistical distribution of the angles is what determines the performance. Note also that the order of the segment vectors is irrelevant, as only the sum of the magnitudes of the vectors is used in calculating the range errors. Thus the specific geometry of the scattered path is not important, so that the vectors can be rearranged in any order if that simplifies the analysis. Indeed for the analysis, any analytical model which obeys the assumptions and constraints can be chosen, so that it is appropriate to choose a simple model to minimize the analytical difficulties. For example, the vectors could be sorted by the segment vector angle, from minimum to maximum. This rearrangement would result in a largely smooth curved abstracted path, rather than the complex zig-zag physical path geometry. Further, in the limit with a large number of nodes (as shown in Fig. 13.8) on the abstracted path, but with the statistical properties defined in assumption (8) above, the curved path will approximate a segment of a circle. The reason why the path is a circular segment is as follows. From assumption (8), the diffraction angle required depends on where the incident ray intercepts the obstacle, and this will be random with a uniform distribution. Thus after the vectors are sorted, the progressive angle of the curved path will increase uniformly (as least in a statistical sense) as a function of the position along the sorted-vector path. As equally-spaced nodes along a circular segment also have the same property (albeit deterministically rather than statistically) a segment of a circle is a good approximation to the sorted-segment path. This approximation improves as the number of path segments averaged increases, and thus in the limit the average characteristics of the path (such as the range excess) will be the same as that of the circular-segment model. Thus the complex zig-zag path in the physical environment has been abstracted to a mathematically simple path, greatly simplifying the analysis.

Thus provided the model used for the analytical calculations has the correct statistical characteristics (as specified in Sect. 13.5.2), the details of the signal scattering can be ignored. Note that although the model and the associated analysis is not rigorous, it does encapsulate the essential features necessary to estimate the delay excesses associated with diffraction around obstacles along the propagation path. With this as the guiding principle, the proposed model as shown in Fig. 13.8 is as follows.

1. With reference to Fig. 13.8 the N diffraction nodes lie equally spaced (by distance D) on a segment of a circle (of radius a) which is symmetrical about the range mid-point. The transmitter and receiver nodes lie on the x-axis, with a separation distance R. The maximum angle subtended by the transmitter/

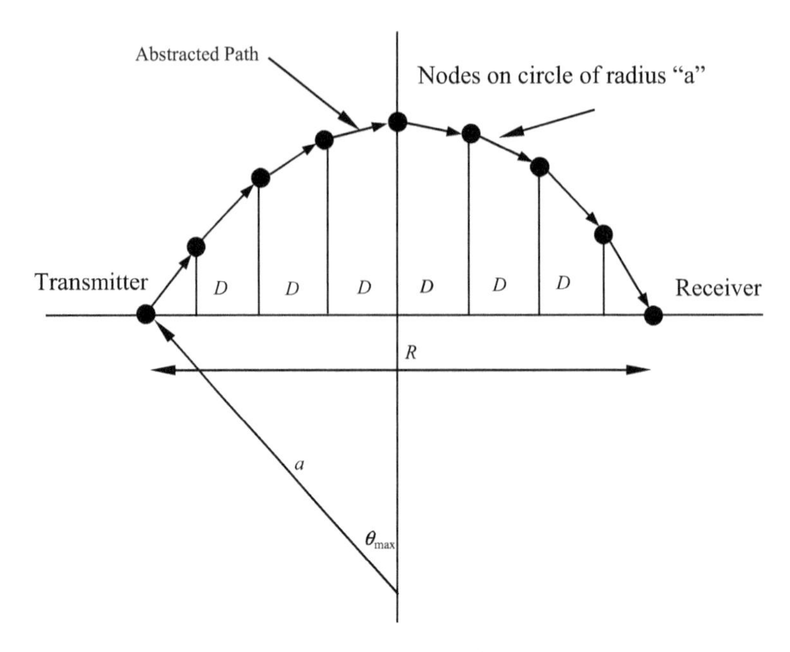

Fig. 13.8 Geometry of the abstracted path used to calculate the range excess

receiver from the centre of the circle is $2\theta_{max}$. The propagation vectors join these nodes. The angular separations are a constant, which is the deterministic equivalent of a uniform statistical distribution. As the number of diffracting nodes increases, the delay excess is simply the difference between the circular segment path length and the straightline path R. This range error can be interpreted as the expected (or mean) range error when averaged over many propagation paths.

2. Observe that the requirements of assumption (8) in Sect. 13.5.2 are valid for this model. Because the nodes are equally spaced and the geometry of the circle, the sum of the vector angles is zero, the mean diffraction angle is zero, and the distribution of the angle is symmetrical about zero (x-axis).

Based on the above-defined geometry shown in Fig. 13.8, the following equation applies

$$R/2 \approx a \sin \theta_{max} \approx a\theta_{max} \tag{13.9}$$

The range error using the circularly arranged segments as the scattered path range is

$$\varepsilon = 2a(\theta_{max} - \sin \theta_{max}) \approx \frac{a}{3}\theta_{max}^3 = \frac{1}{3}\left(\frac{R}{2\theta_{max}}\right)\theta_{max}^3 = \left(\frac{\theta_{max}^2}{6}\right)R \tag{13.10}$$

Comparing (13.10) with (13.2) shows that estimate for the range errors slope parameter is

$$\hat{\lambda} = \frac{\theta_{max}^2}{6} \tag{13.11}$$

Thus this simplified analysis shows that the mean range excess (bias) error is a linear function of range, as observed in the measured data. Further, the analysis provides and estimate of the linear slope parameter λ, which is related to the *maximum* angular size (as measured from the previous obstacle along the path) of an obstacle along the minimum range diffracted path from the transmitter to the receiver. As the angular size of an obstacle depends on the "diameter" of the obstacle and the distance from the adjacent obstacle (or the transmitter for the first obstacle), the linear slope parameter can be estimated from the dimensions of largest objects in the rooms/building, and the separation between the objects in the room. Further, as the objects in rooms are often located along the walls, the slope parameter will be directly related to the mean separation of walls in the building. In this case the mean wall separation distance is a suitable *proxy* for calculating the delay excess, as the actual source of most of the delay excess is not directly related to the properties of walls.

Consider a numerical example based on the building in Fig. 13.4. From Fig. 13.5 for UWB signals (which are not much affected by the errors induced by the TOA algorithm), the slope parameter is $\lambda \approx 0.018$, and the average wall separation for this building is about 4.3 m.[1] The objects in rooms will have a typical separation of less than the wall separation, say 3 m. The size (width) of significant obstacles is in the range 0.5–1 m. Using these data with *maximum* size of 1 m gives $\theta_{max} \approx 1/3 = 0.33$ radians. Using this value in (13.11) the estimated range error slope parameter is $\widehat{\lambda} = 0.018$, which is in agreement with the measured value. While this exact agreement is a coincidence, it is clear that any reasonable values of object separation and object sizes will give values of λ that are in reasonable agreement with measurements. Thus despite the simplicity of (13.11) used to estimate the effects of the complexities of indoor signal scattering, the calculated values using (13.11) gives reasonable estimates of the range bias error slope parameter, at least for the UWB case. Further, this good agreement lends support to the reason why the range bias errors increasing linearly with range, namely it is due the delay excess associated with diffraction around obstacles along the propagation path. Note also that the (13.11) estimate is the minimum possible delay excess, even with very large bandwidths, so that the delay excess cannot be reduced to near zero by increasing the signal bandwidth.

Although (13.11) was not derived using statistical methods, statistical variations in the slope parameter (and hence the range errors associated with the path range

[1]This average is obtained by counting the number is walls along the straight-line paths between pairs of base stations located in the building.

excesses) could be estimated using (13.11) and estimates of the statistical variation angular sizes of obstacles along the propagation path. However, such analysis is not pursued here. The effect on the delay excess with signals with a bandwidth less than UWB is considered in Sect. 13.7.

13.6 NLOS Leading Edge Analysis

From Eq. (13.1) the delay excess associated with a NLOS path has two components. The effect of signal scattering along the path was considered in Sect. 13.5; this section provides an analysis on the errors associated with the TOA detection process of a NLOS received pulse. In particular, the following provides a theoretical analysis of the shape of the leading edge of the received pulse in a NLOS environment. The aim of the analysis is to obtain a more general insight into the characteristics of the leading edge of the NLOS multipath pulse shape. As illustrated in Figs. 13.1, 13.2 and 13.3 there will be multiple scattered signals, which can be characterised by their amplitude, phase and delay excess relative to the straightline path. The following subsections analyse the effect on the shape of the leading edge using various theoretical assumptions about the nature of the multipath signals.

13.6.1 Equi-amplitude with Random Phase Case

The initial (over simplified) case considered is where all the multipath signals are of equal amplitude but with random phase; this simplifying assumption is used in the following analysis, but variation in the amplitude is considered later in Sect. 13.6.2.

In a severe multipath environment there will be a large number of interference signals, the amplitude of which tends to decrease with increasing delay excess (see Fig. 13.2). For theoretical analysis the indoor NLOS multipath signal amplitude distribution can be approximated as an exponential function of delay excess (Ghassemzadeh et al. 2005), with the exponential parameter defining the rate of decay. If the pulse rise-time is much less than this decaying period, then to a first order approximation they can be considered of equal amplitude. For example, the WASP system described in Chapter 4, Sect. 4.2.5.1 has a pulse rise-time of 13.5 ns, which is small compared with the delay spread shown in Fig. 13.2.

Now consider the signal phase characteristics. As the path lengths are very much greater than the wavelength, even small changes in the path length will result in a rapid change in phase. Thus it is reasonable to assume that the interference signals will have random phase, typically assumed to have a uniform distribution over $[0, 2\pi]$.

As only interference signals with delays up to about the pulse rise-time (τ_{pulse}) can affect the performance of a leading edge algorithm, without loss of generality only interference signals with delays up to τ_{pulse} are considered. Further, these signals are assumed to be uniformly distributed in time throughout this period with a separation of $\delta \times \tau_{pulse}$. Thus in the following analysis it is convenient to normalize the time by τ_{pulse}, so the time separation of individual signal components is δ in normalized time. Also it is convenient to normalize the pulse amplitude to unity. While the delay excesses are assumed to be equally spaced, simulations showed that other random distributions (such as statistically uniform distribution of delays) have little effect on the characteristics of the leading edge of the pulse.

First consider that all the signals are in phase and of unit amplitude. As the nominal pulse shape is close to triangular,[2] for this special case the shape of the leading edge of the pulse can be computed by summing up the delayed equi-amplitude signals, so that starting at the leading edge the cumulative pulse amplitude at the normalized times $0, \delta, 2\delta, 3\delta, \ldots$ will be $0, \delta, 2\delta + \delta, 3\delta + 2\delta + \delta, \ldots$ and so on.

Thus if the nth signal has a random phase ϕ_n, the in-phase component of the multipath pulse at pulse time $\tau = N\delta$ is given by

$$C_N = \delta \sum_{n=1}^{N} n \cos \phi_{N-n} \tag{13.12}$$

A similar expression applies for the quadrature component S_N, with sine replacing the cosine. Thus the magnitude of the multipath pulse is given by

$$M_N = \delta \sqrt{\left[\sum_{n=1}^{N} n \cos \phi_{N-n}\right]^2 + \left[\sum_{n=1}^{N} n \sin \phi_{N-n}\right]^2} \tag{13.13}$$

The calculation of the statistics of the magnitude of pulse at sample N using (13.13) is difficult due to the nonlinearity of the square-root function, so an alternative approximate method is used. Because the C_N and S_N components are the summation of a number of random variables, each component will have approximately a Gaussian distribution due to the central limit theorem (CLT). Further, the expected value of the summation will be zero, as the expected values of the cosines and sines are zero. Under such circumstances the magnitude of the multipath pulse at the Nth sample will exhibit approximately Rayleigh statistics, provided N is sufficiently large; this simplification is later confirmed through simulations. At this stage it is also worth noting that because of the consequences of the CLT this result also applies to any statistical (or non-statistical) amplitude distribution, due to the randomizing effect of the random phase.

Using this Rayleigh distribution approximation the mean (μ) and the standard deviation (σ) of the pulse amplitude are sought. These are given by definition

---

[2] Based on the autocorrelation function of a direct-sequence spread-spectrum signal.

$$\begin{aligned}\mu &= E[M_N] \\ \sigma^2 &= E\left[M_N^2\right] - E[M_N]^2\end{aligned} \tag{13.14}$$

By expanding the summations in (13.13), and noting that the expectation of the cross-product terms will be zero due to their random phases, the expected value of the magnitude-squared is given by

$$E\left[M_N^2\right] = \delta^2 \sum_{n-1}^{N} n^2 = \delta^2 S(N) \approx \delta^2 \frac{N^3}{3} \quad (N \gg 1) \tag{13.15}$$

where $S(N) = N(N+1)(2N+1)/6$.

While the expected value of the magnitude cannot be easily calculated, by assuming Rayleigh statistics the variance to the mean-squared ratio has a known value of $4/\pi - 1$. Thus

$$\frac{\sigma^2}{\mu^2} = \frac{4}{\pi} - 1 = \frac{E\left[M_N^2\right]}{E[M_N]^2} - 1 \tag{13.16}$$

Combining Eqs. (13.14) and (13.16) gives the mean of the magnitude as

$$\mu \approx \sqrt{\frac{\pi}{4} E[M_N^2]} = \delta \frac{\sqrt{\pi}}{2} \sqrt{S(N)} \approx \frac{\delta}{2} \sqrt{\frac{\pi}{3}} \sqrt{N^3} \ (N \gg 1) \tag{13.17}$$

If there are N samples in the leading edge, then $\delta = 1/N$ and $\tau_n = n\delta = n/N$. For a unit amplitude pulse the Nth sample is used to normalize the nth sample of the multipath pulse, so the expected value of the magnitude of the normalized pulse at the nth sample is given approximately by

$$\mathrm{E}[M(\tau)] = \mu(\tau) \approx \left[\frac{n}{N}\right]^{3/2} = \tau^{3/2} \tag{13.18}$$

Finally again assuming Rayleigh statistics, the ratio of the standard deviation to the mean can be used to estimate the shape of the standard deviation of the leading edge, namely from (13.16) and (13.18)

$$\sigma(\tau) \approx \sqrt{\frac{4}{\pi} - 1}\, \tau^{3/2} \tag{13.19}$$

Figure 13.9 shows the comparison between a simulation and the above analytical results. The simulation uses 20 multipath sources, and the mean and standard deviation of 2000 random leading edge pulses plotted. Also plotted are the analytical estimate (13.18) of the mean shape of the leading edge, and its associated standard deviation (13.19). As can be observed, there is excellent agreement

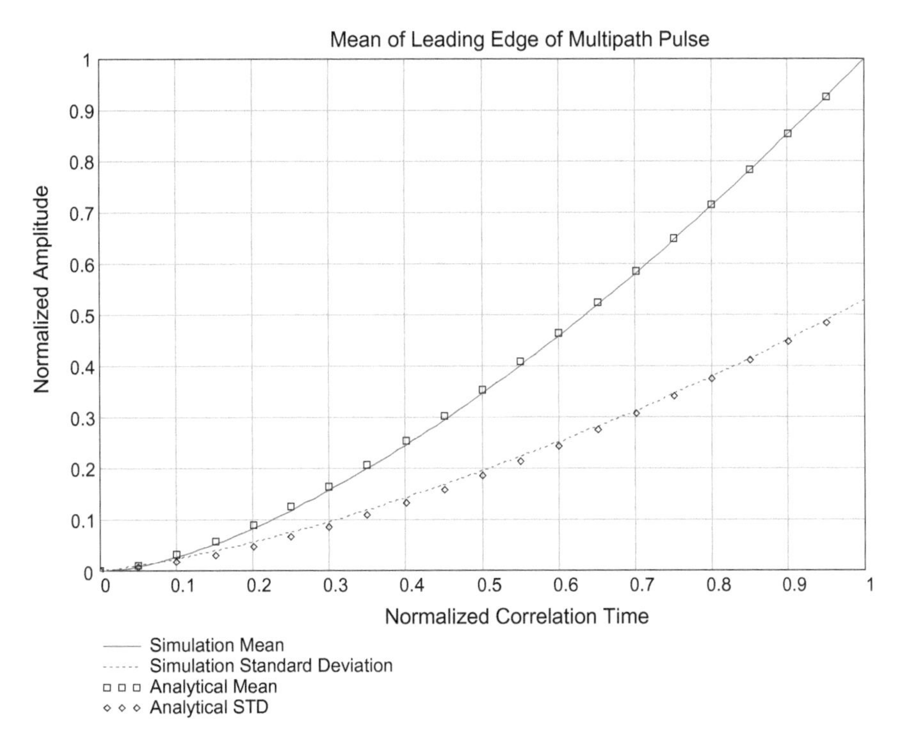

Fig. 13.9 Comparison of the mean and STD of 2000 random leading edge pulses and the analytical estimates of the shape

between the simulation and analytical results; other amplitude distributions give identical normalized results, as will be demonstrated in the next subsection.

An important consequence of the above analysis on the design of a leading edge TOA algorithm is the rapid increase in the standard deviation of the multipath pulse "noise" as a function of delay from the epoch of the pulse. Clearly the best quality measurements are near the epoch of the pulse, and the worst data are near the peak.

13.6.2 Random Amplitude and Phase Case

While the equi-amplitude analysis provides useful estimates of the shape of the leading edge, actual propagation conditions will have varying amplitudes (see Fig. 13.2), so that the statistical characteristics with varying signal amplitudes and phases are analysed in this section.

From the analysis in Sect. 13.6.1, the magnitude-squared of the Nth sample in the pulse leading edge is given by

$$\left[\frac{M_N}{\delta}\right]^2 = \left[\sum_{n=1}^{N} na_{N-n}\cos\phi_{N-n}\right]^2 + \left[\sum_{n=1}^{N} na_{N-n}\sin\phi_{N-n}\right]^2 \tag{13.20}$$

where the nth signal has an amplitude a_n. The expected value of the magnitude-squared is thus

$$E\left[M_N^2\right] = \delta^2 E\left[\sum_{n=1}^{N} n^2 a_{N-n}^2\right] = \delta^2 \sum_{n=1}^{N} n^2 E\left[a^2\right] \tag{13.21}$$

The expected value of the signal amplitude-squared in (13.21) can be evaluated if the statistics of the signal amplitude are known. For example, with a Rayleigh distribution of the amplitude the corresponding magnitude-squared is

$$E\left[M_N^2\right] = \frac{1}{3}N(2N+1)(N+1)(\delta\sigma)^2 \approx \left(\frac{2}{3}(\delta\sigma)^2\right)N^3 \tag{13.22}$$

If (13.22) is compared with (13.15) it can be observed that the only difference is the constant term, which is also true for any other amplitude statistics, albeit with a different constant. Thus if the multipath pulse is *normalized* in the same manner as described in Sect. 13.6.1, the resulting mean shape will be the same, namely as defined by (13.18). As stated previously, this result is expected as consequence of the CLT. Thus provided there is a sufficiently large number of scattering sources such that statistics of large numbers is reasonably valid, the mean shape of the leading edge is *independent* of the signal amplitude statistics.

The standard deviation in the shape of the leading edge can also be calculated in a manner similar to that described in Sect. 13.6.1. Thus the shape of the standard deviation curve for the random amplitude case is given by (13.19), as shown in Fig. 13.9.

13.7 Effect of Signal Bandwidth

The previous analysis has determined the range errors as a result of RF propagation characteristics through materials (in walls) and the diffraction around objects. Such errors are not related to the signal bandwidth, at least not explicitly. The nature of the range errors for different bandwidths is illustrated in Fig. 13.10. In this figure the range error scatter is plotted as a function of range for a pulse rise-time of 0.5 and 25 ns (or signal bandwidths of about 2 GHz and 40 MHz); also plotted are linear least squares fit lines.

The paths shown in Fig. 13.4 are all NLOS, but at short ranges (say less than 5 m) the path will be LOS; in such cases it is expected that the mean error will be approximately zero, as shown in the lines in Fig. 13.10. As can be observed the

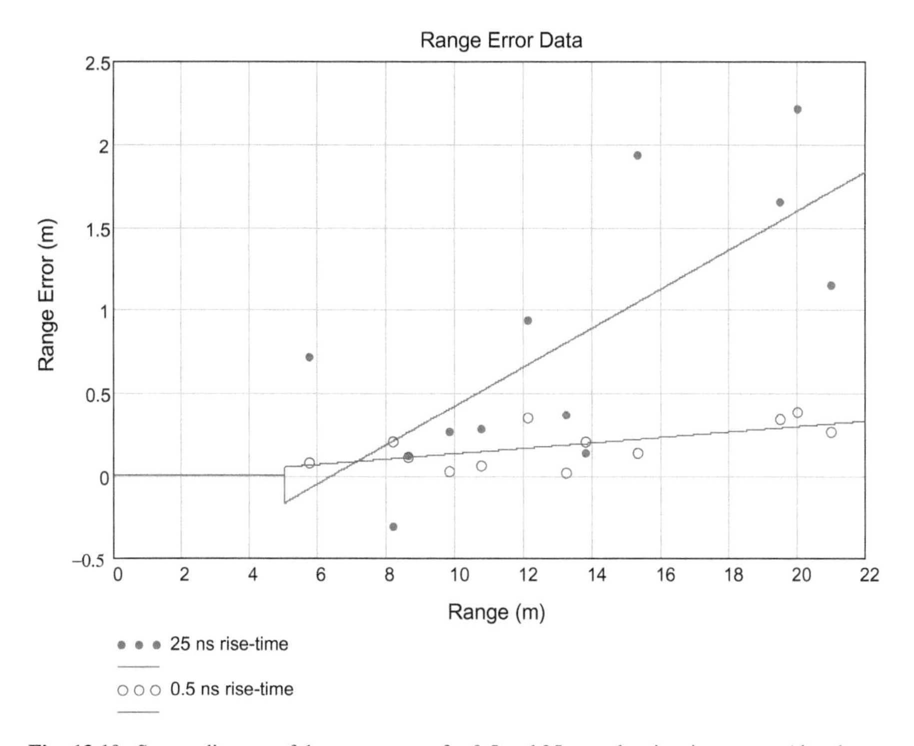

Fig. 13.10 Scatter diagram of the range errors for 0.5 and 25 ns pulse rise-time cases. Also shown is the linear regress plots for ranges greater than 5 m. The slopes of the lines in this example are 0.016 m per meter and 0.118 m per meter respectively. The scatter standard deviations are 0.097 m and 0.52 m respectively

range errors appear to have a bias error which increases linearly with range beyond the LOS range. This linear-range behaviour occurs at both signal bandwidths in Fig. 13.10, but this characteristics is general for all bandwidths, with the slope increasing as the bandwidth decreases.

Figure 13.11 shows an overall summary of the range slope parameter λ (as derived from Fig. 13.10 and other similar plots) plotted as a function of the pulse rise-time. As can be observed for rise-times up to about 5 ns (or 200 MHz bandwidth) the slope parameter is approximately constant, but then increases linearly for greater pulse rise-times. This result is contrary to the error characteristics of the mechanisms described in Sect. 13.5, namely the range errors are independent of the signal bandwidth. Clearly there is another mechanism other than those described in Sect. 13.5 that causes these range error characteristics as a function of the number of walls along the propagation path, as explained below.

An analysis of range error data at various RF frequencies and bandwidths in (Sharp et al. 2009) showed that there are two mechanisms for TOF range errors, namely those associated with the propagation path (as discussed in Sect. 13.2), and

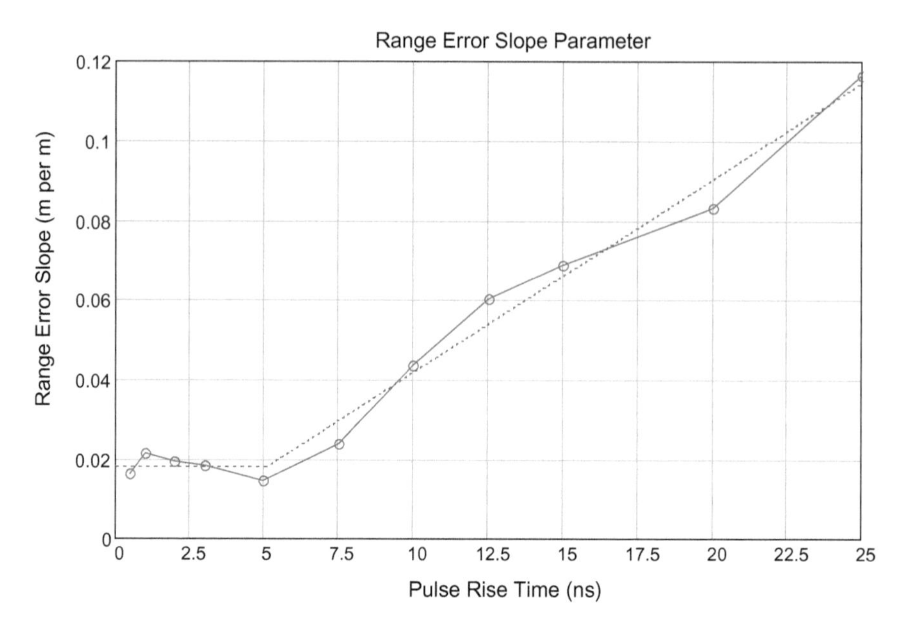

Fig. 13.11 Plot of the slope parameter (λ) associated with range errors plotted as a function of the pulse rise-time. Also shown is the least-squares piecewise-linear fit to the slope data. The slope of the regression line is 0.49 cm per meter per ns. The limiting value for small pulse rise-times is 0.018 m per meter

those associated with the TOA detection algorithm; this is summarized by the (13.1), and repeated here for convenience in the form

$$\varepsilon_{meas}\left(R, \tau_{pulse}; f_c\right) = \varepsilon_{path}(R; f_c) + \varepsilon_{TOA}\left(\tau_{pulse}\right) \tag{13.23}$$

where f_c is the RF carrier frequency. Note in (13.23) the first term is independent of the signal bandwidth, while the second is only a function of the bandwidth (or pulse rise-time). Thus any bandwidth-related phenomenon must be associated with the shape of the leading edge of the received pulse, and how the TOA algorithm determines the epoch of the pulse. As the measured data always couples these two effects summarized by (13.23), decoupling the measured data to determine the bandwidth effects is not easy. However, it is shown in Sharp and Yu (2014) that for UWB signals the errors associated with the TOA algorithm are small (of the order of 1–2 cm for a 4 GHz bandwidth signal). Thus to a good degree of accuracy, by assuming that the RF-related range error characteristics are the same for every bandwidth, the TOA component of the error is approximately given by

$$\varepsilon_{TOA}\left(\tau_{pulse}\right) \approx \varepsilon_{meas}\left(R, \tau_{pulse}; f_c\right) - \varepsilon_{meas}(R, \tau_{uwb}; f_c) \tag{13.24}$$

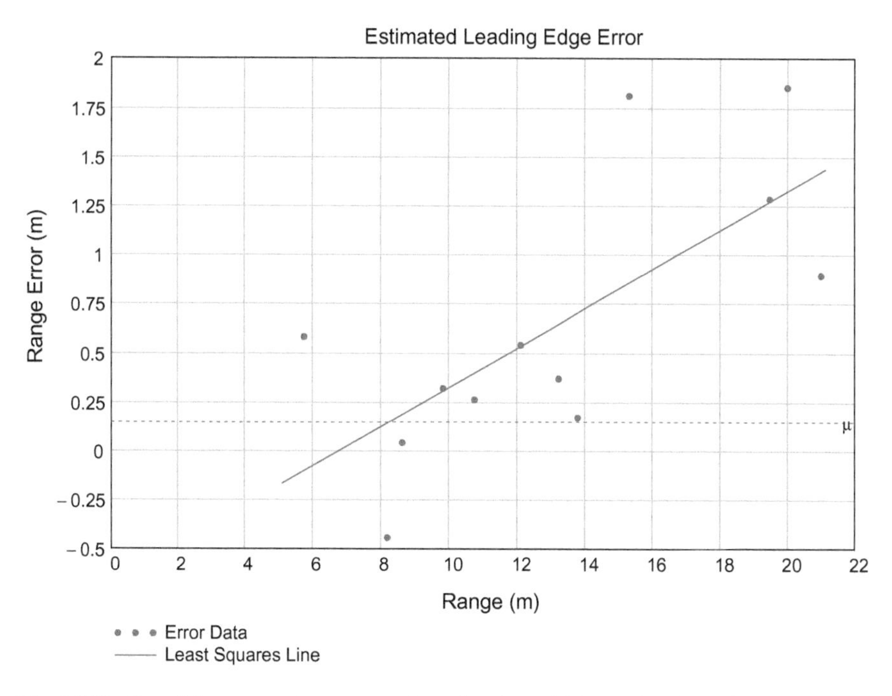

Fig. 13.12 Plot of the corrected range error for a rise-time of 25 ns. Also shown is the mean error (μ) for ranges up to 14 m, and the least-squares fit straightline to all the data

That is, to estimate the range errors associated with the TOA algorithm for a lower bandwidth signal, the measured lower bandwidth range error data are corrected by the corresponding UWB measured data. For example, Fig. 13.12 shows the estimated leading edge error for a bandlimited signal with a rise-time of 25 ns. Observe that there appears to be two groups of measurements. For data with ranges up to 14 m, the range errors are approximately independent of range (with some statistical variation), but for data with ranges greater than 14 m the errors are greater, with a greater spread. The former Group I cases exhibit the characteristic that the TOA measurement errors are independent of range; the second (Group II) points needs further investigation to determine the source of the errors.

For the Group I type measurements at relatively short range and for relatively large pulse rise times the range errors are dominated by the second term in (13.23), namely the errors associated with the shape of the leading edge. At longer ranges the first term in (13.23) becomes increasingly important, so the range errors increase with range.

To determine the reason why Group II points have larger errors, two competing explanations need to be examined—namely that the cause of the errors is due to either extra path delays or due to distortions in the shape of the received pulse. From Fig. 13.10 it can be ascertained that large Group II range errors are not apparent in the UWB data with small pulse rise times, so distortions in the low bandwidth pulse shape seems to be the explanation for the increase in range errors with range. This is consistent with the causes of measured range errors discussed in Chap. 4, Sect. 4.2.3.4.

ANNEX A—Measurements of Diffraction and Associated Delay Excess

This Annex provides the results of experiments of measuring delay excesses where the propagation is through a wall dividing two rooms in a building, and with an intervening obstacle requiring path diffraction. This experiment provides some validation of the theoretical analysis described in Sect. 13.5.

To provide some actual data on the delay excess through walls, some UWB (6–9 GHz) measurements were performed. The effective rise time of the pulse is about 0.5 ns, (equivalent to about 15 cm), with an effective pulse resolution of about 1 cm in the multipath environment. The experimental setup is shown in Fig. 13.13.

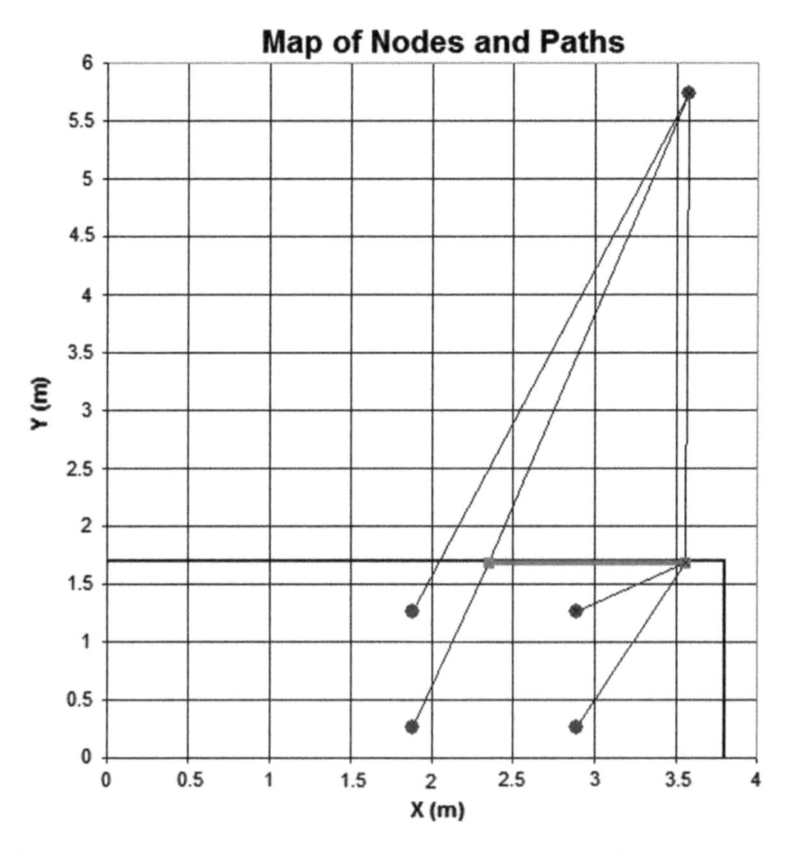

Fig. 13.13 Map of nodes and paths for range measurements through a wall (thick L-shaped line), with an opaque whiteboard (thick line between square dots). The diffraction paths around the whiteboard are also shows

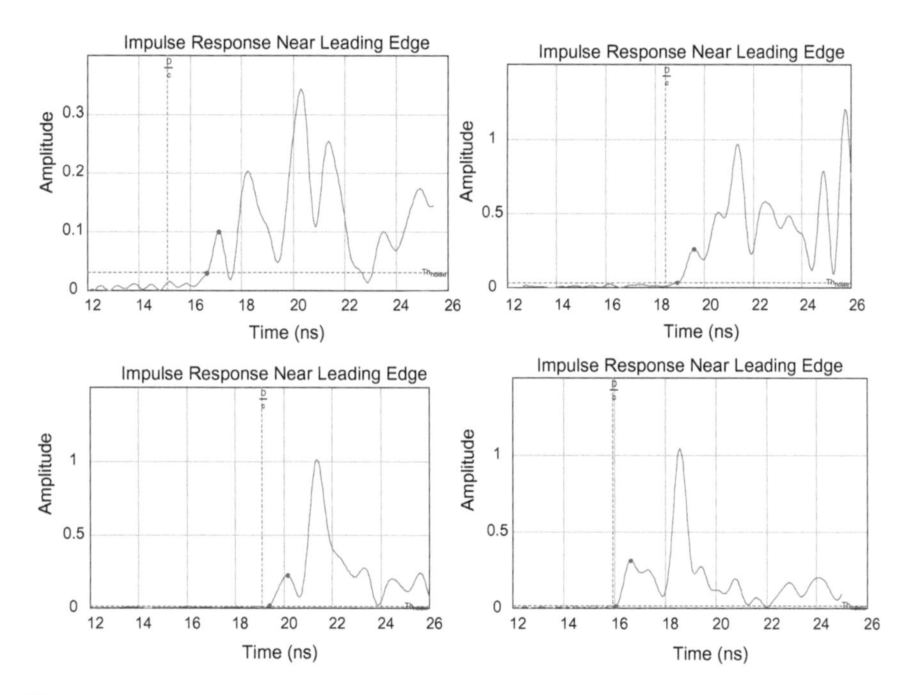

Fig. 13.14 Examples of four measurements of TOA using a Threshold algorithm

The four points on one side of the wall are at the corners of a 1 m square, with the side closest to the wall about 0.4 m from the wall and parallel to the wall. The receiving point is on the other side of the wall at a distance of about 4 m. The wall is made of wood (plywood with wooden studs), but the wall also includes a metal-backed whiteboard which obscures the straight-line path from two of the transmitting points to the receiver. The whiteboard has a steel backing which cannot be penetrated by the radio signal, so there will be diffraction around the whiteboard in these two cases.

Figure 13.14 shows the details of the measured impulse responses of the four measurements through the wall. The signal is UWB 6–9 GHz, with a range of $D = 5 \pm 0.5$ m through a wall, with the path approximately normal to the wall. The signal epoch is measured where the signal exceeds the noise threshold level, and is shown by a dot in the figures. The local peak on the leading edge is also shown by another dot; note that this local peak has a much smaller amplitude than the signal peak, and is considerably delayed relative to the first peak. The straight-line distance propagation delay between the transmitter and the receiver in each case is also shown. The measurement error is the difference between the epoch time and the time D/c.

The figures can be interpreted as follows. The first case (top-left) has a propagation excess of about 1.5 ns, even though the threshold algorithm correctly detects the first (small) signal above the threshold. This result is a consequence of the

Table 13.3 Summary of the measured path delay excesses for the four cases in Fig. 13.14

Path	Range (m)	Measured range (m)	Diffraction range (m)	Range excess (m)	Diffraction excess (m)	Estimated wall excess (cm)
1	4.53	4.85	4.84	0.32	0.31	1
2	5.52	5.69	5.63	0.17	0.11	6
3	5.73	5.81	5.73	0.08	0.00	8
4	4.78	4.83	4.79	0.04	0.01	3

diffraction around the whiteboard. Also observe a second peak with about a delay excess of 2.7 ns is associate with diffraction around the other edge of the whiteboard. The second case (top-right) is similar, although the first signal is smaller due to the greater diffraction losses. In contrast, the last two cases are not obscured by the whiteboard, and the delay excess is small. For a wooden wall (plywood) the delay excess can be just a few centimetres. For example, applying the theory in Sect. 13.4.1 with two plywood sheets of thickness 6 mm gives a mean delay excess of 10 mm, which is about the measurement resolution.

The data for these four cases are summarized in Table 13.3. The results in the table give data on the true range, the diffracted path range, the measured range, and an estimate of the wall excess. The measured range excess and the diffraction excess are generally in reasonable agreement. The difference between these two measurements are nominally the extra delay associated with the wall. Because the expected wall excess is about 1 cm, and the measurement resolution about 1–2 cm, accurate estimates of the wall excess delay is not possible.

In conclusion, the results in this Annex suggest that the delay excess associated with a wall can vary widely, depending on the details of the construction of the wall. For a wooden plywood construction the mean delay excess will be about 1 cm, but a brick wall can be about 20 cm. Further, if there are opaque obstacles such as the whiteboard in the above example, the delay excess will be associated with the diffraction around the obstacle, not the details of the wall construction. Thus in a large building with a variety of wall types and obstacles, deterministic measures of the delay excess are not feasible, but statistical parameters (mean, standard deviation) could be determined.

References

Alavi B, Pahlavan K (2006) Modeling of the distance measure error using UWB indoor radio measurement. IEEE Commun Lett 10(4):275–277

Alavi B, Pahlavan K (2003) Modeling of the distance error for indoor geolocation. In: Proceedings of IEEE wireless communications and networking, pp 668–672, Mar 2003

Alsindi N, Alavi B, Pahlavan K (2009) Measurement and modelling of ultrawideband TOA-based ranging in indoor multipath environments. IEEE Trans Veh Technol 58(3):1046–1058

Bellusci G, Janssen G, Yan J, Tiberius C (2008) Model of distance and bandwidth dependency of TOA-based UWB ranging error. In: Proceedings of IEEE international conference on ultra-wideband, pp 193–196

Chehri A, Fortier P, Tardif PM (2007) On the TOA estimation for UWB ranging in complex confined area. In: Proceedings of international symposium on signals, systems and electronics, pp 533–536

Dardari D, Chong CC, Win MZ (2008) Threshold-based time-of-arrival estimators in UWB dense multipath channels. IEEE Trans Commun 56(8):1366–1378

Gentile C, Kik A (2006) An evaluation of ultra wideband technology for indoor ranging. In: Proceedings of IEEE GLOBECOM, pp 1–6

Ghassemzadeh S, Greenstein L, Kavčić A, Sveinsson T, Tarokh V (2005) UWB indoor delay profile model for residential and commercial environments. IEEE Trans Veh Technol 54 (4):1235–1244

Ibraheem I, Schoebel J (2007) Time of arrival prediction for WLAN systems using prony algorithm. In: Proceedings of workshop on positioning, navigation and communication, pp 29–32

McNamara DA, Pistorius CW, Malherbe JA (1990) Introduction to the uniform geometrical theory of diffraction. Artech House

Prieto J, Bahillo A, Mazuelas S, Lorenzo RM, Blas J, Fernandez P (2009) NLOS mitigation based on range estimation error characterization in an RTT-based IEEE 802.11 indoor location system. In: Proceedings of IEEE international symposium on intelligent signal processing, pp 61–66

Sharp I, Yu K, Guo YJ (2009) Peak and leading edge detection for time-of-arrival estimation in band-limited positioning systems. IET Commun 3(10):1616–1627

Sharp I, Yu K (2013) Enhanced least squares positioning algorithm for indoor positioning. IEEE Trans Mob Comput 12(8):1640–1650

Sharp I, Yu K (2014) Indoor TOA error measurement, modeling and analysis. IEEE Trans Instrum Meas 63(9):2129–2144

Wang W, Jost T, Mensing C, Dammann A (2009) ToA and TDoA error models for NLoS propagation based on outdoor to indoor channel measurement. In: Proceedings of IEEE wireless communication and networking conference, pp 1–6

Yu K, Sharp I, Guo YJ (2009) Ground-based wireless positioning. Wiley-IEEE Press

Chapter 14
GDOP Analysis for Positioning Design

14.1 Introduction

A practical indoor system used for tracking people is typically based on both fixed anchor nodes and mobile nodes attached to people or other objects. In this mode of operation, mobile nodes whose position coordinates have been previously determined can contribute to the position determination of other mobile nodes whose position are sought. While a design can specify the location of the anchor nodes for adequate position performance, the mobile nodes' positions can only be specified in a statistical sense. Further, in the initial design phase, the positions of the fixed nodes within buildings may be difficult to specify, as the exact positioning of nodes can only be determined during installation, and based on detailed aspects of the architecture of the building. For example, an indoor-system designer may decide for performance reasons that the nominal locations of base stations should be on a 10 m square grid. However, if the base stations are to be attached to walls, the actual locations will vary somewhat (in a random fashion) from the plan. Thus the classical analysis (Torieri 1984; Spirito 2001; Miao et al. 2007; Yu 2007; Yu and Guo 2008) based on fixed nodes is not appropriate for such indoor systems. Instead, position performance prediction may have to be based only on statistical information of the spatial density of nodes. As a consequence, the positional accuracy analysis in this chapter is based on a statistically uniform spatial distribution rather than known, deterministic positions of nodes, and the analysis allows the performance degradation due to this random nature to be estimated. Note that the focus of this chapter is on the a priori estimation of performance, particularly related to the required node density, node radio range, and consequential positional errors given a specified propagation environment, rather than algorithms for position determination. Further, the aim is to provide relatively simple analytical expressions that allow design performance estimates to be determined without the need for more complex numerical simulations.

I. Sharp and K. Yu, *Wireless Positioning: Principles and Practice*, Navigation: Science and Technology, https://doi.org/10.1007/978-981-10-8791-2_14

There are a number of metrics for measuring the accuracy of positioning algorithms and systems, including geometric dilution of precision (GDOP) (Sharp et al. 2009; Zhu 1992; Lee 1975), root-mean-square, cumulative distribution probability, and the mean and variance (Bulusu et al. 2001). For positioning indoors and in wireless sensor networks, anchors would be typically deployed in the form of a mesh for both accuracy and robustness (Yu et al. 2009; Yick et al. 2004) Although simulation-based performance evaluation has been performed by researchers, simple analytical solutions for accuracy evaluation and system design in such mesh networks under rich multipath and non-line-of-sight (NLOS) propagation conditions are rare in the literature for indoor operation. Because of the desire to develop analytical expressions and the complexity of the problem with NLOS propagation and statistically defined node positions, complex (and more accurate) positioning algorithms are not the basis for the analysis in this chapter. A GDOP method is particularly attractive for design estimation, as this approach decouples the geometric aspects from the propagation aspects of positional error estimation. For example, a GDOP map of the coverage area can be calculated based solely on the (deterministic) location of nodes. If the TOA error statistics such as the standard deviation are known or can be estimated, the positional accuracy can be expressed as the product of GDOP and the TOA error standard deviation. Thus this chapter adopts this GDOP approach, but modified to account for the random specification of node positions. While the actual system implementation may adopt more sophisticated position determination algorithms, the GDOP method described is closely related to the commonly-used iterative least-squares (Yu et al. 2009, Chap. 7) method, and thus the design estimations using formulas derived in the chapter can be considered as providing conservative estimates of the performance of actual systems.

Previous GDOP analysis in the literature has focused on fixed and known anchor locations and zero-mean measurement errors. On the other hand, this chapter focuses on GDOP and accuracy analysis especially for indoor positioning system design when the measurement errors are biased and the anchor node deployment is random. The characteristics of the bias errors for indoor positioning systems are described and their characteristics analyzed in Chaps. 4 and 13, and these results are adopted in the analysis in this chapter.

14.2 GDOP Concept and Problem Formation

GDOP is a concept of specifying positional errors relative to ranging (or pseudo-ranging) errors; in practical situations GDOP is often greater than unity, so that the ranging errors are "diluted" by the dimensionless GDOP factor. GDOP is related only to the geometrical arrangement of the nodes, and is independent of the radio propagation characteristics, so that it is a useful concept for the design of the layout of nodes in a mesh network. Classical GDOP analysis is based on zero-bias operating environments with deterministic position data for the base stations

(Yu et al. 2009; Sharp et al. 2009; Lee 1975). The fundamental formula of GDOP is defined in terms of the expectations of the squares of the position (x-coordinate and y-coordinate) errors $(\Delta x$ and $\Delta y)$ and the range or pseudorange error $(\Delta r$ or $\Delta p)$, and is determined by

$$GDOP = \sqrt{\frac{E[\Delta x^2] + E[\Delta y^2]}{E[\Delta p^2]}} = \sqrt{\frac{\sigma_x^2 + \sigma_y^2}{\sigma_p^2}} \tag{14.1}$$

where the σ terms are standard deviations of the associated variable. Note that as pseudoranges are equivalent to ranges plus a constant, $\sigma_p = \sigma_r$ where σ_r is the standard deviation of the range measurement errors. Equation (14.1) is completely general, so that given the statistics of pseudorange errors and the consequential position error statistics, GDOP can be calculated using (14.1). In this chapter, the focus is on indoor positioning, which has different statistics compared with the classical outdoor case. The particular model for the indoor case is introduced in Sect. 14.3.

Suppose that there are N_R anchor nodes that are within the radio range of a mobile node. Pseudorange measurements are made between the mobile node and each of these anchor nodes. Due to multipath propagation and other measurement errors, the pseudorange measurement model is defined by

$$\hat{p}_i = p_i + \varepsilon_i, \; i = 1, 2, \ldots N_R \tag{14.2a}$$

where the zero-mean random component denoted by ε_i. Note that this measurement error model is generic, with details of the statistics undefined except for the zero-mean condition; the specific model for indoor positioning systems is somewhat different, and is considered in Sect. 14.3. The pseudorange p_i is defined as

$$p_i = r_i + c\phi_0 \tag{14.2b}$$

where c is the speed of signal propagation, and ϕ_0 is the common time offset between the mobile node and the anchor nodes which are assumed to be time synchronized. The pseudorange offset is commonly assumed to be associated with the clock offset in the unsynchronized mobile device, but it is shown in Yu et al. (2009, Chap. 5) that equipment delays are also a contributing factor even if round-trip-time (RTT) measurements are used, as in the experimental equipment described in Sect. 14.5. The distance r_i between the mobile unit and the ith anchor node in the 2D environment is given by

$$r_i = \sqrt{(x_i - x)^2 + (y_i - y)^2} \tag{14.3}$$

where (x, y) and (x_i, y_i) are the coordinates of the mobile and the ith anchor, respectively.

The following is the fundamental analytical framework that will be used in the remainder of the chapter. As mentioned previously, the position coordinate errors are denoted by Δx and Δy, and the pseudorange offset error as $\Delta\phi_0$, which are a consequence of the pseudorange measurement errors. Assume that these errors are relatively small compared with the overall geometry of the mesh. Then, applying a truncated Taylor series expansion on (14.3) produces the pseudorange error as

$$\begin{aligned}\Delta p_i &\approx \frac{\partial r_i}{\partial x}\Delta x + \frac{\partial r_i}{\partial y}\Delta y + \frac{\partial r_i}{\partial \phi_0}\Delta\phi_0 \\ &= \alpha_i\,\Delta x + \beta_i\,\Delta y + c\,\Delta\phi_0\end{aligned} \tag{14.4}$$

where $\alpha_i = (x - x_i)/r_i = \cos\theta_i$, $\beta_i = (y - y_i)/r_i = \sin\theta_i$, and θ_i is the angular position of the mobile node relative to the ith anchor node. Equation (14.4) can be written in a compact form as

$$\mathbf{A\delta} = \mathbf{p},\quad \mathbf{A} = \begin{bmatrix} \alpha_1 & \beta_1 & c \\ \vdots & \vdots & \vdots \\ \alpha_{N_R} & \beta_{N_R} & c \end{bmatrix},\quad \boldsymbol{\delta} = \begin{bmatrix} \Delta x \\ \Delta y \\ \Delta\phi_0 \end{bmatrix},\quad \mathbf{p} = \begin{bmatrix} \Delta p_1 \\ \vdots \\ \Delta p_{N_R} \end{bmatrix} \tag{14.5}$$

Equation (14.5) represents a set of overly-defined linear equations which can be solved by least-squares (LS) techniques. The optimum LS method depends on the statistics of the random processes associated with the errors Δp in the measured pseudoranges. It will be shown in Sect. 14.3 that based on the statistical model of range errors in an indoor environment the errors Δp are statistically uncorrelated, $E[\Delta p] = 0$, and $\mathrm{var}[\Delta p_i] = \sigma^2$, a constant. In such a case the classical LS solution is known to be the "best linear unbiased estimate" (BLUE) (Teunissen 2002), and is given by

$$\boldsymbol{\delta} = (\mathbf{A}^T\mathbf{A})^{-1}(\mathbf{A}^T\mathbf{p}) = \mathbf{\Phi}^{-1}\mathbf{h} \tag{14.6}$$

where

$$\mathbf{\Phi} = \begin{bmatrix} \sum\limits_i \alpha_i^2 & \sum\limits_i \alpha_i\beta_i & \sum\limits_i \alpha_i \\ \sum\limits_i \alpha_i\beta_i & \sum\limits_i \beta_i^2 & \sum\limits_i \beta_i \\ \sum\limits_i \alpha_i & \sum\limits_i \beta_i & N_R \end{bmatrix},\qquad \mathbf{h} = \begin{bmatrix} \sum\limits_i \alpha_i\Delta p_i \\ \sum\limits_i \beta_i\Delta p_i \\ \sum\limits_i \Delta p_i \end{bmatrix} \tag{14.7}$$

The constant c does not have any effect on the solution, so that it is simply set to unity (Sharp et al. 2009).

In the following sections GDOP as defined by (14.1) and the associated positional accuracy will be studied based on the system model defined by (14.2a), but also under the assumption of random and uniformly deployed anchor nodes. Note that typically indoor anchor locations would be carefully surveyed so that the location errors would be very small, say on the order of one or two centimeters. The impact of such a small error magnitude on the mobile node location estimation would be negligible. On the other hand, in the presence of relatively large anchor location errors, such as due to uncalibrated movement of mobile nodes used as "pseudo-anchor" nodes, the impact of the anchor location errors can be significant. The effect of such random errors in the position of anchor nodes is not directly analyzed in the chapter; rather the focus of the chapter is the effect on GDOP and the positional accuracy assuming the nodes have a uniform statistical spatial distribution typical of nodes attached to people.

14.3 Statistical Analysis with Random Node Positions

In the literature GDOP analysis is based on the assumption that the statistical ranging accuracy (standard deviation) is the same for all nodes in the network, and that the mean of the ranging errors is zero (Sharp et al. 2009). This assumption is approximately valid for line-of-sight (LOS) situations, such as outdoor locations, but is not appropriate for indoor situations, where NLOS propagation conditions typically prevail. In particular, NLOS propagation results in a range measurement errors consisting of a positive bias as well as a random component. Although the situation is complex, measurements made in various indoor environments reported in Gentile and Kik (2006) indicate that the ranging errors can be modeled approximately by

$$\Delta r = \varepsilon + \lambda R \tag{14.8}$$

where ε is a random variable that has a normal distribution with zero mean and standard deviation σ_r, R is the range between the communicating nodes, and λ is a scaling factor, which is usually of unknown value. See also Chap. 4 for more details, and in particular Sect. 4.2.7 and Eq. (4.22). For a more theoretical basis for (14.8), see Chap. 13.

Clearly in this model (14.8), σ_r and λ are the two parameters which characterize a particular indoor environment. Typically the scaling factor λ varies from 5 to 30 cm/m, while the standard deviation parameter σ_r ranges from 0.2 to 3 m (Yu et al. 2009, Gentile and Kik 2006). However, it should be also noted that the model parameters will include both the effects of the radio propagation environment and the method of detecting the TOA in the receiver, so they will be affected by system design parameters such as the signal bandwidth, modulation technique and the signal processing method used to detect the TOA.

The model in (14.8) defines two ranging error mechanisms, and as the measurement equations defined by (14.5) are linear, it is convenient to consider each type of error separately. The first measurement error type in (14.8) is assumed to be random with zero mean. This error is associated with the scattering variability of radio signals along the path from the mobile node to each of the receiving nodes.

Provided these receiving nodes are not too close together, the scattering will be different for each path, and thus the corresponding random errors for each path will be uncorrelated. Further, the model assumes (and the measurements described in Sect. 14.5 largely confirm) that the variance of the errors is independent of range and thus can be modeled as a constant parameter. Thus these conditions are those required for the standard LS procedure described in Sect. 14.2. The second error type defined in (14.8) is a bias error associated with the zig-zag scattering path between nodes, and assumed to be a linear function of range. This error appears to be deterministic rather than a random variable, but for the position determination process the range is not known, except that it is in the interval $[0, R_{\max}]$, where $R_{\max}$ is the maximum radio range for communications between nodes. Thus λR_i is a random variable, but with a non-zero mean. As LS theory requires that the random error has a zero mean, it appears that LS fitting cannot be (optimally) applied for bias errors. However, as shown in (Yu et al. 2009, Sect. 5.4.3) the bias error averaged over all the nodes in range can be effectively subtracted from all the pseudorange measurement; this mean $\lambda\bar{R}$ can incorporated into the pseudorange constant term in (14.2b) without affecting the position determination process. Thus the modified bias error has zero mean and approximately a constant variance, as required by LS theory. Further, as the ranges to the nodes are random, the modified bias errors will be statistically uncorrelated. Thus again the conditions required for the standard LS procedure are met.

The classical analysis is based on deterministic anchor node positions, so that $\mathbf{\Phi}$ matrix in (14.6) is deterministic. However, in the case where the anchor nodes are randomly deployed, the $\mathbf{\Phi}$ matrix is random. To examine the effect of the randomness in anchor node positions, it is necessary to study the statistics of the $\mathbf{\Phi}$ matrix as described below.

14.3.1 Statistics of Matrix $\mathbf{\Phi}$

Since $\mathbf{\Phi}$ is a random matrix, it can be described as the sum of two matrices, namely

$$\mathbf{\Phi} = \mathbf{\Phi}_0 + \mathbf{\Phi}_\varepsilon \tag{14.9}$$

where $\mathbf{\Phi}_0$ is the expected (mean) value of $\mathbf{\Phi}$ which is deterministic, and

$$\mathbf{\Phi}_\varepsilon = \begin{bmatrix} \varepsilon_{\alpha^2} & \varepsilon_{\alpha\beta} & \varepsilon_\alpha \\ \varepsilon_{\alpha\beta} & \varepsilon_{\beta^2} & \varepsilon_\beta \\ \varepsilon_\alpha & \varepsilon_\beta & 0 \end{bmatrix} \tag{14.10}$$

is the random part of the matrix whose components, except for the zero, are random variables with zero means. The mean matrix $\mathbf{\Phi}_0$ and the standard deviation of $\mathbf{\Phi}_0$ can be determined as follows. It is assumed that the nodes have a uniform but random spatial distribution, so their angular distribution relative to the mobile node is statistically uniform with an angular probability density function (PDF) given by

$$p(\theta) = \begin{cases} 1/2\pi, & -\pi \leq \theta \leq \pi \\ 0\,, & \text{otherwise} \end{cases} \tag{14.11}$$

As seen from (14.7) there are five different components in the matrix $\mathbf{\Phi}$. Let us determine the mean and standard deviation of one such component of the matrix $\mathbf{\Phi}$ in (14.7), say the $\sum\alpha$ term. For convenience, the subscript is dropped. As $\alpha = \cos\theta$, the PDF of α can be calculated by

$$q(\alpha) = 2p(\theta)/\left|\frac{d\alpha}{d\theta}\right| = \frac{1}{\pi\sin\theta} = \frac{1}{\pi\sqrt{1-\alpha^2}} \tag{14.12}$$

where the scaling factor 2 comes from the overlapping in the mapping of α. Therefore, the mean and variance of α are given by

$$\mu_\alpha = \int_{-1}^{1} \alpha q(\alpha)\,d\alpha = 0,\ \sigma_\alpha^2 = \int_{-1}^{1} (\alpha-\mu_\alpha)^2 q(\alpha)\,d\alpha = \int_{-1}^{1} \alpha^2 q(\alpha)\,d\alpha = \frac{1}{2} \tag{14.13}$$

Further, if there are N_R nodes within radio range, then the expected value of the sum of these independent random variables will also be zero, and the associated standard deviation will be $\sqrt{N_R/2}$, so that $[\mathbf{\Phi}_0]_{1,3} = 0$ and the standard deviation of $[\mathbf{\Phi}_\varepsilon]_{1,3} = \sqrt{N_R/2}$. Now consider one more component of $\mathbf{\Phi}$, say the $\sum\alpha^2$ component. The mean and variance of α^2 are determined as

$$\mu_{\alpha^2} = \int_{-1}^{1} \alpha^2 q(\alpha)\,d\alpha = \frac{1}{2},\quad \sigma_{\alpha^2}^2 = \int_{-1}^{1} (\alpha^2-\mu_{\alpha^2})^2 q(\alpha)\,d\alpha = \frac{1}{8} \tag{14.14}$$

That is $[\mathbf{\Phi}_0]_{1,1} = N_R/2$ and the standard deviation of $[\mathbf{\Phi}_\varepsilon]_{1,1}$ equals $\sqrt{N_R/8}$. The statistics of the other three components can be similarly calculated. The results show that the matrix $\mathbf{\Phi}_0$ is given by

$$\mathbf{\Phi}_0 = \frac{N_R}{2}\begin{bmatrix} 1 & 0 & 0 \\ 0 & 1 & 0 \\ 0 & 0 & 2 \end{bmatrix} \tag{14.15a}$$

The off-diagonal terms in (14.10) are defined by

$$\varepsilon_{\alpha} = \sum_i \alpha_i,\ \varepsilon_{\beta} = \sum_i \beta_i,\ \varepsilon_{\alpha\beta} = \sum_i \alpha_i\beta_i \tag{14.15b}$$

so that from (14.15a) the mean of these three error variables is zero, as required by the definition for the random variables. However from (14.15a), the mean of the two diagonal variables (α^2 and β^2) both have non-zero means, so that definitions of the associated error terms are

$$\varepsilon_{\alpha^2} = \sum_i \alpha_i^2 - \frac{N_R}{2},\ \varepsilon_{\beta^2} = \sum_i \beta_i^2 - \frac{N_R}{2} \tag{14.15c}$$

and include the term $-N_R/2$ to satisfy the condition that the mean of the random part of the matrix is zero. As a result, the element-wise standard deviation matrix of $\mathbf{\Phi}_\varepsilon$ is given by

$$\boldsymbol{\sigma}_{\mathbf{\Phi}_\varepsilon} = \sqrt{\frac{N_R}{8}} \begin{bmatrix} 1 & 1 & 2 \\ 1 & 1 & 2 \\ 2 & 2 & 0 \end{bmatrix} \tag{14.15d}$$

Note also that the PDF of individual components of matrix $\mathbf{\Phi}$ are not Gaussian (see for example $q(\alpha)$ in (14.12)); however, from the Central Limit Theorem (CLT), their distributions will be close to Gaussian as their derivation involves a summation process. The number of nodes required to approach Gaussian statistics is of the order of 5–10 (Yu et al. 2009, Chap. 2), so in most practical situations the Gaussian assumption will be a good approximation.

Having determined the characteristics of the random $\mathbf{\Phi}$ matrix, the position errors can be computed by using (14.6) with range errors as defined by the model (14.8), that is

$$\boldsymbol{\delta} = \lambda(\mathbf{\Phi}_0 + \mathbf{\Phi}_\varepsilon)^{-1} \left[\sum_i \alpha_i R_i \quad \sum_i \beta_i R_i \quad \sum_i R_i \right]^T \tag{14.16}$$

as the model deterministic measurement error is λR. The task now is to determine the inverse of $(\mathbf{\Phi}_0 + \mathbf{\Phi}_\varepsilon)$. In general there is no closed form for the inverse of the summation of two matrices, so the calculation of the inverse in (14.16) is somewhat involved. From (14.15a) and (14.15b), it can be seen that the diagonal components of the $\mathbf{\Phi}_0$ matrix are generally much larger than those of the $\mathbf{\Phi}_\varepsilon$ matrix, so one possible method to determine an approximate analytical solution for the inverse is to apply a Neumann series (Meyer 2000). Detailed analysis described in Annex A results in the approximate solution

$$(\mathbf{\Phi}_0 + \mathbf{\Phi}_\varepsilon)^{-1} \approx \tilde{\mathbf{\Phi}}_0 + \tilde{\mathbf{\Phi}}_\varepsilon \tag{14.17a}$$

where

$$\tilde{\mathbf{\Phi}}_0 \approx diag\{\,2\zeta,\quad 2\zeta,\quad \zeta\,\},\qquad \zeta = \frac{1}{N_R} + \frac{2}{N_R^2} + \frac{4}{N_R^3}$$

$$\tilde{\mathbf{\Phi}}_\varepsilon = \mathbf{\Phi}_0^{-1}\mathbf{\Phi}_\varepsilon\mathbf{\Phi}_0^{-1} = -\frac{1}{N_R^2}\begin{bmatrix} 4\varepsilon_{\alpha^2} & 4\varepsilon_{\alpha\beta} & 2\varepsilon_\alpha \\ 4\varepsilon_{\alpha\beta} & 4\varepsilon_{\beta^2} & 2\varepsilon_\beta \\ 2\varepsilon_\alpha & 2\varepsilon_\beta & 0 \end{bmatrix} \tag{14.17b}$$

It can be shown that the matrix $\tilde{\mathbf{\Phi}}_\varepsilon$ has an element-variance matrix given by

$$\boldsymbol{\sigma}_\varepsilon^2 = \frac{1}{N_R^3}\begin{bmatrix} 2 & 2 & 2 \\ 2 & 2 & 2 \\ 2 & 2 & 0 \end{bmatrix} \tag{14.18}$$

Based on this approximation, the positioning performance with bias errors will be analyzed in Sect. 14.3.2, the performance with zero-mean random range errors in Sect. 14.3.3, and the performance with the complete range error model in Sect. 14.3.4.

14.3.2 Statistics of Position Errors with Bias Ranging Errors

The analysis in Sect. 14.3.1 provides a general theory for the calculation of position errors when the nodes have a statistically uniform spatial distribution. In this section this theory is applied to the case where the ranging errors are associated with range-related biases only. In this case from (14.16) the position errors are given by

$$\boldsymbol{\delta} = \lambda(\mathbf{\Phi}_0 + \mathbf{\Phi}_\varepsilon)^{-1}\mathbf{s},\qquad \mathbf{s} = \left[\sum_{i=1}^{N_R}(x - x_i)\sum_{i=1}^{N_R}(y - y_i)\sum_{i=1}^{N_R} r_i\right]^{\mathrm{T}} \tag{14.19}$$

For example, from (14.19) the x-coordinate error is given by

$$\Delta x = 2\lambda\left(1 + \frac{2}{N_R}\right)m_x - \lambda(4m_{\alpha^2}m_x + 4m_{\alpha\beta}m_y + 2m_\alpha m_r) \tag{14.20a}$$

where

$$m_{\alpha^2} = \frac{1}{N_R}\varepsilon_{\alpha^2},\quad m_{\alpha\beta} = \frac{1}{N_R}\sum_i c_{\alpha\beta},\quad m_\alpha = \frac{1}{N_R}\sum_i c_\alpha$$
$$m_x = \frac{1}{N_R}\sum_i x_i,\quad m_r = \frac{1}{N_R}\sum_i r_i \tag{14.20b}$$

and for notational simplicity the origin of the coordinate system is set at the mobile node position and only two terms of $\mathbf{\Phi}^{-1}$ are used. Taking the expectation of both sides of (14.20a) results in

$$\begin{aligned}\mu_{\Delta x,bias} &= E[\Delta x] \\ &= 2\lambda\left(1+\frac{2}{N_R}\right)E[m_x] - \lambda(4E[m_{\alpha^2}m_x] + 4E[m_{\alpha\beta}m_y] + 2E[m_\alpha]E[m_r])\end{aligned} \tag{14.21}$$

where "bias" in the subscript indicates that the statistics are associated with the range bias error. Note that by definition the mean of each component of $\mathbf{\Phi}_\varepsilon$ matrix is zero, namely $E[m_{\alpha^2}] = E[m_{\alpha\beta}] = E[m_\alpha] = 0$. Applying these conditions, and the PDF of α and β as given by (14.12), it can also be shown that $E[m_{\alpha^2}m_x] = 0$ and $E[m_{\alpha\beta}m_y] = 0$, and $E[m_x] = m_x = E[r\alpha] = E[r]E[\alpha] = 0$ (a consequence of the uniform statistical distribution of nodes), so that (14.21) become

$$\mu_{\Delta y,bias} = 0 \tag{14.22}$$

The same result applies to the y-coordinate error: $\mu_{\Delta y,bias} = 0$. That is, as the distribution of anchor nodes within radio range circle is assumed to be statistically spatially uniform, the expected value of the x-coordinate error for all positions in the coverage area is zero.[1] However, there will be considerable variation around this mean value, so it is necessary to estimate the variance of the position error to examine position accuracy. The variance of the x-coordinate error is given by

$$\sigma^2_{\Delta x,bias} = E[\Delta x^2] = 4\lambda^2 E\left[\left\{\left(1+\frac{2}{N_R}\right)m_x - 2m_{\alpha^2}m_x - 2m_{\alpha\beta}m_y - m_\alpha m_r\right\}^2\right] \tag{14.23}$$

The expression for Δx has four terms, so that the squared function has ten terms, four for the square of the individual components and six cross-product terms. In determining analytical expressions for the components in (14.23), the following properties of the expectation operation can be exploited in the analysis:

1. The angular and radial position components are independent random variables, so the expectation of the product of functions of the variables is equal to the product of their expectations. Thus, for example, the random variable $x_i = r_i \cos\theta_i = r_i\alpha_i$ is actually the product of two random variables, and the expectation of x_i is the product of the two expectations, namely $E[x] = E[r\alpha] = E[r]E[\alpha]$ as the angular and radial distributions are statistically independent.

[1] This assumes the mobile node position is not near the edge of the coverage area.

2. The summations associated with the means of variables in (14.23) can be split into two groups, one where the summed variables all have the same index, and another where the indices $i, j, k..., i \neq j \neq k...$ are different. In the latter case the expectation of product of variables will be zero if at least one of the variables ε_α, ε_β or $\varepsilon_{\alpha\beta}$ has a zero expectation. This procedure greatly simplifies the calculation of the analytical expressions. An example of the analysis is given in Annex B.2.
3. The calculation of the expectation of some variables, such as x^2, can be performed by an appropriate surface integral, as described in Annex B.1.
4. Using the methods outlined above analytical expressions for the ten terms can be determined. The results are summarized in Table 14.1. Note that all the terms are a function of $R^2_{\max}$ and powers of $1/N_R$, with a weighting term proportional to the error model parameter squared (λ^2). Note that $R_{\max}$ is the maximum radio range of a node.

The combined effect can be obtained by summing the ten components in Table 14.1, resulting in

$$\sigma^2_{\Delta x,bias} = \frac{\lambda^2 R^2_{\max}}{9}\left(\frac{1}{N_R} + \frac{10}{N_R^2}\right) \tag{14.24}$$

The same result can be obtained for $\sigma^2_{\Delta y,bias}$. An example of the values of each of the ten components is shown in Fig. 14.1; it can be observed that the main contributions to the result are from the first, third and sixth components, while all the other components are much smaller. This comes from the fact that these three

Table 14.1 Summary of the terms in the calculation of Eq. (14.23)

No.	Component	Weighting	Analytical expression
1	$E[m_x^2]$	$4\left(1+\frac{2}{N_R}\right)^2\lambda^2$	$\frac{R^2_{\max}}{4N_R}$
2	$E[m_{\alpha^2}^2 m_x^2]$	$16\lambda^2$	$\frac{1}{32}\left(\frac{R_{\max}}{N_R}\right)^2$
3	$E[m_\alpha^2 m_r^2]$	$4\lambda^2$	$\left(\frac{2}{9}+\frac{1}{36N_R}\right)\frac{R^2_{\max}}{N_R}$
4	$E[m_{\alpha\beta}^2 m_y^2]$	$16\lambda^2$	$\frac{1}{32}\left(\frac{R_{\max}}{N_R}\right)^2$
5	$E[m_{\alpha^2}^2 m_x^2]$	$-16\left(1+\frac{2}{N_R}\right)\lambda^2$	$\frac{1}{16}\left(\frac{R_{\max}}{N_R}\right)^2$
6	$E[m_\alpha m_x m_r]$	$-8\left(1+\frac{2}{N_R}\right)\lambda^2$	$\frac{2}{9}\frac{R^2_{\max}}{N_R}+\frac{1}{36}\left(\frac{R_{\max}}{N_R}\right)^2$
7	$E[m_{\alpha^2} m_x m_\alpha m_r]$	$16\lambda^2$	$\frac{1}{18}\left(\frac{R_{\max}}{N_R}\right)^2+\frac{1}{72}\left(\frac{R_{\max}}{N_R^3}\right)^2$
8	$E[m_{\alpha\beta} m_x m_y]$	$-16\left(1+\frac{2}{N_R}\right)\lambda^2$	$\frac{1}{16}\left(\frac{R_{\max}}{N_R}\right)^2$
9	$E[m_{\alpha^2} m_x m_{\alpha\beta} m_y]$	$32\lambda^2$	0
10	$E[m_\alpha m_x m_{\alpha\beta} m_y]$	$16\lambda^2$	$\frac{1}{18}\left(\frac{R_{\max}}{N_R}\right)^2+\frac{1}{72}\frac{R^2_{\max}}{N_R^3}$

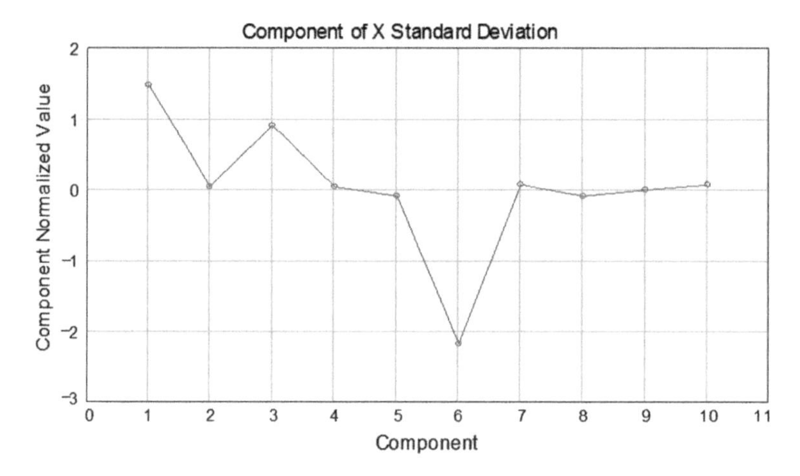

Fig. 14.1 Example of the normalized value of the ten components defined in Table 14.1. The normalization factor is $N_R/(\lambda R_{\max})^2$, which implies the first normalized component for a large number of base stations is unity. The parameters of the example are $N_R = 9$, $\lambda = 0.1$ and $R_{\max} = 25$ m

components are proportional to $R^2_{\max}/N_R$, whereas the other seven terms are proportional to either $R^2_{\max}/N^2_R$ or $R^2_{\max}/N^3_R$. Thus, the combined result will mainly depend on these three components. Using the three main components only, the variance of the x-coordinate error as well as the y-coordinate error can be computed to be

$$\sigma^2_{\Delta x,bias} = \sigma^2_{\Delta y,bias} = \frac{\lambda^2 R^2_{\max}}{9}\left(\frac{1}{N_R} + \frac{3}{N^2_R}\right) \tag{14.25}$$

which only differs from (14.24) by the coefficient of the $1/N^2_R$ term.

The comparison of these two theoretical formulas with numerical calculations is shown in Fig. 14.2 in Sect. 14.4. The numerical results tend to lie between the above two estimates (especially for large mean separations between nodes separation), with (14.24) being the upper estimate and (14.25) the lower estimate. In the event that a multipath mitigation technique is employed, (14.25) would provide a better performance prediction so that it will be used in further analysis in following sections.

14.3.3 Position Accuracy Derivation with Random Ranging Errors

In this section the general theory in Sect. 14.3.1 is used to calculate GDOP for the case where ranging error is a random variable of zero mean, which is defined as ε

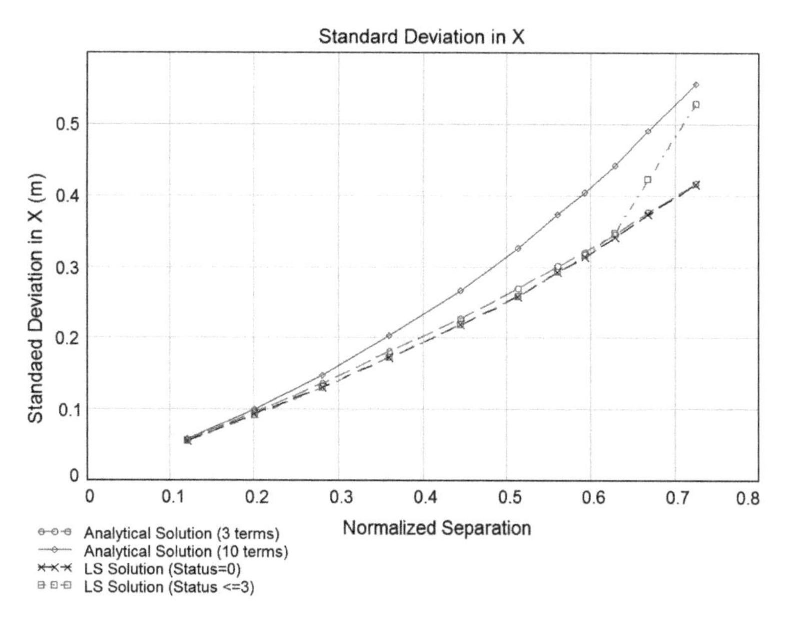

Fig. 14.2 Simulated and analytical ("3 terms" is for (14.25) and "10 terms" is for (14.24)) x–position errors (standard deviation) when using random sets of node positions and only bias ranging errors. The parameters are $\sigma_r = 0$, $\lambda = 0.1$ and $R_{\max} = 25$ m. The effect of two different completions status flags for the simulations is also shown. The "Normalized Separation" is the mean separation between nodes divided by the maximum radio range ($R_{\max}$), as defined in Eq. (14.32)

in (14.8). The analysis is similar to classical GDOP theory, except that it includes the effect of random node distribution. Applying the results in (14.6), (14.7), (14.17a), and (14.17b), the error in the x-coordinate position is described as

$$\begin{aligned}\Delta x = &\left[\frac{2}{N_R} + \frac{4}{N_R^2} + \frac{8}{N_R^3}\right]\left(\sum_i \alpha_i \Delta p_i\right) - \frac{4}{N_R^2}\varepsilon_{\alpha^2}\left(\sum_i \alpha_i \Delta p_i\right) \\ &- \frac{4}{N_R^2}\left(\sum_i \alpha_i \beta_i\right)\left(\sum_i \beta_i \Delta p_i\right) - \frac{2}{N_R^2}\left(\sum_i \alpha_i\right)\left(\sum_i \Delta p_i\right)\end{aligned} \tag{14.26}$$

The expectation of Δx can be obtained by determining the expectations of the individual terms in (14.26). It was shown earlier that $E[\alpha_i] = 0$ and $E[\beta_i] = 0$ and in the absence of range bias, $E[\Delta p_i] = 0$. Further, the mutual statistical independence among α_i, β_i and Δp_i results in

$$E[\alpha_i \Delta p_j] = 0,\ \ E[\alpha_i^2 \alpha_j \Delta p_j] = 0,\ \ E[\alpha_i \beta_i \beta_j \Delta p_j] = 0,\quad i, j = 1,\ 2,\ \ldots,\ N_R \tag{14.27}$$

As a result, the mean of each of the four terms in (14.26) is zero, producing

$$\mu_{\Delta x,ran} = E[\Delta x] = 0 \quad (14.28)$$

where "*ran*" in the subscript indicates that the statistics are associated with the zero-mean random range error. The same result applied to the y-coordinate error: $\mu_{\Delta y,ran} = 0$. These results are the same as the classical case with fixed nodes located symmetrically in 2D space relative to the mobile node. To compute the variance of Δx, (14.26) must be squared, and then the expectation is taken. The calculations are similar to those described for the bias error case in Sect. 14.3.2. As in the previous case there are a total of 10 expectation terms to be calculated; the results are summarized in Table 14.2, where σ_r^2 is the variance of the range error. From the sum of the terms in Table 14.2 the variance of Δx is given by

$$\sigma^2_{\Delta x,ran} = E[\Delta x^2] = \left(\frac{2}{N_R} + \frac{4}{N_R^2} + \frac{12}{N_R^3}\right)\sigma_r^2 \quad (14.29)$$

Note that the first-order approximation to (14.29) is the classical result $2\sigma_r^2/N_R$. Also note that as the number of nodes becomes large the last two terms in the

Table 14.2 Summary of the terms in the calculation of $E[\Delta x^2]$ for random ranging errors

No.	Component	Weighting	Analytical expression
1	$E\left[\left(\sum_i \alpha_i \Delta p_i\right)^2\right]$	$\frac{4}{N_R^2} + \frac{16}{N_R^3} + \frac{48}{N_R^4}$	$\frac{N_R}{2}\sigma_r^2$
2	$E\left[\left(\sum_i \alpha_i^2 - \frac{N_R}{2}\right)^2 \left(\sum_i \alpha_i \Delta p_i\right)^2\right]$	$\frac{16}{N_R^4}$	$\frac{N_R^2}{16}\sigma_r^2$
3	$E\left[\left(\sum_i \alpha_i \beta_i\right)^2 \left(\sum_i \beta_i \Delta p_i\right)^2\right]$	$\frac{16}{N_R^4}$	$\frac{N_R^2}{16}\sigma_r^2$
4	$E\left[\left(\sum_i \alpha_i\right)^2 \left(\sum_i \Delta p_i\right)^2\right]$	$\frac{4}{N_R^4}$	$\frac{N_R^2}{2}\sigma_r^2$
5	$E\left[\left(\sum_i \alpha_i^2 - \frac{N_R}{2}\right)^2 \left(\sum_i \alpha_i \Delta p_i\right)^2\right]$	$-\left(\frac{16}{N_R^3} + \frac{32}{N_R^4}\right)$	$\frac{N_R}{8}\sigma_r^2$
6	$E\left[\left(\sum_i \alpha_i \Delta p_i\right)\left(\sum_i \alpha_i \beta_i\right)\left(\sum_i \beta_i \Delta p_i\right)\right]$	$-\left(\frac{16}{N_R^3} + \frac{32}{N_R^4}\right)$	$\frac{N_R}{8}\sigma_r^2$
7	$E\left[\left(\sum_i \alpha_i \Delta p_i\right)\left(\sum_i \alpha_i\right)\left(\sum_i \Delta p_i\right)\right]$	$-\left(\frac{8}{N_R^3} + \frac{16}{N_R^4}\right)$	$\frac{N_R}{2}\sigma_r^2$
8	$E\left[\left(\sum_i \alpha_i^2 - \frac{N_R}{2}\right)\left(\sum_i \alpha_i \Delta p_i\right) \times \left(\sum_i \alpha_i \beta_i\right)\left(\sum_i \beta_i \Delta p_i\right)\right]$	$\frac{32}{N_R^4}$	0
9	$E\left[\left(\sum_i \alpha_i^2 - \frac{N_R}{2}\right)\left(\sum_i \alpha_i \Delta p_i\right) \times \left(\sum_i \alpha_i\right)\left(\sum_i \Delta p_i\right)\right]$	$\frac{16}{N_R^4}$	$\frac{N_R}{8}\sigma_r^2$
10	$E\left[\left(\sum_i \alpha_i \beta_i\right)\left(\sum_i \beta_i \Delta p_i\right) \times \left(\sum_i \alpha_i\right)\left(\sum_i \Delta p_i\right)\right]$	$\frac{16}{N_R^4}$	$\frac{N_R}{8}\sigma_r^2$

brackets in (14.29) become insignificant, so that the solution approaches the classical result as the number of nodes in range becomes large. On the other hand, as the number of nodes becomes small, the result varies significantly from that in the classical case.

The same result applies to $\sigma^2_{\Delta y,ran}$, so that the GDOP in the presence of uniform random anchor node deployment and zero-mean random range error is given by

$$GDOP_{ran} = \frac{\sqrt{E[\Delta x^2] + E[\Delta y^2]}}{\sigma_r} = \sqrt{\frac{4}{N_R} + \frac{8}{N_R^2} + \frac{24}{N_R^3}} \tag{14.30}$$

Thus the resulting GDOP is solely a function of the number of nodes in range. For the uniform random distribution case there is no deterministic geometry, so from the fundamental definition of GDOP (based on statistical expectations), there can be only a single solution which is a function of the only problem parameter, N_R.

The mean of the radial position errors is commonly used as an indication of positional accuracy, and this measure can be related to the above GDOP calculation. From (14.26), it is seen that functions for Δx and Δy are summations of a number of independent random variables so that their distributions are (approximately) Gaussian according to CLT, especially when involving the summation of a relatively large (5–10 is sufficient Yu et al. 2009, Chap. 2) number of random variables. Accordingly, the radial position error $\sqrt{\Delta x^2 + \Delta y^2}$ will approximately be Rayleigh distributed and thus the mean of the radial position error is

$$\rho_d = \bar{d}/R_{\max} = \sqrt{\pi/N_R} \tag{14.31a}$$

where $E[\Delta x^2] = E[\Delta y^2] = \sigma^2_{\Delta x,ran}$ is employed, producing

$$\mu_{\Delta r,ran} = \sqrt{\frac{\pi}{2}}\sigma_{\Delta x,ran} = \sqrt{\frac{\pi}{N_R}\left(1 + \frac{2}{N_R} + \frac{6}{N_R^2}\right)}\,\sigma_r \tag{14.31b}$$

By defining the mean spacing of anchor nodes as $\bar{d} = \sqrt{A/N_R}$, where A is the area of the coverage region, the normalized mean mesh distance can be described for a circular coverage area of radius $R_{\max}$ by

$$\rho_d = \bar{d}/R_{\max} = \sqrt{\pi/N_R} \tag{14.32}$$

Accordingly, (14.31b) becomes

$$\mu_{\Delta r,ran} = \sqrt{1 + \frac{2}{N_R} + \frac{6}{N_R^2}}\,\rho_d\sigma_r \approx \rho_d\sigma_r, \quad N_R \gg 1 \tag{14.33}$$

Thus the mean of the radial position error has approximately a simple linear relationship with the standard deviation of the range error when the number of anchor nodes within radio range is large. The above formulas can be used to aid system design in practical situations.

14.3.4 Summary of Overall Positioning Performance

In the previous sections the positioning performance was derived for the scenarios where the ranging error has either a positive range-dependent bias or a zero-mean random error, and the anchor nodes have a uniform random spatial distribution. In this section the positioning performance with the simultaneous presence of both these types of ranging errors is studied. First, making use of the results in (14.22) and (14.28) produces

$$\mu_{\Delta x,total} = \mu_{\Delta x,bias} + \mu_{\Delta x,ran} = 0 \tag{14.34}$$

where the same result applies to the y-coordinate positional error, that is $\mu_{\Delta y,total} = 0$. As explained earlier, the two ranging error components are statistically independent. It can also be shown that the position coordinate estimation error in the presence of range bias error is uncorrelated with that in the presence of random range error. Thus the resulting position coordinate error variance will be a superposition of the two coordinate error variances in the presence of range bias and random range errors. Then, combining the results in (14.25) and (14.29) produces the variance of the x-coordinate estimation error as

$$\begin{aligned} \sigma^2_{\Delta x,total} &= \sigma^2_{\Delta x,bias} + \sigma^2_{\Delta x,total} \\ &= \frac{\lambda^2 R^2_{\max}}{9}\left(\frac{1}{N_R} + \frac{3}{N_R^2}\right) + \left(\frac{2}{N_R} + \frac{4}{N_R^2} + \frac{12}{N_R^3}\right)\sigma_r^2 \end{aligned} \tag{14.35}$$

As a consequence, applying the GDOP formula (14.1) produces the GDOP in the presence of uniform random anchor node deployment with both range bias and zero-mean random range errors as

$$\begin{aligned} GDOP_{total} &= \frac{\sqrt{\sigma^2_{\Delta x,total} + \sigma^2_{\Delta y,total}}}{\sigma_r} \\ &= \sqrt{\frac{\lambda^2 R^2_{\max}}{9\sigma_r^2}\left(\frac{1}{N_R} + \frac{3}{N_R^2}\right) + \left(\frac{2}{N_R} + \frac{4}{N_R^2} + \frac{12}{N_R^3}\right)} \\ &= \frac{2}{\sqrt{N_R}}\frac{\sigma_{\Delta r,total}}{\sigma_r} \end{aligned} \tag{14.36}$$

where $\sigma^2_{\Delta y,total} = \sigma^2_{\Delta x,total}$ is used, and

$$\sigma_{\Delta r,total} = \sqrt{\left(1 + \frac{2}{N_R} + \frac{6}{N_R^2}\right)(\sigma_r^2 + \sigma_B^2)}, \quad \sigma_B = \frac{\lambda R_{\max}}{6}\sqrt{\frac{N_R(N_R+3)}{N_R^2+2N_R+6}} \tag{14.37}$$

where σ_B is the quasi-standard deviation associated with the range bias error.

As seen from (14.34) the means of the final x-coordinate and y-coordinate estimation errors are zero. Also, as mentioned earlier, the position coordinate estimation errors Δx and Δy are the sum of a number of random variables, so that they can be approximated as a zero-mean Gaussian random variable. Accordingly, the radial position error $\sqrt{\Delta x^2 + \Delta y^2}$ is approximately a Rayleigh random variable, so that the mean of the radial position error is given by

$$\mu_{\Delta r,total} = \sqrt{\frac{\pi}{2}}\sigma_{\Delta x,total} = \frac{\sqrt{\pi}}{2}\frac{2}{\sqrt{N_R}}\sigma_{\Delta r,total} = \rho_d \sigma_{\Delta r,total} \tag{14.38}$$

which is in the same form as that for the classical case with only random range errors. Note also that the standard deviation of the radial position error is given by $\sqrt{2-\pi/2}\sigma_{\Delta x,total}$ instead of $\sqrt{\sigma^2_{\Delta x,total} + \sigma^2_{\Delta y,total}}$.

Now consider a numerical example with parameters typical for an indoor positioning system, as summarized in Table 14.3. These results show that even though the bias errors are greater than the random ranging errors, the effect of the bias errors on positional accuracy is relatively modest. Further, the theory shows that for a given node spatial density, increasing the radio range will improve the position accuracy, even though this will result in the long-range nodes having large ranging errors; in this respect the performance is similar to the classical case with zero-mean random errors. This observation means that even though indoor radio

Table 14.3 Example of computed values based on theoretical expressions

Param.	Value	Equation	Comment
σ_r	2 m		Random STD, typical for indoor system
λ	0.1		Bias parameter, typical for indoor system
$\bar{d}$	15 m		Assumed mean separation
$R_{\max}$	30 m		Assumed maximum radio range
ρ_d	0.5	(14.31a, 14.31b)	Normalized mean separation
N_R	12	(14.31a, 14.31b)	Expected nodes in range
ε_B	3 m	$\lambda R_{\max}$	Maximum bias error
μ_B	2 m	$2/3\varepsilon_B$	Mean bias error
σ_B	0.72 m	(14.33)	Effective bias error STD
σ_t	2.32 m		Effective total range error STD
$\bar{\mu}_{\Delta r}$	1.16 m		Mean radial error

propagation can result in large ranging measurement errors, a positioning system based on pseudorange measurements could result in acceptable performance accuracy, especially when there are sufficient anchor nodes within radio range of the mobile.

14.4 Positioning Performance Based on Simulations

To assess the performance of the above theoretical analysis, performing Monte Carlo simulations allow a comparison between the analytical formulas and the performance of numerical position determination. In particular, the numerical position error statistics of the iterative least-squares (LS) solution can be compared with analytical solutions as a function of the normalized node density parameter ρ_d defined by (14.32). The LS method is essentially the same as described in Sect. 14.2 for GDOP, except that the solution is iterated until the incremental change in position is below a small threshold (0.01 m in this implementation).

The performance of a position determination algorithm will be influenced by the multipath propagation conditions. Although the employed LS method can compensate for the range bias errors because the mean of the range errors is treated as part of the pseudorange offset, it does not include extra multipath mitigation techniques. Thus the results can be considered as the performance upper bound for systems implemented with multipath mitigation methods. Note that the numerical implementation used does include some enhancements (as summarized in Table 14.4) to reduce the effects of the ranging errors. The position determination

Table 14.4 Summary of the LS solution status flags

Flag	Description	Comment
−6	Poor position accuracy	The distance between the computed position and the mean central point of the anchor nodes in radio range is greater than half the nominal diameter of the covered area
−5	Less than three nodes	No position can be determined
−4	Three nodes but no solution	Matrix inverse is invalid
−3	Iterative solution does not converge	Number of iterations is greater than 10
−2	Iterative solution does not converge	LS root-mean-squared fit error is greater than 10 m
−1	Invalid matrix inverse	More than 3 nodes, but no position determined
0	Valid position fix	Normal completion status
1	Three base stations only	No redundancy in number of nodes
2	GDOP greater than 2	Position determined, but unreliable

algorithm returns a completion "status flag" to indicate valid, invalid and unreliable position fixes.

Examples showed in Fig. 14.2 (bias errors only), Fig. 14.3 (random errors only) and Fig. 14.4 (both bias and random errors) use typical parameters of an indoor positioning system. The results show that the simple analytical formulas provide good estimates of the position accuracy over the complete range of normalized mean node separation distances. For Fig. 14.4, the corresponding LS solution status flag performance data are shown in Table 14.5. The results show that for small normalized node separations there are 100% valid positions, but this drops to 67% for a separation of $\sigma_d = 0.8$. Thus a designer should be cautious in the selection of this parameter when planning a positioning system. For instance, if a practical system requires reliable position fixes 95% of the time, then based on the data in Table 14.5, it is suggested that $\rho_d \leq 0.6$.

14.5 Experimental Results

The theoretical analysis presented in previous sections is intended to aid the design of mesh positioning systems, particularly regarding the tradeoff between radio range and node spatial density when given a particular propagation environment. To test

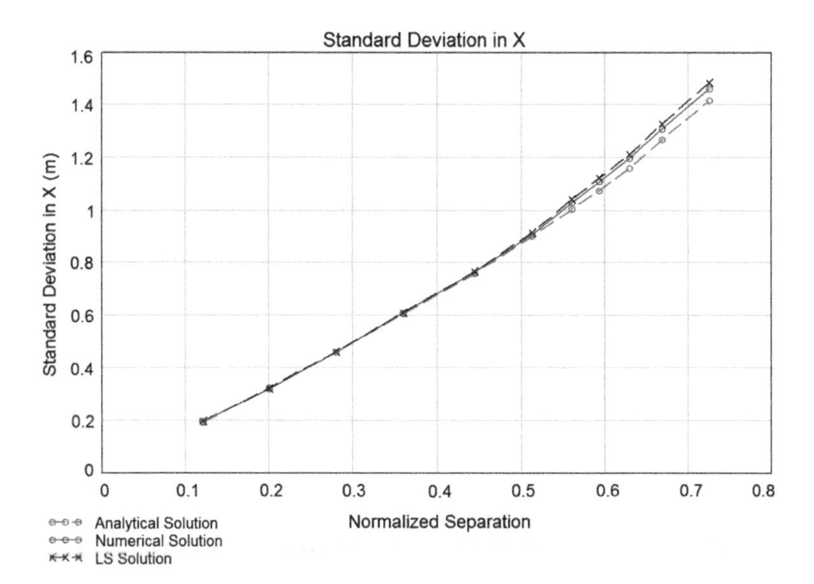

Fig. 14.3 Simulated and analytical position errors when using random sets of node positions and only zero-mean random ranging errors. The parameters are σ_r = 2 m, λ = 0 and R_{max} = 25 m. The simulation solution uses status flag = 0 only

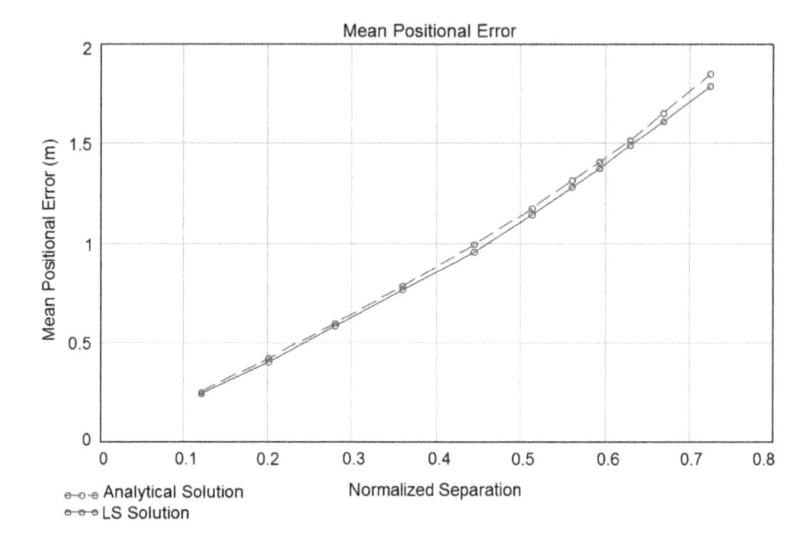

Fig. 14.4 Simulated and analytical mean radial position errors when using random sets of node positions and the ranging error has both biased and random components. The parameters are $\sigma_r = 2$ m, $\lambda = 0.2$ and $R_{\max} = 25$ m. The simulation solution uses status flag = 0 only

Table 14.5 Percentage of the flag status versus the mean normalized separation distance between nodes when the ranging error is both biased and random. The "missing" status flags in the table have zero occurrence

Flag	Mean normalized distance between nodes							
	0.1	0.2	0.3	0.4	0.5	0.6	0.7	0.8
−6	0	0	0	0	0	0.05	0.62	1.53
−5	0	0	0	0	0	0	0.38	3.25
−3	0.02	0.21	0.60	1.22	2.40	3.66	4.15	3.73
0	99.98	99.79	98.40	98.78	97.56	95.26	85.62	67.58
1	0	0	0	0	0	0	0.59	2.96
2	0	0	0	0	0.04	0.97	8.32	20.38

the analysis presented above, measurements using an actual mesh positioning system from which the measurement-based performance data are compared with the theoretical results.

14.5.1 Description of Positioning System and Measurements

The testing of the theoretical results is based on the Wireless Ad hoc System for Positioning (WASP) (Hedley et al. 2008; Sathyan et al. 2011), developed by CSIRO.

This experimental system operates in the 5.8 GHz ISM band with an effective bandwidth of 125 MHz. Position calculations are based on TOA estimation of a reconstructed wideband spread-spectrum signal with an effective leading-edge rise-time period of 13.5 ns; round-trip tine (RTT) between node pairs is determined from the TOA measurements. The TOA is estimated using a leading-edge super-resolution algorithm (Humphrey and Hedley 2008), which has been shown to out-perform other techniques such as the MUSIC algorithm (Schmidt 1986). The radio also supports ad hoc data communications which allows the measured TOA data to be forwarded to a central data logging computer.

The basic measurement procedure is as follows. Time is coarsely synchronized between nodes which use a time division multiple access technique so that each node transmits in each super-frame a spread-spectrum beacon signal. Each node that receives the beacon signals measures the TOA, which is then included in the data field of subsequent beacon signals. From these responses, a round-trip time (RTT) is estimated. The raw RTT is corrected by compensating for both delays in the equipment (using calibrated offset delays) and frequency offset related timing errors associated with the clocks in each node, which are not frequency synchronized. The measurement accuracy under calibration conditions is better than 1 ns. Because of the limited radio range, some nodes do not respond. The average maximum range of the responding nodes can be used to define the maximum range parameter $R_{\max}$. Thus for this 36-node mesh network each time frame will measure up to $36 \times 35 = 1260$ ranges. These measurements are performed 50 times, and the raw range measurements averaged, effectively removing temporal effects such as receiver noise. Thus the resulting measurements only include the effects of multipath propagation and small calibration errors.

The positional accuracy within the coverage area is not measured using a mobile node, but rather using the 36 base stations only. Each base station in turn can be considered to be a "mobile" node and its location is estimated using the remaining base stations that are within radio range. The computed locations are compared to the known surveyed locations of the base stations, so that the position errors within the mesh can be estimated throughout the coverage area. Note that while RTT provides range data, the positions are determined by the pseudorange iterative LS fit procedure (Yu et al. 2009, Chap. 7) to mitigate the effects of the synchronization and the range bias errors.

The measurements using the WASP system were performed within an office-type building with the 36 nodes covering an area of approximately 2100 m^2. The layout of the nodes and the building can be observed in Fig. 14.7. The building includes both office-style rooms (about 2.5×3 m), as well as some larger areas including a lecture theater (top-left), two atriums and some open-plan offices. The building construction is varied, including concrete with some plasterboard, steel and wood. According to the definition introduced previously, the mean node separation is calculated as: $\bar{d} = \sqrt{2100/36} = 7.6$ m.

14.5.2 Accuracy Comparison

The range measurement model defined by (14.8) and the consequential theoretical expected average positional error summarized by (14.38) can be checked by comparisons using measured data, thus providing some verification of the mathematical developments in the chapter. This procedure is thus equivalent to comparing the a priori design estimate and the performance achieved with an actual positioning system.

The ranging errors were determined from the RTT measurements and the surveyed positions of the nodes. The results are summarized in Fig. 14.5. The propagation parameters were determined to be: $\lambda = 0.023$, $\sigma_r = 1.08$ m. As the measured statistical characteristics of the range errors are broadly in agreement with the assumed model (linear bias error with range, constant random standard deviation), the measured positional accuracy can be compared with that predicted from the theoretical formulas.

A comprehensive comparison of the theoretical average positional error with the measurements is difficult as a large number of anchor node configurations would be required. Nevertheless, one specific measured example with one specific layout of nodes can be compared with the statistical expectations of the theory. Thus the test is whether the one measured example (of the overall average radial error) is consistent with the statistical prediction of the theory. The analytical expressions (14.37) and (14.38) involve the number of nodes in radio range (N_R) and the normalized mean

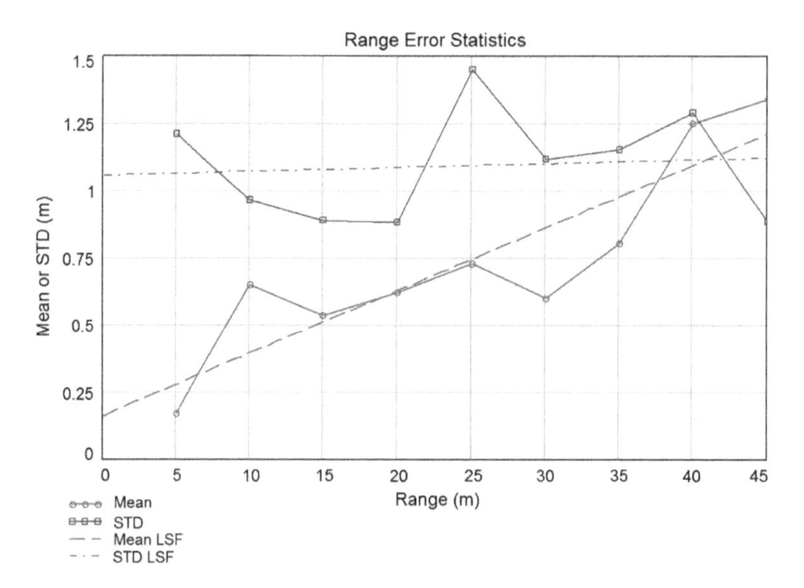

Fig. 14.5 Measured range error data plotted as a function of the range using data grouped into 5 m bins. From the least-squares fit (LSF) regression the range bias error parameter is $\lambda = 0.023$ and the mean of the range error standard deviation is $\sigma_r = 1.08$ m

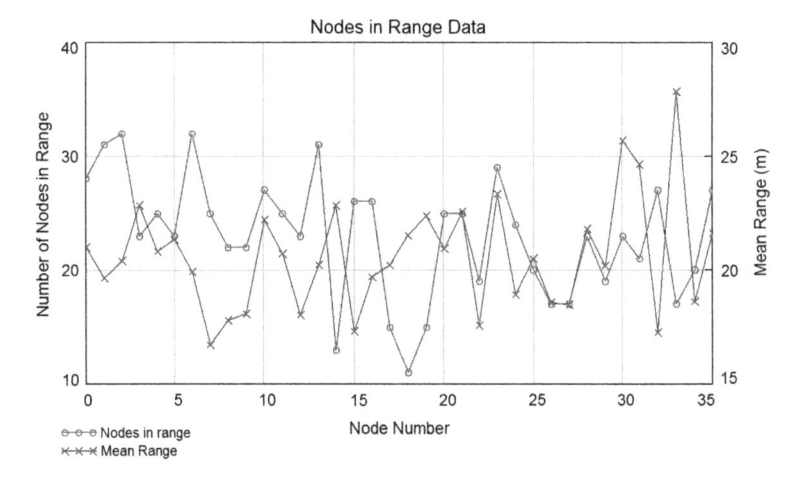

Fig. 14.6 Data for the number of neighboring nodes in range, and the corresponding mean range for each of the 36 nodes in the mesh network. The "Number of Nodes in Range" is the parameter N_R for each node

distance (ρ_d) to surrounding nodes. These parameters can be determined for each node from the measured data, as shown in Fig. 14.6, where the number of nodes in radio range and the mean distance (μ_R) for each node are shown. The number of nodes in radio range depends on the propagation conditions and the location of the node; nodes in the "core" of the building have more neighbors than those on the periphery of the building. The mean distance varies from about 15 to 28 m, with a mean averaged over all nodes of $\mu_R = 20.6$ m. If the distribution of neighboring nodes were circular (which is only approximately true), then the average maximum range parameter can be determined by $R_{\max} = 3/2\mu_R = 31$ m, assuming a uniform random distribution of range over $[0, R_{\max}]$. On this basis, the normalized mean distance between nodes is $\rho_d = 7.6/31 = 0.245$. As a result, the parameters σ_B and $\sigma_{\Delta r,total}$ in (14.37), GDOP and the mean of the radial position error can be calculated for each node using the derived formulas.

Using the measured RTT data and an iterative LS positioning algorithm to estimate the position of each node in turn, as described in Sect. 14.5.1, the positional errors are determined from the known surveyed node positions. A contour map based on interpolating the spatial positional errors at each node is shown in Fig. 14.7.

To compare the analytical estimates with these computed positional errors, only measurements at the "core" of the building are used, as the analytical estimates assume the mobile node is remote from the edge of the coverage area. The mean of the radial position errors of the LS algorithm for the $N_{core} = 13$ core locations is 0.46 m, while the corresponding analytical estimate from (14.37) and (14.38) and using the parameters determined from Fig. 14.5 is 0.40 m, or a difference of 0.06 m compared with the measurements. However, this average of LS mean radial errors is one sample from the population of random node position sets, and has a standard

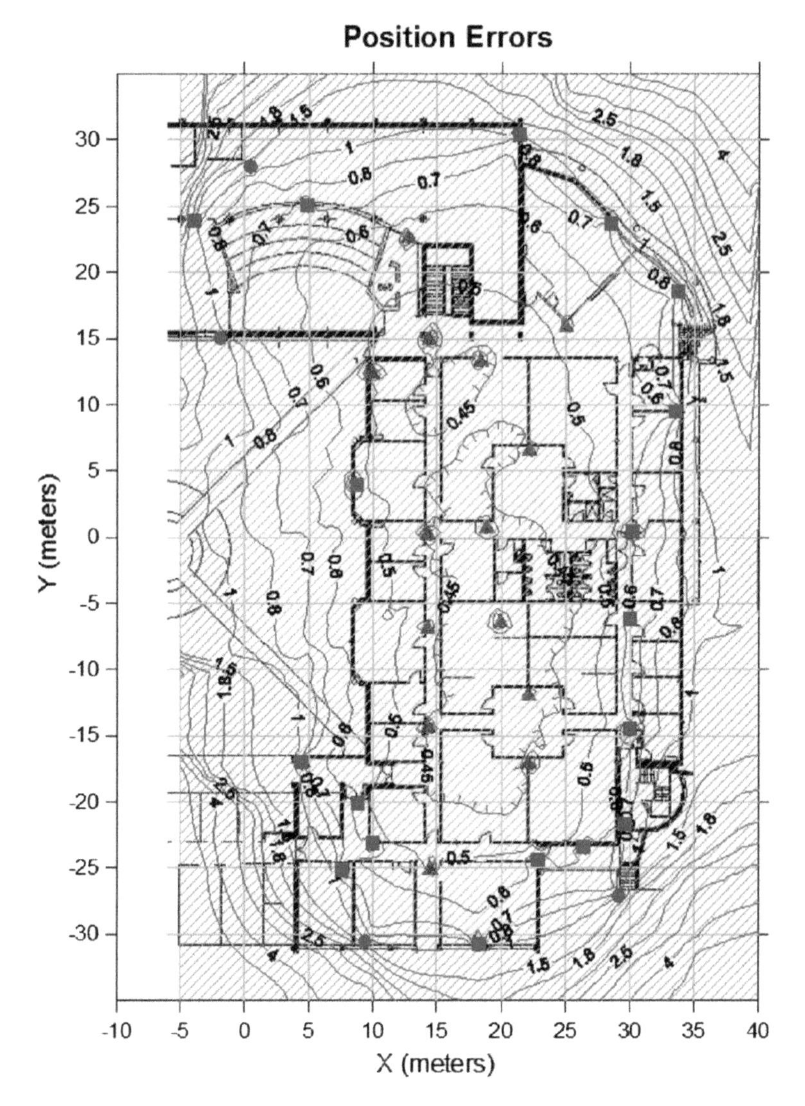

Fig. 14.7 Contour plot of position errors in meters, based on measured ranging data. Also shown are the positions of the 36 base stations, which cover an area of about 2100 m^2 inside a building. Core base stations are indicated by triangles. Note that the contours beyond the coverage area defined by the base stations are extrapolated, and do not represent accurate position error estimates

deviation in the average of $\sigma/\sqrt{N_{core}} = 0.076$ m, where σ is the standard deviation of the position errors; thus the measured and analytical estimates agree to within one standard deviation. That is, without conducting simulations and/or experiments, the derived formulas have reliably predicted the performance of

TOA-based indoor positioning systems. As a consequence, the design process can be greatly simplified.

Annex A: Expectation Calculation for the Inverse of $\mathbf{\Phi}$ Matrix

In this Annex the details of deriving the inverse of the $\mathbf{\Phi}$ matrix in (14.9) are provided. The $\mathbf{\Phi}$ matrix is the sum of two matrixes, a mean component plus a random component which has zero mean. In general, there is no simple expression for the inverse of the sum of two matrices, so that calculating the inverse requires extensive analysis. Ideally the inverse should be expressed in the same form as (14.9), namely a mean component matrix plus a zero-mean random component. The following analysis provides an approximation to such a solution.

Because the mean components of the $\mathbf{\Phi}$ matrix are much larger than the random components as seen from (14.15a) and (14.15b), one possible method involves the application of a Neumann series (Meyer 2000) for the inverse of the sum of two specific matrices, namely

$$(\mathbf{I}-\mathbf{A})^{-1}=\mathbf{I}+\sum_{n=1}^{\infty}\mathbf{A}^{n} \tag{14.39}$$

Applying (14.39) to the inverse of the $\mathbf{\Phi}$ matrix yields

$$\begin{aligned}(\mathbf{\Phi}_0+\mathbf{\Phi}_\varepsilon)^{-1}&=[\mathbf{\Phi}_0(\mathbf{I}-(-\mathbf{\Phi}_0^{-1}\mathbf{\Phi}_\varepsilon))]^{-1}\\&=\mathbf{\Phi}_0^{-1}+\left[\sum_{n=1}^{\infty}(-1)^n(\mathbf{\Phi}_0^{-1}\mathbf{\Phi}_\varepsilon)^n\right]\mathbf{\Phi}_0^{-1}\\&=(\mathbf{\Phi}_0^{-1}+\mathbf{\Delta})-\mathbf{\Phi}_0^{-1}\mathbf{\Phi}_\varepsilon\mathbf{\Phi}_0^{-1}\end{aligned} \tag{14.40a}$$

where

$$\mathbf{\Delta}=\left[\sum_{n=2}^{\infty}(-1)^n(\mathbf{\Phi}_0^{-1}\mathbf{\Phi}_\varepsilon)^n\right]\mathbf{\Phi}_0^{-1} \tag{14.40b}$$

The first order solution ignores the (small) $\mathbf{\Delta}$ term so that

$$(\mathbf{\Phi}_0+\mathbf{\Phi}_\varepsilon)^{-1}\approx\mathbf{\Phi}_0^{-1}-\mathbf{\Phi}_0^{-1}\mathbf{\Phi}_\varepsilon\mathbf{\Phi}_0^{-1} \tag{14.41}$$

Now consider the calculation of the expectation (or mean) of the inverse. Because the expectation of the $\mathbf{\Phi}_\varepsilon$ matrix is a null matrix, it is clear that

$$E[(\mathbf{\Phi}_0 + \mathbf{\Phi}_\varepsilon)^{-1}] = \mathbf{\Phi}_0^{-1} + E[\mathbf{\Delta}] \tag{14.42}$$

However, rather than calculating $E[\mathbf{\Delta}]$ directly from (14.40b) which involves an infinite series, an alternative more direct approach is taken. In particularly, from (14.7) the inverse of the 3×3 $\mathbf{\Phi}$ matrix can be determined directly. Consider one element of this inverse, namely

$$[\mathbf{\Phi}^{-1}]_{1,1} = \frac{N_R \sum_i \beta_i^2 - \left(\sum_i \beta_i\right)^2}{D} \tag{14.43a}$$

where the denominator is given by

$$\begin{aligned} D = {} & N_R \sum_i \alpha_i^2 \sum_i \beta_i^2 - N_R \left(\sum_i \alpha_i \beta_i\right)^2 - \sum_i \alpha_i^2 \left(\sum_i \beta_i\right)^2 \\ & - \sum_i \beta_i^2 \left(\sum_i \alpha_i\right)^2 + 2\left(\sum_i \alpha_i \beta_i\right)\left(\sum_i \alpha_i\right)\left(\sum_i \beta_i\right) \end{aligned} \tag{14.43b}$$

The calculation of the expectation of (14.43a) cannot be performed by calculating the expectation of the numerator and denominator and dividing. However, the problem can be solved by expanding the inverse of the denominator as a Taylor series, so that the expectation can be determined as a summation of the expectation of each of the Taylor series terms. The expectation of (14.43b) can be calculated as

$$E[D] = D_0 = \frac{N_R(N_R - 1)(N_R - 2)}{4} \tag{14.44a}$$

so that

$$D = D_0 + d = D_0\left(1 + \frac{d}{D_0}\right) \tag{14.44b}$$

where

$$d = D - D_0 \tag{14.44c}$$

is the zero-mean random component. Applying (14.44b) to (14.43a) and expanding D^{-1} as a Taylor series yield

$$[\mathbf{\Phi}^{-1}]_{1,1} = D_0^{-1}\left(N_R \sum_i \beta_i^2 - \left(\sum_i \beta_i\right)^2\right)\left(1 - \frac{d}{D_0} + \left(\frac{d}{D_0}\right)^2 + \cdots\right) \tag{14.45}$$

Finally the expected value of (14.43a) can be approximated by the summation of the expectation of the individual terms in (14.45), resulting in

$$\begin{aligned} E[[\mathbf{\Phi}^{-1}]_{1,1}] &= D_0^{-1}\left(N_R E\left[\sum_i \beta_i^2\right] - E\left[\left(\sum_i \beta_i\right)^2\right]\right)\left(1+E\left[\left(\frac{d}{D_0}\right)^2\right]\right) \\ &= \left(\frac{4}{N_R(N_R-1)(N_R-2)}\right)\left(\frac{N_R^2}{2}-\frac{N_R}{2}\right)\left(1+\left(\frac{\sigma_d}{D_0}\right)^2\right) \end{aligned} \tag{14.46}$$

where it has been assumed that d is statistically independent of the other random terms in (14.46), σ_d is the standard deviation of d, and the Taylor series is limited to three terms. While it is possible to calculate σ_d analytically, the number of terms makes this calculation rather cumbersome. As the term involving σ_d in (14.46) is rather small, one approach is to simply ignore the small correction associated with it. In this case the expectation becomes

$$E[[\mathbf{\Phi}^{-1}]_{1,1}] \approx \frac{2}{N_R-2} = \frac{2}{N_R} + \frac{4}{N_R^2} + \frac{8}{N_R^3} + \cdots \tag{14.47}$$

The first term in the series in (14.47) can be recognized as the inverse component for $\mathbf{\Phi}_0^{-1}$, which is first order inverse component from (14.41). The other terms in (14.47) are small corrections which approach zero as the number of nodes in range becomes large. Note also that including σ_d in the calculation only affects the last cubic term in (14.47). The other components of the inverse can be similarly calculated. In particular, the other two diagonal components are

$$\begin{aligned} E[[\mathbf{\Phi}^{-1}]_{2,2}] &\approx \frac{2}{N_R} + \frac{4}{N_R^2} + \frac{8}{N_R^3} + \cdots \\ E[[\mathbf{\Phi}^{-1}]_{3,3}] &\approx \frac{1}{N_R} + \frac{2}{N_R^2} + \frac{4}{N_R^3} + \cdots \end{aligned} \tag{14.48}$$

The remaining components of the expectation can be similarly calculated to be all zero. Thus ignoring the σ_d effect and limiting the series to three terms, the expected value of the inverse of the **Φ** matrix is then as given by (14.17a, 14.17b).

Now consider the random component of the inverse. From (14.41), the first order solution to the random component of the inverse is

$$\tilde{\Phi}_\varepsilon = -\mathbf{\Phi}_0^{-1}\mathbf{\Phi}_\varepsilon\mathbf{\Phi}_0^{-1} \tag{14.49}$$

To calculate higher-order solutions, one needs to consider the random components of **Δ**. However, such analysis results in very complex analytical expressions, and thus only the first order solution (14.49) is used in the analysis.

Annex B: Calculation of Expectations

Expectation Calculation Based on Surface Integrals

This section briefly presents the derivation of the first expectation in Table 14.1, based on the surface integral technique. The required expectation is

$$E[m_x^2] = \frac{1}{N_R^2} E\left[\left(\sum_i x_i\right)^2\right] = \frac{1}{N_R^2} E\left[\sum_i x_i^2\right] \quad (14.50)$$

where the last expression in (14.50) is obtained because the mean of the x-coordinates is zero and the samples are statistically independent, as it is assumed that the distribution of anchor nodes is statistically uniform throughout a circular coverage area whose center is the origin of the coordinate system. The mean of the summation of the squared x-coordinate of the anchor node positions can be determined by use of an appropriate surface integral. That is, integrating over one quadrant of the circle produces

$$\begin{aligned} E\left[\sum_i x_i^2\right] &= \sum_i E[x_i^2] = N_R E[x_i^2] \\ &= \frac{N_R}{\pi R_{\max}^2/4} \int_0^{R_{\max}} \int_0^{\sqrt{R_{\max}^2 - y^2}} x^2 dx\, dy = \frac{N_R R_{\max}^2}{4} \end{aligned} \quad (14.51)$$

Substituting the results in (14.51) into (14.50) produces

$$E[m_x^2] = \frac{1}{N_R^2} E\left[\sum_i x_i^2\right] = \frac{R_{\max}^2}{4N_R} \quad (14.52)$$

Calculation of Expectation of Products of Means of Random Variables

This subsection gives an illustration of the method of calculating the expectation of the product of means of random variables, as described in Sect. 14.3.2, paragraph 2. The method can be applied generally, but will be illustrated with a particular example, namely the calculation of $E[m_x m_\alpha m_r]$—see Table 14.1, item 6. The method is based on splitting the summations associated with the mean operations into two groups, one where all the indices are the same, and one where they are different. From the definition of a mean (m), the above expectation of the product of three means can be written as

$$E[m_x m_\alpha m_r] = \frac{1}{N_R^3} E\left[\sum_i x_i \sum_j \alpha_j \sum_k r_k\right] = \frac{1}{N_R^3} E\left[\sum_i \alpha_i r_i \sum_j \alpha_j \sum_k r_k\right] \quad (14.53)$$

where all the summations are over all nodes in range (N_R), and α, r are statistically independent random variables. First consider the group with all the indices the same. In this case (14.53) becomes

$$\begin{aligned} \frac{1}{N_R^3} E\left[\sum_i \alpha_i r_i \sum_j \alpha_j \sum_k r_k\right] &= \frac{1}{N_R^3} E\left[\sum_i \alpha_i^2 r_i^2\right] = \frac{1}{N_R^2} E\left[\alpha^2\right] E\left[r^2\right] \\ &= \frac{1}{N_R^2}\left(\frac{1}{2}\right)\left(\frac{R_{\max}^2}{2}\right) = \frac{1}{4}\left(\frac{R_{\max}}{N_R}\right)^2 \end{aligned} \quad (14.54)$$

where the expectation of r^2 can be calculated by the surface integral method described in 14.B.1. Now consider the group where some or all the indices are different. In this case with $(i, j, k, i \neq j \neq k)$ (14.53) becomes

$$\begin{aligned} \frac{1}{N_R^3} E\left[\sum_i \alpha_i r_i \sum_j \alpha_j \sum_k r_k\right] &= \frac{1}{N_R^3} E\left[\sum_i \sum_j \sum_k \alpha_i \alpha_j r_i r_k\right] \\ &= \frac{1}{N_R^3} \sum_i \sum_j \sum_k E[\alpha_i] E\left[\alpha_j\right] E[r_i] E[r_k] = 0 \end{aligned} \quad (14.55)$$

as $E[\alpha] = 0$, and as the indices are all different the random variables are statistically independent. Next consider the subgroup when x and α have the same index, but r has a different index. In this case (14.53) becomes

$$\begin{aligned} \frac{1}{N_R^3} E\left[\sum_i \alpha_i r_i \sum_i \alpha_i \sum_j r_j\right] &= \frac{1}{N_R^3} E\left[\sum_i \sum_j \alpha_i^2 r_i r_j\right] = \frac{1}{N_R^3} \sum_i \sum_j E\left[\alpha_i^2\right] E\left[r_i r_j\right] \\ &= \frac{N_R^2 - N_R}{N_R^3}\left(\frac{1}{2}\right) E[r]^2 \end{aligned} \quad (14.56)$$

where the number of elements in the group is $N_R^2 - N_R$. Using the surface integral method described in 14.B.1, it can be shown that $E[r] = \frac{2}{3} R_{\max}$, so that

$$\frac{1}{N_R^3} E\left[\sum_i \alpha_i r_i \sum_i \alpha_i \sum_j r_j\right] = \frac{2}{9}\frac{R_{\max}^2}{N_R} - \frac{2}{9}\left(\frac{R_{\max}}{N_R}\right)^2 \quad (14.57)$$

Similar types of calculations show that the expectation of all other subgroups is zero as $E[\alpha] = 0$. Finally, combining (14.54) and (14.57), the required expectation in (14.53) (see also Table 14.1, item 6) is

$$E[m_x, m_\alpha, m_r] = \frac{2}{9}\frac{R_{\max}^2}{N_R} + \frac{1}{36}\left(\frac{R_{\max}}{N_R}\right)^2 \tag{14.58}$$

The above example illustrates the method by which splitting the summations associated with the mean operation the expectation can be calculated, with most subgroups having a zero expectation.

References

Bulusu N, Heidemann J, Estrin D (2001) Adaptive beacon placement. In: Proceedings of international conference on distributed computing systems, Phoenix, Arizona, USA, April 2001, pp 489–498

Gentile C, Kik A (2006) An evaluation of ultra wideband technology for indoor ranging. In: Proceedings of IEEE global telecommunications conference (GLOBECOM), San Francisco, California, USA, Nov 2006, pp 1–6

Hedley M, Humphrey D, Ho P (2008) System and algorithms for accurate indoor tracking using low-cost hardware. In: Proceedings of IEEE/IOA position, location and navigation symposium, Monterey, California, USA, May 2008, pp 633–640

Humphrey D, Hedley M (2008) Super-resolution time of arrival for indoor localization. In: Proceedings of international conference on communications (ICC), Beijing, China, May 2008, pp 3286–3290

Lee HB (1975) A novel procedure for assessing the accuracy of hyperbolic multilateration systems. IEEE Trans Aerosp Electron Syst 11(1):2–15

Meyer C (2000) Matrix analysis and applied linear algebra. Society for Industrial and Applied Mathematics

Miao H, Yu K, Juntti M (2007) Positioning for NLOS propagation: algorithm derivations and Cramer-Rao bounds. IEEE Trans Veh Technol 56(5):2568–2580

Sathyan T, Humphrey D, Hedley M (2011) A system and algorithms for accurate radio localization using low-cost hardware. IEEE Trans Soc Man Cybern—Part C 41(2):211–222

Sharp I, Yu K, Guo Y (2009) GDOP analysis for positioning system design. IEEE Trans Veh Technol 58(7):3371–3382

Schmidt R (1986) Multiple emitter location and signal parameter estimation. IEEE Trans Antenna Propag 34(3):276–280

Spirito M (2001) On the accuracy of cellular mobile station location estimation. IEEE Trans Veh Technol 50(3):674–685

Teunissen P (2002) Adjustment theory—an introduction. Series on mathematical geodesy and positioning. VSSD

Torieri D (1984) Statistical theory of passive location systems. IEEE Trans Aerosp Electron Syst 20(2):183–198

Yick J, Bharathidasan A, Pasternack G, Mukheriee B, Ghosal D (2004) Optimizing placement of beacons and data loggers in sensor network—a case study. IEEE wireless communications and networking conference (WCNC), Atlanta, Georgia, USA, March 2004, pp 2486–2491

Yu K (2007) 3-D localization error analysis in wireless networks. IEEE Trans Wirel Commun 6 (10):3473–3481

Yu K, Guo Y (2008) Improved positioning algorithms for nonline-of-sight environments. IEEE Trans Veh Technol 57(4):2342–2353

Yu K, Sharp I, Guo Y (2009) Ground-based wireless positioning. Wiley-IEEE Press, Chippenham

Zhu J (1992) Calculation of geometric dilution of precision. IEEE Trans Aerosp Electron Syst 28 (3):893–895

Chapter 15
Signal Strength Positioning

15.1 Introduction

In radio positioning, five basic signal parameters have been widely considered for position determination, namely signal strength, time-of-arrival, frequency difference-of-arrival, carrier phase, and angle-of-arrival. Each of these methods have different strengths and weaknesses, so in general there must be a tradeoff between complexity, the associated cost and positional accuracy. Nearly all modern radio communication devices include a Receiver Signal Strength Indicator (RSSI) functionality. As a consequence, there is a strong motivation to utilize this capability for position determination, as a software-only method can be implemented even in the simplest of devices. Thus there has been a considerable interest in developing appropriate methods, particularly associated with the widespread adoption of WiFi (which mainly operates in the 2.4 GHz ISM band), both in commercial applications and in homes. This chapter provides some new methods for position determination (to an accuracy of a few meters) which only requires the RSSI measurements from several transmitters (such as WiFi access points), and apart from the location data of these transmitters no other information is required for position determination. This means that positioning indoors is often possible due to the widespread adoption of WiFi.

Chapter 8 described methods using WiFi to determine position in situations where multiple WiFi access points (AP) are within radio range. However, these methods are based on the signal strength measured at calibration points throughout the building, and thus this method cannot be used in general where no such calibration data are available. The concept behind such methods is that the complexity of the indoor signal strength map can be exploited to determine position, by matching a measured signal pattern from multiple APs to the calibration database. Further, the accuracy of position determination is directly related to the scale of the calibration grid points, so that meter-accuracy position determination is possible, albeit with a large database of calibration points. Indeed, the methods described in

I. Sharp and K. Yu, *Wireless Positioning: Principles and Practice*, Navigation: Science and Technology, https://doi.org/10.1007/978-981-10-8791-2_15

Chap. 8 show with a calibration grid size in S the maximum position errors are of the order of S, and the RMS errors of the order of $S/2$. Although such methods have been advocated for many years (Caso et al. 2015; Chirakkal et al. 2015; Koo and Cha 2012), this concept has not been commercially successful for indoor applications due to the effort of generating and maintaining databases. However, for more general position determination in urban areas worldwide databases of over one billion WiFi access points (in homes and commercial buildings) have been established commercially (iBeacon 2018; Combain 2018), allowing positions to be estimated to an accuracy of 30–50 m on smart phones and other similar devices using appropriate "Apps". This allows devices without GPS positioning capability to obtain an approximate position in urban areas, including when the mobile device is located indoors. However, these applications do not provide any useful position information for use within buildings, but only a rough geographic position for navigation within urban areas.

Thus in this chapter some methods of determining position based solely on the knowledge of the location of *radio nodes* and their associated RSSI readings are considered, and importantly the accuracy of such methods is analysed and compared with measurements. The radio nodes could be based on WiFi, or any other system such as Zigbee or wireless sensor networks. However, in this chapter the measured data are associated with an ad hoc network of devices which includes a time-of-arrival positioning capability, thus allowing a comparison between RSS and TOA methods.

15.2 Overview of Signal Strength Positioning

In theory Receiver Signal Strength (RSS) positioning is simple to implement and can use existing radio infrastructures such as WiFi access points (AP). However, the variation of the signal strength as a function of distance/position can be complex, especially indoors, so that in general the position accuracy that can be obtain is rather poor, compared with other methods such as signal time-of-arrival. Typically, the implementation of RSS measurements is based on a diode detector in the IF of the radio receiver, and thus is cheap and easy to implement. However, note that RSSI only gives an *indication* of the signal strength, and is not intended to be an accurate measurement of the signal strength at the antenna port of the device. As the measurement is not directly of the RF signal, some form of calibration is required. Specifications for standard common consumer radio systems such as WiFi, Bluetooth and Zigbee only require the measurement within a specified range; for example, the Zigbee IEEE 802.15.4 specification has an accuracy range of ± 5 dB. Such imprecision in raw RSSI measurements (Lui et al. 2011) must be taken into account if RSSI positioning is to achieve a reasonable accuracy.

In many indoor environments such as offices, hospitals, hotels, shopping centres and other similar large buildings, there are often WiFi access points (AP) and Bluetooth devices which can provide short-range wireless communications. Note that these technologies are designed for data communications, with indoors range being limited to less than 10 m for Bluetooth, and perhaps 30–40 m for WiFi. Devices such as mobile phones include a RSSI measurement as a standard part of the API access to the device operating system software. This software access can be exploited by radiolocation applications on the device.

There are two main types of location-based-applications. One is based on using multiple RSSI measurements to determine position using methods based on the propagation loss inferred from the RSSI measurements, a variant of which is the main topic of this chapter. The second method is based on a database of WiFi APs, and is typically the default method of position determination in mobile devices, particularly for indoor applications. This technology is based on providing low-accuracy position information (of the order of 20–100 m) mainly in urban areas, and is based on commercial databases (Combain, Unwired Labs), of the locations of billions of WiFi APs throughout the world. However, the methods described in this chapter are intended for tasks such as tracking and navigating within buildings, with a positional accuracy of (ideally) a few meters.

One common method that can be used for RSS positioning is to estimate the range to base stations such as WiFi access points, and then use standard methods (Liu et al. 2007; Gu et al. 2009) for position determination. The basic methodology is to use the path link budget (see Chap. 9) to estimate the propagation loss, and then estimate the range using a loss model of the form $L(R)$; inverting this function provides an estimate of the range. However, it is shown in (Sharp and Yu 2018) that this simple method results in rather poor and unreliable positioning performance, particularly for ranges beyond about 15 m. At longer ranges the RSSI value in decibels only changes slowly with distance, so that the range estimates are very poor as the RSSI in decibels changes increasingly slowly as the range increases. Further, the link budget parameters are often not known. In particular, the transmitter powers, antenna gains on both the transmitter and the receiver, and the RSSI calibration in the receiver are generally not available, so that the link budget calculation will introduce considerable errors in any calculation. As a result, the accuracy of such methods indoors is at best about 10 m, with some positions errors much greater. As a consequence, applications are limited to identifying the location of the mobile device to a particular building or house, but not within the building.

To overcome these limitations of the standard method, and alternative approach has been suggested that exploits the complex variable nature of indoor radio propagation. Suppose that the RSSI signals received from base stations over the coverage area have a range of 50 dB, and that the RSSI can be read to an accuracy of 2 dB. Then if four base stations are simultaneously in radio range of the mobile, the number of distinct RSSI "fingerprints" will be $25^4 \approx 4 \times 10^5$. Thus even if the base station signals are sampled on 1 m grid within a building, the number of samples will be far less then the maximum number of possible fingerprints. The method is thus to survey the coverage area on a grid of points, measuring and recording the RSSI at each grid point, the spacing of which will be of the order of the required positional accuracy; closer spacing allows greater positional accuracy, but requires more surveying effort. Each point's RSSI data can be stored as a vector in a central database. To obtain a position the mobile device measures the RSSI of the base stations (usually WiFi access points) within range, and uses the radio data communication network to forward the RSSI readings vector to the central database. By finding the closest match to an entry in the database, the position can be determined, and them forwarded back to the mobile device. This is a general description which is common to these fingerprint methods, but more sophisticated algorithms have been devised. For more details, refer to Chap. 8. However, there are major problems in implementing such a system. First, a major survey of the coverage area must be undertaken, and the data stored in a large central database. This central system must also perform the computational search procedure to determine the position, so the computational requirements will be large if these are large numbers of mobile devices updating their position frequently. Further, there is a considerable data traffic requirement on the data network. Thus this concept is complex and expensive to implement, and is not suitable to most situations. In contrast, the method described in this chapter is simple to implement, with all the (minimal) computations being performed in software in the mobile device.

Because of the limitations of these standard RSS positioning methods, this chapter investigates an alternative simple method that only requires the RSSI data from a number of base stations to determine the position. Further, measurements of a person walking around a building show a position accuracy of better than 4 m. The details of this method are described in the remainder of this chapter.

15.3 Indoor Radio Propagation

The characteristics of the radio signal are important in designing an indoor positioning system. To estimate the range from a transmitter to a receiver using the RSSI, the nature of large-scale signal variation with distance is required. While in free space a simple inverse-square law can be used to determine the signal strength as a function of range, in a NLOS environment the complexity of the environment makes calculations of the large-scale signal attenuation with distance difficult. Further, while the average variation with distance is important, the medium-scale variation of the signal strength is also an important parameter in the design of indoor positioning systems. These medium-scale effects are typically on a scale of 5–50 wavelengths (0.5–5 m at 2.4 GHz used by WiFi), and thus are associated with major structural elements of a building such as rooms. Small-scale signal variations also can be considerable, further complicating matters. In particular, constructive and more importantly destructive interference (Rayleigh and Rician statistical fading) occurs on the scale of the order of a wavelength (12.5 cm at 2.4 GHz). These dividing lines are only rough guidelines which are useful for engineering analysis of the complex propagation environment of indoor radio propagation at microwave frequencies. In summary, small-scale variations in signal strength are associated with complex (amplitude and phase) constructive and destructive interference between the signal components, medium-scale effects are associated with signal scattering, diffraction and the absorption of RF energy by obstacles along the propagation path, and large-scale variation is associated with the free-space attenuation in addition to signal losses mainly associate with walls along the propagation path. It is this complicated spatially varying signal that must be utilised for position determination.

15.3.1 Link Budget

The basis of signal strength positioning is the determination of the propagation loss between a number of transmitters (fixed base stations) and a common receiver (mobile device), and then use this information to determine the position of the receiver. The transmission process can be summarised in terms of a link budget (for more details refer to Chap. 9), whereby all the major hardware components and associated parameters are defined, so that the propagation loss can be calculated. For calculations it is convenient to define the link budget in decibels, so that various elements in the budget can be added rather than multiplied. Thus the link budget can be expressed in the form

$$P_{rx} = P_{tx} + G_{tx} - L + G_{rx} \tag{15.1}$$

Where

P_{rx} received power (in dBm, or dB relative to a milliwatt)
P_{tx} transmit power (20 dBm for the nodes used in this chapter)
G_{tx} transmitter antenna gain relative to a hypothetical omnidirectional antenna (5 dBi for nodes in this chapter)
L propagation loss (in dB)
G_{rx} receiver antenna gain (5 dBi)

The basic starting point for loss determination is the free-space loss (also referred to as the line-of-sight (LOS) loss), as this represents the minimum possible loss. The free space path loss for a range R and a wavelength λ is given by

$$\begin{aligned} l_{free\_space}(R) &= \left(4\pi\frac{R}{\lambda}\right)^2 \\ L_{free\_space}(R) &= \left(\frac{20}{\ln(10)}\right)\ln\left(4\pi\frac{R}{\lambda}\right) = \alpha_0 + \beta_0 \ln\left(\frac{R}{\lambda}\right) \quad \text{(in dB)} \end{aligned} \tag{15.2}$$

However, indoors the propagation environment is more complex, and usually there is no LOS, so that propagation occurs through multiple walls along the path from the transmitter to the receiver. The data used in the following analysis are based on measurements from a system with 23 nodes (base stations) in a building (see Fig. 15.1 for a basic map, and Fig. 11.20 in Chap. 11 for a detailed map) covering 2100 m^2, with a average separation of 8 m; note as this is a positioning system requiring reception at multiple nodes for position determination, the node density is considerably greater than for a data-only (WiFi) network. The inter-nodal RSSI measurements were based on each node (base station) in turn transmitting, and the remaining 22 nodes receiving. This arrangement approximates a mobile node located at 23 locations throughout the building.

At a 30 m range, the propagation path loss excess is typically of the order of 35 dB, as shown in Fig. 15.2. The effective node transmitter power used for testing

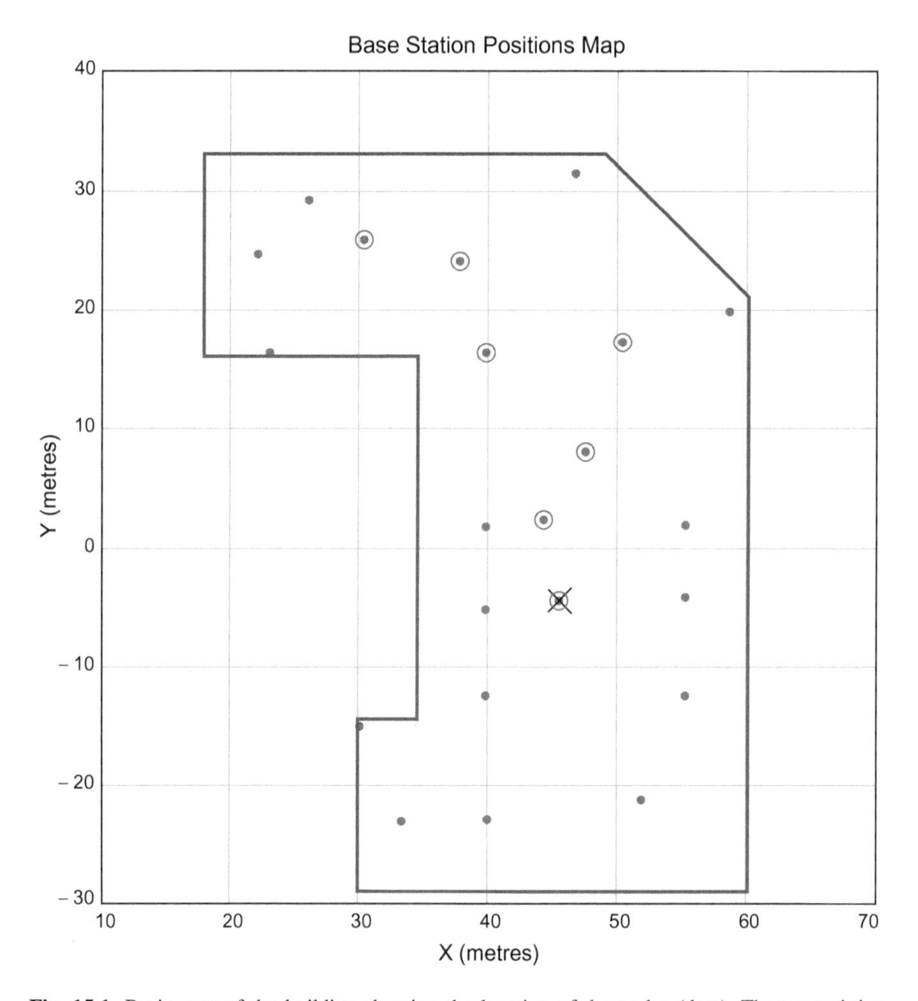

Fig. 15.1 Basic map of the building showing the location of the nodes (dots). The transmitting node is marked by an X. The seven nodes marked by an O are nodes in the core of the building, and are surrounded by other nodes. Other nodes are on the edge of the building, and thus are not completely surrounded by other nodes

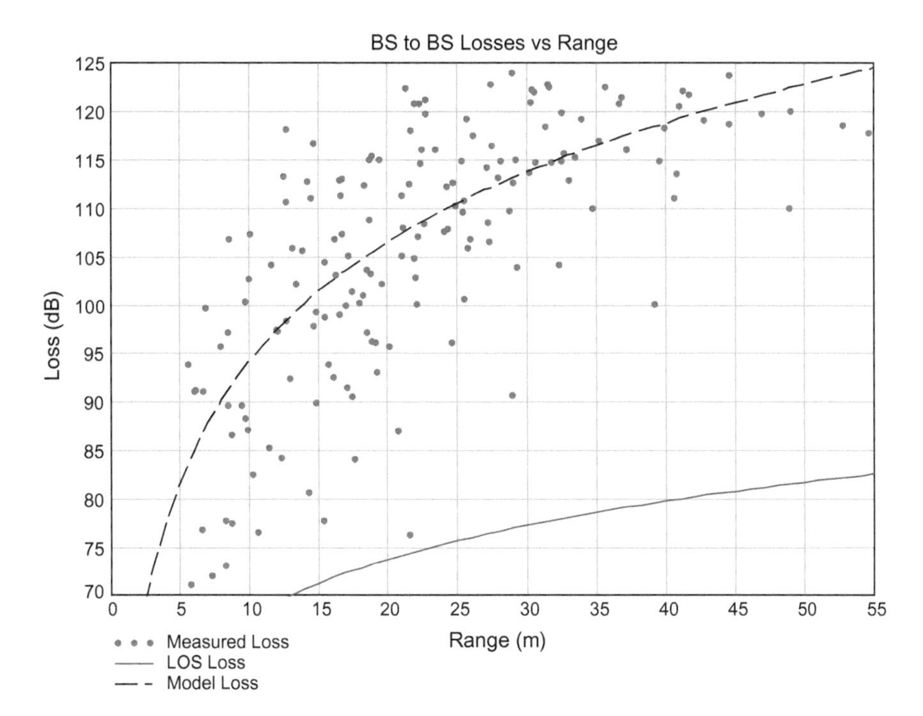

Fig. 15.2 Propagation loss as a function of range based on the raw RSSI signal strength data. The dashed line is a least-squares fit to a model of the form $L(R) = \alpha + \beta \ln(R/\lambda)$, where $\alpha = 0.7$ dB and $\beta = 40.9$ so $\gamma = 4.1$. Also shown is the LOS propagation path loss. The model RMS fit error is 9.1 dB

the this chapter is of 25 dBm, so P_{rx} is about −90 dBm at 30 m. The noise floor for the receiver used for the measurements is −98 dBm, but the receiver has a process gain of 30 dB so that output SNR is a high 38 dB.

15.3.2 Large Scale Propagation Characteristics

Large-scale propagation effects are the basis for determining the distance of transmitting nodes from a receiver, and hence determining the position. Given a radio system parameter such as the transmitter power, the receiver sensitivity and antenna gains, a link budget can be devised to estimate the propagation loss between the transmitter and receiver units. From this estimate and the large-scale attenuation characteristics, the range can be estimated using an appropriate range-loss model. As the propagation characteristics indoors are complex, the typical approach taken is for empirical generic equations to represent the propagation loss, with the parameters determined by appropriate fitting these equations to measured data.

The propagation loss as a function of distance for the ad hoc node network (Sathyan et al. 2011) used in this chapter is shown plotted in Fig. 15.2. One very common indoor propagation model based on an extension of (15.2) is of the form

$$l = l_0 \left(\frac{R}{\lambda}\right)^{\gamma} \tag{15.3}$$

In this power-law model, the attenuation as a function of distance is controlled by the parameter γ with $\gamma > 2$ as the loss is greater than the free-space case, and l_0 the loss at one wavelength. The loss is assumed to be a function of the propagation distance measured in wavelengths, as with the free-space case. However, as in practice loss is measured in decibels, a more appropriate form for indoor NLOS large-scale propagation loss is

$$\begin{aligned} &L(\mathrm{dB}) = \alpha + \beta \ln(R/\lambda) \qquad (L \geq 0 \quad \text{or} \quad R \geq R_{\min}) \\ &\alpha = \frac{10}{\ln(10)} \ln(l_0) \qquad \beta = 10\gamma/\ln(10) \qquad R_{\min} = e^{-\alpha/\beta} \end{aligned} \tag{15.4}$$

Notice that natural logarithms are used in (15.4), as it is more convenient in latter mathematical developments. Figure 15.2 also shows the free-space loss for comparison, as well as the least-squares fit based on model (15.4). As can be observed, the measured loss is considerably greater than the free-space loss, and also exhibits random (medium-scale) variation. From the least-squares fit to these data the power law parameter γ is 4.1 (compared with 2 for free-space). As the power law exponent $\gamma = 4.1$ is much greater than the free space exponent, the losses increase much more rapidly with propagation distance. While this model is perhaps a natural extension to the free-space law (which is based on a rigorous solution to Maxwell's equations), the model defined by Eq. (15.4) is not totally satisfactory. For example, when the propagation range is short LOS conditions will exist, implying that $\gamma = 2$ at short range, but a larger value applies at long ranges, with presumably a transitionary region at intermediate ranges. Further, at medium ranges the NLOS propagation indoors will typically be through a number of walls with a consequential loss at each wall, so this implies a linear increase (in dB) in the loss excess as a function of distance. As the range further increases and the losses through walls increases, there may be an alternative lower loss path along corridors in the building. Thus at long range the propagation loss (in dB) will increase at a slower rate than predicted by a simple wall loss excess model. Thus in summary, all these effects cannot be captured with a constant power law exponent.

Another criticism of the simple power law (15.4) is the difficulty in relating the power law parameters to the physical nature of the building. In practice the power law exponent γ can only be determined by measurements in specific buildings, with reported values in the range of 3–6 for indoor propagation. Because of the difficulty of applying (15.4) in practical situation in various buildings, an alternative approach is taken in the algorithms described in the following sections. While model (15.4) is used, it is not used directly in the position determination, but rather it is used to estimate the *differential* loss between the node with the greatest RSSI signal and other nodes with a smaller RSSI.

15.4 Weighted Closest Nodes Algorithm Positioning

This section describes a simple algorithm for computing positions of mobile modules based on the received signal strength (RSS) from nodes transmitting a radio signal in an indoor environment. Unlike classical RSS positioning which uses the link budget to estimate the propagation loss and hence the ranges, instead the method uses the receiver signal strength indicator (RSSI) to determine a weighting function for each transmission. From this weighting the position of the mobile is determined as a weighted sum of the locations of the base stations transmitting the signal to the receiving device. The method is not intended for accurate position determination, but as a simple quick method of estimating the position based solely on the RSSI data and the known locations of the transmitting nodes. As no other information is required, the method would be appropriate for simple "smart phone" application programs based on using existing access points in a building.

15.4.1 Overview of Algorithm

The basic concept of the algorithm is that the mobile receiver typically will be "surrounded" by transmitting nodes (see "core" nodes in Fig. 15.1), so that the receiver location can be estimated as the "average" of the locations of these nearby transmitters. For example, the mobile node whose position is sought will have an x-coordinate which lies somewhere between the x-coordinates of the surrounding nearby nodes, that is some are greater and some smaller than the mobile's x-coordinate; the same applies for the y-coordinate. Thus by averaging the coordinates of these nearby nodes a rough estimate of the position of the mobile node can be determined, with the error of the order of the standard deviation of the coordinates of the surrounding nodes divided by the square-root of the number of surrounding nodes; this is a standard statistical property of averaging random data. Further, by weighting the averaging process appropriately, this geometric arrangement *always* can result in *exact* coordinates of the mobile node, but the weighting required to achieve this ideal goal is not known. Also observe that the weighting pattern required for zero positional errors in not unique. Thus if an appropriate weighting pattern (it will not be unique) can be independently estimated (based on the RSSI data), then the mobile position can be determined exactly. In practice a "correct" weighting pattern for the averaging process cannot be determined from the available information (RSSI measurements which can only be approximately modeled), so that there will be some residual errors in the computed position. Nevertheless, by applying the weighted averaging to appropriately selected nearby nodes a better estimate than simply averaging the coordinates of the nearby nodes can be obtained.

While in the layout shown in Fig. 15.1 a mobile unit (attached to a person) would almost always be surrounded by the receiving nodes (as required by the algorithm), with base stations simulating a mobile device most of the tested positions are near the edge of the building, and are thus not particularly representative. However, a small number (seven) of nodes marked by a O in Fig. 15.1 are surrounded by other nodes; it is expected positioning accuracy for these nodes will be better than for the nodes on the periphery of the building.

15.4.2 Measured Performance

The algorithm for position determination is based on the observation that the receiver will be located somewhere near the center of the locations of surrounding transmitting nodes. These signals give a rough indication of the range from the transmitter to the receiving node, with the strongest signal probably being the closest transmitting node to the mobile position. Thus as a very rough estimate, the mobile position can be equated to the location of this node. A somewhat better estimate is to use the mean of the positions of the surrounding base stations, as the mobile must be roughly in the "centre" of the surrounding cluster. As a mobile device moves, the base stations in radio communications will change, and thus the computed position will also change. This averaging method weights all base receiving stations equally. The positional performance can be tested as the number of nodes used in the averaging is varied, beginning at one. The order of including nodes is based on the order of signal strength. Note that this is only an *indication* of the order of range, as in some cases distant nodes with a clear line of sight will have a stronger signal than a nearby node but with no line-of-sight. Thus the method uses the signal strength as a *proxy* for determining the order of closeness of receiving nodes to the transmitting nodes. What is of interest is the computed position error, and the effect of increasing the number of transmitting nodes.

The computed performance with four nearby nodes is shown in Fig. 15.3. The mean errors for all locations in the x and y coordinates are respectively −0.2 m and −0.3 m, or essentially zero. This result shows that the above described method of estimating the position results in an unbiased estimate. This result also confirms the theoretical expectation, as described in Sect. 15.4.4.

The RMS error of both x and y coordinates is 7.0 m and the RMS radial error (indicated by the circle) is 9.9 m. However, if the nodes are restricted to those in the core of the building (indicated by O around the point) the RMS errors are 1.6 m and 1.8 m respectively for x and y, and the RMS radial error is 3.9 m. It is also useful to normalise the results based on the mean distance the nearest neighboring

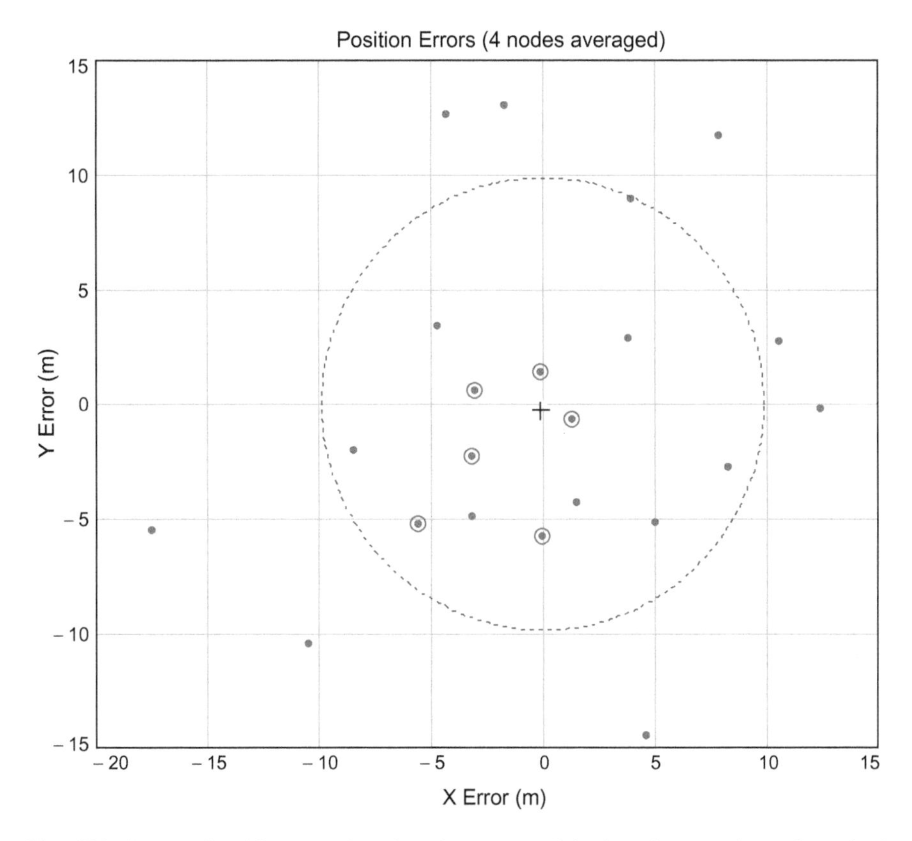

Fig. 15.3 Scatter of position errors based on the average of the four closest nodes as determined by the RSSI. The circle indicates the RMS radial error of 9.9 m, and the cross marks the mean error (−0.2 m, −0.3 m)

transmitting node ($\bar{d} = 8$ m); the overall normalised RMS radial error is 1.2 and for the core nodes 0.49.

The statistical distribution of the radial errors is shown in Fig. 15.4 for two cases: using the closest node only and using the four closest nodes, both based on the RSSI. As can be observed, there is a considerable improvement using four nodes, particularly the maximum error. In particular notice the long "tail" on the "closest node" distribution, where position errors up to 35 m occurs. The reason for the occasional large errors is that a distant node with a LOS path can exist, which results in a bad selection of the nearest neighbour using the RSSI as the indicator. This large error is significantly reduced when four nodes of position data are averaged. The testing showed that the overall accuracy improved as the number of nodes increased to about 3–4, and then there was no further improvement. It is suggested that using four nodes is a good choice based on some basic geometry, as explained in the next subsection. As the node distribution throughout a building would typically be designed based on a square grid of some defined size, the four

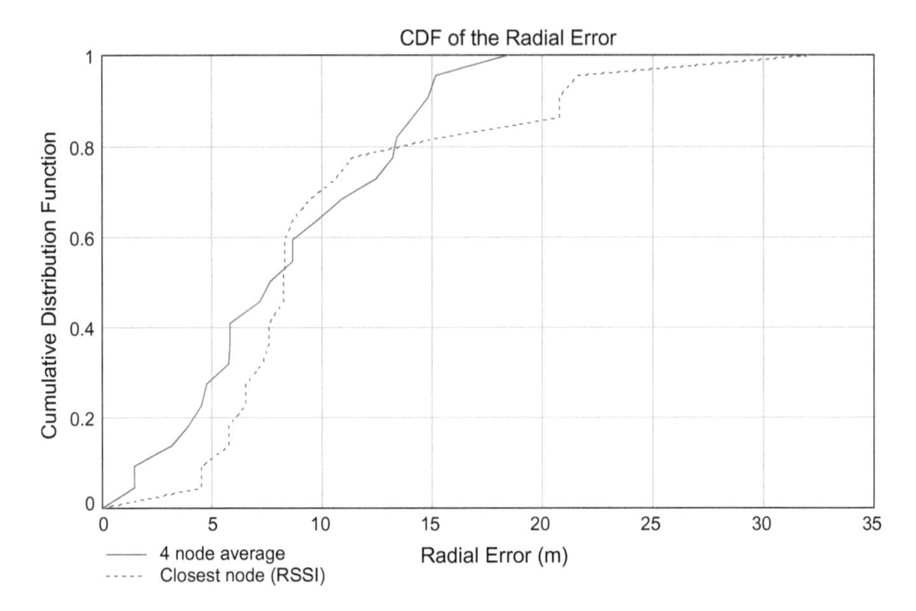

Fig. 15.4 Cumulative distribution functions of the radial error based on the closest node and the average of the four closest nodes based on the RSSI. Note the long "tail" associated with the closest node based on the strongest RSSI measurement

nearest transmitting nodes to a mobile node would be on the corners of a square, and hence the choice of four closest nodes is appropriate.

15.4.3 Closest Nodes Algorithm Performance Overview

The measured performance of the closest nodes position estimation algorithm in Sect. 15.4.2 provided numerical and statistical information on the performance of the algorithm; this section provides an analytical statistical analysis which explains the results and provides useful formula for designing a system based on this method of position determination. In particular, the following needs to be explained:

1. The average position error is unbiased, even though the estimated position is based on the closest node or the average of a number of close nodes.
2. There is a difference in performance for locations in the core of the building, compared with positions more on the periphery of the building.
3. The accuracy statistics needs to be explained in terms of the geometry and the statistical variations associated with the mobile node position and the locations of nearby transmitting nodes.

The starting point in the analysis is the basic geometry of the positioning algorithm. Note that the RSSI measurements are only used in determining the nodes used the positioning algorithm, so in this very basic method only geometric factors are relevant. The most important factor is thus the distribution of nodes throughout the coverage area. As a default in planning a system it can be assumed that the fixed "anchor" nodes are located on a square grid of size S. More generally for tracking mobiles within the coverage area, the mobile nodes can be assumed to be randomly located, and the "size" parameter is the mean distance to the nearest neighbouring anchor node $\bar{S}$ (8 m in the above measurements). Note that the measurements in Sect. 15.4.2 cannot be considered to adhere to random mobile position criteria, as base stations are used as "mobile" measurement points, and because of the shape of the building most of the measurement points (simulating a mobile node) are near the edge of the building, and there are only 6 out of 23 points that are in the "core" of the building—see the map of locations in Fig. 15.1.

Now consider the geometry of the closest nodes in relationship to mobile nodes in real-world applications. In the ideal case the closest nodes (up to four are used in the algorithm) "surround" the mobile node is a random fashion, so that ideally the angular distribution can be considered to have the uniform random distribution $U(-\pi, \pi)$; this distribution would only apply to mobile locations in the core of the building. However, for mobile locations at an edge of the building, the angular range will be restricted to a range of just π radians. For example, consider a mobile on the left edge of a rectangular building (and with the coordinates orientated as in Fig. 15.1), then the angular distribution is $U(-\pi/2, \pi/2)$, and the nearby receiving nodes will generally be in a region to the right of the mobile node. Thus the average estimated position will be to the right of the true position, or a positive x-bias error. Similarly, when the mobile is near the right edge of the building there will be a negative bias x-bias error. The same principle applies to mobile locations near the top or bottom edges of the building, except the bias will be in the y-direction. Thus considering the overall average over mobile positions throughout the building these bias errors will tend to average to zero, which is the measured result in Sect. 15.4.2. Thus the somewhat surprising result is that using a simple average of the locations of nearby nodes as the estimate of the mobile node position results in (overall) an unbiased position estimate. Further, if the mobile is in the core of the building (not near the edges), then the *individual* position estimates are also unbiased. These principles form the basis of this basic positioning technique. However, while the positions are on average unbiased, what is of practical importance is the associate variation and statistical mean (or RMS) radial error; a mathematical analysis is provides in the following subsection.

15.4.4 Analysis of Closest Nodes Algorithm Performance

The principles of the statistical analysis of the position errors was introduced in Sect. 15.4.3; this section provides the mathematical details. The first case analysed is the most common practical situation, namely in the mobile node is located in the core of the building, "surrounded" by transmitting nodes. If the true position of the mobile node is at (x_m, y_m) and the surrounding nodes (n) are located relatively at a range and angle (R_n, θ_n) then the estimated mobile position $(\hat{x}_m, \hat{y}_m)$ is the average of N nodes, namely

$$\hat{x}_m = x_m + \frac{1}{N}\sum_{n=1}^{N} R_n \cos(\theta_n) \qquad \hat{y}_m = y_m + \frac{1}{N}\sum_{n=1}^{N} R_n \sin(\theta_n) \tag{15.5}$$

In (15.5) both R_n and θ_n are random variables; the statistics of R are not accurately known (but are considered further below), but the statistics of θ are assumed to be the uniform angular distribution $U(-\pi, \pi)$. Now consider the expected (mean) position

$$\begin{aligned} E[\hat{x}_m] &= E\left[x_m + \frac{1}{N}\sum_{n=1}^{N} R_n \cos(\theta_n)\right] = x_m + \frac{1}{N}E\left[\sum_{n=1}^{N} R_n \cos(\theta_n)\right] \\ &= x_m + \frac{1}{N}E\left[\sum_{n=1}^{N} R_n\right]E\left[\sum_{n=1}^{N} \cos(\theta_n)\right] \end{aligned} \tag{15.6}$$

where the statistical independence of R_n and θ_n has been invoked. However, as the statistics of θ are assumed to be $U(-\pi, \pi)$ the expected value of the second expectation in the last expression in (15.6) is zero, so overall the expected value is $\hat{x}_m = x_m$; a similar result applies to the y-coordinate. Note that this result applies even though the statistics of the node range R are unspecified. These range statistics will largely be dependent on the geometry of the location of the anchor nodes relative to the mobile node, but also in part on the accuracy of using the RSSI measurements to select the closest N nodes. However, even if this selection process results in some rather poor node selections from the list of closest nodes ranked according to the RSSI, the overall *expectation* of the location of the mobile node will be unaffected, provided the relative angular locations of anchor nodes have approximately a uniform angular distribution. Indeed, later it will be suggested that this angular distribution is used in conjunction with the RSSI in selecting suitable receiving nodes. Thus RSSI measurement errors have a relatively minor effect on the mean position determined by this simple averaging algorithm. This result provides the mathematical confirmation of the measured results shown in Fig. 15.3, namely that the error scatter is centered about the origin of the error plot.

Before proceeding in the estimation of the RMS radial error, the probability density function of the cosine and sine of the radial angle θ needs to be computed. This is analysis has been done in Chap. 14, with the result in Eq. (14.12), namely letting $\cos(\theta) = \alpha$, the probability density function of α with θ assumed to be $U(-\pi, \pi)$ is

$$p(\alpha) = \frac{1}{\pi\sqrt{1 - \alpha^2}} \quad (-1 \leq \alpha \leq 1) \tag{15.7}$$

Now consider the expectation of the square of the error, using the last expression in (15.6)

$$\sigma_x^2 = E\left[(\hat{x}_m - x_m)^2\right] = E\left[\left(\frac{1}{N}\sum_{n=1}^{N} R_n \alpha_n\right)^2\right] = \frac{1}{N} E\left[R^2\right] E\left[\alpha^2\right] = \frac{1}{2N} E\left[R^2\right] \tag{15.8}$$

where from (15.7) $E[\alpha^2] = 1/2$, and the statistical independence of all of the cross-product components in the summation in (15.8) has been invoked, so that the expectation of the cross-product terms is zero as $E[\alpha] = 0$. A similar result applies for the y-coordinate, so that the overall RMS radial error is

$$\sigma_r = \sqrt{\sigma_x^2 + \sigma_y^2} = \sqrt{\frac{1}{N} E[R^2]} \tag{15.9}$$

Thus the RMS radial error is simply related to the number of nodes used in the averaging, and the statistical characteristics of the radial distance to the N closest nodes, as determined by the RSSI measurements and any method designed to ensure the angular distribution is approximately uniform. This analysis is performed in the next subsection. Note that while using more nodes in the averaging reduces the error, it also increases the expectation of R^2, so that overall this strategy of averaging over more nodes for improving the accuracy may not be effective. Indeed, from the practical measurements described in Sect. 15.4.2 it was found that the RMS error defined by (15.9) does not improve when the number of nodes exceeds four, so that $N = 4$ is used in measurements and the analysis in the following subsections. Indeed, this number of nodes is often used (such as in Chap. 8 for WiFi positioning based on signal strength measurements), and the layout of nodes is assumed to be on a square grid with a node separation parameter defined as S.

15.4.5 Analysis of Statistics of the Distance to Nearby Nodes

From (15.9) the RMS radial error depends on the statistics of the radial distance to nearby nodes. For the simplest case where just one node $N = 1$ is used, applying (15.9) gives a simple result, namely

$$\sigma_r^{\langle 1 \rangle} = \sqrt{E[R^2]} = d_{rms} \tag{15.10}$$

where $d_{rms} = 9.3$ m is the RMS distance to the nearest neighbouring node in Fig. 15.1. The actual measured value of the RMS radial error is 12.6 m averaged of all nodes, which is somewhat greater than the theoretical estimate (15.10). However, the expression (15.10) is based on the geometrical closest node, rather than the estimated closest node as determined by the strongest RSSI signal. In practice the RSSI will sometimes select the wrong node, one further distant than the true closest neighbour. The reason for this is that the node further distant may have a LOS path (lower path propagation loss) than a closer node with a NLOS path (greater path loss). Thus in this case the performance is rather poor; averaging more nodes would provide more redundancy and thus better performance. Thus the performance of the algorithm improves initially by increasing the number of nodes averaged, but as previously mentioned this improvement is limited to a maximum of about four nodes averaged.

Now consider the recommended number of nodes to average, namely $N = 4$, where the four nodes are on the corners of a square of side S. The mobile node is inside this square, which is a square in the grid of nodes which defines the coverage area. Thus while there will be a large number of such squares in the coverage area, without loss of generality the analysis needs only to consider one such square. If the transmitting nodes are on the corners of a square, and the RSSI measurements are even moderately accurate, the four closest nodes will be the four nodes on the corners of the square. Thus initially it will be assumed that this simplified geometry is a reasonable approximation to the actual situation. Later in Sect. 15.6.1 a more complex selection process is described.

Based on the above discussion it is assumed that the mobile node is randomly located within a square, and according to (15.9) the expected ranges squared to these four nodes need to be calculated. For simplicity, but without loss of generality, the analysis will be based on a normalized unit square, with the four nodes located at (0,0), (1,0), (1,1) and (0,1), and the mobile located at (x_m, y_m), where $0 \leq x_m \leq 1$ and similarly $0 \leq y_m \leq 1$. Both these are random variables with a uniform probability density function $U(0, 1)$. Then the expectation of the range squared for node #1 is

$$E\left[R_1^2\right] = E\left[x_m^2 + y_m^2\right] = E\left[x_m^2\right] + E\left[y_m^2\right] \tag{15.11}$$

as the two coordinates are statistically independent. Now consider calculating the x-coordinate expectation. With $u = x_m^2$ the associated PDF of u can be calculated using $p(x)\,dx = q(u)\,du$, so that

$$q(u) = 1/2\sqrt{u} \tag{15.12a}$$

and

$$E\left[x_m^2\right] = E[u] = \int_0^1 uq(u)u = 1/3 \tag{15.12b}$$

The same calculation applies to the y-coordinate, so from (15.11) $E\left[R_1^2\right] = 2/3$. A similar calculation can be applied to the other three nodes with the same result. Thus the RMS error averaged over four nodes is thus from (15.9)

$$\sigma_r = \sqrt{\frac{1}{4}E[R^2]}\,S = \sqrt{\frac{1}{4}\left(\frac{2}{3}\right)}S = S/\sqrt{6} = 0.41S \tag{15.13}$$

where the nodes are located on a square of side S. In a real world situation the nodes will not be on a regular square grid, so the mean distance $\bar{d}$ to the closest neighboring node could be used for estimating the RMS error by (15.13).

For the measurements in Sect. 15.3.2 with $\bar{d} = 8$ the mean radial error by applying (15.13) is 3.3 m; the actual measured value for core positions is 3.9 m (or $\sigma_r \approx 0.49S$), which is close to the above simplified theoretical value in (15.13) based on a square grid.

Another useful comparison is with the WiFi positioning performance described in Chap. 8. In particular the results of the more complex methods described in Chap. 8 can be approximately summarised by $\sigma_r \approx 0.5S$, where $S = 4$ m (see Tables 8.1 and 8.6 in Chap. 8). Thus these results are also in broad agreement with (15.13). The conclusion is that the performance of methods which utilize a square grid of reference nodes for position determination is largely a function of the geometry of the nodes and not on the processing algorithm, provided the mobile is surrounded by measurement nodes and the RSSI data are moderately accurate. Thus given a particular required design accuracy, (15.13) gives reasonable preliminary estimate of the required measurement grid node size S. Alternatively, given an existing network of nodes whose locations are known (and hence the mean distance to the closest neighbouring node), then (15.13) gives an estimate of the positional accuracy that will be achieved.

15.5 Weighted Closest Nodes Algorithm

To improve on the node averaging position method described in Sect. 15.4, it is conjectured that the mobile position can be better estimated as a *weighted* sum of the positions of the N nearby transmitting node locations. Thus for example, the x-coordinate of the mobile can be estimated as

$$\begin{aligned} \hat{x}_m &= \sum_{n=1}^{N} w_n x_n^{bs} \\ w_n > 0, &\quad \sum_{n=1}^{N} w_n = 1 \end{aligned} \tag{15.14}$$

with a similar expression for the y-coordinate. Note that from (15.14) the mobile position is constrained to the extremes of the positions of the neighbouring measurement nodes (again as determined by the RSSI measurements) in radio communication with the mobile, namely

$$\min\left(x^{bs}\right) < \hat{x}_m < \max\left(x^{bs}\right) \tag{15.15}$$

Thus if the weighting function can be determined in (15.14) an estimate of the position of the mobile device can be made again solely using the known locations of these nodes. Further, if greater weight is given to nodes closer to the mobile position using the RSSI data, then improved accuracy will be possible compared with the simple averaging method described in Sect. 15.4.

Now consider the location of the base stations relative to the (unknown) location (x_m, y_m) of the mobile device. If the range to the nth base station is R_n and the corresponding direction is θ_n relative to the mobile position, then the x-coordinate of the base station n is given by

$$x_n^{bs} = x_m + R_n \cos(\theta_n) \tag{15.16}$$

Thus from (15.14) and (15.16) the mobile position estimate is

$$\hat{x}_m = \sum_{n=1}^{N} w_n x_n^{bs} = \sum_{n=1}^{N} w_n x_m + \sum_{n=1}^{N} w_n R_n \cos(\theta_n) = x_m + \sum_{n=1}^{N} w_n R_n \cos(\theta_n) \tag{15.17a}$$

so that

$$x_m = \sum_{n=1}^{N} w_n x_n^{bs} - \sum_{n=1}^{N} w_n R_n \cos(\theta_n) \tag{15.17b}$$

Now to get a better estimate using (15.17b) the effects of propagation in the indoor environment on the measured RSSI the model (15.4) can be invoked to estimate the ranges in (15.17a, 15.17b). Inverting (15.4) the range as a function of the propagation loss is given by the form

$$R(L) = R_0 e^{L/\beta} \tag{15.18}$$

where R_0 is an unknown range constant, and the parameter β can be estimated from inter base station signal strength measurements, similar to that shown in Fig. 15.2 where the estimated value is $\beta = 40.9$ dB. Now the propagation loss (L) in (15.18) is not known to this algorithm as only the RSSI is measured. However, assuming that all the base stations transmit a similar power, the *differential* loss between paths can be determined. Using differential loss (actually differential RSSI measurements) has the advantage that calibration offsets in the receiver RSSI measurement subsystem does not affect the performance. If the path with the strongest signal is used as a reference, and designated as node $n = 1$, then the propagation loss for the nth path can be described as

$$L_n = L_1 + \Delta L_n \qquad (\Delta L_n > 0, n > 0) \tag{15.19}$$

Applying (15.19) to (15.18) gives

$$R_n \approx R_0 e^{(L_1 + \Delta L_n)/\beta} = R_1 e^{\Delta L_n/\beta} \equiv R_1 e^{\Delta RSSI_n/\beta} \tag{15.20}$$

The following analysis is for the x-coordinate; similar expressions apply for the y-coordinate. Applying (15.20) to (15.17b) gives

$$x_m \approx \sum_{n=1}^{N} w_n x_n^{bs} - R_1 \sum_{n=1}^{N} w_n e^{\Delta L_n/\beta} \cos(\theta_n) \tag{15.21a}$$

Now consider setting the weighting function such that $w_n = e^{-\Delta L_n/\beta}/W$, where $W = \sum_{n=1}^{N} e^{-\Delta L_n/\beta}$. The is an intuitive choice as the nodes with the smaller propagation loss (and nominally the closer to the mobile) get the largest weighting the averaging process. Then (15.21a) becomes

$$x_m \approx \sum_{n=1}^{N} w_n x_n^{bs} - \frac{R_1}{W} \sum_{n=1}^{N} \cos(\theta_n) \tag{15.21b}$$

Equation (15.21b) can be further simplified if is it assumed that the base stations "surround" the mobile location is a quasi-uniform manner, so that $\sum_{n=1}^{N} \cos(\theta_n) \approx 0$, in which case (15.21b) reduces to

$$x_m \approx \sum_{n=1}^{N} w_n x_n^{bs} \tag{15.21c}$$

which is the original contention defined by (15.14). A similar expression applies to the y-coordinate.

Thus by using an appropriate weighting function the weighted sum of the nearby nodes is an estimate of the mobile location, but with an assumption that the mobile is surrounded in a quasi-uniform manner such that the summation of the cosines of the angles to the nearby nodes used in the estimation is close to zero. For the case of the y-coordinate the requirement is that the summation of the sines of the angles to the nearby nodes is zero, in addition to the cosine constraint. While this is possible with super-symmetry about the x and y axes, in general this will not be possible with the limited number of neighboring nodes. However, nodes can be selected to minimise the expression

$$CS = \sqrt{\left[\sum_{n=1}^{N}\cos(\theta_n)\right]^2 + \left[\sum_{n=1}^{N}\sin(\theta_n)\right]^2} \tag{15.22}$$

so that the computed position error is minimized.

However (15.21b) provides a better estimate of the mobile position than (15.21c), provided the range R_1 to the reference base station 1 is known, and the angles to the base stations as referenced at the mobile position are also known. Thus if the position is first estimated using the simple averaging position algorithm described in Sect. 15.3, an initial estimate of ranges and angles (R_n and θ_n) can be used in the first iteration of this process, which allows a better estimate of the position of the mobile node. This iterative process continues until convergence. Convergence is indicated when the differential position error falls below a specified threshold, or divergence starts to occur indicated by the differential position error starts to increase. Such divergence is associated with poor geometry of the surrounding nodes (such as near the corners of buildings), or where the measured RSSI data vary considerably from the assumed propagation loss model.

Thus the mobile position is calculated as a weighted sum of the surrounding base station positions, where the weighting function is derived from differential RSSI measurements and the propagation loss is modeled as a function of range. Thus unlike other methods using RSSI measurements to determine position the RSSI measurements are only indirectly used in the position determination process. Further any errors in the loss modeling relative to the real-world losses are of less importance compared with traditional methods, as the method relies on a weighted summation of the locations of (four) base stations in radio range, so that random errors in the modeling tend to cancel.

If the mobile position is "surrounded" by the base stations, the summation in (15.21b) will be approximately zero, as the angles will be spread through 360°; more precisely the angles typically have an approximate uniform statistical distribution over $[0, 2\pi]$. Thus the cosine summation in (15.18b) will be a random quantity with approximately Gaussian statistics (due to the Central Limit Theorem) with approximately a zero mean and a standard deviation of $\sigma = \sqrt{N/2}$. More specifically the expected values of the computed mobile coordinates are

$$E[\hat{x}_m] = x_m \quad E[\hat{y}_m] = y_m \tag{15.23}$$

which shows the expected position is an unbiased estimate of the true coordinates of the mobile device. The 2D (x, y) distribution of positional errors associated with the cosine and sine summations will result in the radial error having approximately Rayleigh statistics with a probability distribution function *Rayleigh* $(\varepsilon_r, \sigma_r)$, where from (15.21b) σ_r is given by

$$\sigma_r = \left(\frac{E[R_1]}{W}\right)\sigma \tag{15.23a}$$

For a Rayleigh distribution the mean radial error is given by $\sqrt{\pi/2}\sigma_r$, so that the mean radial position error is

$$\sigma_\varepsilon \approx \sqrt{\frac{\pi}{2}}\frac{E[R_1]}{W}\sqrt{\frac{N}{2}} = \frac{\sqrt{\pi}E[R_1]}{W} \text{ for } N = 4 \tag{15.23b}$$

The expected value of the distance of the mobile to the nearest base station can be approximated for the $N = 4$ case using the square grid approximation given by (15.13), namely $E[R_1] \approx \bar{S}/\sqrt{6}$, so that the mean radial error becomes

$$\sigma_r \approx \sqrt{\pi/6}\frac{\bar{S}}{W} \tag{15.23c}$$

where $\bar{S}$ is the mean separation between base station nodes. The sum of the weights W will be somewhat less than N, typically in the range 2–3.

For example with the case shown Fig. 15.1 where $\bar{S} = 8$ m the expected standard deviation in the radial position given by (15.23c) with $W = 3$ is 1.9 m. However, the assumption that the surrounding nodes have a uniform angular distribution about the mobile node is often not an accurate description, so in practice the mean radial error will exceed that given by (15.23c).

These estimates compare favourably with other methods based on range estimation, where the range error standard deviation associated with propagation loss noise determines the positional accuracy. With the above method noise in the RSSI data does not have a significant impact on the positional accuracy, although geometric factors of the location of base stations can have a more significant impact than the classical approach of position determination using RSSI signal strength.

15.6 Weighted Closest Nodes Algorithm Performance

The performance of the weighted closest nodes algorithm was tested in a similar manner to the closest node algorithm in Sect. 15.4.2, namely each base station in turn acts as a simulated mobile node, with the position determined by using the RSSI data measured at the remaining base stations using the method described in Sect. 15.5. The results for a simple averaging of the position of the four closest nodes as determined by the measured RSSI at the base station resulted in a RMS position error of 9.9 m using all 23 nodes (see Fig. 15.3). The performance of the weighted averaging algorithm is expected to be considerably better.

15.6.1 Details of the Weighted Closest Nodes Algorithm

The weighted algorithm described in this section again uses four nodes, the selection of which is based on both the measured RSSI and the relative location of the nearby neighboring nodes, such that the nodes are approximately symmetrically located around the mobile position. The detection of the angular distribution is based on minimising the cosine-sine measure *CS* defined by (15.22). In particular, the four nodes are selected as follows:

1. The measured RSSI data from each of the base station nodes that are in radio range of the mobile transmitter are listed according to decreasing RSSI signal values, so that (nominally) the closest base station with the highest RSSI is listed first. This is a good, but not perfect, method of determining the receiving node closest to the mobile node. This "closest" node is defined as the reference node (node 1), as in the description of the algorithm in Sect. 15.5.
2. The second node selected is the second in the list, nominally the second closest anchor node to the mobile node.
3. To avoid the potential poor selection of nodes that all have the same or closely similar x or y coordinate, rectangular exclusion zone areas of width $\pm\Delta$ (say with $\Delta = 3$ m) relative to the first two nodes are defined, and the next node in the list outside these exclusion zones is selected as the third node. This procedure is necessary to avoid the situation of selecting three or four nodes in a straight line, as would occur with nodes located along a corridor. Such nodes have LOS paths, and thus are likely to be high up in the RSSI-sorted list, above closer nodes which have a NLOS path.
4. The fourth node is selected to try to maximize the node "surrounding" requirement for good operation of the algorithm. In particular, the node is selected which has the lowest value of *CS*, as defined by (15.22). Minimising this parameter will minimise the positioning error, as shown by (15.21b); a similar equation applies for the y-coordinate. For good-quality positioning the condition $CS < 0.5$ should apply, with smaller values indicating more accurate positions. Indeed, the positional error will be proportional to $R_1 \times CS$, as can be observed from the correction term in (15.21b).

Using the nodes selected by the above-defined procedure, the position is determined by applying (15.21b) and a similar equation for the y-coordinate. However, the procedure is iterative, as the calculation is based on using an estimated position of the mobile node for determining R_1 and θ_n in (15.21b). For initiating the first iteration estimate the simple average method is used for the position of the mobile node. This initial estimate allows a new selection of the four nodes to be made using the above procedure (but typically this is not required), and updated values of R_1 and θ_n calculated, and finally an updated estimate of the position of the mobile node is computed using (15.21b).

This iterative process ideally should converge to close to the true position of the mobile node. However, due to variation in the geometry and the quality of the RSSI data, this may not occur. Quite frequently it was found that the procedure initially converges (iterations becoming closer to the true node position), and thereafter diverges. In some particularly bad geometric situation (such as near a corner of the building), the iterations diverge (slowly) after the first iteration. However, it was found the about four iterations is optimum.

15.6.2 Weighted Closest Nodes Performance

The positional performance of the weighted closest node algorithm was measured using the same data set as used to test the basic averaging algorithm, described in Sect. 15.3.2. However, as the weighted averaging algorithm assumes that the receiving nodes surround the mobile node, at least partially, nodes near the corners of the building (shown in Fig. 15.1) were excluded, so that only 14 mobile position were tested. The resulting error distribution scatter is shown in Fig. 15.5. The mean

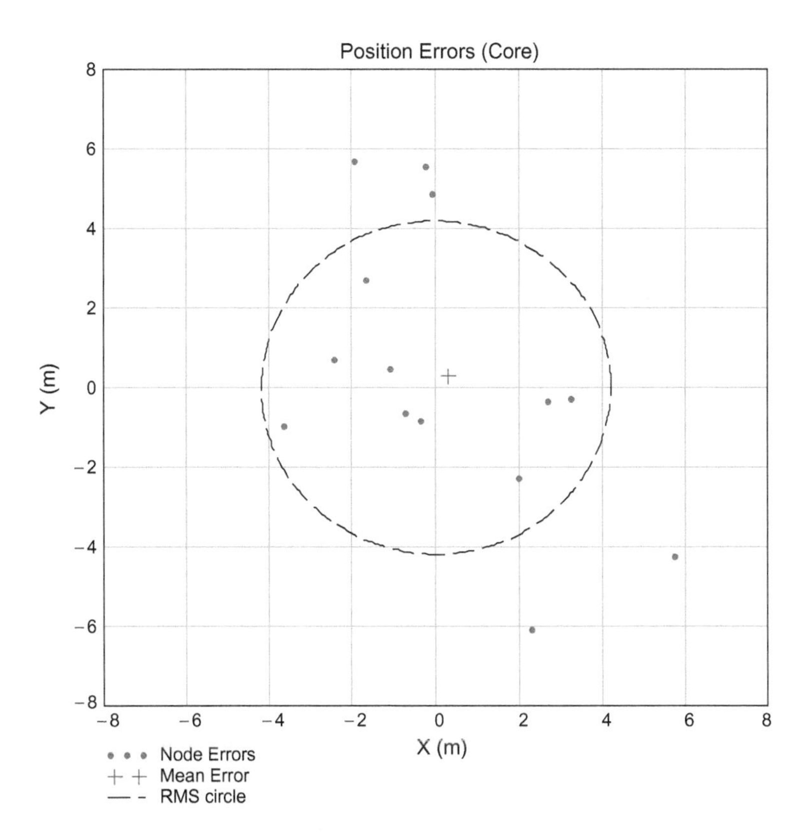

Fig. 15.5 Position errors of all nodes, excluding those near the corner of the building. The mean radial error is 4.2 m. Also shown by a cross is the mean error (0.3, 0.3) m. The circle radius is the mean radial error

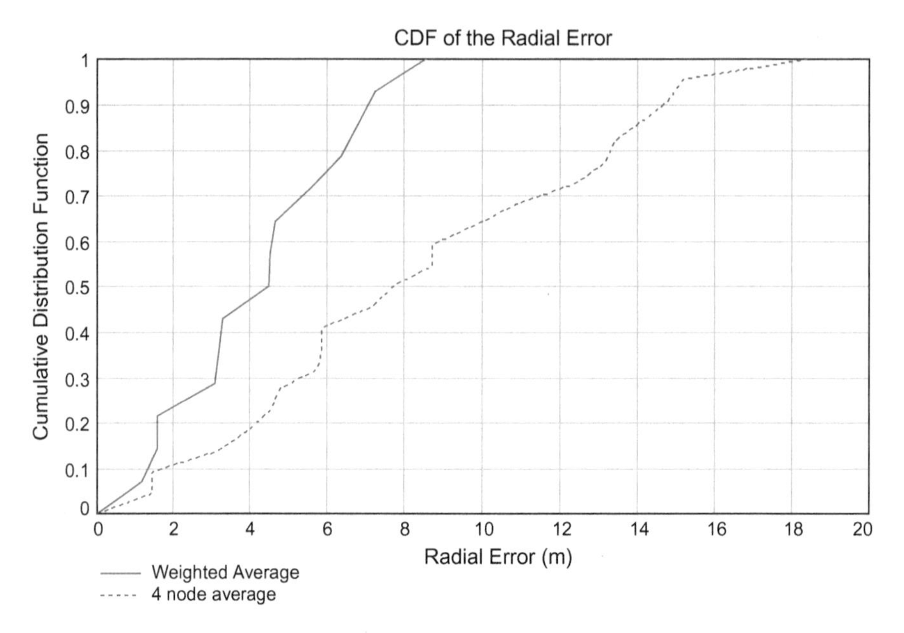

Fig. 15.6 Comparison of the CDF of the radial errors for the weighted average method and the simple position averaging method. Median error is from 8 to 3 m, maximum error from 18 to 7 m

radial error is 4.2 m, which compares with the 9.9 m shown in the similar Fig. 15.3 for the basic averaging of the four "closest" nodes algorithm. Thus there is a considerable improvement in the positional accuracy using the weighted averaging algorithm. The associated cumulative distribution function is shown in Fig. 15.6, together with the CDF for the basic averaging algorithm. As can be observed there is a substantial improvement, particularly with the reduction in the probability of large errors. For example, the peak errors are reduced from 18 to 7 m, so that basic reliability of the position determination is greatly improved. The median error is just 3 m, which with a mean separation of nodes $\bar{S} = 8$ m represent a normalised median error of just 0.38. For comparison theoretical normalise mean radial error given by (15.13) is 0.41 for a square grid of nodes. Thus the measured result for the weighted average algorithm is close to the theoretical expectation, and the normalised result is also similar to (but somewhat better) the performance of the WiFi positioning algorithms described in Chap. 8.

15.7 Weighted Closest Nodes Algorithm Performance with Moving Mobile

The performance of the weighted averaging positioning algorithm for static "mobile" nodes was tested by the procedures described in Sect. 15.6. While this provides useful performance data, most applications of the method would be associated with locating mobile devices such a cellular smart phones. These devices incorporate both the ability to receive signals from nearby WiFi access points (AP), as well as a connecting to the Internet (by WiFi or via the cellular phone network). The requirements for implementing positioning using the methods described in this chapter would thus be available to an appropriate application program (App). In particular, the WiFi functions allow the smart phone to detect the SSID signal of local access points, even if data communications with these APs is not possible. Further, if the App can access a database which defines the location coordinates of these APs (say in a particular building or building complex), then the methods described in this chapter allows moderately accurate positioning (to within a few meters) to be performed. Note that while such location services are currently available in smart phones and other similar devices such a tablets, the accuracy of such services is generally rather poor, typically of the order of 30–50 m when using commercially available data bases.

The testing of the performance of the weighted averaging algorithm with a moving mobile device was performed in the same building as the other testing described in Sect. 15.4, but the mobile device was attached to a person walking around the building (see photo in Fig. 11.19 in Chap. 11). The logged RSSI data was for the same 23 base stations shown in Fig. 15.1. The path consisted of walking along corridors in the building in a figure-8 pattern, with the path repeated in two opposite directions. As the person's body will cause some blockage of the signal (frequency 5.8 GHz, or a wavelength of about 5 cm which is much smaller than the size of the human body), the effective power at receiving node will be dependent on the orientation of the body. By performing a figure-8 path in both directions, the procedures provide a thorough test of the effects of path loss variation due to the mobile device being located on the body of a person.

Data were recorded with a 0.6 s update period for a total of 5 min. This procedure provides 500 sampled positions with a wide variety of signal reception conditions within the coverage area of the base stations. Further, as the area covered was broadly the same as the static tests described in Sect. 15.3, a direct comparison of the results is possible.

Examples of the RSSI data recorded at two base stations (up to 500 samples) is shown in Fig. 15.7a, b. During the walk the (straightline) range to base stations varies depending on the mobile location, so that at a long range base station may not receive the signal above the required threshold of −95 dBm. The Fig. 15.7a, b show examples of the measured RSSI as a function of range to the mobile node. These two examples are base stations with 10 m LOS separation, one in an open reception area, and the second at the entrance of the long corridor of length 40 m. In RSSI data in Fig. 15.7a show the RSSI signal falling rapidly for the first 15 m, and then more slowly out to a distance on 45 m. This RSSI behaviour is broadly in line with a power law loss model as defined by (15.2). The RSSI random noise is about ±5 dB. This scatter results in considerable range uncertainty if the RSSI measurement is used for range estimation. For example, at −80 dBm the range varies from 15 to 30 m.

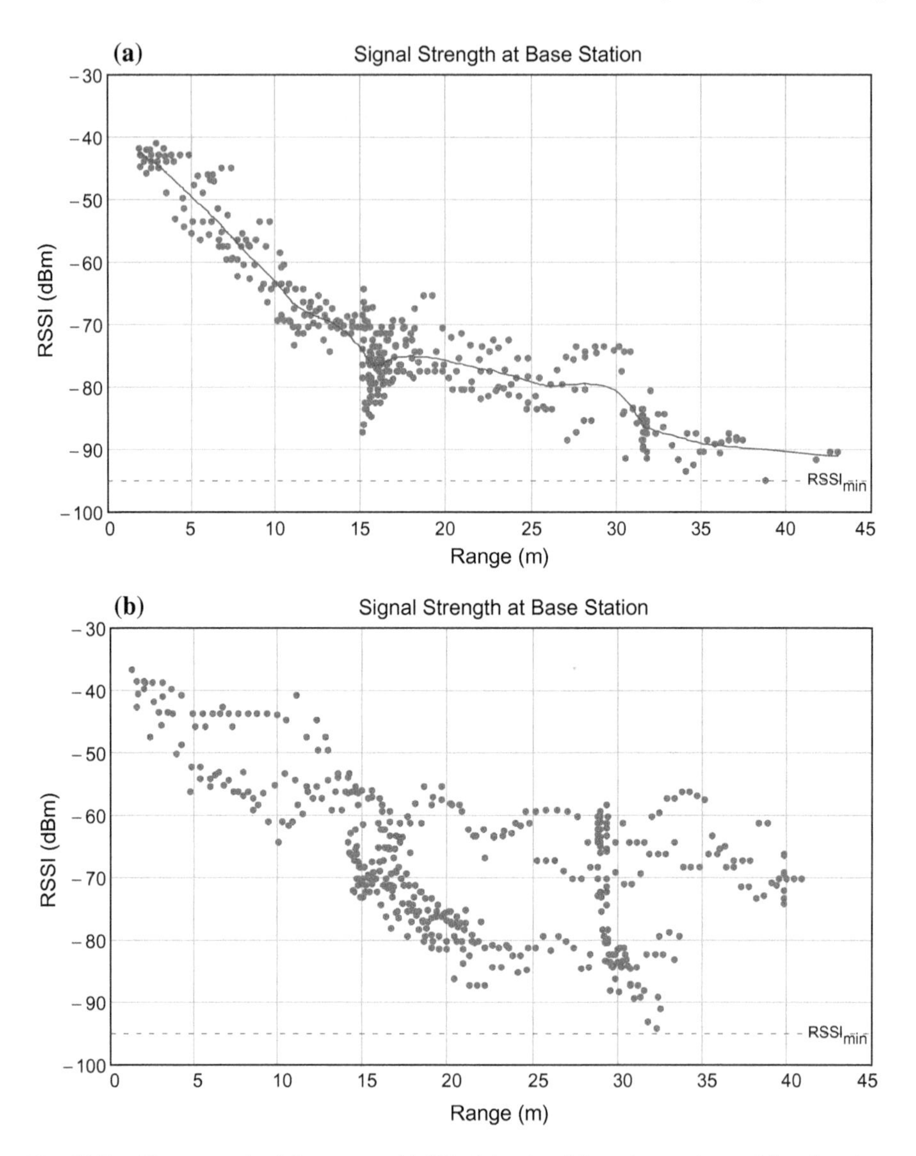

Fig. 15.7 **a** First example of the measured RSSI of the signal from the moving mobile, plotted as a function of the true range. Also shown is a smoothed estimate of the RSSI versus range function. The base station is located in a open reception area at position (51 m, 17 m) in Fig. 15.1. **b** Second example of the measured RSSI of the signal from the moving mobile, plotted as a function of the true range. The base station is located in at the end of a corridor at position (40 m, 16 m) in Fig. 15.1

The RSSI data in Fig. 15.7b shows a very different pattern to that in Fig. 15.7a, even though the base stations are closely adjacent. In particular, a bimodal pattern is evident, with one mode only dropping the RSSI slowly by about 30 dB in 40 m, while the second mode drops 40 dB in 20 m. Both modes show little change in RSSI beyond 20 m. Within each mode the noise is again about ±5. The RSSI data of this case thus has no pattern that can be easily described as a function of propagation distance, and thus using the RSSI measurements for determining range in this case would not be of any practical value. However, as the weighted average algorithm only uses the RSSI data for selecting the four nearby nodes and their associated weighting, the variability in the RSSI data in both Fig. 15.7a, b is expected to have considerably less effect on the measured positional performance.

A summary of the computed position results is given in Fig. 15.8 for both the x (top) and y (bottom) coordinates. The raw measured coordinates are the noisy graph line, the straightline segments the true coordinates, and the thick smoothed line is the Kalman filtered data. For most positioning systems tracking a moving target some form of data smoothing is used, so the smoothed data will be used in the accuracy calculation. (This smoothing only reduces the mean errors by a few tenths on a meter). The RMS radial error for this data set is 4.1 m, compared with 4.2 m for the data in Fig. 15.5. Thus despite the many degrading effects of tracking a moving device described above, the overall performance of the weighted averaging algorithm is essentially the same as the static case. The scatter diagram of the errors in shown in Fig. 15.9. Overall the mean errors in the x and y coordinates is close to zero (and hence the method results in unbiased estimates). Most of the

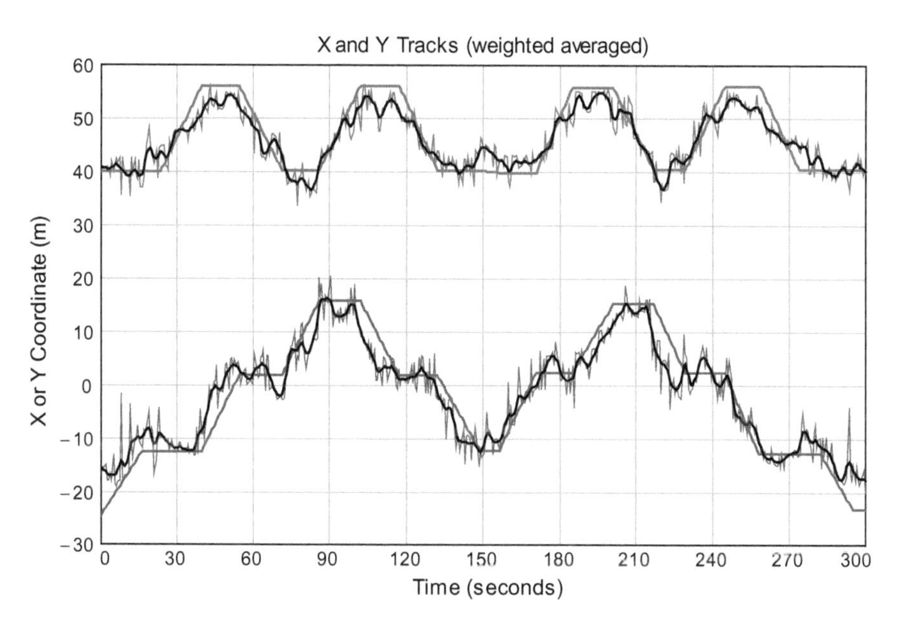

Fig. 15.8 Plot of the x and y coordinates of the track within the building. The straightline segments indicate the true track over time. The "noisy track" is the raw position data derived from the weighted average algorithm, and the associated thick line is the Kalman filtered track. The time update period is 0.6 s. The RMS radial error is 4.1 m

errors lie within the circle of radius equal to the RMS radial error of 4.1 m, but in the y-direction there are a moderate number of errors up to ± 8 m.

The comparison of the cumulative distribution functions of the position errors for the static and moving device is shown in Fig. 15.10. As can be observed the two curves are very similar in nature, with the moving CDF slightly superior. However, as the number of measurements used in the static case is only 14 compared with 500 for the moving case, these differences are not statistically significant. For the moving CDF graph, it can be observed the median radial error for tracking a mobile node is 3.5 m, while the peak errors are 9 m in both cases. As the size of the rooms in the building is about 3 m (see Fig. 11.20 in Chap. 11 for a detailed map of the building), the weighted average method accuracy is sufficient to mostly determine the room in which a tracked person is located.

It is interesting to compare the performance of tracking a person using the weighted averaging RSSI method and a time-of-arrival (TOA) positioning method.

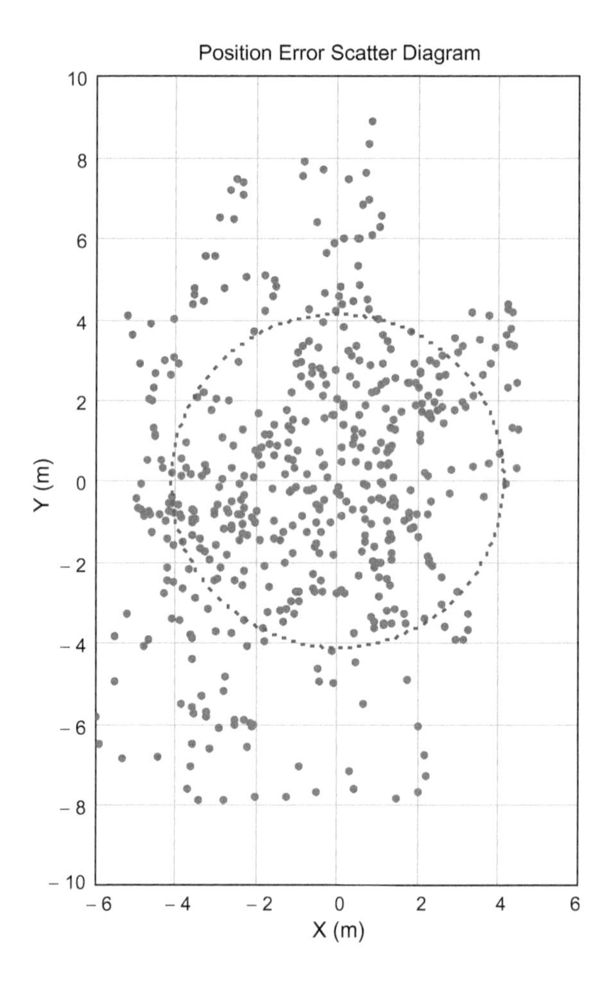

Fig. 15.9 Plot of the x and y coordinates of the track errors for the data in Fig. 15.8. The circle radius is the RMS radial error of 4.1 m. Notice that the spread of errors in the y-direction is about twice that in the x-direction

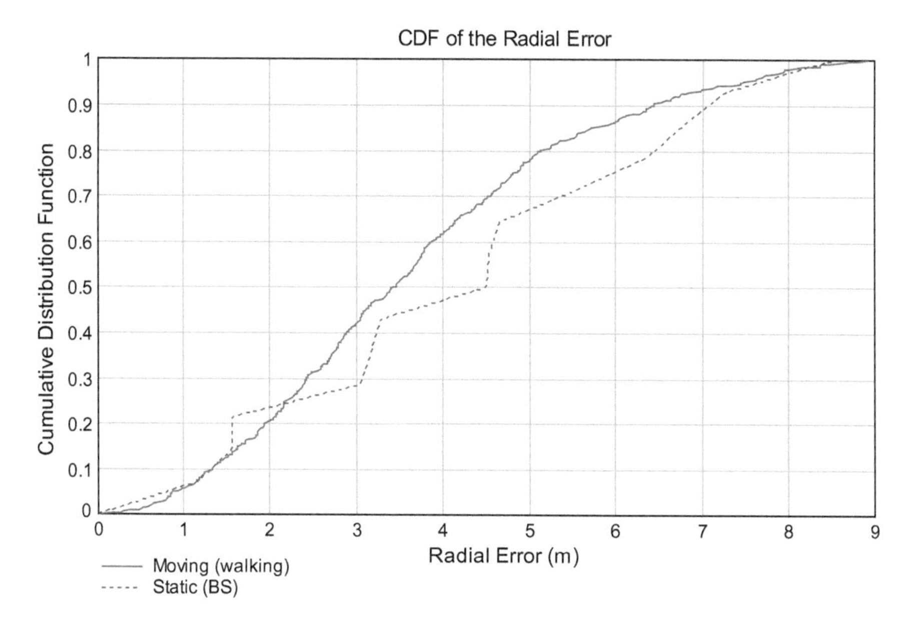

Fig. 15.10 Comparison of the CDFs of the static and moving test error data for the weighted average algorithm

The testing described above for determining the positions using the RSSI data also included measurement of the positions using TOA with the WASP system (Sathyan et al. 2011). The CDF of the TOA results is shown in Fig. 11.22 in Chap. 11. For comparison the WASP RMS radial error for the same walking path and the same base stations is 0.5 m, or about eight times more accurate. However, this performance requires a specially designed tracking system, whereas the RSSI method can be applied to wide range radio systems (WiFi, cellular phone network, wireless sensor networks, Zigbee networks) without any special hardware, and only requires application software in an appropriate mobile device. Thus while the accuracy is much inferior, it is of sufficient accuracy and of sufficient reliability that it could find widespread applications in a large number of practical situations.

References

Caso G, Nardis L, Benedetto M (2015) Frequentist inference for WiFi fingerprinting 3D indoor positioning. In: Proceedings of the IEEE international conference on communication workshop (ICCW), pp 809–814

Chirakkal V, Park M, Han D (2015) Indoor navigation using WiFi for smartphones: an improved Kalman filter based approach. In: Proceedings of the IEEE International conference on consumer electronics (ICCE), pp 82–83

Combain (2018) https://combain.com

Gu Y, Lo A, Niemegeers I (2009) A survey of indoor positioning systems for wireless personal networks. IEEE Commun Surv Tutor 11(1):13–32
iBeacon (2018) https://en.wikipedia.org/wiki/IBeacon
Koo J, Cha H (2012) Unsupervised locating of WiFi access points using smartphones. IEEE Trans Syst Man Cybern Part C (Appl Rev) 42(6):1341–1353
Liu H, Darabi H, Banerjee P, Liu J (2007) Survey of wireless indoor positioning techniques and systems. IEEE Trans Syst Man Cybern C: Appl Rev 37(6):1067–1080
Lui G, Gallagher T, Li B, Dempster A, Rizos C (2011) Differences in RSSI readings made by different Wi-Fi chipsets: a limitation of WLAN localization. In: Proceedings of international conference on localization and GNSS, Tampere, Finland, pp 53–57
Sathyan T, Humphrey D, Hedley M (2011) WASP: a system and algorithms for accurate radio localization using low-cost hardware. IEEE Trans Syst Man Cybern—Part C, 41(2):211–222
Sharp I, Yu K (2018) New methods for improved indoor signal strength positioning. In: Yu K (ed) Positioning and navigation in complex environments. IGI Global

Chapter 16
NLOS Mitigation for Vehicle Tracking

16.1 Introduction

Precisely tracking moving objects in complex propagation scenarios such as deep urban canyons and indoors is a challenging problem. In some cases, GPS repeaters may be used to obtain better performance (Jardak and Samama 2009). However, a more reliable way to provide position information in the harsh propagation environments is to utilize the network-based positioning systems (Torieri 1984; Schmidt 1996; Caffery and Stuber 1998; Hellebrandt and Mathar 1999; Juurakko and Backman 2004; Wang et al. 2006; Mourad et al. 2011). Although network-based systems may provide continuous target position information in general, the position accuracy is usually not satisfactory in severe non-line-of-sight (NLOS) propagation scenarios. To improve the accuracy of network-based positioning systems, a range of methods and techniques have been proposed (Yu et al. 2009). One of these efforts has been focused on mitigating the NLOS propagation effect. For instance, by exerting constraints and/or introducing a NLOS error related parameter into the cost function, optimization algorithms can be developed to mitigate the NLOS effect (Wang et al. 2003; Kim et al. 2006; Yu and Guo 2008). If the statistics of the NLOS errors and measurement noise are known, such as based on field trials, statistical processing can significantly reduce the NLOS effect (Cong and Zhuang 2005). When a database is established in advance, signature matching can be employed to greatly improve positional accuracy in NLOS scenarios (Bahl and Padmanabhan 2000; Nezafat et al. 2005; Nerguizian et al. 2006). Further, using angle measurements at the mobile unit is another efficient method to mitigate NLOS effect provided that it is feasible to obtain angle-of-arrival or angle-of-departure information at the mobile (Miao et al. 2007). When tracking moving objects, Kalman filtering (KF) has been widely considered and different tracking algorithms have been developed. Kalman filtering with consideration of NLOS mitigations has been investigated by many researchers (Woo et al. 2000; Le et al. 2003; Liao and Chen 2006; Lakhzouri et al. 2003; Chen et al. 2009; Huerta et al. 2009). In

I. Sharp and K. Yu, *Wireless Positioning: Principles and Practice*, Navigation: Science and Technology, https://doi.org/10.1007/978-981-10-8791-2_16

(Woo et al. 2000) Kalman filtering is used to reconstruct the NLOS corrupted distance measurements, by introducing a bias state variable. In (Le et al. 2003) both unbiased and biased Kalman filtering of TOA estimates are employed based on the outcome of NLOS identification. The smoothed TOA estimates are then transferred to time-difference-of-arrival (TDOA) estimates which are used for position determination. A KF based interacting-multiple-model smoother is proposed in Liao and Chen (2006) for TOA data smoothing. A two-state Markov process is employed to describe the transition between LOS and NLOS propagation conditions between the mobile and each base station. Speed and heading information is employed for position determination in Yu and Dutkiewicz (2012), basically using the dead-reckoning technique. A Markov-transitioned fuzzy-tuned hybrid framework is proposed in Ho (2013) for modeling the dynamics of mobile station, LOS and NLOS range measurements, and NLOS bias variations for each base station. Based on the framework, a selective fuzzy-tuned interacting multiple-model (SFT-IMM) algorithm is derived. A closed-form NLOS range estimation method is proposed in Abu-Shaban et al. (2016), which makes use of a distance-dependent ranging bias model. More relevant references can be found in the literature.

Although different NLOS mitigation methods do improve position accuracy in cellular networks, the resulting position errors are still relatively large, often many tens of meters and even more than one hundred meters. The performance depends on the propagation environment, signal characteristics such as the modulation technique and the signal bandwidth, and how many base stations are within the radio range of the mobile unit. In this chapter a two-stage cellular-network based positioning approach is studied which consists of two stages. The first stage is filtering of raw distance measurements with NLOS mitigation techniques, and second stage uses fusion of distance, velocity and heading for position determination. At the first stage, a LOS and NLOS identification method is proposed to mitigate the NLOS bias error. This online identification method makes use of online mean and variance estimates which can be replaced with the median and the mean absolute deviation (MAD) to improve identification robustness. The mean of the bias errors is largely removed through subtracting the measurements from the estimated bias mean. At the second stage, the velocity and heading angle information as well as the distance measurements are incorporated through the extended KF. The estimates of the mobile kinematics can be provided by the off-the-shelf velocity and heading sensors.

16.2 Measurement Model

Suppose that the distance measurements are made between the mobile and each of N base stations through measuring the time-of-arrival or the round-trip-time of the signal. The positions of the base stations are known and constant. For notational simplicity, the subscripts related to base stations are dropped. The distance measurement is modeled as

$$r_j = d_j + \varepsilon_j, j \geq 1 \tag{16.1}$$

where d_j is the Euclidean distance between the mobile and the base station at time instant j, which in a two-dimensional scenario is defined as $d_j = \sqrt{(x_{bs} - x_{m,j})^2 + (y_{bs} - y_{m,j})^2}$ where (x_{bs}, y_{bs}) and $(x_{m,j}, y_{m,j})$ are respectively the positions of the base station and the mobile. Note that although a two-dimensional mobile positioning scenario is considered and the application intended here is vehicular, the developed algorithms can be extended to three-dimensional positioning. The measurement error (ε_j) is modeled as

$$\varepsilon_j = \begin{cases} n_j, & \text{LOS condition} \\ b_j + n_j, & \text{NLOS condition} \end{cases} \tag{16.2}$$

where n_j is the Gaussian measurement noise with a zero mean and a variance σ_n^2, and b_j is the positive measurement bias error induced by NLOS propagation. It is assumed that as the vehicle travels the LOS and NLOS propagations change from over time. In addition to distance measurements, mobile velocity and heading angle measurements are assumed to be available, such as provided by a speedometer and a heading sensor, respectively. The purpose is to develop algorithms to enhance positioning accuracy using the range measurements and the velocity and heading angle information.

16.3 Two-Stage Algorithm

A block diagram of the two-stage approach is shown in Fig. 16.1. The details of the approach are described in the following sections.

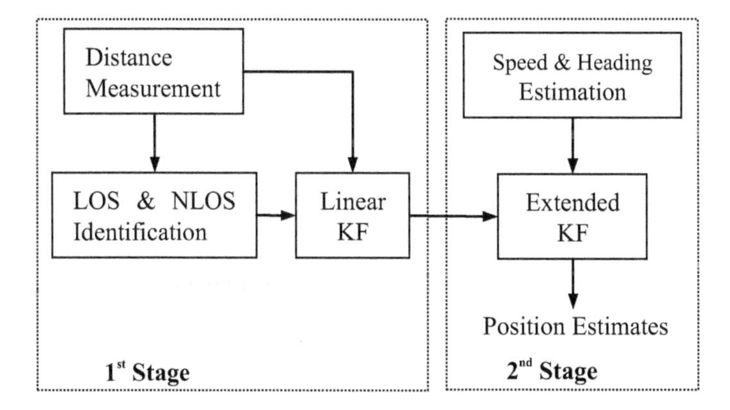

Fig. 16.1 Block diagram of the two-stage approach

16.3.1 Distance Smoothing with NLOS Mitigation

16.3.1.1 LOS and NLOS Identification

The existing distance smoothing algorithms typically use online variance estimation to identify the NLOS propagation. The variance of the observation errors in the KF is then changed accordingly. In addition to using the online variance or standard deviation (STD) estimates, an online mean estimation can also be performed to aid LOS and NLOS identification and to update the KF equations. This identification is performed independently using distance measurements associated with each base station. It is assumed that the sampling frequency for ranging is much greater than one, say 30 samples per second. Such a sampling frequency would be feasible for cellular networks and other wireless systems in general provided that the central processing power is sufficiently high. Figure 16.2 shows the flow chart of this LOS and NLOS identification method and more details of the method are described in following subsections.

Note that higher position update rates have been considered for some practical positioning systems. For instance, in mid-2010 the position solution of the Locata was calculated at a 25 positions per second (Locata 2017, visited). The Locata Corporation had also demonstrated the update rates up to 50 positions per second, and envisaged that the update rate could be further increased by increasing the processing power of the central processing unit. The main drawback of such a relatively high sampling frequency is that the power consumption will increase.

Initial Identification

The initial identification uses the sequence of past distance measurements for hypothesis testing. Specifically, suppose that the current time instant is k. Then the mean and STD of the past sequence of L original distance measurements are computed by

$$m_1 = \frac{1}{L}\sum_{j=k-L}^{k-1} r_j,\ \sigma_1 = \sqrt{\frac{1}{L}\sum_{j=k-L}^{k-1} (r_j - m_1)^2} \tag{16.3}$$

where the length L should be less than the number of samples per second when the vehicle is traveling at a moderate or high speed. For instance, if the sampling rate is 50 samples per second, L may be chosen around 20. If the vehicle is traveling at a speed of 40 km/h, then the maximum distance variation among the L position points will not be greater than 4.4 m. Such a quantity is relatively small compared to the measurement noise STD which can be greater than 100 m for 2G and 3G cellular networks and the mean and STD of the measurement bias error which can be over one hundred meters. Thus, the effect of such a variation on the STD

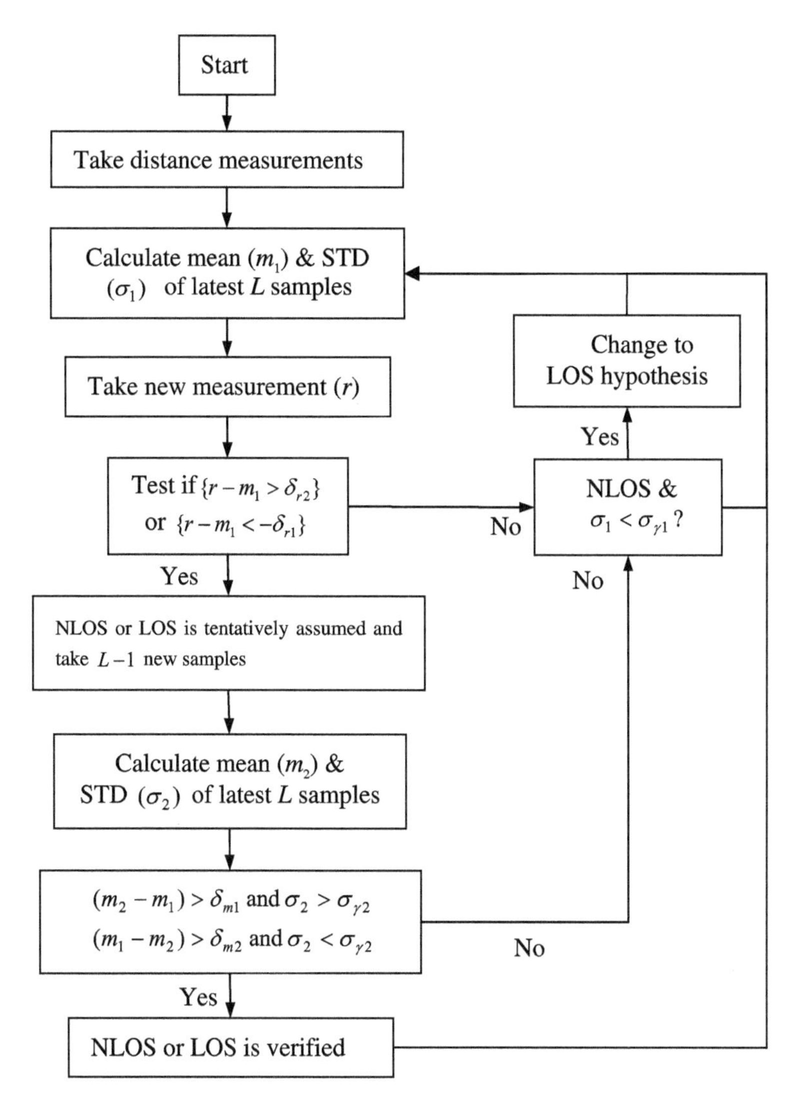

Fig. 16.2 Flow chart of the proposed LOS and NLOS identification method

calculation would be minor. Although the variance or STD can be estimated in other ways, such a method is simple, straightforward and reliable.

Note that the mean and STD calculated according to (16.3) may not be robust when the measurements contain outliers, which can occur when either measurements are corrupted by the NLOS effect, or under LOS conditions, depending on which hypothesis is being tested. To improve the robustness of the algorithm, the mean and the STD can be replaced respectively with the median and the MAD. In the simulation the performance of the mean and STD based method will be compared with that of the median and MAD based method.

The current measurement is then compared with the estimated mean in (16.3). If the difference is within a pre-defined range

$$-\delta_{r1} < r_k - m_1 < \delta_{r2} \tag{16.4}$$

where δ_{r1} and δ_{r2} are two thresholds, then the current measurement is used to update the mean and the STD in (16.3), where the oldest measurement r_{k-L} is excluded from the calculation. That is, a sliding window is applied to constrain the number of the distance measurements. Then if the current propagation hypothesis is NLOS and

$$\sigma_1 < \sigma_{\gamma 1} \tag{16.5}$$

where $\sigma_{\gamma 1}$ is the STD threshold, the hypothesis is changed to LOS. The issue of the threshold selection will be discussed in Sect. 16.4. Next a new measurement is taken and the identification procedure continues. On the other hand, if

$$r_k - m_1 \geq \delta_{r2} \tag{16.6}$$

then the current propagation is assumed to be NLOS. Alternatively, if

$$r_k - m_1 \leq -\delta_{r1} \tag{16.7}$$

then the propagation is assumed to be LOS. The tentative decision made based on (16.6) or (16.7) may not be correct especially when the thresholds δ_{r1} and δ_{r2} are relatively small. To reduce the probability of a wrong decision making, a verification process is necessary, as described in the following subsection.

Hypothesis Verification

When the threshold is crossed in the initial identification, a verification process is required as follows. In the verification phase, $L - 1$ new measurements are taken to form two sequences of measurements as shown in Fig. 16.3.

The mean and STD of the second sequence of measurements are calculated by

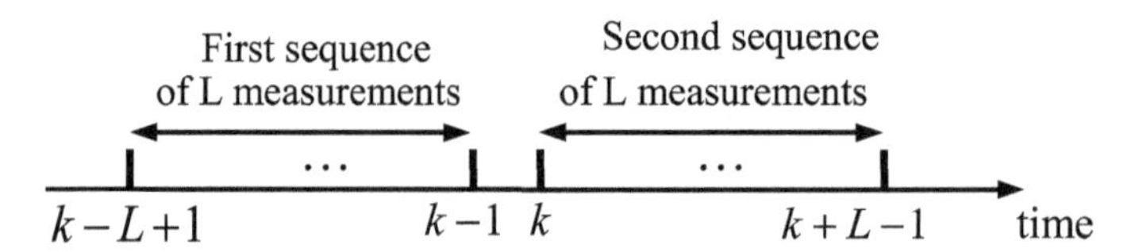

Fig. 16.3 Two sequences of measurements for LOS and NLOS identification

$$m_2 = \frac{1}{L}\sum_{j=k}^{k+L-1} r_j,\ \sigma_2 = \sqrt{\frac{1}{L}\sum_{j=k}^{k+L-1} (r_j - m_2)^2} \tag{16.8}$$

In the case where (16.6) holds, and

$$m_{2,1} = m_2 - m_1 \geq \delta_{m1};\ \sigma_2 \geq \sigma_{\gamma 2} \tag{16.9}$$

where δ_{m1} is the threshold for the mean difference and $\sigma_{\gamma 2}$ is another STD threshold, then all the measurements of the second sequence are categorized as NLOS measurements. The mean and STD calculated by (16.8) now become m_1 and σ_1, respectively. Alternatively, if (16.6) holds, but one or two conditions in (16.9) do not hold, the initial decision of the NLOS measurement at time instant k is rejected. The other $L - 1$ measurements of the second sequence are now examined one by one to see if any of the three Eqs. (16.4), (16.5), and (16.6) is satisfied.

Similarly, if (16.7) holds in the initial identification, and

$$m_{2,1} \leq -\delta_{m2};\ \sigma_2 \leq \sigma_{\gamma 2} \tag{16.10}$$

where δ_{m2} is another threshold for the mean difference, the measurements of the second sequence are all categorized as LOS measurements. The mean and variance are updated accordingly. However, even if (16.7) holds but one or two conditions in (16.10) are not satisfied, the initial LOS hypothesis is rejected. Then starting from r_{k+1}, the identification process continues. Meanwhile, if both (16.9) and (16.10) do not hold, the condition in (16.5) is checked and the NLOS hypothesis is changed to LOS hypothesis if (16.5) is satisfied.

Note that in the hypothesis testing, it is inevitable that some incorrect decisions will be made. However, since the identification is based on a sliding window of measurements, no serious error propagation would occur. Also, the online distance measurement variance estimation would usually strengthen the correct decision and reject the incorrect decision. Further, the online mean and STD estimation can be replaced with the more robust operations such as the median and the mean absolute deviation estimation in the presence of outliers. Therefore, the identification algorithm is robust in the presence of incorrect decisions.

Linear Kalman Filtering

For notational simplicity, the indexes for the base stations are dropped as mentioned earlier. At time instant k the state vector is defined as: $\boldsymbol{\theta}^{(k)} = [\,\theta_d^{(k)} \quad \dot{\theta}_d^{(k)}\,]^T$ where $\theta_d^{(k)}$ and $\dot{\theta}_d^{(k)}$ are respectively the distance state and its variation state. The process equation of the KF is then defined as

$$\boldsymbol{\theta}^{(k)} = \boldsymbol{\Phi}\boldsymbol{\theta}^{(k-1)} + \boldsymbol{\beta}\zeta^{(k)} \tag{16.11}$$

where ζ is the process noise and

$$\boldsymbol{\Phi} = \begin{bmatrix} 1 & \delta t \\ 0 & 1 \end{bmatrix}, \boldsymbol{\beta} = \begin{bmatrix} 0.5\delta t^2 \\ \delta t \end{bmatrix} \tag{16.12}$$

where δt is the distance sampling interval. The observation equation can be written as

$$p^{(k)} = \mathbf{h}\boldsymbol{\theta}^{(k)} + w^{(k)} \tag{16.13}$$

where $\mathbf{h} = [1 \quad 0]^T$ and w is the observation noise (distance measurement error). Then the state prediction ($\hat{\boldsymbol{\theta}}^{(k|k-1)}$), the error covariance matrix prediction ($\mathbf{C}^{(k|k-1)}$), the Kalman gain computation ($\mathbf{k}^{(k)}$), and the state and error covariance correction stages of the linear KF are performed recursively as follows

$$\begin{aligned} \hat{\boldsymbol{\theta}}^{(k|k-1)} &= \boldsymbol{\Phi}\,\hat{\boldsymbol{\theta}}^{(k-1|k-1)} \\ \mathbf{C}^{(k|k-1)} &= \boldsymbol{\Phi}\,\mathbf{C}^{(k-1|k-1)}\boldsymbol{\Phi}^T + \boldsymbol{\beta}\mathbf{Q}\boldsymbol{\beta}^T \\ \mathbf{k}^{(k)} &= \mathbf{C}^{(k|k-1)}\mathbf{h}^T(\mathbf{h}\mathbf{C}^{(k|k-1)}\mathbf{h}^T + R^{(k)}) \\ \hat{\boldsymbol{\theta}}^{(k|k)} &= \hat{\boldsymbol{\theta}}^{(k|k-1)} + \mathbf{k}^{(k)}(p^{(k)} - \mathbf{h}\hat{\boldsymbol{\theta}}^{(k|k-1)}) \\ \mathbf{C}^{(k|k)} &= (1 - \mathbf{k}^{(k)}\mathbf{h})\mathbf{C}^{(k|k-1)} \end{aligned} \tag{16.14}$$

where $\mathbf{Q}$ is the covariance matrix of the process/acceleration noise vector and $R^{(k)}$ is the variance of the observation noise variable. In a simulation these two covariance matrices can be selected manually through a number of tests. For simplicity they can be set as a diagonal matrix. The value of the diagonal element of $\mathbf{Q}$ should be small, whereas that of $R^{(k)}$ should be relatively large, depending on the variance of the measurement noise and error. In practice, field trials are needed to choose the most suitable parameters. The initial values for the error covariance matrix can be simply set as an identity matrix multiplied by a large number such as 10^6. The initial values for the mobile position coordinates can be obtained using a simple single-shot non-iterative least-squares position determination algorithm, such as the one with linearization and introduction of an extra intermediate parameter as described in Yu et al. (2009).

The linear KF is updated in accordance with the updated variance estimates and the hypothesis testing results as discussed earlier. That is, the measurement error variance $R^{(k)}$ in (16.14) is continuously updated based on the current variance estimates of the distance measurements. Also, the fourth equation in (16.14) is updated using the mean estimates according to

$$\hat{\boldsymbol{\theta}}^{(k|k)} = \hat{\boldsymbol{\theta}}^{(k|k-1)} + \mathbf{k}^{(k)}(p^{(k)} - \delta p - \mathbf{h}\hat{\boldsymbol{\theta}}^{(k|k-1)}) \tag{16.15}$$

where δp is the adjustment term to compensate for the mean of the bias errors, which is chosen according to

$$\delta p = \begin{cases} 0, & \text{if current hypothesis is LOS} \\ m_2 - m_1, & \text{if current hypothesis is NLOS} \end{cases} \tag{16.16}$$

where m_1 is the mean of the last L distance measurements under the latest LOS hypothesis and calculated by (16.3), and m_2 is the mean of the first L distance measurements under current NLOS hypothesis and calculated by (16.8). Note that these two sequences of measurements are next to each other, distinguished by the transition from the LOS hypothesis to the NLOS hypothesis.

Clearly this distance smoothing algorithm is suited for scenarios where the NLOS and LOS propagations alternate frequently in an irregular pattern. Such a radio propagation pattern can occur in reality as reported in Woo et al. (2000). It is worth mentioning that the above studied distance smoothing approach is similar to that in Le et al. (2003). That is, online variance estimation results are used to update the KF. The main difference is that in the studied approach online bias error mean estimation is also performed and the estimated mean is used to update the state vector in the KF.

16.3.2 Integrating Distance Measurements with External Velocity and Heading Information

At the second stage the filtered distance measurements are employed to locate the vehicle. Vehicles are usually equipped with speedometers to provide the estimate of the speed. Also the direction of movement of a vehicle can be measured by the off-the-shelf heading sensors such as the Honeywell HMR3200 digital compass which provides heading information with accuracy of 0.5° in the absence of electromagnetic disturbance. In this section the speed and heading angle estimates are integrated with the distance measurements to improve the position estimation accuracy.

The four-element state vector is defined as

$$\boldsymbol{\theta}^{(k)} = \begin{bmatrix} x_{m,k} & y_{m,k} & v_{x,k} & v_{y,k} \end{bmatrix}^T \tag{16.17}$$

where $x_{m,k}$ and $y_{m,k}$ are the coordinate state variables, and $v_{x,k}$ and $v_{y,k}$ are the velocity state variables of the mobile along the x-axis and the y-axis, respectively. Note that a number of similar notations are used in both preceding section and this section, but with different definitions. Similar to (16.11), the process equation can be written as

$$\boldsymbol{\theta}^{(k)} = \boldsymbol{\Phi}\,\boldsymbol{\theta}^{(k-1)} + \mathbf{B}\xi^{(k)} \tag{16.18}$$

where $\boldsymbol{\xi}^{(k)} = [\, \xi_x \quad \xi_y \,]^T$ is the system noise vector, and

$$\boldsymbol{\Phi} = \begin{bmatrix} 1 & 0 & \delta t & 0 \\ 0 & 1 & 0 & \delta t \\ 0 & 0 & 1 & 0 \\ 0 & 0 & 0 & 1 \end{bmatrix}, \quad \mathbf{B} = \begin{bmatrix} 0.5\delta t^2 & 0 \\ 0 & 0.5\delta t^2 \\ \delta t & 0 \\ 0 & \delta t \end{bmatrix} \tag{16.19}$$

In contrast to (16.13), the observation equation is a nonlinear function of the state vector, defined as

$$\mathbf{p}^{(k)} = \mathbf{g}(\boldsymbol{\theta}^{(k)}) + \boldsymbol{\eta}^{(k)} \tag{16.20}$$

where $\boldsymbol{\eta}^{(k)}$ is the observation noise vector (the distance measurement error vector after smoothing) and

$$\begin{aligned} \mathbf{g}(\boldsymbol{\theta}^{(k)}) &= [\, d_1^{(k)} \quad d_2^{(k)} \quad \ldots \quad d_N^{(k)} \,]^T \\ \mathbf{p}^{(k)} &= [\, r_{1,k} \quad r_{2,k} \quad \ldots \quad r_{N,k} \,]^T \end{aligned} \tag{16.21}$$

where the vector $\mathbf{p}^{(k)}$ contains the filtered distance measurements that are obtained from the first stage. In the presence of speed estimate $\hat{v}_k$ and heading estimate $\hat{\phi}_k$, the speed along the x-axis and that along the y-axis can be modeled as

$$\begin{aligned} \hat{v}_{x,k} &= \hat{v}_k \cos \hat{\phi}_k & \hat{v}_{y,k} &= \hat{v}_k \sin \hat{\phi}_k \\ &= v_{x,k} + \vartheta_{x,k}, & &= v_{y,k} + \vartheta_{y,k} \end{aligned} \tag{16.22}$$

where $\vartheta_{x,k}$ and $\vartheta_{y,k}$ are respectively the speed errors along the x-axis and y-axis. These are modeled as Gaussian random variables. Note that the angle between the moving direction and the positive x-axis is defined as the heading angle here. Then the observation equation consists of both (16.20) and (16.22), resulting in the extended format as

$$\tilde{\mathbf{p}}^{(k)} = \tilde{\mathbf{g}}(\boldsymbol{\theta}^{(k)}) + \tilde{\boldsymbol{\eta}}^{(k)} \tag{16.23}$$

where

$$\tilde{\mathbf{p}}^{(k)} = \begin{bmatrix} \mathbf{p}^{(k)} \\ \hat{v}_{x,k} \\ \hat{v}_{y,k} \end{bmatrix}, \; \tilde{\mathbf{g}}(\boldsymbol{\theta}^{(k)}) = \begin{bmatrix} \mathbf{g}(\boldsymbol{\theta}^{(k)}) \\ v_{x,k} \\ v_{y,k} \end{bmatrix}, \; \tilde{\boldsymbol{\eta}}^{(k)} = \begin{bmatrix} \boldsymbol{\eta}^{(k)} \\ \vartheta_{x,k} \\ \vartheta_{y,k} \end{bmatrix} \tag{16.24}$$

Similarly, the extended KF can be implemented as follows

$$\begin{aligned}
\hat{\boldsymbol{\theta}}^{(k|k-1)} &= \boldsymbol{\Phi}\hat{\boldsymbol{\theta}}^{(k-1|k-1)} \\
\mathbf{C}^{(k|k-1)} &= \boldsymbol{\Phi}\mathbf{C}^{(k-1|k-1)}\boldsymbol{\Phi}^T + \mathbf{BQB}^T \\
\mathbf{K}^{(k)} &= \mathbf{C}^{(k|k-1)}\mathbf{H}^T(\mathbf{HC}^{(k|k-1)}\mathbf{H}^T + \mathbf{R}^{(k)}) \\
\hat{\boldsymbol{\theta}}^{(k|k)} &= \hat{\boldsymbol{\theta}}^{(k|k-1)} + \mathbf{K}^{(k)}(\tilde{\mathbf{p}}^{(k)} - \tilde{\mathbf{g}}(\hat{\boldsymbol{\theta}}^{(k|k-1)})) \\
\mathbf{C}^{(k|k)} &= (\mathbf{I} - \mathbf{K}^{(k)}\mathbf{H})\mathbf{C}^{(k|k-1)}
\end{aligned} \tag{16.25}$$

where $\mathbf{Q}$ is the covariance matrix of $\boldsymbol{\xi}^{(k)}$ and $\tilde{\mathbf{R}}^{(k)}$ is the covariance matrix of $\tilde{\eta}^{(k)}$ and

$$\begin{aligned}
[\mathbf{H}^{(k)}]_{i,1} &= \left.\frac{\partial d_i^{(k)}}{\partial x}\right|_{\boldsymbol{\theta}^{(k)}=\hat{\boldsymbol{\theta}}^{(k|k-1)}}, \quad [\mathbf{H}^{(k)}]_{i,2} = \left.\frac{\partial d_i^{(k)}}{\partial y}\right|_{\boldsymbol{\theta}^{(k)}=\hat{\boldsymbol{\theta}}^{(k|k-1)}} \\
[\mathbf{H}^{(k)}]_{i,3} &= \left.\frac{\partial^2 d_i^{(k)}}{\partial x^2}\right|_{\boldsymbol{\theta}^{(k)}=\hat{\boldsymbol{\theta}}^{(k|k-1)}}, \quad [\mathbf{H}^{(k)}]_{i,4} = \left.\frac{\partial^2 d_i^{(k)}}{\partial y^2}\right|_{\boldsymbol{\theta}^{(k)}=\hat{\boldsymbol{\theta}}^{(k|k-1)}} \\
[\mathbf{H}^{(k)}]_{N+1,1} &= [\mathbf{H}^{(k)}]_{N+1,2} = [\mathbf{H}^{(k)}]_{N+1,4} = 0 \\
[\mathbf{H}^{(k)}]_{N+2,1} &= [\mathbf{H}^{(k)}]_{N+2,2} = [\mathbf{H}^{(k)}]_{N+2,3} = 0 \\
[\mathbf{H}^{(k)}]_{N+1,3} &= [\mathbf{H}^{(k)}]_{N+2,4} = 1
\end{aligned} \tag{16.26}$$

where $i = 1, 2, \ldots, N$ and the two speed components in $\tilde{\mathbf{g}}(\hat{\boldsymbol{\theta}}^{(k|k-1)})$ are calculated according to

$$\begin{aligned}
v_{xk} &\approx \frac{[\hat{\boldsymbol{\theta}}^{(k|k-1)}]_{1,1} - [\hat{\boldsymbol{\theta}}^{(k-1|k-1)}]_{1,1}}{\delta t} \\
v_{yk} &\approx \frac{[\hat{\boldsymbol{\theta}}^{(k|k-1)}]_{2,1} - [\hat{\boldsymbol{\theta}}^{(k-1|k-1)}]_{2,1}}{\delta t}
\end{aligned} \tag{16.27}$$

where note that in the case of large system noise variances, the above equations may not hold.

When the hypothesis verification is involved and more samples are taken in the LOS and NLOS identification, some delay will occur in generating position estimates, although the delay may not be significant. To avoid the problem of non-real time tracking, different techniques can be employed. In the case where the speed and heading measurements are available, the mobile positions during this period can be readily calculated using a basic dead-reckoning technique. In the absence of speed and heading sensors, a simple way to resolve the problem is to approximate the positions at these time points with the latest position estimate. Certainly, one can use the Kalman filter in the prediction mode when raw measurement data are not available, which is the normal practice. Alternatively, the positions can be predicted as for the case where the speed and heading sensors are available. In this case the speed and heading are estimated based on the previous position estimates.

Another issue when the sampling rate is large is that the measurement errors in (16.2) may be somewhat correlated. In this case, this correlation should be considered in the design of the KF.

16.4 LOS and NLOS Hypothesis Testing and Threshold Selection

In the LOS and NLOS identification method studied earlier a number of thresholds are required. In this section, the general rules in selecting these thresholds are provided based on the hypothesis testing theory. Specifically, it is assumed that the distributions of the distance measurement noise and bias errors are known a priori, such as based on processing field measurements and modeling the noise and bias errors. The thresholds are associated with the probability of a false alarm (PFA). When the PFAs are given, the thresholds can be determined accordingly. In this case these thresholds are obtained in advance and thus there is no issue of computational time. However, when the distribution parameters of the noise and bias errors are not known a priori, the thresholds may be determined adaptively online. Then the computational time of obtaining the thresholds can be an issue. However, investigation of such an adaptive technique is beyond the scope of the chapter. In the remainder of this section, the details of selecting the thresholds in presence of known distance measurement noise and bias error statistics are described.

In the literature the measurement noise is typically modeled as a Gaussian random variable. On the other hand, since the NLOS-induced bias component is always positive when the measurement is TOA or distance, it is often modeled as a one-parameter Rayleigh or exponential random variable when evaluating the performance of positioning algorithms (Cong and Zhuang 2005; Lay and Chao 2005). Field measurements also demonstrated that the TOA measurement errors had a good match with the Rayleigh distribution (Huerta and Vidal 2005). Thus, a Rayleigh bias variable is assumed for the analysis in this chapter.

The probability density function (PDF) of the sum of a zero-mean Gaussian variable and a Rayleigh variable, which are mutually independent, is equal to the convolution of the PDFs of the two random variables and can be derived as

$$\begin{aligned} p(\varepsilon_j) = & \frac{1}{\sqrt{2\pi\sigma_i^2/\sigma_n}} \exp\left(-\frac{\varepsilon_j^2}{2\sigma_n^2}\right) + \frac{\varepsilon_j}{\sigma_i^3/\sigma_R} \\ & \times \exp\left(-\frac{\varepsilon_j^2}{2\sigma_i^2}\right) Q\left(-\frac{\varepsilon_j}{\sigma_i\sigma_n/\sigma_R}\right) \end{aligned} \tag{16.28}$$

where $Q(\cdot)$ is the standard Q-function, σ_n is the STD of the Gaussian noise variable, σ_R is the Rayleigh distribution parameter, and $\sigma_i = \sqrt{\sigma_n^2 + \sigma_R^2}$.

16.4.1 Analysis for Initial Identification

16.4.1.1 NLOS Identification

First consider the case where the immediate L past distance measurements were taken under LOS propagation. Thus the mean and STD of the average of L past LOS distance measurement errors are respectively zero and $\sigma_n/\sqrt{L}$. Then the variable $u = r - m_1$ where r is the current measurement has a PDF that is dependent on whether the measurement condition is LOS or NLOS propagation. If the current propagation is LOS, then u is still Gaussian with a PDF given by

$$p(u|los : los) = \mathcal{N}(0, \tilde{\sigma}_n^2) \tag{16.29}$$

where $\tilde{\sigma}_n = \sigma_n\sqrt{1 + 1/L}$. Alternatively, if the current propagation is NLOS, then the PDF $(p(u|los : nlos))$ is given by (16.28), but with ε_j replaced by u and σ_n replaced by $\tilde{\sigma}_n$. Figure 16.4 shows an example of these two density functions. Clearly in this example the two density functions overlap significantly. That is, if the probability of detection (POD) is high, then the probability of false alarm (PFA) will not be small. On the other hand, if the PFA is small, then the POD will not be large. Note that in this case the POD is the probability of deciding NLOS as NLOS, whereas the PFA is the probability of deciding LOS as NLOS.

Now consider the two probabilities, the thresholds, and their relationships. Giving a PFA denoted by P_{FA}, the threshold δ_{r1} can be determined by

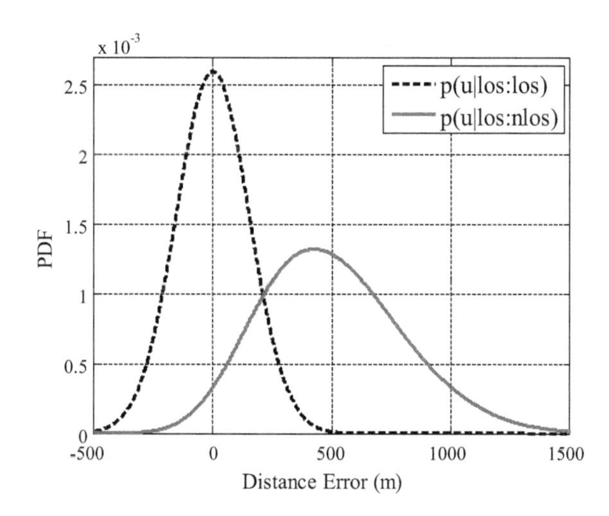

Fig. 16.4 Example of PDFs for NLOS hypothesis testing at the first step

$$P_{FA} = \int_{\delta_{r1}}^{\infty} \frac{1}{\sqrt{2\pi}\tilde{\sigma}_n} \exp\left(-\frac{u^2}{2\tilde{\sigma}_n^2}\right) du = Q\left(\frac{\delta_{r1}}{\tilde{\sigma}_n}\right) \tag{16.30}$$

Then the POD is calculated according to

$$P_D = \int_{\delta_{r1}}^{+\infty} p(u|los : nlos) du \tag{16.31}$$

where numerical integration is needed to calculate the POD. For instance, when $\delta_{r1} = 350$ m, $\sigma_n = 150$ m, and $L = 20$, the PFA and the POD are 1.14% and 67.22%, respectively. In this case, although the PFA is small, the probability of missing (POM) is as high as 32.87%. Note that the POM is the complement of the POD. However in the case of $\delta_{r1} = -100$ m, the PFA and the POD are respectively 74.23% and 99.02%. Although the POD is very high and thus the POM is very small, the PFA is rather large. As discussed later, when NLOS is identified and confirmed, the measurement will be subtracted from the estimate of the bias mean. Thus it is necessary to keep the PFA small.

16.4.1.2 LOS Identification

When the immediate L past distance measurements were taken under NLOS propagation, the mean of the sum of the L Rayleigh bias errors can be approximated as a Gaussian variable with a mean $m_R = \sigma_R\sqrt{\pi/2}$ and a STD $\tilde{\sigma}_R = \sqrt{(4-\pi)/(2L)}\sigma_R$ according to the central limit theorem. Then similarly to (16.29), one of the two density functions now becomes

$$p(u|nlos : los) = \mathcal{N}(-m_R, \tilde{\sigma}_G^2) \tag{16.32}$$

where $\tilde{\sigma}_G = \sqrt{\tilde{\sigma}_n^2 + \tilde{\sigma}_R^2}$. The other density function can be obtained as follows. The variable u now is the sum of a Rayleigh variable and a Gaussian variable of non-zero mean $\{-m_R\}$ and a STD $\tilde{\sigma}_G$. In this case, although (16.28) cannot directly be used to describe the density function of u, following the same derivation procedure produces

$$\begin{aligned} p(u|nlos : nlos) = &\frac{1}{\sqrt{2\pi\tilde{\sigma}_i^2/\tilde{\sigma}_G}} \exp\left(-\frac{(u+m_R)^2}{2\tilde{\sigma}_G^2}\right) \\ &+ \frac{u+m_R}{\tilde{\sigma}_i^3/\sigma_R} \exp\left(-\frac{(u+m_R)^2}{2\tilde{\sigma}_i^2}\right) Q\left(-\frac{u+m_R}{\tilde{\sigma}_i\tilde{\sigma}_G/\sigma_R}\right) \end{aligned} \tag{16.33}$$

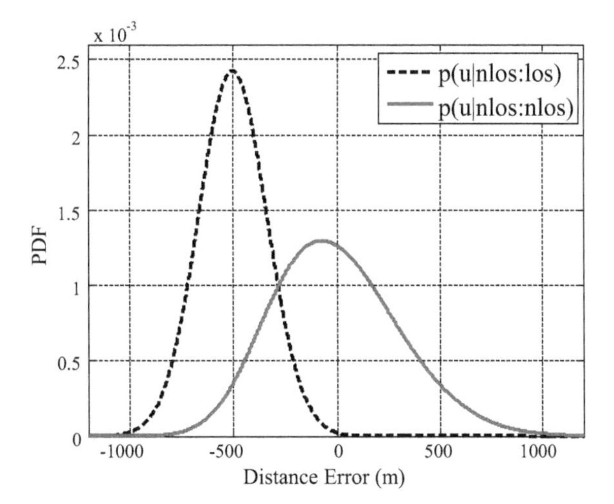

Fig. 16.5 Example of PDFs for LOS hypothesis testing at the first step

where $\tilde{\sigma}_i = \sqrt{\tilde{\sigma}_G^2 + \sigma_R^2}$. Figure 16.5 shows the two density functions when testing the LOS hypothesis. In this case the PFA and the POD are given by

$$\begin{aligned} P_{FA} &= \int_{-\infty}^{-\delta_{r2}} \frac{1}{\sqrt{2\pi}\tilde{\sigma}_G} \exp\left(-\frac{(u+m_R)^2}{2\tilde{\sigma}_G^2}\right) du = Q\left(\frac{\delta_{r2} - m_R}{\tilde{\sigma}_G}\right) \\ P_D &= \int_{-\infty}^{\delta_r} p(u|nlos : nlos) du \end{aligned} \tag{16.34}$$

With the threshold set at −250 m and the other parameters as given earlier, the PFA and POD are 21.53% and 94.21%, respectively. To reduce the probability of deciding LOS as NLOS, the threshold can be set at a larger value so that the POD is larger. However, the PFA which is the probability of deciding NLOS as LOS will simultaneously also be larger. For instance, if the POD is increased to 99.4%, the PFA will be increased to 39.83%. To resolve such a contradiction, a verification stage is required. The following subsection presents the hypothesis testing analysis in the verification phase.

16.4.2 Analysis for Hypothesis Verification

The hypothesis assumed in the initial identification phase is verified by the confirmation process described in this section. The focus is on the dominant cases where the measurements of each of the two sequences are made under the same propagation condition. The analysis is based on the density function of the difference of the means of the two sequences of distance measurements $m_{2,1}$.

16.4.2.1 Verification of NLOS Hypothesis

First consider the case where the first sequence of L measurements is under LOS propagation, whereas the second sequence of L measurements is under either LOS or NLOS propagation, but not both. Note that the two sequences are next to each other as shown in Fig. 16.3. If the second sequence of measurements is under LOS propagation, the difference of the means of the two sequences $m_{2,1}$ is a Gaussian variable which has a PDF given by

$$p(m_{2,1}|los : los) = \mathcal{N}\left(0, 2\sigma_n^2/L\right) \tag{16.35}$$

On the other hand, when the second sequence of measurements is under NLOS propagation, the mean of the sum of L Rayleigh bias errors can be approximated as a Gaussian variable with a mean $m_R = \sigma_R\sqrt{\pi/2}$ and a STD $\tilde{\sigma}_R = \sqrt{(4-\pi)/(2L)}\sigma_R$ as mentioned earlier. Thus the variable $m_{2,1}$ is approximately Gaussian with a PDF given by

$$p(m_{2,1}|los : nlos) = \mathcal{N}(m_R, \sigma_a^2) \tag{16.36}$$

where $\sigma_a = \sqrt{2\sigma_n^2/L + (4-\pi)\sigma_R^2/(2L)}$. Figure 16.6 shows two examples of the two density functions when $\sigma_n = 150$ m, and $\sigma_R = 400$ m and $L = 30$ for the upper figure, whereas the lower figure is produced when $L = 10$. Note that the sum of ten independent and identically distributed Rayleigh variables has a distribution that has a near perfect match with a Gaussian distribution. It can be readily observed that when the mean of the Rayleigh bias error is relatively large, as typically observed in practice, there is nearly no overlapping between the two density functions even for the case of $L = 10$. This comes from the fact that the PDF of the average of L measurements is compared with the PDF of the average of L other measurements. The shapes of the two density functions are squeezed significantly because of the averaging operation. This implies that if the threshold is chosen appropriately, for instance around 150 m, the decision on the hypothesis will nearly always be correct.

The PFA that is the probability of taking the LOS measurement as the NLOS one is given by

$$P_{FA} = \int_{\delta_{m1}}^{+\infty} p(m_{2,1}|los : los)dm_{2,1} = Q\left(\sqrt{\frac{L}{2}}\frac{\delta_{m1}}{\sigma_m}\right) \tag{16.37}$$

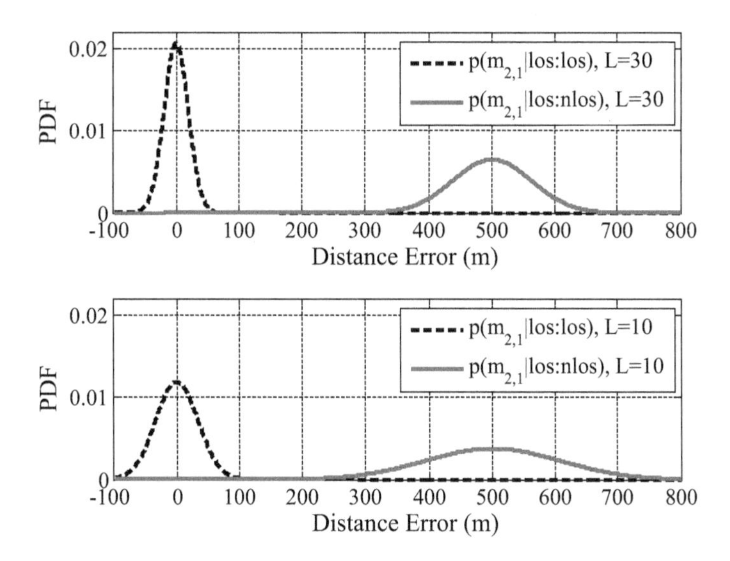

Fig. 16.6 Examples of PDFs for NLOS hypothesis testing at the second step when the mean of the bias error is 500 m

Also the POD that is the probability of correctly deciding the NLOS propagation is given as

$$P_D = \int_{\delta_{m1}}^{+\infty} p(m_{2,1}|los : nlos)dm_{2,1} = Q\left(\frac{\delta_{m1}}{\sigma_a}\right) \tag{16.38}$$

From the density functions shown in Fig. 16.6, the PFA and the POD are respectively virtually zero and one, when the threshold is set around 150 m.

16.4.2.2 Verification of LOS Hypothesis

Now consider the case where the first sequence of L measurements is under NLOS propagation, whereas the second sequence of L measurements is under LOS or NLOS propagation, but not both. Similarly to (16.35) and (16.36), the two density functions of $m_{2,1}$ are respectively given by

$$\begin{aligned} p(m_{2,1}|nlos : los) &= \mathcal{N}(-m_R, \sigma_a^2) \\ p(m_{2,1}|nlos : nlos) &= \mathcal{N}(0, \sigma_b^2) \end{aligned} \tag{16.39}$$

where $\sigma_b = \sqrt{2(\sigma_n^2 + (4-\pi)\sigma_R^2/2)/L}$. Figure 16.7 shows the two PDFs with the same parameters for Fig. 16.6. The distances between the two PDFs are smaller

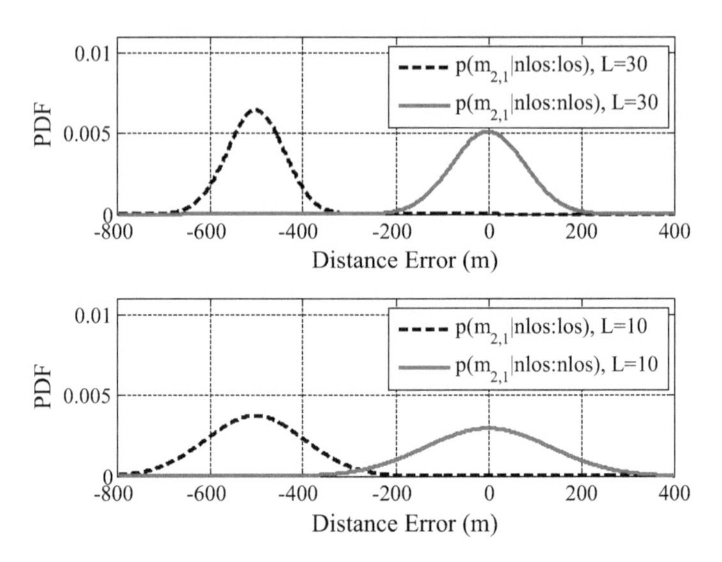

Fig. 16.7 Example of PDFs for LOS hypothesis verification when the mean of the bias error is 500 m

than those in Fig. 16.6. Nevertheless there is only small overlapping between the two density functions when $L = 10$. Thus the decisions made at the first stage can be correctly verified with a very high probability when the threshold is chosen appropriately, such as around −300 m. Specifically the PFA and POD are given by

$$\begin{aligned} P_{FA} &= \int_{-\infty}^{-\delta_{m2}} p(m_{2,1}|nlos : nlos) dm_{2,1} = Q\left(\frac{\delta_{m2}}{\sigma_b}\right) \\ P_D &= \int_{-\infty}^{-\delta_{m2}} p(m_{2,1}|nlos : los) dm_{2,1} = Q\left(\frac{\delta_{m2} - m_R}{\sigma_a}\right) \end{aligned} \tag{16.40}$$

For instance if the threshold is set at −300 m, the length is $L = 10$ and the other parameters are the same as given earlier, the POD and PFA are 98.69% and 2.95%, respectively. That is the wrong decision at the initial identification will be accepted with a probability of 2.95%, whereas the correct decision will be accepted with a probability of 98.69%. Also the hypothesis on the latest $L - 1$ measurements will be tested correctly with a probability of 98.69%.

Note that the thresholds selected based on the PFA are used as a reference when the smoothed distance measurement accuracy is considered. Since there is no formula to describe the relationship between the distance measurement accuracy and the thresholds, the final threshold values are determined manually, and the PFA-based threshold values are only used as a guide.

16.5 Simulation Results

In a hexagonal cellular deployment, the mobile unit will typically be able to communicate with a small number (say four) of base stations, due to the limited radio range of reception of the cellular radio signals. As the mobile moves through a cell it will lose a radio link with one or more base stations that are out of the radio range. Meanwhile, one or more new base stations will come within radio range of the mobile, so that new communication links will be established. For simplicity consider a scenario where the mobile is able to communicate with four base stations denoted by a square and traverses along a path with two linear track segments as shown in Fig. 16.8. While such a scenario is simplistic, the position accuracy calculated under this simulation setup would be approximately similar to a more realistic one.

The mobile is assumed to move along the path from point P_1 (3, 3) through P_2 (3, 0.5) and P_3 (1.5, 0.5) (units in kilometers) at variable speeds. Specifically, the vehicle starts from point P_1 with an initial speed of zero and an acceleration of 2.5 m/s^2. After reaching the speed of 36 km/h, the vehicle travels at this speed until it is close to P_2. Then the vehicle slows down at an deceleration of −2.5 m/s^2 until the speed is reduced to nearly zero at point P_2. Next the vehicle turns right and gradually speeds up to 27 km/h at an acceleration of 2.5 m/s^2 and then keeps moving at this speed along the road segment. The selection of these speeds and accelerations are a bit ad hoc, but these assumed relatively low vehicle speeds are typical of urban driving in large cities. The distance sampling rate is set at 50 samples per second. The distance measurement noise is modeled as a zero-mean Gaussian variable with a STD of 150 m which is known a priori, whereas the bias error is modeled as a Rayleigh variable with the distribution parameter set at 400 m. These parameters are selected based on some early field trials conducted with a

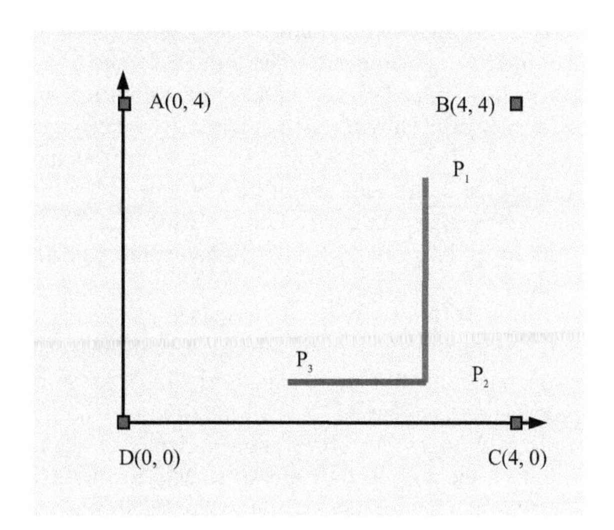

Fig. 16.8 Base station configuration and mobile track for simulation

GSM or 2G cellular network (Silventoinen and Rantalainen 1996; Wylie and Holtaman 1996). Currently, 3G and 4G systems are widely used and the development of 5G network is in progress, so that new field measurements are needed to provide new statistical data for algorithm development and performance evaluation.

The developed algorithm was validated only based on simulation results, without any actual data from a real system. It is assumed that the signal propagation between the mobile and each base station alternates between LOS and NLOS. Intuitively the NLOS or LOS status would usually remain unchanged for a period of time a few seconds. Such a LOS and NLOS propagation pattern in outdoor environments has been observed through field measurements (Woo et al. 2000). Although the duration of either LOS or NLOS propagation would be irregular and dependent on the specific environment, it is modeled as a uniform random variable ranging between 1 and 4 s.

16.5.1 Performance of Distance Estimates Smoothing

The post-smoothing simulated distance measurements of two algorithms versus sampled time is shown in Fig. 16.9. "Proposed" denotes results of the studied smoothing approach. "Biased & Unbiased" denotes results when the LOS and NLOS status and the variance of the bias variable are known perfectly so that the exact variances of the distance measurement errors are used in the KF. Thus these results can be treated as the best performance that can be achieved by the biased and unbiased smoothing algorithm, and the interacting multiple model algorithm (Le et al. 2003; Liao and Chen 2006). For the studied approach the sequence length is set at $L = 30$ and the thresholds are set as $\delta_{r1} = \delta_{r2} = 450$ m, $\delta_{m1} = 170$ m, $\delta_{m2} = 300$ m, and $\sigma_{\gamma 1} = \sigma_{\gamma 2} = 1.1\sigma_n = 165$ m, where σ_n is the STD of the Gaussian noise. These STD thresholds are manually selected and they can be simply set at the STD of the measurement noise if it is difficult to choose better thresholds. Clearly the studied smoothing algorithm produces more accurate distance estimates than the other algorithms. The performance gain comes from the fact that the bias error is reduced significantly through subtracting the estimates of the bias mean, as can be observed in Fig. 16.10. Table 16.1 shows the root mean square error (RMSE) of the original, adjusted, and smoothed distance measurements. The RMSE is reduced about 40% after adjustment and the RMSE of the studied smoothing algorithm is reduced by between 50 and 80% compared with that of the Biased & Unbiased method.

Figure 16.11 shows the performance of the algorithms in terms of the cumulative distribution function (CDF) of the absolute value of the distance estimation errors related to all four base stations. "LOS Assumption" denotes results when the propagation is assumed LOS all the time so that the variance of the measurement noise is used as the observation variance throughout the filtering process. The length of the sequence of measurements for the hypothesis testing is set at three

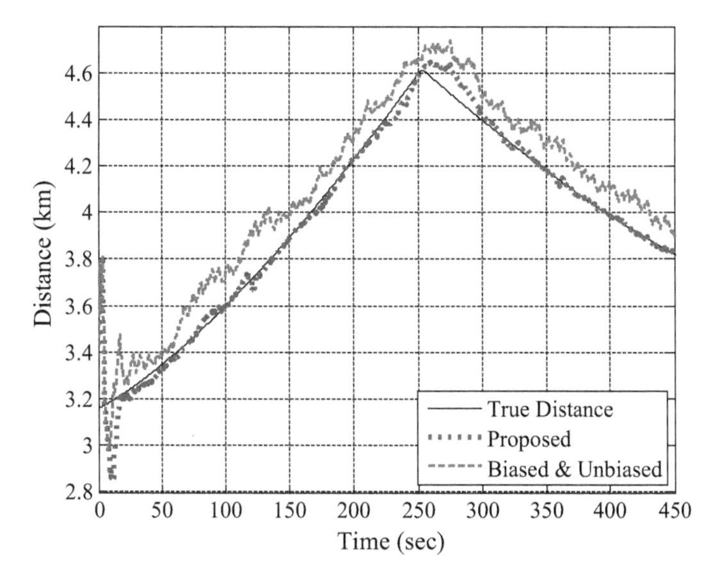

Fig. 16.9 True distances and distance estimates between the mobile and base station A. The sequence length is set at 30 for the proposed approach

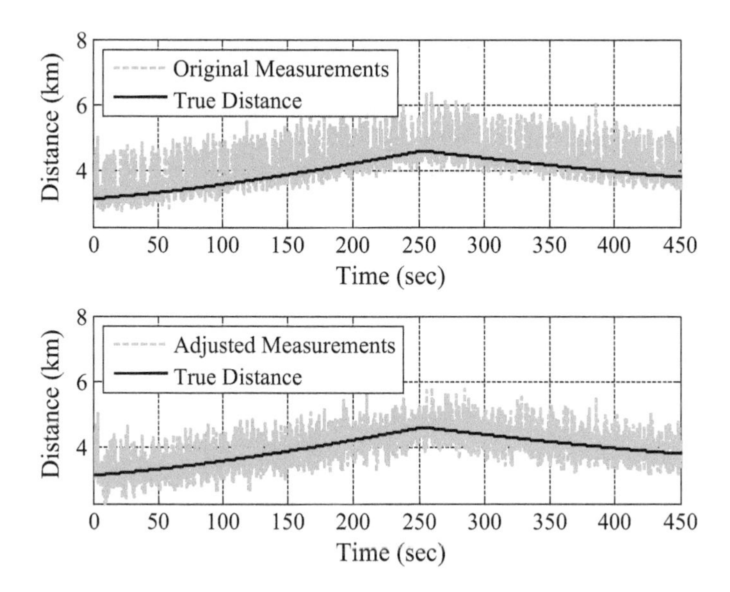

Fig. 16.10 True distance, original distance measurements, and adjusted distance measurements related to one base station

Table 16.1 RMSE (m) of original, adjusted, and smoothed distance measurements

	BS-A	BS-B	BS-C	BS-D
Original	428.95	420.79	420.79	440.77
Adjusted	272.15	259.74	263.02	272.36
Biased and unbiased	113.27	98.99	121.61	133.28
Proposed	34.32	32.49	23.83	68.17

different values: $L = \{10, 20, 30\}$. When $L = 20$ the threshold values are set as $\sigma_{\gamma 1} = \sigma_{\gamma 2} = 0.95\sigma_n = 142.5$ m, whereas it is $\sigma_{\gamma 1} = \sigma_{\gamma 2} = 0.8\sigma_n = 120$ m for $L = 10$. These values are selected manually to achieve better performance; however, the two thresholds may be simply set at the noise STD. Simulation results show that when the two thresholds are set as above, the RMSE of the smoothed distance measurements associated with all four base stations and three sequence lengths is 45.12 m and the RMSE of the corresponding position estimation is 46.62 m. On the other hand, when the two thresholds are set at the noise STD, the RMSE of the smoothed distance measurements and the position estimation are respectively 49.99 m and 48.69 m. The performance degradation is about 10% for distance measurement and about 5% for position estimation.

From Figs. 16.10 and 16.11, it is observed that the distance estimation accuracy can be improved significantly using the studied approach. Also from Fig. 16.11 it can be observed that the effect of the length of the sequence of measurements on the accuracy is marginal for the range of the sequence lengths (10–30).

When the mean and the STD used in the LOS and NLOS identification are replaced respectively with the median and the MAD, the RMSE of the smoothed distance measurements and the position estimation is respectively 41.92 m and 44.07 m, an accuracy improvement by around 6%. It is expected that when the radio propagation alternates more frequently between LOS and NLOS, the accuracy

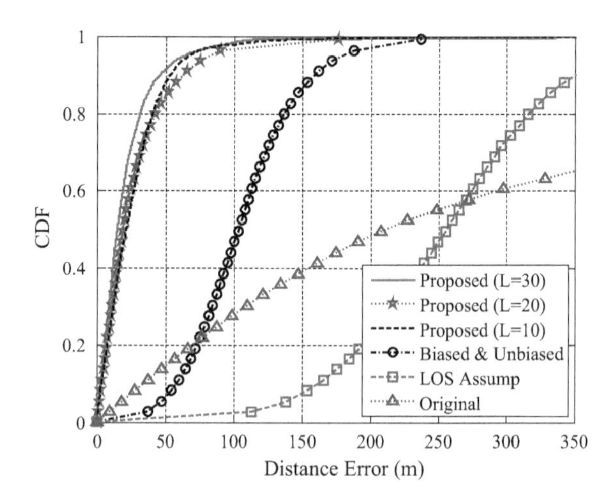

Fig. 16.11 Cumulative distribution function of original distance measurement error and three different algorithms

gain would be higher. Thus, the median and the MAD are preferable in the identification of the LOS and NLOS propagations.

16.5.2 Comparison of Positional Accuracy

In the following simulation, the sampling rate of the speed and heading data is set at 10 samples per second, which is similar to the sampling rates of current off-the-shelf products. This value is much smaller than the sampling rate for ranging data in the above simulations. Thus the speed and heading sample values remain constant for position determination until the next speed and heading sampling time. The speed and heading estimation errors are modeled as Gaussian random variables of zero mean and standard deviations of respectively 2 m/s and 8°. The selection of these values is based on the fact that the speed error of the speedometer is typically less than 10% of the actual value. Although the resolution of the off-the-shelf heading sensors or digital compasses can be as high as ±0.5°, the angle measurement error can be much higher than the resolution in the presence of disturbances.

The CDF of the studied two-stage mobile tracking algorithm is shown in Fig. 16.12, denoted by "Proposed", the algorithm in Le et al. (2003) denoted by "LAT", and the algorithm in Liao and Chen (2006) denoted by "LC". The curve "Proposed-a" denotes the results of the studied approach without using velocity and heading measurements, whereas the curve "Proposed-b" denotes the results using these sensor measurements. The length of the sequence of measurements for hypothesis verification is set at $L = 20$. It can be observed that the studied NLOS mitigation based two-stage approach outperforms the existing algorithms significantly. The use of the speed and heading information in the Kalman filter yields a

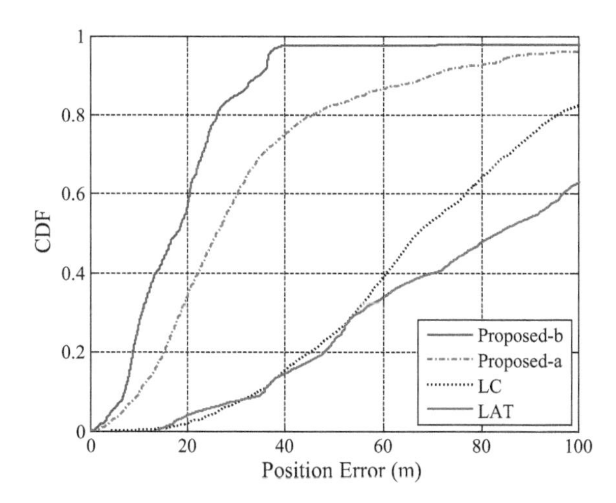

Fig. 16.12 Positional accuracy in terms of CDF of four different algorithms

significant accuracy gain compared to the case where these speed and heading measurements are not exploited.

A similar performance gain would be expected for the case where three-dimensional positioning is considered. However, it needs further investigation on whether or not the vertical accuracy improvement would be similar to the x-coordinate or y-coordinate position accuracy improvement.

16.5.3 Impact of Velocity and Heading Measurement Errors

The impact of the heading measurement errors on the positional accuracy of the studied algorithm is shown in Fig. 16.13. The length of the sequence of measurements for hypothesis verification is set at $L = 20$ and the STD of the velocity measurement error is set at 7.2 km/h. Clearly when the STD of the angular errors is less than 6°, the algorithm is insensitive to the heading measurement error and the impact is insignificant. However, when the STD is equal to or greater than 14° the accuracy degradation can be significant in terms of the CDF for smaller position errors. Thus it would be desirable to keep the STD of the heading measurement error below 10°. As mentioned earlier, a number of the off-the-shelf heading sensors can provide measurements with accuracy better than 10° in normal operating environments.

The effect of the mobile speed measurement errors when the STD of the heading measurement error is set at 6° is shown Fig. 16.14. Results with four different STD values (3.6, 7.2 10.8, and 14.4 km/h) are presented. It can be observed that the proposed tracking algorithm is rather insensitive to the speed measurement errors. The reason for such insensitivity may be explained as follows. The range of the speed errors STD studied is 10.8 km/h, which is equivalently 3 m/s. The

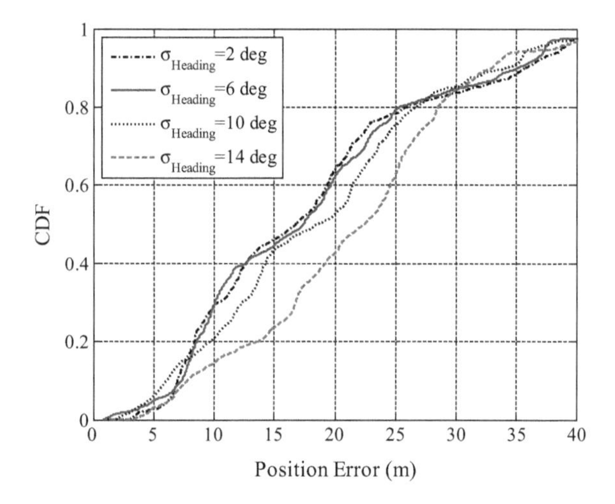

Fig. 16.13 Impact of heading angle measurement error on the positional accuracy of the proposed algorithm

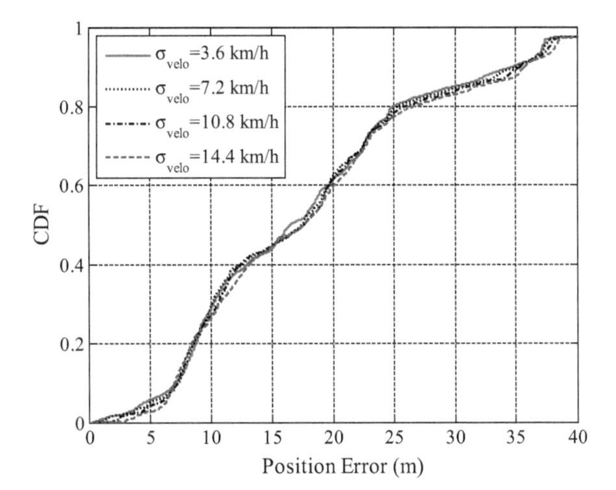

Fig. 16.14 Impact of mobile velocity measurement error on the positional accuracy of the proposed algorithm

corresponding distance error over a small fraction of a second would be on the order of a decimeter. Such a distance error is much smaller than the smoothed distance measurement error, that can be tens of meters.

References

Abu-Shaban Z, Zhou X, Abhayapala TD (2016) A novel TOA-based mobile localization technique under mixed LOS/NLOS conditions for cellular networks. IEEE Trans Veh Technol 65(11):8841–8853

Bahl P, Padmanabhan V (2000) RADAR: an in-building RF-based user location and tracking system. In: Proceedings of IEEE conference on computer communications (INFOCOM), pp 775–784, Tel Aviv, Israel, March 2000

Caffery JJ, Stuber GL (1998) Subscriber location in CDMA cellular networks. IEEE Trans Veh Technol 47(2):406–416

Chen B-S, Yang C-Y, Liao F-K, Liao J-F (2009) Mobile location estimator in a rough wireless environment using extended Kalman-based IMM and data fusion. IEEE Trans Veh Technol 58 (3):1157–1169

Cong L, Zhuang W (2005) Nonline-of-sight error mitigation in mobile location. IEEE Trans Wirel Commun 4(2):560–573

Hellebrandt M, Mathar R (1999) Location tracking of mobiles in cellular radio networks. IEEE Trans Veh Technol 48(5):1558–1562

Ho TJ (2013) Urban localization estimation for mobile cellular networks: a fuzzy-tuned hybrid systems approach. IEEE Trans Wirel Commun 12(5):2389–2399

Huerta JM, Vidal J (2005) Mobile tracking using UKF, time measurements and LOS-NLOS expert knowledge. In: Proceedings of IEEE international conference on acoustics, speech, and signal processing, Philadelphia, pp 901–904, PA, USA, March 2005

Huerta JM, Vidal J, Giremus A, Tourneret J-Y (2009) Joint particle filter and UKF position tracking in severe non-line-of-sight situations. IEEE J. Sel Top Sig Process 3(5):874–888

Jardak M, Samama N (2009) Indoor positioning based on GPS-repeaters: performance enhancement using an open code loop architecture. IEEE Trans Aerosp Electron Syst 45(1): 147–159

Juurakko S, Backman W (2004) Database correlation method with error correlation for emergency location. Wirel Pers Commun 30(2–4):183–194

Kim W, Lee JG, Jee G-I (2006) The interior-point method for an optimal treatment of bias in trilateration location. IEEE Trans Veh Technol 55(4):1291–1301

Lakhzouri A, Lohan ES, Hamila R (2003) M (2003) Extended Kalman filter channel estimation for line-of-sight detection in WCDMA mobile positioning. EURASIP J Appl Sig Process 13:1268–1278

Lay K-T, Chao W-K (2005) Mobile positioning based on TOA/TSOA/TDOA measurements with NLOS error reduction. In: Proceedings of international symposium on intelligent signal processing and communication systems, pp 545–548, Hong Kong, Dec 2005

Le BL, Ahmed K, Tsuji H (2003) Mobile location estimator with NLOS mitigation using Kalman filtering. In: Proceedings of wireless communications and networking, New Orleans, pp 1969–1973, Louisiana, USA, March 2003

Liao J-F, Chen B-S (2006) Robust mobile location estimator with NLOS mitigation using interacting multiple model algorithm. IEEE Trans Wirel Commun 5(11):3002–3006

Locata (2017, visited) FAQs: What is the uprate rate of the Locata position and/or individual LocataLite ranges? http://www.locata.com/technology/faqs/#faq_1553

Miao H, Yu K, Juntti M (2007) Positioning for NLOS propagation: algorithm derivations and Cramer-Rao bounds. IEEE Trans Veh Technol 56(5):2568–2580

Mourad F, Snoussi H, Richard C (2011) Interval-based localization using RSSI comparison in MANETs. IEEE Trans Aerosp Electron Syst 47(4):2897–2910

Nerguizian C, Despins C, Affes S (2006) Geolocation in mines with an impulse response fingerprinting and neural networks. IEEE Trans Wirel Commun 5(3):603–611

Nezafat M, Kaveh M, Tsuji H, Fukagawa T (2005) Statistical performance of subspace matching mobile localization using experimental data. In: Proceedings of IEEE workshop on signal processing advances in wireless communications, pp 645–649, New York, USA, June 2005

Schmidt R (1996) Least squares range difference location. IEEE Trans Aerosp Electron Syst 32 (1):234–242

Silventoinen MI, Rantalainen T (1996) Mobile station emergency locating in GSM. In: Proceedings of IEEE international conference on personal wireless communications, New Delhi, India, pp 232–238, Feb 1996

Torieri DJ (1984) Statistical theory of passive location systems. IEEE Trans Aerosp Electron Syst 20(2):183–198

Wang G, Amin MG, Zhang Y (2006) New approach for target locations in the presence of wall ambiguities. IEEE Trans Aerosp Electron Syst 42(1):301–315

Wang W, Wang Z, O'Dea B (2003) A TOA-based location algorithm reducing the errors due to nonline-of-sight (NLOS) propagation. IEEE Trans Veh Technol 52(1):112–116

Wylie MP, Holtzman J (1996) The non-line of sight problem in mobile location estimation. In: Proceedings of IEEE international conference on universal personal communications, pp 827–831, Cambridge, Massachusetts, Sept–Oct 1996

Woo S-S, You H-R, Koh J-S (2000) The NLOS mitigation technique for position location using IS-95 CDMA networks. In: Proceedings of IEEE vehicular technology conference, pp 2556–2560, Boston, MA, USA, Sept 2000

Yu K, Dutkiewicz E (2012) Geometry and motion based positioning algorithms for mobile tracking in NLOS environments. IEEE Trans Mob Comput 11(2):254–263

Yu K, Guo YJ (2008) Improved positioning algorithms for nonline-of-sight environments. IEEE Trans Veh Technol 57(4):2342–2353

Yu K, Sharp I, Guo YJ (2009) Ground-based wireless positioning. Wiley-IEEE Press, Chippenham

Chapter 17
Data Fusion and Map Matching for Position Accuracy Enhancement

In complex and hostile propagation environments such as in deep urban canyons or underground tunnels, a positioning algorithm based on a single position estimation technology or method, or based on one type of signal measurements, may not achieve satisfactory performance. To improve positioning performance, data fusion can be applied. Specifically, in a positioning system, different signal parameter measurements can be made, such as the received signal strength (RSS), the time-of-arrival (TOA), and the angle-of-arrival (AOA). The different types of measurements are then jointly employed for position determination to achieve enhanced position accuracy. Alternatively giving a set of signal measurements, different position estimation algorithms can be used to produce a set of position estimates. Then the position estimates are combined to produce a position estimate which would be better than any of the results from the individual algorithms. Further, measurements or position estimation results from different technologies or systems can be integrated to achieve a performance gain and provide non-interrupted location services.

This chapter describes methods of using position information from multiple sources, and fusing these data to obtain more accurate and reliable estimates of position data. Data fusion sources considered in this chapter include data described in other chapters of the book, including TOA, RSS, and pseudoranges, as well as position data from other sources including GPS, inertial navigation system, and other dead-reckoning sensors. Further, a number of road map mapping approaches are studied, which can be used to improve vehicle tracking performance for safe driving and intelligent transportation.

I. Sharp and K. Yu, *Wireless Positioning: Principles and Practice*, Navigation: Science and Technology, https://doi.org/10.1007/978-981-10-8791-2_17

17.1 Data Fusion with Different Techniques in Wireless Networks

In this section the focus is on the fusion for position determination in wireless networks. In particular, the various algorithms and techniques for combining two or more types of signal measurements are presented. The term of *hybrid* is also often used for describing these methods.

17.1.1 Accuracy Limits of TOA and TDOA Based Positioning

The TOA and time-difference-of-arrival (TDOA) based methods are often employed to produce accurate position estimates, especially when wideband and ultra-wideband technology is exploited. In this subsection the performance limits of the two positioning approaches are studied, which may help in the selection of the algorithms to achieve better positioning accuracy (Qi and Kobayashi 2003).

Let $\hat{\boldsymbol{\theta}} = [\hat{x} \quad \hat{y}]^T$ be the estimate of the position coordinate vector $\boldsymbol{\theta} = [x \quad y]^T$ of the mobile node. Also denote $p(\mathbf{r}|\boldsymbol{\theta})$ as the conditional probability density function with $\mathbf{r}$ as the observation vector. The Fisher information matrix (FIM) is defined as

$$\mathbf{F}(\boldsymbol{\theta}) = E\left[\frac{\partial}{\partial \boldsymbol{\theta}} \ln p(\mathbf{r}|\boldsymbol{\theta}) \left(\frac{\partial}{\partial \boldsymbol{\theta}} \ln p(\mathbf{r}|\boldsymbol{\theta})\right)^T\right] \tag{17.1}$$

The Cramer-Rao lower bound (CRLB) sets a lower bound on the variance of any unbiased estimates of an unknown parameter, which is defined as the inverse of the FIM, that is $\mathrm{F}^{-1}(\boldsymbol{\theta})$. Accordingly the covariance matrix of $\hat{\boldsymbol{\theta}}$ satisfies

$$\begin{aligned} \operatorname{cov}(\hat{\boldsymbol{\theta}}) &= E\left[(\hat{\boldsymbol{\theta}} - \boldsymbol{\theta})(\hat{\boldsymbol{\theta}} - \boldsymbol{\theta})^T\right] \\ &\geq \mathbf{F}^{-1}(\boldsymbol{\theta}) \end{aligned} \tag{17.2}$$

Suppose that the received signal at the *i*th base station is described as

$$r_i(t) = A_i s(t - \tau_i) + n_i(t),\ i = 1, 2, \ldots, N \tag{17.3}$$

where $s(t)$ is the baseband signal waveform, A_i and τ_i are respectively the signal amplitude and time delay, and $\{n_i(t)\}$ are the independent complex-valued Gaussian noise processes with single-sided spectral density N_0. The time delay τ_i for TOA positioning is given by

$$\tau_i = \frac{1}{c}\sqrt{(x - x_i)^2 + (y - y_i)^2} \tag{17.4}$$

where c is the speed of propagation and (x_i, y_i) are the coordinates of the ith base station. The FIM for the mobile position estimate can be derived as

$$\mathbf{F}_{\mathrm{TOA}} = \mathbf{H}_{\mathrm{TOA}}\Lambda\mathbf{H}_{\mathrm{TOA}}^{\mathrm{T}} \tag{17.5}$$

where

$$\mathbf{H} = \frac{1}{c}\begin{bmatrix} \cos\varphi_1 & \cos\varphi_2 & \cdots & \cos\varphi_N \\ \sin\varphi_1 & \sin\varphi_2 & \cdots & \sin\varphi_N \end{bmatrix}$$

$$\Lambda = \mathrm{diag}\{\lambda_1, \quad \lambda_2, \quad \ldots, \quad \lambda_N\} \tag{17.6}$$

Here angle φ_i is the geometric angle between the positions of the mobile and the ith base station, given by

$$\varphi_i = \tan^{-1}\frac{y - y_i}{x - x_i} \tag{17.7}$$

The diagonal elements of Λ is given by

$$\lambda_i = 8\pi^2\beta^2 \times \gamma_i \tag{17.8}$$

where γ_i is the signal-to-noise ratio (SNR) of the received signal at the ith base station, defined as

$$\gamma_i = \frac{\int |A_i s(t)|^2 dt}{N_0} \tag{17.9}$$

and β is the effective bandwidth of the signal waveform $s(t)$, defined as

$$\beta^2 = \frac{\int f^2 |S(f)|^2 df}{\int |S(f)|^2 df} \tag{17.10}$$

where $S(f)$ is the Fourier transform of $s(t)$.

For TDOA based positioning, there is an unknown time offset between the clock of the mobile and those of the base stations. Thus the time delay becomes

$$\tau_i = \frac{1}{c}\left(\sqrt{(x - x_i)^2 + (y - y_i)^2} + l_0\right) \tag{17.11}$$

where l_0/c is the time offset. The associated FIM can be expressed as

$$\mathbf{F}_{\mathrm{TDOA}} = \mathbf{H}_{\mathrm{TDOA}} \mathbf{\Phi}^{-1} \mathbf{H}_{\mathrm{TDOA}}^{\mathrm{T}} \tag{17.12}$$

where

$$\mathbf{H}_{\mathrm{TDOA}} = \frac{1}{c} \begin{bmatrix} \cos\varphi_1 & \cos\varphi_2 & \cdots & \cos\varphi_{N-1} \\ \sin\varphi_1 & \sin\varphi_2 & \cdots & \sin\varphi_{N-1} \end{bmatrix} - \frac{1}{c} \begin{bmatrix} \cos\varphi_N \\ \sin\varphi_N \end{bmatrix} \times \mathbf{1}^T$$

$$\mathbf{\Phi} = \mathrm{diag}\{ \lambda_1^{-1}, \quad \lambda_2^{-1}, \quad \cdots, \quad \lambda_{N-1}^{-1} \} + \lambda_N^{-1} \mathbf{1} \times \mathbf{1}^T \tag{17.13}$$

Here $\mathbf{1}$ is the column vector of all ones and size $(N-1)$. By defining the unit vector

$$\mathbf{h}_i = \begin{bmatrix} \cos\varphi_i \\ \sin\varphi_i \end{bmatrix} \tag{17.14}$$

The TOA related FIM in (17.5) can be rewritten as

$$\mathbf{F}_{\mathrm{TOA}} = \frac{1}{c^2} \sum_{i=1}^{N} \lambda_i \mathbf{h}_i \mathbf{h}_i^T \tag{17.15}$$

Define weight coefficient w_i as

$$w_i = \frac{\lambda_i}{\lambda} \tag{17.16}$$

where

$$\lambda = \sum_{i=1}^{N} \lambda_i \tag{17.17}$$

Then (17.15) becomes

$$\mathbf{F}_{\mathrm{TOA}} = \frac{\lambda}{c^2} \sum_{i=1}^{N} w_i \mathbf{h}_i \mathbf{h}_i^T \tag{17.18}$$

Similarly, the TDOA related FIM in (17.12) can be rewritten as

$$\mathbf{F}_{\mathrm{TDOA}} = \frac{\lambda}{c^2} \left(\sum_{i=1}^{N} w_i \mathbf{h}_i \mathbf{h}_i^T - \left(\sum_{i=1}^{N} w_i \mathbf{h}_i \right) \left(\sum_{i=1}^{N} w_i \mathbf{h}_i \right)^T \right) \tag{17.19}$$

Define a random vector $\mathbf{h}$ that takes values of

$$\mathbf{h}_1, \mathbf{h}_2, \ldots, \mathbf{h}_N \tag{17.20}$$

with probabilities

$$w_1, w_2, \ldots, w_N \tag{17.21}$$

and its weighted average as

$$\bar{\mathbf{h}} = \sum_{i=1}^{N} w_i \mathbf{h}_i \tag{17.22}$$

Accordingly $\mathbf{F}_{\mathrm{TOA}}$ and $\mathbf{F}_{\mathrm{TDOA}}$ can be expressed in terms of the second moment and covariance of $\mathbf{h}$, respectively. That is

$$\mathbf{F}_{\mathrm{TOA}} = \frac{\lambda}{c^2} E\left[\mathbf{h}\mathbf{h}^T\right] \tag{17.23}$$

and

$$\mathbf{F}_{\mathrm{TDOA}} = \frac{\lambda}{c^2} E\left[(\mathbf{h} - \bar{\mathbf{h}})(\mathbf{h} - \bar{\mathbf{h}})^T\right] \tag{17.24}$$

Consequently, it is seen that

$$\mathbf{F}_{\mathrm{TDOA}}^{-1} \geq \mathbf{F}_{\mathrm{TOA}}^{-1} \tag{17.25}$$

where the equality holds if and only if $\bar{\mathbf{h}} = 0$.

The relationship between the TOA based CRLB and the TDOA based CRLB is reasonable due to the fact that there is one extra unknown parameter l_0 in the TDOA based approach. However, this does not mean that the TOA based approach is superior to the TDOA based approach. In practice, the TOA measurements may be obtained through synchronization so that a synchronization error can be added to the TOA measurement error. In the event that the TOA is obtained by making round-trip-time (RTT) measurements, the time estimation error at the mobile may be larger than that at the base stations. On the other hand, the TDOA approach requires synchronization among all the base stations and the actual synchronization may be imperfect, resulting in synchronization errors. In some circumstances such as in wireless sensor networks, it may be inappropriate to maintain clock synchronization among a group of anchor nodes. Furthermore, the positioning accuracy will also depend on the specific positioning algorithm used. For instance, when the measurements are linearized and the least-squares estimator is applied, the TOA and the TDOA approaches achieve the same positioning accuracy (Shin and Sung 2002). Thus the choice of the two approaches will depend on the application scenarios, the synchronization performance, the positioning algorithms, and the nodes configurations.

17.1.2 Combining TOA and TDOA Data

This subsection describes a data fusion approach based on a sequence of independent TOA and TDOA measurements (Kleine-Ostmann and Bell 2001). Figure 17.1 shows the block diagram of the data fusion process. At the first level of fusion, the raw TOA measurements are converted into the TDOA format which are combined with the raw TDOA measurements. The two sets of independent TDOA data serve as the input to a position estimator, such as the weighted least squares estimator, or the maximum likelihood estimator. Since these estimators are iterative, an initial position estimate is required, which may be obtained by using a non-iterative algorithm as discussed in Yu et al. (2009).

Second level of fusion combines position estimates from the TOA based and the TDOA based position estimators. Suppose that the TOA based position coordinates estimates have means denoted by $(\hat{x}_{\mathrm{TOA}}, \hat{y}_{\mathrm{TOA}})$ and variances denoted by $(\sigma^2_{\hat{x}_{\mathrm{TOA}}}, \sigma^2_{\hat{y}_{\mathrm{TOA}}})$. Also the means and variances of the TDOA based method are respectively denoted by $(\hat{x}_{\mathrm{TDOA}}, \hat{y}_{\mathrm{TDOA}})$ and $(\sigma^2_{\hat{x}_{\mathrm{TDOA}}}, \sigma^2_{\hat{y}_{\mathrm{TDOA}}})$. Then according to Bayesian inference the fused position estimate is given by

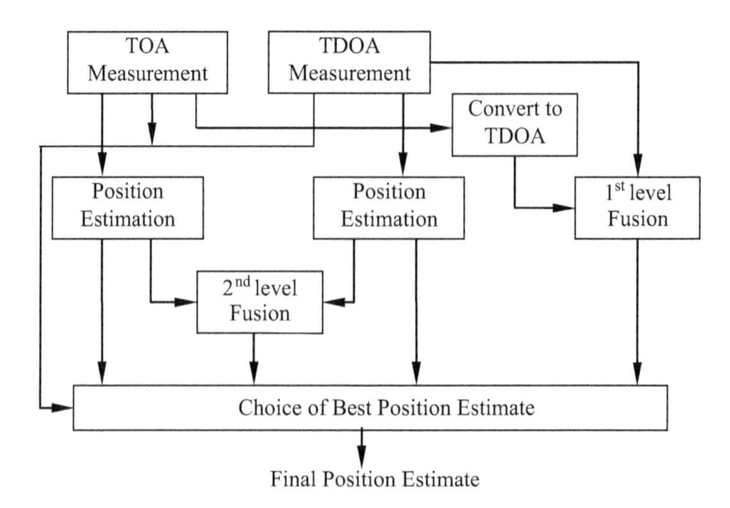

Fig. 17.1 Block diagram of TOA and TDOA based data fusion for position estimation

$$\begin{aligned}\hat{x} &= \hat{x}_{\mathrm{TDOA}} + \frac{\sigma^2_{\hat{x}_{\mathrm{TDOA}}}}{\sigma^2_{\hat{x}_{\mathrm{TDOA}}} + \sigma^2_{\hat{x}_{\mathrm{TOA}}}} (\hat{x}_{\mathrm{TDOA}} - \hat{x}_{\mathrm{TOA}}) \\ \hat{y} &= \hat{y}_{\mathrm{TDOA}} + \frac{\sigma^2_{\hat{y}_{\mathrm{TDOA}}}}{\sigma^2_{\hat{y}_{\mathrm{TDOA}}} + \sigma^2_{\hat{y}_{\mathrm{TOA}}}} (\hat{y}_{\mathrm{TDOA}} - \hat{y}_{\mathrm{TOA}})\end{aligned} \tag{17.26}$$

The variances of the coordinate estimates are respectively given by

$$\sigma^2_{\hat{x}} = \frac{\sigma^2_{\hat{x}_{\mathrm{TDOA}}} \sigma^2_{\hat{x}_{\mathrm{TOA}}}}{\sigma^2_{\hat{x}_{\mathrm{TDOA}}} + \sigma^2_{\hat{x}_{\mathrm{TOA}}}}, \quad \sigma^2_{\hat{y}} = \frac{\sigma^2_{\hat{y}_{\mathrm{TDOA}}} \sigma^2_{\hat{y}_{\mathrm{TOA}}}}{\sigma^2_{\hat{y}_{\mathrm{TDOA}}} + \sigma^2_{\hat{y}_{\mathrm{TOA}}}} \tag{17.27}$$

The data fusion at the third level chooses the appropriate position estimates based on the outputs of the four estimators as well as on a priori knowledge of the geometric base station configuration and the statistics of the TOA and TDOA measurements. Typically, the estimate with the smallest variance is considered the most reliable estimate. However, the choice cannot be based solely on the variance. The geometric dilution of precision (GDOP) should also be considered. When a mobile is close to one or more base stations, the TDOA-based estimate should not be used since it suffers from a large GDOP, resulting in poor position estimation accuracy. Of the remaining estimates, the one with the lowest variance should be chosen. On the other hand, when a mobile is far away from all base stations, the TDOA-based estimate would be the best solution. The TOA-based estimate may be used to decide whether the mobile is close to any of the base stations. When the base stations are on the boundary of a circular shape, the mobile may be considered close to a base station if the distance is less than a tenth of the radius.

17.1.3 Performance Limit Based Method

The TOA, TDOA and RSS based methods are the major radio positioning techniques. As mentioned in early chapters the RSS-based ranging method is simple. Many existing communication systems such as the Bluetooth related systems automatically provide the RSS measurements. Alternatively, unlike the TOA approach that requires a dedicated hardware implementation, a simple device can be utilized to measure the RSS. However, the major limitation for the RSS positioning is its poor performance when the mobile is located in a region with severe multipath. It is useful to study the accuracy limit for the RSS-based ranging as well as the TOA based ranging (Qi and Kobayashi 2003). The purpose is to provide a guideline for the choice of the different methods to achieve better positioning accuracy. When using the log path loss model, the path loss P_L and the distance d are related by

$$P_L = 10\eta \log d + \varepsilon \tag{17.28}$$

where for analytical simplicity the constant terms in the path loss model are dropped, log denotes the base 10 logarithm, η is the path loss factor which is equal to 2 in free space and typically greater than 2 in multipath propagation channels, and ε is the modeling error which may include log-normal shadowing. Suppose ε is a Gaussian variable of zero mean and variance σ_ε^2. Then the probability density function of ε conditioned on d is given by

$$p(\varepsilon|d) = \frac{1}{\sqrt{2\pi}\sigma_\varepsilon}\exp\left(-\frac{1}{2\sigma_\varepsilon^2}(P_L - 10\eta\log d)^2\right) \tag{17.29}$$

The corresponding CRLB for the distance estimate can be readily derived as

$$\mathbf{F}_{\mathrm{d,RSS}}^{-1} = \left(\frac{\ln 10}{10}\right)^2 \frac{\sigma_\varepsilon^2}{\eta^2} d^2 \tag{17.30}$$

Thus the standard deviation of the distance estimate satisfies

$$\sigma_{\mathrm{d,RSS}} \geq \frac{\ln 10}{10}\frac{\sigma_\varepsilon}{\eta} d \tag{17.31}$$

Clearly in the RSS-based ranging, the accuracy limit is completely determined by the characteristics of a communication channel and the distance. Therefore, there is little we can do to change the parameters of the CRLB expression and the resulting positioning accuracy when the optimal position estimator or the maximum likelihood estimator is applied.

For TOA based positioning, consider the case where the time delay estimation is performed based on the matched filter output with the delay estimate described as

$$\hat{\tau} = \tau + \varsigma \tag{17.32}$$

where ς is a Gaussian random variable of mean zero and variance given by

$$\sigma_\varsigma^2 = \frac{1}{8\pi^2\beta^2\gamma} \tag{17.33}$$

It can be easily shown that the associated CRLB is given by

$$\mathbf{F}_{\mathrm{d,TOA}}^{-1} = \frac{c^2}{8\pi^2\beta^2\gamma} \tag{17.34}$$

Accordingly

$$\sigma_{\mathrm{d,TOA}} \geq \frac{c}{2\sqrt{2}\pi\beta\sqrt{\gamma}} \tag{17.35}$$

From (17.35) it is seen that the distance estimation accuracy depends on the effective bandwidth and the SNR. That is the positioning accuracy can be improved by increasing the bandwidth, the SNR, or both. Thus, comparatively, the TOA approach is more suited for long-range positioning. Based on (17.31) and (17.35), a critical distance may be defined as

$$d_c = \frac{5c}{(\sqrt{2}\ln 10)\pi}\frac{\eta}{\sigma_\varepsilon}\frac{1}{\beta\sqrt{\gamma}}. \tag{17.36}$$

That is by comparing the critical distance with the distance estimates for both the RSS and the TOA approaches, a decision can be made on which of the two estimates should be used for position determination. If the distance is greater than d_c, the TOA distance estimate should be used. Alternatively if the distance is less than d_c, the RSS distance estimate should be used. However, the true distance is not known a priori, so both the TOA-based and the RSS-based distance estimates should be considered. Thus when both TOA and RSS data are available, the positioning scheme may take one of three different modes as follows:

1. The RSS mode. If the RSS-based distance estimate $\hat{d}_{RSS}$ is much smaller than d_c, then $\hat{d}_{RSS}$ is accepted as the distance estimate. Only RSS associated data are used for position determination.
2. The TOA mode. If the TOA-based distance estimate $\hat{d}_{TOA}$ is much greater than d_c, then $\hat{d}_{TOA}$ is accepted as the distance estimate. The TOA data are thus used for position determination.
3. The hybrid mode. If both $\hat{d}_{RSS}$ and $\hat{d}_{TOA}$ are around d_c, both RSS and TOA data should be used for position determination.

The main condition for the above positioning scheme is that the two models for describing the TOA error and the path loss are precise. The comparison should be made under the same condition of only multipath propagation and NLOS is not considered. Also, the comparison is based on CRLB which is not the actual ranging accuracy, hence the theory presented may just be used as a reference. In the event that the path loss model is not accurate due to the complexity and variation of the environment, the TOA mode is preferable. In particular, when using the ultra wideband technology, super-resolution TOA estimation can be achieved for short-range positioning.

17.1.4 EK-IMM Based Fusion

Consider fusing RSS data and TOA data for locating mobile stations in urban microcell radio propagation environments (Chen et al. 2009). An extended Kalman-filter based interacting multiple Model (EK-IMM) smoother is studied for smoothing the TOA-based distance and the RSS measurements in hostile propagation environments. The extended Kalman filter is used for nonlinear parameter estimation, whereas the IMM is employed as a switch between the LOS and the NLOS states which are considered as a Markov process with two interactive modes.

The TOA based distance measurement between the kth base station and the mobile at time n under line-of-sight (LOS) condition is modeled as

$$r_k(n) = d_k(n) + \varepsilon_{dk}(n),\ k = 1,\ 2,\ \ldots,\ K \tag{17.37}$$

where $d_k(n)$ is the true distance and $\varepsilon_{dk}(n)$ is the measurement noise modeled as a Gaussian variable of zero mean and variance σ_m^2. In the event of non-line-of-sight (NLOS) propagation, the true propagation distance is the sum of the distance from the base station to the scatterer, $d_{ck}(n)$, and that from the scatterer to the mobile, $d_{rk}(n)$, or vice versa. Thus the NLOS distance measurement is modeled as

$$\begin{aligned} r_k(n) &= d_k(n) + \varepsilon_{dk}(n) \\ &= d_{ck}(n) + d_{rk}(n) + \varepsilon_{dk}(n). \end{aligned} \tag{17.38}$$

Note that the LOS and NLOS propagation conditions can occur alternatively over the observation period. The path loss related to the RSS under LOS propagation is modeled as

$$L_{k,1}(n) = 10\log\left(d_k^a(n)(1 + d_k(n)/g)^b\right) + \varepsilon_{pk,1}(n),\ k = 1,\ 2,\ \ldots,\ K \tag{17.39}$$

where $\varepsilon_{pk,j}(n)$ is the RSS measurement noise modeled as a Gaussian variable of zero mean and variance σ_p^2, g is the break point distance ranging from 150 to 300 m, and a and b are the slope path loss parameters before and after the break point. Under NLOS propagation the path loss is modeled as

$$\begin{aligned} L_{k,2}(n) = 10\log\left(d_{ck}^a(n)\left(1 + \frac{d_{ck}(n)}{g}\right)^b (d_k(n) - d_{ck}(n))^a\left(1 + \frac{d_k(n) - d_{ck}(n)}{g}\right)^b\right) \\ + \varepsilon_{pk,2}(n),\ k = 1,\ 2,\ \ldots,\ K \end{aligned} \tag{17.40}$$

To enhance the position estimation performance, the TOA-based distance and the RSS-based pass loss measurement equations are combined. Define

$$\mathbf{z}_{k,j}(n) = \left[L_{k,j}(n) \quad r_{k,j}(n)\right]^T, j = 1, 2, k = 1, 2, \ldots, K$$
$$\varepsilon_{k,j}(n) = \left[\varepsilon_{pk,j}(n) \quad \varepsilon_{dk,j}(n)\right]^T$$
$$\mathbf{h}_{k,1}(n) = \left[10\log d_k^a(n)\left(1+\frac{d_k(n)}{g}\right)^b \quad d_k(n)\right]^T$$
$$\mathbf{h}_{k,2}(n) = \left[10\log\left(d_{ck}^a(n)\left(1+\frac{d_{ck}(n)}{g}\right)^b (d_k(n)-d_{ck}(n))^a\left(1+\frac{d_k(n)-d_{ck}(n)}{g}\right)^b\right) \quad d_k(n)\right]^T. \tag{17.41}$$

Then the combined observation equation under LOS conditions is given by

$$\mathbf{z}_{k.1}(n) = \mathbf{h}_{k,1}(n) + \varepsilon_{k,1}(n) \tag{17.42}$$

and in NLOS conditions the combined observation equation becomes

$$\mathbf{z}_{k.2}(n) = \mathbf{h}_{k,2}(n) + \varepsilon_{k,2}(n) \tag{17.43}$$

Define the state vector as

$$\boldsymbol{\theta}_k(n) = \left[d_k(n) \quad \dot{d}_k(n)\right]^T \tag{17.44}$$

where $\dot{d}_k(n)$ is the speed of the mobile relative to the kth base station. Then the state equation is given by

$$\boldsymbol{\theta}_k(n+1) = \mathbf{G}\boldsymbol{\theta}_k(n) + \boldsymbol{\beta}\mathbf{v}_k(n) \tag{17.45}$$

where $\mathbf{v}_k(n)$ is the process noise vector and

$$\mathbf{G} = \begin{bmatrix} 1 & T_s \\ 1 & 1 \end{bmatrix}, \quad \boldsymbol{\beta} = \begin{bmatrix} T_s^2/2 \\ T_s \end{bmatrix} \tag{17.46}$$

where T_s is the sample period. The terms $\mathbf{h}_{k,1}(n)$ and $\mathbf{h}_{k,2}(n)$ in the observation equations are associated with the state vector by $d_k(n) = [\boldsymbol{\theta}_k(n)]_{1.1}$. Thus the distance estimation problem is changed to the state estimation problem of the two-mode system, namely the LOS mode described by (17.42) and (17.45) and the NLOS mode represented by (17.43) and (17.45). The LOS/NLOS transition may be modeled as a switching-mode system that can be described as a two-state Markov process as shown in Fig. 17.2.

The block diagram of the EKF-IMM distance smoother is shown in Fig. 17.3. Two parallel extended Kalman filters are employed to smooth the distance measurements under the assumed LOS and NLOS conditions, producing the corresponding mode distance estimates. By using the respective distance estimation

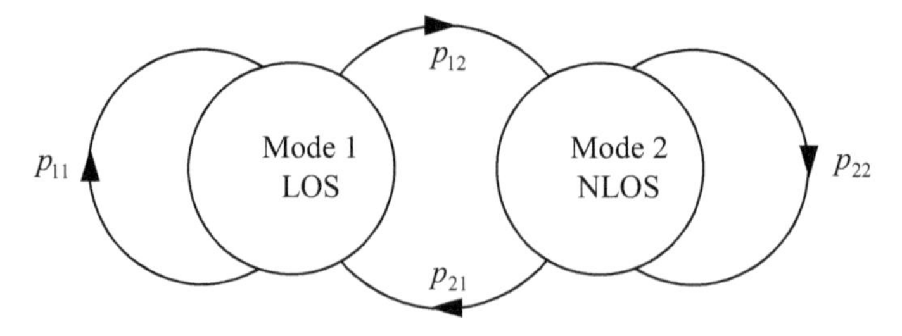

Fig. 17.2 Two-state Markov switching model

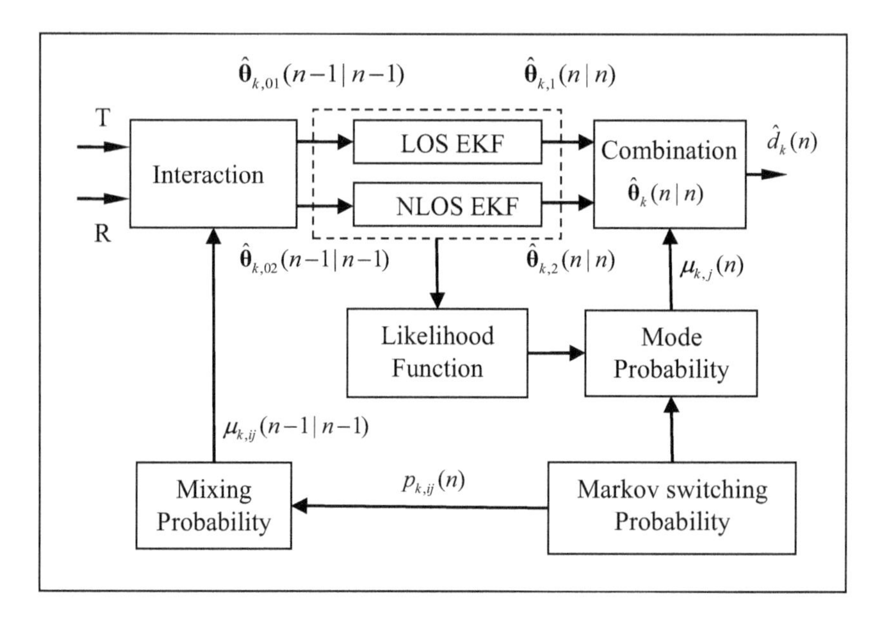

Fig. 17.3 Block diagram of EK-IMM distance estimator

errors and the likelihood function, the mode probabilities of the present environment condition are computed. The results from the two parallel smoothers are then combined based on their mode probabilities. The Markov transition probabilities are assumed known and used for computing the mixing probabilities that are exploited to calculate the priori state and covariance estimation. More details about the algorithm are described below.

Priori State and Covariance Estimation

Denote $p_{k,ij}$ as the transition probability from mode i to mode j for the kth base station. Also denote $\mu_{k,i}(n)$ as the probability of the kth base station in mode i at time n. Then the mixing probability from mode i to mode j at time $n-1$ is defined as

$$\mu_{k,ij}(n-1|n-1) = \frac{1}{\bar{u}_{k,j}} p_{k,ij}\mu_{k,i}(n-1) \tag{17.47}$$

where

$$\bar{u}_{k,j} = \sum_i p_{k,ij}\mu_{k,i}(n-1) \tag{17.48}$$

As for the jth mode (either LOS or NLOS) EKF in the kth base station at time $n-1$, the mixed priori state estimate $\hat{\boldsymbol{\theta}}_{k,0j}(n-1|n-1)$ and covariance matrix $\boldsymbol{\Phi}_{k,0j}(n-1|n-1)$ are computed according to

$$\begin{aligned} \hat{\boldsymbol{\theta}}_{k,0j}(n-1|n-1) &= \sum_i \hat{\boldsymbol{\theta}}_{k,i}(n-1|n-1)\mu_{k,i|j}(n-1|n-1) \\ \boldsymbol{\Phi}_{k,0j}(n-1|n-1) &= \sum_i \Big\{ \boldsymbol{\Phi}_{k,i}(n-1|n-1) + \Big(\hat{\boldsymbol{\theta}}_{k,i}(n-1|n-1) - \hat{\boldsymbol{\theta}}_{k,0i}(n-1|n-1)\Big) \\ &\quad \times \Big(\hat{\boldsymbol{\theta}}_{k,i}(n-1|n-1) - \hat{\boldsymbol{\theta}}_{k,0i}(n-1|n-1)\Big)^T \Big\} \mu_{k,i|j}(n-1|n-1) \end{aligned} \tag{17.49}$$

where $\hat{\boldsymbol{\theta}}_{k,i}(n-1|n-1)$ and $\boldsymbol{\Phi}_{k,i}(n-1|n-1) = E[(\boldsymbol{\theta}_k(n) - \hat{\boldsymbol{\theta}}_{k,j}(n|n-1))(\boldsymbol{\theta}_k(n) - \hat{\boldsymbol{\theta}}_{k,j}(n|n-1))^T]$ are respectively the state estimate and the error covariance matrix of the state estimation of the ith mode EKF.

Mode-Matched EKF

When the priori state estimate and the error covariance matrix are obtained, an individual EKF can be used to smooth the distance estimates. According to the EKF the state estimate is predicted by

$$\hat{\boldsymbol{\theta}}_{k,j}(n|n-1) = \mathbf{G}\,\hat{\boldsymbol{\theta}}_{k,j}(n|n-1) \tag{17.50}$$

where recall that k indexes the base stations and j indexes the LOS and NLOS modes. The minimum prediction of the error covariance matrix is computed as

$$\boldsymbol{\Phi}_{k,j}(n|n-1) = \mathbf{G}\boldsymbol{\Phi}_{k,j}(n-1|n-1)\mathbf{G}^T + \boldsymbol{\beta}\boldsymbol{\Psi}\boldsymbol{\beta}^T \tag{17.51}$$

where Ψ is the covariance matrix of the process error vector. The Kalman gain matrix is then determined by

$$\mathbf{K}_{k,j}(n) = \boldsymbol{\Phi}_{k,j}(n|n-1)\mathbf{H}_{k,j}\Big(\mathbf{H}_{k,j}\boldsymbol{\Phi}_{k,j}(n|n-1)\mathbf{H}_{k,j}^T + \mathbf{Q}_k\Big)^{-1} \tag{17.52}$$

where $\mathbf{Q}_k$ is the covariance matrix of the measurement error vector and

$$\mathbf{H}_{k,j} = \left.\frac{\partial \mathbf{h}_{k,j}}{\partial \boldsymbol{\theta}}\right|_{\boldsymbol{\theta}=\hat{\boldsymbol{\theta}}_{k,j}(n|n-1)}, j = 1, 2 \tag{17.53}$$

The derivations in (17.53) are given by

$$\mathbf{H}_{k,1} = \begin{bmatrix} 10[ad^{-1} + b(1+d/g)^{-1}/\ln 10 & 0 \\ 1 & 0 \end{bmatrix}, \text{ for } d = [\hat{\boldsymbol{\theta}}_{k,1}(n|n-1)]_{1,1}$$
$$\mathbf{H}_{k,2} = \begin{bmatrix} 10[a(d-d_{ck})^{-1} + b(1+(d-d_{ck})/g)^{-1}/\ln 10 & 0 \\ 1 & 0 \end{bmatrix}, \text{ for } d = [\hat{\boldsymbol{\theta}}_{k,2}(n|n-1)]_{1,1} \tag{17.54}$$

Then the state estimate is updated by

$$\hat{\boldsymbol{\theta}}_{k,j}(n|n) = \hat{\boldsymbol{\theta}}_{k,j}(n|n-1) + \mathbf{K}_{k,j}(n)\left(\mathbf{z}_{k,j}(n) - \mathbf{h}_{k,j}(\hat{\boldsymbol{\theta}}_{k,j}(n|n-1))\right), j = 1, 2 \tag{17.55}$$

The covariance matrix is updated according to

$$\boldsymbol{\Phi}_{k,j}(n|n) = \left(\mathbf{I} - \mathbf{K}_{k,j}(n)\mathbf{H}_{k,j}\right)\boldsymbol{\Phi}_{k,j}(n|n-1) \tag{17.56}$$

The mode probabilities are then determined by

$$\mu_{k,j}(n) = \frac{1}{u_k}\Lambda_{k,j}(n)\sum_i p_{k,ij}\mu_{k,i}(n-1) = \frac{1}{u_k}\Lambda_{k,j}(n)\bar{u}_{k,j} \tag{17.57}$$

where u_k is a normalization factor given by

$$u_k = \sum_i \Lambda_{k,j}(n)\bar{u}_{k,j} \tag{17.58}$$

and the likelihood function $\Lambda_{k,j}(n)$ is associated with the probability density function of the error

$$\varepsilon_{k,j}(n) = \mathbf{z}_{k,j}(n) - \mathbf{h}_{k,j}(\hat{\boldsymbol{\theta}}_{k,j}(n|n-1)) \tag{17.59}$$

which has a Gaussian density function of zero mean and covariance matrix given by

$$\mathbf{S}_{k,j}(n) = E[\varepsilon_{k,j}(n)\varepsilon_{k,j}^T(n)] = \mathbf{H}_{k,j}\boldsymbol{\Phi}_{k,j}(n|n-1)\mathbf{H}_{k,j}^T + \mathbf{Q}_k \tag{17.60}$$

Combination

The state vector and the covariance matrix estimates can be produced by combining the corresponding results of the two modes based on their mode probabilities. That is

$$\hat{\boldsymbol{\theta}}_k(n|n) = \sum_j \hat{\boldsymbol{\theta}}_{k,j}(n|n)\mu_{k,j}(n)$$
$$\boldsymbol{\Phi}_k(n|n) = \sum_j \left\{ \boldsymbol{\Phi}_{k,j}(n|n) + \left(\hat{\boldsymbol{\theta}}_{k,j}(n|n) - \hat{\boldsymbol{\theta}}_k(n|n)\right)\left(\hat{\boldsymbol{\theta}}_{k,j}(n|n) - \hat{\boldsymbol{\theta}}_k(n|n)\right)^T \right\} \mu_{k,j}(n) \tag{17.61}$$

Consequently the range between the mobile and the kth base station at time n is simply obtained as the first element of $\hat{\boldsymbol{\theta}}_k(n|n)$, namely

$$\hat{d}_k(n) = [\hat{\boldsymbol{\theta}}_k(n|n)]_{1,1} \tag{17.62}$$

The main advantage of the EKF-IMM distance smoothing method is that the LOS/NLOS conditions are adaptively evaluated without the need of repeatedly checking the conditions by using another identification algorithm. One challenge is that knowledge of the Markov transition probabilities is required. When a map of the urban area is available the transition probabilities can be well determined. On the other hand, when there is no priori information about the transition probabilities, the EKF-IMM algorithm may not be able to achieve a satisfactory performance.

17.1.5 Fusion for Cooperative Localization

In wireless sensor networks the distance between an ordinary node and an anchor node may be determined based on TOA measurements. In general anchor nodes have more computational power than ordinary nodes. Also they usually have the hardware and software to perform TOA estimation. Further it is desirable to measure the distances between an ordinary node and its neighboring ordinary nodes. Typically, the distance between the neighboring ordinary nodes is relatively short so that the RSS distance estimate may satisfy the accuracy requirement, and the more power-hungry TOA approach can be avoided. In particular, ordinary nodes may not have the capability to measure TOA. The TOA and the RSS data are then fused to perform cooperative positioning (or localization) to localize a group of ordinary nodes simultaneously. In general cooperative localization can achieve better performance than localizing nodes individually. In some circumstances cooperative localization is necessary since one or more ordinary nodes do not have sufficient neighbors that are either anchor nodes or localized ordinary nodes. Some useful information on cooperative localization is in, Yu (2018), Chap. 6.

17.1.6 Algorithm Aggregation

Algorithm aggregation refers to making use of the position estimates from a number of algorithms to improve position estimation accuracy when giving certain types of signal measurements. In the case of indoor RSS-based position determination, as reported in Li and Martin (2008), aggregation with just two algorithms can improve the accuracy by 23%, while aggregation with 12 algorithms can increase the accuracy by 51%. It is also noted that the performance gain by using algorithm aggregation becomes saturated at around four algorithms. The studied RSS-based algorithms have similar performance in accuracy and none of them can achieve the good accuracy all of the time. In the following it is intended to exploit the difference in position estimation results at each particular point from different algorithms to achieve better accuracy.

First, simply averaging the results from different algorithms can achieve a significant performance gain. For instance, consider the case where three algorithms are aggregated. Suppose that the position coordinate estimates based on each of the three algorithms have the same mean and variance. Then the averaging-based aggregation yields a position estimate with the same mean, but with one third of the variance if the estimates are uncorrelated. However, there are better aggregation approaches to enhance the performance gain. The best solution may be that the aggregation algorithm simply selects the best estimate with the minimum estimation error as the final position estimate at each individual position. However, it is not clear which estimate is the best a priori. The problem may be partially solved if there is an offline training phase, which can be saved in an online database which is accessible to the nodes in the network. Specifically, at each grid point all the algorithms are tested and a few (say three) algorithms with the best performance chosen. During real-time position estimation an initial position estimate is obtained through the averaging-based aggregation. Then the corresponding chosen algorithms for estimating the position are employed to produce a final position estimate. For instance, the results from the chosen algorithms may be averaged, or simply one of the algorithms is chosen from the database with the minimum estimation error. In the absence of offline training, the GDOP properties of the algorithms also may be used to choose one or more algorithms for each position point.

17.2 Combining GNSS and Other Technologies

When there are LOS propagation paths between a ground GNSS receiver and at least four GNSS satellites, the position of the GNSS receiver can be determined at anytime, anywhere, and in all weather conditions. For this reason, GNSS has been widely used for personal navigation and tracking in outdoor environments. However, in a range of circumstances such as in urban areas where there are dense high-rise buildings, GNSS might not perform well because some or all of the satellite signals are blocked. In this case the GNSS receivers either cannot provide any position information, or the provided position estimate has rather poor accuracy. One way to solve this problem is to combine GNSS with other technologies. In this section, a number of such hybrid positioning approaches are studied.

17.2.1 Integration of GNSS and Inertial Navigation System

In this subsection, the basic concepts of inertial navigation system (INS) and GNSS are studied. The reason why GNSS/INS integration is desirable is also briefly explained. Furthermore, the fundamental integration approaches are described, followed by the utilization of redundant inertial measurement units (IMU).

17.2.1.1 Fundamentals of INS and GNSS

INS equipment consists of inertial sensors: gyroscopes and accelerometers. The output of a gyroscope is a signal proportional to angular rate of change, whereas the output of an accelerometer is a signal proportional to velocity change along the axis of the accelerometer. To have a clear understanding of the operation of INS as well as GNSS, a number of different coordinate frames are first explained, as follows:

1. *Earth-centered inertial (ECI) frame*: coordinate frames with their origins at the center of the earth. They are termed inertial since they are non-rotating frames, contrasting to the earth-centered earth-fixed (ECEF) frame. ECI frames are useful for studying the motion of objects in space. There are different definitions for the ECI frame. One such commonly used frame is referred to as J2000, defined with the Earth's Mean Equator and Equinox at 12:00 Terrestrial Time on 1 January 2000. The x-axis is aligned with the mean equinox, the z-axis is defined by the spin axis of the Earth (North Pole), and the y-axis completes the orthogonal set of the inertial frame by complying with the right-hand rule.
2. *ECEF frame/Earth frame*: a coordinate frame which rotates with the earth at the earth spin rate and has its origin at the center of the earth. The x-axis passes through the equator at the prime meridian; the z-axis passes through the North Pole; and the y-axis is determined by the right-hand rule. It is convenient for representing the positions and velocities of terrestrial objects in ECEF frame.

3. *Local geodetic frame/geographic frame*: For ground-based applications such as land-vehicle navigation or tracking, the local geodetic frame is typically employed, which is defined as a local East, North, Up (ENU) Cartesian coordinate system with its origin at the position of the IMU. Alternatively, for aerospace applications it is more convenient to define the local geodetic frame as a local North, East, Down (NED) Cartesian coordinate system with its origin at the position of the IMU.
4. *Navigation frame*: a coordinate frame which has origin at the position of the IMU and has a principal axis (say z-axis) pointing vertically up or down. The local geodetic frame can be used as a navigation frame. However, in the local geodetic frame, the North at the North Pole and the South at the South Pole are not defined due to a singularity problem. To avoid such a problem, the navigation frame can be defined by rotating the local geodetic frame through the z-axis at an angle known as the wander angle.
5. *Body frame*: a local coordinate frame associated with a structure such as the host vehicle body where the accelerometers and gyroscopes are mounted and are not mechanically moved.
6. *Sensor frame*: a local coordinate frame defined with respect to the orientations of three accelerometers, each of which corresponding to one of the three principal axes. The sensor frame may be different from the body frame since the sensors may not be perfectly aligned mechanically with respect to the body frame. In this case the sensor measurements need to be transformed from the sensor frame to the body frame for further processing.

By applying Newton law of motion and performing coordinate transformation between different frames (Titterton and Weston 2004) the navigation equation of a strapdown INS can be produced as

$$\dot{\mathbf{v}}_e^n = \mathbf{C}_b^n \mathbf{f}^b - (2\omega_{ie}^n + \omega_{en}^n) \times \mathbf{v}_e^n + \mathbf{g}^n \tag{17.63}$$

where the symbol $\times$ denotes the cross product and

- $\mathbf{v}^n$ ground velocity vector of the vehicle expressed in the navigation frame
- $\mathbf{C}_b^n$ transformation matrix (direction cosine matrix) from the body frame to the navigation frame
- $\mathbf{f}^b$ force acceleration vector measured by the accelerometers and expressed in the body frame
- $\boldsymbol{\omega}_{ie}^n$ angular velocity vector of the Earth frame relative to the inertial frame and expressed in the navigation frame
- $\boldsymbol{\omega}_{en}^n$ angular velocity vector of the navigation frame relative to the Earth frame and expressed in the navigation frame
- $\mathbf{g}^n$ local gravity vector expressed in the navigation frame

The matrix $\mathbf{C}_b^n$ propagates according to

$$\dot{\mathbf{C}}_b^n = \mathbf{C}_b^n \Omega_{nb}^b \tag{17.64}$$

where Ω_{nb}^b is the skew-symmetric matrix form of ω_{nb}^b which is the angular velocity vector of the body frame relative to the navigation frame and expressed in the body frame. The skew-symmetric matrix (Ω) form for an angular velocity vector ($\omega = \begin{bmatrix} \omega_x & \omega_y & \omega_z \end{bmatrix}^T$) is defined as

$$\Omega = \begin{bmatrix} 0 & -\omega_z & \omega_y \\ \omega_z & 0 & -\omega_x \\ -\omega_y & \omega_x & 0 \end{bmatrix} \tag{17.65}$$

The angular velocity vector ω_{nb}^b is expressed as

$$\omega_{nb}^b = \omega_{ib}^b - \mathbf{C}_n^b \omega_{in}^n = \omega_{ib}^b - \mathbf{C}_n^b (\omega_{ie}^n + \omega_{en}^n) \tag{17.66}$$

The basic principle of the INS is to set up and then resolve the navigation equation to determine the acceleration and orientation of the vehicle with respect to a reference frame. Then the time integral of the acceleration measurement yields the instantaneous velocity of the vehicle provided that the initial velocity is known. A second integration produces the distance travelled with respect to a known starting point. Figure 17.4 shows the basic structure of a strap-down inertial navigation system, where the dashed arrows indicate possible data inputs for calibration.

The basic principles of GNSS is briefly described as follows. Suppose that a GNSS satellite at a space position (x_s, y_s, z_s) broadcasts a coded signal at time t_e, which is received at time t_r by the GNSS receiver at a ground position (x_r, y_r, z_r). The satellite position varies with time since it orbits the earth about twice a day. The GNSS receiver position also changes with time when it is attached to a mobile unit. The pseudorange measurement between the satellite and the receiver can be described by

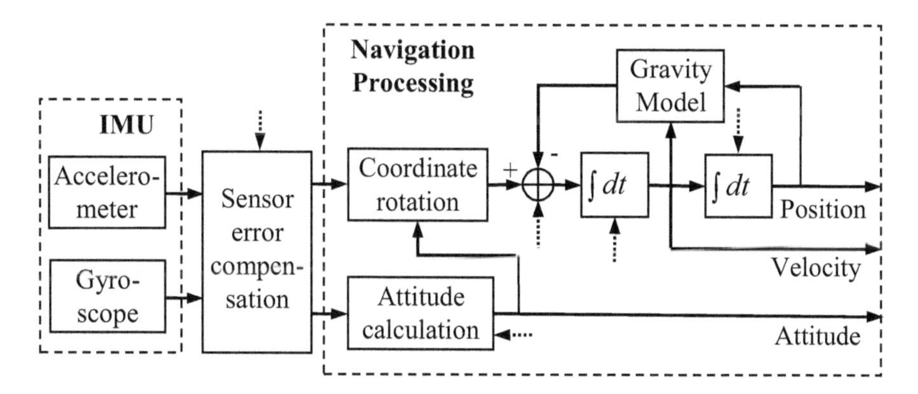

Fig. 17.4 Simplified block diagram of a strap-down inertial navigation system

$$r_r^s(t_r, t_e) = \rho_r^s(t_r, t_e) - (\delta t_r - \delta t_s)c + \delta_{ion} + \delta_{tro} + \delta_{tide} + \delta_{mul} + \varepsilon_{pse} \tag{17.67}$$

where the geometric distance between the satellite and the receiver is given by

$$\rho_r^s(t_r, t_e) = \sqrt{(x_r - x_s)^2 + (y_r - y_s)^2 + (z_r - z_s)^2} \tag{17.68}$$

The clock errors of the receiver and satellite are denoted respectively by δt_r and δt_s, δ_{ion} and δ_{tro} respectively denote the ionospheric and the tropospheric effects; δ_{tide} denotes the effect of the Earth tide and ocean loading tide; δ_{mul} denotes the multipath effect; and ε_{pse} models all the remaining errors (Xu 2007).

The second basic measurement of the GNSS receiver is the carrier phase of the received signal, which can be modeled in the same way as

$$\psi_r^s(t_r) = \frac{\rho_r^s(t_r, t_e)}{\lambda} - f_c(\delta t_r - \delta t_s) + N_r^s + \frac{\delta_{ion}}{\lambda} + \frac{\delta_{tro}}{\lambda} + \frac{\delta_{tide}}{\lambda} + \frac{\delta_{mul}}{\lambda} + \varepsilon_{ph} \tag{17.69}$$

where λ and f_c are respectively the wavelength and nominal frequency of the satellite signal; N_r^s is the integer ambiguity due to the fact that the receiver can only measure the fractional phase; and ε_{ph} accounts for all the other errors.

Doppler frequency is another measurement made by GNSS receivers, which is caused by the relative movement between the satellite and the receiver. The radial velocity of the satellite relative to the receiver is calculated according to

$$v_\rho = v \cos \alpha \tag{17.70}$$

where v is the relative velocity between the satellite and the receiver and α is the angle between the satellite motion direction and the directional vector from the receiver to the satellite. The relative radial frequency is positive when the satellite is moving towards the receiver. Then the Doppler frequency shift is calculated by

$$f_d = \frac{v_\rho}{c} f_c = \frac{v_\rho}{\lambda} = \frac{1}{\lambda} \frac{d\rho}{dt} \tag{17.71}$$

In the presence of measurement errors, the Doppler frequency measurement is modeled as

$$f_d = \frac{1}{\lambda} \frac{d\rho_r^s(t_r, t_e)}{dt} - f_c \frac{d(\delta t_r - \delta t_s)}{dt} + \varepsilon_{Dop} \tag{17.72}$$

where ε_{Dop} accounts for other un-modeled errors.

Based on the three different measurements at the receiver associated with at least four satellites, the position and velocity of the receiver as well as the time information can be estimated. Some of the error terms in the models such as the ionosphere delay and the troposphere delay are provided by the satellites. In differential GNSS (DGNSS), the receiver can be assisted by other terrestrial base stations or reference stations which provide corrections to compensate the un-modeled errors to improve estimation accuracy. In some circumstances, the GNSS receiver is also aided by ground base stations which provide the receiver with the relevant GNSS data to speed up the positioning process.

INS exhibits relatively low short-term (say up to a few seconds) noise, but tends to drift over long time periods, so that errors accumulation due to the integration processes. Small errors in the measurement of acceleration and angular velocity accumulate with time to produce larger errors in velocity, which then result in even a larger error in position. Thus the position must be corrected periodically by some other external reference to maintain positional accuracy. In contrast GNSS have relatively noisy short-term errors, but exhibit no long-term drift. Further, GNSS receivers may suffer from outages, being unable to provide any positional information if the satellite signals are blocked. The integration of GNSS and INS through Kalman filtering can thus significantly enhance the overall positioning performance. When the GNSS is performing well, it can contribution to the INS correcting the INS positional drifts. When GNSS is not performing well, for example when a vehicle passes through a tunnel, the INS will provide mobile position estimates for a limited period of time, and the inertial position and velocity information can reduce the search time required to reacquire the GNSS signals after an outage.

GNSS/INS integration is a well-studied topic and various integration schemes exist. Instead of giving a comprehensive study on all the different types of integration methods, the focus here is on briefly describing the two important broad GNSS/INS integration approaches: loosely coupled and tightly coupled.

17.2.1.2 Use of a Single IMU

In loosely coupled GNSS/INS integration with a single IMU as shown in Fig. 17.5, the internal GNSS-processed measurements of position, velocity and time, derived from raw measurements of pseudorange, carrier phase and Doppler frequency shift,

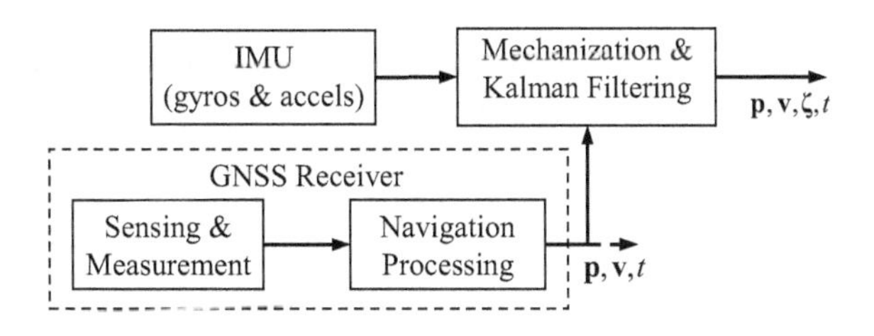

Fig. 17.5 Simplified block diagram of loosely coupled GNSS/INS integration

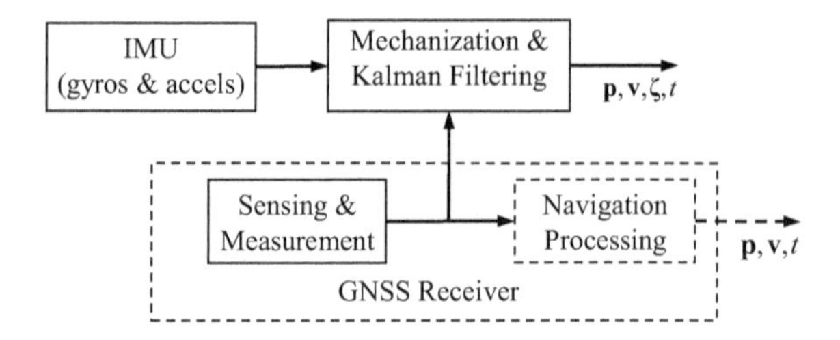

Fig. 17.6 Simplified block diagram of tightly coupled GNSS/INS integration

are used in the integration process with the INS. The main advantage of such an integration approach is its simplicity. Knowledge of the details of the GNSS receiver is not required, and it virtually does not involve any modification to the GNSS receiver. However, when a GNSS outage occurs because the receiver can only receive signals from less than four satellites, no GNSS data are available for calibration of the INS sensors. As a result, the Kalman filtering for the INS integration will either stop immediately or perform reliably only for a short period of time, depending on the specific algorithm.

The alternative fusion approach is the tightly coupled GNSS/INS integration as shown in Fig. 17.6. The raw GNSS measurements are directly used in the data integration. Thus when the GNSS receiver detects signals from two or three satellites, raw measurements can still be integrated with the INS data. This makes the tightly coupled approach more attractive for mobile navigation or tracking in harsh propagation environments such as in urban canyons. Also there is a further potential benefit from fusing raw measurements, namely higher overall positional accuracy. The drawback of this integration approach is that the Kalman filtering will be more complex since extra state variables are introduced. In addition, software and perhaps hardware modifications are required to access the GNSS raw measurements.

17.2.1.3 Use of Redundant IMUs

In the presence of redundant IMUs the positioning performance can be enhanced due to the *diversity gain* associated with multiple IMUs. The multiple IMUs can be configured in different structures; three such structures are illustrated as below.

Synthetic Method

In the synthetic approach as shown in Fig. 17.7, the redundant IMU data are combined before being input to the GNSS/INS fusion algorithm developed for scenarios where only a single IMU is involved. While fusing the IMU data, defective sensors can be detected and realistic noise and covariance terms can be estimated. This synthetic IMU concept is simple since it does not require any modification of the standard GNSS/INS algorithm. One disadvantage of the method is that there is no link to feedback the fusion results to the individual sensors in the IMUs.

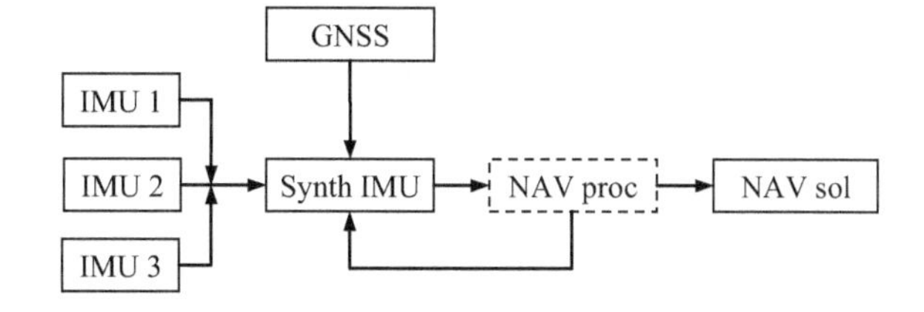

Fig. 17.7 Simplified structure of a synthetic IMU based GNSS/IMU fusion

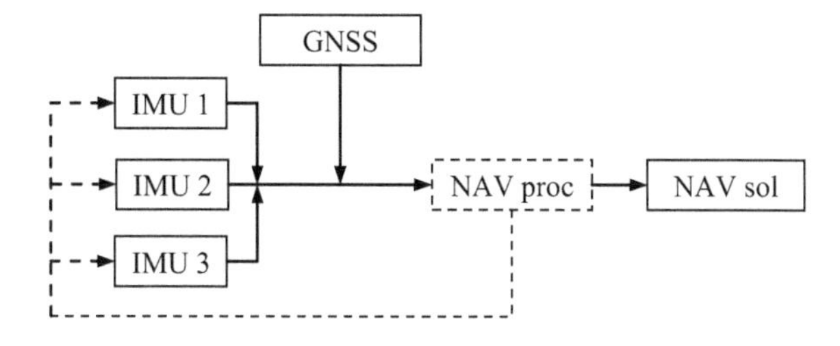

Fig. 17.8 Simplified structure of extended IMU based GNSS/IMU fusion

Extended Method

In the extended method shown in Fig. 17.8, the processed GNSS/INU data are linked to the individual IMU sensors. As a result, the systematic errors of each sensor can be modeled and estimated, in addition to the possible detection of defective sensors, and the estimation of noise terms during data fusion. One drawback of the method is that it requires the modification of the GNSS/INS software to accommodate the hardware structural change.

Geometrically-Constrained Method

The geometrically-constrained approach shown in Fig. 17.9 provides an option for system calibration if the relative sensor geometry is insufficiently known. In this approach, multiple navigation solutions are provided (one for each IMU) and compared at regular time intervals. Similar to the extended method, the geometrically constrained method allows estimating the individual sensor errors. However, there are a few disadvantages associated with this method. First, the computational effort is increased considerably compared to the other two methods so that it requires a significant modification to the GNSS/INS software. Second, it is more

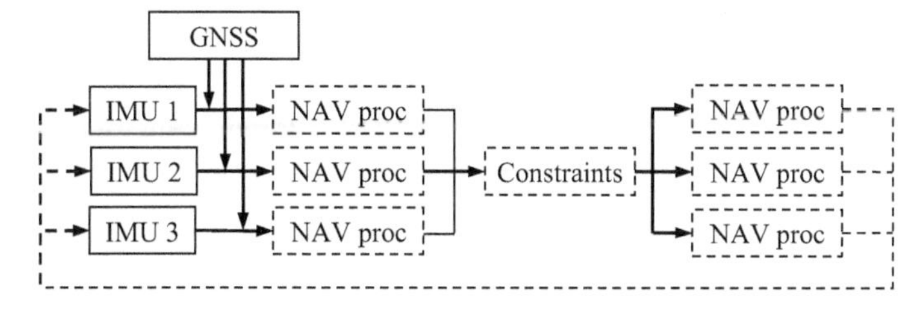

Fig. 17.9 Simplified structure of Geometrically-constrained IMU based GNSS/IMU fusion

sensitive to sensor failures because defects can only be noticed at the update stage and the measurement faults cannot generally be isolated.

17.2.2 Fusion of GNSS and Wireless Network Data

In recent years more wireless communication devices such as mobile phones have embedded GNSS receivers to provide navigation and tracking facility. Consequently, the GNSS technology and the cellular network system can be combined to produce better positioning services. Here a number of hybrid GNSS and network technology based positioning methods are briefly described.

17.2.2.1 A-GNSS

In a range of situations such as indoors and in urban canyons GNSS may have very poor performance since the GNSS receiver cannot adequately detect the weak satellite signals. A-GNSS (assisted GNSS) is one of the solutions to solve the problem, which uses network services to improve the startup performance of GNSS-based positioning systems. The assistance can be realized in two different ways. One is that a base station with a GNSS receiver and an appropriate antenna can quickly and reliably detect GNSS data such as almanac and then sends the information via internet connection to a mobile handset to acquire satellites more quickly. The other is that the handset captures a snapshot of the GNSS signal which is processed by a server at the base station to generate a position estimate for the handset. A base station also has more accurate information about the local ionospheric conditions and other conditions that affect the GNSS signal, so that better corrections can be used. Also, because the position calculation is carried out by the base station, the complexity of the positioning related hardware and software in the mobile handset can be reduced significantly. Thus the main advantage of A-GNSS is either to speed up the mobile position acquisition, or to improve position accuracy and reduce complexity significantly. The main drawback is the extra cost due to the increased data communication/access.

17.2.2.2 Differential GNSS

In a basic GNSS receiver compensation for variable parameters in the pseudorange measurement equations, such as the ionospheric and tropospheric delays, are typically imperfectly known or unmodeled; this is the main reason the poor accuracy in such basic GNSS receivers. Differential GNSS (DGNSS) is an enhancement to GNSS by using ground-based reference stations. To assist with providing better estimates of these parameters for mobile receivers, fixed reference stations at known geographic locations are equipped with GNSS receivers. The position estimates of the reference stations measured by thsee GNSS receivers are compared with their actual known positions, so that the difference between the pseudorange measurements and the actual pseudoranges can be calculated. These pseudorange differences are then forwarded to other mobile GNSS receivers so that their pseudorange measurements can be corrected, at least better than the default values used in basic receivers. The distribution of these DGNSS data can be realized by any suitable wireless link, in particular with Internet access.

17.2.2.3 System-Level Combining

The simplest way to generate a joint position estimate from different positioning systems and/or technologies is the weighted sum of the individual position estimates. Consider the scenario where position estimates are generated by GNSS, the network-based received signal strength method, or the network-based TOA method, denoted respectively by $(\hat{x}_{GPS}, \hat{y}_{GPS}, \hat{z}_{GPS})$, $(\hat{x}_{RSS}, \hat{y}_{RSS}, \hat{z}_{RSS})$, and $(\hat{x}_{TOA}, \hat{y}_{TOA}, \hat{z}_{TOA})$. Without loss of generality, combining of these x-coordinate estimates is considered. If the variances of these three different estimates of the x-coordinate are known, either based on priori information, or from an online variance estimation database, the minimum-variance combination method can be employed. Specifically, the weighted sum of the three x-coordinate estimates ia

$$\hat{x} = a_1\hat{x}_{GPS} + a_2\hat{x}_{RSS} + a_3\hat{x}_{TOA} \tag{17.73}$$

where $\{a_i\}$ are the weighting coefficients which are positive and satisfy

$$a_1 + a_2 + a_3 = 1 \tag{17.74}$$

With estimates of the variances of the three x-coordinate donated as σ_{GPS}^2, σ_{RSS}^2, and σ_{TOA}^2, the variance of the weighted-sum estimate $\hat{x}$ is given by

$$\sigma_{\hat{x}}^2 = a_1^2\sigma_{GPS}^2 + a_2^2\sigma_{RSS}^2 + a_3^2\sigma_{TOA}^2 \tag{17.75}$$

where it is assumed that the three x-coordinate estimates $\{\hat{x}_{GPS}, \hat{x}_{RSS}, \hat{x}_{TOA}\}$ are mutually statically independent. This assumption is reasonable due to the fact that they are produced by three different technologies. Substituting (17.74) into (17.75) yields

$$\sigma_{\hat{x}}^2 = a_1^2\sigma_{GPS}^2 + a_2^2\sigma_{RSS}^2 + (1 - a_1 - a_2)^2\sigma_{TOA}^2 \tag{17.76}$$

Differentiating (17.76) with respect to a_1 and a_2 produces

$$\begin{aligned} \frac{\partial \sigma_{\hat{x}}^2}{\partial a_1} &= 2a_1\sigma_{GPS}^2 + 2(a_1 + a_2 - 1)\sigma_{TOA}^2 \\ \frac{\partial \sigma_{\hat{x}}^2}{\partial a_2} &= 2a_2\sigma_{RSS}^2 + 2(a_1 + a_2 - 1)\sigma_{TOA}^2 \end{aligned} \tag{17.77}$$

Setting the above derivatives to zero and solving the two equations yields

$$\begin{aligned} \hat{a}_1 &= \frac{\sigma_{RSS}^2\sigma_{TOA}^2}{\sigma_{GPS}^2\sigma_{RSS}^2 + \sigma_{GPS}^2\sigma_{TOA}^2 + \sigma_{RSS}^2\sigma_{TOA}^2} \\ \hat{a}_2 &= \frac{\sigma_{GPS}^2\sigma_{TOA}^2}{\sigma_{GPS}^2\sigma_{RSS}^2 + \sigma_{GPS}^2\sigma_{TOA}^2 + \sigma_{RSS}^2\sigma_{TOA}^2} \end{aligned} \tag{17.78}$$

Replacing a_1 and a_2 in (17.74) by $\hat{a}_1$ and $\hat{a}_2$ in (17.78) produces

$$\hat{a}_3 = \frac{\sigma_{GPS}^2\sigma_{RSS}^2}{\sigma_{GPS}^2\sigma_{RSS}^2 + \sigma_{GPS}^2\sigma_{TOA}^2 + \sigma_{RSS}^2\sigma_{TOA}^2} \tag{17.79}$$

Thus at each time instant the x-coordinate estimate is equal to the weighted sum of the three x-coordinate estimates as given by (17.73), with the weighting coefficients determined by (17.78) and (17.79). The y-coordinate and z-coordinate estimates can be handled in the same way.

17.2.3 Fusion of GNSS/INS and Other Sensors

To further enhance positioning performance in harsh propagation environments, extra sensors can be added to GNSS/INS integration. One technique is the fusion of a Lidar sensor with a tightly-coupled GNSS/INS assembly (Chap. 10, Yu 2018). Lidar sensors provide relative motion estimation and such a sensor fusion is able to enhance the dead-reckoning accuracy of INS when GNSS signal is absent or of poor quality. Another example of an integrated navigation system beyond GNSS/INS integration is the GNSS//Locata/INS integration. CKF (centralized Kalman filter), FKF, and GOF (global optimal fusion) can be used to realize such integration of triple systems (Chap. 11, Yu 2018). Locata is a ground-based local positioning system which makes use of a signal design similar to the GNSS signal, but in different signal frequency band. Another triple integration system is GNSS/WiFi/INS integration which is particularly suited for indoor pedestrian navigation (Chap. 11, Yu 2018). The GNSS/WiFi/INS integration can significantly improve the WiFi stand-alone positioning performance. The positioning and attitude solutions from the INS will be aided with the WiFi positioning solutions, and the WiFi positioning will stabilize any INS positional drifts. In addition, step detection can be implemented to reduce INS errors during the indoor navigation phase.

17.3 Map Matching for Vehicle Navigation and Tracking

While the motion of people is largely unrestricted, vehicles are almost exclusively driven on streets or highways which form a road network of a specific city or a larger region; such a road network has available published maps. Accordingly, digital road maps can be used to improve positioning performance through matching the position estimation results obtained from GNSS and/or other positioning technologies with the road map. In this section a number of key map matching approaches are described. Although the focus is on outdoor vehicle applications, the basic principles can also be applied to indoor applications.

17.3.1 Digital Road Map

A digital road map is a database consisting of topological (connectivity properties) and metrical (coordinates) information, and attributes such as road class, street names, speed limits, and turn restrictions (Skog and Handel 2009). The road network is typically represented by a planar model on a digital map. The street system is represented by a set of curves/arcs, each of which represents the centerline of a road segment. An arc can be described by a finite set of points and the segment between each pair of neighboring points is commonly assumed to be linear. Thus a digital road map can be readily handled by a computer. As shown in Fig. 17.10, the road centerlines are denoted by dashed lines, called arcs. The black circles denote nodes, whereas the white circle denotes shape points. There are also other different types of nodes, including intersection nodes and dead-end nodes.

In the creation of a road map, a number of procedures are involved, including map scaling, map projection, and map datum. These procedures inevitably introduce errors and thus affect the map quality. In addition, digitalization and identification processes are required to generate a digital road map, which introduce further errors to digital road maps. For instance, in creating digital road maps some features of the real-world road networks such as roundabouts, junctions, and curves

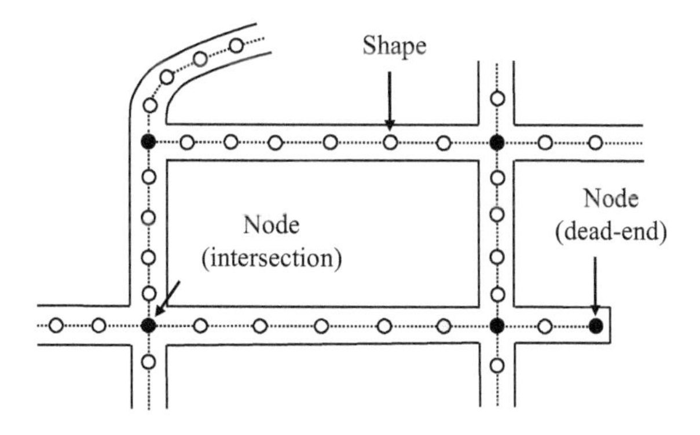

Fig. 17.10 Basic Components of a digital road map

are often omitted or simplified, producing the topological errors. Further, in the creation of the digital road maps, road centerlines or junctions can be displaced, resulting in the geometrical errors. As a consequence, the digital road maps do not perfectly represent the actual road networks. Thus in analyzing the performance of the map matching, the map-caused errors needs to be considered in any integration procedures.

17.3.2 Geometric Map Matching

Geometric *map matching* makes use of the geometric information of the spatial road network, namely the *shapes* of the links, not the way in which they are connected (Bernstein and Kornhauser 1996). Three basic geometric map matching methods are briefly described as below.

17.3.2.1 Geometric Point-to-Point Matching

In this method, the position estimate is matched to the closest node or shape point in the road network. The "closest" is usually measured in terms of the Euclidean distance between the position estimate examined and the nodes or shape points. As shown in Fig. 17.11 there are four streets, four nodes and a number of shapes in the road network. The vehicle traverses on street 1, then turns right at the node, and traverses on street 3. The position estimates of the vehicle are denoted by triangles. The Euclidean distance based point-to-point matching is paired by a dashed line. This point-to-point matching method is easy to implement and fast; however, it also has problems in practice. In particular, the matching performance depends critically on the density of shape points of the road segments in the road network. Segments with a higher density of shape point are more likely to be accurately matched to.

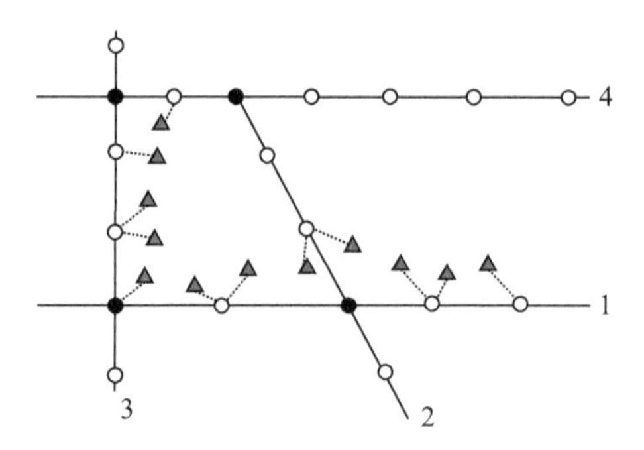

Fig. 17.11 Illustration of geometric point-to-point matching

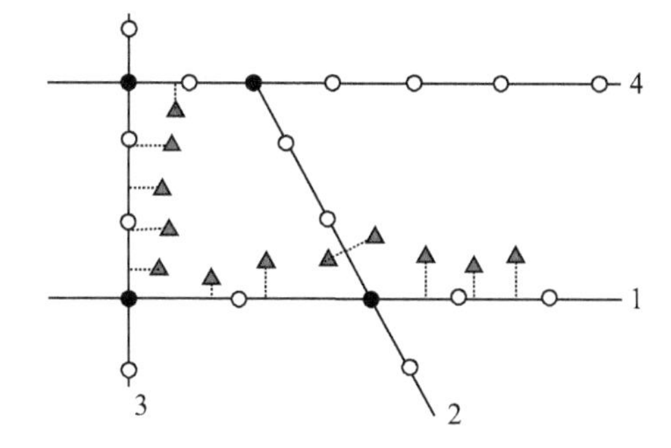

Fig. 17.12 Illustration of geometric point-to-curve matching

17.3.2.2 Geometric Point-to-Curve Matching

In the point-to-curve matching, the examined position point is matched to the arc/road segment which is closest to the examined point. Again the "closest" means the distance from the point to the road segment is the minimum distance compared to the distances to other road segments. The distance from a point (x_0, y_0) to a line $ax + by + c = 0$ can be calculated according to

$$d = \frac{|ax_0 + by_0 + c|}{\sqrt{a^2 + b^2}} \tag{17.80}$$

where $|z|$ is the absolute value of z and the projection of the point on the line. The matched position point on the line is calculated by

$$x = \frac{b^2x_0 - aby_0 - ac}{a^2 + b^2}, \qquad y = \frac{a^2y_0 - abx_0 - bc}{a^2 + b^2} \tag{17.81}$$

Figure 17.12 shows an example of point-to-curve matching with the same road network used in Fig. 17.11. The minimum distances are represented by dashed lines. The point-to-curve matching is equivalent to the point-to-point matching when the shape point density goes to infinity. Thus the performance of point-to-curve matching is better than that of point-to-point matching. The drawback of the point-to-curve matching as well as the point-to-point matching is that it produces unstable matching results, oscillating from one road segment to another segment, especially due to high road density in dense urban areas.

17.3.2.3 Geometric Curve-to-Curve Matching

As mentioned earlier, the point-to-point and the point-to-curve matching methods are unstable in dense road networks. Curve-to-curve matching can be used to reduce the instability, through comparing the estimated vehicle trajectory against known road segments. The estimated trajectory can be simply constructed by connecting each pair of neighboring estimated position points with a straight line. The road segment is selected provided that the distance between this segment and the estimated trajectory is the minimum. To speed up the search process, the road segment candidates are only those identified by the point-to-point or the point-to-curve matching methods. Although there are different ways to define the distance between the two curves, one simple and efficient way is to make use of the results from the point-to-curve matching. That is the distance is simply defined as the average of the minimum distances from the estimated position points to the road segments. The main disadvantage of this matching method is that it is sensitive to outliers.

17.3.3 Geometric and Topological Map Matching

The geometric map matching methods do not consider the ways about how the different parts or shapes of the road network are connected. In the geometric and topological map matching, the connectivity, adjacency and proximity characteristics of the road network and the location sequence of the mobile unit are employed. In this section the map matching procedure proposed in Greenfeld (2002) is described.

17.3.3.1 Basic Matching Procedure

This map matching procedure is composed of two separate algorithms, the initial matching algorithm and the main matching algorithm, each of which is applied at a specific stage of the process. The initial algorithm determines the initial match of an observed position on an arc based a geometric matching process as follows:

1. Find the closest street node to the observed position point.
2. Determine all the arcs in the road network, which are connected to the selected street node.
3. Map the next observed position point onto one of the selected arcs.

The initial algorithm is applied in the following circumstances:

1. When the first observed position is received.
2. When the distance between the new observed point and the previous observed point exceeds a pre-defined distance tolerance.
3. When the main algorithm cannot successfully map a particular observed position point. In this case the procedure is reset to start a new matching task.

The initial algorithm is followed by the main algorithm which is composed of the following steps:

1. Obtain the new observed position point.
2. Form an observed line segment between the current and the previous observed points.
3. Evaluate the proximity (distance) and orientation (direction or azimuth) of the observed line to the currently matched street segment.
4. If the new point does not map onto the current segment, then find another segment which is also selected based on the same proximity and orientation evaluation scheme to be discussed later.
5. Repeat from step 1 to step 4 unless the initial algorithm is required.

The above described method only uses coordinate information of the observed positions of the mobile. It does not use any heading and/or speed information from GNSS or other sensors. However, heading and speed information could be used to improve map matching; this topic is addressed in later subsections.

17.3.3.2 Similarity Assessment

Different measures can be used to determine the proximity of an observed point to an arc. The first one is to compute the shortest distance from the point to the arc as already discussed in the geometric map matching. The second proximity measure is to determine whether the line between the latest two observed position points and the road segment intersect. If the two lines intersect and basically follow the same direction, it is likely that a correct match was found. On the other hand, if the two lines are close to being perpendicular to each other, it is very unlikely that the match is correct. As shown in Fig. 17.13 the lines P_1-P_2, P_2-P_3, and P_3-P_4 intersect with arc A_1 and in general follow the same direction. Thus the points P_1 through P_4 should be mapped onto arc A_1. The line P_3-P_4 is nearly perpendicular to arc A_2, so that the observed point P_4 should not be mapped onto arc A_2.

Another measure of likelihood for a correct match is the orientation or the direction of the arcs and the observed lines. The direction of the observed line and the direction of the road segment should be similar. The azimuth similarity criterion can be used to determine whether or not an arc can be an appropriate candidate for a match. Since the coordinates of the observed positions are assumed available, the azimuth of a line from P_1 (x_1, y_1) to P_2 (x_2, y_2) can be calculated according to

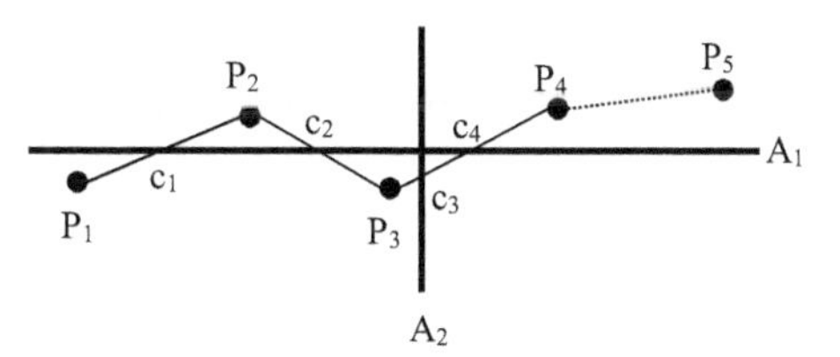

Fig. 17.13 Intersection between observed lines and road segments

$$\beta_{Az} = \begin{cases} \tan^{-1}\frac{y_2-y_1}{x_2-x_1}, & y_2 > y_1,\ x_2 > x_1 \\ \pi - \tan^{-1}\frac{y_2-y_1}{|x_2-x_1|}, & y_2 > y_1,\ x_2 < x_1 \\ \pi + \tan^{-1}\frac{|y_2-y_1|}{|x_2-x_1|}, & y_2 < y_1,\ x_2 < x_1 \\ 2\pi - \tan^{-1}\frac{|y_2-y_1|}{x_2-x_1}, & y_2 < y_1,\ x_2 > x_1 \end{cases} \tag{17.82}$$

17.3.3.3 Similarity Evaluation

To determine which arc the observed position should be matched with, a weighted similarity criterion can be utilized. That is a likelihood score is computed for each of the candidate arcs, which is defined as

$$S = S_{Az} + S_D + S_I \tag{17.83}$$

where S is the total score, S_{Az} is the score for the orientation, S_D is the score for the distance based proximity, and S_I is the score for the case where an intersection occurs. The individual scores can be calculated based on some specific formulas such as

$$\begin{aligned} S_{Az} &= w_{Az} \cos^{n_{Az}} \beta_{Az} \\ S_D &= w_D(1 - ad^{n_D}) \\ S_I &= w_I \cos^{n_I} \beta_{Az} \end{aligned} \tag{17.84}$$

where $\{n_i\}$ are the power parameters, $\{w_i\}$ and a are the coefficients. The power parameters and the coefficients need to be tuned in practice to maximize the correct match.

Note that other similarity criteria, if available, can also be developed in the same way for performance enhancement. For instance, when a vehicle travels from one street to another, the two streets must be connected. Thus the road continuity criterion can be exploited. Another criterion would be based on the speed limit change which would be typically provided by a road network.

17.3.3.4 Filtering Outliers

Sometimes abnormal observed position estimates may be produced due to sensor failure or severe NLOS propagation. Matching mistakes would be produced when such an abnormal observation or outlier exists. As shown in Fig. 17.14, a spike happens at the observed position P_6 which is an outlier. The path joining points P_1, P_2, P_3, P_5, and P_6 are generally parallel to arc 1, whereas point P_4 does not fit that pattern. If simply using the computed direction for matching, point P_4 would be matched to arc 2. To avoid such a mistake, the matching of point P_4 will be delayed

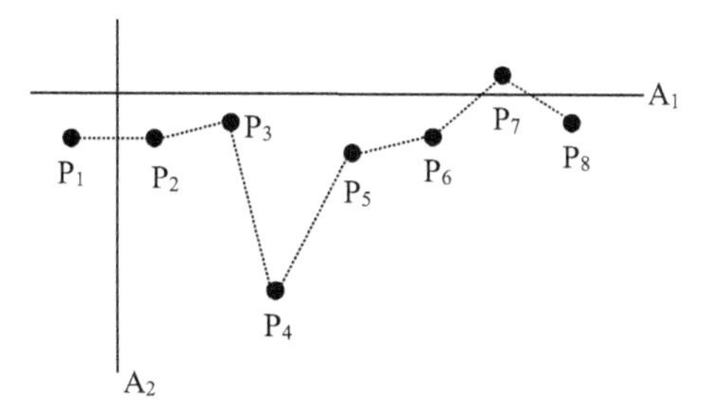

Fig. 17.14 Spike caused by abnormal observed positions

by two observation-time periods. That is point P_4 will be mapped after the directions for the lines P_4-P_5 and P_5-P_6 are computed. The delay allows the algorithm to determine whether the computed direction of line P_3-P_4 is consistent with the directions of the lines P_4-P_5 and P_5-P_6. If the consistency does not hold, then the observed position P_4 is considered to be an outlier. The observed position point which is identified as an outlier should be excluded from the matching process. The corresponding actual position on the road segment can be simply computed as the average of the two neighboring mapped points on the segment, one behind and the other in front.

17.3.3.5 Skipping Arcs

In practice the land-vehicles travel from one road segment to another, which are connected to each other. In the topological map matching such continuity is also assumed, and thus the vehicle will only travel onto a new road segment (arc) that is connected to the previous one. However, when an arc is very short and the vehicle travels very fast, arc-skipping may happen; that is there may be no sampling of a position during the time interval when the vehicle travels on the arc. This event can also occur when there is an outage of the positioning system, producing no position estimate over a short period of time. To ensure correct and efficient map matching, arc-skipping should be taken into account. This issue can be addressed in different ways. When a new observed position cannot be matched on the current arc, or on any other arc which is connected to the current arc, then matching process should be restarted. The drawback of this simple solution is that the already known mobile location information is lost in the initialization process. Another solution to the arc-skipping problem is to extend the search area. Specifically, instead of looking for a match only on arcs which are connected to the nodes of the current arc, one can look for a match on arcs that are connected to the nodes of the arcs that are connected to the nodes of the current arc. The drawback of such an extended matching technique is the increased complexity since more information about the neighboring arcs and nodes has to be stored and the match finding is more intensive.

17.3.4 Road Selection Based on Belief Theory

In Sect. 17.3.3.3 a likelihood-score based geometric and topological map-matching method is described. In this section a belief theory based road selection approach described in El Najjar and Bonnifait (2007) is introduced. The belief theoretical concepts are first studied and then a belief-function based similarity criteria determination and criteria fusion are described.

17.3.4.1 Belief Theory Concepts

The belief theory, also known as the Dempster-Shafer theory, is a generalization of the Bayesian theory of subjective probability (Dempster 1967; Shafer 1976). It allows one to combine evidence from different sources and arrive at a degree of belief, represented by a *belief function*. Although the Bayesian theory requires probabilities for each question of interest, belief functions allow degrees of belief for one question to be based on probabilities for a related question.

A *basic entity* is a set of all possible answers, also called hypotheses, to a specific question. All the hypotheses in a set denoted as θ must be exclusive and exhaustive, and each subset can be a possible answer to the question. The power set denoted by 2^{θ} is defined as all the subsets of θ, including the empty set. The number of elements of the power set is thus equal to 2^N where N is the size of θ. The degree of belief of each hypothesis is represented by a real number in the range [0, 1], which is called a *mass function*, $m(\cdot)$ satisfying

$$\begin{aligned} m(\varphi) &= 0 \\ \sum_{A\subseteq\theta} m(A) &= 1 \end{aligned} \tag{17.85}$$

where φ is an empty set and A is a given member of the power set. Associated with the mass assignments, two measures namely *belief* denoted by $Bel(\cdot)$ and *plausibility* denoted by $Pl(\cdot)$ are defined as follows:

$$\begin{aligned} Bel(A) &= \sum_{B\subseteq A} m(B) \\ Pl(A) &= \sum_{B\cap A\neq\varphi} m(B) \end{aligned} \tag{17.86}$$

These two measures are related to each other by the following relationship

$$Pl(A) = 1 - Bel(\bar{A}) \tag{17.87}$$

where $\bar{A}$ denotes the complement of set A. Another important issue is the combination of a number of independent sets of mass assignments. That is, one needs to combine evidence from difference sources. The evidence combination of two different sources can be determined by the Dempster-Shafer rule, whereas the evidence combination of more than two different sources is calculated according to the Yager's rule (Yager 1987). Suppose that there are K basic mass assignments, $\{m_i\}_{i=1}^{K}$, for K different belief structures. Then based on the Yager's rule, the combination of the K mass assignments is calculated by

$$\begin{aligned} m(\varphi) &= 0 \\ m(A) &= \frac{1}{1-q} \sum_{\cap_{i=1}^{K} A_i = A} m_1(A_1) m_2(A_2) \ldots m_K(A_K) \end{aligned} \tag{17.88}$$

where

$$q = \sum_{\cap_{i=1}^{K} A_i = \varphi} m_1(A_1) m_2(A_2) \ldots m_K(A_K) \tag{17.89}$$

Note that in the case of the combination of two different sources, the Yager's rule becomes the Dempster-Shafer rule. In the following subsections, the *belief theory* is applied to road selection for vehicle tracking.

17.3.4.2 Proximity Criterion

As mentioned earlier the proximity criterion is based on the measurement of the Euclidean distance between the estimated position and each segment in the vicinity of the estimated position. When GNSS or other technologies such as INS are used for two-dimensional positioning, the position estimation error vector $\Delta = [\Delta x \quad \Delta y]^T$ can be modeled to have a bivariate Gaussian distribution. Alternatively when the position estimate is defined as $\hat{\boldsymbol{\theta}} = [\hat{x} \quad \hat{y}]^T$, the true position $\boldsymbol{\theta} = [x \quad y]^T$ can also be assumed to have a bivariate Gaussian density. That is the conditional PDF is given by

$$p(\boldsymbol{\theta}|\hat{\boldsymbol{\theta}}) = \frac{1}{2\pi \det(\mathbf{Q})} \exp\left(-\frac{1}{2} a\right) \tag{17.90}$$

where $\mathbf{Q}$ is the covariance matrix of the position estimation error vector defined as

$$\mathbf{Q} = \begin{bmatrix} \sigma_x^2 & \sigma_{xy}^2 \\ \sigma_{xy}^2 & \sigma_y^2 \end{bmatrix} \tag{17.91}$$

and

$$a = \Delta^T \mathbf{Q}^{-1} \Delta \tag{17.92}$$

These values of Δx and Δy define the points on an ellipse in the $(\Delta x, \Delta y)$ plane, namely

$$a = \frac{1}{1-\rho^2}\left(\frac{\Delta x^2}{\sigma_x^2} - 2\rho\frac{\Delta x \Delta y}{\sigma_x \sigma_y} + \frac{\Delta y^2}{\sigma_y^2}\right) \tag{17.93}$$

where $\rho = \frac{\sigma_{xy}^2}{\sigma_x \sigma_y}$. The major and minor axes of the ellipse given by (17.93) are not aligned with the coordinate axes, but rotated by an angle: $\frac{1}{2}\arctan\left(\frac{2\rho\sigma_x\sigma_y}{\sigma_x^2 - \sigma_y^2}\right)$ (Chamlou 2004). The covariance matrix $\mathbf{Q}$ is typically a nonsingular matrix, so applying eigenvalue decomposition produces

$$\mathbf{Q} = \mathbf{A}\,\Lambda\,\mathbf{A}^{-1} \tag{17.94}$$

where $\mathbf{A}$ is formed from the eigenvectors of $\mathbf{Q}$ associated with the two eigenvalues λ_1 and λ_2, and Λ is the diagonal matrix with the two eigenvalues as the diagonal elements. These two eigenvalues can be determined by solving the equation

$$\det\left(\begin{bmatrix} \sigma_x^2 & \sigma_{xy}^2 \\ \sigma_{xy}^2 & \sigma_y^2 \end{bmatrix} - \begin{bmatrix} \lambda & 0 \\ 0 & \lambda \end{bmatrix}\right) = 0 \tag{17.95}$$

producing

$$\lambda_{1,2} = \frac{(\sigma_x^2 + \sigma_y^2) \pm \sqrt{(\sigma_x^2 + \sigma_y^2)^2 + 4\sigma_{xy}^4}}{2} \tag{17.96}$$

Substituting (17.94) into (17.93) produces

$$\begin{aligned} a &= (\mathbf{A}^T\Delta)^T \Lambda^{-1} (\mathbf{A}^T\Delta) \\ &= \frac{\Delta x_r^2}{\lambda_1} + \frac{\Delta y_r^2}{\lambda_2} \end{aligned} \tag{17.97}$$

where

$$\Delta_r = [\,\Delta x_r \quad \Delta y_r\,]^T = \mathbf{A}^T \Delta \tag{17.98}$$

describes a rotational change only, and thus $\det(\mathbf{A}^T)$ equals 1 or −1. It can be seen from (17.97) for $a = 1$, the transformed ellipse has the standard form as

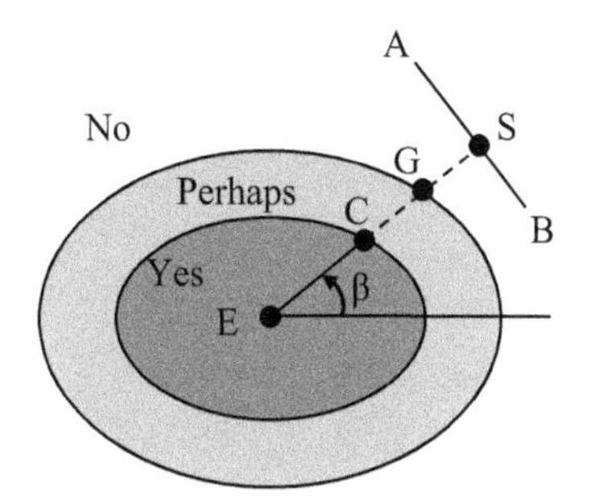

Fig. 17.15 Illustration of estimated position, road segment and elliptical error

$$\frac{x^2}{axis_{major}^2} + \frac{y^2}{axis_{\min or}^2} = 1 \tag{17.99}$$

where

$$axis_{major} = \sigma_{x_r} = \max(\sqrt{\lambda_1},\ \sqrt{\lambda_2}), \qquad axis_{\min or} = \sigma_{y_r} = \min(\sqrt{\lambda_1},\ \sqrt{\lambda_2}) \tag{17.100}$$

Accordingly, the PDF in (17.100) becomes

$$p(\boldsymbol{\theta}|\hat{\boldsymbol{\theta}}) = \frac{1}{2\pi\sigma_{x_r}\sigma_{y_r}} \exp\left(-\left(\frac{\Delta x_r^2}{2\sigma_{x_r}^2} + \frac{\Delta x_r^2}{2\sigma_{y_r}^2}\right)\right) \tag{17.101}$$

The desired probability is generated by integrating the above PDF over the ellipse bounded by $k\sigma_{x_r}$ and $k\sigma_{y_r}$. That is

$$\begin{aligned} P(\boldsymbol{\theta}|\hat{\boldsymbol{\theta}}) &= \iint_{\frac{\Delta x_r^2}{\sigma_{x_r}^2} + \frac{\Delta x_r^2}{\sigma_{y_r}^2} < k^2} p(\boldsymbol{\theta}|\hat{\boldsymbol{\theta}}) d\Delta x_r d\Delta y_r \\ &= \int_0^k \int_0^{2\pi} \frac{1}{2\pi} \exp\left(-\frac{u^2}{2}\right) u d\alpha du \\ &= 1 - \exp\left(-\frac{k^2}{2}\right) \end{aligned} \tag{17.102}$$

Alternatively, the last equation in (17.102) can be written as

$$k = \sqrt{-2\ln(1 - P(\boldsymbol{\theta}|\hat{\boldsymbol{\theta}}))} \tag{17.103}$$

Thus when giving a value of k, the probability that the true target position is in the ellipse bounded by $k\sigma_{x_r}$ and $k\sigma_{y_r}$ can be calculated, or vice versus. For instance,

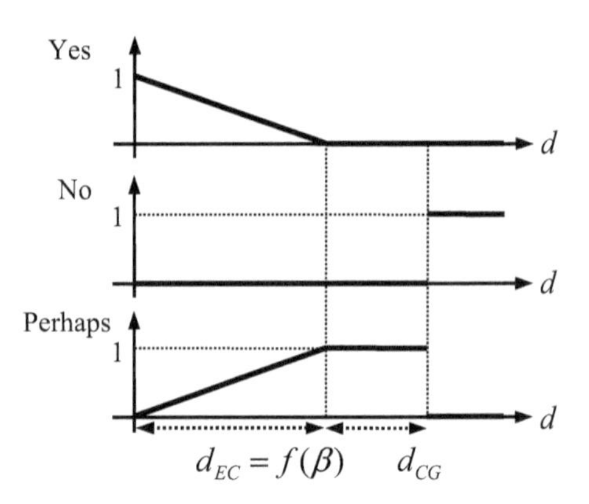

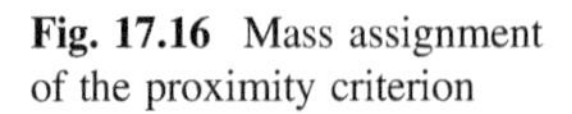
Fig. 17.16 Mass assignment of the proximity criterion

when $k = 2.5$, the probability is $P(\boldsymbol{\theta}|\hat{\boldsymbol{\theta}}) = 95.61\%$. Figure 17.15 shows the elliptical error density of equitable probability related to a specific road segment AB. Point S is the projection of the estimated position point E onto the segment. To apply the belief theory, the set is defined as $\theta = \{Yes, \ No\}$, corresponding to the response to the question: is this the right segment? Accordingly the power set can be defined as $2^{\theta} = \{\varphi, \ Yes, \ No, \ Perhaps\}$, where the empty set φ can be ignored. The inner ellipse is obtained by assigning a value to the probability $P(\boldsymbol{\theta}|\hat{\boldsymbol{\theta}})$, calculating the parameter k, and replacing a in (17.97) by k^2. The angle between the major axis of the ellipse and the line ES is denoted by β.

One method for the mass assignment of the proximity criterion is shown in Fig. 17.16. Let d denote the distance from point E to the segment AB. The mass assigned for the set member “Yes” linearly decreases from one at point E to zero at point C. The distance d_{EC} is a function of the angle β. The allocated mass for the “Perhaps” hypothesis complements that of the “Yes” hypothesis, namely $m(Perhaps) = 1 - m(Yes)$ when $d < d_{EG}$. The mass assigned for the “No” hypothesis is the step function starting from distance d_{EG}. The selection of the distance d_{CG} would depend on the specific application.

17.3.4.3 Angular Criterion

In the presence of heading angle measurements such as provided by heading sensors, the heading information can be used for road selection. Intuitively the most credible segment is the one that has an angle closest to the heading of the vehicle. Let the headings of the road segment and the vehicle be denoted by α_S and α_V. Then the heading difference can be defined as

$$\Delta\alpha = \min(|\alpha_S - \alpha_V|, \ |\alpha_S - \alpha_V + \pi|), \quad \alpha_S \in [0, \ \pi], \ \alpha_V \in [0, \ \pi] \tag{17.104}$$

where the segments are assumed to be two-way streets. When the segment is one-way, the heading difference can be defined as

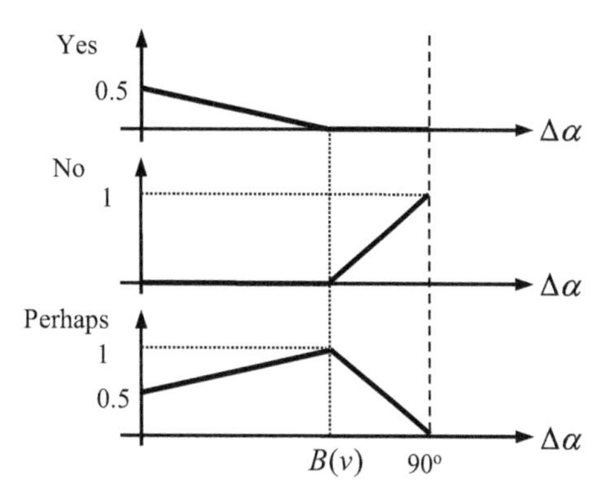

Fig. 17.17 Heading based mass assignment when giving a velocity and $\sigma_{\alpha_V} = \pi/12$

$$\Delta\alpha = |\alpha_S - \alpha_V|, \quad \alpha_S \in [0,\ 2\pi), \quad \alpha_V \in [0,\ 2\pi) \tag{17.105}$$

The mass assignment for the "Yes" hypothesis versus the heading difference is adaptive with respect to the speed (v) of the vehicle and the standard deviation (σ_{α_V}) of the estimation error of the vehicle heading angle. Specifically, the maximum mass assigned to the "Yes" hypothesis is defined by

$$m(\sigma_{\alpha_V}) = \max\left\{1 - \frac{6}{\pi}\sigma_{\alpha_V},\ 0\right\} \tag{17.106}$$

The parameter B is used to fix the angular difference limit for a given speed, defined as

$$B(v) = \pi/2 - \lambda v \tag{17.107}$$

where $\lambda = (90° - 10°)\pi/(v_{\max} \cdot 180°)$ and $v_{\max}$ is the maximum speed of the vehicle. An example of the mass assignments for the three hypotheses is illustrated in Fig. 17.17.

17.3.4.4 Criteria Fusion

Once the mass assignments are accomplished for all the different criteria, combination of the mass assignments can be performed for each of the hypotheses or power set members, which are the "Yes", "No", and "Perhaps" hypothesis, for the vehicle tracking problem. The combination is performed according to the formula given by (17.105). Although only two criteria, proximity and heading direction are examined, there are other possible criteria such as speed limit and road continuity that could be used to further improve the road selection performance.

After the combination step several decision rules can be used to produce the final result. Based on the set of combined masses, the *belief* and *plausibility* are calculated. If an optimistic decision is desired, the maximum of plausibility should be

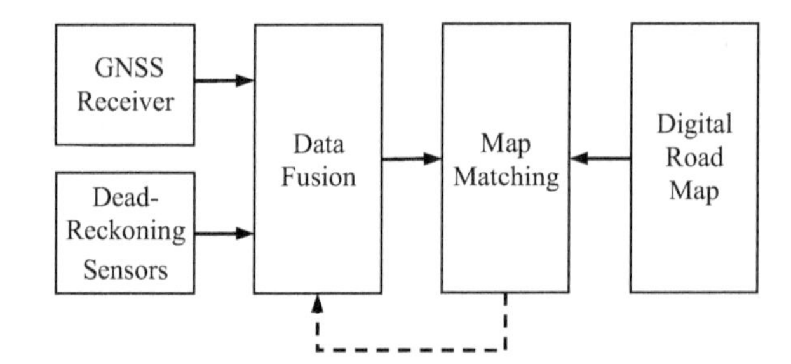

Fig. 17.18 Illustration of unidirectional (solid line connection) and bidirectional (both solid and dashed line connection) map-matching

used. On the other hand, the maximum of belief should be employed to produce a pessimistic decision. The belief and plausibility correspond respectively to the minimum probability and the maximum probability that an assumption is true. It is clear that when a hypothesis has a belief that is higher than the plausibility of any other hypothesis, this hypothesis is selected without any ambiguity.

17.3.5 Unidirectional Method Versus Bidirectional Method

Traditionally the map-matching process is unidirectional or open loop. That is the position and trajectory estimated by the GNSS receiver and dead-reckoning sensors have been used as input to the map-matching process. The map-matching outcome does not have any effect on the fusion of the measurements from the GNSS receiver and dead-reckoning sensors to produce position and trajectory estimates. On the other hand, the map-matching can be bidirectional. Specifically, the map-matching results such as the heading and length of the road segment identified can be compared to the results of the sensor data fusion such that the sensors can be calibrated online. Figure 17.18 shows the simplified block diagram of the unidirectional and bidirectional matching approaches.

References

Bernstein D, Kornhauser, A (1996) An introduction to map matching for personal navigation assistants. pp 1–16, Aug 1996. http://www.njtide.org/reports/mapmatchintro.pdf

Chamlou R (2004) Calculation of navigation accuracy category for position and velocity parameters. In: 2004 Proceedings of IEEE digital avionics systems conference, pp 1.D.3-1-13

Chen BS, Yang CY, Liao FK, Liao JF (2009) Mobile location estimator in rough wireless environment using extended Kalman-based IMM and data fusion. IEEE Trans Veh Technol 58 (3):1157–1169

Dempster AP (1967) Upper and lower probabilities induced by a multivalued mapping. Ann Math Stat 38(2):325–339

El Najjar MEB, Bonnifait P (2007) Road selection using multicriteria fusion for the road-matching problem. IEEE Trans Intell Transp Syst 8(2):279–291

Greenfeld JS (2002) Matching GPS observations to locations on a digital map. In: 2002 Proceedings of 81st annual meeting of the transportation research board, Washington, D.C.
Kleine-Ostmann T, Bell AE (2001) A data fusion architecture for enhanced position estimation in wireless networks. IEEE Commun Lett 5(8):343–345
Li X, Martin, RP (2008) Simple algorithm aggregation improves signal strength based localization. In Proceedings of international symposium on wireless pervasive computing, pp 540–544, May 2008
Qi Y, Kobayashi, H (2003) On relation among time delay and signal strength based geolocation methods. In: 2003 Proceedings of IEEE global communications conference (GLOBECOM), pp 4079–4083
Shafer G (1976) A mathematical theory of evidence. Princeton University Press
Shin D-H, Sung T-K (2002) Comparisons of error characteristics between TOA and TDOA positioning. IEEE Trans Aerosp Electron Syst 38(1):307–311
Skog I, Handel P (2009) In-car positioning and navigation technologies – a survey. IEEE Trans Intell Transp Syst 10(1):4–21
Titterton DH, Weston JL (2004) Strapdown inertial navigation technology. IET Press
Xu G (2007) GPS Theory, Algorithms and Applications. Springer
Yager R (1987) On the Dempster-Shafer framework and new combination rules. Inf Sci 41(2): 93–137
Yu K (2018) Positioning and navigation in complex environments. IGI Global
Yu K, Sharp I, Guo Y (2009) Ground based wireless positioning. Wiley-IEEE Press

Chapter 18
Utilization of Multiple Antennas for Positioning

Radio-based positioning performance is affected by a range of factors such as radio propagation conditions, node configurations, location determination algorithms, signal design, and signal parameter estimation. To improve positioning accuracy various algorithms and techniques have been developed, among which is the utilization of multiple antennas. If transmitter nodes, receiver nodes (or both) are equipped with multiple antennas, appropriate signal processing algorithms can be exploited to determine the location of a mobile node. This chapter provides a comprehensive coverage on the utilization of multiple antennas for positioning.

The topic of multiple antennas has been extensively studied for high data-rate wireless communications especially for coping with the adverse fading channels. There exist numerous multi-input multi-output (MIMO) schemes and methodologies. Naturally the existing multiple-antenna techniques in wireless communications can be properly exploited for radio positioning and to enhance positioning accuracy. In some circumstances where the physical size of the mobile nodes is rather small, implementation of multiple antennas poses a challenge and would be impractical. However, when the constraint on the physical size and complexity is relaxed especially at fixed base stations, implementation of multiple antennas is feasible and practical, as evidenced by a range of positioning technologies.

Multiple antennas can be utilized in different ways to assist positioning. One well-known aspect is the angle-of-arrival (AOA) estimation. Antenna array based AOA estimation has been widely studied especially in relation to signal processing and data communications. Good AOA measurements can be employed to locate the target of interest by developing AOA-only positioning algorithms or hybrid positioning approaches which may incorporate range determination. Another important usage of antenna array is beamforming of the signals received at or transmitted from multiple antenna elements, so that multiple access interference and multipath

I. Sharp and K. Yu, *Wireless Positioning: Principles and Practice*, Navigation: Science and Technology, https://doi.org/10.1007/978-981-10-8791-2_18

interference can be greatly mitigated, thus producing more accurate position estimates. This chapter includes description and analysis of the following topics:

- Array-based AOA estimation methods
- Basic beamforming approaches
- Two AOA-based positioning methods
- Application examples of multiple antennas

18.1 Array-Based AOA Estimation

Multi-antenna based AOA estimation has been extensively investigated and a wide range of angle estimation methods and techniques have been proposed in the literature. This section is not intended to provide a thorough study on all the existing methodologies related to AOA estimation. Instead a number of typical algorithms are described.

18.1.1 Least-Squares Method

The least-squares (LS) method provides a closed-form solution for 2-D angle estimation of a single narrowband source with a uniform circular array (UCA) which is centrosymmetric (Wu and So 2008). Note that UCA has been extensively studied and employed for 2-D angle estimation because it has some attractive characteristics such as providing the full 360° azimuthal coverage, elevation angle information, and a smooth directional pattern. By exploiting the phase of an autocorrelation sequence, a set of linear equations can be obtained for determining the 2-D angles. This algorithm is computationally efficient since it does not require a 2-D search and/or eignenvalue decomposition.

A simplified representation of the geometry of a centrosymmetric spatial sensor configuration is shown in Fig. 18.1. The K sensors are uniformly deployed on a ring in the XY plane. The first sensor is on the x-coordinate so that the positional angle of the kth sensor is given by

$$\alpha_k = 2\pi(k-1)/K \tag{18.1}$$

The narrowband signal impinges on the circular array with a 2-D angle pair (θ, ϕ) where $\theta \in [0, 2\pi)$ and $\phi \in [0, \pi)$ are respectively the azimuth and elevation angles.

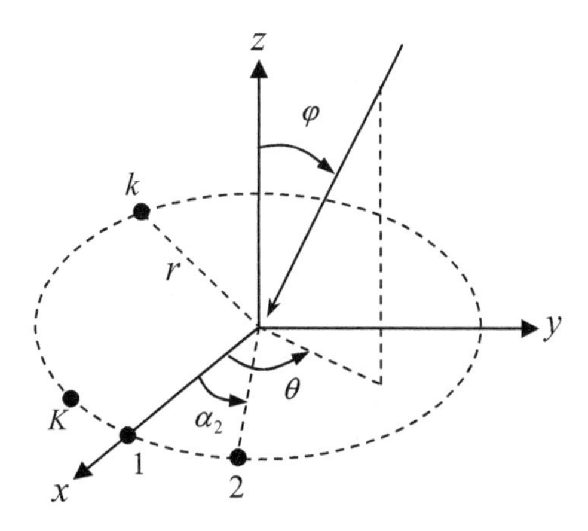

Fig. 18.1 Centrosymmetric circular array configuration

The output of the kth sensor of the circular array at discrete time t is given by

$$x_k(t) = \exp\left(j\frac{2\pi r}{\lambda}\cos(\theta - \alpha_k)\sin\phi\right)s(t) + n_k(t),\\ k = 1, 2, \ldots, K,\ t = 1, 2, \ldots, L \tag{18.2}$$

where at each sensor L data samples are received, $s(t)$ is the zero-mean and complex narrowband signal source, $\{n_k(t)\}$ are assumed to be independent zero-mean complex random processes which are independent from the signal, λ is the wavelength of the signal, defined as the speed of propagation divided by carrier frequency, and r is the radius of the circle. The purpose is to estimate the angle pair by making use of the outputs $\{x_k(t)\}$.

Construct a sequence of time average of $\{x_k(t)\}$ as

$$u(k) = \frac{1}{L}\sum_{t=1}^{L} x_k(t)x^*_{k+K/2}(t),\ k = 1, 2, \ldots, K/2 \tag{18.3}$$

The array geometric symmetry results in

$$\alpha_{k+K/2} = \alpha_k + \pi \tag{18.4}$$

and thus

$$\cos(\theta - \alpha_{k+K/2}) = -\cos(\theta - \alpha_k) \tag{18.5}$$

Then when both K and signal-to-ratio (SNR) are sufficiently large, the signal can be approximated to be ergodic, namely the ensemble average is approximate to the time average, resulting in

$$u(k) \approx E[x_k(t)x^*_{k+K/2}(t)] = \exp\left(j\frac{4\pi r}{\lambda}\cos(\theta - \alpha_k)\sin\phi\right)\sigma_s^2 + \sigma_n^2 \tag{18.6}$$

where σ_s^2 is the power of the signal $s(t)$ and σ_n^2 is the noise power. When the noise power is known, the phase of $u(k) - \sigma_n^2$ can be computed by (18.3) and denoted by ω_k. Note that in the case of high SNR, the noise power may be neglected in calculating the phase. Then

$$\frac{4\pi r}{\lambda}\cos(\phi - \alpha_k)\sin\theta \approx \omega_k + 2m_k\pi,\ k = 1, 2, \ldots, K/2 \tag{18.7}$$

where m_k is a non-negative integer and the phase ambiguity $2m_k\pi$ comes from the fact that the computed phase ω_k is typically limited to $-\pi < \omega_k < \pi$, whereas the left-hand side of (18.7) can be beyond this region depending on the angle pair (θ, ϕ), the angles $\{\alpha_k\}$, and the ratio between the radius and the wavelength as well. Assuming $m_k = 0$, then, (18.7) can be written in a compact form as

$$\mathbf{Ab} \approx \boldsymbol{\omega} \tag{18.8}$$

where

$$\mathbf{A} = \begin{vmatrix} \cos\alpha_1 & \sin\alpha_1 \\ \vdots & \vdots \\ \cos\alpha_{K/2} & \sin\alpha_{K/2} \end{vmatrix}$$
$$\mathbf{b} = \frac{4\pi r}{\lambda}[\sin\theta\cos\phi \quad \sin\theta\sin\phi]^T \tag{18.9}$$
$$\boldsymbol{\omega} = [\omega_1 \ \cdots \ \omega_{K/2}]^T$$

The ordinary LS estimate of $\mathbf{b}$ is given by

$$\hat{\mathbf{b}} = [\hat{b}_1 \quad \hat{b}_2]^T \approx (\mathbf{A}^T\mathbf{A})^{-1}\mathbf{A}^T\boldsymbol{\omega} \tag{18.10}$$

From $\hat{\mathbf{b}}$ the estimate of the angle pair (θ, ϕ), denoted by $(\hat{\theta}, \hat{\phi})$, can be determined as

$$\hat{\theta} = \sin^{-1}\left(\frac{\lambda}{4\pi r}\sqrt{\hat{b}_1^2 + \hat{b}_2^2}\right)$$
$$\hat{\phi} = \tan^{-1}\left(\frac{\hat{b}_2}{\hat{b}_1}\right) \tag{18.11}$$

The following example illustrates the accuracy of this LS angle estimation algorithm. The angle pair (θ, ϕ) is set at $(26°, 325°)$ and the radius is set at half the wavelength. The accuracy is evaluated with respect to the SNR, the number of samples, and the number of array elements. Figure 18.2 shows the distribution pattern of the azimuth and elevation angle estimation errors. The SNR is equal to 15 dB, 80 data samples from each array element are used, there are six array elements, and 2000 simulation runs are conducted. Figure 18.3 shows the root-mean-square error (RMSE) of the azimuth angle estimation versus the number of array elements and number of samples when the SNR is set at 15 dB. As expected, the accuracy increases with the number of samples and the number of array elements. Figure 18.4 shows the RMSE of the azimuth angle estimation versus the number of array elements and SNR when the number of samples is set at 80. As long as the SNR is above 15 dB, the RMSE of the azimuth angle can be kept below 0.5°.

18.1.2 MUSIC Method

The MUSIC (multiple signal classification) method was originally proposed by Schmidt (1986). It is one of the signal-subspace methods, and has been widely studied for high-resolution signal parameter estimation. The MUSIC method is applicable to arrays of arbitrary but known configuration and response, and can be employed to estimate multiple parameters per source, such as the number of

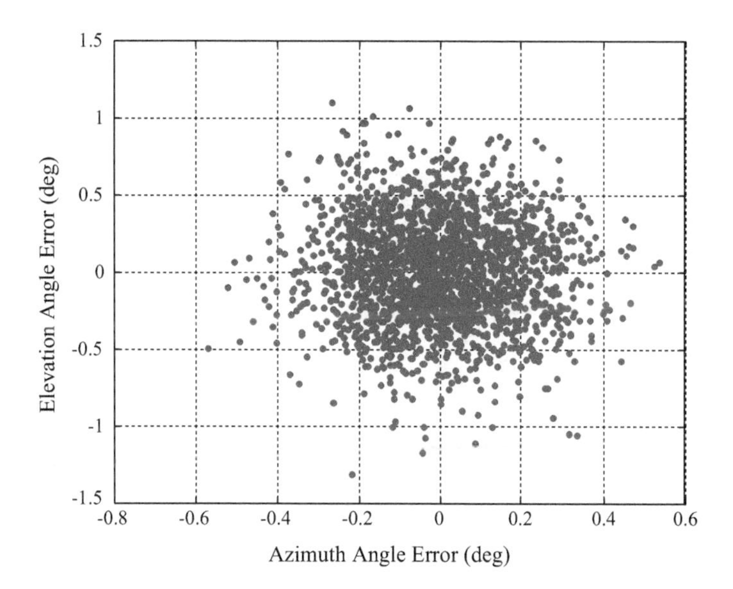

Fig. 18.2 Distribution pattern of azimuth and elevation angle estimation errors

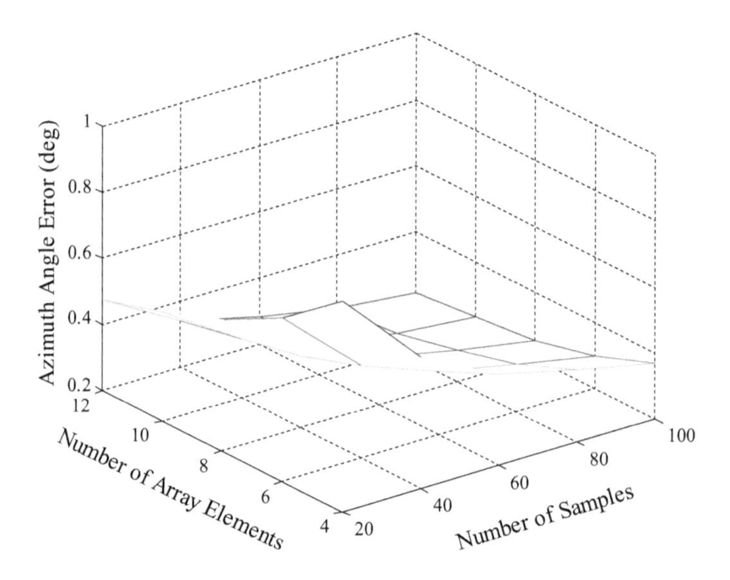

Fig. 18.3 Azimuth angle errors with respect to number of array elements and number of samples

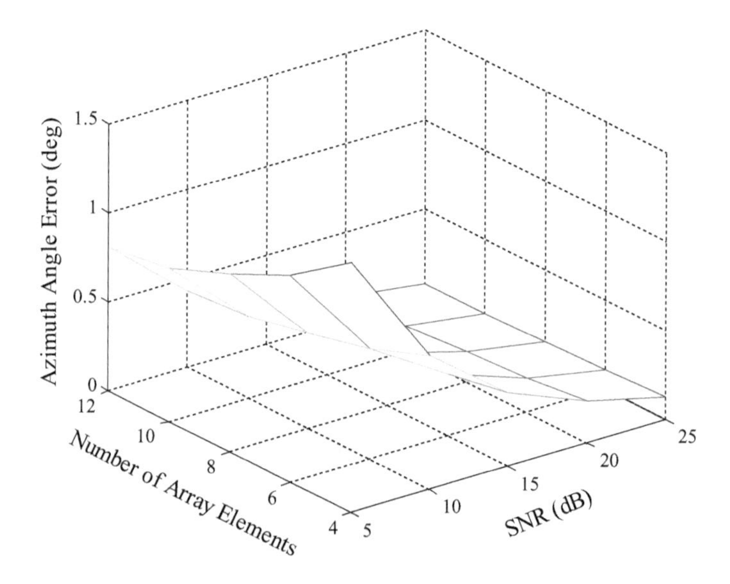

Fig. 18.4 Azimuth angle errors with respect to number of array elements and SNR

incident wavefronts, azimuth and elevation angles, range, and polarization. In this subsection the classic MUSIC algorithm and the root MUSIC algorithm are described.

18.1.2.1 Basic MUSIC

The basic MUSIC algorithm for one-dimensional angle estimation is analyzed first. Assume that the wavefronts of N narrowband far-field signal sources impinge on an M-element antenna array. Then the array output can be described as

$$\mathbf{x}(t) = \mathbf{A}\mathbf{s}(t) + \mathbf{n}(t) \tag{18.12}$$

where

$$\mathbf{s}(t) = \begin{bmatrix} s_1(t) & s_2(t) & \cdots & s_N(t) \end{bmatrix}^T \tag{18.13}$$

is the complex source signal vector at time t,

$$\mathbf{n}(t) = \begin{bmatrix} n_1(t) & n_2(t) & \cdots & n_N(t) \end{bmatrix}^T \tag{18.14}$$

is the additive complex noise vector, and $\mathbf{A}$ is the array manifold matrix defined by

$$\mathbf{A} = \begin{bmatrix} \mathbf{a}(\theta_1) & \mathbf{a}(\theta_2) & \cdots & \mathbf{a}(\theta_N) \end{bmatrix} \in R^{M \times N} \tag{18.15}$$

where

$$\mathbf{a}(\theta_i) = \begin{bmatrix} a_1(\theta_i) & a_2(\theta_i) & \cdots & a_M(\theta_i) \end{bmatrix}^T \tag{18.16}$$

If the noise and signals are uncorrelated, and for the moment the noise is spatially white, then the covariance matrix of the received signals is given by

$$\begin{aligned} \mathbf{R_x} &= E[\mathbf{x}(t)\mathbf{x}^H(t)] \\ &= \mathbf{A}\mathbf{R_s}\mathbf{A}^H + \sigma^2\mathbf{I} \end{aligned} \tag{18.17}$$

where

$$\mathbf{R_s} = E[\mathbf{s}(t)\mathbf{s}^H(t)] \tag{18.18}$$

is the covariance matrix of the source signals, σ^2 is the noise power in each channel, and $\mathbf{I}$ is an identity matrix of size $M \times M$.

Like most of the other subspace-based methods, the basic MUSIC algorithm first estimates the dominant subspace of the observations, and then finds the elements of the manifold matrix, which are in some sense closest to this subspace. The subspace estimation is realized by performing an *eigendecomposition* on the covariance matrix $\mathbf{R_x}$, resulting in

$$\begin{aligned} \mathbf{R_x} &= \sum_{i=1}^{M} \lambda_i \mathbf{e}_i \mathbf{e}_i^H \\ &= \mathbf{E_s}\mathbf{\Lambda_s}\mathbf{E_s}^H + \sigma^2\mathbf{E_n}\mathbf{E_n}^H \end{aligned} \tag{18.19}$$

where $\{\mathbf{e}_i\}$ are the column eigenvectors and $\{\lambda_i\}$ are the eigenvalues which satisfy

$$\lambda_1 \geq \lambda_2 \geq \cdots \geq \lambda_N > \lambda_{N+1} = \lambda_{N+2} = \cdots = \lambda_M = \sigma^2 \tag{18.20}$$

and

$$\begin{aligned} \mathbf{E_s} &= [\mathbf{e}_1 \quad \mathbf{e}_2 \quad \cdots \quad \mathbf{e}_N] \\ \mathbf{E_n} &= [\mathbf{e}_{N+1} \quad \mathbf{e}_{N+2} \quad \cdots \quad \mathbf{e}_M] \end{aligned} \tag{18.21}$$

The span of the N vectors in $\mathbf{E_s}$ defines the signal subspace, whereas the orthogonal complement spanned by the vectors in $\mathbf{E_n}$ defines the noise subspace. The orthogonality of the two subspaces can be exploited to estimate the AOAs. That is one finds vectors on the array manifold $\{\mathbf{a}(\theta)\}$ that have zero projection in the noise subspace. Specifically, one finds the zeros of the function

$$\begin{aligned} f(\theta) &= \frac{\mathbf{a}^H(\theta)\mathbf{E_n}\mathbf{E_n}^H\mathbf{a}(\theta)}{\mathbf{a}^H(\theta)\mathbf{a}(\theta)} \\ &= \mathrm{Tr}(\mathbf{P}_\theta \mathbf{E_n}\mathbf{E_n}^H) \end{aligned} \tag{18.22}$$

where $\mathrm{Tr}(\mathbf{C})$ denotes taking trace of matrix $\mathbf{C}$ and $\mathbf{P}_\theta$ is the projection matrix, defined as

$$\mathbf{P}_\theta = \mathbf{a}(\theta)[\mathbf{a}^H(\theta)\mathbf{a}(\theta)]^{-1}\mathbf{a}^H(\theta) \tag{18.23}$$

In practice, the method of finding the zeros of the function is not appropriate since only an estimated noise subspace can be obtained due to various errors. Replacing the noise subspace in (18.22) by its estimate $\hat{\mathbf{E}}_\mathbf{n}$ produces

$$\hat{f}(\theta) = \mathrm{Tr}(\mathbf{P}_\theta \hat{\mathbf{E}}_\mathbf{n}\hat{\mathbf{E}}_\mathbf{n}^H) \tag{18.24}$$

The approach of the MUSIC algorithms to try to minimize the function to find out the angle estimates, that is

$$\hat{\theta} = \arg\min_\theta \{\hat{f}(\theta)\} \tag{18.25}$$

When the noise is not spatially white, its covariance matrix may be written as

$$E[\mathbf{n}(t)\mathbf{n}^H(t)] = \sigma^2 \mathbf{\Sigma} \tag{18.26}$$

where $\mathbf{\Sigma}$ is a known square non-identity matrix (Swindlehurst and Kailath 1992). Then the minimization can be performed via an eigendecomposition of the pre-whitened covariance, namely

$$\begin{aligned}\boldsymbol{\Sigma}^{-\frac{1}{2}}\mathbf{R_x}\boldsymbol{\Sigma}^{-\frac{1}{2}} &= \boldsymbol{\Sigma}^{-\frac{1}{2}}\mathbf{ASA}^H\boldsymbol{\Sigma}^{-\frac{1}{2}} + \sigma^2\mathbf{I} \\ &= \sum_{i=1}^{M}\lambda_i\mathbf{e}_i\mathbf{e}_i^H \\ &= \mathbf{E_s}\boldsymbol{\Lambda}_\mathbf{s}\mathbf{E_s}^H + \sigma^2\mathbf{E_n}\mathbf{E_n}^H\end{aligned} \tag{18.27}$$

The MUSIC function is now given by

$$\hat{f}(\theta) = \mathrm{Tr}(\mathbf{P}_{\Sigma,\theta}\hat{\mathbf{E}}_\mathbf{n}\hat{\mathbf{E}}_\mathbf{n}^H) \tag{18.28}$$

where

$$\mathbf{P}_{\Sigma,\theta} = \boldsymbol{\Sigma}^{-\frac{1}{2}}\mathbf{a}(\boldsymbol{\theta})\left[\mathbf{a}^H(\boldsymbol{\theta})\boldsymbol{\Sigma}^{-\frac{1}{2}}\mathbf{a}(\boldsymbol{\theta})\right]^{-1}\mathbf{a}^H(\boldsymbol{\theta})\boldsymbol{\Sigma}^{-\frac{1}{2}} \tag{18.29}$$

In the special case of a uniform linear array (ULA) of identical sensors, a variation of the MUSIC algorithm, the root-MUSIC algorithm is often considered. The array response of a ULA is a Vandermonde vector, given by

$$\mathbf{a}(\boldsymbol{\theta}) = \mathbf{a}_v(\boldsymbol{\theta}) = \begin{bmatrix}1 & z & z^2 & \cdots & z^{M-1}\end{bmatrix}^T \tag{18.30}$$

where

$$z = \exp\left(-j2\pi\frac{d}{\lambda}\sin\theta\right) \tag{18.31}$$

Accordingly, the function $f(\boldsymbol{\theta})$ may be written as a polynomial in z of order $2M-2$, that is,

$$\begin{aligned}f(\boldsymbol{\theta}) &= \mathbf{a}_v^H(z)\boldsymbol{\Sigma}^{-\frac{1}{2}}\mathbf{E_n}\mathbf{E_n}^H\boldsymbol{\Sigma}^{-\frac{1}{2}}\mathbf{a}_v(z) \\ &= \sum_{i=N+1}^{M}\mathbf{e}_i(z^{-1})\mathbf{e}_i^H(z)\end{aligned} \tag{18.32}$$

where

$$\begin{aligned}\mathbf{e}_i(z^{-1}) &= \mathbf{a}_v^T(z^{-1})\boldsymbol{\Sigma}^{-\frac{1}{2}}\mathbf{e}_i \\ \mathbf{e}_i^H(z) &= \mathbf{e}_i^H\boldsymbol{\Sigma}^{-\frac{1}{2}}\mathbf{a}_v(z)\end{aligned} \tag{18.33}$$

When the covariance matrix is perfectly known, N out of $2M-2$ roots of function $f(\boldsymbol{\theta})$ will be on the unit circle at values of z that correspond to the true AOAs. In the presence of noise, the root-MUSIC algorithm chooses the roots with modulus that are closest to unity and only roots inside the unit circle need be considered.

18.1.2.2 Real Beamspace MUSIC for Circular Arrays

The real-valued beamspace MUSIC (RB-MUSIC) operates in beamspace and employs a phase mode excitation-based beamformer, which offers advantages over element-space operation, including reduced computation, enhanced performance in correlated source scenarios, and the applicability of root-MUSIC (Mathews and Zoltowski 1994)

Beamforming Matrices and Manifold Vectors

The element space is transformed into beamspace through

$$\mathbf{F}_r^H \mathbf{a}(\boldsymbol{\theta}) = \mathbf{b}(\boldsymbol{\theta}) \tag{18.34}$$

where $\mathbf{F}_r^H$ is the transforming matrix, the beamspace manifold $\mathbf{b}(\boldsymbol{\theta})$ is real-valued, and

$$\begin{aligned} \mathbf{a}(\boldsymbol{\theta}) &= \mathbf{a}(\varsigma, \phi) = \left[e^{j\varsigma\cos(\phi-\alpha_1)} \quad e^{j\varsigma\cos(\phi-\alpha_2)} \quad \dots \quad e^{j\varsigma\cos(\phi-\alpha_N)} \right]^T \\ \varsigma &= \frac{2\pi}{\lambda} d \sin\theta \end{aligned} \tag{18.35}$$

where the definition of the angles $\{\phi, \ \theta, \ \alpha_i\}$ can be found in Fig. 18.1. Thus the transformation enables real-valued beamspace processing. Also define another transformation which is solely based on phase mode excitation, namely

$$\mathbf{F}_e^H \mathbf{a}(\boldsymbol{\theta}) = \mathbf{a}_e(\boldsymbol{\theta}) \tag{18.36}$$

where the beamspace manifold $\mathbf{a}_e(\boldsymbol{\theta})$ is centro-Hermitian, and the beamforming matrix $\mathbf{F}_e^H$ is defined as

$$\mathbf{F}_e^H = \mathbf{C}_v \mathbf{V}^H \tag{18.37}$$

here

$$\begin{aligned} \mathbf{C}_v &= \mathrm{diag}\{ j^{-M}, \ \dots, \ j^{-1} \ \ j^0 \ \ j^{-1} \ \ \dots \ \ j^{-M} \} \\ \mathbf{V} &= \sqrt{N} [\, \mathbf{w}_{-M} \quad \cdots \quad \mathbf{w}_0 \quad \cdots \quad \mathbf{w}_M \,] \\ \mathbf{w}_m &= \frac{1}{N} [\, 1 \quad e^{-j2\pi m/N} \quad \dots \quad e^{-j2\pi(N-1)/N} \,]^T \end{aligned} \tag{18.38}$$

The vector $\mathbf{w}_m$ excites the UCA with phase mode m, producing the UCA array pattern for mode m as

$$\begin{aligned} f_m &= \mathbf{w}_m^H \mathbf{a}(\boldsymbol{\theta}) \\ &= \frac{1}{N} \sum_{n=0}^{N-1} e^{jm\alpha_{n+1}} e^{j\varsigma \cos(\phi - \alpha_{n+1})} \\ &\approx j^m J_m(\varsigma) e^{jm\phi} \\ &= j^{|m|} J_{|m|}(\varsigma) e^{jm\phi} \end{aligned} \tag{18.39}$$

where $J_m(\varsigma)$ is the Bessel function of the first kind of order m and in the last equation the property of the Bessel function $J_{-m}(\varsigma) = (-1)^m J_m(\varsigma)$ is applied. Substituting (18.37) into (18.36) produces

$$\begin{aligned} \mathbf{a}_e(\boldsymbol{\theta}) &= \mathbf{C}_v \mathbf{V}^H \mathbf{a}(\boldsymbol{\theta}) \\ &= \sqrt{N} \mathbf{J}_\varsigma \mathbf{v}(\varphi) \end{aligned} \tag{18.40}$$

where

$$\begin{aligned} \mathbf{J}_\varsigma &= \mathrm{diag}\{J_M(\varsigma), \ldots, J_1(\varsigma), J_0(\varsigma), J_1(\varsigma), \ldots, J_M(\varsigma)\} \\ \mathbf{v}(\varphi) &= \left[e^{-jM\phi}, \ldots, e^{-j\phi}, e^0, e^{j\phi}, \ldots, e^{jM\phi}\right]^T \end{aligned} \tag{18.41}$$

It is clear that the beamspace manifold has been decoupled into two parts, one related to the azimuth angle and the other depending on the elevation angle. The beamspace vectors are centro-Hermitian, satisfying

$$\tilde{\mathbf{I}} \mathbf{a}_e(\boldsymbol{\theta}) = \mathbf{a}_e^*(\boldsymbol{\theta}) \tag{18.42}$$

where $\tilde{\mathbf{I}}$ is the reverse permutation matrix with ones on the anti-diagonal and zeros elsewhere. It can easily be seen that the transforming matrix $\mathbf{F}_e$ is orthogonal, namely

$$\mathbf{F}_e^H \mathbf{F}_e = \mathbf{I} \tag{18.43}$$

Now construct the transforming matrix $\mathbf{F}_r^H$, which is defined as the product of a matrix $\mathbf{W}^H$ with centro-Hermitian rows and $\mathbf{F}_e^H$, namely

$$\begin{aligned} \mathbf{F}_r^H &= \mathbf{W}^H \mathbf{F}_e^H \\ &= \mathbf{W}^H \mathbf{C}_v \mathbf{V}^H \end{aligned} \tag{18.44}$$

Accordingly the real-valued beamspace manifold $\mathbf{b}(\boldsymbol{\theta})$ becomes

$$\begin{aligned} \mathbf{b}(\boldsymbol{\theta}) &= \mathbf{F}_r^H \mathbf{a}(\boldsymbol{\theta}) \\ &= \sqrt{N} \mathbf{W}^H \mathbf{J}_\varsigma \mathbf{v}(\varphi) \end{aligned} \tag{18.45}$$

Note that any matrix W satisfying $\tilde{\mathbf{I}}\mathbf{W} = \mathbf{W}^*$ will enable $\mathbf{b}(\boldsymbol{\theta})$ to be a real-valued beamspace manifold. Thus one may construct $\mathbf{W}$ according to

$$\mathbf{W} = \frac{1}{\sqrt{M'}}\left[\mathbf{v}(\alpha_{-M}) \quad \cdots \quad \mathbf{v}(\alpha_0) \quad \cdots \quad \mathbf{v}(\alpha_M)\right] \tag{18.46}$$

where $\mathbf{v}(\alpha_0)$ is defined in (18.41) and

$$\begin{aligned} M' &= 2M + 1 \\ \alpha_i &= \frac{2\pi i}{M'} \end{aligned} \tag{18.47}$$

RB-MUSIC Algorithm

This subsection describes and analyses the RB-MUSIC algorithm using the transforming matrices constructed previously. Let $\mathbf{x}(t)$ be the output vector of the array elements, $\mathbf{A}$ be the array manifold matrix, $\mathbf{s}(t)$ be the signal vector of the M signal sources, and $\mathbf{n}(t)$ be the additive noise vector. Assume that the signals and the noise are zero-mean uncorrelated random processes. The UCA-based RB-MUSIC algorithm employs the orthogonal beamforming matrix $\mathbf{F}_r^H$ constructed in the preceding subsection to make the transformation from the element space to the beamspace, producing

$$\begin{aligned} \mathbf{y}(t) &= \mathbf{F}_r^H \mathbf{x}(t) \\ &= \mathbf{B}\mathbf{s}(t) + \mathbf{F}_r^H \mathbf{n}(t) \end{aligned} \tag{18.48}$$

where

$$\mathbf{B} = \mathbf{F}_r^H \mathbf{A} \tag{18.49}$$

is a real-valued beamspace AOA matrix with column vectors given by $\mathbf{b}(\boldsymbol{\theta}_i)$. The covariance matrix of $\mathbf{y}(t)$ is given by

$$\begin{aligned} \mathbf{R}_y &= E[\mathbf{y}(t)\mathbf{y}^H(t)] \\ &= \mathbf{B}\mathbf{R}_x\mathbf{B}^T + \sigma\mathbf{I} \end{aligned} \tag{18.50}$$

Let $\mathbf{R}$ be real part of $\mathbf{R}_y$. Then

$$\begin{aligned} \mathbf{R} &= \mathrm{Re}\{\mathbf{R}_y\} \\ &= \mathbf{B}\mathrm{Re}\{\mathbf{R}_x\}\mathbf{B}^T + \sigma\mathbf{I} \end{aligned} \tag{18.51}$$

Performing real-valued eigenvalue decomposition of the matrix $\mathbf{R}$ produces bases for the beamspace signal and noise subspaces. Let $\mathbf{S}$ and $\mathbf{G}$ denote the two orthogonal matrices that respectively span the two subspaces, namely

$$\begin{aligned} \mathbf{S} &= \begin{bmatrix} \mathbf{s}_1 & \mathbf{s}_2 & \cdots & \mathbf{s}_K \end{bmatrix} \\ \mathbf{G} &= \begin{bmatrix} \mathbf{g}_{K+1} & \mathbf{g}_{K+2} & \cdots & \mathbf{g}_{M'} \end{bmatrix} \end{aligned} \tag{18.52}$$

where $\{\mathbf{s}_i\}$ and $\{\mathbf{g}_i\}$ are the eigenvectors. The RB-MUSIC spectrum is given by

$$\frac{1}{\mathbf{b}^T(\boldsymbol{\theta})\mathbf{G}\mathbf{G}^T\mathbf{b}(\boldsymbol{\theta})} \propto \frac{1}{\mathbf{v}^H(\varphi)\mathbf{J}_\varsigma(\mathbf{W}\mathbf{G}\mathbf{G}^T\mathbf{W}^H)\mathbf{J}_\varsigma\mathbf{v}(\varphi)} \tag{18.53}$$

which has peaks at $\boldsymbol{\theta} = \boldsymbol{\theta}_i$, $i = 1, 2, \ldots, K$ corresponding to the arrival angles of the source signals. Clearly two-dimensional searching is required to obtain the arrival angle estimates. Initial angle estimates are needed to start the iterative process.

18.1.2.3 Spatial Averaging

As mentioned in Sect. 18.1.1 by utilizing the symmetric configuration of the circular array, a centrosymmetric array can be produced. Due to the centrosymmetric property, the covariance matrix of the received data can be constructed in the form of a Hermitian persymmetric matrix. Then the spatial averaging approach can be applied to estimate the direction-of-arrival (DOA) of the incoming signal (Ye et al. 2007).

Consider the situation where N narrowband far-field incoherent signal sources impinge on the UCA. The output of the array can be described as

$$\mathbf{x}(t) = \mathbf{A}\mathbf{s}(t) + \mathbf{n}(t) \tag{18.54}$$

where $\mathbf{s}(t)$ is the source signal vector, $\mathbf{n}(t)$ is the zero-mean Gaussian noise vector, and $\mathbf{A}$ is the array manifold matrix defined by

$$\mathbf{A} = \begin{bmatrix} \mathbf{a}(\phi_1, \theta_1) & \mathbf{a}(\phi_2, \theta_2) & \cdots & \mathbf{a}(\phi_N, \theta_N) \end{bmatrix} \in R^{M \times N} \tag{18.55}$$

where

$$\mathbf{a}(\phi_i, \theta_i) = \begin{bmatrix} a_1(\phi_i, \theta_i) & a_2(\phi_i, \theta_i) & \cdots & a_M(\phi_i, \theta_i) \end{bmatrix}^T \tag{18.56}$$

is the complex array response for a source impinging from the angle pair (ϕ_i, θ_i), with its components given by

$$a_m(\phi_i, \theta_i) = \exp\left(j\frac{2\pi r}{\lambda}\cos(\phi_i - \alpha_m)\sin\theta_i\right) \tag{18.57}$$

Similar to the definitions in Sect. 18.1.1 and shown in Fig. 18.1, $\phi_i \in [0, 2\pi)$ is the azimuth angle and $\theta_i \in [0, \pi)$ is the elevation angle of the ith source signal. The position angle of the mth sensor of the array, α_m, is given by (18.1).

The covariance matrix of the output of the array is then given by

$$\begin{aligned}\mathbf{R_x} &= E[\mathbf{x}(t)\mathbf{x}^H(t)] \\ &= \mathbf{A}\mathbf{R_s}\mathbf{A}^H + \sigma^2\mathbf{I}\end{aligned} \tag{18.58}$$

where

$$\mathbf{R_s} = E[\mathbf{s}(t)\mathbf{s}^H(t)] \tag{18.59}$$

is the covariance matrix of the source signals, σ^2 is the noise power, and $\mathbf{I}$ is an identity matrix of size $M \times M$. Due to the centrosymmetry of the sensors positions

$$\alpha_{M/2+m} = \alpha_m + \pi \tag{18.60}$$

and thus

$$a_m(\phi_i, \theta_i) = a^*_{M/2+m}(\phi_i, \theta_i),\ m = 1, 2, \ldots, M/2 \tag{18.61}$$

Reordering these sensors as

$$\{1\ \ 2 \ldots M/2 \ldots M/2+2 \quad M/2+1\} \tag{18.62}$$

the directional vectors in (18.56) can be rewritten as

$$\begin{aligned}\mathbf{a}(\phi_i, \theta_i) = [\,a_1(\phi_i, \theta_i) \quad a_2(\phi_i, \theta_i) \quad \cdots \quad a_{M/2}(\phi_i, \theta_i) \quad a_M(\phi_i, \theta_i) \\ \cdots \quad a_{M/2+2}(\phi_i, \theta_i) \quad a_{M/2+1}(\phi_i, \theta_i)\,]^T\end{aligned} \tag{18.63}$$

Then it can be observed that

$$\mathbf{a}(\phi_i, \theta_i) = \mathbf{J}\mathbf{a}^*(\phi_i, \theta_i) \tag{18.64}$$

where $\mathbf{J}$ is the antidiagonal matrix with $[\mathbf{J}]_{i,M-i+1} = 1,\ i = 1, 2, \ldots, M$. Note that $\mathbf{a}^*$ denotes the conjugate of $\mathbf{a}$, whereas $\mathbf{a}^H$ is the conjugate transpose of $\mathbf{a}$. Equation (18.64) is equivalent to

$$\mathbf{A}(\phi_i, \theta_i) = \mathbf{J}\mathbf{A}^*(\phi_i, \theta_i) \tag{18.65}$$

Substituting (18.65) into (18.58) produces

$$\mathbf{R_x} = \mathbf{J}\mathbf{R_x^*}\mathbf{J} \tag{18.66}$$

In practice the covariance of the received signal is usually computed from a finite number of samples, namely

$$\begin{aligned}\hat{\mathbf{R}}_\mathbf{x} &= \frac{1}{L}\sum_{\ell=1}^{L}\mathbf{x}(\ell)\mathbf{x}^H(\ell) \\ &= \mathbf{R_x} + \boldsymbol{\Delta}\end{aligned} \tag{18.67}$$

where $\boldsymbol{\Delta}$ is the error matrix resulting from by using a finite number (L) of samples to calculate the covariance. From (18.58) and the first equation in (18.67), it is clear that both $\hat{\mathbf{R}}_\mathbf{x}$ and $\mathbf{R_x}$ are Hermitian matrices. Therefore $\boldsymbol{\Delta}$ is also a Hermitian matrix, that is, $\delta_{k,\ell} = [\boldsymbol{\Delta}]_{k,\ell} = \delta^*_{\ell,k}$. The estimation of the covariance matrix may be improved by using the spatial averaging, that is combining (18.66) and (18.67) yields

$$\begin{aligned}\hat{\mathbf{R}} &= \frac{1}{2}(\hat{\mathbf{R}}_\mathbf{x} + \mathbf{J}\hat{\mathbf{R}}_\mathbf{x}^*\mathbf{J}) \\ &= \mathbf{R_x} + \frac{1}{2}(\boldsymbol{\Delta} + \mathbf{J}\Delta^*\mathbf{J})\end{aligned} \tag{18.68}$$

Suppose that $\{\delta_{k,\ell}\}$ are independent identically distributed with zero-mean and variance σ_δ^2. Then each element of matrix $\frac{1}{2}(\boldsymbol{\Delta} + \mathbf{J}\Delta^*\mathbf{J})$ is a random variable of zero-mean and variance $\sigma_\delta^2/2$ except for the antidiagonal elements which have a variance equal to σ_δ^2. Thus the estimate of the covariance matrix using (18.68) is more accurate than that using (18.67).

18.1.3 ESPIRIT Method

Estimation of signal parameters via rotational invariance technique (ESPRIT) is another subspace-based approach for high-resolution signal parameter estimation. In this subsection two ESPRIT algorithms are described.

18.1.3.1 TLS Esprit

First, the *total least squares* ESPRIT (TLS EPRIT) method is described, with a focus on estimation of one-dimensional angles (Ottersten et al. 1991). One of the primary assumptions of the ESPRIT method is that the sensor array consists of two identical sub-arrays, which are displaced by a known translation vector $\boldsymbol{\Delta}_\mathbf{t}$. Each sub-array has m elements and can have arbitrary geometry and sensor responses.

The response of each sub-array to a unit wavefront from direction θ is still modeled by the complex column vector $\mathbf{a}(\theta) = [a_1(\theta) \quad a_2(\theta) \quad \cdots \quad a_N(\theta)]^T$ which is termed *steering vector*. The signals from the N sources at time t are again modeled by $\mathbf{s}(t)$ which is a column vector of length N. The signals are narrowband and have the same known center frequency. Then the outputs of the two sub-arrays can be described as

$$\begin{aligned}\mathbf{x}_1(t) &= [\mathbf{a}(\theta_1) \quad \mathbf{a}(\theta_2) \quad \cdots \quad \mathbf{a}(\theta_N)]\mathbf{s}(t) + \mathbf{n}_1(t) \\ &= \mathbf{A}\mathbf{s}(t) + \mathbf{n}_1(t) \\ \mathbf{x}_2(t) &= [\mathbf{a}(\theta_1)e^{j\phi_1} \quad \mathbf{a}(\theta_2)e^{j\phi_2} \quad \cdots \quad \mathbf{a}(\theta_N)e^{j\phi_N}]\mathbf{s}(t) + \mathbf{n}_2(t) \\ &= \mathbf{A}\mathbf{\Phi}\,\mathbf{s}(t) + \mathbf{n}_2(t)\end{aligned} \tag{18.69}$$

where $\mathbf{n}_1(t)$ and $\mathbf{n}_2(t)$ are the complex noise vectors and

$$\begin{aligned}\mathbf{A} &= [\mathbf{a}(\theta_1) \quad \mathbf{a}(\theta_2) \quad \cdots \quad \mathbf{a}(\theta_N)] \\ \mathbf{\Phi} &= \mathrm{diag}\{e^{j\phi_1}, \quad e^{j\phi_2}, \quad \ldots, \quad e^{j\phi_N}\} \\ \phi_k &= 2\pi|\mathbf{\Delta}_t|\sin\theta_k\end{aligned} \tag{18.70}$$

where $|\mathbf{\Delta}_t|$ is normalized to the wavelength. Equation (18.69) can be written in a compact form as

$$\begin{aligned}&\mathbf{x}(t) = \mathbf{G}\,\mathbf{s}(t) + \mathbf{n}(t) \\ &\mathbf{x}(t) = \begin{bmatrix}\mathbf{x}_1(t) \\ \mathbf{x}_2(t)\end{bmatrix}, \ \mathbf{G} = \begin{bmatrix}\mathbf{A} \\ \mathbf{A}\mathbf{\Phi}\end{bmatrix}, \ \mathbf{n}(t) = \begin{bmatrix}\mathbf{n}_1(t) \\ \mathbf{n}_2(t)\end{bmatrix}\end{aligned} \tag{18.71}$$

The purpose of the analysis is to determine the angles of arrival of the N source signals when giving L snapshots $\mathbf{x}(1)$, $\mathbf{x}(2)$, …, $\mathbf{x}(L)$. Then the sample covariance matrix is first computed by

$$\hat{\mathbf{R}}_\mathbf{x} = \frac{1}{L}\sum_{t=1}^{L}\mathbf{x}(t)\mathbf{x}^H(t). \tag{18.72}$$

Performing eigendecomposition of the sample covariance matrix produces

$$\begin{aligned}\hat{\mathbf{R}}_\mathbf{x} &= [\hat{\mathbf{E}}_\mathrm{s} \quad \hat{\mathbf{E}}_\mathbf{n}]\begin{bmatrix}\hat{\Lambda}_\mathrm{s} & \mathbf{0} \\ \mathbf{0} & \hat{\Lambda}_\mathbf{n}\end{bmatrix}\begin{bmatrix}\hat{\mathbf{E}}_\mathrm{s}^H \\ \hat{\mathbf{E}}_\mathbf{n}^H\end{bmatrix} \\ &= \hat{\mathbf{E}}_\mathrm{s}\hat{\Lambda}_\mathrm{s}\hat{\mathbf{E}}_\mathrm{s}^H + \hat{\mathbf{E}}_\mathbf{n}\hat{\Lambda}_\mathbf{n}\hat{\mathbf{E}}_\mathbf{n}^H\end{aligned} \tag{18.73}$$

Define the cost function as

$$||\hat{\mathbf{E}}_\mathbf{s} - \mathbf{GT}||_F^2 \tag{18.74}$$

where $||.||_F$ refers to the Frobenius norm which is defined as the square root of the sum of the absolute squares of its elements, and $\mathbf{T}$ is some nonsingular matrix. The angle estimates are then determined by minimizing the cost function with respect to $\mathbf{G}$ and $\mathbf{T}$. By defining

$$\begin{aligned} \mathbf{B} &= \mathbf{AT} \\ \mathbf{\Psi} &= \mathbf{T}^{-1}\mathbf{\Phi}\mathbf{T} \end{aligned} \tag{18.77}$$

the minimization problem becomes

$$\min_{\mathbf{B},\mathbf{\Psi}} \left\| \begin{bmatrix} \hat{\mathbf{E}}_{s1} \\ \hat{\mathbf{E}}_{s2} \end{bmatrix} - \begin{bmatrix} \mathbf{B} \\ \mathbf{B}\mathbf{\Psi} \end{bmatrix} \right\|_F^2 \tag{18.76}$$

where

$$\begin{bmatrix} \hat{\mathbf{E}}_{s1} \\ \hat{\mathbf{E}}_{s2} \end{bmatrix} = \hat{\mathbf{E}}_s \tag{18.77}$$

Clearly $\{\mathbf{\Phi}\}_{i,i}$, $i = 1, 2, \ldots, N$ are the eigenvalues of $\mathbf{\Psi}$ and $\mathbf{T}$ is the eigenvector matrix of $\mathbf{\Psi}$. The TLS estimate of $\mathbf{\Psi}$ is given by

$$\mathbf{\Psi}_{\text{TLS}} = -\mathbf{V}_{12}\mathbf{V}_{22}^{-1} \tag{18.78}$$

where $\mathbf{V}_{12}$ and $\mathbf{V}_{22}$ are obtained by the eigendecomposition

$$\begin{bmatrix} \hat{\mathbf{E}}_{s1}^H \\ \hat{\mathbf{E}}_{s2}^H \end{bmatrix} [\hat{\mathbf{E}}_{s1} \quad \hat{\mathbf{E}}_{s2}] = \begin{bmatrix} \mathbf{V}_{11} & \mathbf{V}_{12} \\ \mathbf{V}_{21} & \mathbf{V}_{22} \end{bmatrix} \mathbf{Q} \begin{bmatrix} \mathbf{V}_{11}^H & \mathbf{V}_{21}^H \\ \mathbf{V}_{12}^H & \mathbf{V}_{22}^H \end{bmatrix} \tag{18.79}$$

where $\mathbf{Q}$ is a diagonal matrix. Denoting the phase of $\{\mathbf{\Phi}\}_{i,i}$ as $\angle\{\mathbf{\Phi}\}_{i,i}$, the estimates of the angles are given by

$$\hat{\theta}_i = \sin^{-1}\frac{\angle\{\mathbf{\Phi}\}_{i,i}}{2\pi|\mathbf{\Delta}_t|}, \; i = 1, 2, \cdots, N \tag{18.80}$$

In the case of uniform linear arrays, the array manifold is a Vandermonde matrix

$$\mathbf{A}_{\text{ULA}} = \begin{bmatrix} a_1 & a_2 & \cdots & a_N \\ a_1 e^{j\phi_1} & a_2 e^{j\phi_2} & \cdots & a_N e^{j\phi_N} \\ \vdots & \vdots & \ddots & \vdots \\ a_1 e^{j(M-1)\phi_1} & a_2 e^{j(M-1)\phi_2} & \cdots & a_N e^{j(M-1)\phi_N} \end{bmatrix} \tag{18.81}$$

which manifests multiple invariances so that different choices of sub-arrays can be made. Let the distance between the two identical sub-arrays in a ULA be k times the element spacing. The selection of a particular sub-array can be described by a selection matrix. For instance, when choosing m rows of $\mathbf{A}_{\mathrm{ULA}}$ for the first sub-array, the selection matrix becomes $\begin{bmatrix} \mathbf{J}_{m\times(M-k)} & \mathbf{0}_{m\times k} \end{bmatrix}$. Then the selection matrix for the second sub-array is given by $\begin{bmatrix} \mathbf{0}_{m\times k} & \mathbf{J}_{m\times(M-k)} \end{bmatrix}$. Accordingly a matrix similar to the matrix $\mathbf{G}$ in (18.71) can be produced as

$$\tilde{\mathbf{J}}\mathbf{A}_{\mathrm{ULA}} = \begin{bmatrix} \mathbf{A} \\ \mathbf{A}\bar{\Phi}^k \end{bmatrix} = \begin{bmatrix} \mathbf{A} \\ \mathbf{A}\Phi \end{bmatrix} \tag{18.82}$$

where the matrix $\mathbf{A}$ contains rows of $\mathbf{A}_{\mathrm{ULA}}$ that are selected by the selection matrix $\begin{bmatrix} \mathbf{J}_{m\times(M-k)} & \mathbf{0}_{m\times k} \end{bmatrix}$ and

$$\tilde{\mathbf{J}} = \begin{bmatrix} \mathbf{J}_{m\times(M-k)} & \mathbf{0}_{m\times k} \\ \mathbf{0}_{m\times k} & \mathbf{J}_{m\times(M-k)} \end{bmatrix}$$

$$\bar{\Phi} = \mathrm{diag}\left\{ e^{j\phi_1}, \quad e^{j\phi_2}, \quad \cdots, \quad e^{j\phi_N} \right\}.$$

One simple and natural way of configuring two sub-arrays is the overlapping sub-arrays as shown in Fig. 18.5. In this case $m = M - k$ and the sub-matrix $\mathbf{J}_{m\times(M-k)}$ is an identity matrix. Clearly the two sub-arrays share no elements if $m \geq M/2$.

Thus giving the selection matrices the angle estimation for a ULA follows the same procedure as from (18.72) to (18.80). However, it is worth noting some differences. The output from a ULA can be described as

$$\mathbf{x}_{\mathrm{ULA}}(t) = \mathbf{A}_{\mathrm{ULA}}\,\mathbf{s}(t) + \mathbf{n}(t)$$

A sufficient number of output samples is first collected and then applying (18.72) produces the sample covariance matrix. The eigendecomposition of the covariance matrix is then performed according to (18.73) to generate $\hat{\mathbf{E}}_s$. Since the TLS ESPRIT problem becomes $\min_{\mathbf{G},\mathbf{T}} ||\tilde{\mathbf{J}}\hat{\mathbf{E}}_s - \mathbf{G}\mathbf{T}||_F^2 \hat{\mathbf{E}}_{s1}$ and $\hat{\mathbf{E}}_{s2}$ in (18.76) and (18.79) are replaced by $\begin{bmatrix} \mathbf{J}_{m\times(M-k)} & \mathbf{0}_{m\times k} \end{bmatrix}$ and $\begin{bmatrix} \mathbf{0}_{m\times k} & \mathbf{J}_{m\times(M-k)} \end{bmatrix}$, respectively.

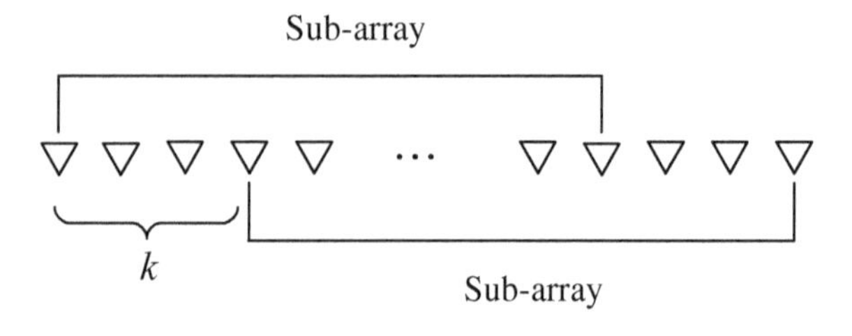

Fig. 18.5 Two overlapping sub-arrays formed from a uniform linear array. The two sub-arrays are displaced by k times the element spacing

18.1.3.2 UCA-ESPRIT

The UCA-ESPRIT algorithm is not based on the displacement invariance array structure, but makes use of a recursive relationship between Bessel functions (Mathews and Zoltoski 1991). The method provides paired source azimuth and elevation angle estimates in closed form through eigenvalue decomposition. This algorithm has low computational complexity since it does not require complicated spectral searches, iterative multidimensional optimization, or a pairing procedure for associating azimuth estimates with elevation estimates. In some circumstance the results from the UCA-ESPRIT algorithm can be employed as initial estimates of the arrival angles for more complex methods such as the RB-MUSIC described in Sect. 18.1.2.2.

Like the other subspace based algorithms the UCA-ESPRIT uses the covariance matrix of the output of the array elements, based on a certain number of sample data snapshots as described previously. Let $\hat{\mathbf{R}}$ denote the array sample covariance matrix. Then the sample beamspace covariance matrix is formed by

$$\hat{\mathbf{R}}_y = \mathbf{F}_r^H \hat{\mathbf{R}} \mathbf{F}_r \tag{18.83}$$

where $\mathbf{F}_r^H$ is the transforming matrix as defined in section "Beamforming Matrices and Manifold Vectors". The real-valued eigenvalue decomposition is then carried out for the real part of $\hat{\mathbf{R}}_y$, namely $\mathrm{Re}\{\hat{\mathbf{R}}_y\}$, to generate $M' = 2M - 1$ ordered eigenvalues as

$$\hat{\lambda}_1 \geq \hat{\lambda}_2 \geq \cdots \geq \hat{\lambda}_{M'} \tag{18.84}$$

and the corresponding eigenvectors, $\hat{\mathbf{s}}_1, \ldots, \hat{\mathbf{s}}_\eta, \hat{\mathbf{g}}_{\eta+1}, \ldots, \hat{\mathbf{g}}_{M'}$. The eigenvectors based matrices

$$\begin{aligned} \hat{\mathbf{S}} &= [\hat{\mathbf{s}}_1 \quad \hat{\mathbf{s}}_2 \quad \cdots \quad \hat{\mathbf{s}}_\eta] \\ \hat{\mathbf{G}} &= [\hat{\mathbf{g}}_{\eta+1} \quad \hat{\mathbf{g}}_{\eta+2} \quad \cdots \quad \hat{\mathbf{g}}_{M'}] \end{aligned} \tag{18.85}$$

span the estimated signal and noise subspaces, respectively. A phase mode excitation based beamforming matrix is generated according to

$$\mathbf{F}_o^H = \mathbf{C}_o \mathbf{F}_e^H \tag{18.86}$$

where $\mathbf{F}_e^H$ is defined in section "Beamforming Matrices and Manifold Vectors" and

$$\mathbf{C}_o = \mathrm{diag}\left\{(-1)^M, \ldots, \quad (-1)^1, \quad 1, \quad 1, \ldots, 1\right\} \tag{18.87}$$

The beamspace manifold is then produced by

$$\begin{aligned} \mathbf{a}_o(\boldsymbol{\theta}) &= \mathbf{F}_o^H \mathbf{a}(\boldsymbol{\theta}) \\ &= \sqrt{N} \mathbf{J}_{\varsigma -} \mathbf{v}(\varphi) \end{aligned} \tag{18.88}$$

where

$$\mathbf{J}_{\varsigma -} = \operatorname{diag}\{J_{-M}(\varsigma), \ldots, J_{-1}(\varsigma), J_0(\varsigma), J_1(\varsigma), \ldots, J_M(\varsigma)\} \tag{18.89}$$

Three vectors of size $M_e = M' - 2$ are extracted from the beamspace manifold as

$$\mathbf{a}^{(i)} = \Delta^{(i)} \mathbf{a}_o(\boldsymbol{\theta}),\ i = -1,\ 0,\ 1 \tag{18.90}$$

where $\{\Delta^{(i)}\}$ are the selection matrices which pick out the first, middle, and last M_e elements from $\mathbf{a}_o(\boldsymbol{\theta})$. Using the recursive relationship of the Bessel function

$$J_{m-1}(\varsigma) + J_{m+1}(\varsigma) = \frac{2m}{\varsigma} J_m(\varsigma) \tag{18.91}$$

the three vectors have the relationship as

$$\Gamma \mathbf{a}^{(0)} = \mu \mathbf{a}^{(-1)} + \mu^* \mathbf{a}^{(1)} \tag{18.92}$$

where

$$\begin{aligned} \mu &= \sin\theta\, e^{j\phi} \\ \boldsymbol{\Gamma} &= \frac{\lambda}{\pi r} \operatorname{diag}\{\ -(M-1),\ \ \ldots,\ \ -1,\ \ 0,\ \ 1,\ \ \ldots,\ \ M-1\,\} \end{aligned} \tag{18.93}$$

where r is the radius of the circular array. Let $\mathbf{A}_o = \mathbf{F}_o^H \mathbf{A}$ denote the beamspace AOA matrix, and let $\mathbf{S}_0$ span $\mathrm{Re}\{\mathbf{A}_o\}$. Since $\mathbf{S}$ spans $\mathrm{Re}\{\mathbf{B}\}$ such that $\mathbf{S} = \mathbf{BT}$ where $\mathbf{T}$ is real nonsingular matrix, $\mathbf{B} = \mathbf{F}_r^H \mathbf{A}$ and $\mathbf{F}_o^H = \mathbf{C}_o \mathbf{W} \mathbf{F}_r^H$, it can be observed that

$$\begin{aligned} \mathbf{S}_o &= \mathbf{C}_o \mathbf{WS} \\ &= \mathbf{C}_o \mathbf{WBT} \\ &= \mathbf{A}_o \mathbf{T} \end{aligned} \tag{18.94}$$

Similarly to forming $\mathbf{a}^{(i)}$ from $\mathbf{a}_o(\boldsymbol{\theta})$, the AOA matrix $\mathbf{A}_o$ and the subspace matrix $\mathbf{S}_0$ can be portioned as

$$\begin{aligned} \mathbf{A}^{(i)} &= \Delta^{(i)} \mathbf{A}_o,\ i = -1,\ 0,\ 1 \\ \mathbf{S}^{(i)} &= \Delta^{(i)} \mathbf{S}_o \end{aligned} \tag{18.95}$$

It can be shown that

$$\mathbf{A}^{(1)} = \mathbf{D}\tilde{\mathbf{I}}\mathbf{A}^{(-1)*} \tag{18.96}$$

where

$$\mathbf{D} = \mathrm{diag}\{\, (-1)^{M-2}, \quad \ldots, \quad (-1)^1, \quad (-1)^0, \quad (-1)^1, \quad \ldots, \quad (-1)^M \} \tag{18.97}$$

From (18.94) and (18.95), it can be seen that

$$\mathbf{S}^{(i)} = \mathbf{A}^{(i)}\mathbf{T} \tag{18.98}$$

so that

$$\mathbf{S}^{(1)} = \mathbf{D}\tilde{\mathbf{I}}\mathbf{S}^{(-1)*} \tag{18.99}$$

From (18.92), it is easy to show that

$$\Gamma\mathbf{A}^{(0)} = \mathbf{A}^{(-1)}\mathbf{\Phi} + \mathbf{A}^{(1)}\mathbf{\Phi}^* \tag{18.100}$$

where

$$\mathbf{\Phi} = \mathrm{diag}\{\ \mu_1, \quad \mu_2, \quad \cdots, \quad \mu_\eta\,\} \tag{18.101}$$

Equation (18.100) can be written in terms of signal subspace matrices as

$$\Gamma\mathbf{S}^{(0)} = \mathbf{S}^{(-1)}\mathbf{\Psi} + \mathbf{D}\tilde{\mathbf{I}}\mathbf{S}^{(-1)*}\mathbf{\Psi}^* \tag{18.102}$$

where

$$\mathbf{\Psi} = \mathbf{T}^{-1}\mathbf{\Phi}\mathbf{T} \tag{18.103}$$

Writing (18.102) in block matrix yields

$$\tilde{\mathbf{S}}\tilde{\Psi} = \Gamma\mathbf{S}^{(0)} \tag{18.104}$$

where

$$\begin{aligned} \tilde{\mathbf{S}} &= [\,\mathbf{S}^{(-1)} \quad \mathbf{D}\tilde{\mathbf{I}}\mathbf{S}^{(-1)*}\,] \\ \tilde{\Psi} &= [\,\mathbf{\Psi}^T \quad \mathbf{\Psi}^H\,]^T \end{aligned} \tag{18.105}$$

If $M_e > 2\eta$, the system given by (18.104) is over-determined so that there exists a unique solution for $\tilde{\Psi}$, or equivalently a unique solution for $\mathbf{\Psi}$. Since

$$\mathbf{\Phi} = \mathbf{T}\,\mathbf{\Psi}\,\mathbf{T}^{-1} \tag{18.106}$$

the eigenvalues of $\mathbf{\Psi}$ corresponds to the diagonal elements of $\mathbf{\Phi}$, which are

$$\mu_i = \sin\theta_i\, e^{j\phi_i},\ i = 1, 2, \ldots, \eta \tag{18.107}$$

Consequently, the paired source azimuth and elevation angle estimates are automatically produced.

18.2 Beamforming for AOA Estimation

As mentioned at the beginning of the chapter, positioning accuracy is significantly affected by the signal parameter estimation, such as the TOA or pseudorange. Parameter estimation accuracy is often dominated by the signal-to-noise ratio (SNR). Certainly increasing transmission power will usually improve parameter estimation accuracy and thus the position determination accuracy. However, the transmission power level is constrained by regulations to limit interference to other systems in general. In addition, it is not a wise option to increase power level for a SNR gain when the transmitter device is battery-powered. In many circumstances such as in sensor networks, recharging or changing the battery of hundreds of or even thousands of sensor nodes is costly or impractical, lower power consumption is crucial for battery-powered sensor nodes.

A different way to increase the SNR is to apply beamforming at the transmitter, receiver or both. At the transmitter beamforming focuses most of the signal power into a beam which points in the desired direction, whereas the power in side lobes pointing at other directions is minimized. At the receiver the desired incoming signal at a specific angle is maximized, whereas the interference coming from other directions is dramatically suppressed through beamforming. As a result, the accuracy of the parameter estimation can be greatly enhanced. Also by employing beamforming the effect of non-line-of-sight (NLOS) propagation conditions on position determination in indoor environments can be considerably mitigated, because a direct-path signal can be more readily detected even under NLOS conditions.

Beamforming can be broadly divided into three different categories: analog, digital, and hybrid beamforming. Analog beamforming typically makes use of phase shifters to compensate for the phase differences between the antenna elements. This usually takes place in the radio-frequency (RF) part of the system, including the intermediate-frequency (IF) components. Digital beamforming is generally applied to baseband signals and the element signal weightings are typically generated adaptively based on some specific criterion. Hybrid beamforming is realized by performing both analog and digital beamforming. In this section basic beamforming concepts and joint beamforming and AOA estimation approaches are presented and analyzed.

18.3 Basic Concepts

In wireless communications when the bandwidth of a signal does not considerably exceed the channel coherence bandwidth, the signal is referred to as narrowband. Otherwise the signal is classified as wideband. Channel fading for narrowband signals is treated as *flat fading*, namely there is uniform fading across the signal bandwidth. Alternatively, for wideband signals it is termed *frequency-selective fading*, as the signal fades in only part of the signal bandwidth. Further, the *fractional bandwidth* of a signal is defined as the bandwidth divided by the center frequency, namely

$$\mathrm{FBW} = \frac{f_H - f_L}{f_0} \tag{18.108}$$

where f_H is the upper frequency, f_L is the lower frequency (both measured at the −3 dB points of the signal spectrum), and f_0 is the center frequency. Alternatively, a signal is referred to as narrowband if its fractional bandwidth is less than 1%, and wideband if greater than 10%.

A simplified diagram of the typical digital beamformer for narrowband signals is shown in Fig. 18.6. The signal impinges on the antenna elements at an angle θ with respect to the array normal. Note that the angle is treated as the same when the signal comes from the far field source, whereas the angles are different when

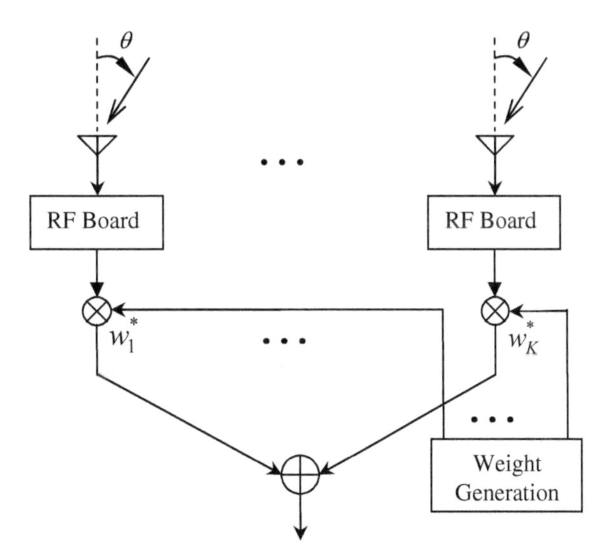

Fig. 18.6 Simplified block diagram of adaptive digital beamformer

coming from a near field source. In the case of far-field signals, the output of the kth element can be described as

$$x_k(t) = s(t)\exp\left(j\frac{2\pi}{\lambda}kd\sin\theta\right) + v_k(t), \quad 0 \le k < K \tag{18.109}$$

where $s(t)$ is the signal input at the antenna array, K is the number of antenna elements of the array, d is the inter-element spacing, λ is the wavelength of the signal, and $v_m(t)$ is the additive noise. After processed through the RF components and converted to baseband, the signal from each antenna element is multiplied by a complex coefficient of the adaptive filter.

There are a variety of adaptive beamforming algorithms which can be used to generate the weighting coefficients. A simple adaptive beamforming algorithm is analyzed in the following subsection.

18.3.1 Simple Beamforming Method

Beamforming algorithms can be divided into two broad categories: *temporal* and *spatial* methods. The advantage of the temporal algorithms is that they are simple, whereas the drawback is that a training sequence or pilot symbols are required. On the other hand, spatial algorithms are intended to estimate the AOA of the incoming signal. Comparatively the spatial reference algorithms are more complex, but the AOA information is explicitly provided. However, it would be desirable to have an algorithm which has the advantages of both the temporal reference and the spatial reference algorithms. The iterative beam steering (IBS) algorithm achieves the objective, that is low computational complexity and providing AOA information (Guo 2004). In the IBS algorithm the weighting coefficient vector $\mathbf{w}(n) = \left[w_0(n) \quad w_1(n) \quad \cdots \quad w_{K-1}(n)\right]^T$ is iteratively produced according to

$$\mathbf{w}(n) = \alpha\mathbf{w}(n-1) + (1-\alpha)s^H(n)\mathbf{x}(n) \tag{18.110}$$

where α is the *forgetting factor* that is used to control the effective window size of the averaging process, $s(n)$ is the reference signal, and

$$\mathbf{x}(n) = \left[x_0(n) \quad x_1(n) \quad \cdots \quad x_{K-1}(n)\right]^T \tag{18.111}$$

is the signal vector input to the adaptive filter. It is assumed that a perfect reference signal is obtainable. This assumption is reasonable since a training sequence is usually used in practice. The training sequence may simply be the header in a data packet, as the bit error rate of a data receiver is typically very low, for instance 10^{-4}.

In practice to reduce the sensitivity of the beamformer to the magnitude of the input signal, a normalized IBS algorithm can be employed. That is in (18.110) the input signal vector is normalized, resulting in

$$\mathbf{w}(n) = \alpha \mathbf{w}(n-1) + (1-\alpha)s^*(n)\frac{\mathbf{x}(n)}{||\mathbf{x}(n)||^2}. \tag{18.112}$$

The output of the digital beamformer is then given by

$$\mathbf{z}(n) = \mathbf{w}^H(n)\mathbf{x}(n) \tag{18.113}$$

where $\mathbf{w}^H(n)$ denotes the conjugate transpose of $\mathbf{w}(n)$.

Suppose that a plane wave impinges on a linear array with an angle denoted by θ with respect to the array normal as shown in Fig. 18.6. The antenna array factor can be described as

$$F(\theta) = \sum_{n=0}^{N-1} \exp\left(j\frac{2\pi f}{c} nd \sin\theta\right) \times w_n^* \tag{18.114}$$

where w_n^* is the conjugate of w_n which is adaptively generated under one specific AOA. Figures 18.7 and 18.8 show two examples of the beam pattern in terms of the amplitude of the array factor when the number of elements of the linear array is set at 3 and 5, respectively. The coefficients are generated by using the IBS algorithm. The true AOA is equal to 20° and the SNR is set at 20 dB. It is evident that the transmit power will be focused on the desired direction, or the desired signal coming along the main beam will be strengthened while the interfering signals from other directions will be suppressed. As the number of elements increases, the antenna gain is more (narrower beamwidth). The *antenna gain* is computed for these two cases. The antenna gain is defined as the multiplication of the *directive gain* and the transmit pattern of each element. When considering an omni-directional antenna, the transmit pattern is uniform (pattern gain of unity). The directive gain of a linear array is calculated by

$$D(\theta_0) = \frac{4\pi|F(\theta_0)|^2}{\int_0^{2\pi}|F(\theta)|^2 d\theta} \tag{18.115}$$

where θ_0 is the desired angle. The computed directive gain is 26.5 dB and 28.3 dB respectively for the 3-element and the 5-element linear arrays. Therefore, the 5-element array produces 1.8 dB more directive gain than the 3-element array.

18.3.2 *Joint Beamforming and AOA Estimation*

When the IBS algorithm is applied for the adaptation of the coefficient, the AOA information can be directly extracted from the coefficients of the adaptive filter.

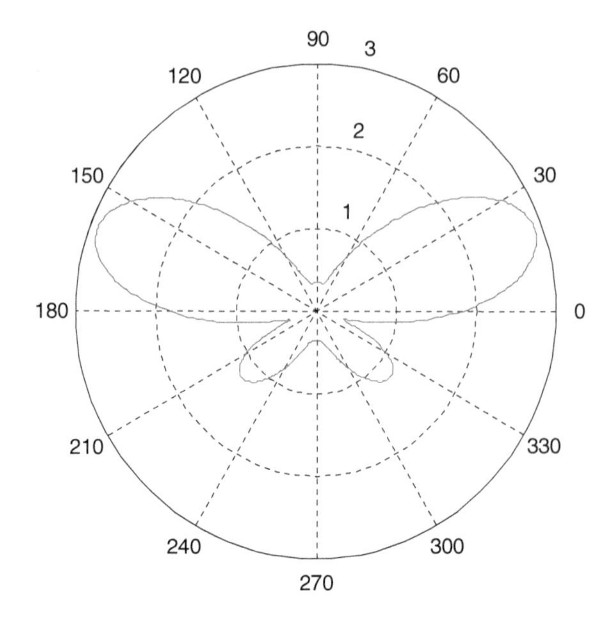

Fig. 18.7 Beam pattern of the adaptive array with three elements

A detailed description on how to estimate the AOA based on the coefficients is as follows. Let the complex coefficient vector be

$$w = [a_1 + jb_1 \quad a_2 + jb_2 \quad \cdots \quad a_K + jb_K]^T \tag{18.116}$$

where a_i and b_i are the real and imaginary components of the ith coefficient of the adaptive filter, and the number of the sub-arrays is set at K. Take the first element as the reference, so that the phase of the signal impinging on the center of the sub-array is equal to zero. Then the distance induced phase of the signal impinging on the ith element is given by

$$\alpha_i = \frac{2\pi}{\lambda}(i - 1)d \sin\theta \tag{18.117}$$

As mentioned previously the purpose of the digital beamformer is to remove these phases, that is the phases of the coefficients of the adaptive filter are designed to balance the phases $\{\alpha_i\}$. Generally speaking, the coefficients are associated with the AOA by

$$\frac{a_i + jb_i}{\sqrt{a_i^2 + b_i^2}} = \exp(j\alpha_i) + \varepsilon_i \tag{18.118}$$

where ε_i is the coefficient error caused by noise and the error incurred by adaptation and processing. Let $\phi_i \in [0, 2\pi)$ be the phase of the complex number $a_i + jb_i$. Then we have

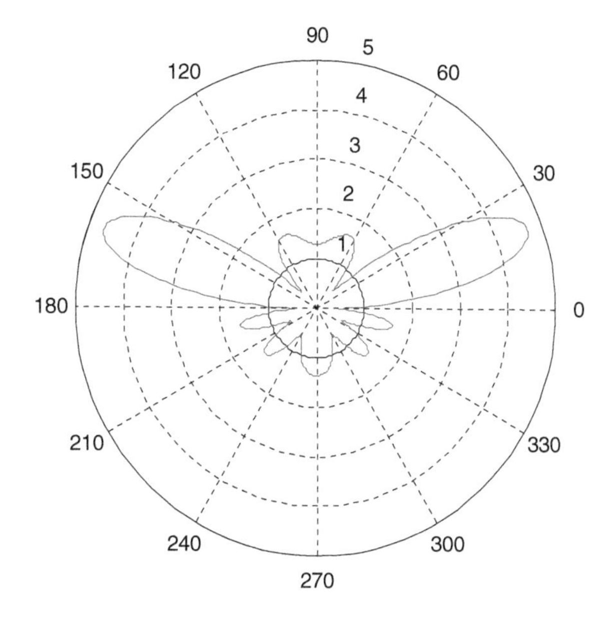

Fig. 18.8 Beam pattern of the adaptive array with five elements

$$\phi_i = \begin{cases} \tan^{-1}\frac{b_i}{a_i}, & a_i > 0,\ b_i \geq 0 \\ \pi - \tan^{-1}\frac{b_i}{|a_i|}, & a_i < 0,\ b_i \geq 0 \\ \pi + \tan^{-1}\frac{|b_i|}{|a_i|}, & a_i < 0,\ b_i \leq 0 \\ 2\pi - \tan^{-1}\frac{|b_i|}{a_i}, & a_i > 0,\ b_i \leq 0 \end{cases} \tag{18.119}$$

where $|x|$ denotes taking the absolute value of x. By ignoring any measurement error, it can be shown that

$$\phi_i \approx \text{mod}\left(\frac{2\pi}{\lambda}(i-1)d\sin\theta,\ 2\pi\right) \tag{18.120}$$

where $\text{mod}(x, y)$ denotes the modulus after dividing x by y. To prevent grating lobes, the inter-element spacing must be constrained to be less than or equal to half the wavelength of the signal, resulting in

$$0 \leq \frac{2\pi}{\lambda} d\sin\theta \leq \pi \tag{18.121}$$

The AOA can be estimated from the coefficients of the adaptive receiver, which contain the AOA information as shown in (18.119) and (18.120). Note that ϕ_1 does not provide useful angle information since it is set as the reference. More specifically ϕ_2 remains unchanged, whereas ϕ_i, $i > 2$ obtained from (18.119) need to be corrected as indicated by (18.120); this correction is made as follows. Starting from ϕ_3, perform

$$\phi_i + 2\pi \Rightarrow \phi_i, \text{ if } \phi_i < \phi_{i-1} \tag{18.122}$$

This process repeats until $\phi_i > \phi_{i-1}$. It can be observed that the phase difference between any two neighboring elements is the same, given by $\frac{2\pi}{\lambda} d \sin\theta$. In the presence of noise and errors, the differential phases would also be similar. The purpose is to make use of the differential phases to determine the AOA. The sequence of the differential phases is calculated according to

$$\delta\phi_{i,i-1} = \phi_i - \phi_{i-1}, i \geq 2 \tag{18.123}$$

In the case of $i = 2$, $\delta\varphi_{2,1} = \phi_2$. The next step is to remove those differential phases which are much greater than the mean differential phase, which is given by

$$\delta\phi = \sum_{i=2}^{M} \delta\phi_{i,i-1} \tag{18.124}$$

These abnormal differential phases are usually produced by the operation in (18.122) when the true phases are very close to $2i\pi, i \geq 0$. In this case, the neighboring phases are very similar and close to zero. In the presence of noise, the operation by (18.122) may mistakenly add 2π to the phases. To reduce the error caused by such a wrong operation, the abnormal differential phases should be identified and then discarded. The differential phase is treated as abnormal if

$$\delta\varphi_{i,i-1} - \delta\phi > \rho \times \delta\phi, 0 < \rho < 1 \tag{18.125}$$

The constant ρ can be chosen in advance such as based on a simulation or experimental results, but simulations suggest a suitable value is 0.4. After removal the abnormal differential phases, a new mean of the differential phases is then computed by

$$\delta\phi = \frac{1}{J} \sum_{j} \delta\varphi_{j,j-1} \tag{18.126}$$

where $J \leq M$ is the number of the remaining differential phases after discarding the abnormal ones. As a consequence, the AOA estimate is obtained as

$$\hat{\theta} = \sin\left(\frac{\delta\phi}{\frac{2\pi}{\lambda} d}\right) \tag{18.127}$$

This angle estimate is then employed for beamforming and the procedure repeats until the difference between two neighboring angle estimates is sufficiently small or the number of iterations reaches a pre-defined value.

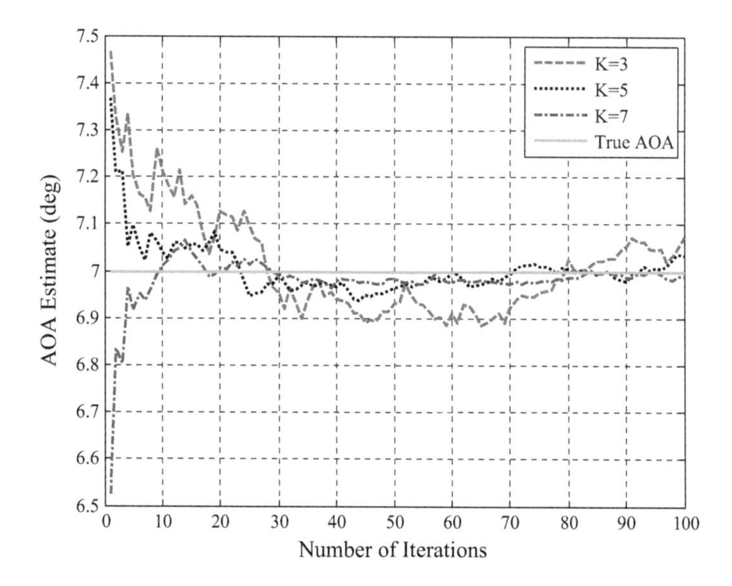

Fig. 18.9 AOA estimates versus number of iterations. The number of the elements of the array is set at 3, 5, and 7. The true AOA equals 7° and the forgetting factor is set at 0.98

To further reduce the estimation error when the true AOA is equal or very close to zero, one more step may be required. That is all the coefficients of the adaptive filter are set to unity and used to calculate the output of the beamformer. If the real part of the output with unit coefficients is greater than the estimated angle in (18.127), the angle estimate of zero degree is accepted. Otherwise the angle estimate given by (18.127) is selected.

To illustrate the above procedure a simulation is set up as follows. The carrier frequency is set at 2.4 GHz and the antenna inter-element spacing is set at half the wavelength 6.2 cm to avoid both correlation and grating lobes. The SNR is set at 20 dB and three different numbers (3, 5, and 7) of antenna array elements are used in the simulation. Figures 18.9 and 18.10 show the results of the AOA estimation when the true AOA is 7 and 70°. The forgetting factor is set at 0.98 and the initial values of the complex coefficients of the adaptive filter are simply set at zero, except for the first coefficient which is set at unity. It can be observed that the speed of convergence is better as the number of elements increases. When the number of elements is greater than three, the process approaches steady-state in around 10 iterations. The accuracy also generally increases with the number of elements. In the case of five elements, the estimation error is limited below 0.2°.

Figure 18.11shows the results when the forgetting factor is set at 0.9. Compared with the results in Fig. 18.10, it can be observed that a larger forgetting factor performs better than a smaller factor. This is usually true for adaptive filtering with forgetting factors. A larger forgetting factor is more sensitive to the input data, causing larger variations in the estimation as the signal varies. Therefore, when the transmitter and the receiver are static or moving very slowly, a forgetting factor closer to unity could

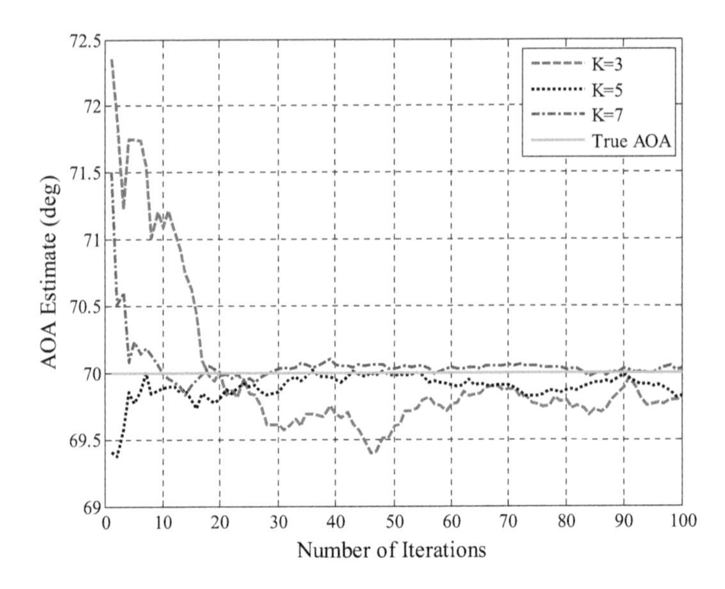

Fig. 18.10 AOA estimates versus number of iterations. The number of the elements of the array is set at 3, 5, and 7. The true AOA equals 70° and the forgetting factor is set at 0.98

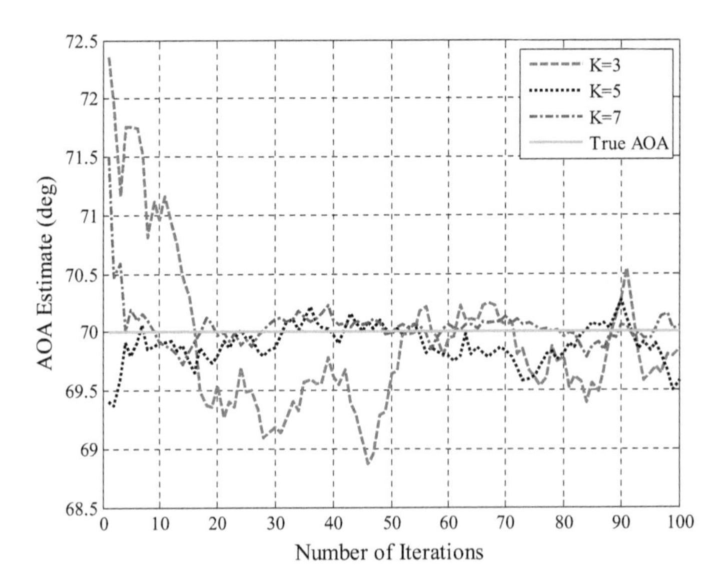

Fig. 18.11 AOA estimates versus number of iterations. The number of the elements of the array is set at 3, 5, and 7. The true AOA equals 70° and the forgetting factor is set at 0.9

be selected. On the other hand, when the target is moving quickly so that the AOA varies substantially with time, the forgetting factor should be chosen smaller such as between 0.9 and 0.98, depending on the variation rate of the angle.

In reality radio signal propagation environments can be complex, such as inside buildings. In such adverse environments, multipath fading and non-line-of-sight (NLOS) propagation can be significant, resulting in accuracy degradation compared to the results in Figs. 18.9 and 18.10.

18.4 AOA-Assisted Position Determination

In this section the basic idea of using AOA estimates to determine the target position is described. The first basic technique makes use of AOA estimates alone, whereas the second approach utilizes both AOA and TOA estimates.

18.4.1 AOA-Only Based Method

When the AOA of the target is measured at base stations, the position of the target can be determined purely by using the angle information. In the case of two-dimensional positioning, the minimum requirement is the azimuth angle measurements from two base stations, subject to the constraint that the two base stations and the target do not lie in a straight line. Figure 18.12 illustrates how to make use of two angle measurements to locate the target.

For three-dimensional positioning using AOA, two-dimensional angle information is required, namely the azimuth and elevation angles. Figure 18.13 shows spatial relationship between one base station and the target for three-dimensional positioning. To obtain a unique solution for the target position, a minimum three base stations are required. In the special case where the base station antenna is far above the mobile antenna, the minimum number of base stations for a unique solution becomes two. The basic way to determine the target position is as follows. Denote the unknown mobile position as (x, y, z) and the known position of the ith

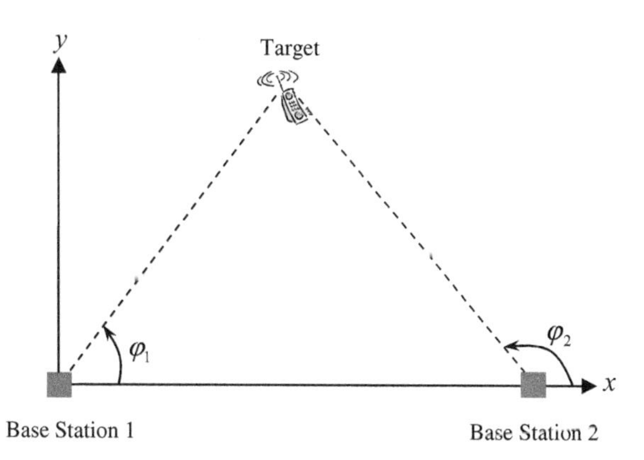

Fig. 18.12 Illustration of 2-D positioning by using AOA

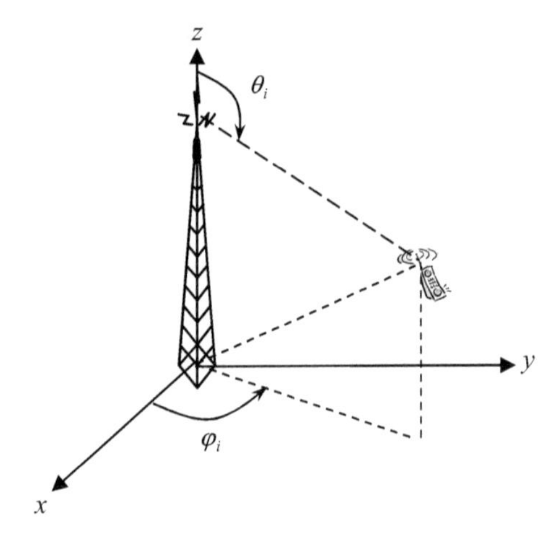

Fig. 18.13 Spatial relationship between a base station and the mobile station (target)

base station as (x_i, y_i, z_i). Let ϕ_i and θ_i be the azimuth and elevation angles of the signal transmitted by the mobile and impinging on the antenna of the ith base station. Then the angles and the known and unknown positions are related by

$$\begin{aligned} y - x\tan\hat{\phi}_i &\approx y_i - x_i\tan\hat{\phi}_i,\ i = 1, 2, 3 \\ (z - z_i)\sin\hat{\theta}_i &\approx \sqrt{(x - x_i)^2 + (y - y_i)^2}\cos\hat{\theta}_i \end{aligned} \tag{18.128}$$

where $\hat{\phi}_i$ and $\hat{\theta}_i$ are the estimates respectively of ϕ_i and θ_i, and the approximation comes from dropping angle estimation errors. In the case of $z_i \gg z$,, the x-coordinate and y-coordinate of the mobile position can be determined by angle measurements at two base stations (say #1 and #2) as

$$\begin{aligned} \hat{x} &= \frac{y_2 - y_1 + x_1\tan\hat{\phi}_1 - x_2\tan\hat{\phi}_2}{\tan\hat{\phi}_1 - \tan\hat{\phi}_2} \\ \hat{y} &= \frac{\tan\hat{\phi}_1(y_2 - x_2\tan\hat{\phi}_2) - \tan\hat{\phi}_2(y_1 - x_1\tan\hat{\phi}_1)}{\tan\hat{\phi}_1 - \tan\hat{\phi}_2} \end{aligned} \tag{18.129}$$

Then the z-coordinate is calculated by

$$\hat{z}^{(i)} = z_i + \frac{\sqrt{(\hat{x} - x_i)^2 + (\hat{y} - y_i)^2}}{\tan\hat{\theta}_i},\ i = 1, 2 \tag{18.130}$$

The two solutions from two base stations in (18.130) can be averaged to produce the final z-coordinate estimate

$$\hat{z} = \frac{1}{2}(\hat{z}^{(1)} + \hat{z}^{(2)}) \tag{18.131}$$

If there is uncertainty about whether the mobile antenna position is lower than the base station antennas, then at least three base stations are required. In this case, the x-coordinate and y-coordinate of the mobile position is computed by

$$[\hat{x} \quad \hat{y}]^T = (\mathbf{A}^T\mathbf{A})^{-1}\mathbf{A}^T\mathbf{h} \tag{18.132}$$

where

$$\mathbf{A} = \begin{bmatrix} \tan\hat{\phi}_1 & -1 \\ \tan\hat{\phi}_2 & -1 \\ \tan\hat{\phi}_2 & -1 \end{bmatrix}, \mathbf{h} = \begin{bmatrix} x_1 \tan\phi_1 - y_1 \\ x_2 \tan\phi_2 - y_2 \\ x_3 \tan\phi_3 - y_3 \end{bmatrix} \tag{18.133}$$

Note that the solution given by (18.132) is an ordinary least-squares (OLS) solution. A weighted LS solution also may be obtained if a confidence measure on the reliability of the angle measurements at each base station is available. The z-coordinate estimate can be simply calculated according to

$$\hat{z} = \frac{1}{3}\sum_{i=1}^{3}\left(z_i + \frac{\sqrt{(\hat{x} - x_i)^2 + (\hat{y} - y_i)^2}}{\tan\hat{\theta}_i}\right) \tag{18.134}$$

If there are more than three base stations, the unknown mobile position can be determined in the same way.

The performance of such an AOA-based positioning can be illustrated through the following example. Three base station antennas are located (data in meters) at (0, 0, 40), (800, 0, 30), and (400, 500, 50), and the mobile is located at (500, 200, 3). The angle estimation error is assumed as a Gaussian random variable of mean 1.5° and standard deviation (STD) respectively equal to 2, 4, and 6°. 2000 angle samples (both azimuth and elevation angles) at each base station are processed. The x-coordinate and y-coordinate estimation errors are shown in Figs. 18.14, and 18.15 shows the z-coordinate estimation errors when the STD is 4°. The RMS errors of the three coordinate estimates are listed in Table 18.1. Also shown is the position error which is defined as the average distance between the true and estimated mobile positions. Observed that the estimated z-coordinate error for the mobile is relatively large. However, in typical real-world situations the accuracy of the x-coordinate and y-coordinate estimations is more important than the z-coordinate, especially for vehicle and people tracking.

The cumulative distribution of the position errors under the three STDs is shown in Fig. 18.16. When the STD is 6°, the probability of the position errors below 60 m is only about 60%. This positioning accuracy can be improved when other measurements such as TOA or TDOA are jointly used for position determination. In the following subsection a hybrid AOA/TDOA-based method is analysed.

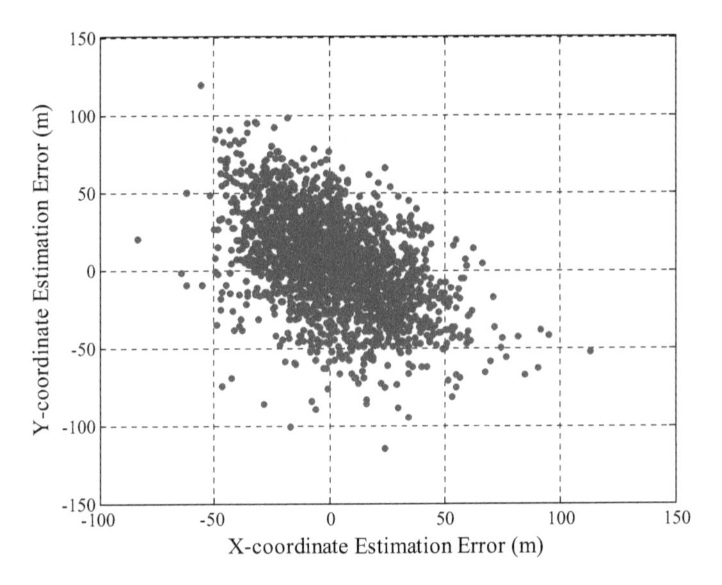

Fig. 18.14 X-coordinate and Y-coordinate estimation errors of the AOA-based method

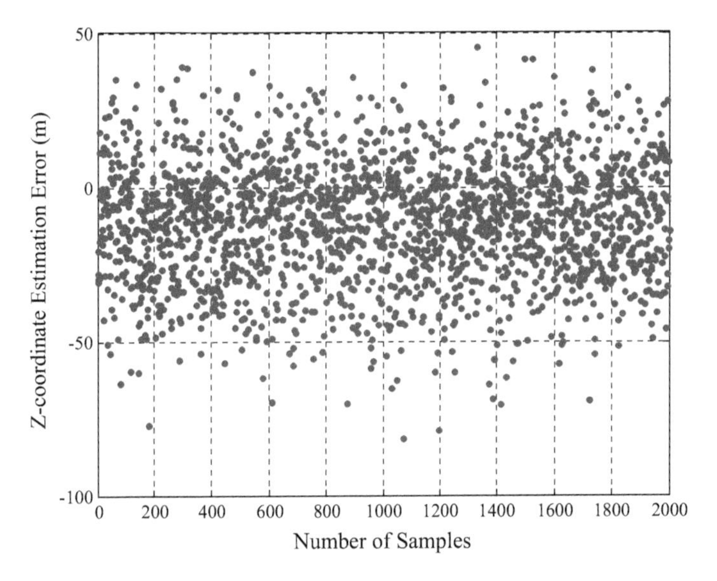

Fig. 18.15 Z-coordinate estimation errors of AOA-based method

18.4.2 Hybrid AOA and TOA Based Method

Typically, when more parameter measurements and/or more position determination techniques are jointly employed, more accurate position estimation can be achieved.

Table 18.1 RMS errors of the three coordinate estimation and the averaged position error

Angle STD (deg)	x-coordinate RMSE (m)	y-coordinate RMSE (m)	z-coordinate RMSE (m)	Position error (m)
2	12.5	14.9	13.8	22.2
4	24.2	28.8	19.7	38.7
6	35.7	43.4	27.5	56.4

In this subsection dual AOA and TOA measurements are used to enhance the positioning performance.

It is obvious that a mobile position can be determined using only one base station when both AOA and TOA (range) measurements are made at the base station. For example, the following analyses how TOA measurements from three base stations allow position determination, and the effect on positioning accuracy. The TOA-based distance measurement at the ith base station can be described as

$$\hat{d}_i = c\hat{\tau}_i = d_i + n_i \tag{18.135}$$

where c is the speed of radio propagation, $\hat{\tau}_i$ is the estimate of the time-of-flight (TOF) of the radio wave traveling from the base station to the mobile or vice versa, n_i is the distance estimation error, and

$$d_i = \sqrt{(x - x_i)^2 + (y - y_i)^2 + (z - z_i)^2} \tag{18.136}$$

Note that the TOF, TOA, and propagation time are treated in this case as equivalent. However, when the transmitter and receiver are not time synchronized, the TOA will be different from the TOF due to the time offset between the transmitter and the receiver. Also the measured TOA at the receiver incorporates the TOF and the system internal delays due to electronic processing in the transceiver; additionally, measurement errors occur even if the transmitter and receiver are synchronized. More details about this issue refer to Chap. 5.

When using the distance measurements, additional measurement equations can be obtained, namely

$$\begin{aligned} x &\approx x_i + \hat{d}_i \sin\hat{\theta}_i \cos\hat{\phi}_i,\ i = 1, 2, \ldots, N \\ y &\approx y_i + \hat{d}_i \sin\hat{\theta}_i \sin\hat{\phi}_i \\ z &\approx z_i + \hat{d}_i \cos\hat{\theta}_i \end{aligned} \tag{18.137}$$

where it is assumed that there are N base stations. Combining the first equation in (18.128) and the equations in (18.137), the ordinary LS solution for the position estimation is given by

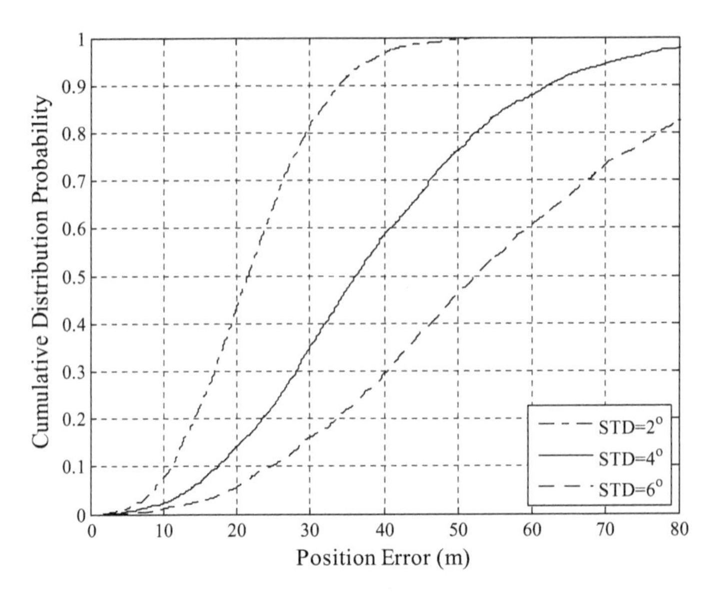

Fig. 18.16 Cumulative distribution probability of position error of AOA based method

$$[\hat{x} \quad \hat{y} \quad \hat{z}]^T = (\mathbf{A}^T\mathbf{A})^{-1}\mathbf{A}^T\mathbf{h} \tag{18.138}$$

where

$$\mathbf{A} = \begin{bmatrix} \tan\hat{\phi}_1 & -1 & 0 \\ \vdots & \vdots & \vdots \\ \tan\hat{\phi}_N & -1 & 0 \\ 1 & 0 & 0 \\ \vdots & \vdots & \vdots \\ 1 & 0 & 0 \\ 0 & 1 & 0 \\ \vdots & \vdots & \vdots \\ 0 & 1 & 0 \\ 0 & 0 & 1 \\ \vdots & \vdots & \vdots \\ 0 & 0 & 1 \end{bmatrix} \in R^{4N\times 3}, \ \mathbf{h} = \begin{bmatrix} x_1\tan\phi_1 - y_1 \\ \vdots \\ x_N\tan\phi_N - y_N \\ x_1 + \hat{d}_1\sin\hat{\theta}_1\cos\hat{\phi}_1 \\ \vdots \\ x_N + \hat{d}_N\sin\hat{\theta}_N\cos\hat{\phi}_N \\ y_1 + \hat{d}_1\sin\hat{\theta}_1\sin\hat{\phi}_1 \\ \vdots \\ y_N + \hat{d}_N\sin\hat{\theta}_N\sin\hat{\phi}_N \\ z_1 + \hat{d}_1\cos\hat{\theta}_1 \\ \vdots \\ z_N + \hat{d}_N\cos\hat{\theta}_N \end{bmatrix} \in R^{4N\times 1} \tag{18.139}$$

Consider using three base stations to examine the performance of a joint AOA/TOA based LS method. The positions of the three base stations and the mobile are the same as in the preceding subsection. The distance measurement error is modeled

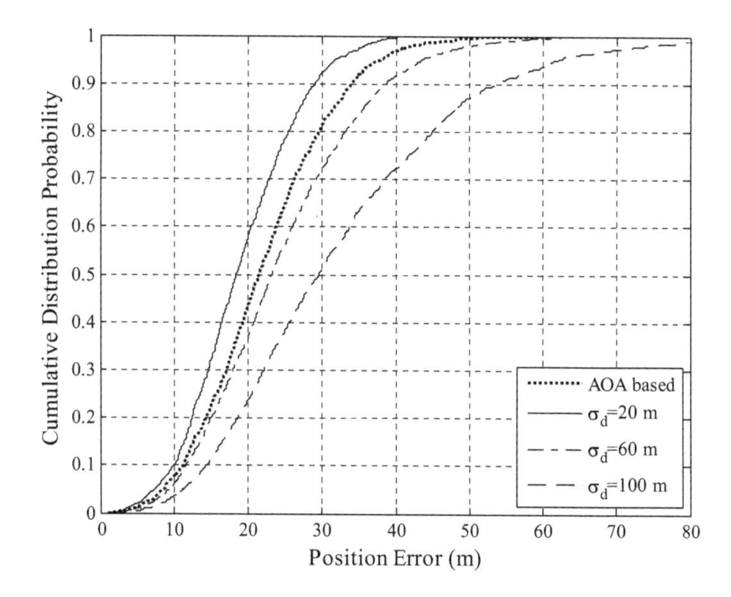

Fig. 18.17 Cumulative distribution probability of position error of joint AOA/TOA based method. The STD of the AOA error is equal to 2°

as a random variable which consists of two parts: one is Gaussian distributed and the other exponential distributed. The mean of the exponential component is set at 30 m. Figures 18.17 and 18.18 show the cumulative distribution of the position errors when the STD of the Gaussian component of the distance estimation error is set at 20 and 100 m. Also shown are the results using AOA-only methods. It can be observed that the AOA-based method can outperform the joint AOA/TOA system when the AOA error is relatively small and the TOA error is relatively large. On the other hand, when the AOA error is relatively large, the joint TOA/AOA system outperforms AOA-only system. Therefore, in practice it is important to obtain some knowledge of the confidence and reliability of parameter estimates. However, the weighted measurements can improve positioning accuracy, but this is not analyzed here.

18.5 Application Examples

Antenna array based AOA estimation and/or beamforming have been widely used in practice in the areas of communications, signal processing, and positioning. In this section, four specific application examples associated with antenna array based positioning are briefly described.

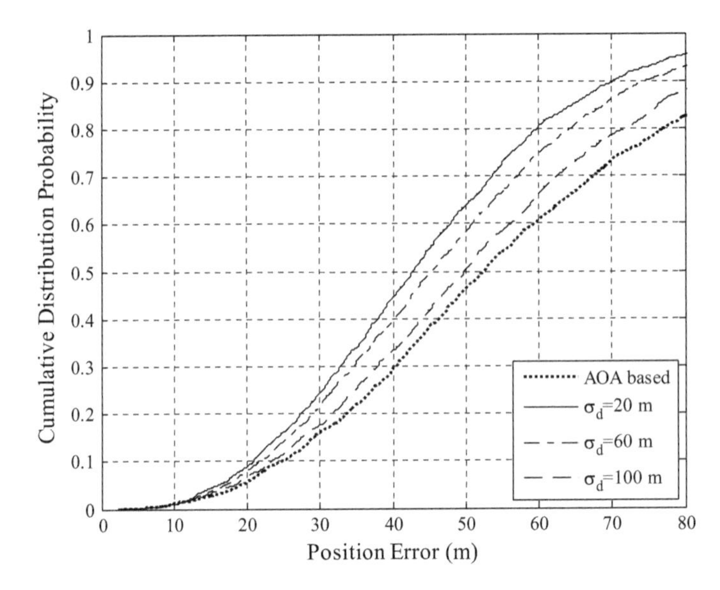

Fig. 18.18 Cumulative distribution probability of position error of joint AOA/TOA based method. The STD of the AOA error is equal to 6°

18.5.1 Animal Tracking

Animal tracking is often an important part of research on animal behaviour and associated fields such as the impact of climate change. The classical approach to radio tracking animals makes use of a multiple-element Yagi antenna to find the direction of the incoming signal emitted by a transmitter mounted on the animal. In addition to conventional Yagi-antennas, a group of ecologists from Germany and the United States made use of a phased antenna array to track Molossid bats in Panama to study their sociality (Dechmann et al. 2010). After the Molossid bats were captured, they were measured and weighed and their sex and reproductive status determined. Finally a 0.35-g radio transmitter was glued to the bats, and also tagged for identification with a subcutaneous transponder. The average mass of individual bats was 10-g, so that the transmitters remained well under the recommended 5% upper weight limit. After processing, all bats were released at the site of capture.

To track groups of bats AR8200 telemetry receivers and 3-element Yagi antennas were used. At two of the three chosen locations, the simple Yagi antennas were used to determine the angle of arrival of signal emitted from the transmitter, with estimated accuracy better than 15°. A third location was located on a canopy tower, where the two Yagi antennas were fixed in parallel on a wooden pole two wavelengths apart, to form a simple two-element phased array. The pole was then placed on a tripod to allow for simple field operations and quick directional scanning. The antenna beam is formed with the aid of a precise directional compass so that the direction can be determined. This simple antenna array enabled the scientists to obtain direction measurements with errors less than 3°.

An alternative tracking scheme was to attach a device consisting of a GPS receiver and a transmitter to the body of the animal of interest. The GPS data are forwarded to a processing center via a radio link. There are a number of issues related to this GPS option, including the power consumption, the weight and dimensional size of the device, and environmental conditions necessary for satellite signal reception. Thus which tracking scheme should be used will depend on the size of the animals to be tracked, and the radio propagation conditions in the monitored area. For instance, it would be impractical to attach an off-the-shelf GPS receiver to a 10-g Molossid bat. On the other hand, device weight would not be an issue at all when attaching a GPS receiver onto a large bear or a large marine mammal.

18.5.2 Antenna Array Based GPS Receivers

In a range of circumstances GPS receivers are required to reject multiple access radio frequency interference and multipath interference for accurate GPS positioning. To achieve an accuracy goal a multi-element antenna array can be introduced in addition to the GPS receiver to enable adaptive beamforming. One such application example is the American navy and air force Joint Precision Approach and Landing System (JPALS), which is an all-weather landing system making use of differential GPS technology. JPALS consists of modular avionics and ground/shipboard components to provide a range of landing minima and system configurations. Aircrafts receive ranging and navigation data from the satellite constellation and differential ranging data or corrections from a ground/shipboard station via a data link (Stanford 2017). Since the JPASL-equipped aircrafts are expected to operate in scenarios where the received GPS satellite signal is weak and there is strong interference, controlled reception pattern antenna technology is employed to make the system more robust in a hostile operating environment.

In (Lorenzo et al. 2006) it is noted that a multi-element antenna array may introduce additional biases in the pseudorange and carrier-phase estimates. This is caused by mutual element electronic coupling as well as distortion that may be introduced by the spatial and temporal weighting in forming the composite array output signal. Thus in the design of antenna array based GPS receivers, it is necessary to consider the trade-offs between pseudorange and carrier phase bias errors and interference rejection. Also it is desirable to develop techniques such as frequency-domain equalization to compensate the pseudorange and carrier-phase biases.

18.5.3 US Air Force Space Surveillance System

The US Air Force Space Surveillance System is a multistatic radar system that detects orbital objects passing over America, and is envisioned to be able to detect space orbiting objects at heights up to 30,000 km. This system was formerly operated by the U.S. Navy and known as Space Command Surveillance (NAVSPASUR) (Fas 2017). The system consists of one large master transmitter and two supplementary transmitters. The master transmitter is able to transmit continuous radiowaves with an average radiated power of 770 kW, and operates at a frequency of 216.98 MHz. There are six receiving stations located in six different cities in the United States.

The basic concept of NAVSPASUR is rather simple. The master transmitter generates a large fan beam of energy which reflects signals from an orbiting object (either a satellite or space debris) back to ground receiving stations. These receiving stations use large antenna arrays to form an interferometer to determine the angle of arrival and angle rates of arrival from the reflected signals. By observing the orbiting target from several stations, the object position can be determined. Also the object's orbit can be inferred by using multiple angular data over time. Although the NAVSPASUR system performs the functions of detection and satellite orbit determination very well, there are limitations on coverage and time required to determine an orbit with the CW signalling approach. Later the system was updated to transmit ranging signals as well as the primary CW signal, so that not only could the angles of arrival at the receiving sites be measured, but also the distance to the target.

Recently the Space Fence is intended to replace the Air Force Space Surveillance System, which was transferred from American Navy to American Air Force in 2004. The higher radiowave frequency of the Space Fence allows for detection of much smaller satellites and debris. The intention is to deploy the most precise radar network for the space surveillance, and to provide the high accuracy detection of even the smallest space objects. As a consequence, in addition to maximizing national security from a space point of view, but to provide data to minimize collisions of even small objects orbiting the Earth. All these objects present potential threats for communication or GPS satellites or even the International Space Station.

18.5.4 Shooter and Explosion Location

Boomerang is an acoustics system developed by BBN Technologies and funded by U.S. Defence Advanced Research Projects Agency (DARPA), which is intended to locate shooters or snippers, and thus reduce casualties. Since 2005 more than 1000 units of the snipper locator have been deployed in Iraq and Afghanistan by the U.S. military (Cherry 2008). The system consists of seven microphones arranged on a mast as a sensor array, which can be mounted on a vehicle moving at up to 100 km/h on either a highway or rough terrain. The system is able to deliver alert information by a voice message or on a screen. There are three different versions or generations of Boomerang, namely original Boomerang, Boomerang II, and Boomerang III.

Two acoustic waves are produced when firing a bullet, the shock wave generated by the air pressure caused by the supersonic flight of the bullet, and the muzzle blast generated by the explosion in the weapon's barrel (Hengy et al. 2016). The muzzle blast can directly travel from the muzzle to the acoustic sensor array in the absence of any obstruction. By estimating the time-of-arrival of the blast signal and processing the TOA measurements using methods such as the time-difference-of-arrival (TDOA), the direction of the signal (azimuth and elevation) can be determined (Borzino et al. 2016). The direction of a bomb explosion can be estimated in a similar way (Cui et al. 2015). A shock wave is not generated at the muzzle, but at the detach point during the flight of the bullet. The exact location of the detach point depends on a number of factors such as bullet dimensions and speed. Typically, the two waves will arrive at the sensor array at different time instants. By measuring the TDOA between the shock wave and the muzzle blast, the distance between the shooter and the array can be estimated. It is claimed that the Boomerang system is able to yield information about the azimuth and elevation angles and distance in less than 1.5 s.

References

Borzino AMCR, Apolinario JA Jr, de Campos MLR (2016) Consistent DOA estimation of heavoisy gunshot signals using a microphone array. IET Radar Sonar Navig 10(9):1519–1527

Cherry S (2008) Spotting snippers with sound. IEEE Spectr 45(12):14–14

Cui XX, Yu K, Lu S (2015) Evolutionary TDOA-based direction finding methods with 3-D acoustic array. IEEE Trans Instrum Meas 64(9):2347–2359

Dechmann DKN, Kranstauber B, Gibbs D, Wikelski M (2010) Group hunting—a reason for sociality in Molossid bats. PLoS ONE 5(2):1–7

Fas (2017) http://www.fas.org/spp/military/program/track/spasur_at.htm

Guo YJ (2004) Advances in mobile access networks. Artech House, Boston

Hengy S, Duffner P, DeMezzo S, Heck S, Gross L, Naz P (2016) Acoustic shooter localization using a network of asynchronous acoustic nodes. IET Radar Sonar Navig 10(9):1528–1535

Lorenzo DSD, Rife J, Enge PK, Akos D (2006) Navigation accuracy and interference rejection for an adaptive GPS antenna array. In: Proceedings of ION institute of navigation global navigation satellite systems conference, Fort Worth, TX, USA, Sept 2006

Mathews CP, Zoltowski MD (1994) Eigenstructure techniques for 2-D angle estimation with uniform circular arrays. IEEE Trans Signal Process 42(9):2395–2407

Ottersten B, Viberg M, Kailath T (1991) Performance analysis of the total least squares ESPRIT algorithm. IEEE Trans Signal Process 39(5):1122–1135

Schmidt RO (1986) Multiple emitter location and signal parameter estimation. IEEE Trans Antenna Propag 34(3):276–280

Stanford (2017). http://waas.stanford.edu/research/jpals.htm

Swindlehurst AL, Kailath T (1992) A performance analysis of subspace-based methods in the presence of model errors, part I: the MUSIC algorithm. IEEE Trans Signal Process 40 (7):1758–1774

Wu Y, So HC (2008) Simple and accurate two-dimensional angle estimation for a single source with uniform circular array. IEEE Antennas Wirel Propag Lett 7(99):78–80

Ye Z, Xiang L, Xu X (2007) DOA estimation with circular array via spatial averaging algorithm. IEEE Antennas Wirel Propag Lett 6:74–76

Appendix A
Introduction to Kalman Filtering of Time-Series Data

A.1 Introduction

The topic of Kalman filtering has a vast associated literature (Brookner 1998; Brown and Hwang 1997), so the simple overview in this Appendix can only touch on some on the more important aspects used in this book. Thus the purpose of this Appendix is to present the basic mathematical underpinnings of Kalman filtering, and to present the method of Kalman filter design appropriate for particular applications.

Because location systems have inherent errors in determining positions, methods of improving the accuracy are an important part in the overall design of a system. One of the simplest methods of removing errors in positional data is to simply smooth the raw positional data in real time. This concept is based on the observation that moving objects have finite dynamics, while positional (particularly radiolocation) data can result in large changes in position between successive position updates. Under these circumstances it is intuitive that some form of data smoothing will result in a measured track which more closely represents the actual track than the raw data. However, smoothing the raw data to minimize the noise will also distort the desired positional data track of a moving object being tracked. Thus there needs to be a compromise between reducing the noise component while having minimal effect on the tracking of positions. Clearly the optimum solution will depend on the dynamics of the object being tracked—a tracked person will have much more agile dynamics then (say) a motor vehicle. The most popular method of such data smoothing is the Kalman Filter (the subject of this Appendix) which provides the optimum compromise solution for the given specified dynamics and noise characteristics.

The data associated with position fixing will be corrupted with random errors. To minimise the effect of these errors, a Kalman filter is applied to the measured or computed data. Reducing the bandwidth of the filter will certainly reduce the noise

I. Sharp and K. Yu, *Wireless Positioning: Principles and Practice*, Navigation: Science and Technology, https://doi.org/10.1007/978-981-10-8791-2

(variance), but due to the dynamics of the data (changes in the position as well as velocity and acceleration changes) the filtering will also introduce errors in noise-free data. To minimise the overall errors, the parameters of the Kalman filter must be optimised for the dynamics appropriate to the particular time-series data of the application. The order of the filter is also important in minimising the errors. For a third-order state vector with $[x, x', x'']$, the filter will track with zero nominal error both the position x and speed x' components for a constant acceleration x''. As both position and speed estimates are important for position location applications, a third order (or g-h-k) filter will be described in a later section.

The following Sect. A.2 provides details of the implementation of the Kalman filter. The method closely follows that given in the book "Tracking and Kalman Filtering Made Easy", by Eli Brookner. The Kalman filter algorithms are presented in matrix form without proof. The elegant nature of this form is that the basic algorithm is the same for all filters, regardless of the details of the particular application. However, the matrices must be configured for each case. The details for the typical application are given in Sect. A.3 following.

A.2 Matrix Representation

The matrix representation of the Kalman filter is summarised in this section. The aim of the algorithm is to optimally filter a state vector **X**, based on measured input data **Y**. Estimated variables are indicated by an asterisk (*) superscript. It is assumed that the measurements and the state vector are updated at a regular fixed period T (s), and samples are indicated by a subscript n. Typical update periods for tracking vary from 0.1 to 1 s for people, and 1 to 5 s for vehicles.

There are two types of uncertainty in the measured input data **Y**. The first type of uncertainty (process noise) is associated with the dynamics of the particular application. In the case of a tracking system, this uncertainty is associated with the motion of the mobile device; such motion is modelled as a random process. The process noise, specified by the matrix **Q**, is defined as an input (often assumed invariant) to the Kalman filter. The nature of the matrix is dependent on the particular application and its dynamics. See Sect. A.3 for more details.

The second type of uncertainty is associate with **Y** measurement errors (measurement noise). Such noise is assumed to be random (typically Gaussian), and are usually uncorrelated with the process noise. As well as a smoothed estimate of the state vector, the Kalman filter also outputs an estimate of the measurement noise, based on the input data. In a positioning system this can be roughly interpreted as the "accuracy" of the position fixes.

The Kalman filter uses two types of estimates, namely a priori estimates of the state vector (and its associated covariance matrix), and a posteriori estimates of the same parameters. The a priori estimate occurs just before the new data are available at time period n, and can be considered as a prediction of the state vector for time n. The a posteriori estimate at time n is based on the latest data (at time n), and can be

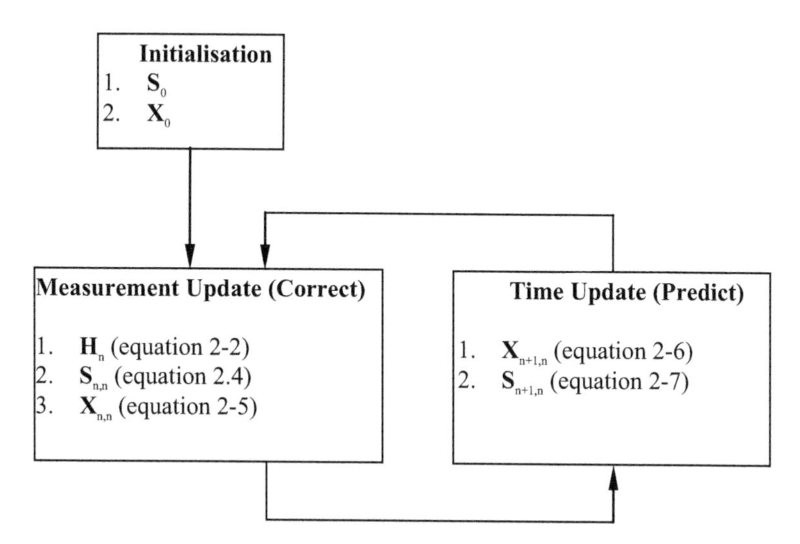

Fig. A.1 Block diagram of the Kalman filter process, showing the cyclical loop used for calculations

considered as a update to the estimate. These two groups of calculations are performance in a cyclical fashion, with in input of one set coming from the output of the second set. This cyclical process is shown in more detail in Fig. A.1.

A.2.1 Initialization

The filter is initialized with an estimate of the state vector $\mathbf{X_0}$ (a priori estimate). This initial value is not too critical, as the filter dynamics will soon track the error to their steady-state values. In typical practical cases, the measurement vector and the state vector may not be of the same order. For example, a positioning system measures the position only, whereas the state vector may also include the speed and acceleration. In some cases, with other sensors, these two elements of the state vector may also be measured by other sensors (for example accelerometers), but this is not essential for implementation of the Kalman filter.

Thus for time period n, the measurements $\mathbf{Y_n}$ are related to the state vector by

$$\mathbf{Y}_n = \mathbf{M}\mathbf{X}_n + \mathbf{N}_n \tag{A.1}$$

In (A.1), $\mathbf{N}$ is the random measurement noise component which corrupts the measurements. The $\mathbf{M}$ vector is applied (if necessary) to convert the measurement data to match the state vector; for example if the measurements are just the position x but the state vector is $[x, x', x'']^T$, then $M = [1, 0, 0]$.

The other a priori input is the initial estimate of the error matrix $\mathbf{S}^*$. The matrix $\mathbf{S}^*$ is associated with the accuracy in predicting the state vector. Mathematically $\mathbf{S}^*$ is related to the covariance of the predicted state vector $\mathbf{X}^*$, and is updated by the Kalman filter during the filtering process. When initialising the $\mathbf{S}$ matrix, if there is no other information, all the elements of the matrix can be set to zero. The Kalman filter will update the matrix as measured data become available. However, typically initial values of the matrix can be estimated from approximate knowledge of the errors associated with the measurements and the state vector dynamics.

A.2.2 Measurement Update (Correction)

The first set of calculations is associated with the update of the Kalman loop parameters, based on the a priori data. The first step is to define the filter. These parameters are similar in nature to the parameters of a g-h-k filter (see Sect. A.3), except that in the Kalman filter these parameters vary according to the characteristics of the input measured data (and the associated noise). These parameters are defined by a $\mathbf{H}$ matrix. The $\mathbf{H}$ parameter vector can now be updated as defined in (A.2)

$$\mathbf{H}_n = \mathbf{S}^*_{n,n-1}\mathbf{M}^T[\mathbf{R} + \mathbf{M}\mathbf{S}^*_{n,n-1}\mathbf{M}^T]^{-1} \tag{A.2}$$

The dual subscripts have the following interpretation: the first index indicates the time index for which the calculated value applies, and the second index defines the time index of the data used for the calculation. Equation (A.2) introduces a new matrix associated with the Kalman filter. The $\mathbf{R}$ matrix is associated with the measurement noise, and is defined by the following covariance matrix

$$\mathbf{R}_n = \text{cov}[\mathbf{N}_n] = E[\mathbf{N}_n\mathbf{N}_n^T] \tag{A.3}$$

The next update is associated with determining the a-posteriori estimate of the $\mathbf{S}$ matrix. This updating process is described mathematically by the matrix equation

$$\mathbf{S}^*_{n,n} = [\mathbf{I} - \mathbf{H}_n\mathbf{M}]\ \mathbf{S}^*_{n,n-1} \tag{A.4}$$

The filtered state vector is now updated, based on the updated $\mathbf{H}$ matrix and the input measured data $\mathbf{Y}$

$$\mathbf{X}^*_{n,n} = \mathbf{X}^*_{n,n-1} + \mathbf{H}_n[\mathbf{Y}_n - \mathbf{M}\mathbf{X}^*_{n,n-1}] \tag{A.5}$$

A.2.3 *Time Update (Prediction)*

The time update processing (or predictions) use the measurement updates to determine the a priori (predicted) parameters. First, the predicted state vector is given by

$$\mathbf{X}_{n+1,n} = \mathbf{\Phi}\,\mathbf{X}_{n,n} \tag{A.6}$$

The $\mathbf{\Phi}$ matrix is the predictor matrix, and is based on the system dynamics model appropriate for the particular case. The $\mathbf{\Phi}$ matrix allows the state vector to be updated (predicted) one time period.

The second a priori update is associated with the $\mathbf{S}$ matrix.

$$\mathbf{S}^{*}_{n+1,n} = \Phi\mathbf{S}^{*}_{n,n}\Phi^{T} + \mathbf{Q} \tag{A.7}$$

The $\mathbf{Q}$ matrix is defined as the covariance matrix of the random system dynamics noise (process noise), and is given by

$$\mathbf{Q} = \mathrm{cov}[\mathbf{U}_n] = E[\mathbf{U}_n\mathbf{U}_n^{T}] \tag{A.8}$$

The filter algorithm is looped repeatedly by processing Eqs. (A.2)–(A.8) to obtain a prediction of the state vector $\mathbf{X}$ at time $n+1$, based on data $\mathbf{Y}$ available up to time n.

A.3 Kalman Filter for Position Determination Applications

The general theory presented in Sect. A.2 is now applied for a typical position determination application. The main requirement is usually associated with filtering the estimated position (x, y) of the mobile device. It is assumed that the x and y coordinates are statistically independent, so that the Kalman filter is applied to x and y separately. Each tracking application will have well defined dynamics. For example, tracking people will have a time constant of the order of a second, as the stride rate is typically about 1.5 per second, and the direction can change in a couple of strides. The main period of acceleration is at the beginning of walking, with low acceleration when walking in a straight line. However, direction change can occur at unpredictable times, so that the accelerations should be modelled as a random process with some maximum associated acceleration. For accurate tracking and predictions, it is essential that the position and speed are determined accurately without lag. Thus a third-order filter is appropriate, as this filter type will have zero steady-state lag for constant acceleration.

When tracking the individual x and y coordinates the situation is a little more complicated. In this case, even if the speed is constant, there will be accelerations in both x and y around bends. These accelerations are termed "pseudo-accelerations", as they are a consequence of the geometry and the selected variables to be tracked, and are related only indirectly to the real mobile accelerations. (An alternative set would be in polar coordinates, where both r and θ vary relatively slow, even when in a bend).

For the x or y coordinates, the pseudo-accelerations for a oval track[1] are readily calculated. In particular, assuming an oval track with semi-circular bends, the pseudo-speed and pseudo-acceleration are of the form

$$\begin{aligned} V(t) &= -V_0 \cos\left(\frac{V_0 t}{r}\right) \\ A(t) &= \frac{V_0^2}{r} \sin\left(\frac{V_0 t}{r}\right) \end{aligned} \tag{A.9}$$

where r is the radius of the bend, and V_0 is the mobile speed, assumed constant. As both these parameters can be approximately defined for a given application, the peak acceleration can be specified for the Kalman filter dynamics. For example, if $r = 1$ m and $V = 1.5$ m/s (typical for a person walking), then the peak pseudo-acceleration is 2.25 m/s^2.

The proposed implementation of the Kalman filter is a simplification of the Singer g-h-k Kalman filter described in Sect. 2.9 of the above-referenced book. This filter has a number of components in defining the process noise statistics (probability density function) that can be tuned to a wide variety of applications, including walking. The PDF is illustrated in Fig. A.2, which shows that there are three delta functions plus a uniform distribution. The delta function at zero acceleration is associated with no movement or moving at a constant speed and direction. The two delta functions at the maximum acceleration are associated with moving around a bend, as described above. The uniform distribution is associated with other random motions which cannot be described explicitly, but are assumed to cover a wide range of accelerations up to the maximum associated with turning a tight bend. The relative magnitude of there components can be estimated by simulation or actual measurements. Figure A.3 shows the PDF of movement around an oval track at constant speed; as can be observed, the resulting PDF of the acceleration approximates the assumed model of the PDF. The PDF has a dominant peak at the origin (acceleration is assumed zero on straight-line segments), and approximately a constant distribution up to the maximum acceleration defined by (A.9). The Singer-Kalman filter parameters for the human walking application are assumed to be as follows: $P_0 \approx 0.4$, $P_{\max} = 0$, $A_{\max} = 2.25$ m/s^2. Note that it is assumed that people spend a large proportion of time when walking with zero acceleration.

---

[1] An oval track is an appropriate model as it includes both straight-line segments and constant radius bends, which approximate the types of path segments for people walking around a building.

This filter has an exponential autocorrelation function given by

$$E\left[\ddot{x}(t)\ddot{x}(t+t')\right] = \sigma_a^2 e^{-|t'/\tau|} \quad (A.10)$$

The time constant τ of the acceleration can be computed from the autocorrelation function of the pseudo-acceleration around a path. For the Kalman filter to be effective the data sampling period (T) should be much smaller than the dynamics time constant. As a guide it is recommended that $T < \tau/10$. For this Singer distribution, the variance can be computed to be

$$\sigma_a^2 = \frac{A_{\max}}{3}\left[1 + 4P_{\max} - P_0\right] \quad (A.11)$$

For any given application, the above parameters can be computed, thus optimising the solution for each track. For example, while the above parameters are appropriate for people walking, the same model could be used for tracking vehicles, but with different parameters. However, race cars on a track will be different from cars on suburban streets, but the same model is broadly appropriate.

The measured data input to the filter are (separately) the computed x and y positions from the position location system. The corresponding state vector includes the position, speed and acceleration. (In the following, only the x coordinate is used; the y coordinate analysis is identical).

Thus the state noise and observation matrices are

$$\mathbf{X}_n = \begin{bmatrix} x_n \\ \dot{x}_n \\ \ddot{x}_n \end{bmatrix} \qquad \mathbf{U}_n = \begin{bmatrix} u_{x_n} \\ u_{v_n} \\ u_{a_n} \end{bmatrix} \qquad \mathbf{M} = \begin{bmatrix} 1 & 0 & 0 \end{bmatrix} \quad (A.12)$$

As the system dynamics parameter satisfies $\tau \gg T$, the transition matrix $\mathbf{\Phi}$ reduces to that associated with constant acceleration, namely

$$\mathbf{\Phi} = \begin{bmatrix} 1 & T & T^2/2 \\ 0 & 1 & T \\ 0 & 0 & 1 \end{bmatrix} \quad (A.13)$$

The covariance of the white-noise manoeuvre excitation vector $\mathbf{U}_n$ can be shown to be

$$\mathbf{Q} = \frac{2\sigma_a^2}{\tau}\begin{bmatrix} \frac{T^5}{20} & \frac{T^4}{8} & \frac{T^3}{6} \\ \frac{T^4}{8} & \frac{T^3}{3} & \frac{T^2}{2} \\ \frac{T^3}{3} & \frac{T^2}{2} & T \end{bmatrix} \quad (A.14)$$

The predictor covariance matrix $\mathbf{S}$ is initialised as follows

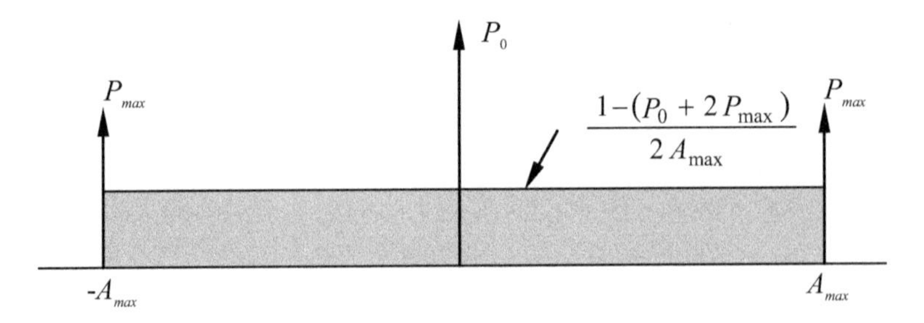

Fig. A.2 Generic PDF for Singer-Kalman filter

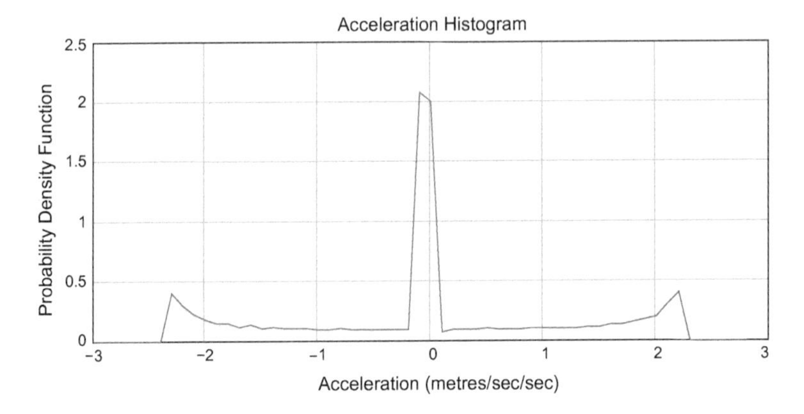

Fig. A.3 Probability Density Function of the acceleration for constant speed around an oval path

$$\mathbf{S}_0 = \begin{bmatrix} \sigma_x^2 & \frac{\sigma_x^2}{T} & 0 \\ \frac{\sigma_x^2}{T} & \frac{2\sigma_x^2}{T^2} & \tau\sigma_a^2 \\ 0 & \tau\sigma_a^2 & \sigma_a^2 \end{bmatrix} \tag{A.15}$$

where σ_x is the measurement accuracy (standard deviation) of the x measurements. By applying these Kalman filter matrix definitions to the generic algorithm given in Sect. A.2, the optimum filtering of the positional data can be obtained. The Kalman filter is implemented using the generic matrix equations described in Sect. A.2, and the specific definitions of the matrices described above.

A.4 G-H-K Filter Equivalent

The Kalman filter implementation defined in Sect. A.2 uses matrix-based algorithm to implement the filter. While the algorithm is relatively simple when expressed in matrix algebra form, the requirement for floating-point matrix calculations can be intensive in computation, particularly if the simple processors in Wireless Sensor Networks are used. Thus an alternative method is sought which requires less computational resources, yet retains the benefits of the Kalman filter. In this section, the Kalman filter is shown to be equivalent (with some minor constraints) to a g–h–k filter, which can be implemented as a classical finite impulse response (FIR) filter, which only requires a few multiplications and additions for its implementation.

A g–h–k filter is based on defining the motion differential equations assuming constant acceleration. It is shown in the cited reference that this filter is defined by three parameters (g, h and k), and three differential equations for the position, speed and acceleration. Importantly, the (g, h, k) parameters can be defined in terms of the Kalman **H** matrix, and thus the Kalman filter design can be transformed into a g–h–k filter. An overview of the key results will now be presented.

The Kalman filter described in Sect. A.3 approaches a steady-state g-h-k filter, provided the process noise (**Q**) and the measurement noise (**R**) are constant matrices (which is normally the case in the typical applications). In this case the third-order filter parameters (g, h, k) are defined by the **H** matrix as follows

$$\mathbf{H} = \begin{bmatrix} g \\ h/T \\ 2k/T^2 \end{bmatrix} \tag{A.16}$$

Additionally, the filter parameters obey the following conditions

$$\begin{aligned} g &= \sqrt{2h} - h/2 \\ h &= 4 - 2g - 4\sqrt{1-g} \\ k &= \frac{h^2}{4g} \end{aligned} \tag{A.17}$$

Note that equations (A.17) are in fact only two conditions, as the first two are two different forms of the one condition. Thus these equations are not sufficient to fully define these parameters. One parameter (typically g) is determined from the standard deviation of the acceleration (σ_a) and the standard deviation of x (σ_x). In practice, the following condition always applies: $g \gg h \gg k$.

The prediction equations for the g-h-k filter with constant acceleration are given by

$$\begin{aligned} x^*_{n+1} &= x^*_n + g\,(y_n - x_n) + T\dot{x}^*_n + h\,(y_n - x_n) + \frac{T^2}{2}\ddot{x}^*_n + k\,(y_n - x_n) \\ \dot{x}^*_{n+1} &= \dot{x}^*_n + \frac{h}{T}(y_n - x_n) + T\ddot{x}^*_n + \frac{2k}{T}(y_n - x_n) \\ \ddot{x}^*_{n+1} &= \ddot{x}^*_n + \frac{2k}{T}(y_n - x_n) \end{aligned} \tag{A.18}$$

Equations (A.18) can be converted into the more convenient form of the z-transform of the filter transfer function by taking the z-transform of the equations. Taking the z-transform of equations (A.18) yields

$$
\begin{aligned}
zX_n &= X_n + g\,(Y_n - X_n) + T\dot{X}_n + h\,(Y_n - X_n) + \frac{T^2}{2}\ddot{X}_n + k\,(Y_n - X_n) \\
z\dot{X}_n &= \dot{X}_n + \frac{h}{T}(Y_n - X_n) + T\ddot{X}_n + \frac{2k}{T}(Y_n - X_n) \\
z\ddot{X}_n &= \ddot{X}_n + \frac{2k}{T}(Y_n - X_n)
\end{aligned} \tag{A.19}
$$

By substituting for $\dot{X}_n$ and $\ddot{X}_n$ in the first equation in (A.19), the z-transform of the transfer function can be determined after much algebraic manipulation

$$
H(z) = \frac{X_n}{Y_n} = \frac{(g+h+k)\,z^2 + (k-2g-h)\,z + g}{z^3 + (g+h+k-3)\,z^2 + (k-2g-h+3)\,z + (g-1)} \tag{A.20}
$$

Now that the Kalman filter implementation has been reduced to a z-transform transfer function, the filter can be implemented using the standard techniques for FIR filters. As the polynomial in the denominator of the transfer function is a cubic, the FIR filter will require samples with a delay of up to three samples.

The frequency response defined by Eq. (A.20) can be determined by substituting $z = \exp(j\omega)$. Thus at $\omega = 0$, it is easy to show that $H(0) = 1$. Similarly, at the maximum scaled frequency of $f = 0.5$, the magnitude of the transfer function is given by

$$
|\,H(0.5)\,| = a = \left|\frac{2g+h}{2g+h-4}\right| \approx \frac{g}{2} \tag{A.21}
$$

which is the attenuation (a) of the filter at high frequencies, and thus defines the degree of attenuation of the high-frequency noise. Thus for good attenuation g should be small, but g also largely determines the filter bandwidth. This compromise is one of the essential characteristics of the design of Kalman filters.

The filter delay is another important parameter, and ideally for accurate tracking the delay should be zero for all input signals. A filter group delay is given by

$$
\Gamma(\omega) = \frac{d\Phi(\omega)}{d\omega} \qquad \Phi(\omega) = \arg\left[H\left(e^{j\omega}\right)\right] \tag{A.22}
$$

where $\Phi(\omega)$ is the phase of the transfer function. With the transfer function given by Eq. (A.21), Eq. (A.22) can be evaluated at $\omega = 0$ (or $z = 1$), which shows that the group delay is indeed zero at zero frequency. Thus input signals with low spectral bandwidth can be tracked with essential zero error.

A typical example is shown in Fig. A.4, based on the values determined by the Singer-Kalman filter (standard deviation in range 1 m. Note that the initial filter gain of 0 dB increases to a peak near the motion dynamics bandwidth parameter $(1/\tau)$, and then falls to about −20 dB at the maximum frequency of 5 Hz (Nyquist frequency for sampling rate of 10 per second assumed in this example). Thus as expected noise with frequencies greater than the motion dynamics bandwidth are significantly filtered, while frequencies below the motion dynamics bandwidth are largely unfiltered.

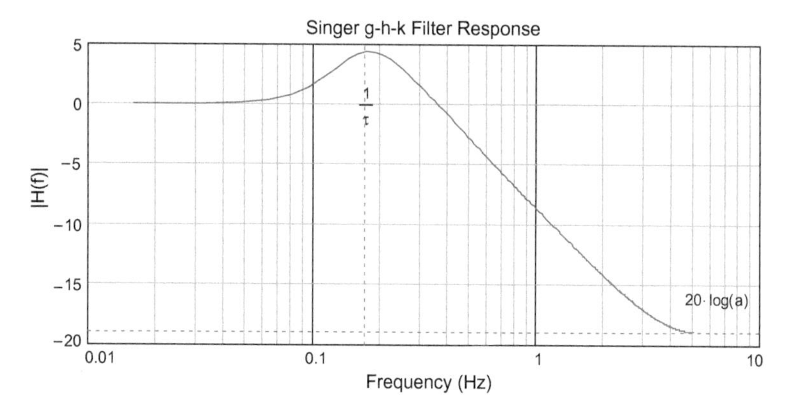

Figure A.4. Example of the transfer function of the g-h-k tracking filter

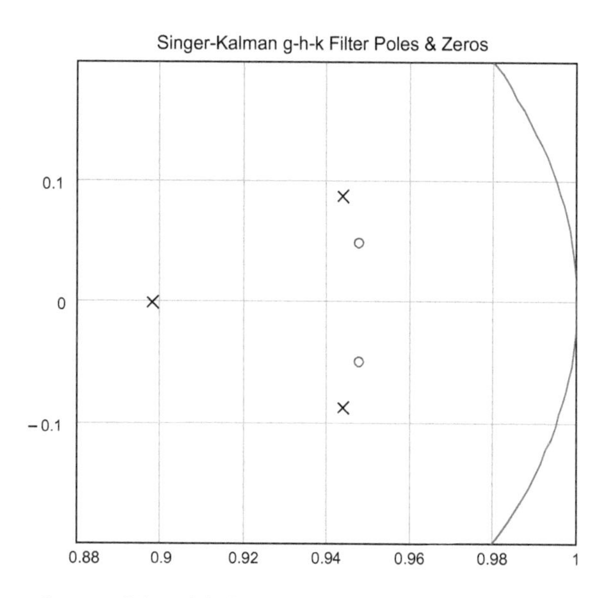

Fig. A.5 Poles and zeros of the g-h-k filter

The corresponding poles and zeros of the transfer functions are shown in Fig. A.5. The pairs of poles and zeros near $z = 1$ define the filter shape (bandwidth) at low frequencies. As the frequency response is determined by the product magnitude of the zero vectors (zero to point on unit circle) divided by the magnitude of the pole vectors (pole to point on unit circle), it is clear that the effect of the poles and zeros approximately cancel at low frequencies, as the pole-zero pairs are close together.

References

Brookner E (1998) Tracking and Kalman filtering made easy. Wiley. ISBN 0-471-18407-1.

Brown R, Hwang P (1997) Introduction to random signals and applied Kalman filtering. Wiley. ISBN 0-0471-12839-2.

CPSIA information can be obtained
at www.ICGtesting.com
Printed in the USA
LVHW082203090519
617272LV00002B/19/P